MASSIVE STARS AS COSMIC ENGINES

IAU SYMPOSIUM No. 250

Cover photo of ghostly Hawaiian beach taken through an IR filter by Fabio Bresolin.

IAU SYMPOSIUM PROCEEDINGS SERIES

2007 EDITORIAL BOARD

Chairman

I.F. CORBETT, IAU Assistant General Secretary
European Southern Observatory
Karl-Schwarzschild-Strasse 2
D-85748 Garching bei München
Germany
icorbett@eso.org

Advisers

K.A. VAN DER HUCHT, IAU General Secretary,
SRON Netherlands Institute for Space Research, Utrecht, the Netherlands
E.J. DE GEUS, *Dynamic Systems Intelligence B.V., Assen, the Netherlands*
U. GROTHKOPF, *European Southern Observatory, Germany*
M.C. STOREY, *Australia Telescope National Facility, Australia*

Members

IAUS242
J.M. CHAPMAN, *Australia Telescope National Facility, Sydney, NSW, Australia*
IAUS243
J. BOUVIER, *Laboratoire Astrophysique Observatoire de Grenoble, Grenoble, France*
IAUS244
J.I. DAVIES, *Department of Physics and Astronomy, Cardiff University, Cardiff, UK*
IAUS245
M. BUREAU, *Sub-Department of Astrophysics, University of Oxford, Oxford, UK*
IAUS246
E. VESPERINI, *Department of Physics, Drexel University, Philadelphia, PA, USA*
IAUS247
C.A. MENDOZA-BRICEÑO, *Universidad de los Andes, Merida, Venezuela*
IAUS248
WENJING JIN, *Shanghai Astronomical Obeservatory, Shanghai, China*
IAUS249
S. FERRAZ MELLO, *IAG, Universidade de São Paulo, São Paulo, SP, Brasil*
IAUS250
F. BRESOLIN, *Institute for Astronomy, Honolulu, HI, USA*

INTERNATIONAL ASTRONOMICAL UNION
UNION ASTRONOMIQUE INTERNATIONALE

MASSIVE STARS AS COSMIC ENGINES

PROCEEDINGS OF THE 250th SYMPOSIUM OF THE
INTERNATIONAL ASTRONOMICAL UNION
HELD IN KAUAI, HAWAII, USA
DECEMBER 10–14, 2007

Edited by

FABIO BRESOLIN
Institute for Astronomy, University of Hawaii, Honolulu, HI, USA

PAUL A. CROWTHER
Department of Physics and Astronomy, University of Sheffield, UK

and

JOACHIM PULS
Universitätssternwarte, München, Germany

CAMBRIDGE UNIVERSITY PRESS
The Edinburgh Building, Cambridge CB2 8RU, United Kingdom
32 Avenue of the Americas, New York, NY 10013-2473, USA
477 Williamstown Road, Port Melbourne, VIC 3207, Australia
Ruiz de Alarcón 13, 28014 Madrid, Spain
Dock House, The Waterfront, Cape Town 8001, South Africa

© International Astronomical Union 2008

This book is in copyright. Subject to statutory exception
and to the provisions of relevant collective licensing agreements,
no reproduction of any part may take place without
the written permission of the International Astronomical Union.

First published 2008

Printed in the United Kingdom at the University Press, Cambridge

Typeset in System LaTeX 2_ε

A catalogue record for this book is available from the British Library

Library of Congress Cataloguing in Publication data

ISBN 9780521874724 hardback
ISSN 1743-9213

Table of Contents

Preface ix

Organizing committee xi

Conference photograph xii

Conference participants xiii

1. Atmospheres of Massive Stars

Massive Stars as Cosmic Engines Through the Ages 3
    A. Maeder, G. Meynet, S. Ekström, R. Hirschi, & C. Georgy

X-ray Emission from O Stars 17
    D. H. Cohen

Physical and Wind Properties of OB-Stars 25
    J. Puls

The Metallicity Dependence of the Mass Loss of Early-Type Massive Stars 39
    A. de Koter

Properties of Wolf-Rayet Stars 47
    P. A. Crowther

Wolf-Rayet Wind Models from Hydrodynamic Model Atmospheres 63
    G. Gräfener & W.-R. Hamann

Luminous Blue Variables & Mass Loss near the Eddington Limit 71
    S. Owocki & A.-J. van Marle

Pulsation-Initiated Mass Loss in Luminous Blue Variables: A Parameter Study . 83
    A. J. Onifer & J. A. Guzik

What Do We Really Know About the Winds of Massive Stars? 89
    D. J. Hillier

The Physical Properties of Red Supergiants: Comparing Theory and Observations 97
    P. Massey, E. M. Levesque, B. Plez, & K. A. G. Olsen

The Evolutionary State of the Cool Hypergiants – Episodic Mass Loss, Convective
    Activity and Magnetic Fields 111
    R. M. Humphreys

Massive Binaries 119
    A. F. J. Moffat

3-D SPH Simulations of Colliding Winds in η Carinae 133
    A. T. Okazaki, S. P. Owocki, C. M. P. Russell, & M. F. Corcoran

The First Determination of the Rotation Rates of Wolf-Rayet Stars 139
    A-N. Chené & N. St-Louis

2. Physics and Evolution of Massive Stars

Developments in Physics of Massive Stars 147
    G. Meynet, S. Ekström, A. Maeder, R. Hirschi, C. Georgy, & C. Beffa

Can Pulsational Instabilities Impact a Massive Star's Rotational Evolution? ... 161
    R. Townsend & J. MacDonald

Rotation and Massive Close Binary Evolution 167
    N. Langer, M. Cantiello, S.-C. Yoon, I. Hunter, I. Brott, D. Lennon, S. de Mink, & M. Verheijdt

The Effect of Massive Binaries on Stellar Populations and Supernova Progenitors 179
    J. J. Eldridge, R. G. Izzard, & C. A. Tout

Thoughts on Core-Collapse Supernova Theory 185
    A. Burrows, L. Dessart, C. D. Ott, E. Livne, & J. Murphy

Episodic Mass Loss and Pre-SN Circumstellar Envelopes 193
    N. Smith

The Progenitor Stars of Core-Collapse Supernovae 201
    S. J. Smartt, R. M. Crockett, J. J. Eldridge, & J. R. Maund

Can Very Massive Stars Avoid Pair-Instability Supernovae? 209
    S. Ekström, G. Meynet, & A. Maeder

Stellar Evolution at Low Metallicity 217
    R. Hirschi, C. Chiappini, G. Meynet, A. Maeder, & S. Ekström

Evolution of Progenitor Stars of Type Ibc Supernovae and Long Gamma-Ray Bursts 231
    S.-C. Yoon, N. Langer, M. Cantiello, S. E. Woosley, & G. A. Glatzmaier

Core Overshoot and Nonrigid Internal Rotation of Massive Stars: Current Status from Asteroseismology 237
    C. Aerts

3. Massive Star Populations in the Nearby Universe

Young Massive Clusters 247
    D. F. Figer

Massive Stars in the Galactic Center 257
    F. Martins, D. J. Hillier, R. Genzel, F. Eisenhauer, T. Ott, S. Gillessen, & S. Trippe

Metallicity Studies in the IR: Unveiling Obscured Clusters of Our Galaxy 265
    F. Najarro

Massive Stars in the Nuclei and Arms of Spirals 273
    F. Bresolin

UCHII Regions and Newly Born O-type Stars 285
    P. S. Conti, J. Rho, J. Furness, & P. A. Crowther

Binary Populations and Stellar Dynamics in Young Clusters.................. 293
   D. Vanbeveren, H. Belkus, J. Van Bever, & N. Mennekens

Westerlund 1 as a Template for Massive Star Evolution 301
   I. Negueruela, J. S. Clark, L. J. Hadfield, & P. A. Crowther

One Hundred 30 Dors? 307
   M. Hanson & B. Popescu

Extragalactic Stellar Astronomy with the Brightest Stars in the Universe...... 313
   R. Kudritzki, M. A. Urbaneja, F. Bresolin, & N. Przybilla

VLT/FORS Surveys of Wolf-Rayet Stars in the Nearby Universe............. 327
   L. J. Hadfield & P. A. Crowther

LBT Discovery of a Yellow Supergiant Eclipsing Binary in the Dwarf Galaxy Holmberg IX 333
   J. L. Prieto, K. Z. Stanek, C. S. Kochanek, & D. R. Weisz

4. Hydrodynamics and Feedback from Massive Stars in Galaxy Evolution

Bubbles and Superbubbles: Observations and Theory 341
   Y.-H. Chu

The Evolution of the Circumstellar and Interstellar Medium Around Massive Stars 355
   S. J. Arthur

Infrared Tracers of Mass-Loss Histories and Wind-ISM Interactions in Hot Star Nebulae... 361
   P. Morris & the Spitzer WRRINGS team

Stellar Feedback Through Cosmic Time: Starbursts & Superwinds............ 367
   M. A. Dopita

Gemini/IFU Observations of Galactic Outflows in Starburst Galaxies 379
   L. J. Smith & M. S. Westmoquette

Radiative Feedback in Galaxies 385
   M. S. Oey, E. S. Voges, R. A. M. Walterbos, G. R. Meurer, S. Yelda, & E. Furst

The Role of Massive Stars in Galactic Chemical Evolution 391
   F. Matteucci

Detailed Nucleosynthesis Yields from the Explosion of Massive Stars.......... 401
   C. Fröhlich, T. Fischer, M. Liebendörfer, F.-K. Thielemann, & J. W. Truran

Evidence for a Mass Outflow from Our Galactic Center..................... 407
   C. Law

Part 5. Massive Stars as Probes of the Early Universe

Massive Stars at High Redshifts 415
   M. Pettini

Star Forming Galaxies at $z > 5$ 429
  Y. Taniguchi

Core-Collapse Supernovae as Dust Producers 437
  R. Kotak

GRBs as Probes of Massive Stars Near and Far 443
  J. P. U. Fynbo & D. Malesani

Probing the Interstellar Medium and Stellar Environments of Long-Duration GRBs   457
  M. Dessauges-Zavadsky, J. X. Prochaska, & H.-W. Chen

The Connection between Gamma-Ray Bursts and Extremely Metal-Poor Stars as
  Nucleosynthetic Probes of the Early Universe 463
  K. Nomoto, N. Tominaga, M. Tanaka, K. Maeda, & H. Umeda

The First Stars 471
  J. L. Johnson, T. H. Greif, & V. Bromm

Imprint of First Stars Era in the Cosmic Infrared Background Fluctuations ... 483
  A. Kashlinsky

Imaging and Spectroscopy with the James Webb Space Telescope 491
  G. Sonneborn

The Impact of Extremely Large Telescopes on the Study of the Most Luminous
  Stellar Objects 495
  S. D'Odorico

Metallicities at the Sites of Nearby SN and Implications for the SN-GRB Connection 503
  M. Modjaz, L. Kewley, R. P. Kirshner, K. Z. Stanek, P. Challis, P. M. Garnavich, J. E. Greene, P. L. Kelly, & J. L. Prieto

Abstracts of additional oral talks 509

6. Conclusion

Symposium Summary 513
  C. Leitherer

7. Posters

Poster Abstracts 525
  F. Bresolin, P. A. Crowther, & J. Puls

8. Reports on Special Sessions

Evolution of Massive Stars at Low Metallicity 571
  G. Meynet, N. R. Walborn, I. Hunter, C. Martayan, A. J. van Marle, S. Marchenko, J. S. Vink, M. Limongi, E. M. Levesque, & M. Modjaz

Magnetic Massive Stars 577
  R. Townsend, D. H. Cohen, L. Dessart, S. Hubrig, Y. Nazé, V. Petit, A. ud-Doula, & N. R. Walborn

Author index 587

Preface

Within the past few years, great progress has been made towards our understanding of the astrophysical role played by massive stars. From an observational perspective, temperatures of OB stars have been revised downward based on the most recent observations with FUSE, HST and ground-based facilities; the role of clumping in stellar winds has been recognized, with potentially dramatic consequences for stellar evolution, due to its influence on derived mass-loss rates; close binaries with masses of up to 80 solar masses have been identified and studied visually and with exquisite detail using Chandra, XMM, VLA; visibly obscured young massive clusters have been identified at our Galactic Centre, elsewhere in our own Milky Way and in external galaxies. These have been studied with HST, VLT, Gemini and Subaru, exploiting natural guide star Adaptive Optics (AO) techniques from the ground. Increasingly the use of AO with laser guide stars is expected to revolutionise the study of massive star forming regions.

Quantitative spectroscopy of massive stars beyond the Local Group has been undertaken with VLT and Keck to disentangle chemical evolution of galaxies in the nearby Universe and to determine independent distances; star formation histories have been inferred from population/spectrum synthesis of resolved/unresolved populations of nearby star forming galaxies; nearby starbursts – templates for high redshift counterparts – have been studied with FUSE, HST, GALEX and Spitzer. Large surveys for star forming galaxies from redshifts 1 to 6, making use of colour selection techniques at optical, infrared and sub-mm wavelengths, have provided quantitative measures of their massive stellar populations over most of the age of the universe, including their past history of star formation, the IMF, assembled stellar masses, metallicities and chemical yields; from space, HETE-2 and SWIFT have allowed an increasing number of Gamma Ray Bursts (GRBs) to be studied in detail, with rapid follow-up from ground-based facilities permitting chemical information on their host galaxies to be obtained. These are all tremendously exciting topics, at the forefront of present-day astrophysical research and providing some of the core scientific cases for the next generation of extremely large telescopes currently under development.

Theoretically, great advances have been made towards improved evolutionary and atmospheric models for massive stars allowing for rotation and magnetic fields, and towards the evolution of massive binary systems; the impact of internal waves generated at the boundary of the convective core on the transport of angular momentum and chemical species in the stellar interior; important developments have taken place with respect to spectral synthesis of starbursts, improved spectral energy distributions of young stellar populations, hydrodynamic simulations of GRB explosions, and notably numerical simulations of star formation at the earliest epochs, including very massive Population III stars which are thought to play the dominant role in the reionization of the universe at redshift $z > 6$.

The present 'beach' Symposium follows in a long line of successful meetings, held between 1971 (IAU Symposium 49, Argentina) and 2002 (IAU Symposium 212, Spain). IAU Symposium 250 was held at the Grant Hyatt hotel in Kauai between 10–14 December 2007, where once again the massive star community has gathered along sun-drenched beaches to discuss the state of the art in an exciting field of astronomical research. The participants have focused on how massive stars shape the Universe, from our immediate neighborhood to high redshift galaxies and the first generation of stars. Despite their rarity, these 'cosmic engines' are most effective in drawing our attention, with explosions, bold winds, and bolometric extravagances, from all corners of the observable Universe.

These proceedings collect summaries of 24 invited review talks and 33 oral contributions. Two special sessions, on 'massive stars at low metallicity', and on 'magnetic massive stars' were organized on December 9, by Georges Meynet and Richard Townsend, respectively. We have collected here their reports on these special sessions, as well as the abstracts for the approximately 130 posters presented at the Symposium. One poster from each session was selected by SOC members for a brief oral presentation, namely André-Nicolas Chené, Conny Aerts, Jose Prieto, Casey Law and Maryan Modjaz, with extended reports from these incorporated in the proceedings at the end of each scientific session. The overall poster competition winner was Maryam Modjaz, selected by SOC co-chairs. The photographs in this book have been taken by Karen Teramura, Véronique Petit, Jesús Maíz Apellániz, Joachim Puls and Fabio Bresolin. The Hawaiian Island chain evolutionary concept in this page was provided by Tony Moffat. A well-attended reception took place on the evening of Sunday 9 December, while an open-air banquet (luau) took place on the evening of Thursday 13 December.

Paul A. Crowther and Joachim Puls, co-chairs SOC
Fabio Bresolin, chair LOC
Sheffield, Munich, Honolulu, March 1, 2008

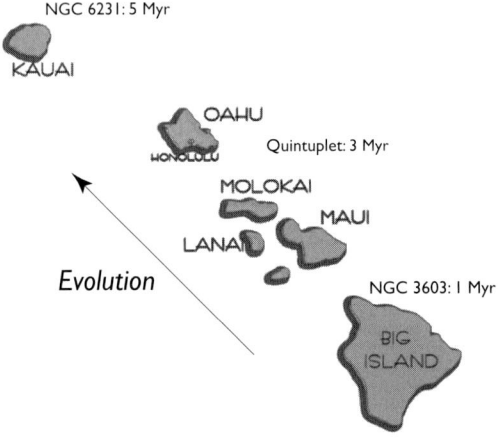

THE ORGANIZING COMMITTEE

Scientific

P.A. Crowther (co-chair, UK)
J. Fynbo (Denmark)
T. Heckman (USA)
G. Koenigsberger (México)
N. Langer (The Netherlands)
F. Matteucci (Italy)
A.F.J. Moffat (Canada)
M. Pettini (UK)

M. Dopita (Australia)
E. Grebel (Germany)
D. Hunter (USA)
R.-P. Kudritzki (USA)
A. MacFadyen (USA)
G. Meynet (Switzerland)
K. Nomoto (Japan)
J. Puls (co-chair, Germany)

Local

F. Bresolin (chair)
L. Good
A. Miyashiro
K. Toyama
M.A. Urbaneja
C.G. Wynn-Williams

L. Clark
E. Levesque
K. Teramura
V. U
K. Uyehara

Acknowledgements

The symposium has been sponsored and supported by the IAU Divisions IV (Stars), VI (Interstellar Matter), VII (Galactic System), VIII (Galaxies) and XI (Space and High Energy Astrophysics); and by the IAU Commissions No. 28 (Galaxies), No. 29 (Stellar Spectra), No. 34 (Interstellar Matter), No. 35 (Stellar Constitution), No. 36 (Theory of Stellar Atmospheres), No. 37 (Star Clusters and Associations) and No. 44 (Space and High Energy Astrophysics).

Financial support was provided by the International Astronomical Union, providing travel grants for 31 participants, consisting of speakers, PhD students and scientists from less favoured countries. In addition, we appreciate a generous contribution by R. Kudritzki, the director of the Institute for Astronomy. The Local Organizing Committee operated under the auspices of the Institute for Astronomy of the University of Hawaii.

Conference photograph

Participants

Conny **Aerts**, Institute for Astronomy, Leuven University, Belgium	conny@ster.kuleuven.be
Nurdan **Anilmis**, University of Delaware, USA	nurdan@udel.edu
Jane **Arthur**, Centro de Radioastronomía y Astrofísica, UNAM, Morelia, Mexico	j.arthur@astrosmo.unam.mx
Matthew **Austin**, University College London, UK	mja@star.ucl.ac.uk
Cassio **Barbosa**, IP&D, Universidade do Vale do Paraíba, Brazil	cassio@univap.br
Sara **Beck**, Tel Aviv University, Israel	sara@wise.tau.ac.il
Joanne **Bibby**, University of Sheffield, UK	j.bibby@sheffield.ac.uk
Thomas **Bisbas**, Cardiff University, UK	spxtb@astro.cf.ac.uk
Jon **Bjorkman**, University of Toledo, USA	Jon.Bjorkman@utoledo.edu
Karen **Bjorkman**, University of Toledo, USA	Karen.Bjorkman@utoledo.edu
Ronny **Blomme**, Royal Observatory of Belgium, Belgium	Ronny.Blomme@oma.be
Ann **Boesgaard**, Institute for Astronomy, University of Hawaii, USA	boes@ifa.hawaii.edu
Alceste **Bonanos**, Carnegie Institution of Washington, USA	bonanos@dtm.ciw.edu
Jean-Claude **Bouret**, Laboratoire d'Astrophysique de Marseille, France	jean-claude.bouret@oamp.fr
Fabio **Bresolin**, Institute for Astronomy, University of Hawaii, USA	bresolin@ifa.hawaii.edu
Volker **Bromm**, University of Texas, USA	vbromm@astro.as.utexas.edu
Kate **Brooks**, Australia Telescope National Facility, Australia	Kate.Brooks@csiro.au
Geoffrey **Burbidge**, CASS, University of California, San Diego, USA	gburbidge@ucsd.edu
Adam **Burrows**, Steward Observatory, University of Arizona, USA	burrows@as.arizona.edu
Francesco **Calura**, Osservatorio Astronomico di Trieste, Italy	fcalura@oats.inaf.it
Matteo **Cantiello**, Astronomical Institute, Utrecht University, The Netherlands	M.Cantiello@astro.uu.nl
Norberto **Castro Rodriguez**, Instituto de Astrofísica de Canarias, Spain	norberto@iac.es
Pascal **Chardonnet**, Université de Savoie, France	chardonnet@lapp.in2p3.fr
Jane **Charlton**, Pennsylvania State University, USA	charlton@astro.psu.edu
André-Nicolas **Chené**, CGO, Herzberg Institute of Astrophysics, Canada	andre-nicolas.chene@nrc-cnrc.gc.ca
Sabina **Chita**, Astronomical Institute, Utrecht University, The Netherlands	s.chita@astro.uu.nl
Minho **Choi**, Korea Astronomy and Space Science Institute, Korea	minho@kasi.re.kr
You-Hua **Chu**, University of Illinois, USA	chu@astro.uiuc.edu
Candace **Church Joggerst**, University of California, Santa Cruz, USA	cchurch@ucolick.org
Laurie **Clark**, Institute for Astronomy, University of Hawaii/NRAO, USA	laurieclark5@comcast.net
David **Cohen**, Swarthmore College, USA	cohen@astro.swarthmore.edu
Peter **Conti**, JILA, University of Colorado, USA	pconti@jila.colorado.edu
Jeff **Cooke**, University of California, Irvine, USA	cooke@uci.edu
Jackie **Cooper**, RSAA, The Australian National University, Australia	jcooper@mso.anu.edu.au
Michael **Corcoran**, CRESST/USRA/GSFC, USA	Michael.F.Corcoran@nasa.gov
Steven **Cranmer**, Harvard-Smithsonian Center for Astrophysics, USA	scranmer@cfa.harvard.edu
Paul **Crowther**, University of Sheffield, UK	Paul.crowther@sheffield.ac.uk
Michel **Curé**, Universidad de Valparaíso, Chile	michel.cure@uv.cl
Augusto **Damineli**, IAGUSP, Universidade de São Paulo, Brazil	damineli@astro.iag.usp.br
Kris **Davidson**, University of Minnesota, USA	kd@aps.umn.edu
Ben **Davies**, Rochester Institute of Technology, USA	davies@cis.rit.edu
Michael **De Becker**, Université de Liège, Belgium	debecker@astro.ulg.ac.be
Alex **de Koter**, University of Amsterdam, The Netherlands	dekoter@science.uva.nl
Antoine **de la Chevrotière**, Université de Montréal, Canada	antoine@astro.umontreal.ca
Selma **de Mink**, Astronomical Institute, Utrecht University, The Netherlands	mink@astro.uu.nl
Massimo **Della Valle**, Osservatorio Astronomico di Arcetri, Italy	massimo@arcetri.astro.it
Luc **Dessart**, Steward Observatory, University of Arizona, USA	luc@as.arizona.edu
Miroslava **Dessauges**, Geneva Observatory, Switzerland	miroslava.dessauges@obs.unige.ch
Sandro **D'Odorico**, European Southern Observatory, Germany	sdodoric@eso.org
Michael **Dopita**, RSAA, The Australian National University, Australia	Michael.Dopita@anu.edu.au
Laurent **Drissen**, Université Laval, Canada	ldrissen@phy.ulaval.ca
Philippe **Eenens**, Universidad de Guanajuato, Mexico	eenens@gmail.com
Sylvia **Ekström**, Geneva Observatory, Switzerland	sylvia.ekstrom@obs.unige.ch
John **Eldridge**, Institute of Astronomy, Cambridge, UK	jje@ast.cam.ac.uk
Cesar **Esteban**, Instituto de Astrofísica de Canarias, Spain	cel@iac.es
Chris **Evans**, Royal Observatory Edinburgh, UK	cje@roe.ac.uk
Remi **Fahed**, Université de Montréal, Canada	fahed@astro.umontreal.ca
Robert **Fesen**, Dartmouth College, USA	fesen@snr.dartmouth.edu
Donald **Figer**, Rochester Institute of Technology, USA	figer@cis.rit.edu
Carla **Fröhlich**, Enrico Fermi Institute, University of Chicago, USA	frohlich@uchicago.edu
Alexander **Fullerton**, Space Telescope Science Institute, USA	fullerton@stsci.edu
Johan **Fynbo**, Dark Cosmology Centre, Copenhagen University, Denmark	jfynbo@astro.ku.dk
Bryan **Gaensler**, The University of Sydney, Australia	bgaensler@usyd.edu.au
Christa **Gall**, Dark Cosmology Centre, Copenhagen University, Denmark	christa@astro.ku.dk
John **Gallagher III**, University of Wisconsin-Madison, USA	jsg@astro.wisc.edu
Miriam **Garcia**, Instituto de Astrofísica de Canarias, Spain	mgg@iac.es
Kenneth **Gayley**, University of Iowa, USA	kenneth-gayley@uiowa.edu
Mélanie **Godart**, Université de Liège, Belgium	melanie.godart@ulg.ac.be
Götz **Gräfener**, Universität Potsdam, Germany	goetz@astro.physik.uni-potsdam.de
Eva **Grebel**, University of Heidelberg, Germany	grebel@ari.uni-heidelberg.de
Jose **Groh**, Max-Planck Institute for Radio Astronomy, Germany	jgroh@mpifr-bonn.mpg.de
Theodore **Gull**, NASA Goddard Space Flight Center, USA	gull@milkyway.gsfc.nasa.gov
Lucy **Hadfield**, Rochester Institute of Technology, USA	hadfield@cis.rit.edu
Margaret **Hanson**, University of Cincinnati, USA	margaret.hanson@uc.edu
Peter **Hargrave**, Cardiff University, UK	p.hargrave@astro.cf.ac.uk
Takashi **Hattori**, Subaru Telescope, National Astronomical Observatory of Japan	hattori@subaru.naoj.org
Vincent **Hénault-Brunet**, Université de Montréal, Canada	henault@astro.umontreal.ca
Artemio **Herrero**, Instituto de Astrofísica de Canarias, Spain	ahd@iac.es
Grant **Hill**, W. M. Keck Observatory, USA	ghill@keck.hawaii.edu
John **Hillier**, University of Pittsburgh, USA	hillier@pitt.edu
Raphael **Hirschi**, Keele University, UK	r.hirschi@epsam.keele.ac.uk
Jennifer **Hoffman**, University of Denver, USA	jennifer.hoffman@du.edu
Ian **Howarth**, University College London, UK	idh@star.ucl.ac.uk
Swetlana **Hubrig**, European Southern Observatory, Chile	shubrig@eso.org
Roberta **Humphreys**, University of Minnesota, USA	roberta@umn.edu
Deidre **Hunter**, Lowell Observatory, USA	dah@lowell.edu
Ian **Hunter**, Queen's University Belfast, UK	i.hunter@qub.ac.uk
Rosina **Iping**, NASA Goddard Space Flight Center/Catholic University of America, USA	Rosina.C.Iping@nasa.gov
Eric **Josselin**, GRAAL, Université de Montpellier, France	josselin@graal.univ-montp2.fr
Vicky **Kalogera**, Northwestern University, USA	vicky@northwestern.edu
Alexander **Kashlinsky**, NASA Goddard Space Flight Center, USA	kashlinsky@milkyway.gsfc.nasa.gov
Daniel **Kiminki**, University of Wyoming, USA	nawade@uwyo.edu
Chip **Kobulnicky**, University of Wyoming, USA	chipk@uwyo.edu
Gloria **Koenigsberger**, Instituto de Ciencias Físicas, UNAM, Cuernavaca, Mexico	gloria@fis.unam.mx
Rubina **Kotak**, Queen's University Belfast, UK	r.kotak@qub.ac.uk
Chryssa **Kouveliotou**, NASA Marshall Space Flight Center, USA	chryssa.kouveliotou@nasa.gov

Chael **Kruip**, Sterrewacht Leiden, The Netherlands	kruip@strw.leidenuniv.nl
Rolf **Kudritzki**, Institute for Astronomy, University of Hawaii, USA	kud@ifa.hawaii.edu
Cornelia **Lang**, University of Iowa, USA	cornelia-lang@uiowa.edu
Norbert **Langer**, Astronomical Institute, Utrecht University, The Netherlands	n.langer@astro.uu.nl
Casey **Law**, University of Amsterdam, The Netherlands	claw@science.uva.nl
Claus **Leitherer**, Space Telescope Science Institute, USA	leitherer@stsci.edu
Maurice **Leutenegger**, NASA Goddard Space Flight Center, USA	maurice@astro.columbia.edu
Emily **Levesque**, Institute for Astronomy, University of Hawaii, USA	emsque@ifa.hawaii.edu
Adriane **Liermann**, Universität Potsdam, Germany	adriane@astro.physik.uni-potsdam.de
Marco **Limongi**, Osservatorio Astronomico di Roma, Italy	marco@oa-roma.inaf.it
Cesar **Lopez Solano**, Sieltec Canarias S.L., Spain	cesarls@sieltec.es
Siegfried **Luehrs**, Former University of Muenster, Germany	siegfriedluehrs@compuserve.de
Andrew **MacFadyen**, New York University, USA	macfadyen@nyu.edu
Thomas **Madura**, University of Delaware, USA	tmadura@udel.edu
André **Maeder**, Geneva Observatory, Switzerland	andre.maeder@obs.unige.ch
Jesús **Maíz Apellániz**, Instituto de Astrofísica de Andalucía, CSIC, Spain	jmaiz@iaa.es
Daniele **Malesani**, Dark Cosmology Centre, Copenhagen University, Denmark	malesani@astro.ku.dk
Sergey **Marchenko**, Western Kentucky University, USA	sergey.marchenko@wku.edu
Amparo **Marco**, Universidad de Alicante, Spain	tobarra@dfists.ua.es
Christophe **Martayan**, Royal Observatory of Belgium/Observatoire de Paris, France	Martayan@oma.be
Fabrice **Martins**, GRAAL, Université de Montpellier, France	martins@graal.univ-montp2.fr
Philip **Massey**, Lowell Observatory, USA	Phil.Massey@lowell.edu
Francesca **Matteucci**, Università di Trieste, Italy	matteucci@ts.astro.it
Maria **Messineo**, Rochester Institute of Technology, USA	messineo@cis.rit.edu
Georges **Meynet**, Geneva Observatory, Switzerland	georges.meynet@obs.unige.ch
Dan **Milisavljevic**, Dartmouth College, USA	danmil@dartmouth.edu
Maryam **Modjaz**, University of California, Berkeley, USA	mmodjaz@astro.berkeley.edu
Anthony **Moffat**, Université de Montréal, Canada	moffat@astro.umontreal.ca
Sarah **Moll**, University of Sheffield, UK	s.moll@shef.ac.uk
Gabriela **Montes**, Instituto de Astrofísica de Andalucía, CSIC, Spain	g.montes@astrosmo.unam.mx
Pat **Morris**, NASA Herschel Science Center, IPAC, Caltech, USA	pmorris@ipac.caltech.edu
Jeremiah **Murphy**, Steward Observatory, University of Arizona, USA	jmurphy@as.arizona.edu
Francisco **Najarro**, DAMIR, Instituto de Estructura de la Materia, Spain	najarro@damir.iem.csic.es
Yaël **Nazé**, Université de Liège, Belgium	naze@astro.ulg.ac.be
Ignacio **Negueruela**, Universidad de Alicante, Spain	ignacio@dfists.ua.es
Maria **Nieva**, Dr. Reimes Observatory Bamberg, Germany	nieva@sternwarte.uni-erlangen.de
Ken'ichi **Nomoto**, University of Tokyo, Japan	nomoto@astron.s.u-tokyo.ac.jp
Dieter **Nürnberger**, European Southern Observatory, Chile	dnuernbe@eso.org
Sally **Oey**, University of Michigan, USA	msoey@umich.edu
Atsuo **Okazaki**, Hokkai-Gakuen University, Japan	okazaki@elsa.hokkai-s-u.ac.jp
Mary **Oksala**, University of Delaware, USA	meo@udel.edu
Andrew **Onifer**, Los Alamos National Laboratory, USA	aonifer@lanl.gov
Stanley **Owocki**, University of Delaware, USA	owocki@bartol.udel.edu
Elliot **Parkin**, University of Leeds, UK	phy1erp@leeds.ac.uk
Véronique **Petit**, Université Laval, Canada	veronique.petit.1@ulaval.ca
Max **Pettini**, Institute of Astronomy, Cambridge, UK	pettini@ast.cam.ac.uk
Julian **Pittard**, University of Leeds, UK	jmp@ast.leeds.ac.uk
Jose **Prieto**, Ohio State University, USA	prieto@astronomy.ohio-state.edu
Norbert **Przybilla**, Dr. Remeis Observatory Bamberg, Germany	przybilla@sternwarte.uni-erlangen.de
Joachim **Puls**, University Observatory Munich, Germany	uh101aw@usm.uni-muenchen.de
Frederic **Rasio**, Northwestern University, USA	rasio@northwestern.edu
Gregor **Rauw**, Université de Liège, Belgium	rauw@astro.ulg.ac.be
Jeonghee **Rho**, Spitzer Science Center, Caltech, USA	rho@ipac.caltech.edu
Ricardo **Rizzo**, Laboratorio de Astrofísica Espacial y Física Fundamental, Spain	ricardo.rizzo@laeff.inta.es
Margarita **Rosado**, Instituto de Astronomia, UNAM, Mexico	margarit@astroscu.unam.mx
Christopher **Russell**, University of Delaware, USA	crussell@udel.edu
Hugues **Sana**, European Southern Observatory, Chile	hsana@eso.org
Florian **Schiller**, Dr. Remeis Observatory Bamberg, Germany	schiller@sternwarte.uni-erlangen.de
Olivier **Schnurr**, University of Sheffield, UK	o.schnurr@sheffield.ac.uk
Markus **Schoeller**, European Southern Observatory, Chile	mschoell@eso.org
Sergio **Simón-Díaz**, Geneva Observatory, Switzerland	sergio.simon-diaz@obs.unige.ch
Stephen **Skinner**, University of Colorado, USA	skinners@casa.colorado.edu
Stephen **Smartt**, Queen's University Belfast, UK	s.smartt@qub.ac.uk
Linda **Smith**, Space Telescope Science Institute, USA	lsmith@stsci.edu
Lindsey **Smith**, Sydney University, Australia	l.smith@physics.usyd.edu.au
Nathan **Smith**, University of California, Berkeley, USA	nathans@astro.berkeley.edu
George **Sonneborn**, NASA Goddard Space Flight Center, USA	george.sonneborn@nasa.gov
Alfredo **Sota**, Universidad Autónoma de Madrid, Spain	alfredo.sota@gmail.com
Krzysztof **Stanek**, Ohio State University, USA	kstanek@astronomy.ohio-state.edu
Robert **Stencel**, Denver University, USA	rstencel@du.edu
Nicole **St-Louis**, Université de Montréal, Canada	stlouis@astro.umontreal.ca
Susan **Stolovy**, Spitzer Science Center, Caltech, USA	stolovy@ipac.caltech.edu
Vanessa **Stroud**, LCOGTN/The Open University, UK	vstroud@lcogt.net
Yoshiaki **Taniguchi**, Ehime University, Japan	tani@cosmos.phys.sci.ehime-u.ac.jp
Karen **Teramura**, Institute for Astronomy, University of Hawaii, USA	teramura@ifa.hawaii.edu
Christina **Thöne**, Dark Cosmology Centre, Copenhagen University, Denmark	cthoene@dark-cosmology.dk
Jose **Torrejon-Vazquez**, Massachusetts Institute of Technology, Boston, USA	torrejon@mit.edu
Richard **Townsend**, University of Delaware, USA	rhdt@bartol.udel.edu
Karen **Toyama**, Institute for Astronomy, University of Hawaii, USA	toyama@ifa.hawaii.edu
Carrie **Trundle**, Queen's University Belfast, UK	c.trundle@qub.ac.uk
Vivian **U**, Institute for Astronomy, University of Hawaii, USA	vivian@ifa.hawaii.edu
Leonardo **Ubeda**, Université Laval, Canada	leonard.ubeda@mac.com
Miguel A. **Urbaneja**, Institute for Astronomy, University of Hawaii, USA	urbaneja@ifa.hawaii.edu
Ana **Ursúa**, Universidad de Alicante, Spain	anaul@dfists.ua.es
Allard Jan **van Marle**, University of Delaware, USA	marle@udel.edu
Dany **Vanbeveren**, Vrije Universiteit Brussel/Association K.U Leuven, Belgium	dvbevere@vub.ac.be
Watson **Varricatt**, Joint Astronomy Centre, USA	w.varricatt@jach.hawaii.edu
Jorick **Vink**, Armagh Observatory, UK	jsv@arm.ac.uk
Nolan **Walborn**, Space Telescope Science Institute, USA	walborn@stsci.edu
Debra **Wallace**, College of Charleston, USA	wallaced@cofc.edu
George **Williams**, MMT Observatory, USA	gwilliams@as.arizona.edu
Stephen **Williams**, Georgia State University, USA	swilliams@chara.gsu.edu
Allan **Willis**, University College London, UK	ajw@star.ucl.ac.uk
John **Wisniewski**, NASA Goddard Space Flight Center, USA	John.P.Wisniewski@nasa.gov
Sung-Chul **Yoon**, University of California, Santa Cruz, USA	scyoon@ucolick.org
Hans **Zinnecker**, Astrophysical Institute Potsdam, Germany	hzinnecker@aip.de
Janos **Zsargo**, University of Pittsburgh, USA	jaz8@pitt.edu

Session I
Atmospheres of Massive Stars

Massive Stars as Cosmic Engines Through the Ages

André Maeder[1], Georges Meynet[1], Sylvia Ekström[1],
Raphael Hirschi[2] and Cyril Georgy[1]

[1]Geneva Observatory, University of Geneva, CH–1290 Sauverny, Switzerland
email: andre.maeder@obs.unige.ch, georges.meynet@obs.unige.ch,
sylvia.ekstrom@obs.unige.ch, cyril.georgy@obs.unige.ch
[2]Astrophysics, EPSAM, University of Keele
email: r.hirschi@epsam.keele.ac.uk

Abstract. Some useful developments in the model physics are briefly presented, followed by model results on chemical enrichments and WR stars. We discuss the expected rotation velocities of WR stars. We emphasize that the (C+O)/He ratio is a better chemical indicator of evolution for WC stars than the C/He ratios. With or without rotation, at a given luminosity the (C+O)/He ratios should be higher in regions of lower metallicity Z. Also, for a given (C+O)/He ratio the WC stars in lower Z regions have higher luminosities. The WO stars, which are likely the progenitors of supernovae SNIc and of some GRBs, should preferentially be found in regions of low Z and be the descendants of very high initial masses. Finally, we emphasize the physical reasons why massive rotating low Z stars may also experience heavy mass loss.

Keywords. stars: early-type – stars: Wolf-Rayet – stars: rotation – stars: mass loss

1. Introduction

It may be worth to quote a few of the important findings which have led to the development of our field. Fifty years ago, Peter Conti *et al.* (1967) found that metal deficient stars have higher O/Fe ratios than the solar ratio. This was the first finding concerning differences of abundance ratios as a function of metallicity, due to a different nucleosynthesis in early stages of galactic evolution. A year after, Lindsey Smith (1968) remarkably found that the various subtypes of WR stars are differently distributed in the Galaxy and that some subtypes are missing in the LMC. The outer galactic regions and the LMC showing the same kind of WC stars (early WC). This was the first evidences of different distributions of massive objects in galaxies. A great discovery from Copernicus and from IUE is the mass loss from O–type stars by Morton (1976, see also Henny Lamers & Donald Morton 1976). The interpretation of WR stars as post-MS stars resulting from mass loss in OB stars was proposed by Peter Conti (1976).

Of course, there are many other big steps which have contributed to our knowledge about massive stars. It would be misleading to believe that these findings were smoothly accepted as such. For at least a decade, there were people claiming that mass loss is an artifact or disputing the status of WR stars, even considering them as pre–MS stars. These controversies, on the whole, contributed to further checks and investigations which eventually resulted in an increased strength of the initial discoveries.

2. Improved Model Physics in Massive Star Evolution

The physics and evolution of massive stars is dominated by mass loss and by rotational mixing. At the origin of both effects, we find the large ratio T/ρ of temperature to density

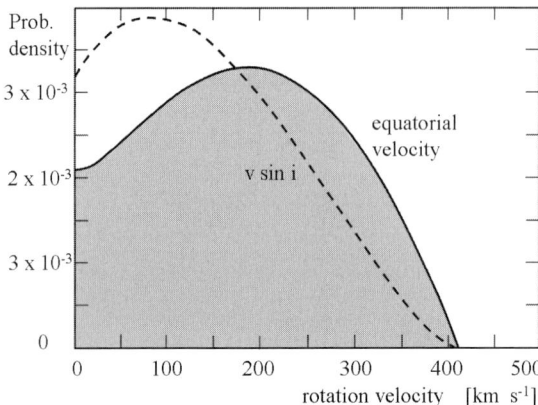

Figure 1. Probability density by km s^{-1} of rotation velocities for 496 stars with types O9.5 to B8. Adapted from Huang and Gies (2006a)

in massive stars. This enhances the ratio of radiation to gas pressure, which goes like

$$\frac{P_{\rm rad}}{P_{\rm gas}} \sim \frac{T^3}{\rho}. \tag{2.1}$$

The high T/ρ favors mass loss by stellar winds. A large fraction of OB stars have high rotational velocities (Fig. 1). A high T/ρ also enhances rotational mixing either by shear diffusion or meridional circulation, since the coefficient of mixing for a vertical shear dv/dz behaves as

$$D_{\rm shear} = 2\,\mathcal{R}i_{\rm crit}\,K\,\frac{(dv/dz)^2}{N_{\rm ad}^2}. \tag{2.2}$$

$\mathcal{R}i_{\rm crit}$ is the critical Reynolds number and $N_{\rm ad}$ is the adiabatic Brunt–Väisälä frequency. The diffusion coefficient scales as the thermal diffusivity $K = 4acT^3/(3C_{\rm P}\,\kappa\,\rho^2)$. Similarly, the velocity of meridional circulation scales as the ratio L/M of the luminosity to mass. Thus, the high T/ϱ favors both mixing and mass loss, the account of both effects brings major revisions of the model results.

For meridional circulation, self–consistent solutions were proposed by Zahn (1992) and Maeder and Zahn (1998). The transport of chemical elements obeys a classical diffusion equation. More critical is the equation for the transport of the angular momentum

$$\frac{\partial}{\partial t}(\varrho r^2 \sin^2\vartheta\,\Omega)_r + \frac{1}{r^2}\frac{\partial}{\partial r}(\varrho r^4 \sin^2\vartheta\,U_r\Omega) + \frac{1}{r\sin\vartheta}\frac{\partial}{\partial\vartheta}(\varrho r^2 \sin^3\vartheta\,U_\vartheta\Omega)$$
$$= \frac{\sin^2\vartheta}{r^2}\frac{\partial}{\partial r}\left(\varrho D_{\rm shear} r^4 \frac{\partial\Omega}{\partial r}\right) + \frac{1}{\sin\vartheta}\frac{\partial}{\partial\vartheta}\left(\varrho D_{\rm h}\sin^3\vartheta\frac{\partial\Omega}{\partial\vartheta}\right). \tag{2.3}$$

It contains both advection terms depending on U_r and U_ϑ the radial and horizontal components of the velocity of meridional circulation and diffusion terms. $D_{\rm h}$ is the diffusion coefficient by the horizontal turbulence. The self–consistent solutions allow us to follow the evolution of the angular velocity $\Omega(r)$ at each level. Many authors ignore the advection terms or represent them by diffusion terms. This is incorrect, since the circulation currents may turn in different ways (Fig. 2 left).

A big question is whether there is a dynamo working in radiative zones, because a magnetic field would have great consequences on the evolution of rotation by exerting

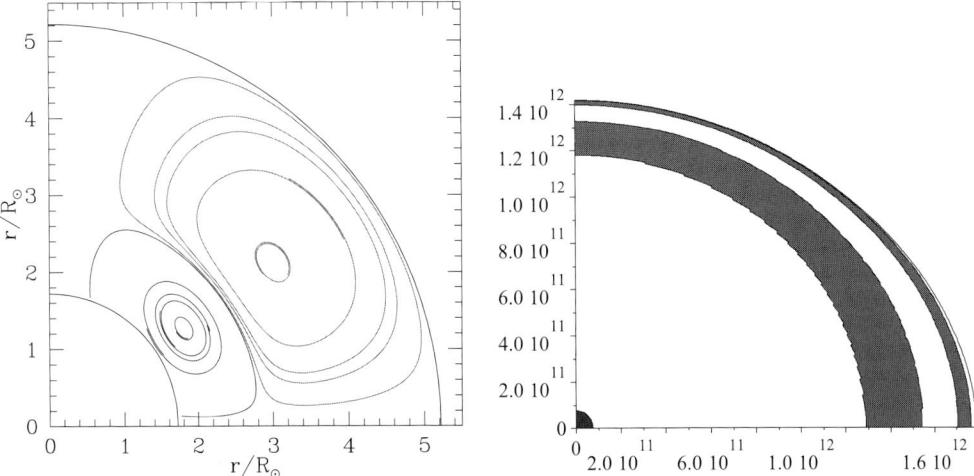

Figure 2. Left: circulation currents in a 20 M$_\odot$ star in the middle of the H–burning phase. The initial rotation velocity is 300 km s^{-1}. The inner loop is raising along the polar axis, while the outer loop, the Gratton–Öpik circulation cell, is going up in the equatorial plane. Right: 2–D representation of the convective zones (dark areas) as a function of radius in a model of 20 M$_\odot$ with $X = 0.70$ and $Z = 0.020$ at the end of MS evolution with $(\Omega/\Omega_{\mathrm{crit}} = 0.94)$.

an efficient torque able to reduce the differential rotation or even to impose uniform rotation. A dynamo can operate in the radiative zone of a differentially rotating star thanks to a magnetic instability as shown by Spruit (2002). Let us consider a star with a shellular rotation law $\Omega(r)$ with an initially weak poloidal magnetic field B_r, so that the magnetic forces are negligible (strong initial field leads to solid body rotation). The radial component is wound up by differential rotation. After a few differential turns, an azimuthal field of component B_φ is present, its strength grows linearly in time and the component B_φ dominates over B_r. At some stage, the field B_φ becomes unstable, due to Tayler's instability which is the first instability encountered. The instabilities mainly have horizontal components, but there is also a small vertical component l_r, limited by the action of buoyancy forces. This small radial component of the field is further wound up by differential rotation, which then amplifies the toroidal component of the field up to a stage where dissipation effects would limit its amplitude. In this way, a strong toroidal field develops together with a limited radial field. The horizontal component enforces shellular rotation, while the vertical field component favors solid body rotation. Numerical models by Maeder and Meynet (2005) show that the field is most effective for transporting angular momentum. The displacement due to the magnetic instability also contribute to enhance the transport of the chemical elements.

In addition to the effects of metallicity on the mass loss rates (see contributions by Vink and Crowther, this volume), the interactions of rotation and stellar winds have many consequences. – Rotation introduces large anisotropies in the stellar winds, the polar regions being hotter than the equatorial ones. – The global mass loss rates are increased by rotation. – The anisotropies of the stellar winds allow a star with strong polar winds to lose lots of mass without losing too much angular momentum. At the opposite, equatorial mass loss removes a lot of angular momentum.

Let us consider a rotating star with angular velocity Ω and a non–rotating star of the same mass M at the same location in the HR diagram. The ratio of their mass loss rates can be written, see Maeder (1999),

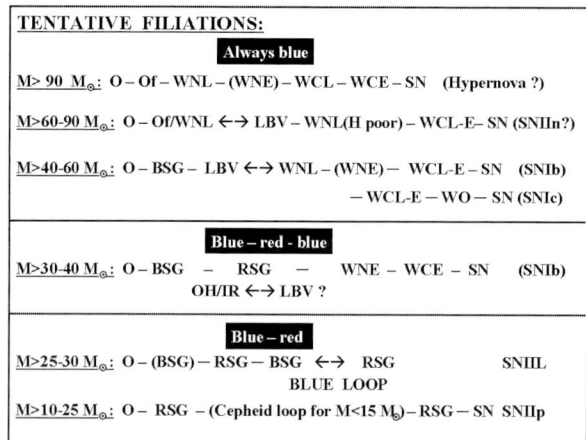

Figure 3. The filiations between Pop. I massive stars for different mass domains.

$$\frac{\dot{M}(\Omega)}{\dot{M}(0)} \approx \frac{(1-\Gamma)^{\frac{1}{\alpha}-1}}{\left[1 - \frac{4}{9}\left(\frac{v}{v_{\text{crit},1}}\right)^2 - \Gamma\right]^{\frac{1}{\alpha}-1}} \ . \qquad (2.4)$$

The ratio $v/v_{\text{crit},1}$ is the ratio of the rotational velocity v to the critical velocity. If $\Omega = 0$, $v/v_{\text{crit},1}$ is equal to 1. For a star with a small Eddington factor Γ, we can neglect Γ with respect to unity. This equation shows that the effects of rotation on the $\dot{M}$ rates remain moderate in general. However, for stars close to the Eddington limit, rotation may drastically increase the mass loss rates, in particularly for low values of the force multiplier α, i.e. for $\log T_{\text{eff}} \leqslant 4.30$. In cases where $\Gamma > 0.639$, a moderate rotation may make the denominator of (2.4) to vanish, indicating large mass loss and instability.

Massive O stars have a small external convective envelope due to their high luminosity. Rotation amplifies these external convective regions (Fig. 2 right). This occurs despite the inhibiting effect of the Solberg–Hoiland criterion, because another more important effect is present in envelopes: the rotational increase of the radiative gradient ∇_{rad} as shown by Maeder *et al.* (2008). These convective zones are likely to play a large role in the origin of the clumping of stellar winds. The matter accelerated in the wind continuously crosses the convective zones in a dynamical process. Convection in the outer layers of O–type stars generates acoustic waves with periods of several hours to a few days. These waves propagate and are amplified in the winds, which have a lower density.

Most remarkably several of these developments lead to the result that the first stars at very low Z have a very different behavior from the present day massive star evolution.

3. Evolution with Mass Loss and Rotation

3.1. *Filiations*

Fig. 3 indicates the possible filiations of massive stars of Pop. I, which can be established from the continuity in the evolution of the chemical abundances, as well as from their properties in star clusters. Globally, one has three main cases.

– For $M > (60-40) M_\odot$: the high mass loss rates remove enough mass so that stars lose their envelopes on the MS or in the blue supergiant stage as LBV. The stars never become red supergiants.

– For about 40 to 30 $M_\odot$: the stars only loose a fraction of their envelopes on the MS. They further evolve to the red supergiant stage, where mass loss is sufficient to remove their envelope, they become bare cores and are observed as WR stars.

– Below about 25 to 30 $M_\odot$: the stars still experience mass loss, however it is not sufficient to alter the global evolution. The mass loss and rotation may nevertheless still modify the lifetimes and the chemical compositions.

The mass limits are uncertain and depend on metallicity Z. At different Z, some sequences may be absent. The last indicated stage before supernovae (SN) are usually reached near the end of central He burning. After this stage, the stellar envelopes do not further evolve and their properties determine the nature of the SN progenitors.

Table 1. The largest [N/H] values observed for different types of stars in the Galaxy, LMC and SMC. The average is equal to about the half of the indicated values. See text.

Types of stars	[N/H] in Galaxy	[N/H] in LMC	[N/H] in SMC
O stars	0.8 - 1.0	–	1.5 - 1.7
B dwarfs $M < 20\ M_\odot$	0.5	0.7 - 0.9	1.1
B giants, supg. $M < 20\ M_\odot$	–	1.1 - 1.2	1.5
B giants, supg. $M > 20\ M_\odot$	0.5 - 0.7	1.3	1.9

3.2. Chemical Abundances in OB Stars and Supergiants

The chemical abundances offer tests of stellar physics and evolution. Mass loss, mixing and mass exchange in binaries affect surface compositions. The removal of the outer layers by stellar winds reveal the inner layers with a composition modified by the nuclear reactions in a beautiful illustration of the effects of the CNO cycles and He–burning reactions. Simultaneously, the internal mixing modifies the surface abundances. Without mixing, there would be no nitrogen enrichment during the MS phase for stars with $M < 60\ M_\odot$. It is only above this value that mass loss can make the products of the CNO cycle to appear at the stellar surface.

The amplitudes of the enrichments of N/H or N/C at the end of the MS phase in massive stars is a reference point telling us the importance of mixing. The main observations in the Galaxy ($Z \approx 0.02$), in the LMC ($Z \approx 0.008$) and in the SMC ($Z \approx 0.004$) at different Z are summarized in Table 1, based on Herrero (2003), Heap *et al.* (2006), Hunter *et al.* (2007), Trundle *et al.* (2007). We notice several facts:

• The N enrichments increase with mass and evolution.

• The N enrichments are larger at lower Z.

• Close to the ZAMS there are both stars with and without N enrichments.

• Away from the ZAMS, but still in the Main Sequence, the N enrichments are larger and they are even larger in the supergiant stages. In B stars, the He excesses are larger as evolution proceeds on the MS and the excesses are greater among the faster rotators as shown by Huang & Gies (2006a, 2006b).

The best credit should be given to the sets of data, where the authors carefully distinguish the mass domains and do not mix in a single plot stars of very different masses. Also, if $\log g$ is taken as an indicator of evolution, the rotational effect on the gravity should be accounted for. Binaries where effects of tidal interactions and mass loss enhancements are possible should be separated from single stars. Finally, error bars should be indicated. Unfortunately, the non–respect of such wise prescriptions is often giving some confusing results.

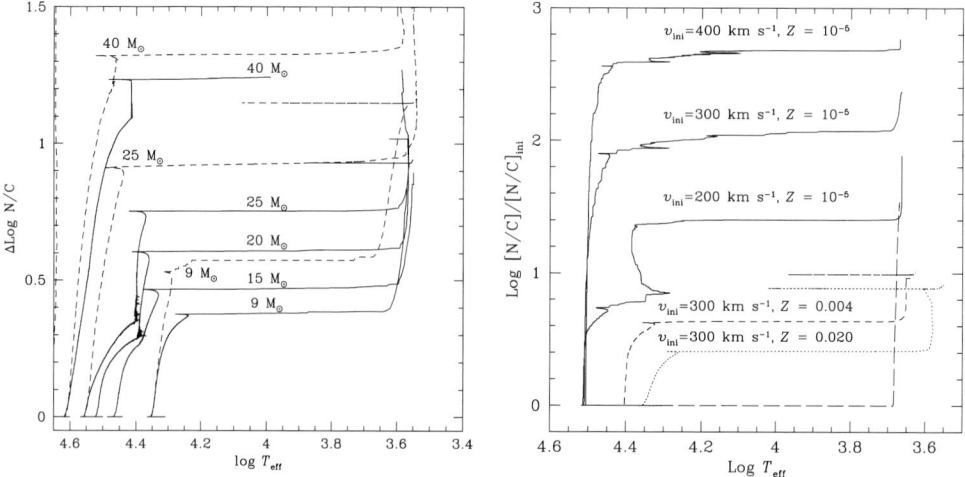

Figure 4. Left: evolution tracks in the plot $\Delta \log(N/C)$ (change with respect to the initial N/C ratio) vs. $\log T_{\mathrm{eff}}$ for various initial masses with $Z = 0.02$ with initial velocities 300 km s^{-1} (continuous lines). The dashed lines show the tracks with different model assumptions (Meynet & Maeder 2000). Right: excesses N/C in log scale for a 9 M$_\odot$ star at different metallicities and rotation velocities. The long–dashed line, at the bottom, corresponds to a non–rotating 9 M$_\odot$ stellar model at $Z = 10^{-5}$. From Meynet & Maeder (2002, 2003).

Fig. 4 (left) shows the predicted changes of the $\log(N/C)$ ratios with respect to the initial ratio. Without rotational mixing (dotted lines), there would be no enrichment until the red supergiant stage. Rotation rapidly increases the N/C ratios on the Main Sequence, with a level depending on the velocities. This results from the steep Ω gradients which produce shear diffusion. The predicted N enrichments are in general agreement with the observed effects. The models consistently predict larger N enrichments with increasing stellar masses and more advanced evolutionary stages. If the stars experience blue loops, they have on the blue side of the HR diagram high N/C ratios typical of red supergiants. Thus, B supergiants at the same location in the HR diagram and with the same rotation may have very different enrichments. The $v \sin i$ converge toward low values during the red phase whatever their initial velocities, thus in the yellow and red phases stars of almost identical $v \sin i$ may exhibit different N/C enrichments.

Fig. 4 (right) shows the N/C ratios in models of rotating stars with 9 M$_\odot$ for $Z = 0.02$, 0.004 and 10^{-5}. At $Z = 10^{-5}$ for the 9 M$_\odot$ model and other masses, there is a large N/C increase by one to two orders of magnitude as shown by Meynet & Maeder (2002). These very large enhancements originate from the very steep Ω gradients in rotating stars at low Z, which drive a strong turbulent shear diffusion. Consistently with observations in Table 1, the lower Z models show larger N enhancements. However, the observations in the SMC show larger N/C ratios than the $Z = 0.004$ model of Fig. 4 right and are in better agreement with relatively lower Z models.

4. Rotation and Chemistry of WR Stars

4.1. *Rotation of WR Stars*

Little is known on the rotation of WR stars. Some information has been recently obtained from the co–rotating regions generating periodic variations in spectral lines (see Chené & St-Louis, this volume). The velocities are typically lower than about 50 km s^{-1} in very

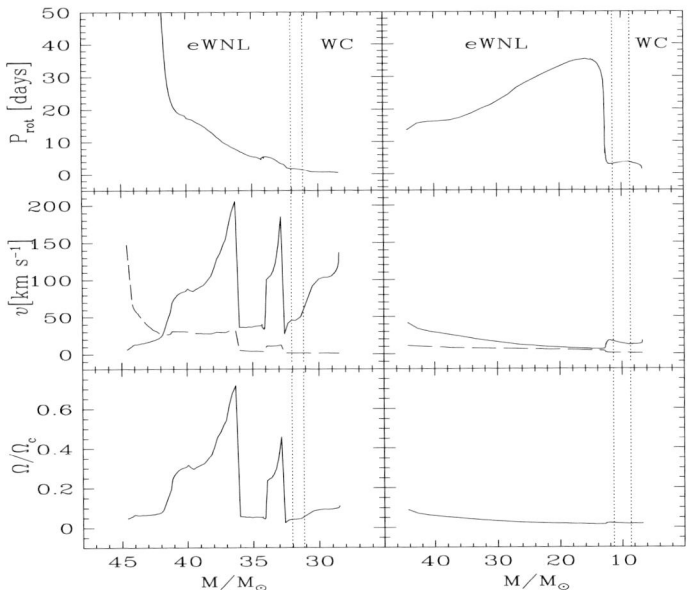

Figure 5. Evolution as a function of the actual mass of the rotation period, of the surface equatorial velocity and of the ratio of the angular velocity to the critical value during the WR stage of rotating stars. The long–dashed lines in the panels for the velocities show the evolution of the radius in solar units. Left: the WR phase of a star with an initial mass of 60 $M_\odot$ with $v_{\rm ini} = 300$ km s^{-1} and $Z = 0.004$. Right: for an initial mass of 60 $M_\odot$ with $v_{\rm ini} = 300$ km s^{-1} and $Z = 0.040$. From Meynet & Maeder (2005).

good agreement with the model predictions. Fig. 5 shows the evolution during the WR stages of the rotation periods $P = (2\pi/\Omega)$, of the rotation velocities v at the equator and of the fractions Ω/Ω_c of the angular velocity to the critical angular velocity at the surface of star models with an initial mass of 60 $M_\odot$ and $v_{\rm ini} = 300$ km s^{-1} at $Z = 0.004$ and 0.040. The evolution in the WR stages is fast and the transfer of angular momentum by meridional circulation is small, thus at this stage the evolution of rotation is dominated by the local conservation of angular momentum unless there is a magnetic field. The variations of v and Ω/Ω_c may nevertheless be fast due to the rapid changes of radius in particular when the star loses its last H layers which makes an opacity decrease. The changes of periods are smoother because v and R both decrease at the same pace.

At solar or higher Z, the expected velocities v are small with $v < 50$ km s^{-1}. The reason is that a large part of the WNL phase occurs during the core H–burning phase, where the high mass loss has time to pump the whole internal angular momentum, so that when the star contracts to the WC stage there is almost no rotation left.

At lower Z, the velocities of WR stars are predicted to be higher, e.g. between 30 and 200 km s^{-1} at $Z = 0.004$. The variations of v and Ω/Ω_c are also greater and more rapid when the radius is changing and the break–up limit might be encountered. The reason is that, at low Z, the WR stage is not entered during the H–burning phase. Despite mass loss, the inner rotation is not killed due to the lack of time.

4.2. WR Star Chemistry

Late WN stars (WNL) generally have H present, with an average value at the surface $X_{\rm s} \approx 0.15$, while early WN stars (WNE) have no H left (see Crowther 2007). In the Galaxy, some WNL stars with weak emission lines have $X_{\rm s} \approx 0.50$. Other abundance

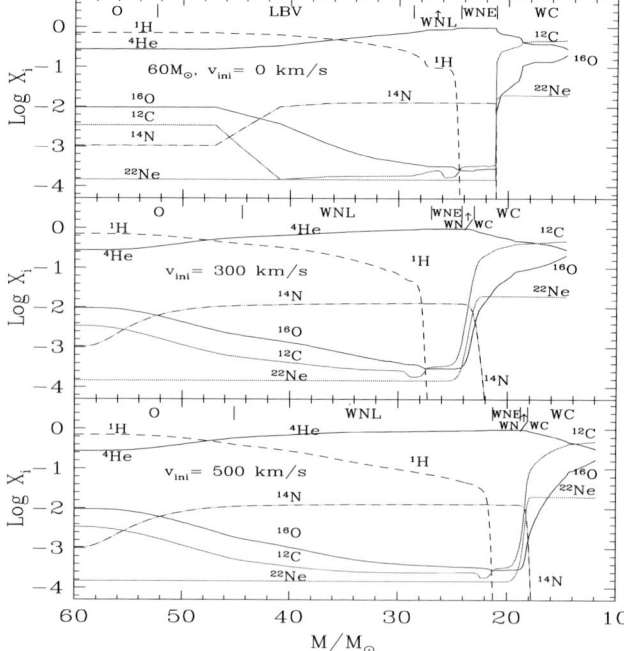

Figure 6. Evolution of the surface abundance for 60 M$_\odot$ models at $Z = 0.02$ for 3 values of the initial velocities. From Meynet & Maeder (2003).

ratios in mass fraction are typically N/He= $(0.035 - 1.4) \times 10^{-2}$, C/He= $(0.21 - 8) \times 10^{-4}$ and C/N= $(0.6 - 6) \times 10^{-2}$. These values are very different from the cosmic values in agreement with model predictions (Fig. 6). The WN abundances are the values of the CNO cycle at equilibrium, they are independent of rotation and are a test of the nuclear cross–sections. At the transition from WN to WC, the rotating models permit the simultaneous presence of ^{14}N, ^{12}C and ^{22}Ne enrichments for a short period of time. This corresponds to the transition WN/C stars, which show mass fractions of N $\sim 1\%$ and C $\sim 5\%$. They represent $\sim$ 4–5 % of the WR stars. Without rotational mixing, there would be no WN/C stars, because of the strong chemical discontinuity at the edge of the convective core in the He–burning phase (Fig. 6). A smooth chemical transition is needed to produce them in the process of peeling–off as shown by Langer (1991).

4.3. A Fundamental Diagram for WC Stars

WC stars have mass fractions of C between about 10 and 60% and of about 5–10 % for O, the rest being helium, e.g. Crowther (2007). The variations are smoother in rotating models (Fig. 6). In the WC stage, rotation broadens the range of possible C/He and O/He ratios, permitting the products of He burning to appear at the surface at an earlier stage of nuclear processing with much lower C/He and O/He ratios as suggested by observations.

The destruction of ^{14}N in the He–burning phase leads to the production of ^{22}Ne, which appears at the stellar surface in the WC stage. The models predict Ne enhancements by a factor 20-30. However, the abundance of the CNO elements have been reduced by a factor of ~ 2 and the Ne abundance has been revised upward by Asplund et al. (2004). Thus, a new estimate has to be made. ^{22}Ne is the daughter of ^{14}N, which is itself the daughter of CNO elements. The sum of CNO elements is $X(\mathrm{CNO})=0.00868$, which essentially

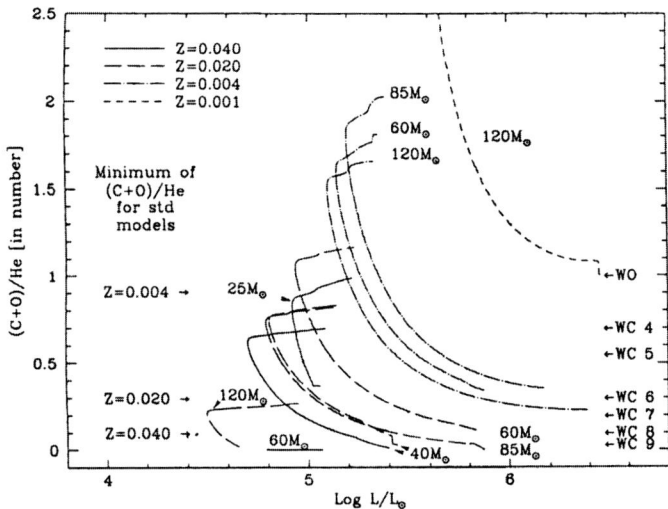

Figure 7. The (C+O)/He ratios in WC stars as a function of L for different Z and initial masses, cf. Maeder & Meynet (1994) and Meynet & Maeder (2005).

becomes ^{14}N. Since two α particles are added to ^{14}N to form ^{22}Ne, the abundance of ^{22}Ne in WC stars should be

$$X(^{22}\text{Ne}) = \frac{22}{14} X(^{14}\text{N}) \quad \text{num.} \quad X(^{22}\text{Ne}) = 1.57 \times 0.00868 = 0.0136\,. \tag{4.1}$$

We get a sum of Ne isotopes of about $X(\text{Ne}) = 0.0156$ compared to $X(\text{Ne})_\odot = 0.0020$. This gives a relative Ne enhancement by a factor of 8, instead of 20 to 30 with the old abundances. The factor of 8 is in excellent agreement with the observations by Ignace et al. (2007), who find excesses of 9.

Fig. 7 is a fundamental diagram for the chemical abundances of WC stars. There are versions of this figure with only mass loss and also including rotation by Meynet & Maeder (2005). It shows the (C+O)/He ratios as a function of the luminosity for WC stars of different initial masses and metallicities. At low Z, since mass loss is low, only the extremely massive stars enter the WC stage, thus their luminosities are high, as well as their (C+O)/He ratios. The reason for the high (C+O)/He ratios is that the rare stars which enter the WC stage enter it very late in the process of central He burning or even they do it after He exhaustion.

At higher Z, such as $Z = 0.02$ or $Z = 0.04$, due to the higher mass loss rates less massive stars may become WR stars, thus they have lower luminosities (Fig. 7). As the mass loss rates are higher, the products of He burning appear at an earlier stage of nuclear processing, i.e. with much lower (C+O)/He ratios. These properties seem unavoidable and they are also present in models with rotation (rotation does not alter the relations illustrated by Fig. 7). The coupling between L, Z and the (C+O)/He ratios produces the following consequences, see Smith & Maeder (1991).

- At a given luminosity, the (C+O)/He ratios are higher in regions of lower Z.
- For a given (C+O)/He ratio, the WC stars in lower Z regions have much higher luminosities.
- WO stars, which correspond to an advanced stage of evolution, should according to these predictions preferentially be found in regions of low Z and from very high initial

masses. This is a particularly important aspect since WO stars (with little or no He left) are likely the progenitors of supernovae SNIc, a small fraction of which are accompanied by GRBs.

These model predictions are awaiting observational confirmation. Some of the above trends have been discussed by Smith & Maeder (1991). However, they rest on the early results by Smith & Hummer (1988), who suggested an increase of C/He from WCL to WCE stars. Further studies have put doubts on this relation as shown by Crowther (2007). However, nothing is settled in view of the scarcity of the data on the O abundances. We emphasize it is essential not to just consider the C/He ratios, because during the He burning, C is first going up and then going down. Thus, by just considering the C data one may get misleading results. The (C+O)/He ratios vary in a monotonic way and should be preferred as a test of the abundances of WC stars.

5. Toward the First Stars

It is usually considered that mass loss should be small at very low Z, such as $Z < 10^{-3}$. This is not necessarily true as shown by Meynet & Maeder (2002) and Meynet et al. (2006). There are three different effects intervening.

• During the MS phase, the internal coupling of the angular momentum resulting from Eq. (2.3) is sufficient to transmit some of the fast rotation of the contracting core to the stellar surface. Thus, for a large range of initial masses and Ω values the low–Z stars reach the critical velocity during their MS phase (Fig. 8). This produces some moderate mass loss during the MS phase.

• The second effect is due to the self–enrichment of the stellar surface in CNO elements due to internal mixing. There is a remarkable interplay between rotation, mass loss and chemical enrichments in low–Z stars. Low Z implies a weak Gratton–Öpik circulation, which favors high internal Ω–gradients and in turn strong mixing of the chemical elements. The surface enrichments are very important, mainly due to the stellar radii being small (the diffusion timescales vary like R^2). Then, the high surface enrichments in heavy elements, particularly CNO elements, permits radiative winds and mass loss in the He–burning phase of massive and AGB stars.

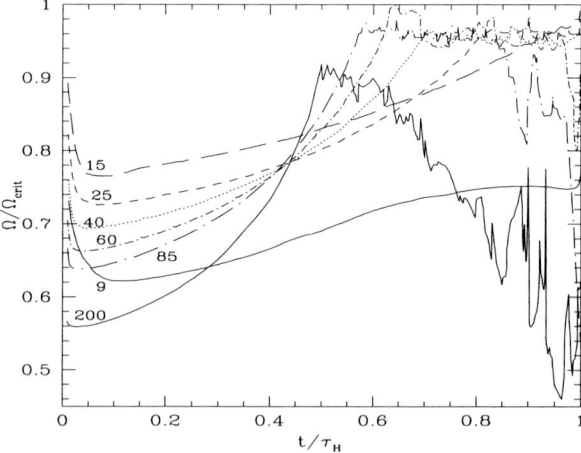

Figure 8. Examples of the evolution of the ratio of the angular velocity Ω to the critical values as a function of the fraction of the MS lifetime for different initial masses at $Z = 0$. From Ekström (2008).

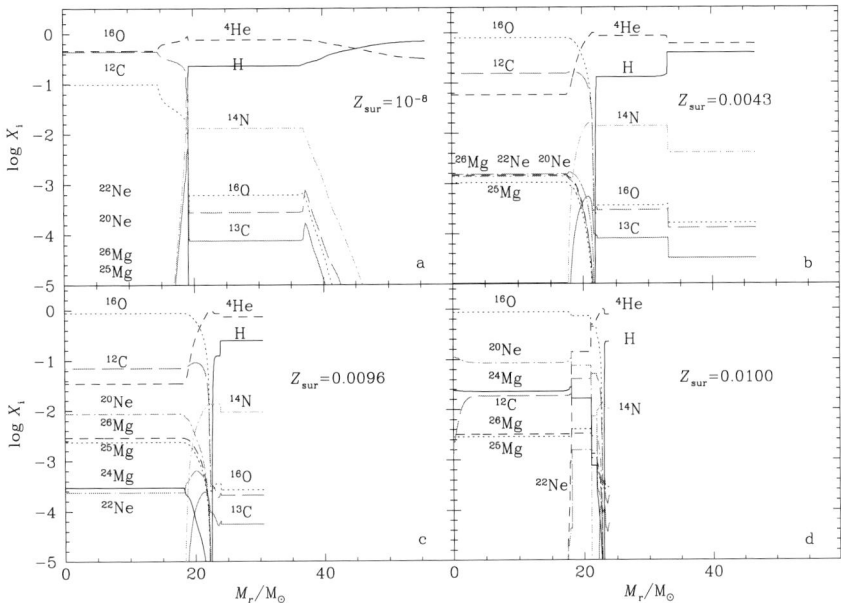

Figure 9. Variations of the abundances (in mass fraction) as a function of the Lagrangian mass within a 60 M$_\odot$ star with $v_{\rm ini}$= 800 km s^{-1} and $Z = 10^{-8}$. The four panels show the chemical composition at four different stages at the end of the core He-burning phase: in panel **a)** the model has a mass fraction of helium at the centre, $Y_{\rm c}$= 0.11 and an actual mass M= 54.8 M$_\odot$ - **b)** $Y_{\rm c}$= 0.06, M= 48.3 M$_\odot$ - **c)** $Y_{\rm c}$= 0.04, M= 31.5 M$_\odot$ - **d)** End of the core C-burning phase, M= 23.8 M$_\odot$. The actual surface metallicity $Z_{\rm surf}$ is indicated in each panel. From Meynet *et al.* (2006).

- The stars with $M < 40$ M$_\odot$, depending on rotation, may make blue trips in the HR diagram. If so, the contraction of the convective envelope brings the surface velocity to critical value and mass loss is enhanced according to Eq. (2.4).

Fig. 9 illustrates the very strong enrichments in CNO during the He–burning phase of 60 M$_\odot$ stars with an initial $Z = 10^{-8}$ and an initial rotation velocity of 800 km s^{-1}. From an initial $Z = 10^{-8}$, the surface metallicity is brought to $Z = 0.01$. This permits the stellar winds of red supergiants and AGB stars to remove the stellar envelopes, especially as the surface is enriched in C which may produce lots of dust. The low–Z AGB and massive stars may lose a large fraction of their mass.

These stellar winds produce very peculiar chemical enrichments of the early galaxies. The chemical composition of the rotationally enhanced winds of very low Z stars show large CNO enhancements by factors of 10^3 to 10^7, together with large excesses of ^{13}C and ^{17}O and moderate amounts of Na and Al. The excesses of primary N are particularly striking. When these ejecta from the rotationally enhanced winds are diluted with the supernova ejecta from the corresponding CO cores, we find [C/Fe], [N/Fe],[O/Fe] abundance ratios that are very similar to those observed in the C–rich, extremely metal–poor stars as shown by Meynet *et al.* (2006). Rotating AGB stars and rotating massive stars have about the same effects on the CNO enhancements. Nevertheless, abundances of s-process elements and the ^{12}C/^{13}C ratios could help us to distinguish between contributions from AGB and massive stars. As shown by Chiappini *et al.* (2006), these peculiar enrichments remarkably well account for the initial chemical evolution of the C/O, N/O and O/Fe ratios in the Galaxy.

References

Asplund, M., Grevesse, N., Sauval, A. J., et al. 2004, *A&A*, 417, 751
Chiappini, C., Hirschi, R., Meynet, G., et al. 2006, *A&A*, 449, 27
Conti, P. S. 1976, *Bull. Soc. Roy. Sci. Liege*, 9, 193
Conti, P. S., Greenstein, J. L., Spinrad, et al. 1967, *ApJ*, 148, 105
Crowther, P. 2007 *ARA&A*, 45, 177
Ekström, S. 2008, *Thesis, University of Geneva*, in prep.
Heap, S. R., Lanz, T., & Hubeny, I. 2006, *ApJ*, 638, 409
Herrero, A. 2003, in: C. Charbonnel, D. Schaerer, & G. Meynet (eds.), *CNO in the Universe*, (San Francisco: ASP), *ASP Conf. Ser.*, 304, 10
Huang, W. & Gies, D. R. 2006a, *ApJ*, 648, 580
Huang, W. & Gies, D. R. 2006b, *ApJ*, 648, 591
Hunter, I., Dufton, P. L., Smartt, S. J., et al. 2007, *A&A*, 466, 277
Ignace, R., Cassinelli, J. P., Tracy, G., et al. 2007, *ApJ*, 669, 600
Lamers, H. J. G. L. M. & Morton, D. C. 1976, *ApJS*, 32, 715
Langer, N. 1991, *A&A*, 248, 531
Maeder, A. 1999 *A&A*, 347, 185
Maeder, A. & Zahn, J. P. 1998,*A&A*, 334, 1000
Maeder, A. & Meynet, G. 2005, *A&A*, 440, 1041
Maeder, A., Georgy, C., & Meynet, G. 2008, *A&A*, 479, L37
Meynet, G. & Maeder, A. 2000, *A&A*, 361, 101
Meynet, G. & Maeder, A. 2002, *A&A*, 390, 561
Meynet, G. & Maeder, A. 2003, *A&A*, 404, 975
Meynet, G. & Maeder, A. 2005, *A&A*, 429, 581
Meynet, G., Ekström, S., & Maeder, A. 2006, *A&A*, 447, 623
Morton, D. C., 1976, *Bull. American Astr. Soc.* 8, 138
Smith, L. F. 1968, *MNRAS*, 141, 317
Smith, L. F. & Hummer, D. G. 1988, *MNRAS*, 230, 511
Smith, L. F. & Maeder, A. 1991, *A&A* 241, 77
Spruit, H. C. 2002, *A&A*, 381, 923
Trundle, C., Dufton, P. L., Hunter, I. et al. 2007, *A&A*, 471, 625
Zahn, J. P. 1992, *A&A*, 265, 115

Discussion

SANA: Thanks for this very nice summary. I have a comment rather than a question. I would like to warn against the use of García & Mermilliod (2001, *A&A*, 368, 122) on the O–star binary fraction, and particularly against the very large fraction (up to 80 %) proposed for some clusters. Several authors mainly from the Liege group (M. De Becker, G. Rauw) and myself, have reinvestigated the question and could not confirm the extreme binary fraction proposed for some clusters. Regarding Tr 14 for which you quote Penny et al. (1993, *PASP*, 105, 588) showing that none of the 9 stars are binaries, it is interesting to keep in mind that there is 2 good binary candidates: one of them is a RV–variable and the other shows NT radio emission. Though this needs to be confirmed, the binary fraction in Tr 14 might be significantly different from zero. I have proposed a poster revising the binary fraction in nearby clusters and my conclusion is that the current available data do not allow to prove that the O-stars binary fraction is varying from one cluster to another.

MAEDER: Thank you for these comments. Thus, you have presented your poster...

LEITHERER: Most the effects you discussed kick in only at really low Z. Observationally this makes many of these predictions hard to test in individual stars because the lowest oxygen abundance in the local universe is just a couple of %. Nevertheless, what can we learn from observations of individual stars in galaxies like I Zw 18?

MAEDER: I think the observations of individual stars in I Zw 18 is the next major step toward the direct observations of very low Z stars, i.e. toward the "First Stars". The large telescopes will open magnificent new possibilities. For now, we nevertheless have most useful infos from the chemical abundances in very low–Z halo stars and in globular cluster stars. Although these are low mass stars, their abundances provide direct information on the massive stars at very low Z.

VANBEVEREN: The Be components in X–ray binaries are probably formed via mass transfer during Roche lobe overflow. Accounting for the fact that a supernova explosion of one of the components of a binary, disrupts the binary, it can be expected that for every Be star in an X–ray binary there are 10 Be stars that are single but are formed via mass transfer as well. So, I do not understand how one can use Be-star statistics in order to defend the role of rotation on the evolution of single stars, by denying the binary formation mechanism for this type of stars.

MAEDER: Maybe you should notice that the issue was not the hypothetical binary formation of Be stars, but whether on the average rotation is higher at lower Z. In this respect, new results by Martayan *et al.* (poster, this volume) beautifully confirm the claim that I made in 1999 with Eva Grebel and J.C. Mermilliod (*A&A* 346, 459), i.e. that low Z stars have on the average a higher rotation.

LIMONGI: A comment on the primary N production in low Z stars. In non rotating models, primary N production is common because of the mixing between the H convective shell and the H rich envelope. Such a phenomenon is common and it was already recognized by Woosley and Weaver in 1982. But as soon as Z increases this phenomenon disappears. In rotating models, on the contrary, you can have primary N production also at higher metallicities.

MAEDER: I agree with your remarks. However, rotation enhances the primary N production. We get primary N production only for metallicities below that of the SMC.

André Maeder.

Part of the Local Organizing Committee resting at Waimea Canyon.
From left to right: Vivian U, Miguel Urbaneja, Laurie Clark and Fabio Bresolin.

X-ray Emission from O Stars

David H. Cohen

Swarthmore College, Department of Physics and Astronomy, 500 College Ave., Swarthmore, Pennsylvania 19081, USA

Abstract. Young O stars are strong, hard, and variable X-ray sources; properties that strongly affect their circumstellar and galactic environments. After ≈ 1 Myr, these stars settle down to become steady sources of soft X-rays. I will use high-resolution X-ray spectroscopy and MHD modeling to show that young O stars like θ^1 Ori C are well explained by the magnetically channeled wind shock scenario. After their magnetic fields dissipate, older O stars produce X-rays via shock heating in their unstable stellar winds. Here too I will use X-ray spectroscopy and numerical modeling to confirm this scenario. In addition to elucidating the nature and cause of the O star X-ray emission, modeling of the high-resolution X-ray spectra of O supergiants provides strong evidence that mass-loss rates of these O stars have been overestimated.

Keywords. hydrodynamics – instabilities – line: profiles – magnetic fields – MHD – shock waves – stars: early type – stars: mass loss – stars: winds, outflows – X-rays: stars

---

O stars dominate the X-ray emission from young clusters, with X-ray luminosities up to $L_\mathrm{x} = 10^{34}$ ergs s^{-1} and emission that is hard (typically several keV) and often variable. This strong X-ray emission has an effect on these O stars' environments, including nearby sites of star formation and protoplanetary disks surrounding nearby low-mass pre-main-sequence stars. The X-ray emission is also interesting in its own right, as it traces important high-energy processes in the extended atmospheres of O stars. In this paper, I will show how the spectral properties of the X-rays from the prototypical young, magnetized O star, θ^1 Ori C are in line with the predictions of the Magnetically Channeled Wind Shock (MCWS) model, but how this process seems to dissipate as O stars age, with weaker line-driven instability wind shocks explaining the X-ray emission in older O stars. I also will show how the X-ray emission can be used as a probe of the conditions in the bulk stellar winds of these objects. Specifically, the resolved X-ray line profiles in normal O supergiants provide an independent line of evidence for reduced mass-loss rates.

Now, it is certainly the case that many young O stars do not show the X-ray signatures of the MCWS mechanism. And indeed, only a handful of O and early B stars have had direct detections of magnetic fields. Of course, highly structured, non-dipole fields will be very difficult to detect on hot stars, even if their local strength is quite high. But one should keep in mind that wind-wind interactions in close binaries can also produce the hard, strong, and variable X-rays seen in many young O stars.

Because θ^1 Ori C has a well established, predominantly dipole field, and because many of its properties are explained by this field, I treat it here as a potential prototype. I also note that the incidence of hard, strong X-ray emission from O stars diminishes rapidly as one looks from young (< 1 Myr) clusters to older (2 to 5 Myr) clusters. This fact can be explained if the fossil fields in young O stars dissipate as the stars age. If wind-wind binaries account for many of these sources, then only the very earliest O stars with very short lifetimes are involved.

We begin the comparison of young and old O stars and their X-rays by showing in Fig. 1 the *Chandra* MEG spectra of two representative stars: θ^1 Ori C (O4-7 V), with an age of ≈ 1 Myr, and ζ Pup (O4 If), with an age of several Myr, and already evolved well off the main sequence. The ζ Pup X-ray spectrum is typical of those measured for most O stars.

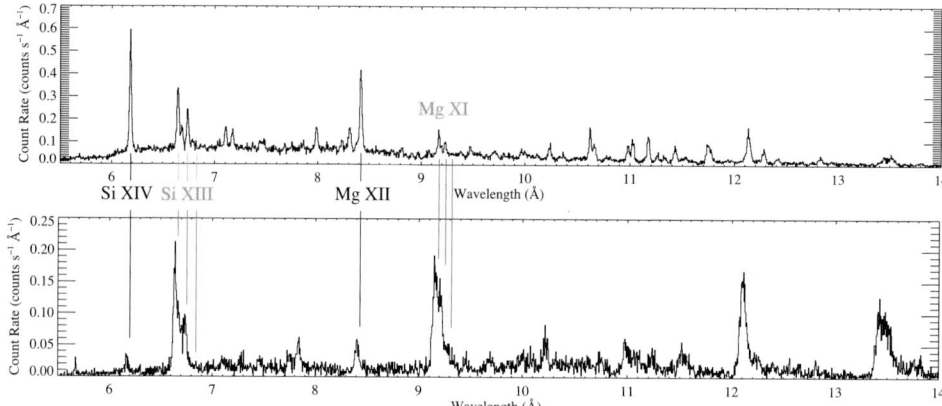

Figure 1. *Chandra* MEG spectra of θ^1 Ori C (top) and ζ Pup (bottom). The hydrogen-like Lyman alpha lines of Si and Mg are indicated in black, while the helium-like resonance-intercombination-forbidden complexes of the same elements are indicated in gray.

Two obvious differences between these spectra are the hardness of the X-ray emission from θ^1 Ori C and the small line widths in that star's spectrum. The hardness implies a much higher plasma temperature in the young O star, and this is best seen in the data when one compares resonance lines of hydrogen-like and helium-like ionization states of abundant elements. In Fig. 1 I have labeled these lines for silicon and magnesium. In θ^1 Ori C the hydrogen-like lines are much stronger, whereas in ζ Pup the helium-like lines are stronger. This reflects a significantly different ionization balance in these two stars which is a direct effect of their different plasma temperatures. The plasma temperature distribution, based on the analysis of high-resolution *Chandra* spectra of several O stars, has been determined by Wojdowski & Schulz (2005). These authors assume a continuous Differential Emission Measure (DEM), which can be thought of as a density-squared weighting of the plasma temperature distribution. Their results are shown in Fig. 2, where it can easily be seen that the DEM for θ^1 Ori C is the only one (of seven O stars) with a positive slope. Its peak is near $T = 30$ million K, whereas the DEMs of the more evolved O stars (including ζ Pup) all peak near one or two million K.

We also show in Fig. 2 the DEM from a snapshot of an MHD simulation of the magnetically confined wind of θ^1 Ori C. The agreement is quite good, both in terms of the overall emission measure and the shape of the DEM. The simulation shows a peak at 15 to 20 million K, modestly lower than that seen in the data, but the main property – a rising DEM up to and beyond 10 million K – matches the data well. These MHD simulations confirm the predictions of Babel & Montmerle (1997) that strong shock fronts near the magnetic equator where the wind from the northern and southern hemispheres meets can heat a significant amount of wind plasma to the observed high temperatures. Snapshots of temperature and emission measure from an MHD simulation are shown in Fig. 3. It can be seen that the bulk of the shock-heated plasma is in the magnetically confined region near $r \approx 2$ R$_*$. Due to this confinement, the speed and line-of-sight velocity of this material is low, and thus the emission lines from the thermal X-ray emission are relatively narrow.

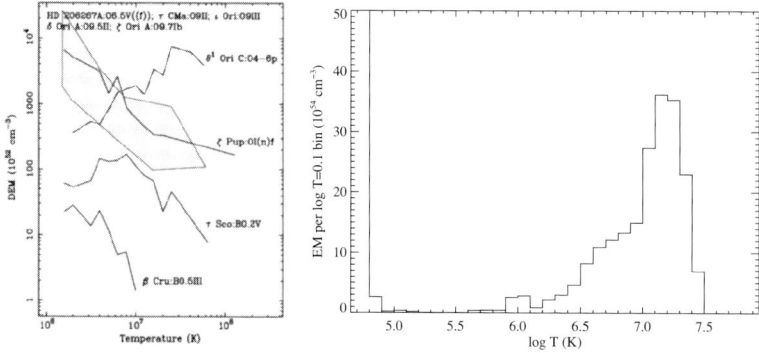

Figure 2. Differential emission measures of seven O stars and two B stars, derived from thermal spectral model fits to *Chandra* spectra (left; taken from Wojdowski & Schulz, 2005). θ^1 Ori C is the only star whose DEM has a positive slope. The panel on the right shows a DEM predicted by the MHD simulations of θ^1 Ori C (taken from Gagné et al., 2005).

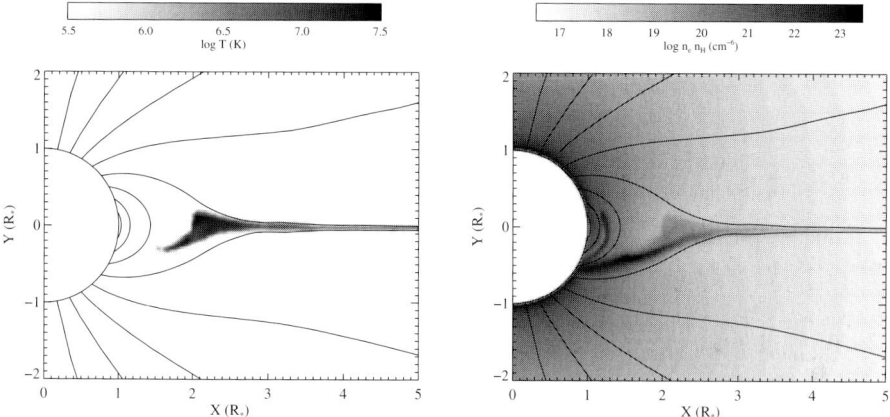

Figure 3. Contour plots of temperature (left) and emission measure per unit volume (right) from a 2-D MHD simulation of the magnetically confined wind of θ^1 Ori C (taken from Gagné et al., 2005). Magnetic field lines are displayed as contours. The wind flow up each closed field line encounters a strong shock due to the ram pressure of the wind flow from the opposite hemisphere, which heats the plasma according to $T_{shock} \approx 10^7 (v_{shock}/1000 \text{ km s}^{-1})^2$ K. The head-on nature of the wind shocks leads to high shock velocities and temperatures. In these MHD models, the field configuration is self-consistently solved for along with the wind dynamics. Note that another difference between the MHD simulations of the MCWS model and the initial analysis of Babel & Montmerle (1997) is the dynamical infall of material from the magnetic equator. Evidence of this can be seen in the snake-like structure visible in the emission measure panel, just above the star's surface, slightly below the equator.

In contrast, the X-ray emission lines of mature O stars, like ζ Pup, are quite broad, as can be seen in Fig. 4, where I show the neon Lyα lines for the two stars. The X-ray spectrum of ζ Pup is also soft, as I have already shown. The X-ray emission from these older, presumably non-magnetized, O stars is thought to arise in much milder wind shocks, embedded in the outflowing, highly supersonic line-driven winds. The Line-Driven Instability (LDI) is generally thought to produce these shocks, although models have difficulty reproducing the overall level of X-ray emission unless the instability is seeded, perhaps by sound waves injected at the base of the wind (Feldmeier et al., 1997).

The softness of the X-ray spectra, along with the large line widths from the high velocity of the shock-heated wind, is well explained by this LDI wind-shock scenario, as I show in Fig. 5. This figure shows a snapshot from a 1-D radiation hydrodynamics simulation of the wind of an O supergiant like ζ Pup, accounting for non-local line radiation transport. The instability grows rapidly beyond about half a stellar radius (in height; $r = 1.5$ R$_*$). Shock fronts can be seen in this snapshot, but they typically have velocity jumps of only a few hundred km s^{-1}, leading to heating of only a few million K. The soft spectrum seen in Fig. 1 and the DEMs weighted to low plasma temperatures, shown in the left-hand panel of Fig. 2, are in line with the results of this simulation.

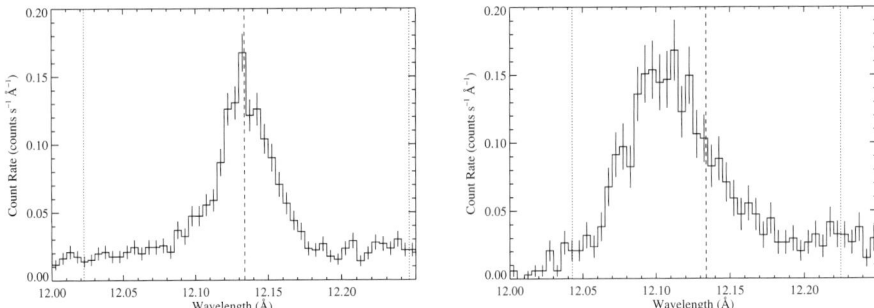

Figure 4. Ne X Lyman alpha lines in the *Chandra* MEG spectra of θ^1 Ori C (left) and ζ Pup (right). The vertical dashed line represents the laboratory rest wavelength of this transition, and the vertical dotted lines represent the blue and red shifts associated with the UV wind terminal velocity in each star. Note that the profile in the ζ Pup spectrum is shifted and skewed as well as being broadened. Error bars are from Poisson photon-counting statistics.

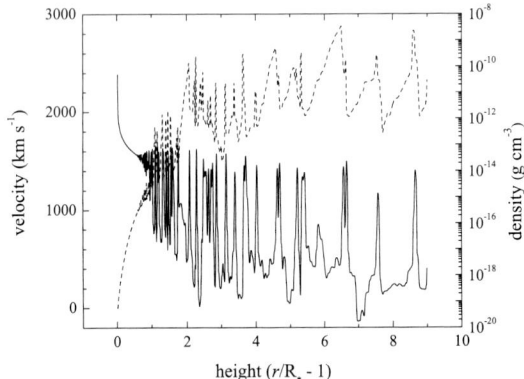

Figure 5. A snapshot showing the velocity (dashed line; left-hand axis) and density (solid line; right-hand axis) as a function of height above the photosphere in a radiation hydrodynamics simulation of the wind of ζ Pup. Shock-heated plasma cools rapidly; the bulk of the wind in any given snapshot is cold.

The predictions of this model can be further tested by examining the X-ray emission line profiles, which have a characteristic asymmetric, skewed shape, as can be seen in the right-hand panel of Fig. 4. This shape, with a deficit of red-shifted photons, arises in a spherically expanding wind with hot, line-emitting material intermixed with warm, continuum-absorbing material, as motivated by the global structure seen in Fig. 5. If the wind is optically thick (in the continuum; $\kappa \approx$ constant across a line), then there is significantly more attenuation of emission from the back of the wind, which is the red shifted portion. And there is comparatively more emission from the front, blue shifted, less attenuated, side.

I fit an empirical model (Owocki & Cohen, 2001) to emission lines in the spectrum of ζ Pup and get good fits by adjusting only three parameters: the normalization, the inner radius below which there is assumed to be no emission (R_o), and the wind optical depth (parameterized by the quantity $\tau_* \equiv \dot{M}\kappa/4\pi R_* v_\infty$). The fit to the Fe XVII line at 15.014 Å is typical and is shown in Fig. 6 as the dashed histogram. The fit is formally good, and the best-fit values with joint 68% parameter confidence limits are $R_o = 1.53^{+0.12}_{-0.15}$ R$_*$ and $\tau_* = 2.0 \pm 0.4$.

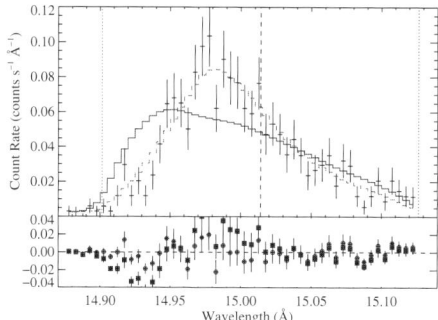

Figure 6. Best-fit wind profile model, for a non-porous wind, (dashed histogram) fit to the Fe XVII line in the *Chandra* spectrum of ζ Pup. The solid histogram is the best-fit non-porous model for which the optical depth parameter is fixed at the value implied by the literature mass-loss rate ($\tau_* = 8$). Fit residuals for the two models are shown in the lower panel, with the circles corresponding to the low optical depth model represented by the dashed histogram, and the squares corresponding to the higher optical depth model.

The R_o derived from the profile is consistent with the onset of the self-excited instability seen in the hydrodynamics simulation shown in Fig. 5. The optical depth consistent with the data is actually quite small, which can be seen qualitatively in the relatively modest asymmetry. The mass-loss rate of 6×10^{-6} M$_\odot$ yr^{-1} derived from Hα emission (Puls et al., 1996) implies $\tau_* = 8$. The best-fit model with that value fixed is shown in Fig. 6 as the solid histogram. The fit is formally very poor. Thus it would appear that the X-ray line profiles provide independent evidence that mass-loss rates of O stars must be revised downward by a factor of several; a factor of 4 for this star, according to this particular line ($\tau_* = 2.0 \pm 0.4$ vs. $\tau_* = 8$).

It has been suggested that porosity associated with large-scale clumping – rather than reduced mass-loss rates – can account for the surprisingly small degree of asymmetry in the observed X-ray emission line profiles in O stars (Oskinova et al., 2006). By fitting the line profile model, modified for the effects of porosity produced by spherical clumps (Owocki & Cohen, 2006), I can quantify the trade-off between atomic opacity and porosity (see also Cohen et al., 2008). In Fig. 7 I show the best-fit porous model with $\tau_* = 8$, the value implied by the literature mass-loss rate, and with the porosity length, $h \equiv \ell/f$, free to vary. Here ℓ is the clump size and f is the volume filling factor of the clumps. The porosity length, h, completely describes the effects of porosity on line profiles and in the limit of small clumps it is equivalent to the interclump spacing. The best-fit porous model with the literature mass-loss rate is nearly indistinguishable from the best-fit non-porous model, although the fit quality is formally not as good. More importantly, it requires a terminal porosity length (the value of h in the outer portion of the wind) of at least $h_\infty = 2.5$ R$_*$ (68% confidence), as shown in the right-hand panel of Fig. 7. Even ignoring the worse quality of the porous model fits, the very high values of the porosity length required to fit the data are vastly larger than the porosity lengths seen in state-of-the-art 2-D radiation hydrodynamics simulations, which show LDI-induced structure down to the grid scale (Dessart & Owocki, 2005).Therefore, I conclude that there is no compelling evidence that porosity explains the modestly asymmetric X-ray line profiles in ζ Pup. Rather, these profiles provide independent evidence for a reduced mass-loss rate.

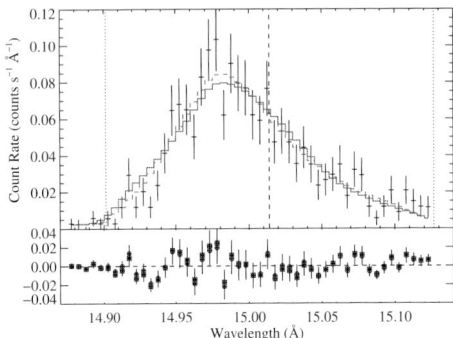 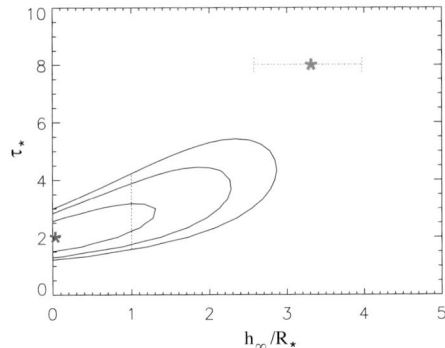

Figure 7. Best-fit wind profile model, for a non-porous wind (dashed histogram), fit to the Fe XVII line in the *Chandra* spectrum of ζ Pup (left). The solid histogram is the best-fit porous model for which the optical depth parameter is fixed at the value implied by the literature mass-loss rate ($\tau_* = 8$). We also show confidence limits (68, 90, 95%) in h_∞ vs. τ_* parameter space (right). The global best-fit model is indicated by the star at $h_\infty = 0, \tau_* = 2$. The star at $h_\infty = 3.3, \tau_* = 8$ represents the best-fit model with the wind optical depth fixed at the value implied by the literature mass-loss rate. The horizontal bar centered on it represents the 68% confidence limit on the value of h_∞, given $\tau_* = 8$ ($2.5 < h_\infty/R_* < 4.0$). The vertical line at $h_* = 1$ emphasizes that even that large porosity length cannot bring the optical depth close to the value associated with the literature mass-loss rate.

Acknowledgements

I acknowledge support from NASA/CXC grant AR7-8002X and from the Hungerford Faculty Support Endowment of Swarthmore College's provost's office.

References

Babel, J. & Montmerle, T. 1997, *ApJ*, 485, L29
Cohen, D. H., Leutenegger, M., & Townsend, R. H. D. 2008, in: W.-R. Hamann, A. Feldmeier, L. M. Oskinova (eds.), *Clumping in Hot Star Winds* (Potsdam: Universitäts-Verlag)
Dessart, L. & Owocki, S. P. 2005, *ApJ*, 437, 657
Feldmeier, A., Puls, J., & Pauldrach 1997, *A&A*, 322, 878
Gagné, M., Oksala, M., Cohen, D. H., *et al.* 2005, *ApJ*, 628, 986
Oskinova, L. M., Feldmeier, A., & Hamann, W.-R. 2006, *MNRAS*, 372, 313
Owocki, S. P. & Cohen, D. H. 2001, *ApJ*, 559, 1108
Owocki, S. P. & Cohen, D. H. 2006, *ApJ*, 648, 565
Puls J., Kudritzki, R.-P., Herrero, A., *et al.* 1996, *A&A*, 305, 171
Wojdowski, P. & Schulz, N. S. 2005, *ApJ*, 627, 953

Discussion

KUDRITZKI: It is very important to repeat the observing experiment that you have done for θ^1 Ori C for young O stars, which are more massive and luminous than θ^1 Ori C. I speculate that for those objects, the magnetic focusing of the stellar winds will be less effective, because the winds are stronger and the ratio of magnetic to mechanical wind energy is lower. I think it is really crucial to do such observations.

COHEN: I agree. The degree of channeling and confinement, however, goes as B^2 but only as $1/\dot{M}$, so the extent of channeling and confinement in any given star is more likely to be dominated by trends in magnetic field strength, which we don't understand, than by trends in mass-loss rate associated with stellar mass and luminosity. I think the key measurements to make in order to test the idea that the MCWS mechanism on θ^1 Ori C is a paradigm for X-ray emission in young O stars would be Zeeman measurements of fields on the massive cluster stars that are known to be strong, hard X-ray sources. High-resolution X-ray spectroscopy would also be very useful, obviously, if it's feasible. As I discussed in my talk in the special session on magnetic massive stars on Sunday, line widths and helium-like forbidden-to-intercombination line strength ratios provide information beyond what's provided by CCD-based (e.g. Chandra ACIS or XMM EPIC) X-ray data.

ZINNECKER: My question refers to θ^1 Ori C and the origin of the obliquity between its rotational axis and magnetic field axis. From star formation theory it would seem an *aligned* magnetic rotator would be expected. Any suggestions to explain the misalignment?

COHEN: I don't have any special expertise in star formation theory, but if models predict aligned rotation and magnetic axes, then there must be some important physics missing from them. Many of the magnetic massive stars have highly misaligned axes: τ Sco, β Cep, and σ Ori E, for example, are all close to $\beta = 90$ degrees.

SKINNER: A new paper by M. Güdel et al. (2008 Sci 319, 309) reports the first detections of hot (1 to 2 MK) *diffuse* X-ray emission in the extended Orion Nebula. This article argues that massive Trapezium O stars (and their shocked winds) are ultimately responsible for the diffuse X-ray emission detected by XMM-Newton.

COHEN: I think that the problem of heating the diffuse, X-ray emitting plasma in massive star clusters is a hard one. The shocked wind (such as ζ Pup's) will adiabatically cool over distances much less than 1 pc. And the morphology of the X-rays isn't what you'd expect from the wind slamming into dense interstellar gas at the boundaries of these cavities.

WALBORN: There are currently only two O stars with observed magnetic fields, θ^1 Ori C and HD 191612. The latter can be understood as a spun-down version of the former, with a rotational period of 538 d (vs. 15 d for θ^1 Ori C), a soft X-ray spectrum, and an age of 3 to 4 Myr. Nevertheless, it is a very unusual object, with a very peculiar spectrum and extreme, periodic spectral variations. So these two stars are inconsistent with your hypothesis that θ^1 Ori C is a typical very young O star, and that magnetic fields have disappeared at ~ 5 Myr. I think that both of these objects are unusual areas of large fossil fields. An alternate interpretation of your cluster X-ray differences might be different frequencies of wind-wind collision binaries.

COHEN: While I agree than wind-wind collision binaries may make a significant contribution to the observed population of hard, strong X-ray sources in young clusters, I think we need more information about the highly unusual O star, HD 191612, not to mention more positive detections or strong upper limits to magnetic field strengths in other O stars, both young and old. The X-ray spectrum of HD 191612 looks like that of a typical, older O star, with very broad lines and an SED that is quite soft. Perhaps the fields of young O stars like θ^1 Ori C become more spatially structured as they evolve. It's possible that HD 191612 has a field that looks more like that of tau Sco and is not dominated by a large scale dipole. If that's the case, then there may not be any large-scale confinement and channeling of the stellar wind, and no substantial MCWS mechanism. The X-ray emission may instead arise in open field regions, in a loose analogy to the solar wind and coronal holes.

David Cohen.

Joachim Puls.

Physical and Wind Properties of OB-Stars

Joachim Puls

Universitätssternwarte München, Scheinerstr. 1, D-81679 München, Germany
email: uh101aw@usm.uni-muenchen.de

Abstract. In this review, the physical and wind properties of OB-stars are discussed, with particular emphasis on metallicity dependence and recent results from the FLAMES survey of massive stars. We summarize the relation between spectral type and $T_{\rm eff}$, discuss the status quo of the "mass-discrepancy", refer to the problem of "macro-turbulence" and comment on the distribution of rotational velocities. Observational constraints on the efficiency of rotational mixing are presented, and magnetic field measurements summarized. Wind properties are reviewed, and problems related to weak winds and wind-clumping highlighted.

Keywords. stars: early-type – stars: atmospheres – stars: fundamental parameters – stars: winds, outflows – stars: mass loss – stars: rotation – techniques: spectroscopic – surveys – magnetic fields

---

1. Atmospheric models for hot stars

Most of our knowledge about the physical parameters of hot stars (effective temperatures, gravities, wind-properties, chemical composition of the outer layers) has (and will be) obtained by *quantitative spectroscopy*, i.e., the analysis of stellar *spectra* by means of atmospheric models. These have to be calculated in NLTE due to the intense radiation field and low densities in the line-forming regions. With respect to such models, the last IAU-symposium on massive stars in Lanzarote (2002) saw the begin of two important developments which are "standard" nowadays, namely the incorporation of metal-line blanketing and the inclusion and diagnostics of wind-clumping (see Sect. 7).

As a brief reminder, line blanketing summarizes the effects of the multitude of (E)UV metal lines which act as a "blanket", in such a way that the corresponding flux in the *outer* atmosphere is reduced, whilst both the mean radiation field and the electron temperature in the *inner* atmosphere increase due to backscattering and thermalization. Consequently, the degree of ionization increases as well (see Fig. 1), and diagnostic lines, such as from He, become weaker. Thus, in comparison to unblanketed models, *lower effective temperatures,* $T_{\rm eff}$ *(and gravities,* $\log g$*), are needed to fit the observations for a given spectral type.*

In Tab. 1, we have summarized the basic features and domains of applications of present state-of-the-art, NLTE, line-blanketed model atmosphere codes which have been/are used to analyze OB-stars (or are suitable to do so). There are two classes of codes: (i) those which refrain from (almost) any approximation within the model assumptions and thus need a substantial amount of computational time to calculate one model atmosphere (TLUSTY, CMFGEN, POWR, PHOENIX); (ii) those which use certain approximations (mostly regarding the treatment of line-blanketing) such that the computational effort becomes considerably reduced (Detail/Surface, WM-basic, FASTWIND). Note that the applied approximations have been carefully tested within the corresponding domains of application, and that the agreement between most of the different codes is satisfactory, except for specific problems such as the strength of the He I singlet lines in later O-types (Najarro *et al.* 2006) or the ionizing fluxes below roughly 400 Å.

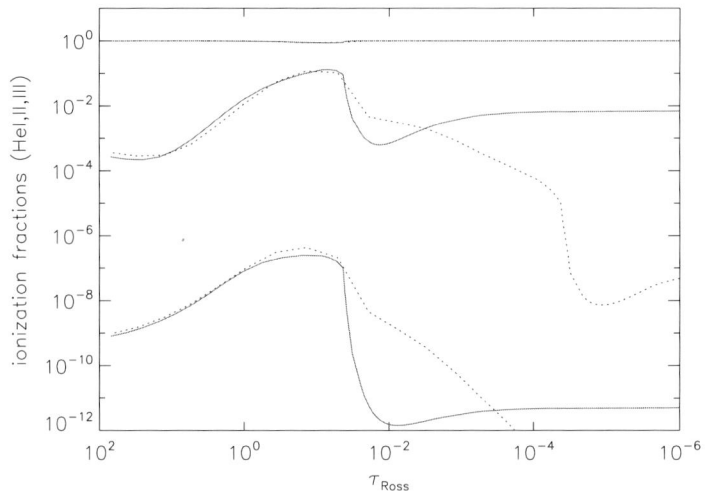

Figure 1. Ionization fractions of He I, He II and He III (from bottom to top) for an unblanketed model atmosphere at $T_{\rm eff}$=45kK and $\log g$=3.9 (dashed) and a blanketed one at $T_{\rm eff}$=40.5kK and $\log g$=3.7 (solid). In the region of photospheric line formation ($\tau_{\rm Ross} > 10^{-2}$), both models predict almost identical occupation numbers and synthetic line profiles, thus requiring a reduction of $T_{\rm eff} = 4{,}500$ K due to the effects of line-blanketing. Adapted from Repolust *et al.* (2004).

2. Effective temperatures of Galactic OB-stars

Since 2002, some 20 spectroscopic analyses of *Galactic* OB stars (excluding Galactic Centre objects, see Martins *et al.* this volume) of various luminosity classes have been

Table 1. Basic features and domains of applications of present state-of-the-art, NLTE, line-blanketed model atmosphere codes. Responsible authors in brackets.

	Detail/Surf. (Butler)	TLUSTY (Hubeny)	CMFGEN (Hillier)	WM-basic (Pauldrach)	FASTWIND (Puls)	POWR (Hamann)	PHOENIX (Hauschildt)
geometry	plane-parallel	plane-parallel	spherical	spherical	spherical	spherical	spherical/pl.-parallel
blanketing	LTE	yes	yes	yes	approx.	yes	yes
line transfer	observer's frame	observer's frame	comoving frame (CMF)	Sobolev	CMF	CMF	CMF/obs.frame
temperature structure	radiative equilibrium	radiative equilibrium	radiative equilibrium	e^- thermal balance	e^- thermal balance	radiative equilibrium	radiative equilibrium
photosphere	yes	yes	from TLUSTY	approx.	yes	yes	yes
diagnostic range	no limitations	no limitations	no limitations	UV	optical/IR	no limitations	no limitations
major application	hot stars with negl. winds	hot stars with negl. winds	OB(A)-stars, WRs, SNe	hot stars w. dense winds, ion. fluxes, SNe	OB-stars, early A-sgs	WRs	stars below 10 kK, SNe
comments	no wind	no wind	start model required	no clumping	explicit/backgr. elements		molecules included, no clump.
execution time	few minutes	hours	hours	1 to 2 h	few min. to 0.5 h	hours	hours

Table 2. Spectroscopic NLTE analyses of Galactic (without Centre) OB-stars since 2002

from	reference	objects	code	$\dot{M}$
UV	Bianchi & Garcia (2002)	O-stars	WM-basic	yes
	Garcia & Bianchi (2004)	O-stars	"	yes
UV + optical	Bouret et al. (2005)	O-stars	CMFGEN	yes
	Martins et al. (2005)	O-dwarfs	"	yes
	Bianchi et al. (2006)	O-stars	WM-basic + CMFGEN	yes
optical	Herrero et al. (2002)	OB-stars (Cyg-OB2)	FASTWIND	yes
	Repolust et al. (2004)	O-stars	"	yes
	Mokiem et al. (2005)	O-stars (genetic algorithm)	"	yes
	Simon-Diaz et al. (2006)	OB-stars (Trapezium)	"	yes
	Urbaneja (2004)	B-supergiants	FASTWIND	yes
	Crowther et al. (2006)	B-supergiants	CMFGEN	yes
	Przybilla et al. (2006)	AB-supergiants	Detail/surface	no
	Lefever et al. (2007)	B-supergiants (period. variable)	FASTWIND	yes
	Trundle et al. (2007)	B-stars (FLAMES: MW + MCs)	TLUSTY	no
	Hunter et al. (2007)	B-stars (FLAMES: MW + MCs)	"	no
	Markova & Puls (2008)	B-supergiants	FASTWIND	yes
NIR	Repolust et al. (2005)	O(B)-stars	FASTWIND	yes

performed, using different wavelength ranges and different atmosphere codes (see Tab. 2, 'MW' = Milky Way, 'MC' = Magellanic Clouds). Combining the results from several investigations, Martins et al. (2005), Trundle et al. (2007) and Markova & Puls (2008, see also Lefever et al. 2007) derived new spectral-type-$T_{\rm eff}$ calibrations for O and B stars, which are displayed in Fig. 2. Whereas for O-stars a linear relation provides a reasonable representation, for B-stars the relation can be described by a 3rd order polynomial, at least for supergiants. Compared to the previous scales by Vacca et al. (1996) and

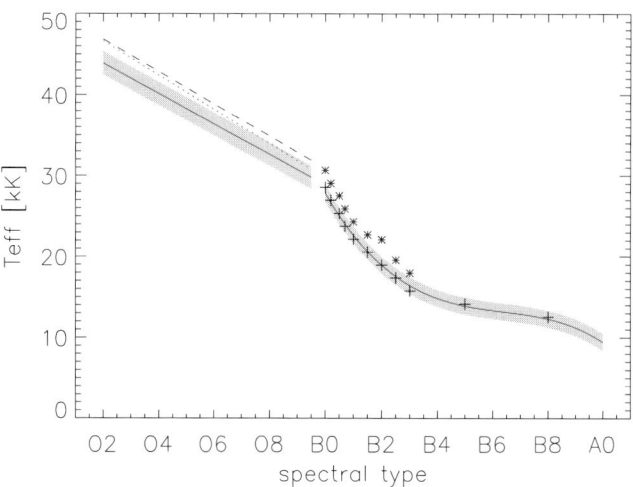

Figure 2. Spectral-type-$T_{\rm eff}$ calibrations for OB-stars. *O-star relations* from Martins et al. (2005) based on own data and data from Herrero et al. (2002); Repolust et al. (2004). Dashed, dotted and solid lines refer to dwarfs, giants and supergiants, respectively. *B-star relations* from Trundle et al. (2007): asterisks and plus-signs refer to dwarfs and supergiants. Overplotted is the corresponding relation (solid) as suggested by Markova & Puls (2008), using own data and data from Crowther et al. (2006); Lefever et al. (2007); Urbaneja (2004); Przybilla et al. (2006). The hatched area refers to the typical uncertainty of the calibrations.

Table 3. NLTE analyses of MC OB-stars since 2002 (without FLAMES). 'dw' = dwarfs, 'sg' = supergiants.

from	reference	objects	code	$\dot{M}$
UV	Martins et al. (2004)	SMC O-dw	CMFGEN	yes
UV + optical	Crowther et al. (2002)	MC O-sg	CMFGEN	yes
	Hillier et al. (2003)	SMC O-sg	"	yes
	Bouret et al. (2003)	SMC O-dw	"	yes
	Evans et al. (2004)	MC OB-sg	"	yes
	Heap et al. (2006)	SMC O-stars	TLUSTY	no
optical	Massey et al. (2004)	MC O-stars	FASTWIND	yes
	Massey et al. (2005)	"	"	yes
	Trundle et al. (2004)	SMC B-sg	FASTWIND	yes
	Trundle & Lennon (2005)	"	"	yes
	Dufton et al. (2006a)	SMC B-sg	TLUSTY	no

Table 4. Targets of the VLT-FLAMES survey of massive stars. Ages from publications as cited in Tab. 5.

Gal.	Cluster	Age	#O	#B
MW	NGC 3293	10-20 Myr	-	99
MW	NGC 4755	10-15 Myr	-	98
MW	NGC 6611	1-4 Myr	13	40
LMC	NGC 2004	10-25 Myr	4	107
LMC	LH9/10	1-5 Myr	44	76
SMC	NGC 330	10-25 Myr	6	109
SMC	NGC 346	1-3 Myr	19	85
total			86	615

McErlean et al. (1999) based on *unblanketed* and "wind-free" models, the new scale is cooler, by 6,000 K for the earliest O-dwarfs to 2,000 K for late O-/early B-supergiants, due to the inclusion of line-blanketing and wind effects.

3. OB-stars in the LMC/SMC and the FLAMES project

Similar investigations have been performed for OB-stars in the Magellanic Clouds, in particular to study metallicity effects on, e.g., effective temperatures and mass-loss rates ($Z(\text{LMC}) \approx 0.5 Z_\odot$, $Z(\text{SMC}) \approx 0.2 Z_\odot$, see Mokiem et al. 2007b and references therein). Tab. 3 summarizes important contributions since 2002, leaving aside the most recent results from the FLAMES survey (see below). Massey et al. (2004, 2005) investigated a large sample of MC O-stars by means of FASTWIND, and provided a spectral-type-T_eff calibration for the SMC. For the LMC, the situation remained unclear, since their sample was concentrated towards the hottest objects, O2-O4. Overall, it turned out that for a given spectral sub-type, $T_\text{eff}(\text{SMC dw}) > T_\text{eff}(\text{MW dw}) \approx T_\text{eff}(\text{SMC sg}) > T_\text{eff}(\text{MW sg})$, where the T_eff-scale for SMC O-stars differs much less from the unblanketed Vacca et al. (1996) calibration than the scale for their Galactic counterparts ('dw' = dwarfs, 'sg' = supergiants). This finding has been attributed to less blanketing and weaker winds due to the lower metallicity. For dwarfs (and also giants), the derived results are in good agreement with investigations of *different* samples performed with CMFGEN, whereas MC supergiants analyzed by means of CMFGEN turned out to be significantly cooler, even cooler than implied by the Galactic scale. Note, however, that a large number of these discrepant objects are somewhat extreme, thus implying significant mass-loss effects (lower T_eff due to strong *wind*-blanketing and wind-emission). Heap et al. (2006) analyzed a sample of SMC O dwarfs and giants, and compared their results (using TLUSTY) with other investigations. In summary, they found a fair agreement (though they stress the large scatter of T_eff within individual sub-types), except for data derived by WM-basic, which seem to suggest systematically cooler temperatures than all other studies.

One of the most important recent projects on OB-stars was the **VLT-FLAMES survey of massive stars** (FLAMES = Fibre Large Array Multi-Element Spectrograph). By means of this ESO *Large Programme* (PI: Stephen Smartt, Belfast), the massive stellar content of 8 Galactic/MC clusters (young and old, see Tab. 4) has been spectroscopically investigated, in order to answer urgent questions regarding (i) rotation and abundances (rotational mixing), (ii) stellar mass-loss as a function of metallicity and (iii) fraction and impact of

Table 5. FLAMES-related investigations

Evans et al. (2005)	overview Galactic clusters
Evans et al. (2006)	overview MC clusters
Mokiem et al. (2006)	SMC O-/early B-stars (parameters, rotation,...)
Mokiem et al. (2007a)	LMC O-/early B-stars (parameters, evolution,...)
Mokiem et al. (2007b)	empirical metallicity dependence of mass-loss rates/ WLR for O-/early B-stars
Dufton et al. (2006b)	Galactic OB stars (parameters, distribution of rotational velocities, cluster ages)
Hunter et al. (2007)	early B-stars in "young" clusters: surface chemical composition
Trundle et al. (2007)	early B-stars in "older" clusters: surface chemical composition, $T_{\rm eff}$-scale
Hunter et al. (2008a)	rotation and evolution of O/ early B-stars in the MCs
Hunter et al. (2008b)	rotation and N-enrichment: the role of rotation

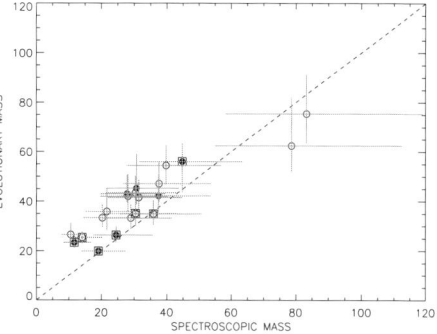

Figure 3. Spectroscopic and evolutionary masses for Galactic O-stars. Squares: rapid rotators; filled: objects with strongly enhanced He. A "real" discrepancy (accounting for errors) is present only for three low-mass stars, but a "mild" discrepancy can still be observed. From Repolust et al. (2004).

binarity. Tab. 5 gives an overview on the various publications which have appeared until now, and we will illuminate important findings in the remainder of this review.

Using a genetic algorithm on top of FASTWIND model atmospheres to obtain the stellar/wind parameters from profile fitting in an objective way (for details, see Mokiem et al. 2005), Mokiem et al. (2006, 2007a) studied the O-/early B-star targets of the FLAMES survey in the SMC and LMC, respectively. Among other results, they confirmed the basic results from Massey et al. (2004, 2005), but refined the spectral-type-$T_{\rm eff}$ scale, particularly with respect to the LMC objects. They showed that at least for O-dwarfs (which are not "contaminated" by additional wind-effects), the effective temperatures for a given spectral sub-type decrease as a function of increasing metallicity, i.e., $T_{\rm eff}$(SMC) > $T_{\rm eff}$(LMC) > $T_{\rm eff}$(MW). A similar result was derived by Trundle et al. (2007) for the FLAMES B-type dwarfs, in this case based on TLUSTY model atmospheres. Interestingly, their $T_{\rm eff}$-scale for B-*supergiants* (which, for Galactic objects, coincides with the calibration provided by Markova & Puls 2008, see Fig. 2) seems to be independent on metallicity. This was interpreted by Markova & Puls (2008) as a consequence of applying, for SMC B-sgs, the (re-)classification scheme by Lennon (1997) which already accounts for the lower metallicity.

4. Masses – still a mass-discrepancy?

An important question concerns the present status of the so-called "mass-discrepancy". Herrero et al. (1992) noted a large discrepancy between masses of *evolved* (Galactic) O-stars derived either spectroscopically (via $\log g$) or via evolutionary calculations, where the latter method resulted in systematically higher values, by roughly a factor of two. Since the spectroscopic analyses had been performed using plane-parallel, unblanketed H/He model-atmospheres (standard at that time), the question arises whether the discrepancy would still be present when the objects were analyzed with more "modern" tools. (Note that even though the inclusion of stellar rotation in recent calculations affects the evolutionary masses, this modification lies well below a factor of two, at least if rotation is not fast enough to result in chemically homogeneous evolution). Fig. 3 shows the outcome of such a re-investigation of the original sample, via blanketed spherical models with winds (Repolust et al. 2004). Indeed, the disagreement has decreased in all cases, and a "real" discrepancy, i.e., non-overlapping error bars, is present only for three low mass objects. However, a milder form seems to have survived, namely that a large fraction of the sample (mostly dwarfs) is located in parallel to the one-to-one relation, shifted by roughly 10 $M_\odot$ upwards in the vertical direction. Similar findings have been reported for LMC and SMC O-/early B-stars (see Massey et al. 2005; Mokiem et al. 2006, 2007a; Heap et al. 2006), and the common result is that the spectroscopic/evolutionary masses of O-*supergiants* seem to agree now, whereas discrepancies are found for dwarfs and giants.

A comprehensive illustration of the (present) problem has been provided by Mokiem et al. (2007a, their Fig. 10) who summarize the mass-discrepancy for Galactic, LMC and SMC O-stars *as a function of Helium content*, i.e, evolutionary status. This figure clearly shows no indication of any discrepancy for all supergiants and bright giants, independent of their Helium enrichment. Insofar, the improvements in atmospheric modeling, evolutionary calculations and spectral analysis techniques seem to have been successful, and the authors argue that the evolution of class I-II objects appears to be "well understood". (These objects are found in that region of the HR-diagram where they have evolved along relatively simple evolutionary tracks). On the other hand, Mokiem et al. find at least a trend that the mass-discrepancy seen for dwarfs and giants increases with increasing He-content, i.e., evolution. They argue that this might be an indication of efficient mixing in the main sequence phase, leading to (near-)chemically homogeneous evolution. Let us point out, however, that there is still a significant number of objects with a normal (or even depleted) He-content and a large discrepancy, which has to be explained in the near future.

5. Macro-turbulence, rotation and rotational mixing

Macro-turbulence. Investigating the efficiency of rotational mixing requires the knowledge of stellar rotational speeds, which can be obtained (at least statistically) from the projected rotational velocities, $v \sin i$, derived primarily from metal lines. For quite a while, however, there have been certain indications of significant broadening processes in addition to rotation (Conti & Ebbets 1977; Lennon et al. 1993; Howarth et al. 1997). Using high resolution, high S/N spectroscopy of early-type B-supergiants, Ryans et al. (2002) showed that this broadening can be described by a *Gaussian* (or similarly shaped) profile, with typical velocities of the order of 50 $\mathrm{km\,s^{-1}}$ for the considered objects. In analogy with solar terminology, this process has been denoted by "macro-turbulence", v_mac.

Fig. 4 clearly shows the presence of such a process, though it turns out that the derivation of unique values for v_mac and $v \sin i$ *in parallel* is difficult using profile fitting methods alone. Fortunately, it is possible to derive an independent estimate of $v \sin i$ from the "first" minimum appearing in the *Fourier transform* of the spectrum. This method was firstly suggested by Carroll (1933), and recent implementations and applications to hot stars have been presented by Simon-Diaz et al. (2006) and Simon-Diaz & Herrero (2007). (For a rigorous discussion, see Gray 2005). Having derived $v \sin i$, the macro-turbulence can be obtained from profile fitting or from the *shape* of the Fourier transform.

Using such an approach, large values of v_mac (of the same order as $v \sin i$) have been derived for OB-supergiants (Dufton et al. 2006a; Lefever et al. 2007; Markova & Puls 2008). Simon-Diaz & Herrero (2007) showed that previous estimates of $v \sin i$ for O-supergiants have been overestimated, by roughly 20-40 $\mathrm{km\,s^{-1}}$. Regarding dwarfs, the

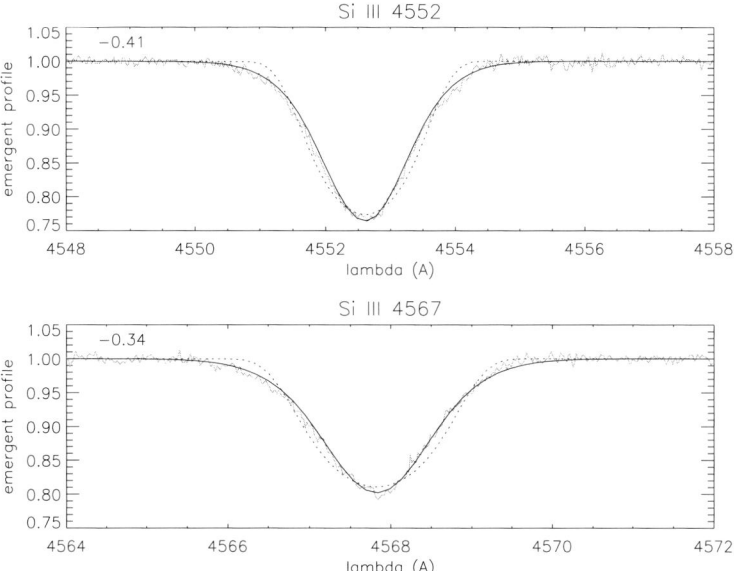

Figure 4. High resolution Si III spectra from HD 89767 (B0Ia), observed with CES@CAT (R = 70,000, S/N ≈ 250). Dotted: Best fitting profile with rotational broadening alone ($v \sin i = 80$ km s^{-1}). Large discrepancies are visible in the wings and cores. Solid: Perfect fit using $v \sin i = 47$ km s^{-1} and $v_{\mathrm{mac}} = 80$ km s^{-1} in parallel. Observations and data from Lefever *et al.* (2007).

present situation remains unclear. Irrespectively, the presence of a Gaussian shaped broadening strongly points to the **presence of symmetrically distributed, deep seated and highly supersonic velocity fields**, in stark contrast to our present understanding of stellar photospheres. This finding needs urgent clarification, and one might speculate about a relation to stellar pulsations and/or winds.

Rotation. Having derived the observed distribution of $v \sin i$ for a large sample of stars, constraints on the intrinsic distribution of rotational speeds, v_{rot}, can be drawn (for the methodology, see Chandrasekhar & Münch 1950), which, for typical sample sizes, is not unique though. Dufton *et al.* (2006b), Mokiem *et al.* (2006) and Hunter *et al.* (2008a) have analyzed the $v \sin i$ distribution of the Galactic, SMC and LMC/SMC FLAMES clusters, respectively. From the B-star sample, Hunter *et al.* (2008a) find a clear difference for LMC and SMC objects, namely that SMC objects (lower metallicity) rotate faster. Obvious reasons are weaker winds (less braking) and smaller stellar radii. Translated to the *intrinsic* v_{rot} distribution, this can be approximated by a Gaussian, with peaks at 100 km s^{-1} for the LMC and 175 km s^{-1} for the SMC (Hunter *et al.* 2008a). The latter value is consistent with the O-/early B-star results from Mokiem *et al.* (2006).

Comparison with evolutionary models: efficiency of rotational mixing. In addition to measuring $v \sin i$, Hunter *et al.* (2008b) have derived surface abundances, in particular those of nitrogen, which is a key-element for stellar evolution. Combining these data with similar data from slow rotators (Hunter *et al.* 2007; Trundle *et al.* 2007) obtained by identical methods/codes (i.e., the complete dataset is internally consistent), they were able to investigate the efficiency of rotational mixing for LMC B-stars, by plotting nitrogen abundance vs. $v \sin i$, and overplotting corresponding evolutionary tracks for different v_{rot} based on recent models from Yoon & Langer (2005). These models have been calibrated with respect to the efficiency of rotational mixing and overshooting parameter to reproduce the observed behaviour of the *bulk* of the objects in this

"Hunter-diagram" (at and slightly above the LMC nitrogen baseline abundance, with $v_{\rm rot}$ not larger than roughly 200 km s^{-1}). In the following, we will concentrate on the results of this comparison for core-hydrogen burning objects alone, which have been found to be located at gravities $\log g \geqslant 3.2$, where objects with $\log g \geqslant 3.7$ are less evolved and the rest is evolved. By comparing the observed positions with the evolutionary tracks, two groups of problematic objects have become visible:

Group 1 consists of rapidly rotating, evolved objects with rather *weak* nitrogen enhancement, i.e., little mixing. According to theory, such objects should be much more enriched, at least if they were single stars. Though the observed low degree of N-enhancement might be the outcome of close binary evolution (Petrovic *et al.* 2005), no indication of binarity (from radial velocity variability) has been found at least for three objects.

Group 2 objects are slow rotators with significant N-enrichment. Since this group comprises a large number of objects, it is very unlikely to see all of them pole-on; consequently, these objects provide a real challenge for stellar evolution. Again, they might be the product of binary evolution, but Hunter *et al.* invoke also a possible correlation with magnetic fields.

The reason for doing so is based on the analysis of slowly rotating Galactic β Cep stars by Morel *et al.* (2006), who found significant nitrogen-enhancement in four out of their ten sample objects, i.e., a similar situation as for the *Group 2* objects. The authors argue that an origin due to pulsations is unlikely. Interestingly now, three of these four objects have strong magnetic fields, of order of several hundred Gauss (see Tab. 6). For the non-enhanced objects, on the other hand, no indication for *B*-fields has been found.

Thus, Hunter *et al.* (2008b) conclude that rotational mixing seems to be not the only mixing process, since (i) it is not efficient at low $v_{\rm rot}$ and (ii) it is unclear if it is the main mixing process at high $v_{\rm rot}$, due to the presence of *Group 1* objects. Binarity and/or magnetic fields might help to explain the observed distribution of nitrogen abundances. Finally, note that Mokiem *et al.* (2006) in their SMC O-star analysis found three slowly rotating OVz stars with strongly enhanced Helium content, which might originate from the same process as present in the *Group 2*-stars.

6. Magnetic fields

So, what is the probability to find a sufficient number of massive stars with strong magnetic fields? Until 2006, only *eight* magnetic massive stars (including the nitrogen enriched β Cep stars from Morel *et al.* 2006) were known, excluding Ap/Bp stars (see Tab. 6, where B_p is the magnetic field strength in Gauss at the magnetic pole of the approximately dipolar field). Interestingly, all of these stars display certain peculiarities in their spectra. The listed field strengths have been derived by means of Stokes-V spectropolarimetry, collected with the MuSiCoS polarimeter (Donati *et al.* 1999) mounted at the Telescope Bernard Lyot(TLB), Pic du Midi and the AAT, with ESPaDOnS@CHFT and with FORS1@VLT.

Recently, the number of analyzed objects considerably increased due to the thesis work by Schnerr (2007) and co-workers. This team has surveyed 25 OB-stars at various phases with MuSiCoS@TBL, and additional 11 O-stars at three different phases with FORS1@VLT. No evidence for magnetic fields has been found in any target, with 1-σ upper limits of $\sim$40-100 Gauss for the longitudinal field averaged over the stellar disk. A similar result has been obtained for 12 A-supergiants by Verdugo *et al.* (2003).

Table 6. Compilation of magnetic massive stars known until 2006, from Schnerr (2007). For references, see same work.

Star	Sp. Type	B_p
θ^1 Ori C	O4-6V	1100 ± 100
HD 191612	O6-8	~ 1500
τ Sco	B0.2V	~ 500
ξ^1 CMa †	B1III	~ 500
β Cep †	B1IV	360 ± 40
V2052 Oph †	B1V	250 ± 190
ζ Cas	B2IV	340 ± 90
ω Ori	B2IVe	530 ± 200

Table 7. Mass-loss rates from H_α alone.

reference	objects	method
Lamers & Leitherer (1993)	Gal. O-stars	approx.
Puls et al. (1996)	Gal./MC O-stars	approx.
Kudritzki et al. (1999)	Gal. BA-sg	unblanketed model atm.
Markova et al. (2004)	Gal. O-stars	approx.

† nitrogen enriched β Cep stars from Morel et al. (2006)

In conclusion, for non-peculiar hot stars, B is either weak and/or acts only on small scales (spots). To cite Donati et al. (2006), "magnetic fields (at least those of moderate to high intensity) are not a common feature of most hot stars, but rather a rare occurrence."

Nevertheless, even weak fields can have some impact, at least on the stellar winds from massive stars. As outlined by ud-Doula & Owocki (2002), the decisive quantity is the so-called confinement parameter, η_*, which measures the ratio of the magnetic to the wind energy at the magnetic equator,

$$\eta_* = \frac{E_B}{E_{\mathrm{wind}}} \ (\mathrm{magn.\ equator}) \approx 0.19 \frac{B_{100}^2(\mathrm{polar}) R_{10}^2}{\dot{M}_{-6} v_8},$$

when B_p is measured in units of 100 Gauss, the stellar radius, R, in units of $10\,R_\odot$, the mass-loss rate, $\dot{M}$, in units of $10^{-6}\,M_\odot/\mathrm{yr}$ and the terminal velocity, v, in units of $1000\,\mathrm{km\,s^{-1}}$. If $\eta_* \geq 1$, magnetic fields have a significant or strong effect on the wind, leading to the formation of a disk and strong shocks. But even if $\eta_* \approx 0.1$, B still has some impact, leading to density enhancements at the magnetic equator. Two examples: to reach $\eta_* = 1$ for the strong wind of ζ Pup ($R_{10} \approx 2$, $\dot{M}_{-6} \approx 4$, $v_8 \approx 2$), a considerable field strength of $B_p \approx 320$ Gauss is required. For a rather weak wind with $\dot{M} = 10^{-8}\,M_\odot/\mathrm{yr}$ and $v_\infty = 2000\,\mathrm{km\,s^{-1}}$, on the other hand, $\eta_* = 1$ is already reached for $B_p \approx 32$ Gauss, i.e., well below present detection capabilities!

7. Wind properties of OB stars at different metallicities

Overview. Wind-properties of Galactic/MC OB-stars (primarily mass-loss rates, $\dot{M}$, and velocity field parameters, β) have been determined by numerous investigations, from H_α alone (Tab. 7) and other spectral ranges (Tab. 2 and 3). In most cases, terminal velocities, v_∞, have been adopted from UV-measurements and/or calibrations (see Kudritzki & Puls 2000 and references therein). Wind-momentum luminosity relations (WLR) have been inferred and compared with theoretical predictions, mostly from Vink et al. (2000, 2001). Remember that radiation driven wind theory predicts a power-law relation between modified wind-momentum rate, D_{mom}, and stellar luminosity,

$$\log D_{\mathrm{mom}} = \log(\dot{M} v_\infty R_\star^{1/2}) \approx x \log(L/L_\odot) + D(\mathrm{metallicity,\ spectral\ type})$$

(see Kudritzki & Puls 2000 and references therein). The most important results of these studies can be summarized as follows: (i) for a given luminosity, mass-loss rates of SMC-stars are lower than for their Galactic counterparts, consistent with theory. (ii) For O-/early B-stars, the theoretical WLR is met, except for O-sgs with rather dense winds, where the "observed" wind-momenta appear as "too large" (explained by wind-clumping, see below), and for low luminosity O-dwarfs, where the "observed" wind-momenta are considerably lower than predicted (denoted as the "weak wind problem"). (iii) B-sgs located *below* the so-called bi-stability jump (i.e., with $T_{\text{eff}} <$ 22 kK, Vink *et al.* 2000) show lower wind-momenta than predicted (Crowther *et al.* 2006; Markova & Puls 2008).

The **metallicity dependence** of the winds from O-/early B-stars could be quantified due to the large sample provided by the FLAMES survey. From the analysis of the SMC/LMC objects by Mokiem *et al.* (2006) and Mokiem *et al.* (2007a), respectively, and in combination with data from previous investigations, Mokiem *et al.* (2007b) derived the WLRs for Galactic, LMC and SMC objects, with rather small 1-σ confidence intervals, and showed that the wind-momenta strictly increase with metallicity Z (i.e., the WLR of the LMC lies in between the corresponding relations for the MW and the SMC). Using $Z(\text{LMC}) \approx 0.5 Z_\odot$, $Z(\text{SMC}) \approx 0.2 Z_\odot$, and allowing for a "clumping correction", they obtained $\dot{M} \propto Z^{0.72 \pm 0.15}$, which is in very good agreement with theory. Further details can be found in de Koter *et al.*, this volume.

The weak wind problem (see also Hillier, this volume). Bouret *et al.* (2003) were the first to note a significant discrepancy between observed and predicted mass-loss rates for late-type O-dwarfs (the observed ones being lower), in their sample of SMC/NGC346 objects. Martins *et al.* (2004) observed a similar disagreement, in this case for extremely young O-dwarfs in SMC/N81, where the differences turned out to be larger than a factor of 10. Though investigating various reasons for this failure of theory, none turned out to be conclusive. The same problem was found by Martins *et al.* (2005) for Galactic O-dwarfs with $\log L/L_\odot < 5.2$. To date, this discrepancy still lacks any explanation, but one might speculate about a relation to magnetic fields, since only low field strengths are necessary to induce significant effects on the wind topology (see above). Note also that τ Sco (B0.2V) with a very large B-field (Tab. 6) is probably affected by the weak wind problem.

Clumping in hot star winds was the objective of an international workshop held in Potsdam 2007. The following is only a *very* brief summary of the status quo. For details, the reader is referred to the corresponding proceedings (Hamann *et al.*, eds., 2008).

During recent years, there have been various direct and indirect indications that hot star winds are not smooth, but clumpy, i.e., that there are small scale density inhomogeneities which redistribute the matter into over-dense clumps and an almost void inter-clump medium. Theoretically, such inhomogeneities have been expected from the first hydrodynamical wind simulations on (Owocki *et al.* 1988), due to the presence of a strong instability inherent to radiative line-driving. This can lead to the development of strong reverse shocks, separating over-dense clumps from fast, low-density wind material. Interestingly, however, the column-depth averaged densities and velocities remain very close to the predictions of stationary theory. (For more recent results, see Runacres & Owocki 2002, 2005 (1-D) and Dessart & Owocki 2003, 2005 (2-D)).

In order to treat wind-clumping in present atmospheric codes, the standard assumption (so far) relates to the presence of optically *thin* clumps and a void inter-clump medium. A consistent treatment of the disturbed velocity field is still missing. The over-density (w.r.t. average density) inside the clumps is described by a "clumping factor", f_{cl}. The most important consequence of such a structure is that any $\dot{M}$ derived from ρ^2-diagnostics (H$_\alpha$, radio) using *homogeneous models* needs to be scaled down by a factor of $\sqrt{f_{\text{cl}}}$.

Based on this approach, Crowther *et al.* (2002); Hillier *et al.* (2003); Bouret *et al.* (2003, 2005) derived clumping factors of the order of 10...50, with clumping starting at or close to the wind base. From these values, a reduction of (unclumped) mass-loss rates by factors 3...7 seems to be necessary. The *radial* stratification of the clumping factor has been studied by Puls *et al.* (2006), from a simultaneous modeling of H_α, IR, mm and radio observations. They found that, at least in dense winds, clumping is stronger in the lower wind than in the outer part, by factors of 4...6, and that unclumped mass-loss rates need to be reduced *at least* by factors 2...3.

Even worse, the analysis of the FUV P v-lines by Fullerton *et al.* (2006) seems to imply factors of 10 or larger, which would have an enormous impact on massive star evolution. However, as suggested by Oskinova *et al.* (2007), the analyses of such optically *thick* lines might require the consideration of wind "porosity", which reduces the effective opacity at optically thick frequencies (Owocki *et al.* 2004, Cohen, this volume). Consequently, the reduction of $\dot{M}$ as implied by the work from Fullerton *et al.* might be overestimated, and factors similar to those cited above (around three) are more likely.

8. Summary

In this review, we have highlighted important findings and conclusions regarding the physical and wind properties of OB-stars, which have been obtained since the last massive star symposium in 2002:
1. The $T_{\rm eff}$-scale of OB-stars has become lower due to the effects of line- and wind-blanketing accounted for in present-day atmospheric models. **2.** Particularly for supergiants, there is a significant spread in $T_{\rm eff}$ for a given spectral sub-type, probably related to different degrees of mass-loss. **3.** For a given spectral sub-type, the effective temperatures of OB-dwarfs increase with decreasing metallicity. **4.** The "mass-discrepancy" has been solved for O-sgs, but there are still problems for dwarfs and giants. **5.** The physics (and consequences) of *supersonic* macro-turbulence detected in OB-sgs needs to be understood. **6.** The $v_{\rm rot}$-distribution of OB-stars peaks at higher values for samples with lower metallicities. **7.** Rotational mixing alone might not be able to explain the observed nitrogen-abundances in OB-stars. Binarity and/or magnetic fields might help. **8.** But: Magnetic fields of *significant* strength seem to be absent in *normal* OB stars, though weak fields can affect weak winds. **9.** Mass-loss rates scale with metallicity as $\dot{M} \propto Z^{0.72\pm0.15}$. **10.** The weak-wind problem needs to be clarified. **11.** Mass-loss rates derived from homogeneous models need to be scaled down, due to the presence of wind-clumping. Factors of three are rather likely.

Acknowledgements

Many thanks to Artemio Herrero and Jon Sundqvist for carefully reading the manuscript. The author gratefully acknowledges a travel grant by the *Deutsche Forschungsgemeinschaft*, under grant Pu117/5-1.

References

Bianchi, L., Herald, J., & Garcia, M. 2006, in: *The Ultraviolet Universe: Stars from Birth to Death, 26th meeting of the IAU, Joint Discussion 4* (JD04, #36)
Bianchi, L. & Garcia, M. 2002, *ApJ*, 581, 610
Bouret, J.-C., Lanz, T., Hillier, D. J., *et al.* 2003, *ApJ*, 595, 1182
Bouret, J.-C., Lanz, T., & Hillier, D. J. 2005, *A&A*, 438, 301
Carroll, J. A. 1933, *MNRAS*, 93, 478
Chandrasekhar, S. & Münch, G. 1950, *ApJ*, 111, 142
Conti, P. S. & Ebbets, D. 1977, *ApJ*, 213, 438
Crowther, P. A., Hillier, D. J., Evans, C. J., *et al.* 2002, *ApJ*, 579, 774
Crowther, P. A., Lennon, D. J., & Walborn, N. R. 2006, *A&A*, 446, 279
Dessart, L. & Owocki, S. P. 2003, *A&A*, 406, L1

Dessart, L. & Owocki, S. P., 2005 *A&A*, 437, 657
Donati, J.-F., Catala, C., Wade, G. A., et al. 1999, *A&AS*, 134, 149
Donati, J.-F., Howarth, I. D., Jardine, M. M., et al. 2006, *MNRAS*, 370, 629
Dufton, P. L., Ryans, R. S. I., Simon-Diaz, S., et al. 2006a, *A&A*, 451, 603
Dufton, P. L., Smartt, S. J., Lee, J. K., et al. 2006b, *A&A*, 457, 265
Evans, C. J., Lennon, D. J., Trundle, C., et al. 2004, *ApJ*, 607, 451
Evans, C. J., Smartt, S. J., Lee, J.-K., et al. 2005, *A&A* 437, 467
Evans, C. J., Lennon, D. J., Smartt, S. J., et al. 2006, *A&A* 456, 623
Fullerton, A. W., Massa, D. L., & Prinja, R. K. 2006, *ApJ*, 637, 1025
Garcia, M. & Bianchi, L. 2004, *ApJ*, 606, 497
Gray, D. F. 2005, *The observation and analysis of stellar photospheres, 3rd edition* (Cambridge: Cambridge University Press)
Hamann, W.-R., Feldmeier, A., & Oskinova, L. M. 2008, *Proc. International Workshop on Clumping in Hot-Star Winds* (Potsdam: Universitätsverlag Potsdam)
Heap, S. R., Lanz, T., & Hubeny, I. 2006, *ApJ*, 638, 409
Herrero, A., Kudritzki, R. P., Vilchez, J. M., et al. 1992, *A&A*, 261, 209
Herrero, A., Puls, J., & Najarro, F. 2002, *A&A*, 396, 949
Howarth, I. D., Siebert, K. W., Hussain, G. A. J., et al. 1997, *MNRAS*, 284, 265
Hillier, D. J., Lanz, T., Heap, S. R., et al. 2003, *A&A*, 588, 1039
Hunter, I., Dufton, P. L., Smartt, S. J., et al. 2007, *A&A*, 466, 277
Hunter, I., Lennon, D. J., Dufton, P. L., et al. 2008a, *A&A*, 479, 541
Hunter, I., Brott, I., Lennon, D. J., et al. 2008b, *ApJ*, 676, L29
Kudritzki, R. P. & Puls, J. 2000, *ARA&A*, 38, 613
Kudritzki, R. P., Puls, J., Lennon, D. J., et al. 1999, *A&A*, 350, 970
Lamers, H. J. G. L. M. & Leitherer, C. 1993, *A&A*, 412, 771
Lefever, K., Puls, J., & Aerts, C. 2007, *A&A*, 463, 1093
Lennon, D. J. 1997, *A&A*, 317, 871
Lennon, D. J., Dufton, P. L, & Fitzsimmons, A. 1993, *A&AS*, 97, 559
Markova, N. & Puls, J. 2008, *A&A*, 478, 823
Markova, N., Puls, J., Repolust, T., et al. 2004, *A&A*, 413, 693
Martins, F., Schaerer, D., Hillier, D. J., et al. 2004, *A&A*, 420, 1087
Martins, F., Schaerer, D., & Hillier, D. J. 2005, *A&A*, 436, 1049
McErlean, N. D., Lennon, D. J., & Dufton, P. L. 1999, *A&A*, 349, 553
Massey, P., Kudritzki, R. P., Bresolin, F., et al. 2004, *ApJ*, 608, 1001
Massey, P., Puls, J., Pauldrach, A. W. A., et al. 2005, *ApJ*, 627, 477
Mokiem, M. R., de Koter, A., Puls, J., et al. 2005, *A&A*, 441, 711
Mokiem, M. R., de Koter, A., Evans, C. J., et al. 2006, *A&A*, 456, 1131
Mokiem, M. R., de Koter, A., Evans, C. J., et al. 2007a, *A&A*, 465, 1003
Mokiem, M. R., de Koter, A., Vink, J. S., et al. 2007b, *A&A*, 473, 603
Morel, T., Butler, K., Aerts, C., et al. 2006, *A&A*, 457, 651
Najarro, F., Hillier, D. J., Puls, J., et al. 2006, *A&A*, 456, 659
Oskinova, L. M., Hamann, W.-R., & Feldmeier, A. 2007, *A&A*, 476, 1331
Owocki, S. P., Castor, J. I., & Rybicki, G. B. 1988, *ApJ*, 335, 914
Owocki, S. P., Gayley, K. G., & Shaviv, N. J. 2004, *ApJ*, 616, 525
Petrovic, J., Langer, N., & van der Hucht, K. A. 2005, *A&A*, 435, 1013
Przybilla, N., Butler, K, Becker, S. R., et al. 2006, *A&A*, 445, 1099
Puls, J., Kudritzki, R. P., Herrero, A., et al. 1996, *A&A*, 305, 171
Puls, J., Markova, N., Scuderi, S., et al. 2006, *A&A*, 454, 625
Repolust, T., Puls, J., & Herrero, A. 2004, *A&A*, 415, 349
Repolust, T., Puls, J., Hanson, M. M., et al. 2005, *A&A*, 440, 261
Runacres, M. C. & Owocki, S. P. 2002, *A&A*, 381, 1015
Runacres, M. C., & Owocki, S. P. 2005, *A&A*, 429, 323
Ryans, R. S. I., Dufton, P. L., Rolleston, W. R. J., et al. 2002, *MNRAS*, 336, 577
Schnerr, R. S. 2007, *PhD Thesis* (University of Amsterdam, The Netherlands)

Simon-Diaz, S. & Herrero, A. 2007, *A&A*, 468, 1063
Simon-Diaz, S., Herrero, A., Esteban, C., et al. 2006, *A&A*, 448, 351
Trundle, C. & Lennon, D. J. 2005, *A&A*, 434, 677
Trundle, C., Lennon, D. J., Puls, J., et al. 2004, *A&A*, 417, 217
Trundle, C., Dufton, P. L., Hunter, I., et al. 2007, *A&A*, 471, 625
ud-Doula, A. & Owocki, S. P. 2002, *ApJ*, 576, 413
Urbaneja, M. A. 2004, *PhD Thesis* (University of La Laguna, Spain)
Vacca, W. D., Garmany, C. D., & Shull, J. M. 1996, *ApJ*, 460, 914
Verdugo, E., Talavera, A., Gómez de Castro, A. I., et al. 2003, in: K. van der Hucht, A. Herrero & C. Esteban (eds.), *A Massive Star Odyssey: From Main Sequence to Supernova* (San Francisco: ASP) *Proc. IAU Symp 212*, p. 255
Vink, J. S., de Koter, A., & Lamers, H. J. G. L. M. 2000, *A&A*, 362, 295
Vink, J. S., de Koter, A., & Lamers, H. J. G. L. M. 2001, *A&A*, 369, 574
Yoon, S.-C. & Langer, N. 2005, *A&A*, 443, 643

Discussion

MEYNET: Regarding the Hunter diagram. Two Remarks:
1. It would have been surprising that present rotating models would successfully reproduce all the observations in all details. My feeling is that rotating models can account for the general trends and the bulk of the observed data.
2. The Hunter diagram is not very easy to analyze. It contains stars of different initial mass, of different ages, the inclination makes the things also complicate to interpret. Thus I would be more careful in the conclusions about the difficulties of rotational mixing based on this diagram.

PULS: I've not established the diagram, I have only reported on this result which is, in my opinion, a very important one. I agree that regarding the stars with high vsini and low Nitrogen abundance, the conclusions are difficult, due to the low number of objects. Certainly, however, there *is* a problem with stars with low vsini which are strongly enriched. Note that due to the large number of stars showing this problem, the probability that all have been observed pole-on is very low, and that rotational mixing for slowly rotating stars cannot be efficient. Thus, I would conclude that indeed there is a class of objects with abundances which cannot be explained by rotational mixing alone.

KUDRITZKI: How do you explain the difference between the new $T_{\rm eff}$-scales for O-stars derived from the He I/II equilibrium and the work by Garcia/Bianchi and Sally Heap and collaborators who use UV lines?

PULS: In my talk, I haven't outlined this problem. Note that also Sally Heap and co-workers state that the $T_{\rm eff}$-scale by Garcia/Bianchi is significantly lower than "all" other results, including their own work. This discrepancy has not been solved yet and urgently requires further effort. But note also that the UV-temperature scale by Garcia/Bianchi results from analyses performed with Adi Pauldrach's code WM-basic, which has not been designed for the analysis of photospheric conditions (e.g., the photospheric line-acceleration is far from being perfect, due to the use of the mCAK formalism to calculate this quantity), so that such discrepancies might also result from the used approximations, at least in part.

STANEK: Why don't you compare the evolutionary masses and the spectroscopic masses with the actual masses derived from eclipsing binaries (EB's)? There are now samples of well observed EB's in the SMC, LMC and Milky Way.

PULS: Indeed, there have been attempts to do this, particularly for B-dwarfs where the present answer is not unique. The major problem is that there are very few EBs (particularly in the O-star range) which have been analyzed by spectroscopic tools, and for those few cases the EB results lie mostly in between the spectroscopic and evolutionary results. Certainly, a systematic comparison has to be performed in the near future. But note also that the "mass-discrepancy" paradigm has somewhat changed: Previously, before using line-/wind-blanketed models, the discrepancy was located in the supergiant range, whereas now (after improving on the models) the dwarfs are most affected.

DOPITA: In turbulent atmospheres with multiple shocks the clumping distribution is more likely to be log-normal rather than the two-phase as assumed in the X-ray opacity models. Have more sophisticated clumping distributions such as log-normal been modelled yet?

PULS: Indeed, analyses of the mass-spectrum of the clumps in Wolf-Rayet winds by Moffat, St-Louis and co-workers point to the presence of a power-law distribution, consistent with the assumption of supersonic turbulence. These results, however, refer mostly to the structure in the outer wind (the inner one is optically thick). With respect to the analyses of the more thinner O-star winds, where the present working hypothesis on the origin of the clumps is related to the line-driven instability, we are just in the beginning of analyses and modeling. Note that for optically thin clumps (regarding, e.g., H_α) the actual distribution plays a minor role, and only the average, effective overdensity (or volume filling factor) is of importance. Here, "we" are just at the beginning to obtain results for the corresponding spatial distribution. For optically thick processes such as UV and X-ray line emission, there is the alternative "porosity" approach by Owocki and co-workers, which defines the porosity length (a combination of blob cross-section and separation) as the photons' mean free path in a porous medium. At least for such models, the distribution of the optical thickness of the clumps has been assumed by a power-law as well.

NIEVA: Concerning the "Hunter-plot", you said something I don't agree very much: "This is a challenge to theory". It is also a challenge to the quantitative spectral analysis. Systematic errors up to 1 dex are not taken into account in most spectral analyses for abundance determinations. I would consider these results with caution.

PULS: Since I did not perform this work myself, I don't know about the individual errors. But note that the Hunter plot compares the derived abundances in a differential way, where the theoretical predictions have been gauged to match the "observed" bulk of the observations. Since the discrepancy (for the low vsini stars) is of the order of one dex and since many objects are affected (and there are many more with "normal" abundances), I think that the discrepancy is real.

The Metallicity Dependence of the Mass Loss of Early-Type Massive Stars

Alex de Koter

Astronomical Institute Anton Pannekoek, University of Amsterdam,
Kruislaan 403, NL-1098SJ, Amsterdam, the Netherlands
email: dekoter@science.uva.nl

Abstract. We report on a comprehensive study of the wind properties of 115 O- and early B-type stars in the Galaxy and the Large Magellanic Clouds. This work is part of the VLT/FLAMES Survey of Massive Stars. The data is used to construct the empirical dependence of the mass-loss in stellar winds on the metal content of their atmospheres. The metal content of early-type stars in the Magellanic Clouds is discussed. Assuming a power-law dependence of mass loss on metal content, $\dot{M} \propto Z^m$, we find $m = 0.83 \pm 0.16$ from an analysis of the wind momentum luminosity relation (Mokiem *et al.* 2007b). This result is in good agreement with the prediction $m = 0.69 \pm 0.10$ by Vink *et al.* (2001). Though the scaling agrees, the absolute empirical value of mass loss is found to be a factor of two higher than predictions. This may be explained by a modest amount of clumping in the outflows of the objects studied.

Keywords. techniques: spectroscopic – stars: atmospheres – stars: early-type – stars: mass loss – galaxies: abundances

1. Introduction

One of the fundamental aims of the *VLT/FLAMES Survey of Massive Stars* (Evans *et al.* 2005, 2006) is to empirically determine the relation between the mass loss rate of early-type massive stars as a result of radiation pressure on spectral lines and their surface chemical composition. The *Fibre Large Array Multi-Element Spectrograph* (FLAMES), the first wide-field, multi-object spectrograph instrument on an 8-m class telescope allowed to collect an unprecedented number of (over) 800 high-quality spectra of stars in the Galaxy and Magellanic Clouds in only ∼100 hours of *Very Large Telescope* (VLT) time. A total of seven clusters was observed. For the main sequence objects the wind strengths are a strong function of spectral type, therefore only the sub-sample of O and early-B stars can be used to determine the relation between mass loss $\dot{M}$ and metal content Z. The young cluster N11 in the Large Magellanic Cloud (LMC) and NGC 346 in the Small Magellanic Cloud (SMC) are particularly rich in these objects (see Fig. 1).

Here we present an overview of the main results of this part of the FLAMES project. Specifically, we discuss the chemical composition of the Magellanic Clouds (Sect. 2) and the modified wind momentum diagram from which the $\dot{M}(Z)$ relation is derived (Sect. 3). We end with an outlook on extending the mass loss vs. metallicity relation towards metallicities below that of the SMC.

2. The metallicity of the Magellanic Clouds

The winds of massive stars are driven by spectral lines of heavy elements, in particular iron. Unfortunately, the optical spectra of O stars show few features of these species. B stars, however, are relatively rich in absorption lines due to carbon, nitrogen, oxygen,

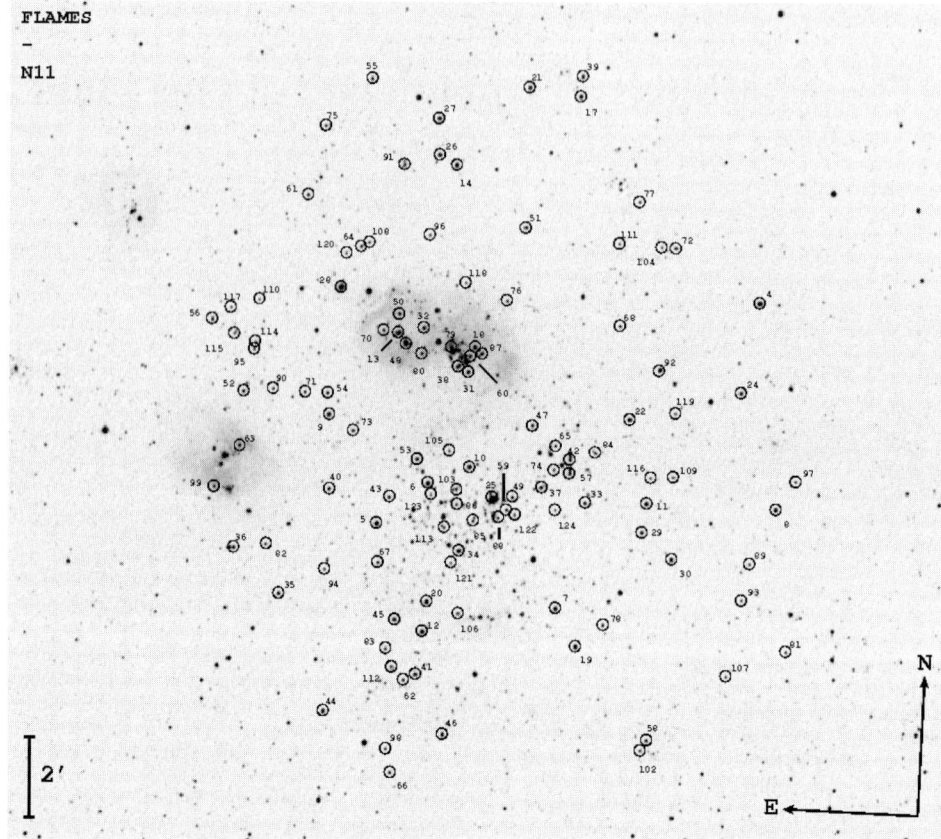

Figure 1. V-band wide field image of FLAMES targets in N11 in the LMC. Our observations sample the central associations of LH9 (south of center) and LH10 (north of center), and the surrounding regions. Stellar parameters of 22 O and early-B stars have been determined using an automated fitting method. Image from Evans *et al.* (2006).

magnesium, aluminum, silicon and sulphur. Iron lines are also present though these are intrinsically quite weak (especially so in the SMC; Rolleston *et al.* 2003). The present-day composition for over 100 slowly-rotating early-B stars in the LMC and SMC was studied by Hunter *et al.* (2007) and Trundle *et al.* (2007) using the TLUSTY model atmosphere code. In Table 1 the derived abundances are given, supplemented with results for Al and S from the literature.

Trundle *et al.* (2007) determine a mean relative to solar iron abundance for 13 stars of their sample in NGC 2004 in the LMC of $\Delta[\text{Fe}/\text{H}] = -0.29 \pm 0.13$, in good agreement with the depletions of O, Mg and S. Notice that C and N are significantly underabundant in the clouds. We adopt $\Delta[Z/\text{H}] = -0.3 \pm 0.1$ for the LMC, but note that the carbon and nitrogen depletion may have a small effect on the wind strengths of the hottest stars where these elements (mainly C) are a significant contributor to the line force.

In the SMC the mean relative to solar of the α elements (O, Mg, Si) is $\Delta[\text{Fe}/\text{H}] = -0.7 \pm 0.1$. The differential result of Al is in good agreement with this value. The iron abundance $\Delta[\text{Fe}/\text{H}] = -0.57 \pm 0.16$ is somewhat higher, but is in agreement within the uncertainties. Studies of AFGK supergiants, which have more and stronger metal lines, tend to yield an iron abundance that is in good agreement with this value (Venn 1999).

Table 1. Present-day chemical composition of the LMC and SMC from B stars. The solar abundances of Asplund *et al.* (2005) are given for reference. References: Hunter *et al.* (2007), Trundle *et al.* (2007), Rolleston *et al.* (2002), Rolleston *et al.* (2003). The latter two references pertain to aluminum and sulfur and reflect LTE results. From Mokiem *et al.* (2007b).

Element	Solar	LMC $12+\log X/H$	LMC $\Delta[X/H]$	SMC $12+\log X/H$	SMC $\Delta[X/H]$
C	8.39	7.73	−0.66	7.37	−1.02
N	7.78	6.88	−0.90	6.50	−1.28
O	8.66	8.35	−0.31	7.98	−0.68
Mg	7.53	7.06	−0.47	6.72	−0.81
Al	6.37	...	...	5.43	−0.72
Si	7.51	7.19	−0.32	6.79	−0.72
S	7.14	...	...	6.44	−0.42
Fe	7.45	7.23	−0.29	6.93	−0.57

We adopt $\Delta[Z/H] = -0.7 \pm 0.1$, but will point out the effect of a higher SMC metal content on the mass loss vs. metallicity relation in Sect. 3.

3. Mass loss as a function of environment

The mass loss vs. metallicity relation presented in the paper is based on 115 objects. As the metallicity differences in the three galaxies considered are modest, it is important that the photospheric and wind parameters of this sample are derived in as homogeneous a way as is possible. To cope with the large dataset provided by the FLAMES survey, to improve the objectivity of the analysis, and to strive towards a homogeneous analysis we have developed an automated fitting method of spectral lines based on genetic algorithms (Mokiem *et al.* 2005). The method uses synthetic line profiles generated by FASTWIND (Puls *et al.* 2005), was successfully tested using well studied Galactic objects (Herrero *et al.* 2002; Mokiem *et al.* 2005) and applied to FLAMES LMC (Mokiem *et al.* 2006) and SMC (Mokiem *et al.* 2007a) targets. An important advantage of our approach is that the derived uncertainties in the model parameters reflect possible degeneracies (in combinations of parameters). Moreover, if the mass loss rate is so low that its prime diagnostic (in our case Hα) is no longer significantly affected by wind emission the method will signal this by being unable to quantify a lower limit to $\dot{M}$.

Though sizeable, the total number of objects does not allow to establish the relation between mass loss and chemical composition on the basis of a comparison of "identical" (except for Z) objects in the Galaxy, LMC and SMC. Actually, this would not even be an appealing approach given the possibility that stars in different parts of the Hertzsprung-Russell diagram and/or of disparate mass may not obey the same $\dot{M}(Z)$. At present the only way to get more insight in the universality of the mass loss-metallicity relation is to turn to predictions of radiation-driven wind theory. A powerful way to proceed is through the use of the *modified wind momentum - luminosity relation*, (WLR; e.g. Kudritzki & Puls 2000)

$$\log D_{\rm mom} \equiv \log(\dot{M} v_\infty \sqrt{R}) \simeq x \log(L/L_\odot) + \log D_\circ, \qquad (3.1)$$

where the slope x and the constant $D_\circ$ may vary as a function of spectral type and metal content (see e.g. Puls *et al.* 2000). The WLR expresses that the mechanical momentum of the stellar wind (the product of mass loss $\dot{M}$ and terminal flow velocity v_∞) is primarily a function of photon momentum (the ratio of the luminosity L and the speed of light c).

The uniqueness of the WLR for specific ranges in spectral type and metallicity has been confirmed by e.g. Vink *et al.* (2000).

Assuming that mass loss and terminal velocity are power laws of metallicity, i.e.

$$\dot{M} \propto Z^m \quad \text{and} \quad v_\infty \propto Z^n, \tag{3.2}$$

it follows that

$$(m + n) = \Delta \log D_{\mathrm{mom}} / \Delta \log Z. \tag{3.3}$$

It is found that for O and early-B stars the slope x is quite similar, though not identical, for the Galaxy and Magellanic Clouds. Because of the slightly varying slopes, a fixed luminosity is picked at which the respective WLRs are compared. We have used $L = 10^{5.75} L_\odot$. For the dependence of v_∞ on metallicity we adopt the theoretical value $n = 0.13$ (Leitherer *et al.* 1992).

The Galactic WLR consists of 49 objects, ranging in spectral type from O2 to B1 and includes dwarfs, giants and supergiants, of which 24 percent was analyzed with the automated method. The properties of the bulk of this sample are collected from the literature. In doing this, we restricted ourselves to using results based on state-of-the-art unified non-LTE line-blanketed model atmospheres only (Hillier & Miller 1998; Puls *et al.* 2005). Stars for which only upper limits to the mass loss rate could be determined were discarded. This was done also for the LMC and SMC case. The LMC sample has 38 objects, of which 58 percent is analyzed using the automated method. Coverage of spectral classes is similar as for the Galactic sample. The 22 targets observed in the FLAMES program have more than doubled the statistics of LMC stars (of these types). It is this improvement in numbers that has allowed us to construct the first robust WLR for this galaxy. Anticipating the outcome, the FLAMES results have established for the first time that the wind strengths of LMC stars are intermediate between those of Galactic and SMC stars. Finally, 28 objects in the SMC were analyzed. Again, coverage of spectral classes was similar as for the Galactic case. For this galaxy 43 percent of the objects were modeled using the automated method. For further details we refer to Mokiem *et al.* (2007b).

The three empirical WLRs are presented in Fig. 2. The top, middle and bottom relations (solid lines), respectively, correspond to Galactic, LMC and SMC observations. One sigma confidence intervals are shown as gray areas. A linear regression using the three D_{mom} values at $L = 10^{5.75} L_\odot$, accounting for both the errors in D_{mom} and Z, yields

$$\dot{M}_{\mathrm{empirical}} \propto Z^{0.83 \pm 0.16}. \tag{3.4}$$

How does this result compare with theoretical expectations? The dashed lines in Fig. 2 show predictions by Vink *et al.* (2000, 2001). The top, middle and bottom relations, respectively, correspond to Galactic, LMC and SMC predictions. They find

$$\dot{M}_{\mathrm{predicted}} \propto Z^{0.69 \pm 0.10}. \tag{3.5}$$

The quoted error in this prediction accounts for random errors in the Monte Carlo method that is applied (see de Koter *et al.* 1997). An independent study by Krtička (2006) confirms the result. We conclude that the power-law behavior of the empirical and predicted $\dot{M}(Z)$ agree within these random error limits. Systematic uncertainties are not accounted for in the value of m. To illustrate this: if the metal content of the SMC is higher than adopted (see Sect. 2), say, $\Delta[Z/\mathrm{H}] = -0.5 \pm 0.1$, both the empirical and theoretical dependence would be stronger: $m \simeq 1.1$.

Notice that the theoretical results underpredict the empirical results by some 0.24–0.35 dex in the logarithm of D_{mom}. This corresponds to about a factor of two difference in

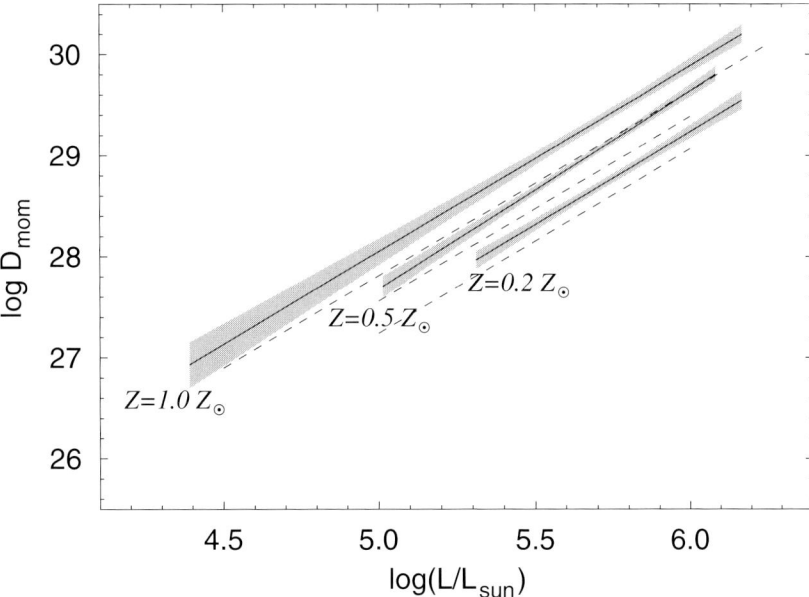

Figure 2. Comparison of the observed wind momentum – luminosity relations for O and early-B stars (solid lines) with the predicted relations of Vink *et al.* (2000, 2001) (dotted lines). Top, middle and bottom lines of each line style, respectively, correspond to Galactic, LMC, and SMC observed and predicted WLRs. One sigma confidence intervals for the empirical relations are shown as gray areas. Among others, this figure for the first time shows that the wind strengths of LMC stars are intermediate between those of Galactic and SMC stars. From Mokiem *et al.* (2007b).

the mass loss rate. The physical reason for this offset is not known, however, if the stellar outflows are clumped on small spatial scales in the region of Hα line formation this discrepancy may be resolved. Owing to the fact that Hα (the most important wind diagnostic in the optical) is the result of recombination, the strength of this line scales with the square of the density. A local overdensity in clumps thus overcompensates the presence of void interclump regions. Introducing the clumping factor $f_{\rm cl} = <\rho^2>/<\rho>^2 \geqslant 1$ where angle brackets denote (temporal) average values (e.g. Puls *et al.* 2006), the mass loss required to fit the Hα profile in case of a structured wind will be less than that assuming a smooth ($f_{\rm cl} = 1$) outflow. It follows that $\dot{M}$(clumped) = $\dot{M}$(smooth wind)/$\sqrt{f_{\rm cl}}$. Given the size of the offset only a modest clumping of $f_{\rm cl} \sim 3$–5 is required to bring the empirical and predicted WLR in agreement. Notice that this assumes that clumping has no significant effect on the predictions of mass loss. Notice also that if the properties of small scale structure are independent of metallicity, *the $\dot{M}(Z)$ scaling is unaffected by the presence of clumping, regardless the absolute value of the clumping factor.*

4. Outlook

The WLR in the SMC is derived from objects more luminous than $10^{5.2} L_\odot$ (see Fig. 2). Therefore, the mass loss vs. metallicity scaling is strictly speaking only valid for such bright objects, objects that show strong winds. For intrinsically dimmer stars, having weaker winds, the WLR relation is being discussed (see e.g. Fullerton *et al.* 2006; Mokiem *et al.* 2007b). Also, $\dot{M}(Z)$ has only been derived for metallicities down to 1/5th the solar value. Though predictions claim a constant power-law for all Z values between

1/30th and 3 $Z_\odot$ (Vink et al. 2001), this remains to be verified. In the foreseeable future the only possibility of studying stars in environments with significantly lower metallicity than the SMC is the young massive population of Local Group dwarf galaxies such as GR8, Leo A, Sextans A, WLM, IC1613, and perhaps IZw18. These systems have metallicities between 1/10th and 1/35th of solar. So far, focussed spectroscopic attempts to characterize the massive star population have been limited. The studies that have addressed blue objects typically focus on supergiants of spectral type mid-B to A (e.g. Kaufer et al. 2004) with the aim to establish abundances and use them as distance indicators, though observations of earlier spectral types including late-O stars are becoming feasible (Bresolin et al. 2006, 2007). The situation may improve once the *X-Shooter* spectrograph is installed on the VLT. Anticipating that X-Shooter will deliver the gains in efficiency that it expects, high-quality spectroscopic observations of mid-O stars should be possible. This will allow for a first confrontation between observations and theory at $Z \leqslant 1/5 Z_\odot$.

References

Asplund, M., Grevesse, N., & Sauval, A. J. 2005, in T. G. Barnes III & F. N. Bash (eds.), *Cosmic Abundances as Records of Stellar Evolution and Nucleosynthesis* (San Francisco: ASP) *ASP Conf. Ser.*, 336, 25
Bresolin, F., Pietrzyński, G., Urbaneja, M. A., et al. 2006, *ApJ*, 648, 1007
Bresolin, F., Urbaneja, M. A., Gieren, W., et al. 2007, *ApJ*, 671, 2028
de Koter, A., Heap, S. R., & Hubeny, I. 1997, *ApJ*, 477, 792
Evans, C. J., Smartt, S. J., Lee, J.-K., et al. 2005, *A&A*, 437, 467
Evans, C. J., Lennon, D. J., Smartt, S. J., & Trundle, C. 2006, *A&A*, 456, 623
Fullerton, A. W., Massa, D. L., & Prinja, R. K. 2006, *ApJ*, 637, 1025
Herrero, A., Puls, J., & Najarro, F. 2002, *A&A*, 396, 949
Hillier, D. J. & Miller, D. L. 1998, *ApJ*, 496, 407
Hunter, I., Dufton, P. L., Smartt, S. J., et al. 2007, *A&A*, 466, 277
Kaufer, A., Venn, K. A., Tolstoy, E., et al. 2004, *AJ*, 127, 2723
Krtička, J. 2006, *MNRAS*, 367, 1282
Kudritzki, R.-P. & Puls, J. 2000, *ARA&A*, 38, 613
Leitherer, C., Robert, C., & Drissen, L. 1992, *ApJ*, 401, 596
Mokiem, M. R., de Koter, A., Puls, J., et al. 2005, *A&A*, 441, 711
Mokiem, M. R., de Koter, A., Evans, C. J., et al. 2006, *A&A*, 456, 1131
Mokiem, M. R., de Koter, A., Evans, C. J., et al. 2007a, *A&A*, 465, 1003
Mokiem, M. R., de Koter, A., Vink, J. S., et al. 2007b, *A&A*, 473, 603
Puls, J., Springmann, U., & Lennon, M. 2000, *A&AS*, 141, 23
Puls, J., Urbaneja, M. A., Venero, R., et al. 2005, *A&A*, 435, 669
Puls, J., Markova, N., Scuderi, S., et al. 2006, *A&A*, 454, 625
Rolleston, W. R. J., Trundle, C., & Dufton, P. L. 2002, *A&A*, 396, 53
Rolleston, W. R. J., Venn, K., Tolstoy, E., & Dufton, P. L. 2003, *A&A*, 400, 21
Trundle, C., Dufton, P. L., Hunter, I., et al. 2007, *A&A*, 471, 625
Venn, K. A. 1999, *ApJ*, 518, 405
Vink, J. S., de Koter, A., & Lamers, H. J. G. L. M. 2000, *A&A*, 362, 295
Vink, J. S., de Koter, A., & Lamers, H. J. G. L. M. 2001, *A&A*, 369, 574

Discussion

WALBORN: I'm concerned that there are three categories of O stars with 'weak winds' that are sometimes not clearly distinguished in analytical studies: 1) SMC stars with weak wind lines due to metal deficiency; 2) Very young O stars that may be on or near the ZAMS and subluminous, with weak wind lines for their spectral types (my review at STScI May 2006 symposium, still astro-ph/0701573 only); 3) Stars with normal wind line strengths for their spectral types, for which current model analyses derive anomalously low mass-loss rates, e.g. 10 Lac.

DE KOTER: Your group 2 and 3 refer to the 'weak wind problem'. I didn't discuss it today, but its good that you mention it. Your group 1 stars have weaker winds, as I've quantified today. They can be understood in the framework of radiation driven wind theory, as I've shown. Concerning your group 2 and 3 stars I think there are two broad avenues to search for an explanation: a) their winds are really weaker, or b) we miss some understanding of the diagnostics used to determine mass loss, but in reality their winds aren't as weak as they seem. Concerning the latter possibility, a recent paper by Oskinova *et al.* (2007, *A&A* 476, 1331) makes the point that porosity effects may lead to underestimating the mass loss when using ultraviolet resonance lines. A thing that I'm always worried about is that the discontinuity between "weak winds" and "strong winds" is at about the point where Hα looses it sensitivity (for weaker winds one has to rely on ultraviolet lines only). Mokiem *et al.*, using their genetic algorithm method did manage to derive the mass loss for three stars in the "weak wind" regime using Hα, ζ Oph, Cyg OB2 #2, and HD 217086. Interestingly, they did recover values that seem consistent with theory.

KUDRITZKI: I guess the Fe abundances that you showed are from the optical spectra from your O and B stars. They must be very uncertain, because you have only a very few Fe lines in those optical spectra. I also wonder whether these are based on NLTE calculations.

DE KOTER: In the context of the FLAMES project, Trundle *et al.* (2007, *A&A*, 471, 625) determined the Fe abundance of B stars in the LMC and SMC. I do agree with you that the Fe abundance, more in general the SMC metallicity, is a very important quantity. Actually, the SMC metallicity might be the largest potential source of systematic error in the power-law of the empirical $\dot{M}(Z)$ dependence. Carrie, do you want to comment on Rolf's question?

TRUNDLE: Only few iron lines are present in the optical spectra of B stars, the best ones being Fe III λ4419 and 4430. For the LMC we used these line to derive an abundance for 13 targets. For our SMC stars Fe III λ4430 is too weak. We used the 4419 line to derive the Fe abundance of five objects only. A second problem is that the iron model atom in our TLUSTY models is too simplistic. Therefore, the line transfer in iron is treated in LTE (though the model atmospheres are NLTE). However, Thompson et al. (2007, *MNRAS* 383, 729) have shown that NLTE effects in Fe appear to be small. Typical errors in our results for iron are $\sim$0.2 dex.

STANEK: The table you show with abundances has many abundances below the average you adopt for both the LMC and SMC. Could you explain that?

DE KOTER: Indeed. In particular the table shows that the abundances of carbon and nitrogen are lower, reflecting that the LMC and SMC chemical composition is not simply a scaled Galactic case. However, C and N are not the most important elements driving the outflow (though C is significant for the hottest stars). The wind is driven by iron group elements, with significant contributions from silicon and sulphur. This is why we focus on these elements is settling on metallicity.

KONIGSBERGER: How does the presence of magnetic fields effect the mass-loss rates?

DE KOTER: I see Stan jumping up and down and raising his finger. I think he would like to address this question.

OWOCKI: Well, the simplest answer is, "not much". But the fraction of the surface that is covered by closed loops can trap the wind and force it to fall back on the star, effectively reducing the global mass loss rate. The reduction can be ca. 50% for a strong field without rotation. With rotation, loops extending above the Kepler co-rotation radius eventually have centrifugal ejection of trapped wind, and so the net reduction in global mass loss is less. Further details can be found in a recent paper with Asif ud-Doula (*MNRAS*, in press; arXiv:0712.2780). There can also be a modest reduction (ca. 10%) of mass flux in open field regions due to the field tilt away from the surface normal. (see Owocki & ud-Doula, *ApJ* 600, 1004).

STANEK: Do you have any "mutants" in your analysis? (i.e. stars that don't fit)

DE KOTER: Our method did not signal clear cases of mutants, i.e. cases in which the derived parameters clearly signaled a non-physical solution. In part this is the result of preselection.

MASSEY: This is a followup to your statement that of these 86 O stars you got good fits for all of them. When we were analyzing our samples of LMC and SMC stars (Massey *et al.* 2004, *ApJ* 608, 1001; 2005, *ApJ* 627, 477) we failed to get satisfactory fits (by eye) to about a third of the stars. Presumably these were the composites (binaries) – not to sound like Dany. Are you concerned that your automatic fitting routines were always satisfied with their fits? Where are the spectral composites, then?

DE KOTER: To expand on my previous answer. There are two reasons why the automatic analyses we performed do not result in unsatisfactory fits in such a high fraction of the targets: 1) From the outset we excluded all known confirmed binaries and radial velocity variables. Therefore, one could say that in this respect our sample is preselected; 2) The automated method is much better in scanning all of parameters space as is a 'by eye' inspection of fits. Still, to be clear about this: in one case (star N11-048 in the LMC) the fit is very poor. We suspect the system to be a binary. In some cases the error bars are quite large (e.g. NGC 330-052) because of low signal-to-noise data, and in several cases one of the (set of potentially ten) diagnostic lines was fitted poorly.

Alex de Koter.

Properties of Wolf-Rayet Stars

Paul A. Crowther

Department of Physics and Astronomy, University of Sheffield, S3 7RH, UK
email: Paul.Crowther@sheffield.ac.uk

Abstract. A review of recent progress relating to Wolf-Rayet (WR) stars is presented. Topics include improved Milky Way statistics from near-IR surveys, different flavours of hydrogen-rich and hydrogen-poor WN stars, WR masses from binary orbits, plus spectroscopic analysis of WR stars resulting in stellar temperatures, luminosities, ionizing fluxes, plus wind properties accounting for clumping. Chemical abundances of WN and WC stars are presented, including a discussion of neon abundances in WC and WO stars from *Spitzer* observations. Empirical evidence supporting metallicity-dependent winds is also presented, including its effect on subtype distributions in different environments. Finally, difficulties in comparisons between evolutionary models and observations are highlighted, plus outstanding issues with predictions from continuous star formation and instantaneous bursts in the Milky Way.

Keywords. stars: Wolf-Rayet – stars: fundamental parameters – stars: atmospheres – stars: evolution – stars: abundances – stars: mass loss – stars: winds, outflows

1. Introduction

Spectroscopically, WR stars are spectacular in appearance, with strong, broad emission lines, due to dense ($\sim 10^{-5} M_\odot \mathrm{yr}^{-1}$), fast ($\sim$2000 km s^{-1}) outflows. Specifically, WN and WC stars show the products of the CNO cycle (H-burning) and the triple-α (He-burning), respectively, although in reality, there is a continuity of physical and chemical properties between O supergiants and WN subtypes. Gamov (1943) first suggested that the anomalous composition of WR stars was the result of nuclear processed material being visible on their surfaces, although this was not universally accepted until the final decade of the 20th Century (Lamers *et al.* 1991). WR stars possess lifetimes of typically a few 10^5 yr, i.e. 10% of the main-sequence O phase. Hundreds are known individually within Local Group galaxies, with evidence for thousands within star forming regions of nearby galaxies, and many more seen in the ultraviolet spectrum of Lyman-break galaxies at high-redshift. WR stars are presumed to be the progenitors of at least some Type Ib/c supernovae and long Gamma-Ray Bursts (GRBs).

2. Inventory and Spectral Classification

A catalogue of 227 Galactic Wolf-Rayet stars has been provided by van der Hucht (2001), to which an extra 72 were added by van der Hucht (2006). The rapid increase in numbers has been driven by infra-red surveys, both within massive clusters within the Galactic Centre itself and other high mass clusters in the inner Milky Way, such as Westerlund 1, which alone hosts 24 WR stars (Crowther *et al.* 2006b). Near-IR surveys have been conducted based on either WR emission line diagnostics (Homeier *et al.* 2003), mid-IR excesses (Hadfield *et al.* 2007) or IR colours consistent with young clusters (Kurtev *et al.* 2007). Dusty, late-type WC stars, which are believed to be close WC+O binaries (e.g. Tuthill *et al.* 2006) exhibit quite different near- and mid-IR colours. It is expected that many more WR stars await discovery in the Milky Way, due to increases in sensitivity and the fraction of the Milky Way disk covered by surveys.

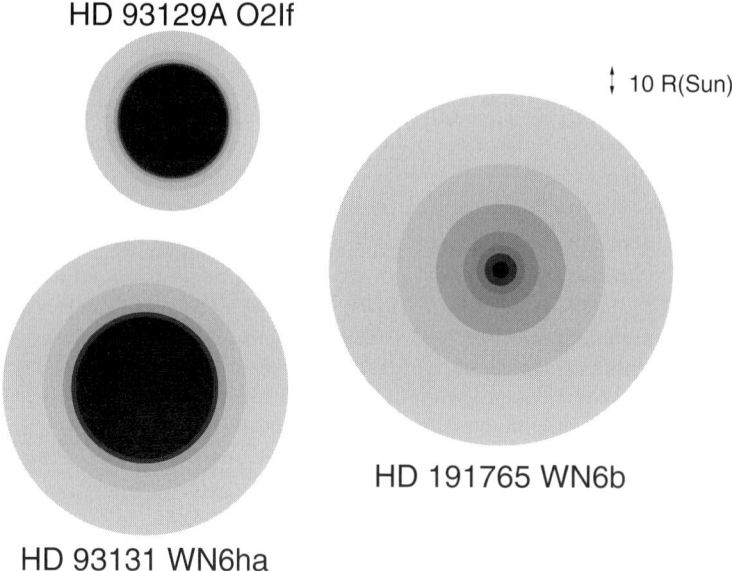

Figure 1. Schematic illustrating the extended atmospheres of Of supergiants (HD 93129A, O2 If), H-rich mid-type WN subtypes (HD 93131, WN6ha) and H-deficient classical WN stars (HD 191765, WN6b). Regions represent (outer to inner) electron densities of $10^{11}, 10^{11.5}, 10^{12}, 10^{12.5}, 10^{13}$ cm^{-3} plus stellar radii at Rosseland optical depths of 2/3 ($= R_{2/3}$) and 20 ($= R_*$).

Visual spectral classification of WR stars is based on emission line strengths and line ratios (Smith 1968). To date, high-ionization WN2 to WN5 are known as 'early WN' (WNE) stars, and low-ionization WN7 to WN9(–11) stars known as 'late WN' (WNL) stars, and WN6 stars either early or late-type. Complications arise for WN stars with intrinsically weak emission lines. For example, HD 93131 (WN6ha) has a He II λ4686 emission equivalent width which is an *order of magnitude* smaller than that observed in other WN6 stars; the 'ha' nomenclature indicates that hydrogen is seen both in absorption and emission. From a standard spectroscopic viewpoint, such stars possess mid to late WN spectral classifications. However, their appearance is more reminiscent of Of stars than classic WN stars (there exists a continuity of properties between normal O stars and some WN stars). Such stars are widely believed to be massive O stars with relatively strong stellar winds at a rather early evolutionary stage, not the more mature, classic He-burning WN stars. A schematic illustrating how such luminous emission-line stars fit between Of and He-burning WN stars is illustrated in Fig. 1. Here, we shall newly identify these as *mid-type* WN stars (Smith & Conti 2008 propose 'WNH' for luminous WN stars with hydrogen).

In contrast, WC stars are a much more uniform group, for which high-ionization WC4–6 subtypes are 'early' (WCE) and low-ionization WC7–9 subtypes are 'late' (WCL). Rare, oxygen-rich WO stars form a yet-higher ionization extension of the WCE sequence, exhibiting strong O VI $\lambda\lambda$3811-34 emission (Kingsburgh *et al.* 1995). Finally, C IV λ5801-12 appears unusually strong in an otherwise normal WN star in a few cases, leading to an intermediate WN/C classification (Conti & Massey 1989). WN/C stars are indeed considered to be at an intermediate evolutionary phase between the WN and WC stages.

3. Binarity and WR masses

The observed binary fraction amongst Milky Way WR stars is 40% (van der Hucht 2001), either from spectroscopic or indirect techniques. Within the low metallicity Magellanic Clouds, close binary evolution would be anticipated to play a greater role, due to the reduced role of mass-loss at earlier phases in producing single WR stars. However, where detailed studies have been carried out, a similar binary fraction to the Milky Way has been obtained (e.g. Schnurr et al. in prep).

Binary derived masses for Galactic WR stars are included in the Crowther (2007) review. WC masses cover a narrow range of 9–16 $M_\odot$. WN stars span a very wide range of ~10–83 $M_\odot$, and exceed their OB companion for some mid-type WN stars. WR20a (SMSP2) currently sets the record for the highest orbital-derived mass of any star, with ~83 $M_\odot$ for each WN6ha component (Rauw et al. 2005), albeit a factor of two lower in mass than the apparent ~ 150 $M_\odot$ stellar mass limit (Figer 2005). More extreme cases may await discovery or confirmation (see Moffat, this volume). Spectroscopic measurement of masses via surface gravities using photospheric lines is not possible for WR stars due to their dense stellar winds.

4. Spectroscopic analysis

Our interpretation of hot, luminous stars via radiative transfer codes is hindered with respect to normal stars by several effects. *First*, the routine assumption of LTE breaks down for high-temperature stars. *Second*, the problem of accounting for the effect of millions of spectral lines upon the emergent atmospheric structure and emergent spectrum – known as line blanketing – remains challenging. *Third*, spherical geometry, rather than plane-parallel geometry must be considered, since the scale height of their atmospheres is non-negligible with respect to their stellar radii.

Radiative transfer for WR atmospheres is either solved in the co-moving frame, as applied by CMFGEN (Hillier & Miller 1998) and PoWR (Gräfener et al. 2002) or via the Sobolev approximation, as used by ISA-wind (de Koter et al. 1993). See Puls (this volume) for a more extensive discussion of stellar atmosphere codes. Regarding the typical assumption of spherical symmetry, spectropolarimetry does not support significant departures from spherical geometry for Milky Way WC stars, with a few exceptions among WN stars (Harries et al. 1998; Williams et al., this volume). Vink (2007) has obtained spectropolarimetry of some LMC WR stars, with similar results, although this sample was based towards bright systems (binaries and late-type WN stars). One of the two non-spherical cases identified, R99 = HDE269445, has long been known to possess unusual spectroscopic properties (e.g. Crowther & Smith 1997).

4.1. Stellar temperatures

Stellar temperatures for WR stars are difficult to characterize, because their geometric extension is comparable with their stellar radii. Atmospheric models for WR stars are typically parameterized by the radius of the inner boundary R_* at high Rosseland optical depth $\tau_{\rm Ross}(\sim 10)$. However, only the optically thin part of the atmosphere is seen by the observer (recall Fig. 1). To obtain R_*, it is necessary to assume that the same velocity law holds for the optically thin and the optically thick part of the atmosphere. The optical continuum radiation originates from a 'photosphere' where $\tau_{\rm Ross} \sim 2/3$. Typical WN and WC winds have reached a significant fraction of their terminal velocity before they become optically thin in the continuum. $R_{2/3}$, the radius at $\tau_{\rm Ross} = 2/3$ lies at highly supersonic velocities, well beyond the hydrostatic domain. For example, Morris et al. (2004) obtain $R_* = 2.9\,R_\odot$ and $R_{2/3} = 7.7\,R_\odot$ for HD 50896 (WN4b), corresponding

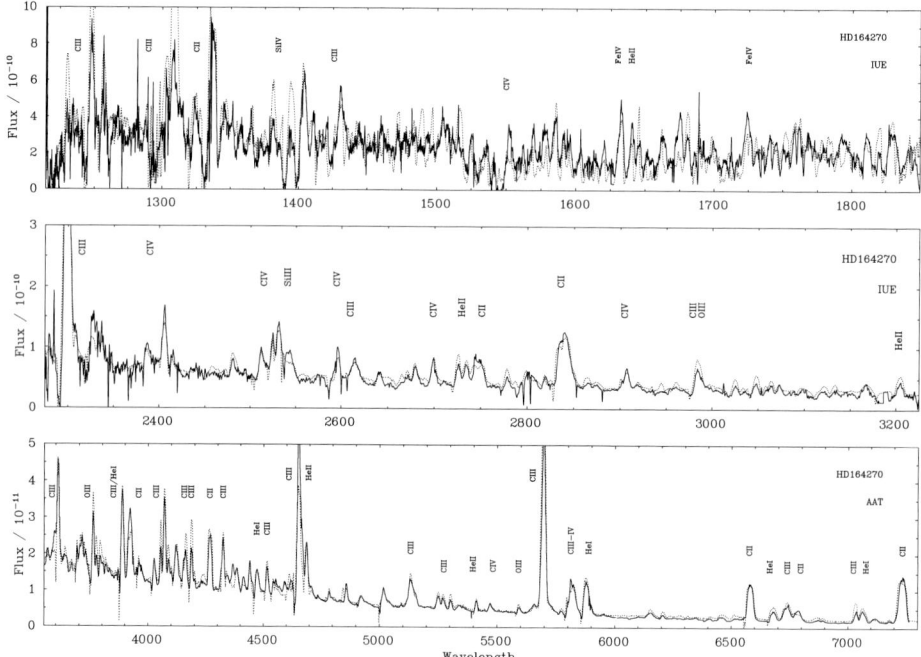

Figure 2. Fit (dotted) to de-reddened UV and optical spectrophotometry (solid) of HD 164270 (WR103, WC9) from Crowther *et al.* (2006a)

to $T_* = 85\,\mathrm{kK}$ and $T_{2/3} = 52\,\mathrm{kK}$, respectively. This is not strictly true for WN stars in possession of weak winds, since their spherical extinction is modest, i.e. $R_* \sim R_{2/3}$ (e.g. HD 9974, Marchenko *et al.* 2004).

Stellar temperatures of WR stars are derived from lines from adjacent ionization stages of helium or nitrogen for WN stars or lines of carbon for WC stars. The agreement between the spectral features and continuum is now generally excellent. By way of example, we compare a synthetic model of HD 164270 (WC9) to UV and optical spectroscopic observations in Fig. 2. Metals such as C, N and O provide efficient coolants, such that the outer wind electron temperature is typically 10±2 kK (Hillier 1989). Hamann *et al.* (2006) have applied their grid of line-blanketed WR models to the analysis of most Galactic WN stars. Temperatures range from 30 kK amongst the latest subtypes to 40 kK at WN8 and approach 100 kK for early-type WN stars. Few recent studies of WC stars have been carried out (e.g. Crowther *et al.* 2002, 2006a), with spectroscopic temperatures are rather higher on average, i.e. 50 kK for WC9 stars, increasing to 70 kK at WC8 and ⩾100 kK for early-type WC stars and WO stars.

4.2. Stellar Luminosities

Bolometric corrections from which stellar luminosities are obtained also depend upon detailed metal line-blanketing (Schmutz 1997; Hillier & Miller 1999). Inferred bolometric corrections range from $M_{\rm bol} - M_v = -2.7$ mag amongst very late-type WN stars (Crowther & Smith 1997) to approximately –6 mag for weak-lined, early-type WN stars and WO stars (Crowther *et al.* 1995b, 2000).

Stellar luminosities of Milky Way WN stars range from 200,000 $L_\odot$ in early-type stars to 500,000 $L_\odot$ in late-type stars. Hydrogen-burning O stars with strong stellar winds, spectroscopically identified as mid-type WNha stars, have luminosities in excess of 10^6 $L_\odot$. For Milky Way WC stars, inferred stellar luminosities are ~150,000 $L_\odot$, increasing by a factor of 2–3 for LMC WC stars. Systematically higher spectroscopic luminosities have recently been determined by Hamann et al. (2006) for Galactic mid- to late-type WN stars† for which a uniformly high absolute magnitude was adopted for all non-cluster member WN6–9h stars. Absolute magnitudes for 'normal' late-type WN stars are subject to large uncertainties since such stars ordinarily shy away from clusters. As a consequence, their results suggest a, perhaps unrealistic, bi-modal distribution around 300,000 $L_\odot$ for early-type WN stars, and 1–2×10^6 $L_\odot$ for all late-type WN stars (solely mid-type WN stars likely occupy the latter values).

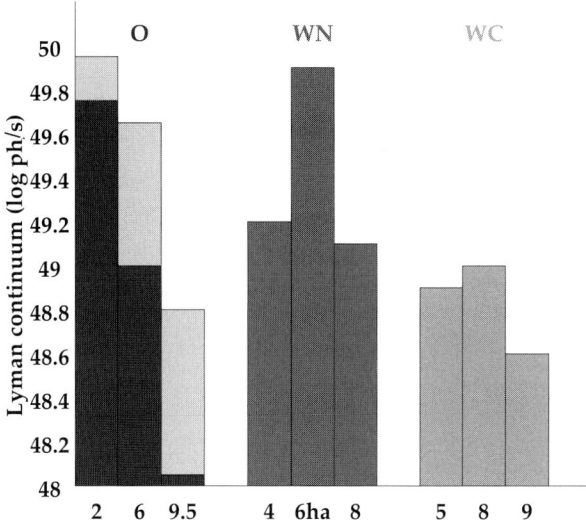

Figure 3. Comparison between Lyman continuum ionizing fluxes for O dwarfs (left, dark) and O supergiants (left, pale) from Conti et al. (2008), plus WN and WC stars from Crowther (2007).

4.3. Ionizing fluxes

Lyman continuum ionizing fluxes, N(LyC), of WR stars are typical of mid-O stars in general (Fig. 3). As such, the low number of WR stars with respect to O stars would suggest that they play only a minor role in the Lyman continuum ionization budget of young star-forming regions. Mid-type WN stars provide a notable exception, since their ionizing output compares closely to O2 stars (Walborn et al. 2004). Crowther & Dessart (1998) showed that the mid-type WN stars in NGC 3603 provided ∼20% of the Lyman continuum ionizing photons, based upon calibrations of non-blanketed models for O (WR) stars, whose temperature scales have since decreased (increased), so such extreme WN stars may provide a yet much greater fraction of the ionizing photons from such young clusters.

WR stars represent an extension of O stars to higher temperatures, so significant He I continuum photons are emitted, plus strong He II continua are predicted for a few high-temperature, low-density stars. The primary effect of metal-line blanketing is to redistribute extreme UV flux to longer wavelengths, reducing the ionization balance in the wind, such that higher temperatures and luminosities are required to match the observed WR emission line profile diagnostics relative to unblanketed models. Atmospheric models for WR stars with dense winds produce relatively soft ionizing flux distributions, in

† Absolute magnitudes of $M_v = -7.2$ mag were adopted from van der Hucht (2001), based in part upon a high 3.2 kpc distance to the mid-type WN stars in Car OB1

which extreme UV photons are redistributed to longer wavelength by the opaque stellar wind (Smith et al. 2002). In contrast, for the low wind-density case, a hard ionizing flux distribution is predicted, in which extreme UV photons pass through the relatively transparent wind unimpeded. Consequently, the shape of the ionizing flux distribution of WR stars depends on both the wind density and the stellar temperature.

4.4. Wind velocities

The wavelength of the blue edge of saturated P Cygni absorption profiles provides a measure of the asymptotic wind velocity, from which accurate wind velocities, v_∞, of WR stars can readily be obtained (e.g. Willis et al. 2004). Alternatively, optical and near-IR He I P Cygni profiles or mid-IR fine-structure metal lines may be used to derive reliable wind velocities (Howarth & Schmutz 1992; Dessart et al. 2000), for which the radial dependence is assumed to follow $v(r) = v_\infty (1 - r/R_*)^\beta$ with $\beta \sim 1$.

In principle, optical recombination lines of He II and C III-IV may *also* be used to estimate wind velocities, since these are formed close to the asymptotic flow velocity. However, velocities obtained from spectral line modelling are preferable. For WR stars exhibiting weak winds – whose lines are formed interior to the asymptotic flow velocity – only lower velocity limits may be obtained. Observational evidence suggests lower wind velocities at later subtypes, by up to a factor of ten, with respect to early-types. Individual WO stars have been identified in a number of external galaxies. One observes a reduction in line width (and so wind velocity) for stars of progressively lower metallicity, by a factor of up to two between the Milky Way and IC 1613 (Crowther & Hadfield 2006). Although numbers are small, this downward trend in wind velocity with decreasing metallicity is believed to occur for other O and WR spectral types.

4.5. Mass-loss rates

The mass-loss rate relates to the velocity field $v(r)$ and density $\rho(r)$ via the equation of continuity $\dot{M} = 4\pi r^2 \rho(r) v(r)$ for a spherical, stationary wind. WR winds may be observed at IR-mm-radio wavelengths via the free-free continuum excess caused by the stellar wind or via UV, optical or near-IR emission lines. Mass-loss rates (e.g. Leitherer et al. 1997) follow from radio *continuum* observations using relatively simple analytical relations, under the assumption of homogeneity and spherical symmetry (Wright & Barlow 1975). Determinations of radio WR mass-loss rates depend upon knowledge of composition and ionization balance at $\sim 100 - 1000\, R_*$.

Alternatively, spectroscopic analysis of UV/optical/IR spectral *lines* observed in WR stars may be used to obtain mass-loss rates. The majority of these can be considered as recombination lines, although line formation is rather more complex in reality (Hillier 1988, 1989). Since recombination involves the combination of ion and electron density, the strength of wind lines scales with the square of the density.

There is overwhelming evidence in favour of highly clumped winds for WR stars. Individual spectral lines, formed at $\sim 10\, R_*$, can be used to estimate volume filling factors f in WR winds – alternatively expressed as a clumping factor $f_{cl} = 1/f$. This technique permits an estimate of f for the inner wind, from a comparison between line electron scattering wings ($\propto \rho$) and recombination lines ($\propto \rho^2$) although it suffers from an approximate radial density dependence and is imprecise due to severe line blending effects (Hillier 1991). Fits to UV/optical/IR line profiles suggest $f \sim 0.05 - 0.25$ ($f_{cl} \sim 4 - 20$). As a consequence, global WR mass-loss rates are reduced by a factor of $\sim 2 - 4$ relative to homogeneous models (dM/dt $\propto f^{-1/2}$). Spectroscopically derived mass-loss rates of Milky Way WN stars span a wide range of $10^{-5.6}$ to $10^{-4.4} M_\odot$ yr^{-1}. In contrast, Galactic WC stars cover a much narrower range in mass-loss rate, from $10^{-5.0}$ to $10^{-4.4} M_\odot$ yr^{-1}.

Independent methods support clumping-corrected WR mass-loss rates. Binary systems permit use of the variation of linear polarization with orbital phase. The modulation of linear polarization originates from Thomson scattering of free electrons due to the relative motion of the companion with respect to the WR star. This technique has been recently applied by Kurosawa et al. (2002) to V444 Cyg using a Monte Carlo approach, from which polarization results suggest $f \sim 0.06$ ($f_{cl} \sim 15$).

5. Chemical abundances

For WR stars, it has long been suspected that abundances represented the products of core nucleosynthesis, although it has taken the development of non-LTE model atmospheres for these to have been empirically supported.

5.1. WN stars

For Galactic WN stars, a clear subtype effect is observed regarding the hydrogen content, with late-type WN stars generally showing some hydrogen (typically $X_H \sim 15 \pm 10\%$), and early-type WN stars being hydrogen-free, although exceptions do exist. This trend breaks down within the lower metallicity environment of the Magellanic Clouds, notably the SMC (Foellmi et al 2003). Milky Way mid-type WN stars are universally H-rich with $X_H \sim 50\%$ (Crowther et al. 1995a; Crowther & Dessart 1998).

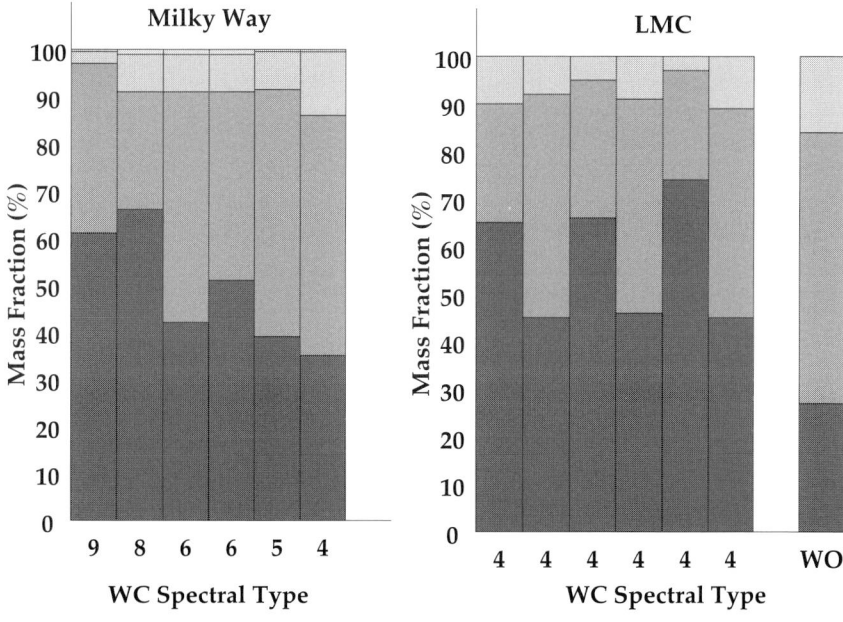

Figure 4. *Left* Distribution of helium (lower), carbon (middle) and oxygen (upper) mass fraction (in %) derived by Crowther et al. (2006a) and unpublished work for Milky Way WC stars; *right* Distribution of helium (lower), carbon (middle) and oxygen (upper) mass fraction (in %) derived by Crowther et al. (2000, 2002) for LMC WC4 and WO stars

Non-LTE analyses confirm that WN abundance patterns are consistent with material processed by the CNO cycle in which these elements are used as catalysts, with $X_N \sim 1\%$ by mass is observed in Milky Way WN stars. Carbon is highly depleted, with typically $X_C \sim 0.05\%$. Oxygen suffers from fewer readily accessible line diagnostics, but probably exhibits a similarly low mass fraction as carbon (e.g. Herald et al. 2001). Non-LTE analysis of transition WN/C stars reveals elemental abundances (e.g. $X_C \sim 5\%$, $X_N \sim 1\%$ by mass) that are in good agreement with the hypothesis that these stars are in a brief transition stage between WN and WC (Langer 1991; Crowther *et al.* 1995c).

5.2. *WC stars*

Detailed abundance determinations for WC stars require non-LTE model atmosphere analyses. Koesterke & Hamann (1995) obtained C/He=0.1–0.5 by number ($20\% \leqslant X_C \leqslant 55\%$) for a large sample of WC5–8 stars, with no systematic WC subtype dependence. Their results were contrary to suggestions by Smith & Maeder (1991) who had followed a recombination line method approach to conclude that C/He increases from late to early-type WC stars. Indeed, LMC WC4 stars possess similar surface abundances to Milky Way WC stars (Crowther *et al.* 2002), as shown in Fig. 4 for which common diagnostics of He II λ5412 and C IV λ5471 are used (Hillier 1989). These recombination lines are formed at high densities of 10^{11} to 10^{12} cm^{-3} at radii of 3–30 R_* (recall Figure 1). Oxygen diagnostics in WC stars lie in the near-UV, such that derived oxygen abundances are rather unreliable unless space-based spectroscopy is available. Where they have been derived, one finds $X_O \sim 5$–10% for WC stars (e.g. Crowther *et al.* 2002).

5.3. *Neon abundances*

Core He burning in massive stars has the effect of transforming ^{14}N (produced in the CNO cycle) to Ne and Mg. Ne lines are extremely weak in the UV/optical spectrum of WC stars (Crowther *et al.* 2002). Ground-state fine-structure lines at [Ne II] 12.8μm and [Ne III] 15.5μm provide abundance indicators if Ne$^+$ or Ne^{2+} is the dominant ionization stage in WR stars at several hundred stellar radii, where these lines originate. Barlow *et al.* (1988) came to the conclusion that Ne was not greatly enhanced in γ Vel with respect to the Solar case ($\sim$0.1% by mass primarily in the form of ^{20}Ne) from their analysis of fine-structure lines. This was a surprising result, since the above reaction is expected to produce $\sim$2% by mass of ^{22}Ne at Solar metallicity.

Once the clumped nature of WR winds is taken into consideration - *assuming* similar wind clumping factors for the inner and outer wind - Ne was found to be enhanced in γ Vel and other WC stars from *ISO*/SWS observations (e.g. Dessart et al. 2000) with an inferred Ne mass fraction of $\sim$1%. Meynet & Maeder (2003) note that the ^{22}Ne enrichment depends upon nuclear reaction rates rather than stellar models, so any remaining disagreement may suggest a problem with the relevant reaction rates. More likely, a lower metal content is inferred from the neon abundance than standard 'Solar metallicity' evolutionary models (Z=0.020). Indeed, if the Solar oxygen abundance from Asplund *et al.* (2004) is taken into account, a revised metal content of Z=0.012 for the Sun is implied. Allowance for depletion of heavy elements due to diffusion in the 4.5 Gyr old Sun suggests a Solar neighbourhood metallicity of Z=0.014 (Meynet 2007).

Recent high-quality *Spitzer*/IRS spectroscopy (e.g. Crowther *et al.* 2006a) - as illustrated in Fig. 5 - allows line fluxes from fine-structure lines to be obtained plus establish the radial dependence of the clumping factor from the mid-IR continuum slope (cf Nugis *et al.* 1998). Analysis supports *similar* inner and outer clumping factors - which, together with reduced CNO total abundances, brings predicted and measured ^{22}Ne abundances into good agreement (Schnurr *et al.*, this volume). *Spitzer*/IRS studies are in progress to determine neon abundances in WO stars, with provisional results supporting similar enrichments in Ne to WC stars (i.e. no evidence for enhanced ^{20}Ne resulting from α–capture of oxygen).

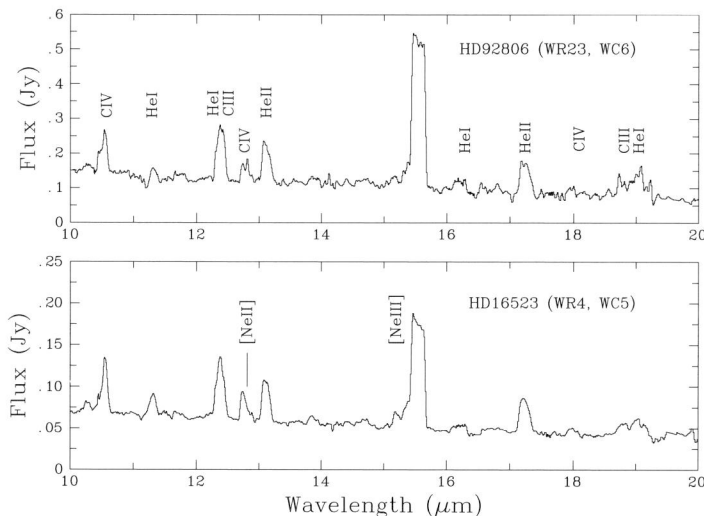

Figure 5. Mid-IR spectroscopy of HD 92806 (WR23, WC6) and HD 16523 (WR4, WC5) from *Spitzer*/IRS revealing strong [Ne III] fine-structure lines (formed at low densities in the outer wind), plus weaker He and C recombination lines (formed at high densities in the inner wind).

6. Metallicity dependent winds

A theoretical framework for mass-loss in normal hot, luminous stars has been developed by Castor *et al.* (1975), known as CAK theory, via line-driven radiation pressure. Historically, it has not been clear whether radiation pressure alone is sufficient to drive the high mass-loss rates of WR stars, whose strength was considered to be mass dependent, but metallicity-independent (Langer 1989). Observationally, Nugis & Lamers (2000) provided empirical mass-loss scaling relations for WR stars by adopting physical parameters derived from spectroscopic analysis and/or evolutionary predictions. As we shall show, observational evidence now favours metallicity-dependent WR winds, with a dependence of $dM/dt \propto Z^m$, with $m \sim 0.8$ for WN stars, and $m \sim 0.6$ for WC stars. O star winds are also driven by radiation pressure, with a metallicity dependence that is similar to WN stars (Mokiem *et al.* 2007). Overall, the notion that WR winds are radiatively driven is supported by observations and theory (see Vink & de Koter 2005 and Gräfener & Hamann, this volume).

6.1. WN stars

Figure 6 (left panel) compares the mass-loss rates of cluster or association member WN stars in the Milky Way with Magellanic Cloud counterparts. Mass-loss estimates are obtained from their near-IR helium lines (Crowther 2006, following Howarth & Schmutz 1992). The substantial scatter in mass-loss rates is in line with the heterogeneity of line strengths within WN subtypes. Stronger winds are measured for WN stars without surface H, in agreement with recent results of Hamann *et al.* (2006).

Measured mass-loss rates of hydrogen-rich early-type WN stars in the SMC (1/5 $Z_\odot$) are 0.4 dex weaker than equivalent stars in the Milky Way and LMC (1/2 to 1 $Z_\odot$). This suggests a metallicity dependence of $dM/dt \propto Z^m$ for WN stars, with $m \sim 0.8 \pm 0.2$. The exponent is comparable to that measured from Hα observations of Milky Way, LMC and SMC O-type stars (de Koter *et al.*, this volume).

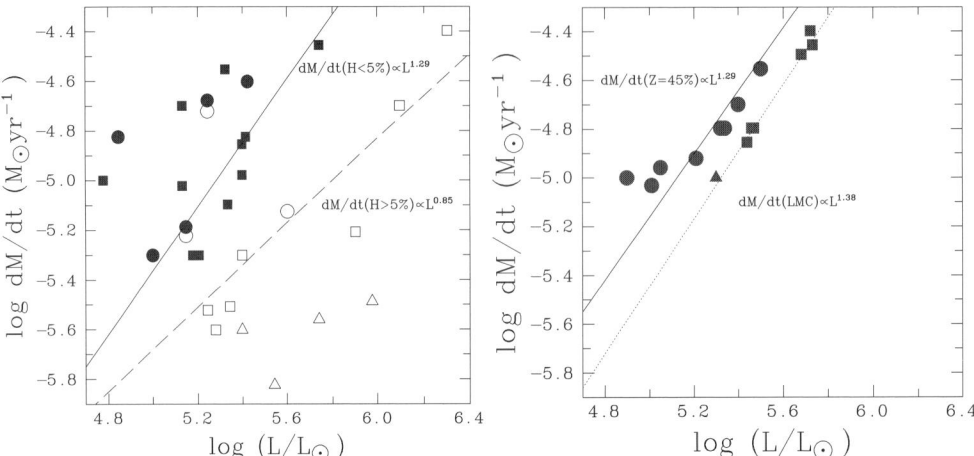

Figure 6. *Left:* Comparison between the mass-loss rates and luminosities of WN3–6 in the Galaxy (circles), LMC (squares) and SMC (triangles), for which open/filled symbols refer to WN stars with/without surface hydrogen, based upon analysis of near-IR helium lines (Crowther 2006). The Nugis & Lamers (2000) relation for H-deficient WN stars (solid line) is show, together with a fit to H-rich WN stars (dotted line). Mass-loss rates are universally high if hydrogen is absent; *Right:* Identical plot for WC4–9 stars in the Galaxy (circles) and LMC (squares), plus a WO star in the LMC (triangle), together with the Nugis & Lamers (2000) relation (solid, for $Z=45\%$) and a fit to the LMC results (dotted line).

6.2. WC stars

It is well established that WC stars in the inner Milky Way, and indeed all metal-rich environments, possess later spectral types than those in the outer Galaxy, LMC and other metal-poor environments (Hadfield & Crowther, this volume). This observational trend led Smith & Maeder (1991) to suggest that early-type WC stars are richer in carbon than late-type WC stars, on the basis of tentative results from recombination line analyses. In this scenario, typical Milky Way WC5–9 stars exhibit reduced carbon abundances than WC4 counterparts in the LMC. However, quantitative analysis of WC stars (e.g. Koesterke & Hamann 1995) do not support a subtype dependence of elemental abundances in WC stars (recall Fig. 3).

If differences of carbon content are not responsible for the observed WC subtype distribution in galaxies, what is its origin? Essentially, C III $\lambda 5696$ only becomes strong (producing a late WC subtype) when $\lambda 574$ is optically thick, i.e. if the stellar temperature is low *or* the wind density is sufficiently high. Since the temperatures of Galactic WC5–7 stars and LMC WC4 stars are comparable, the observed subtype distribution argues that the wind densities of Galactic WC stars must be higher than the LMC stars. Figure 6 (right panel) compares clumping corrected mass-loss rates of WC stars in the Milky Way and LMC, as derived from optical studies, for which the former agree well with Nugis & Lamers (2000).

Crowther *et al.* (2002) obtain a similar mass-loss dependence for WC4 stars in the LMC, albeit offset by –0.25 dex. Crowther *et al.* (2002) argued that the WC subtype distributions in the LMC and Milky Way resulted from this metallicity dependence. C III $\lambda 5696$ emission is very sensitive to mass-loss rate, so weak winds for LMC WC stars would produce negligible C III $\lambda 5696$ emission (WC4 subtypes) and strong winds interior to the Solar circle would produce strong C III $\lambda 5696$ emission (WC8-9 subtypes), in agreement with the observed subtype distributions.

7. Stellar evolution

From membership of WR stars in open clusters, their initial masses may be estimated empirically, for which a revised compilation is provided by Crowther *et al.* (2006b). Overall, hydrogen-rich mid-type WN stars are observed in young, massive clusters; their main-sequence turn-off masses (based on Meynet *et al.* 1994 isochrones) suggest initial masses of $65 - 110\,M_\odot$, and are believed to be core-H burning (Crowther *et al.* 1995a). Lower-mass progenitors of 40–$50\,M_\odot$ are suggested for late WN, late WC, and WO subtypes. Progenitors of some early WN stars appear to be less massive still, suggesting an initial-mass cutoff for WR stars at Solar metallicity around $25\,M_\odot$.

From an evolutionary perspective, the absence of red supergiants (RSGs) at high luminosity together with the presence of H-rich WN stars in young massive clusters suggests the following variation of the 'Conti scenario' in the Milky Way, i.e. for stars initially more massive than $\sim 75\,M_\odot$

$$O \to \mathrm{WN(H-rich)} \to \mathrm{LBV} \to \mathrm{WN(H-poor)} \to \mathrm{WC} \to \mathrm{SN\,Ic},$$

whereas for stars of initial mass from $\sim 40-75\,M_\odot$,

$$O \to \mathrm{LBV} \to \mathrm{WN(H-poor)} \to \mathrm{WC} \to \mathrm{SN\,Ic},$$

and for stars of initial mass in the range 25–$40\,M_\odot$,

$$O \to \mathrm{LBV/RSG} \to \mathrm{WN(H-poor)} \to \mathrm{SN\,Ib}.$$

Indeed, the role of the LBV phase is not yet settled – it may be circumvented entirely in some cases; it may follow the RSG stage, or it may even dominate pre-WR mass-loss for the most massive stars (Smith & Owocki 2006). The presence of dense, circumstellar shells around Type IIn SN indicates that some massive stars might even undergo core-collapse during the LBV phase (Smith *et al.* 2007). Remarkably few Milky Way clusters host both RSG and WR stars, with the notable exception of Westerlund 1 (Clark *et al.* 2005) which suggests that the mass range common to both populations is fairly narrow.

Turning to stellar evolutionary models, their various inputs originate from either laboratory experiments (e.g. opacities) or astronomical observations (e.g. mass-loss properties). Mass-loss and rotation are intimately linked for the evolution of massive stars. Stellar winds will lead to spin-down for the case of an efficient internal angular momentum transport mechanism. A subset of evolutionary models also now consider magnetic fields. Alas, very little is known about the field strengths of WR stars, with the sole exception of HD 50896 (WR6, de la Chevrotière *et al.*, this volume).

Rotation favours the evolution into the WR phase at earlier stages, increasing the WR lifetime, plus lower initial mass stars also enter the WR phase. For an assumed initial rotational velocity of $300\,\mathrm{km\,s^{-1}}$, the minimum initial mass star entering the WR phase is $22\,M_\odot$, versus $37\,M_\odot$ for non-rotating models at Solar metallicity (Meynet & Maeder 2003). Evolutionary models allowing for rotational mixing do predict a better agreement with the observed ratio of WR to O stars at low metallicity, the existence of intermediate WN/C stars (though see Langer 1991), and the ratio of blue to red supergiants in galaxies.

It is possible to predict the number ratio of WR to O stars for regions of constant star formation from evolutionary models, weighted over the Initial Mass Function (IMF). For an assumed Salpeter IMF slope for massive stars, the N(WR)/N(O) ratio predicted from rotating models are in much better agreement with the observed distribution at Solar metallicity (Meynet & Maeder 2003). Since the O star population is relatively imprecise, the predicted WR subtype distributions are often used instead for comparisons with observations. Alas, comparison between massive stellar populations obtained from evolutionary predictions and observations is prone to difficulties. For example, a core H-burning star with $T_{\rm eff}$ = 40 kK, $\log L/L_\odot$ = 6.0 and a surface hydrogen content of X_H = 50% may be defined as an O star from an evolutionary perspective, yet could be a H-rich mid-type WN star from an observational view, if its wind is sufficiently strong. Alternatively, a He-burning star with $T_{\rm eff}$ = 20 kK, $\log L/L_\odot$ = 5.5 and X_H = 20% would be defined as a WN star from evolutionary models versus a B supergiant from an observational perspective.

Table 1. Comparison between expected WR populations from single, Solar metallicity models of Meynet et al. (1994, M94) and binary models of Eldridge et al. (2008, E08) for an instantaneous burst at ~4.5 Myr with the observed WR population of Westerlund 1 (Crowther et al. 2006b)

	M94	E08	Westerlund 1
N(H-rich WN)/N(H-poor WN)	1.2	1.1	0.25
N(WC)/N(WN)	2.4	1.4	0.5

These issues aside, the Solar Neighbourhood WR subtype distribution contains similar numbers of WC and WN stars, also with an equal number of early-type (H-free) and late-type (H-rich) WN stars. From comparison with evolutionary models, the agreement is reasonable, except for the brevity of the H-deficient WN phase in interior models at Solar metallicity. This aspect has been quantified by Hamann et al. (2006). Synthetic WR populations from the Meynet & Maeder (2003) evolutionary tracks, together with a Salpeter IMF, predict that only 20% of WN stars be hydrogen-free, in contrast to over 50% of the observed sample. Non-rotating models provide better statistics, although low luminosity early-type WN stars are absent in such synthetic populations.

We may also take the specific case of predictions from an instantaneous burst with the empirical results for the young massive cluster Westerlund 1, whose age is ~4.5 Myr. Alas, predictions from both single and binary star models, fail to reproduce the observed subtype distribution as illustrated in Table 1. Similar difficulties are experienced at lower metallicities (Hadfield & Crowther 2006), of relevance to He II 1640 as a tracer of high-z galaxies (Brinchmann et al. 2008). The ratio of WC to WN stars is observed to increase with metallicity for nearby galaxies, whose WR content has been studied in detail (Hadfield, this volume).

Evolutionary models, in which convective overshooting and a metallicity scaling for WR stars is included (Eldridge & Vink 2006) - though not rotational mixing - agree fairly well with the observed N(WC)/N(WN) at high metallicity, *unlike* those for which rotational mixing is allowed for, but not a WR metallicity scaling (Meynet & Maeder 2005). A significant WR population formed via a close binary channel is required to reproduce the observed N(WR)/N(O) ratio across the full metallicity range in the Eldridge & Vink (2006) models (see Van Bever & Vanbeveren 2003). A significant binary channel is not required for the Meynet & Maeder (2005) rotating evolutionary models. Each approach resolve some issues with respect to earlier comparisons to observations, although problems do persist.

The end states of massive stars have been studied from a theoretical perspective by Heger et al. (2003). Type II SN typically arising from lower mass red supergiants, while WN and WC stars are the likely progenitors of (at least some) Type Ib and Type Ic core-collapse SN, respectively. This arises because, hydrogen and hydrogen/helium, respectively, are absent in such SNe (Woosley & Bloom 2006). Direct empirical evidence connecting single WR stars to Type Ib/c SN is lacking, for which lower mass interacting binaries represent alternative progenitors. Observations of tens of thousands of WR stars would be required to firmly establish a connection on a time frame of a few years, since WR lifetimes are a few 10^5 yr (Meynet & Maeder 2005) – see Hadfield (this volume) for how this could be achieved. Narrow-band optical surveys of a dozen other high star-forming spiral galaxies within ~10 Mpc would likely provide the necessary statistics. However, ground-based surveys would be hindered by the relatively low spatial resolution of 20 pc per arcsec at 5 Mpc.

The light curves of bright, broad-lined Type Ic supernovae (e.g. SN 2003dh) suggest that the ejected core masses are of order $10 M_\odot$ (Mazzali et al. 2003) which rather well with the masses of LMC WC4 stars inferred by Crowther et al. (2002), if we additionally consider several solar masses which remain as a compact (black hole) remnant. Such supernovae were associated with long GRBs (eg. GRB 030329 Hjorth et al. 2003) in support of the 'collapsar' model (see Woosley & Bloom 2006). WR populations have been detected in the host galaxy of GRB 980425 (SN 1998bw), albeit offset from the location of the GRB by several hundred pc (Hammer et al. 2006). Fruchter et al. (2006) showed that the location of GRBs within their host galaxies is more concentrated on the brightest (youngest) regions than (Type II) SN. Curiously, the location of Type Ib mimic that of Type II SN in their host galaxies, while Type Ic SN trace that of long GRBs (Kelly et al. 2008), suggesting grossly different stellar populations. Fryer et al (2007) provide evidence in favour of binary progenitors of Type Ib/c SN and GRBs.

References

Asplund, M., Grevesse, N., Sauval, A. J., et al. 2004, A&A, 417, 751
Barlow, M. J., Roche, P. F., & Aitken, D. A. 1988, MNRAS, 232, 821
Brinchmann, J., Pettini, M., & Charlot, S. 2008, MNRAS, 385, 769
Cantiello, M., Yoon, S.-C., Langer, N., & Livio, M. 2007, A&A, 465, L29
Castor, J. I., Abbott, D. C., & Klein, R. I., 1975, ApJ, 195, 157
Clark, J. S., Negueruela, I., Crowther, P. A., & Goodwin, S. P. 2005, A&A, 434, 949
Conti, P. S., & Massey, P. 1989, ApJ, 337, 251
Conti, P. S., Crowther, P. A., & Leitherer, C., 2008, From Luminous Hot Stars to Starburst Galaxies, (Cambridge: CUP), Camb. Astrophys. Ser. 45, in press
Crowther, P. A. 2006, in: H. J. G. L. M. Lamers, N. Langer, T. Nugis & K. Annuk (eds.), Stellar Evolution at Low Metallicity: Mass Loss, Explosions, Cosmology (San Francisco: ASP), ASP Conf. Ser., 353, 157
Crowther, P. A. 2007, ARA&A, 45, 177
Crowther, P. A., & Dessart, L. 1998, MNRAS, 296, 622
Crowther, P. A., & Hadfield, L. J. 2006, A&A, 449, 711
Crowther, P. A., & Smith, L. J. 1997, A&A, 320, 500
Crowther, P. A., Smith, L. J., Hillier, D. J., & Schmutz, W. 1995a, A&A, 293, 427
Crowther, P. A., Smith, L. J., & Hillier, D. J. 1995b, A&A, 302, 457
Crowther, P. A., Smith, L. J., & Willis, A. J. 1995c, A&A, 304, 269
Crowther, P. A., Fullerton, A. W., Hillier, D. J., et al. 2000, ApJ, 538, L51
Crowther, P. A., Dessart, L., Hillier, D. J., et al. 2002, A&A, 392, 653
Crowther, P. A., Morris, P. W., & Smith, J. D. 2006a, ApJ, 636, 1033
Crowther, P. A., Hadfield, L. J., Clark, J. S., et al. 2006b, MNRAS, 372, 1407
de Koter, A., Schmutz, W., & Lamers, H. J. G. L. M., 1993, A&A, 277, 561
Dessart, L., Crowther, P. A., Hillier, D. J., et al. 2000, MNRAS, 315, 407
Eldridge, J. J. & Vink, J. S. 2006, A&A, 452, 295
Eldridge, J. J., Izzard, R. G., & Tout, C. A. 2008, MNRAS, 384, 1109
Figer, D. F. 2005, Nat, 434, 192
Foellmi, C., Moffat, A. F. J., & Guerrero, M. A., 2003, MNRAS, 338, 360
Fruchter, A. S., Levan, A. J., Strolger, L., et al. 2006, Nat, 441, 463
Fryer, C. L., Mazzali, P. A., Prochaska, J., et al. 2007, PASP, 119, 1211

Gamov, G. 1943, *ApJ*, 98, 500
Gräfener, G., Koesterke, L., & Hamann, W.-R. 2002, *A&A*, 387, 244
Hadfield, L. J., & Crowther, P. A. 2006, *MNRAS*, 368, 1822
Hadfield, L. J., Van Dyk, S., Morris, P. W., et al. 2007, *MNRAS*, 376, 248
Hamann, W.-R., Gräfener, & G., Liermann, A., 2006, *A&A*, 457, 1015
Hammer, F., Flores, H., Schaerer, D., et al. 2006, *A&A*, 454, 103
Harries, T. J., Hillier, D. J., & Howarth, I. D. 1998, *MNRAS*, 296, 1072
Heger, A., Fryer, C. L., Woosley, S. E., et al. 2003, *ApJ*, 591, 288
Herald, J. E., Hillier, D. J., & Schulte-Ladbeck, R. E. 2001, *ApJ*, 548, 932
Hillier, D. J. 1988, *ApJ*, 327, 822
Hillier, D. J. 1989, *ApJ*, 347, 392
Hillier, D. J. 1991, *A&A*, 247, 455
Hillier, D. J., & Miller, D. L., 1998, *ApJ*, 496, 407
Hillier, D. J., & Miller, D. L., 1999, *ApJ*, 519, 354
Hjorth, J., Sollerman, J., Moller, P., et al. 2003, *Nat*, 423, 847
Homeier, N., Blum, R. D., Pasquali, A., et al. 2003, *A&A*, 408, 153
Howarth, I. D., & Schmutz, W., 1992, *A&A*, 261, 503
Kelly, P. L., Kirshner, R. P., & Pahre, M. 2008, *ApJ*, in press (arXiv:0712.0430)
Kingsburgh, R. L., Barlow, M. J., & Storey, P. J. 1995, *A&A*, 295, 75
Koesterke, L., & Hamann, W.-R., 1995, *A&A*, 299, 503
Kurosawa, R., Hillier, D. J., & Pittard, J. M., 2002, *A&A*, 388, 957
Kurtev, R., Borissova, J., Georgiev, L., et al. 2007, *A&A*, 475, 209
Lamers, H. J. G. L. M., Maeder, A., Schmutz, W., & Cassinelli, J. P. 1991, *ApJ*, 368, 538
Langer, N. 1989, *A&A*, 220, 135
Langer, N. 1991, *A&A*, 248, 531
Leitherer, C., Chapman, J. M., & Koribalski, B. 1997, *ApJ*, 481, 898
Marchenko, S. V., Moffat, A. F. J., Crowther, P. A., et al. 2004, *MNRAS*, 353, 153
Mazzali, P. A., Deng, J., Tominaga, N., et al. 2003, *ApJ*, 599, L95
Meynet, G., 2007, in: *European Phys. Journal: Special Topics* (arXiv:0709.3808)
Meynet, G., & Maeder, A. 2003, *A&A*, 404, 975
Meynet, G., & Maeder, A. 2005, *A&A*, 429, 581
Meynet, G., Maeder, A., Schaller, G., et al. 1994, *A&AS*, 103, 97
Mokiem, M. R., de Koter, A., Vink, J., et al. 2007, *A&A*, 473, 603
Morris, P. W., Crowther, P. A., & Houck, J. R. 2004, *ApJS*, 154, 413
Nugis, T., Crowther, P. A., & Willis, A. J. 1998, *A&A*, 333, 956
Nugis, T., & Lamers, H. J. G. L. M. 2000, *A&A*, 360, 227
Rauw, G., Crowther, P. A., De Becker, M., et al. 2005, *A&A*, 432, 985
Schmutz, W. 1997, *A&A*, 321, 268
Smith, L. F. 1968, *MNRAS*, 138, 109
Smith, L. F., & Maeder, A. 1991, *A&A*, 241, 77
Smith, L. J., Norris, R. P. F., & Crowther, P. A. 2002, *MNRAS*, 337, 1309
Smith, N., & Conti, P. S. 2008, *ApJ*, 679, 1467
Smith, N. & Owocki, S. P., 2006 *ApJ*, 645, L45
Smith, N., Li, W., Foley, R. J., et al. 2007, *ApJ*, 666, 1116
Tuthill, P., Monnier, J., Tanner, A., et al. 2006, *Sci*, 313, 935
Van Bever, J. & Vanbeveren, D. 2003, *A&A*, 400, 63
van der Hucht, K. A. 2001, *New A*, 45, 135
van der Hucht, K. A. 2006, *A&A*, 458, 453
Vink, J. S. 2007, *A&A*, 469, 707
Vink, J. S., & de Koter, A. 2005, *A&A*, 442, 587
Walborn, N. R., Morrell, N. I., Howarth, I. D., et al. 2004, *ApJ*, 608, 1028
Willis, A. J., Crowther, P. A., Fullerton, A. W., et al. 2004, *ApJS*, 154, 651
Woosley, S. E. & Heger, A., 2006, *ApJ*, 637, 914
Wright, A. E., & Barlow, M. J. 1975, *MNRAS*, 170, 41

Discussion

MOFFAT: If C/He ia approximately constant for all WC subtypes, then this means that there is no evolution along the WC sequence from WC9 to WC8 ... to WC4, as suspected years ago. This then would imply that what determines the WC subtype is the ambient metallicity. How is this reconciled with constant C/He? And L and M of WC stars must be on average lower at lower Z, right?

CROWTHER: Yes, ambient metallicity dictates WC subtypes via wind strength, so individual subtypes scan a wide range of carbon (and oxygen) abundances. On average WC masses and luminosities are *higher* at low metallicities (cf. Crowther *et al.* 2002).

WILLIS: In connection with your Spitzer neon abundance for the WO star, having the same Ne abundance as WC stars: there may be a way to reconcile this and yet still have the WO star more advanced in He-burning than WC stars. This is if in fact the ^{20}Ne has been further processed (quickly) to ^{25}Mg. To test this we need to get the Mg abundance in a WO star. Unfortunately, there are no suitable Mg lines in the optical/IR spectral range. However, it would be possible to determine the Mg abundance from X-ray spectra, so we would love to have an X-ray spectrum of a WO star, to determine its Mg abundance and test this possibility.

CROWTHER: Mg V is included in the IRS Spitzer band but is not seen in the two WO stars observed to date. So either there is no Mg enhancement or Mg is hidden in lower ionization stages.

VANBEVEREN: In the early days (1960's) it was indeed believed that most (all) WR stars formed in binaries. However, you may be surprised to hear that together with Peter Conti, I published a paper in 1979-1980 where for the first time convincing arguments were presented that not all WR stars are binary components, and in this paper we proposed a WR binary frequency of 40-50%, a value that still stands after more than 25 years.

Paul Crowther.

Alceste Bonanos.

Jorick Vink.

Massive Stars as Cosmic Engines
Proceedings IAU Symposium No. 250, 2007
F. Bresolin, P.A. Crowther & J. Puls, eds.

© 2008 International Astronomical Union
doi:10.1017/S1743921308020346

Wolf-Rayet Wind Models from Hydrodynamic Model Atmospheres

Götz Gräfener and Wolf-Rainer Hamann

Institut für Physik, Universität Potsdam, Am Neuen Palais 10, 14469 Potsdam, Germany

Abstract. We present a parameter study of WR-type mass loss, based on the PoWR hydrodynamic model atmospheres. These new models imply that optically thick WR-type winds are generally formed close to the Eddington limit. This is demonstrated for the case of hydrogen rich WNL stars, which turn out to be extremely massive, luminous stars with progenitor masses above $\approx 80\,M_\odot$. We investigate the dependence of WR-type mass loss on various stellar parameters, including the metallicity Z. The results depend strongly on the L/M ratio, the stellar temperature $T_\star$, and the assumed wind clumping. For high L/M ratios, strong WR-type winds can be maintained down to very low Z. Even for primordial massive stars we predict considerable mass loss if their surfaces are self-enriched by primary elements.

Keywords. stars: abundances – stars: atmospheres – stars: mass loss – stars: Wolf-Rayet

1. Optically thick stellar winds

A characteristic property of Wolf-Rayet (WR) stars is their strong mass loss. Their wind momenta typically lie above the single scattering limit ($\dot{M} v_\infty > L/c$), raising the question if WR winds are driven by radiation.

In case of radiative driving, photons need to be absorbed and re-emitted more than once to achieve the large wind momenta. A direct consequence of this is that the flux-mean optical depth τ_W of the wind needs to be large. Photons are scattered on average τ_W^2 times before they leave the wind. Because they perform a random walk they transfer a momentum of $\tau_W L/c$ per time interval to the wind. The resulting wind momentum is of the same order of magnitude, i.e., $\dot{M} v_\infty \approx \tau_W L/c$. Actually the wind momentum is slightly lower because part of the radiative momentum is used to overcome the gravitational potential of the star (see, e.g., Sect. 7.2.2 in Lamers & Cassinelli 1999, for an exact derivation).

The large wind optical depth τ_W is also responsible for the spectral appearance of WR stars. For large τ_W the ionizing radiation from the static layers is absorbed within the wind. This leads to strong recombination, and via recombination cascades, to the observed WR emission line spectra. The fact that WR emission lines are observed together with large wind momenta is thus a hint that τ_W is large for WR stars, and that their winds are radiatively driven.

To generate large τ_W a large flux-mean opacity is needed directly above the sonic point, in combination with a large density scale height H_ρ. We thus expect WR-type mass loss to occur 1) close to the Eddington limit when H_ρ becomes large, and 2) in temperature regimes where the sonic point is located close to the Fe-opacity peaks (see also Nugis & Lamers 2002).

2. WR wind models from hydrodynamic model atmospheres

The Potsdam Wolf-Rayet (PoWR) models combine fully line-blanketed non-LTE model atmospheres with the equations of hydrodynamics (for details see Gräfener & Hamann 2005; Hamann & Gräfener 2003; Koesterke et al. 2002; Gräfener et al. 2002). The wind structure ($\rho(r)$ and $v(r)$) and the temperature structure $T(r)$ are computed consistently with the full set of non-LTE populations, and the radiation field in the co-moving frame (CMF). In contrast to other approaches, the radiative wind acceleration $a_{\rm rad}$ is obtained by direct integration

$$a_{\rm rad} = \frac{1}{c} \int \chi_\nu F_\nu {\rm d}\nu, \qquad (2.1)$$

instead of making use of the Sobolev approximation. In this way, complex processes, e.g., due to the strong line overlap or the redistribution of radiation, are automatically included. Moreover, the models take small-scale wind clumping into account. The models describe the conditions in optically thick WR atmospheres in a realistic manner, and provide synthetic spectra, i.e. they allow for a direct comparison with observations.

2.1. Results for different WR subtypes

Utilizing the PoWR models, we have obtained the first fully self-consistent WR wind model for a WC star of early subtype (Gräfener & Hamann 2005). Moreover, we have examined the mass loss of late-type WN stars and its dependence on metallicity (Gräfener & Hamann 2006, 2008).

According to our models, WR-type mass loss is triggered by the proximity to the Eddington limit. The sonic-point temperatures lie in the expected range for optically thick winds ($\approx 200\,{\rm kK}$ for early WC subtypes, corresponding to the hot Fe-peak; 30–45 kK for late WN subtypes, corresponding to the cool Fe-peak). Because of this, we predict two distinct regimes of stellar core temperatures for which strong WR-type mass loss occurs. For early WC subtypes very high core temperatures of the order of $T_\star \approx 140\,{\rm kK}$ are required (Gräfener & Hamann 2005). WNL subtypes can be reproduced successfully for cooler temperatures, in the range $T_\star = 30$–$50\,{\rm kK}$ (Gräfener & Hamann 2006, 2008). For intermediate values of $T_\star$ our models do not provide enough radiative force in the deep atmospheric layers, i.e., we do not obtain a stationary wind solution. In this regime pulsations might play an important role (e.g., strange mode pulsations as proposed by Glatzel & Kaltschmidt 2002).

2.2. The role of wind clumping

An important feature of WR-type winds is their dependence on the wind clumping factor $D(r)$ (see Hamann & Koesterke 1998). The wind acceleration roughly scales as $a_{\rm rad} \propto \sqrt{D}$. This effect is caused by the dominance of recombination processes in WR winds. An increase of D increases the mean $\langle \rho^2 \rangle$, and thus mimics a denser wind with a larger mean opacity. For models of early-type WR stars we usually have to choose relatively large clumping factors around $D = 50$ to reproduce the observed terminal wind velocities. As observational estimates lie more in the range of $D = 10$, our models thus might underestimate $a_{\rm rad}$ by roughly a factor of two. This could possibly reflect the lack of trace elements like Ne, Ar, Mg, or Ca in our models.

3. The most massive stars: H-rich WNL stars

The high L/M ratios which are required to maintain WR-type winds can in principle be reached in different ways. He-burning stars have high L/M ratios because of their large

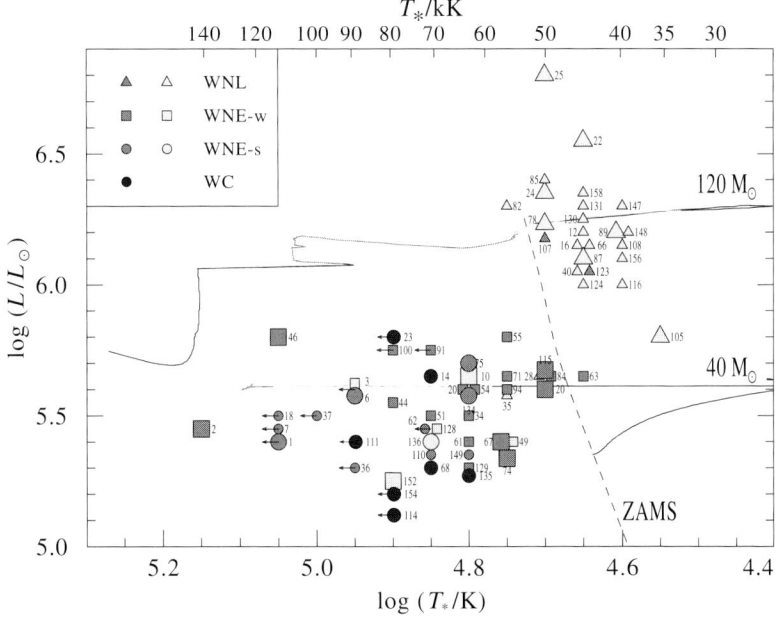

Figure 1. Spectral analyses of galactic WR stars with line-blanketed PoWR models, according to Hamann et al. (2006) and Barniske et al. (2006): symbols in light grey denote H-rich WN stars, whereas H-free objects are indicated in dark grey. WC stars are indicated in black. For objects with large symbols distance estimates are available (van der Hucht 2001), whereas objects with small symbols are calibrated by their spectral subtype. Evolutionary tracks for non-rotating massive stars (Meynet & Maeder 2003) are shown for comparison.

mean molecular weight. H-burning stars, on the other hand, can in principle reach high Eddington factors if their masses are extremely high. In the following we will show that this is indeed the case for H-rich WNL stars in young stellar clusters. Our models imply very high masses for these objects, in line with measurements of very high stellar masses in WNL binary systems (e.g., Rauw et al. 1996; Schweickhardt et al. 1999; Bonanos et al. 2004; Moffat et al. 2007).

3.1. Spectral analyses of galactic WR stars

A comprehensive study of galactic WR stars, based on spectral analyses with line-blanketed PoWR models (Hamann et al. 2006; Barniske et al. 2006, respectively for WN and WC stas) reveal a bimodal WR subtype distribution in the HRD (Fig. 1). H-rich WNL stars are located to the right of the ZAMS with luminosities above $10^6 \, L_\odot$, while the mostly H-free early- to intermediate WN subtypes, as well as the WC stars, have lower luminosities and hotter temperatures.

Although a part of the WNL stars with unknown distances (small symbols in Fig. 1) might actually belong to the group with lower luminosities, the dichotomy implies that H-rich WNL stars are the descendants of very massive stars, possibly still in the phase of central H-burning, whereas the earlier subtypes (including WC stars) are more evolved, less massive, He-burning objects.

3.2. Wind models for H-rich WNL stars

Gräfener & Hamann (2008) investigate the properties of luminous, H-rich WNL stars by means of grid computations with hydrodynamic PoWR models. The most important conclusion from this work is that WR-type mass loss is triggered by the proximity to the

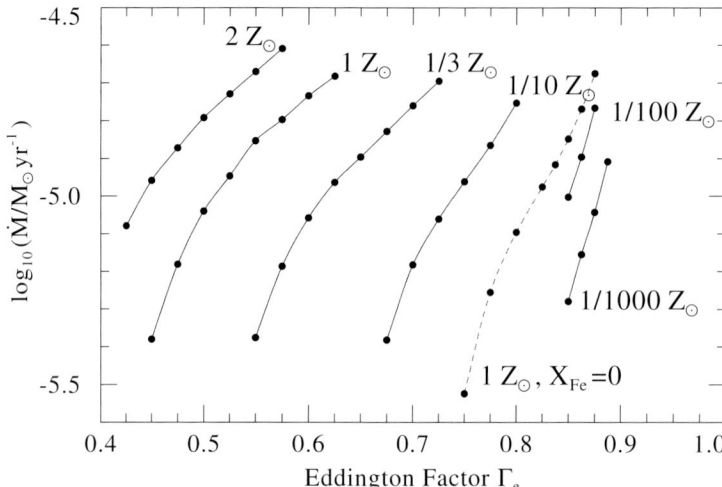

Figure 2. Wind models for WNL stars: mass loss rates from our hydrodynamic models are plotted vs. the Eddington factor $\Gamma_{\rm e}$. The stellar parameters are $L_\star = 10^{6.3}\,L_\odot$, $T_\star = 45\,{\rm kK}$, and $X_{\rm H} = 0.4$. The variation of $\Gamma_{\rm e}$ thus reflects a variation of the stellar mass. Solid curves indicate model series for different metallicities Z. The dashed line indicates models with zero Fe-abundance but solar-like CNO, resembling the composition of self-enriched very metal-poor stars (see Meynet *et al.* 2006).

Eddington limit (i.e., $\Gamma_{\rm e} \equiv \chi_{\rm e} L_\star / 4\pi c G M_\star$ approaching unity), or equivalently, by high L/M ratios. In Fig. 2 we show our results for H-rich WNL stars with a fixed luminosity of $10^{6.3}\,L_\odot$, illustrating the strong dependence of the mass loss on $\Gamma_{\rm e}$ and Z.

H-rich WNL stars typically have stellar temperatures in the range of 30–50 kK and moderate wind optical depths ($\tau_W \approx 1$). Their winds are initiated by the cool Fe-peak opacities. As can be seen in Fig. 2, the mass loss increases very rapidly when the stars approach the Eddington limit. However, the models are still away from the classical Eddington limit (i.e., $\Gamma_{\rm e} = 1$), dependent on the adopted metallicity Z. This effect is caused by the influence of metal lines in the hydrostatic layers. The metal lines increase the flux mean opacity, and shift the 'true' Eddington limit, where the atmosphere actually becomes instable, towards lower values of $\Gamma_{\rm e}$, i.e., towards higher masses. Our solar Z WNL models thus become WR stars already for $\Gamma_{\rm e} \approx 0.5$, a value which can be reached by very massive stars at the end of their main sequence evolution.

A more detailed analysis of the WN 7 component in WR 22, an eclipsing WR+O binary system in Car OB1, supports our results. Our models give a luminosity of $10^{6.3}\,L_\odot$, and a mass of $M_{\rm WR} = 78\,M_\odot$ for this object, in agreement with Rauw *et al.* (1996) who find $M_{\rm WR} \sin^3 i = 72 \pm 3\,M_\odot$ from the binary orbit (see Gräfener & Hamann 2006, 2008). Such high masses imply that H-rich WNL stars are still in the phase of central H-burning, supporting an evolutionary sequence of the form O → WNL (H-rich) → LBV → WN (H-poor) → WC for very massive stars, as originally proposed by Langer *et al.* (1994).

3.3. WNL stars: mass loss properties

In Fig. 3 we show the temperature dependence of our WNL models. With $\dot{M} \propto T_\star^{-3.5}$ the mass loss nearly scales with the size of the stellar surface, a behavior which is not expected from the standard theory of radiatively driven winds (Castor *et al.* 1975, CAK). Moreover, we find a remarkably weak dependence of $\dot{M}$ on the luminosity (for fixed values $\Gamma_{\rm e}$ and $T_\star$). As pointed out by Gräfener & Hamann (2008) these peculiar properties can

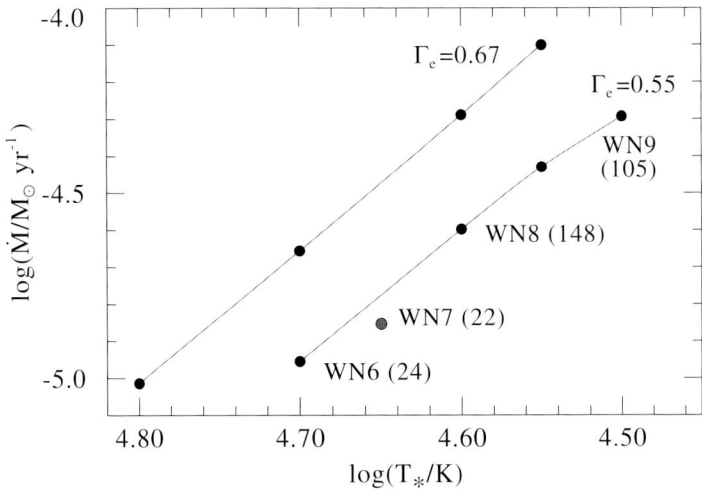

Figure 3. Wind models for galactic WNL stars: mass loss rates for different stellar temperatures $T_\star$ and Eddington parameters Γ_e. The models reproduce the observed spectral sequence from WN 6 to WN 9. The corresponding spectral subtypes are indicated together with WR numbers of specific galactic objects (according to van der Hucht 2001, in brackets).

be understood within the theory of optically thick stellar winds (e.g., Nugis & Lamers 2002). They are caused by the necessity to reach a specific temperatures at the sonic point, which demand for specific wind optical depths τ_W.

3.4. Metallicity dependence & the first WN stars

Our models in Fig. 2 show a Z-dependence where the onset of WR-type mass loss, in terms of the Eddington factor Γ_e, depends strongly on Z. This means that, on average, WR mass loss declines for lower Z. However, for larger Eddington factors our models predict equally strong WR mass loss rates over the whole range of metallicities, from 1/1000–$2 Z_\odot$. Notably, we even obtain strong WR mass loss for iron-free models with solar-like CNO abundances (the dashed red line in Fig. 2). The composition of these models resembles extremely metal-poor stars which are self-enriched with primary nitrogen. The occurrence of such stars has been discussed, e.g., by Meynet et al. (2006). Our models thus predict an efficient mass loss mechanism for initially metal-poor objects (the first WN stars) which could play an important role in the enrichment of the early ISM.

3.5. Mass determinations: the stellar population in the Arches cluster

From IR photometry of the brightest stars in the Arches cluster, Figer (2005) has deduced a general upper mass limit for stars of $\sim 150\,M_\odot$. In a recent work Martins et al. (2008) have determined the stellar parameters of these objects by means of detailed spectral analyses. The stars turned out to be extremely luminous WNL and Of stars, exactly matching the properties of our WNL grid models. Here we apply our results in the form of a parameterized mass loss recipe (of the form $\dot{M}(L_\star, T_\star, \Gamma_e, Z, X_H)$, see Gräfener & Hamann 2008), to estimate the present masses of these objects.

In Fig. 4 we show our results for an adopted solar metallicity (we have performed computations for $Z_\odot$ and $2\,Z_\odot$, giving qualitatively similar results). Notably, we find two stellar populations, namely objects with present masses below $\sim 100\,M_\odot$ which are still on the main sequence, and chemically more evolved objects in the range of 30–100 $M_\odot$ which might partly already burn helium, and seem to originate from more massive stars. The initial masses of the first group seem to lie in a range up to $130\,M_\odot$, lifting the

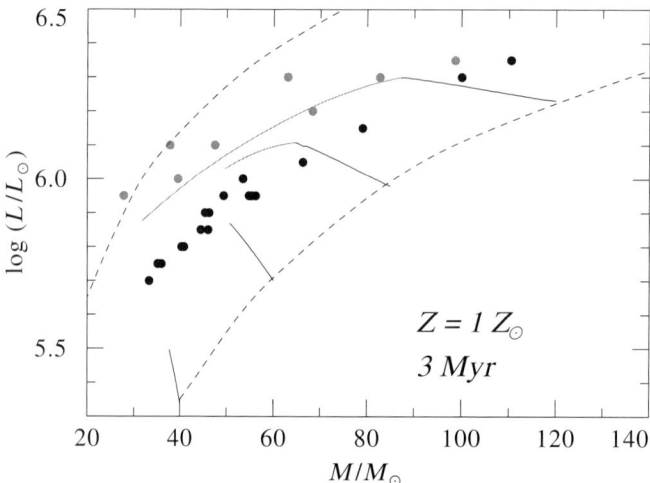

Figure 4. The most massive stars in the Arches cluster: plotted are luminosities taken from Martins *et al.* (2008), vs. present stellar masses, as derived from our wind models, for the brightest stars in the Arches cluster. We detect two populations with different L/M ratios, namely main sequence objects with solar hydrogen content (black dots), and more evolved He-rich objects (grey dots). We compare our results with rotating stellar models with an age of 3 Myr, according to Meynet & Maeder (2003). The dashed lines indicate the ZAMS, and the He-MS for very massive stars (Ishii *et al.* 1999).

second group significantly above this value. If we adopt a larger value for Z our mass estimates would even rise.

References

Barniske, A., Hamann, W.-R., & Gräfener, G. 2006, in: H. J. G. L. M. Lamers, N. Langer, T. Nugis, & K. Annuk (eds.), *Stellar Evolution at Low Metallicity: Mass Loss, Explosions, Cosmology* (San Francisco: ASP) *ASP Conf. Ser.*, 353, 243
Bonanos, A. Z., Stanek, K. Z., Udalski, A., *et al.* 2004, *ApJL*, 611, L33
Castor, J. I., Abbott, D. C., & Klein, R. I. 1975, *ApJ*, 195, 157
Figer, D. F. 2005, *Nature*, 434, 192
Glatzel, W., & Kaltschmidt, H. O. 2002, *MNRAS*, 337, 743
Gräfener, G., & Hamann, W.-R. 2005, *A&A*, 432, 633
Gräfener, G., & Hamann, W.-R. 2006, in: H. J. G. L.M. Lamers, N. Langer, T. Nugis, & K. Annuk (eds.), *Stellar Evolution at Low Metallicity: Mass Loss, Explosions, Cosmology* (San Francisco: ASP) *ASP Conf. Ser.*, 353, 171
Gräfener, G., & Hamann, W.-R. 2008, *A&A*, 482, 945
Gräfener, G., Koesterke, L., & Hamann, W.-R. 2002, *A&A*, 387, 244
Hamann, W.-R., & Gräfener, G. 2003, *A&A*, 410, 993
Hamann, W.-R., Gräfener, G., & Liermann, A. 2006, *A&A*, 457, 1015
Hamann, W.-R., & Koesterke, L. 1998, *A&A*, 335, 1003
Ishii, M., Ueno, M., & Kato, M. 1999, *PASJ*, 51, 417
Koesterke, L., Hamann, W.-R., & Gräfener, G. 2002, *A&A*, 384, 562
Lamers, H. J. G. L. M., & Cassinelli, J. P. 1999, *Introduction to Stellar Winds* (Cambridge: Cambridge University Press)
Langer, N., Hamann, W.-R., Lennon, M., *et al.* 1994, *A&A*, 290, 819
Martins, F., Hillier, D. J., Paumard, T., *et al.* 2008, *A&A*, 478, 219
Meynet, G., Ekström, S., & Maeder, A. 2006, *A&A*, 447, 623
Meynet, G., & Maeder, A. 2003, *A&A*, 404, 975
Moffat, A. F. J., Schnurr, O., Chené, A.-N., *et al.* 2007, *Highlights of Astronomy*, 14, 197

Nugis, T., & Lamers, H. J. G. L. M. 2002, *A&A*, 389, 162
Rauw, G., Vreux, J.-M., Gosset, E., et al. 1996, *A&A*, 306, 771
Schweickhardt, J., Schmutz, W., Stahl, O., et al. 1999, *A&A*, 347, 127
van der Hucht, K. A. 2001, *New Astronomy Review*, 45, 135

Discussion

NAJARRO: The Geneva models with $2\,Z_\odot$ you have used to explain the Arches cluster have indeed four times the Asplund's oxygen abundance and not twice, which is basically the value derived in the literature. Therefore the use of $2\,Z_\odot$ models is not coherent with current estimates of the CNO content in the Arches.

GRÄFENER: The role of Z in evolutionary models is that it controls the stellar mass loss rates. These are dominated by the Fe abundance. The Fe abundance however changed only slightly in Asplund's revision. So, concerning the tracks in the HRD, $2\,Z_\odot$ remains $2\,Z_\odot$. Moreover I have shown that my derived masses are in agreement with $2\,Z_\odot$ tracks at an age of 2 Myr, and with $1\,Z_\odot$ tracks at 3 Myr.

GAYLEY: Using Fe bumps to explain WR mass-loss rates seems to suffer somewhat from the problem of "having your cake and eating it too". If extreme temperature sensitivity in the opacity is crucial for understanding high mass-loss, why don't you have a problem with the wind stagnating as the temperature drops with radius in an optically thick wind?

GRÄFENER: This problem in fact exists for the winds of hot He-stars, but not for the H-rich WNL stars. The latter are driven by the cool Fe-peak alone. Our hot WCE models, on the other hand, with core temperatures of 140kK are initially driven by the hot Fe-peak, and there are difficulties to overcome the gap between the two Fe-peaks. In our models we do this by increasing the clumping factor in this region. For slightly cooler stars the problem might be more severe. In reality the WCL stars even seem to form a second pseudo-photosphere because their initial wind stalls (see Petrovic et al., 2006, *A&A*, 450, 219).

Götz Gräfener.

Philippe Eenens.

Steven Cranmer.

Luminous Blue Variables & Mass Loss near the Eddington Limit

Stan Owocki and Allard Jan van Marle

Bartol Research Institute, Department of Physics & Astronomy
University of Delaware, Newark, DE 19350 USA
email: owocki@bartol.udel.edu, marle@udel.edu

Abstract. During the course of their evolution, massive stars lose a substantial fraction of their initial mass, both through steady winds and through relatively brief eruptions during their Luminous Blue Variable (LBV) phase. This talk reviews the dynamical driving of this mass loss, contrasting the line-driving of steady winds to the potential role of continuum driving for eruptions during LBV episodes when the star exceeds the Eddington limit. A key theme is to emphasize the inherent limits that self-shadowing places on line-driven mass loss rates, whereas continuum driving can in principle drive mass up to the "photon-tiring" limit, for which the energy to lift the wind becomes equal to the stellar luminosity. We review how the "porosity" of a highly clumped atmosphere can regulate continuum-driven mass loss, but also discuss recent time-dependent simulations of how base mass flux that exceeds the tiring limit can lead to flow stagnation and a complex, time-dependent combination of inflow and outflow regions. A general result is thus that porosity-mediated continuum driving in super-Eddington phases can explain the large, near tiring-limit mass loss inferred for LBV giant eruptions.

Keywords. stars: early-type – stars: winds, outflows – stars: mass loss – stars: activity

1. Introduction

Two key properties in making massive stars "cosmic engines" are their high luminosity, and their extensive mass loss. Indeed the momentum of this radiative luminosity is a key factor in driving massive-star mass loss, for example through the coupling with bound-bound opacity that is the basis of their more or less continuous line-driven stellar winds. Among the most luminous hot stars there appears a class of "Luminous Blue Variables" (LBVs) for which the winds are particularly strong, and exhibit irregular variability on time scales ranging from days to years. Contemporary observations generally suggest modest variations in net mass loss, occuring with nearly constant bolometric luminosity, and which might readily be explained by, e.g., opacity instabilities within the standard line-driving mechanism. But historical records, together with the extensive nebulae around many LBVs, suggest there are also much more dramatic eruptions, marked by substantial increases in the already extreme radiative luminosity, and lasting for several years, over which the net mass loss, $0.1 - 10$ $M_\odot$, far exceeds what could be explained by line-driving. Rather, the closeness of such stars to the Eddington limit, for which the radiative force from just the electron scattering continuum would balance the force of gravity, suggests that such "giant eruptions" might instead arise from *continuum* driving, resulting in much higher mass loss, perhaps triggered by interior instabilities that increase the stellar luminosity above the Eddington limit.

The review here focusses on the underlying physical issues behind such historical LBV mass loss. One particular theme is whether such eruptions are best characterized as *explosions*, or as episodes of an enhanced quasi-steady *wind*. Key distinctions to be made include timescale (dynamic vs. diffusive), driving mechanism (gas vs. radiation pressure), and degree of confinement (free expansion vs. gravitationally bound). As detailed below, it seems the characteristics of LBV giant eruptions require a combination of each, i.e. a quasi-steady wind driven by the enhanced luminosity associated with a relatively sudden (perhaps even explosive) release of energy in the interior. But even once an enhanced, super-Eddington luminosity is established, there remain fundamental issues of how the continuum driving can be regulated, e.g. by the spatial "porosity" of the medium, and thus lead to a mass loss that in some cases is inferred to have an energy comparable to the radiative luminosity, representing a "photon-tiring" limit.

2. The Key to Stellar Mass Loss: Overcoming Gravity

2.1. *Basic Momentum and Energy Requirements for Steady Wind*

Gravity is, of course, the essential force that keeps a star together as a bound entity, and so any discussion of stellar mass loss must necessarily focus on what mechanism(s) might be able to overcome this gravity. There are two aspects of this, namely to provide the momentum needed to reverse the inward pull of the gravitational force, but then also to have this outward driving sustained by tapping into a reservoir of energy that is sufficient to lift the material completely out of the star's gravitational potential.

For a steady radial wind flow, momentum balance requires that any acceleration in speed v with radius r must result from a combination of the gradient of gas pressure with any other outward force to overcome the inward pull of gravity,

$$v \frac{dv}{dr} = -\frac{GM}{r^2} - \frac{1}{\rho} \frac{dP}{dr} + g_{out}, \qquad (2.1)$$

with standard notation for, e.g., mass density ρ and stellar mass M. The discussion below focuses on radiative forces as a key to providing the required outward driving term g_{out}, but for now, let us just consider some general properties of such steady wind models.

First, at the base of any such wind outflow this momentum equation reduces to a hydrostatic equilibrium between the inward gravity and outward pressure,

$$-\frac{1}{P} \frac{dP}{dr} \equiv \frac{1}{H_P} = \frac{GM}{a^2 r^2}. \qquad (2.2)$$

Here $a = \sqrt{kT/\mu}$ is the isothermal sound speed, with k Boltzmann's constant and μ the mean molecular weight, and we have used the ideal gas law $P = \rho a^2$ to obtain an expression for the required local pressure scale height H_P.

The transition to a wind outflow occurs at some radius R where the flow speed becomes supersonic, i.e. $v(R) = a$. In massive-star winds, for which the temperature is typically close to the stellar effective temperature, the sound speed $a \approx 20$ km s^{-1}, which is much less than the surface escape speed, $v_{esc} = \sqrt{2GM/R} \approx 600 - 1000$ km s^{-1}. This implies that from the sonic point outward, i.e. from $r > R$, gas pressure plays almost no role in maintaining the outward acceleration against gravity, reducing the momentum equation to

$$v \frac{dv}{dr} \approx -\frac{GM}{r^2} + g_{out} \quad ; \quad r \geqslant R. \qquad (2.3)$$

Integration from this surface radius to infinity then immediately gives an expression for the required work per unit mass,

$$\int_R^\infty g_{out}\, dr \approx \frac{v(\infty)^2}{2} + \frac{GM}{R} = \frac{v_\infty^2}{2} + \frac{v_{esc}^2}{2}. \qquad (2.4)$$

this ignores both the internal and kinetic energy at the sonic point, since these each are only of order $a^2/v_{\rm esc}^2 \approx 10^{-3}$ relative to the terms retained. For a wind with mass loss rate $\dot{M}$, the global rate of energy expended is then

$$L_{wind} = \dot{M}\left[\frac{v_\infty^2}{2} + \frac{GM}{R}\right]. \qquad (2.5)$$

The marginal case in which the wind escapes with vanishing terminal flow speed, $v_\infty = 0$, defines a minimum energy rate for lifting material to escape, $L_{min} = \dot{M}GM/R$. For a given available interior luminosity L, this thus implies a maximum possible, energy-limited mass loss rate

$$\dot{M}_{\rm tir} = \frac{L}{GM/R} = 3.3 \times 10^{-8} \frac{\rm M_\odot}{\rm yr}\left[\frac{L}{M/R}\right], \qquad (2.6)$$

where the latter expression provides a convenient evaluation when the quantities in square brackets are written in solar units. The subscript here refers to reduction or "tiring" of the radiative luminosity as a result of the work done to sustain the outflow against gravity (Owocki & Gayley 1997). Even the most extreme massive-star steady winds, e.g. from WR stars, are typically no more than a few percent of this energy limit; but, as discussed further below, the mass loss during LBV giant eruptions can approach this order.

2.2. *Internal Energy and Virial Temperature*

Although gas pressure is not well-suited to driving a large steady mass loss from the stellar surface, it is generally the key to supporting the star against the inward pull of gravity. The associated pressure scale height is given locally by eqn. (2.2), which applied at the surface radius $r = R$ gives

$$\frac{H_p}{R} = \frac{2a_{\rm eff}^2}{v_{\rm esc}^2} \approx 10^{-3}, \qquad (2.7)$$

where the latter scaling applies for a surface sound speed set by the stellar effective temperature, $a_{\rm eff} \approx \sqrt{kT_{\rm eff}/\mu}$.

However, for the stellar interior, the pressure drops from its central value to nearly zero at the surface, representing an average scale length $H_p \approx R$. This thus implies a characteristic interior sound speed $a_{int} \approx v_{\rm esc}$, and a characteristic interior temperature

$$T_{int} \approx \frac{GM\mu}{kR} \approx 1.4 \times 10^7 \, K \, \frac{M}{R}, \qquad (2.8)$$

where the ratio M/R is in solar units, with $\mu \approx 10^{-24}$ g, roughly appropriate for fully ionized material of solar composition. This characteristic interior temperature can also be derived from the standard "virial theorem" result that the stellar internal thermal energy is half the gravitational binding energy, implying a negative net energy that keeps a star gravitationally bound.

But in the context of mass loss, it means that for gas pressure to have a sufficient internal energy to overcome gravity requires a temperature that is only a factor two larger than the typical interior value. In the solar corona, maintaining temperatures near this escape value does allow a pressure-driven solar wind, but this is only possible because the low density keeps the wind optically thin, with thus limited radiative cooling. For the much higher mass loss rates inferred for hot-star winds, the much higher density means that radiative cooling would prohibit ever reaching temperatures much above $T_{\rm eff} \approx T_{int}/1000$. Thus, as noted above, gas pressure is simply not a viable mechanism for driving a dense, steady surface wind.

2.3. *Gas-Pressure-Driven Expansion in Dynamical Explosions*

On the other hand, gas pressure is indeed the primary driving mechanism for propelling the expansion from supernovae explosions. In this case, dynamical collapse of the stellar core of mass $M_{core} \approx M_\odot$ down to a radius characteristic of a neutron star or black hole, i.e. $R_{ns} \approx 10 \text{ km s}^{-1}$, releases an energy

$$\Delta E \approx \frac{GM_c^2}{R_{ns}} \approx 10^{53} \text{erg} \qquad (2.9)$$

which is of order 10^4 higher than the entire binding energy of the entire stellar envelope, $E_g \approx GM^2/R$. Transfer of just ca. 1% of this collapse energy can thus suddenly heat the surrounding stellar envelope to a temperature up to *hundred times* the equilibrium (virial) value, with an associated sound speed a_{sn} up to ten times the gravitational escape speed. On a short dynamical time scale, R/a_{sn}, of order a few minutes, the associated large gas pressure then drives an acceleration of the full envelope mass ($\sim 10 M_\odot$) to a free expansion at speeds $v_{exp} \approx a_{sn}$, typically several thousand km s^{-1}.

The radiation generated by such SN explosions escapes on a somewhat longer time, with light curves typically peaking a few days after the initial explosion. But this is still significantly shorter than the characteristic time for LBV giant eruptions, which apparently can last for several years. Moreover, the expansion speeds inferred for LBV ejecta are typically a few hundred km s^{-1}, comparable to stellar escape speeds, and much less than the thousands of km s^{-1} typical for the initial expansion of supernovae. Thus, rather than a dynamical explosion wherein the gas overpressure simply overwhelms the binding from stellar gravity, it seems more likely that LBV eruptions may represent a quasi-controlled outburst, induced perhaps by an enhanced radiative brightening that leads to an outward radiative acceleration that exceeds gravity.

3. Radiatively Driven Mass Loss

3.1. *Radiative Acceleration and the Eddington Limit*

The force-per-unit mass imparted to material from interaction with radiation depends on an integration of the opacity and radiative flux over photon frequency ν,

$$\mathbf{g}_{rad} = \int_0^\infty d\nu \, \kappa_\nu \mathbf{F}_\nu / c \equiv \kappa_F \mathbf{F}/c, \qquad (3.1)$$

with c the speed of light, and the latter equality defining the *flux-weighted* opacity κ_F in terms of the bolometric radiative flux $\mathbf{F}$.

In general the opacity κ_ν includes both broad-band continuum processes – e.g. Thomson scattering of electrons, and bound-free or free-free absorption – and bound-bound transitions associated with line absorption and/or scattering. As discussed in §3.3, bound-bound opacity is most effective in near-surface layers where expansion from a not-too-dense wind can partially desaturate the strongest lines. But in a static envelope and atmosphere, the reduction in flux F_ν in such saturated lines keeps the associated line-force small, and so in most regions of a stellar envelope the overall radiative acceleration is set by continuum processes like electron scattering and bound-free or free-free absorption.

In spherical symmetry, both the radial flux $F = L/4\pi r^2$ and gravity $g = GM/r^2$ have similar inverse-square dependence on radius r, which thus cancels in the ratio of radiative acceleration to gravity. In terms of the electron scattering opacity, $\kappa_e \approx 0.34 \text{ cm}^2/g$, this ratio has the scaling

$$\Gamma \equiv \frac{g_{\rm rad}}{g} = \frac{\kappa_F L}{4\pi GMc} = 2.6 \times 10^{-5} \frac{\kappa_F}{\kappa_e} \frac{L}{L_\odot} \frac{M_\odot}{M}. \quad (3.2)$$

When the opacity κ, radiative luminosity L, and mass M are all fixed, then Γ is constant. But, as discussed below, there are various circumstances in which this is not the case.

For pure electron scattering, with $\kappa_F = \kappa_e$, eqn. (3.2) just gives the classical Eddington parameter $\Gamma_e = \kappa_e L/4\pi GMc$. Because stellar luminosity generally scales with a high power of the stellar mass, i.e. $L \propto M^{3-4}$ (see §3.2), massive stars with $M > 10 M_\odot$ generally have electron Eddington parameters of order $\Gamma_e \approx 0.1 - 1$. Indeed, $\Gamma_e \equiv 1$ defines the *Eddington limit*, for which the entire star would formally become unbound, at least in this idealized model of 1-D, spherically symmetric, radiative envelope.

However, because the reversal of gravity formally extends to arbitrarily deep, dense layers of the stellar envelope, any outward mass flux that might be initiated would require a very large mechanical luminosity, and thus would be well above the energy, photon-tiring limit given in eqn. (2.6). As such, exceeding the Eddington limit does not represent an appropriate condition for the steady-state mass loss characteristic of a stellar wind, since that requires an outwardly increasing radiative force that goes from being less than gravity in a bound stellar envelope to exceeding gravity in the outflowing stellar wind. The discussion below summarizes how the necessary force regulation can still occur through line-desaturation for line driving (§3.3), and through porosity of spatial structure for continuum driving (§3.5).

But first let us briefly review the key scalings of stellar structure that lead massive stars to be so close to this fundamental Eddington limit.

3.2. Stellar Structure Scaling for Luminosity vs. Mass

The structure of a stellar envelope is set by the dual requirements for momentum balance and energy transport. The former is described through the equation for hydrostatic equilibrium (cf. eqn. 2.2), modified now to account for a factor $1 - \Gamma$ reduction in the effective gravity, due to the radiation force. Following the same approach as in §2.2, this thus now implies a characteristic interior temperature that scales as

$$T \sim \frac{M(1-\Gamma)}{R}. \quad (3.3)$$

Through most of the stellar envelope, the energy flux $F = L/4\pi r^2$ is transported by diffusion of radiative energy density $U_{\rm rad} \sim T^4$,

$$F = -\frac{1}{\kappa\rho c} \frac{dU_{\rm rad}}{dr}, \quad (3.4)$$

which implies the dimensional scaling

$$L \sim \frac{R^4 T^4}{M}. \quad (3.5)$$

When combined with eqn. (3.3) for the interior temperature, we see that the *radius cancels* in the scaling of luminosity, yielding

$$L \sim M^3 (1-\Gamma)^4. \quad (3.6)$$

Quite remarkably, this scaling does not depend explicitly on the nature of energy generation in the stellar core, but is strictly a property of the envelope structure†.

† Of course, this simple one-point scaling relation does have to be modified to accommodate

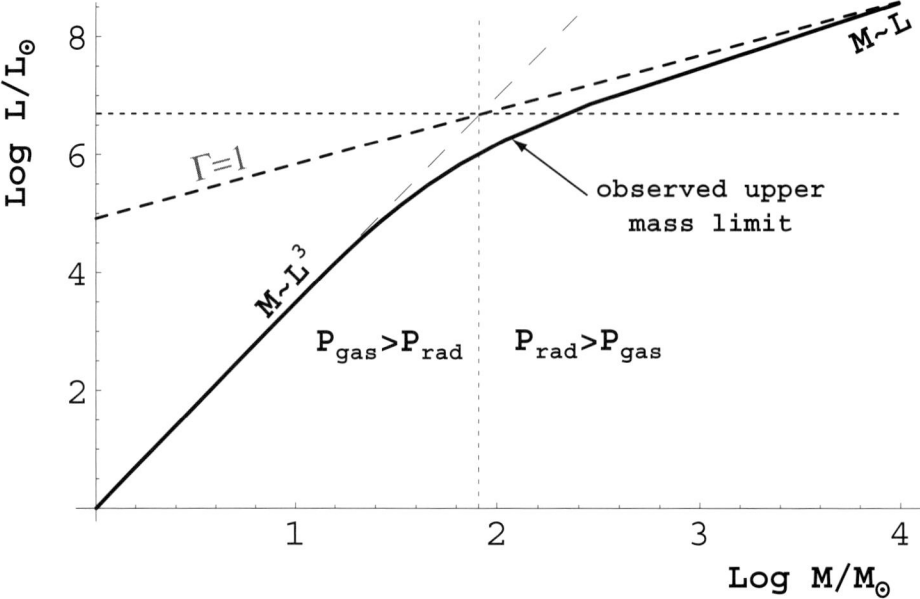

Figure 1. Log-log plot of the scaling of stellar luminosity L vs. mass M implied by the simple relation (3.6).

Figure 1 shows a log-log plot of the resulting variation of luminosity vs. mass. For low-mass stars, it implies a strong $L \sim M^3$ scaling, but as this forces stars to approach the Eddington limit, the $1-\Gamma$ term acts as a strong repeller away from that limit, causing a broad bend toward a linear asymptotic scaling, $L \sim M$.

Formally, this scaling suggests it is in principle possible to have stars with arbitrarily large mass, approaching arbitrarily close to the Eddington limit. But surveys of dense young clusters are providing increasingly strong evidence for a sharp cutoff in the stellar mass distribution at about $M \approx 150 - 200$ M$_\odot$ (Oey and Clarke 2005; Kim et al. 2006).

Note that this inferred upper mass limit corresponds closely to the center of the bend region in fig. 1. This is just somewhat beyond the transition, at $\Gamma \approx 1/2$, to where radiation plays the dominant role in supporting the star against gravity, implying a radiation pressure that is greater than gas pressure, $P_{\rm rad} > P_{\rm gas}$. Somewhat analogous to having a heavier fluid support a lighter one, such a configuration may be subject to various kinds on intrinsic instabilities, leading to spatial clumping and/or the brightness variations that trigger LBV eruptions (Spiegel & Tau 1999; Shaviv 1998, 2000, 2001). The large associated LBV mass loss of such near Eddington stars thus could play a key role in setting the stellar upper mass limit.

3.3. Line-Driven Stellar Winds

The resonant nature of line (bound-bound) absorption leads to an opacity that is inherently much stronger than from free electrons. For example, in the somewhat idealized, optically thin limit that all the line opacity could be illuminated with a flat, unattenuated continuum spectrum with the full stellar luminosity, the total line-force would exceed the free-electron force by a huge factor, of order $Q \approx 2000$ (Gayley 1995). For massive stars

gradients in the molecular weight as a star evolves from the zero-age main sequence, and it breaks down altogether in the coolest stars (both giants and dwarfs), for which convection dominates the envelope energy transport.

with typical electron Eddington parameters within a factor two of unity, $\Gamma_e \approx 1/2$, this implies a net outward line acceleration that could be as high as $\Gamma_{lines} \approx Q\Gamma_e \approx 1000$ times the acceleration of gravity!

Of course, this does not generally occur in practice because of the self-absorption of the lines. For a single line with frequency-integrated opacity $\kappa_q = q\kappa_e$, the reduction in the optically thin line-acceleration $q\Gamma_e$ can be written as

$$\Gamma_{line} \approx q\Gamma_e \frac{1-e^{-qt}}{qt}, \qquad (3.7)$$

where $t \equiv \kappa_e \rho c/(dv/dr)$ is the Sobolev optical depth of a line with unit strength, $q = 1$ (Sobolev 1960; Castor, Abbott & Klein 1975, hereafter CAK). Within the standard CAK line-driven wind theory, the number distribution N of spectral lines is approximated as a power law in line strength $q\, dN/dq = [1/\Gamma(\alpha)](q/Q)^{\alpha-1}$, where the CAK power index $\alpha \approx 0.5 - 0.7$ (and $\Gamma(\alpha)$ here represents the complete Gamma function). The associated line-ensemble-integrated radiation force is then reduced by a factor $1/(Qt)^\alpha$ from the optically thin value,

$$\Gamma_{lines} = \frac{Q\Gamma_e}{(1-\alpha)(Qt)^\alpha} \propto \left(\frac{1}{\rho}\frac{dv}{dr}\right)^\alpha. \qquad (3.8)$$

The latter proportionality emphasizes the key scaling of the line-force with the velocity gradient dv/dr and *inverse* of the density, $1/\rho$. This keeps the line acceleration less than gravity in the dense, nearly static atmosphere, but also allows its outward increase above gravity to drive the outflowing wind. The CAK mass loss rate is set by the associated critical density that allows the outward line acceleration to be just sufficient to overcome the (electron-scattering-reduced) gravity, i.e. with $\Gamma_{lines} \approx 1 - \Gamma_e$,

$$\dot{M}_{CAK} = \frac{\alpha}{1-\alpha}\frac{L}{c^2}\left[\frac{Q\Gamma_e}{1-\Gamma_e}\right]^{-1+1/\alpha}, \qquad (3.9)$$

where we have used the definition of the mass loss rate $\dot{M} \equiv 4\pi\rho v r^2$ and the fact that for such a CAK solution, $v dv/dr \approx g(1-\Gamma_e)$.

This last property further yields the characteristic CAK velocity law scaling $v(r) \approx v_\infty(1-R/r)^{1/2}$, with the wind terminal speed being proportional to the effective surface escape speed,

$$v_\infty \propto v_{\text{eff}} \equiv \sqrt{GM(1-\Gamma_e)/R}. \qquad (3.10)$$

As a star approaches the classical Eddington limit $\Gamma_e \to 1$, these standard CAK scalings formally predict the mass loss rate to diverge as $\dot{M} \propto 1/(1-\Gamma_e)^{(1-\alpha)/\alpha}$, but with a vanishing terminal flow speed $v_\infty \propto \sqrt{1-\Gamma_e}$. The former might appear to provide an explanation for the large mass losses inferred in LBV's, but the latter fails to explain the moderately high inferred ejection speeds, e.g. the 500-800 km s^{-1} kinematic expansion inferred for the Homunculus nebula of η Carinae (Smith 2002, Smith et al. 2003).

But one essential point is that line-driving could never explain the extremely large mass loss rates needed to explain the Homunculus nebulae. To maintain the moderately high terminal speeds, the $\Gamma_e/(1-\Gamma_e)$ factor would have to be of order unity. Then for optimal realistic values $\alpha = 1/2$ and $Q \approx 2000$ for the line opacity parameters (Gayley 1995), the maximum mass loss from line driving is given by (Smith & Owocki 2006),

$$\dot{M} \approx 1.4 \times 10^{-4} L_6 \, M_\odot \text{yr}^{-1}, \qquad (3.11)$$

where $L_6 \equiv L/10^6 L_\odot$. Even for peak luminosities of a few times $10^7 L_\odot$ during η Carinae's eruption, this limit is still several orders of magnitude below the mass loss needed to form the Homunculus. Thus, if mass loss during these eruptions occurs via a wind, it must be a super-Eddington wind driven by continuum radiation force (e.g., electron scattering opacity) and not lines (Owocki, Gayley & Shaviv 2004, hereafter OGS; Belyanin 1999; Quinn & Paczynski 1985).

3.4. *Convective Instability of a Super-Eddington Stellar Interior*

Before discussing such continuum-driven winds during periods of super-Eddington luminosity, it should first be emphasized that locally exceeding the Eddington limit need *not* necessarily lead to initiation of a mass outflow. As first shown by Joss, Salpeter, and Ostriker (1972), in the stellar envelope allowing the Eddington parameter $\Gamma \to 1$ generally implies through the Schwarzschild criterion that material becomes *convectively unstable*. Since convection in such deep layers is highly efficient, the radiative luminosity is reduced, thereby lowering the associated radiative Eddington factor away from unity.

This suggests that a radiatively driven outflow should only be initiated *outside* the region where convection is *efficient*. An upper bound to the convective energy flux is set by

$$F_{\rm conv} \approx v_{\rm conv}\, l\, dU/dr \lesssim a\, H\, dP/dr \approx a^3 \rho, \tag{3.12}$$

where v_{conv}, l, and U are the convective velocity, mixing length, and internal energy density, and a, H, P, and ρ are the sound speed, pressure scale height, pressure, and mass density. Setting this maximum convective flux equal to the total stellar energy flux $L/4\pi r^2$ yields an estimate for the maximum mass loss rate that can be initiated by radiative driving,

$$\dot{M} \leqslant \frac{L}{a^2} \equiv \dot{M}_{\rm max,conv} = \frac{v_{\rm esc}^2}{2a^2} \dot{M}_{\rm tir}, \tag{3.13}$$

where the last equality emphasizes that, for the usual case of a sound speed much smaller than the local escape speed, $a \ll v_{\rm esc}$, such a mass loss would generally be well in excess of the photon-tiring limit set by the energy available to lift the material out of the star's gravitational potential (see eqn. 2.6). In other words, if a wind were to originate from where convection becomes inefficient, the mass loss would be so large that it would use all the available luminosity to accelerate out of the gravitational potential, implying that any such outflow would necessarily stagnate at some finite radius. One can imagine that the subsequent infall of material would likely form a complex spatial pattern, consisting of a mixture of both downdrafts and upflows, perhaps even resembling the 3D cells of thermally driven convection.

Overall, it seems that a star that exceeds the Eddington limit is likely to develop a complex spatial structure, whether due to local instability to convection, to global instability of flow stagnation, or to intrinsic compressive instabilities arising from the dominance of radiation pressure.

3.5. *Super-Eddington Outflow Moderated by Porous Opacity*

Shaviv (1998; 2000) has applied these notions of a spatially structured, radiatively supported atmosphere to suggest an innovative paradigm for how quasi-stationary wind outflows could be maintained from objects that formally exceed the Eddington limit. A key insight regards the fact that, in a laterally inhomogeneous atmosphere, the radiative transport should selectively avoid regions of enhanced density in favor of relatively low-density, "porous" channels between them. This stands in contrast to the usual picture of simple 1D, gray-atmosphere models, wherein the requirements of radiative equilibrium ensure that the radiative flux must be maintained independent of the medium's optical thickness. In 2D or 3D porous media, even a gray opacity will lead to a flux avoidance of the most optically thick regions, much as in frequency-dependent radiative transfer in 1D atmosphere, wherein the flux avoids spectral lines or bound-free edges that represent a localized spectral regions of non-gray enhancement in opacity.

A simple description of the effect is to consider a medium in which material has coagulated into discrete blobs of individual optical thickness $\tau_b = \kappa \rho_b l$, where l is the blob scale, and the blob density is enhanced compared to the mean density of the medium by a volume filling factor $\rho_b/\rho = (L/l)^3$, where L is the interblob spacing. The effective overall opacity of this medium can then be approximated as

$$\kappa_{\rm eff} \approx \kappa \frac{1 - e^{-\tau_b}}{\tau_b}. \qquad (3.14)$$

Note that in the limit of optically thin blobs ($\tau_b \ll 1$) this reproduces the usual microscopic opacity ($\kappa_{\rm eff} \approx \kappa$); but in the optically thick limit ($\tau_b \gg 1$), the effective opacity is reduced by a factor of $1/\tau_b$, thus yielding a medium with opacity characterized instead by the blob cross section divided by the blob mass ($\kappa_{\rm eff} = \kappa/\tau_b = l^2/m_b$). The critical mean density at which the blobs become optically thin is given by $\rho_o = 1/\kappa h$, where $h = L^3/l^2$ is characteristic "porosity length" parameter. A key upshot of this is that the radiative acceleration in such a gray, but spatially porous medium would likewise be reduced by a factor that depends on the mean density.

More realistically, it seems likely that structure should occur with a range of compression strengths and length scales. Noting the similarity of the single-scale and single-line correction factors (cf. eqns. 3.7 and 3.14), let us draw upon an analogy with the power-law distribution of line-opacity in the standard CAK model of line-driven winds, and thereby consider a *power-law-porosity* model in which the associated structure has a broad range of porosity length h. As detailed by OGS, this leads to an effective Eddington parameter that scales as

$$\Gamma_{\rm eff} \approx \Gamma \left(\frac{\rho_o}{\rho}\right)^{\alpha_p} \quad ; \quad \rho > \rho_o, \qquad (3.15)$$

where α_p is the porosity power index (analogous to the CAK line-distribution power index α), and $\rho_o \equiv 1/\kappa h_o$, with h_o now the porosity-length associated with the *strongest* (i.e. most optically thick) clump.

In rough analogy with the "mixing length" formalism of stellar convection, let us assume this porosity length h_o scales with gravitational scale height $H \equiv a^2/g$. Then the requirement that $\Gamma_{\rm eff} = 1$ at the wind sonic point yields a scaling for the mass loss rate scaling with luminosity. For the canonical case $\alpha_p = 1/2$, this takes the form (OGS),

$$\dot{M}_{\rm por} \approx 4(\Gamma - 1) \frac{L}{ac} \frac{H}{h_o} \qquad (3.16)$$

$$\approx 0.004(\Gamma - 1) \frac{{\rm M}_\odot}{{\rm yr}} \frac{L_6}{a_{20}} \frac{H}{h}. \qquad (3.17)$$

The second equality gives numerical evaluation in terms of characteristic values for the sound speed $a_{20} \equiv a/20 \, {\rm km\,s^{-1}}$ and luminosity $L_6 \equiv L/10^6 {\rm L}_\odot$. Comparison with the CAK scalings (3.9) for a line-driven wind shows that the mass loss can be substantially higher from a super-Eddington star with porosity-moderated, continuum driving. Applying the extreme luminosity $L \approx 20 \times 10^6 \, {\rm L}_\odot$ estimated for the 1840-60 outburst of eta Carinae, which implies an Eddington parameter $\Gamma \approx 5$, the derived mass loss rate for a canonical porosity length of $h = H$ is $\dot{M}_{\rm por} \approx 0.32 {\rm M}_\odot \, {\rm yr}^{-1}$, quite comparable to the inferred average $\sim 0.5 {\rm M}_\odot \, {\rm yr}^{-1}$ during this epoch.

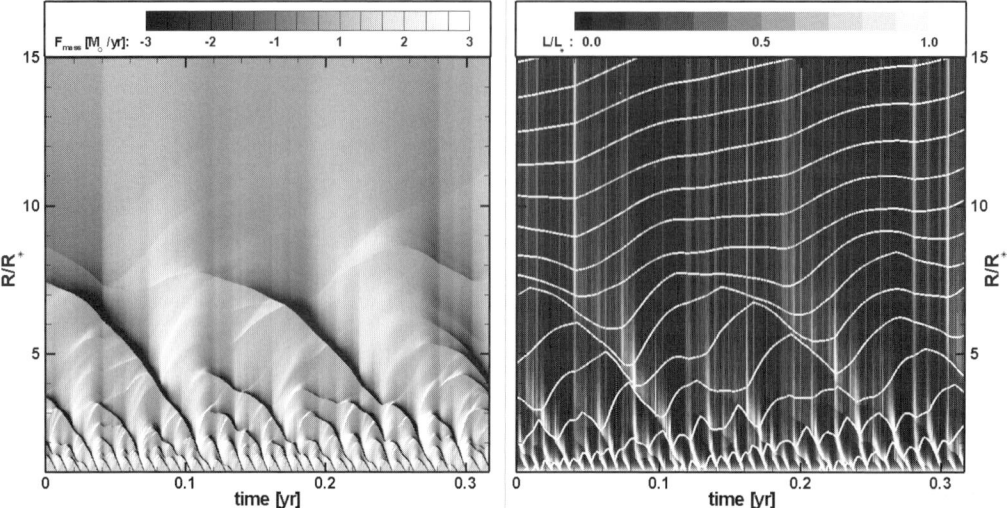

Figure 2. Grayscale plot of radius and time variation of mass flux (left) and luminosity (right) in time-dependent simulation of super-Eddington wind with porosity-mediated base mass flux above the photon tiring limit. The light contours on the right trace the height progression of fixed mass shells.

Overall, it seems that, together with the ability to drive quite fast outflow speeds (of order the surface escape speed), the extended porosity formalism provides a promising basis for self-consistent dynamical modeling of even the most extreme mass loss outbursts of Luminous Blue Variables, namely those that, like the giant eruption of η Carinae, approach the photon tiring limit.

3.6. *1D Simulation of Continuum-Driven Winds above the Photon-Tiring Limit*

For porosity models in which the base mass flux *exceeds* the photon tiring limit, recent numerical simulations (van Marle *et al.* 2008; see also poster in this volume) have explored the nature of the resulting complex pattern of infall and outflow. Despite the likely 3D nature of such flow patterns, to keep the computation tractable, this initial exploration assumes 1D spherical symmetry, though now allowing a fully time-dependent density and flow speed. The total rate of work done by the radiation on the outflow (or vice versa in regions of inflow) is accounted for by a radial change of the radiative luminosity with radius,

$$\frac{dL}{dr} = -\dot{m}g_{\rm rad} = -\kappa_{\rm eff}\,\rho v L/c\,, \qquad (3.18)$$

where $\dot{m} \equiv 4\pi\rho v r^2$ is the local mass-flux at radius r, which is no longer a constant, or even monotonically positive, in such a time-dependent flow. The latter equality then follows from the definition (3.1) of the radiative acceleration $g_{\rm rad}$ for a gray opacity $\kappa_{\rm eff}$, set here by porosity-modified electron scattering. At each time step, eqn. (3.18) is integrated from an assumed lower boundary luminosity $L(R)$ to give the local radiative luminosity $L(r)$ at all radii $r > R$. Using this to compute the local radiative acceleration, the time-dependent equations for mass and momentum conservation are evolved forward to obtain the time and radial variation of density $\rho(r,t)$ and flow speed $v(r,t)$. (For simplicity, the temperature is fixed at the stellar effective temperature.) The base Eddington parameter is $\Gamma = 10$, and the analytic porosity mass flux is 2.3 times the tiring limit.

Figure 2 illustrates the flow structure as a function of radius (for $r = 1-15\,R$) and time (over an arbitrary interval long after the initial condition, set to analytic steady porosity model ignoring photon tiring). The left panel grayscale shows the local mass flux, in $M_\odot\,\text{yr}^{-1}$, with dark shades representing inflow, and light shades outflow. In the right panel, the shading represents the local luminosity in units of the base value, $L(r)/L(R)$, ranging from zero (black) to one (white); in addition, the superposed lines represent the radius and time variation of selected mass shells.

Both panels show the remarkably complex nature of the flow, with positive mass flux from the base overtaken by a hierarchy of infall from stagnated flow above. However, the re-energization of the radiative luminosity from this infall makes the region above have an outward impulse. The shell tracks thus show that, once material reaches a radius $r \approx 5R$, its infall intervals become ever shorter, allowing it eventually to drift outward. The overall result is a net, time-averaged mass loss through the outer that is very close to the photon-tiring limit, with however a terminal flow speed $v_\infty \approx 50\,\text{km s}^{-1}$ that is substantially below the surface escape speed $v_\text{esc} \approx 600\,\text{km s}^{-1}$.

These initial 1D simulation thus provide an interesting glimpse into this competition below inflow and outflow. Of course, the structure in more realistic 2D and 3D models may be even more complex, and even lead itself to a highly porous medium. But overall, it seems that one robust property of such super-Eddington models may well be mass loss that is of the order of the photon tiring limit.

4. Conclusion

The basic conclusion of this review is that the extreme mass loss in giant eruptions of LBV stars seems best explained by quasi-steady, porosity-moderated, continuum-driven stellar wind during episodes of super-Eddington luminosity. The cause or trigger of this enhanced luminosity is unknown, but may be related to the dominance of radiation pressure over gas pressure in the envelopes of massive stars. The mass loss rate in such LBV eruptions is far greater than can be explained by the standard line-driving for hot-star winds in more quiescent phases. In the most massive stars, the cumulative mass loss in such eruptions may also dominate over the quiescent wind, and might even be a key factor in setting the stellar upper mass limit. Moreover, since driving by continuum scattering by free electrons does not directly depend on metallicity, mass loss by LBV eruptions may remain important in low-metallicity environments, including in the early universe. A key outstanding issue, however, is to determine the cause or trigger of the luminosity brightenings, including, for example, whether this might itself depend on metallicity.

Acknowledgements

This research was supported in part by NSF grant AST-0507581. We acknowledge numerous insightful discussions with N. Shaviv and R. Townsend.

References

Belyanin, A. A. 1999, *A&A*, 344, 199
Castor, J., Abbott, D., & Klein, R. 1975, *ApJ*, 195, 157 (CAK)
Gayley, K. 1995, *ApJ*, 454, 410
Glatzel, W. 1994, *MNRAS*, 271, 66.
Joss, P., Salpeter, E., & Ostriker, J. 1973, *ApJ*, 181, 429
Kim, S. S., Figer, D. F., Kudritzki, R. P., & Najarro, F. 2006, *ApJ*, 653, L113
Oey, M. S. & Clarke, C. J. 2005, *ApJ*, 620, L43
Owocki, S. & Gayley, K. 1997, in: A. Nota & H. Lamers (eds.), *Luminous Blue Variables: Massive Stars in Transition* (San Francisco: ASP), *ASP Conf. Ser.*, 120, 121

Owocki, S., Gayley, K., & Shaviv, N. 2004, *ApJ*, 558, 802
Quinn, T. & Paczynski, B. 1985, *ApJ*, 289, 634
Shaviv, N. 1998, *ApJ*, 494, L193
Shaviv, N. 2000, *ApJ*, 529, L137
Shaviv, N. 2001, *ApJ*, 549, 1093
Smith, N. 2002, *MNRAS*, 337, 1252
Smith, N., Gehrz, R. D., Hinz, P. M., *et al.* 2003, *AJ*, 125, 1458
Smith, N. & Owocki, S. 2006 *ApJ*, 645, L45
Sobolev, V. V. 1960, *Moving Envelopes of Stars* (Cambridge: Harvard University Press)
Spiegel, E., & Tao, L. 1999, *Phys. Rep.* 311, 163
van Marle, A. J., Owocki, S. P., & Shaviv, N. 2008, in: B. O'Shea, A. Heger & T. Abel (eds.), *First Stars III*, (New York: AIP) in press (arXiv:0708.4207)

Discussion

ZINNECKER: I completely agree with you that the term "radiation pressure" is ill-conceived, and we should better use a term like "radiative acceleration". I disagree with you, however, on another point: you were writing L/M proportional to M^2 for very massive stars, when in reality it should be proportional to M or even $M^{0.6}$ and approaching a constant near the Eddington limit; see my poster or Zinnecker & Yorke (2007, *ARA&A* 45, 481). Or did I misunderstand you?

OWOCKI: Yes, I think there was some misunderstanding. The luminosity scalings you describe agree well with my simple envelope structure analysis. See eqn. 3.6 and fig. 1.

Stan Owocki talking (left) and in action (right).

Pulsation-Initiated Mass Loss in Luminous Blue Variables: A Parameter Study

Andrew J. Onifer[1] and Joyce A. Guzik[2]

[1]University of Florida, Department of Physics, PO Box 118440, Gainesville, FL 32611 USA
email: onifer@phys.ufl.edu

[2]Los Alamos National Laboratory, Thermonuclear Applications Group (X-2),
PO Box 1663, MS T085, Los Alamos, NM 87545 USA
email: joy@lanl.gov

Abstract. Luminous blue variables (LBVs) are characterized by semi-periodic episodes of enhanced mass-loss, or outburst. The cause of these outbursts has thus far been a mystery. One explanation is that they are initiated by κ-effect pulsations in the atmosphere caused by a blocking of luminosity at temperatures near the so-called "iron bump" ($T \sim 200{,}000$ K), where the Fe opacity suddenly increases and blocks the luminosity. Due to a lag in the onset of convection, the luminosity can build until it exceeds the Eddington limit locally, seeding pulsations and possibly driving some mass from the star. We present preliminary results from a parameter study focusing on the conditions necessary to trigger normal S-Dor type (as opposed to extreme η-Car type) outbursts. We find that as Y increases or Z decreases, the pulsational amplitude decreases and outburst-like behavior, indicated by a large, sudden increase in photospheric velocity, becomes less likely.

Keywords. stars: atmospheres – stars: mass loss – stars: oscillations – stars: variables – hydrodynamics – convection – radiative transfer – instabilities

1. Introduction

Luminous blue variables (LBVs) represent a short-lived phase of massive stellar evolution characterized by episodes of intense mass loss. There are two main types of this behavior. The rare eruptions, such as happened with η-Car and P-Cyg, occur every few decades to centuries. They last as long as a few tens of years and can eject $\geqslant 1 M_\odot$ of material in a single episode. The star also experiences an increase in luminosity during the eruption phase.

In contrast, the more normal S-Dor-type episodes occur every few months to few years. They eject $\sim 10^{-4} M_\odot$ per event. The LBV does *not* become more luminous during these episodes. For the purposes of this paper, we will call the more extreme η-Car-like episodes eruptions and the less extreme S-Dor-like episodes outbursts.

Vink & de Koter (2002) determined that the outburst phase results from changes in the dominant ionization states of Fe in the wind which result from changes in the stellar radius and effective temperature, but what causes the changes in the radius and temperature is not clear. In this paper we explore the hypothesis that it is due to seeding from κ-effect pulsations arising from luminosity blocking due to the increase in Fe opacity from the so-called "hot iron bump" at temperatures near 200,000 K.

This work is based upon the work of Guzik et al. (2005) (see also Guzik (2005), which discusses a similar mechanism for η-Car-like eruptions). We will discuss the possibility that a super-Eddington luminosity in regions of the star near iron-bump temperatures could cause the envelope of the star to become unstable and drive mass loss.

Table 1. Properties of analyzed modes

Y	Z	Mode	Fractional Energy Gain / Period
29	0.01	1H	1.11
29	0.02	1H	2.07
49	0.01	1H	0.552
49	0.02	1H	2.22

2. Models

We start with massive main sequence models in the mass range $50 M_\odot \leqslant M_* \leqslant 80 M_\odot$. We evolve the stars using the Iben evolution code (Iben 1965). Important additions to the code are the Swenson SIREFF analytical EOS (Guzik & Swenson 1997), OPAL opacities using the Grevesse and Noels 93 mixture (Iglesias & Rogers 1996), and the empirical mass-loss prescription of Nieuwenhuijzen & de Jager (1990). We vary the mass-loss rate to allow different surface abundances at the end of the evolution. More details of this code and the other codes used in this analysis are found in Guzik *et al.* (2005a).

Once we have a model with properties similar to an LBV star, we run it through a pulsation analysis code. This code determines the harmonics of the κ-effect pulsations. We choose the the most unstable of the first three harmonics (as defined by the growth rate per period); the fundamental F, first harmonic 1H, or second harmonic 2H for further analysis. The modes chosen for this analysis are shown in Table 1. We analyze this unstable mode using the 1D Lagrangian hydrodynamics code Dynstar (Cox & Ostlie 1993). This code includes the time-dependent convection treatment formulated by Ostlie (1990), which is described below.

The models have $\log L/L_\odot \approx 6.1$ and $T_{\rm eff} = 1.6 \times 10^4$ K. We vary the metallicity Z and the helium abundance Y. Since the pulsations are driven by iron opacity, they are Z-dependent, with lower Z resulting in smaller amplitudes. As Y is increased, the star effectively becomes older and closer to the stable Wolf-Rayet phase, also resulting in smaller pulsation amplitudes.

3. Time-Dependent Convection

Our time-dependent convection model is described in detail in Ostlie (1990). Here are the main highlights.

The time-dependent convection model is built upon standard mixing length theory (Böhm-Vitense 1958; Cox & Giuli 1968). In the standard theory the convective luminosity is

$$L_c = 4\pi r^2 \frac{4T\rho C_P}{g\ell Q} v_c^{0\ 3}, \qquad (3.1)$$

where r, T, C_P, ρ, and g are the local radius, temperature, specific heat, density, and gravitational acceleration, respectively. The mixing length ℓ is assumed to be equal to the pressure scale height. The parameter Q is defined as

$$Q \equiv -\frac{\partial \log \rho}{\partial \log T}\bigg|_P. \qquad (3.2)$$

$Q = 1$ for an ideal gas with a constant mean molecular weight. The convective eddy velocity in the static case is

$$v_c^0 = \frac{1}{4}\sqrt{\frac{gQ\ell^2}{2H_P}}f, \tag{3.3}$$

where H_P is the pressure scale height,

$$f \equiv [\sqrt{1 + 4A^2(\Delta - \Delta_{ad})} - 1]/A, \tag{3.4}$$

$$\Delta \equiv \frac{d\log T}{d\log P}, \tag{3.5}$$

$$A \equiv \frac{Q^{1/2}\, C_P \kappa g \rho^{5/2}\, \ell^2}{12\sqrt{2}acP^{1/2}\, T^3}, \tag{3.6}$$

$\Delta - \Delta_{ad}$ is the super-adiabatic gradient, and κ is the local opacity.

The static case, however, does not account for the effect of convective lag due to the inertia of the material. This lag is included in our model using a quadratic Lagrange interpolation polynomial of v_c over the current and previous two time steps.

For zone i and time step n the time-dependent value of v_c is

$$v_{c,\,i}^n \equiv \frac{(t'-t_{n-1})(t'-t_{n-2})}{(t_n-t_{n-1})(t_n-t_{n-2})} v_{c,\,i}^0 + \frac{(t'-t_n)(t'-t_{n-2})}{(t_{n-1}-t_n)(t_{n-1}-t_{n-2})} v_{c,\,i}^{n-1} \tag{3.7}$$
$$+ \frac{(t'-t_n)(t'-t_{n-1})}{(t_{n-2}-t_n)(t_{n-2}-t_{n-1})} v_{c,\,i}^{n-2},$$

where $v_{c\,i}^0$ is the value determined from equation 3.3,

$$t' \equiv t_{n-1} + \tau(t_n - t_{n-1}) \tag{3.8}$$

and τ is the fraction of a mixing length the convective eddy can travel in a time step:

$$\tau \equiv \frac{v_{c,\,i}^n (t_n - t_{n-1})}{\ell} l_{\text{fac}}. \tag{3.9}$$

The lag factor $l_{\text{fac}} \equiv 1$ in our models.

4. Results

Figure 1 shows how the photospheric velocity v_{phot} is affected by changing the metallicity. The figure shows the v_{phot} as a function of time for a model at solar metallicity (defined here as $Z = 0.02$) and at half-solar, or LMC-like Z. Both models exhibit "outburst"-like behavior, in that extreme super-Eddington zones deep in the star near 200,000 K (see Figure 2) result in a large, sudden increase in the surface velocity. As expected the model with the larger metallicity experienced a larger velocity amplitude.

The $Z = 0.01$ model with Y increased to 0.49 is shown in Figure 3. In contrast to the "outburst" models, this model settles into steady pulsations, and Figure 4 shows that this model only becomes super-Eddington by a few percent. Thus the star seems able to recover and settle into steady, albeit very large amplitude ($v_{\text{phot}} = 100$ km s^{-1}) pulsations.

5. Conclusions and Future Work

We have preliminary results from a study on κ-effect pulsations in LBV stars, with possible implications for S-Dor-like mass loss. We have found that regions in the star near $T \sim 200{,}000$ K experience large super-Eddington luminosities due to the increase of

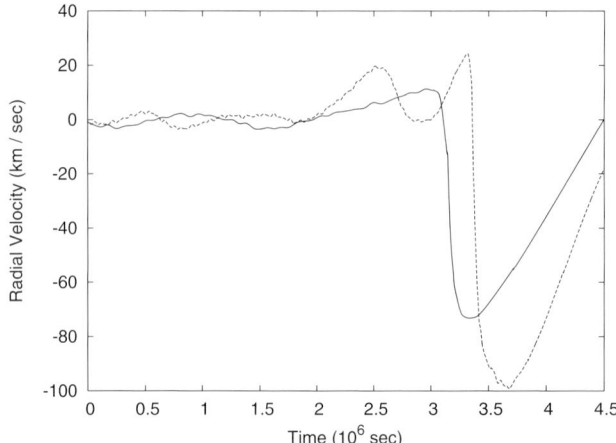

Figure 1. Photospheric velocity as a function of time. Solid line is $Z = 0.02$, dashed line is $Z = 0.01$. Negative velocity is outward.

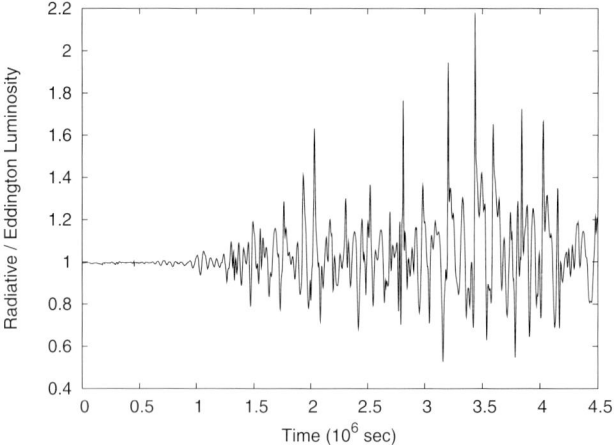

Figure 2. Radiative luminosity at $T \sim 210{,}000$ K relative to the Eddington luminosity.

radiation from iron-bump opacities. We have found that this can cause large amplitude pulsations at the surface. At lower Y and higher Z, a large sudden increase in radial velocity is seen at the surface. Since the Dynstar code does not include outflow, we cannot follow the evolution of outbursts and therefore cannot determine whether the pulsations or the sudden velocity increases evolve into S-Dor-like outbursts.

The results presented here are still very preliminary. Much more work is needed to confirm our results, not the least of which is a zoning study. We have used 60-zone models to improve the throughput, but it is not clear that this is sufficient. Runs at 120 zones and higher should be run in order to confirm that the results are robust.

In addition it is not clear that our "outbursts" result in much, if any mass loss directly. As Owocki, *et al.* showed in these proceedings, the most likely scenario for mass loss in a super-Eddington atmosphere is one in which the material separates into clumps with the clumps essentially stationary and some small amount of material driven above the escape speed between the clumps. As we have only a 1D analysis thus far, it is impossible for us to determine what the clump size distribution would be.

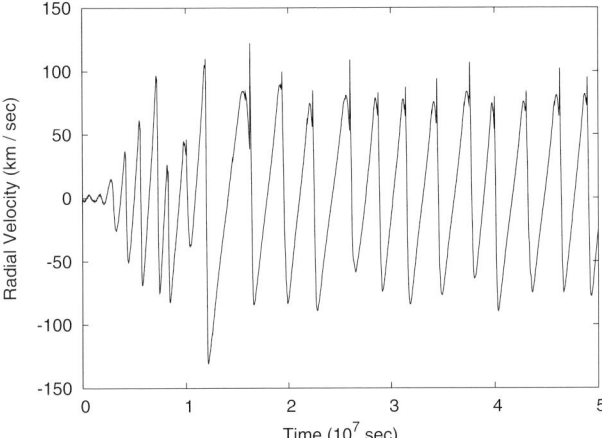

Figure 3. Photospheric velocity as a function of time for the case $Y = 0.49, Z = 0.01$.

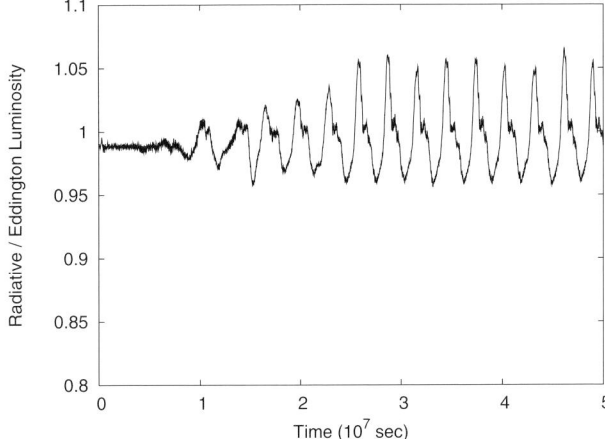

Figure 4. Radiative luminosity at $T \sim 210{,}000$ K relative to Eddington for $Y = 0.49, Z = 0.01$.

This is the beginning of a more formal parameter study on the nature of these pulsations and their connections, if any, to S-Dor behavior. Interesting parameters to study are the zoning, Y, Z, the time-dependent convection parameters, particularly ℓ, which at larger values should result in more "outburst"-like behavior, and evolution parameters like the initial mass and $\dot{M}$, which determine the initial conditions of the LBV. In addition to more coverage of the parameter space, we eventually plan to extend the hydrodynamic analysis to 2D to determine whether the super-Eddington nature of the pulsations leads to enhanced mass loss through pores in the stellar atmosphere.

Acknowledgements

A. Onifer would like to that the IAU for partial support for this presentation. This work was performed under the auspices of the U. S. Department of Energy by the Los Alamos National Security (LANS), LLC under contract No. DE-AC52-06NA25396.

References

Böhm-Vitense, E. 1958, *ZfA*, 46, 108
Cox, J. P. & Giuli, R. T. 1968, *Principles of stellar structure* (New York: Gordon and Breach)
Cox, A. N. & Ostlie, D. A. 1993, *Ap&SS*, 210, 311
Guzik, J. A. 2005, in: R. Humphreys & K. Stanek (eds.), *The Fate of the Most Massive Stars* (San Francisco: ASP), *ASP Conf. Ser.*, 332, 208
Guzik, J. A., Cox, A. N., & Despain, K. M. 2005, in: R. Humphreys & K. Stanek (eds.), *The Fate of the Most Massive Stars* (San Francisco: ASP), *ASP Conf. Ser.*, 332, 263
Guzik, J. A. & Swenson, F. J. 1997, *ApJ*, 491, 967
Guzik, J. A., Watson, L. S., & Cox, A. N. 2005, *ApJ*, 627, 1049
Iben, I., Jr. 1965, *ApJ*, 142, 1447
Iglesias, C. A. & Rogers, F. J. 1996, *ApJ*, 464, 943
Nieuwenhuijzen, H. & de Jager, C. 1990, *A&A*, 231, 134
Ostlie, D. A. 1990, in: J. R. Buchler (ed.), *Numerical Modeling of Nonlinear Stellar Pulsations: Problems and Prospects* (Dordrecht: Kluwer), 89
Vink, J. S. & de Koter, A. 2002, *A&A*, 393, 543

Discussion

LANGER: How many pulsation cycles does it take to lose the pulsating layer in a stationary wind, assuming standard radiation-driven wind mass-loss rates?

ONIFER: The pulsation cycle for the model I showed was about 1 month, and the Fe bump region is $\sim 10^{-3} M_\odot$ below the surface, so it would take about 100 years or ~ 1000 pulsation cycles to strip away the layer.

HUMPHREYS: Thanks for making the distinction between the normal or classical LBVs like S Dor & AG Car and the η Car-like giant eruptions. η Car & its rare relatives actually increase their luminosity during their giant eruptions. The classical LBVs maintain constant luminosity during their optically thick wind stage. Owocki was talking about η-Car-like eruptions. It may only be a difference of degree, and the mechanism may be the same, but what we observe is very different. It is important to make the distinction.

ONIFER: Thank you, this is a very good point.

HIRSCHI: If pulsations are driven by iron opacity, pulsations would be strongly metallicity dependent. Are there other sources of opacity at very low metallicities to drive pulsations?

ONIFER: I'm not aware of any sources of opacity at similar temperatures that could take over for iron at low metallicity.

What Do We Really Know About the Winds of Massive Stars?

D. John Hillier

Department of Physics & Astronomy, University of Pittsburgh, 3941 O'Hara Street,
Pittsburgh, PA, 15260, USA
email: hillier@pitt.edu

Abstract. The standard theory of radiation driven winds has provided a useful framework to understand stellar winds arising from massive stars (O stars, Wolf-Rayet stars, and luminous blue variables). However, with new diagnostics, and advances in spectral modeling, deficiencies in our understanding of stellar winds have been thrust to the forefront of our research efforts. Spectroscopic observations and analyses have shown the importance of inhomogeneities in stellar winds, and revealed that there are fundamental discrepancies between predicted and theoretical mass-loss rates. For late O stars, spectroscopic analyses derive mass-loss rates significantly lower than predicted. For all O stars, observed X-ray fluxes are difficult to reproduce using standard shock theory, while observed X-ray profiles indicate lower mass-loss rates, the potential importance of porosity effects, and an origin surprisingly close to the stellar photosphere. In O stars with weak winds, X-rays play a crucial role in determining the ionization balance, and must be taken into account.

Keywords. stars: atmospheres – stars: mass loss – stars: winds, outflows – stars: individual (10 Lac, τ Sco, ζ Pup) – stars: early-type – line: formation – line: profiles

1. Introduction

Although our understanding of massive star winds has improved dramatically numerous problems remain. There are inconsistencies between theoretical mass-loss rates and terminal velocities and those observed. Wind profiles are variable, and show structure not predicted by the standard model. Atmospheric lines reveal evidence for photospheric motions of unknown origin. While there is consensus that X-rays generally arise from shocks generated by instabilities in the wind we still do not qualitatively understand their production (e.g, Feldmeier 1997). A more detailed account of recent results obtained from X-ray observations of massive stars is given by Cohen (this volume).

The reasons for the discrepancies are not hard to find. To model the mass-loss process, and the spectra, we generally make standard assumptions. In particular we assume (a) stationarity, (b) spherical symmetry and (c) homogeneity. Further we assume that the winds are driven by radiation pressure, and that the momentum, generally only imparted directly to a few species, is shared among all species. In all early-type stars many of these assumptions are invalid.

The reasons why the assumptions are likely to be invalid are clear. First, and foremost, stellar winds are variable. This is expected since radiation-driven winds are intrinsically unstable (Lucy & Solomon 1970, Owocki et al 1988). On a local scale dX/dt, at any given location, is non-zero (X is an arbitrary variable) although on a global scale the statistical properties of the wind may be constant. Emission profiles in some Wolf-Rayet (W-R) stars, for example, often show surprising constancy, and exhibit only small-scale perturbations. Thus for these stars the statistical properties of the wind are fairly constant in time. In such cases it may be possible to ignore the time variability and use a statistical approach to model the spectra.

A direct manifestation of the instability is that winds cannot be homogeneous — they must be clumped. Evidence for clumping is widespread. It manifests itself through line-profile variability in both O (Eversberg et al. 1998) and W-R stars (Lèpine & Moffat 1999). Clumping also influences the relative strength of density squared diagnostics (Hα in O stars, most emission lines in W-R stars, infrared and radio free-free continua) and density diagnostics (electron scattering wings on W-R stars and luminous blue variables (LBVs) [e.g., Hillier 1991]), X-ray profile shapes (e.g., Kramer et al. 2003), and the strength of some resonance lines (e.g., P v) in O stars (e.g., Crowther et al. 2002; Hillier 2003; Hillier et al. 2003).

Clumping is usually incorporated into spectral modeling using the volume-filling factor approach. This approach explicitly assumes that the clumps are small relative to a photon mean-free path — an excellent assumption for continua but less valid for lines. It ignores the fact that at any given radial location clumps will have a range of densities and sizes, and consequently different ionizations. It ignores the possibility that a clumped wind can be porous — that opacity can be hidden in dense clumps – and hence that photons can escape more freely. Porosity comes in two forms — spatial (or classical) porosity and velocity porosity (vorosity). Classical porosity refers to the porosity induced by spatial variations in density — it affects both line and continuum photon transfer. Oskinova et al. have argued that porosity has important influences on X-ray (Oskinova et al. 2007)(but see Owocki & Cohen 2006 and Cohen, this volume, for an alternative view) and UV line profiles (Oskinova 2007). Vorosity (a term introduced by Owocki 2008) refers to the porosity induced by spatial variations in velocity, and affects only lines. In a smooth wind, all velocities are present and properties vary smoothly. In a clumped wind, velocities may change abruptly along a given sight line, and some velocities (because of their low density/spatial extent) may have little influence on line emission/absorption.

Other effects, often neglected, but which can have a substantial influence on spectra, spectral analysis, and winds include rotation and magnetic-fields. Further, the treatment of photospheric velocity fields (microturbulence and macroturbulence) is simplistic.

2. Microturbulence & Macroturbulence

Microturbulence refers to small scale velocity variations, while macroturbulence refers to large scale motions. Unfortunately in early-type stars, because of their large rotation velocity, microturbulence cannot be measured directly. Instead it is determined by using it as a free-parameter to achieve consistency between abundances derived from weak and strong lines. Macroturbulence can be inferred from profile shapes — in particular profiles tend to depart from the classic parabolic rotation profile, and show more Gaussian-like profiles. In O stars, microturbulent velocities are of order the sound speed — in dwarfs they tend to be lower than the sound speed but they can be larger than the sound speed in supergiants. Of course, macroturbulence and microturbulence are just limiting forms. In reality, spectra should be computed using the full 3D photospheric velocity field. When this is done for the Sun, where photospheric motions are related to convection, there is no need for turbulence (Asplund et al. 2000).

Work by Simón-Díaz and Herrero (2007) has shown that the Fourier technique provides a reliable means to separate rotational broadening from other broadening mechanisms in early-type stars. In agreement with earlier work, they find that macroturbulence is generally negligible in dwarfs (compared with rotational broadening) but very important in supergiants which tend to have, on average, lower rotational velocities than dwarfs.

Based on Gaussian broadening, macroturbulence velocities of 5 to nearly $100\,\mathrm{km\,s^{-1}}$ can be inferred to exist in supergiants.

The origins of turbulence in O stars is unknown. It might be related to pulsational instabilities, although evidence for pulsation is not seen in all O stars. However, evidence for the influence of pulsations has been inferred, for example, from variability studies of A and B supergiants by Kaufer *et al.* (1997, 2007). A second alternative is that the turbulence is related to instabilities in the stellar wind — indeed it is possible that there might be a strong coupling between the turbulence and the wind. Another possibility is that the turbulence is related to instabilities — instabilities that can occur when a star is close to the Eddington limit (e.g., Shaviv 2001, Stothers 2003). A final alternative is that turbulence might be related to weak surface convection zones known to exist in some massive stars (e.g., Stothers 2003, Maeder *et al.* 2008).

3. Wind dynamics and theoretical mass-loss rates

It is now generally accepted that O & W-R winds are clumped, with mass-loss estimates from density squared diagnostics overestimating the true mass-loss rates by factors of 2 to 10, with more moderate values (factor of 3) preferred. Recently, Bouret *et al.* (2008a, 2008b) have analyzed ζ Pup. Using a combination of diagnostics they find a volume-filling factor of 0.05, together with a mass-loss rate of $1.7 \times 10^{-6}\,\mathrm{M_\odot\,yr^{-1}}$, which is a factor of 2.8 lower than the predictions of Vink *et al.* (2000; abundances from Allen 1973). Calculations using CMFGEN (with non-CNO abundances from Cox 2000) show that the momentum deposition is reasonably consistent with that needed to drive the flow — there is slightly too much force in the outer regions, and near-consistency in the inner regions can only be achieved with a fast velocity law ($\beta < 1$)(see Bouret *et al.* 2008b for more details).

Two major groups have derived theoretical mass-loss rates for O stars. The Munich group (e.g., Pauldrach *et al.* 1990 & references therein) solves for the momentum-balance equation at selected radii, while Vink *et al.* (2000) use a global momentum argument to deduce the mass-loss rate. Reasonable agreement between the groups is obtained. The level of agreement between theoretical and observed mass-loss rates is unclear. Comparison is made difficult by uncertainties in distances, uncertainties in stellar parameters and abundances, and the unknown correction for clumping (e.g., Markova *et al.* 2004, Puls *et al.* 2006). Work by Puls *et al.* (2006) shows that clumping is less in the radio region than in the Hα formation region, but the actual clumping in the radio region is unknown. Suggestions have also been made that clumping only affects mass-loss rates derived from Hα when it is in emission (i.e., supergiants; Markova *et al.* 2004, Mokiem *et al.* 2007).

Microturbulence can have a substantial influence on the wind dynamics (Poe *et al.* 1990) and line force (Hillier *et al.* 2003) around the sonic point. Recently, a study of the influence of microturbulence on mass-loss rates has been undertaken by Lucy (2007a, 2007b). For the single case studied, Lucy found a mass-loss rate a factor of a few lower than earlier predictions. Lucy's work also highlights many of the uncertainties in understanding wind dynamics, even in a time-steady situation. What is the meaning of the CAK critical point (Castor *et al.* 1975)†? Is it really the CAK critical point that sets the mass-loss rate? Can information from the CAK critical point be communicated, by Abbott waves, back to the photosphere?

† The critical point that comes from an analysis of the momentum equation for radiation driven winds is commonly referred to as the CAK critical point, after its discoverers Castor, Abbott, & Klein (1975). For basic insights into radiation driven stellar winds see Lamers and Cassinelli (1999).

Owocki & Puls (1999) have emphasized the key role of gradients in the diffuse, scattered radiation in setting the mass loss driven through the sonic point. Lucy (2007b) claims that it is really the sonic point that is the critical point in a line driven flow. Practically this appears not to be the case — the velocity gradient can generally be adjusted at the sonic point (where the dynamical and pressure terms balance) so that the line force balances gravity. It is above the sonic point that difficulties appear — the line force now has to balance both gravity and the dynamical term since the pressure term rapidly becomes negligible.

The role of the sound speed in CAK theory is complicated. In models where the sound speed is neglected, multiple solutions to the wind momentum equation exist. In this situation, the mass loss is taken as the critical solution which provides a unique velocity law. In this case, the deduced mass-loss rate is the maximum mass loss that can be driven by the flow. When sound terms are included, this maximum mass-loss solution is essentially equivalent to that found from the critical point analysis. Owocki and ud-Doula (2004; see their Appendix A) applied a perturbation expansion approach to examine how inclusion of a finite sound speed affects the mass-loss rate and terminal flow speed in a CAK wind. The results show the relative changes scale with the ratio of sound speed to escape speed, with mass loss increasing and flow speed decreasing, both typically by about 10% relative to a zero-sound-speed model.

4. Low Mass Rates

For stars with $\log L/L_\odot < 5.2$, mass-loss rates derived from UV analysis are often much lower than predicted by standard radiation driven wind theory (Bouret *et al.* 2003; Martins *et al.* 2004, 2005). The weak wind problem is actually two problems — stars which have weak winds for their spectral type (Vz stars, Walborn 2000) and stars which have low mass-loss rates in comparison with that predicted using the standard relationships between the modified wind momentum and luminosity. The reasons for the discrepancies are not well understood, but there are several plausible explanations.

First, all single O stars appear to be X-ray emitters with $L_X/L_{Bol} \sim 10^{-7}$ (Sana *et al.* 2006, and references therein), with evidence that L_X/L_{Bol} increases for $L_{Bol} < 10^{38}$. Surprisingly, L_X does not correlate with wind properties in the expected manner. Since $\dot{M} \sim L^2$ (Vink *et al.* 2000), X-rays will have a larger influence on the wind ionization at low luminosities and must be allowed for (e.g., Martins *et al.* 2005). Second, since X-ray emission (naively) scales with $\dot{M}^2$, the fraction of X-ray emitting gas must be larger in stars with weak winds — indeed it is possible that for stars with very weak winds a significant fraction of the wind is in the hot state ($T > 10^6$ K). The later conclusion follows from simple cooling arguments. In stars such as ζ Pup, the shocked gas will cool rapidly. In stars with low mass-loss rates and hence wind densities, the cooling length can be comparable to the spatial scale length ($1/r$). In addition, conduction effects between the hot and cool gas can be important (Lucy & White 1980). Given these effects, low density winds will be far from homogeneous, even on large scales.

Weak wind stars also have few wind diagnostics. Hα responds only weakly, if at all, to the mass-loss rate. Indeed if there is any sensitivity, it is only in the core, which is problematical for mass-loss determinations since core intensities are difficult to model accurately. In the recent study by Mokiem *et al.* (2005), 10 Lac was found to have a Hα mass-loss rate less than that predicted by radiation driven wind theory. While marginally consistent (2σ) with predictions of radiation driven wind theory it is only an upper limit — the data is also consistent with a mass-loss rate an order of magnitude lower than predicted theoretically.

In the UV we only have a few resonance lines to study the wind — Si IV, C IV, N V, O VI. In general, none of these lines belong to the dominant ionization stage in the wind, and hence accurate mass-loss rates require accurate abundances, accurate atomic models, and an accurate description of the UV and X-ray radiation fields. The later is problematical since we don't have an adequate theory for the origin of X-rays in O stars.

4.1. *10 Lac*

Recently, Lanz *et al.* (2008) have undertaken a detailed optical/UV study of 10 Lac using TLUSTY (Hubeny & Lanz 1995) and CMFGEN (Hillier & Miller 1998). The difficulty of determining accurate mass-loss rates in low luminosity stars can be observationally illustrated by comparing resonance profiles in 10 Lac (O9 V) and τ Sco (B0 V). The C IV and N V profiles are very similar in the two stars. While wind absorption due to O VI is strongest in 10 Lac it is also present in τ Sco (Lamers & Rogerson 1978). Conversely, Si IV wind absorption can be seen in τ Sco, but it is absent in 10 Lac. The latter property is consistent with the stars' spectral types, while the presence of O VI requires, in both stars, super-ionization caused by X-rays. The similarity of the C IV and N V profiles seems highly fortuitous, especially since τ Sco is known to possess a strong magnetic field (B~500 Gauss, Donati 2006), it has a higher $L_{\rm X}/L_{\rm Bol}$ (Berghöfer *et al.* 1996) than 10 Lac, and its narrow line X-ray spectrum does not fit the standard wind shock model (Cohen *et al.* 2003).

From a theoretical point of view, the derived UV mass-loss rates appear to be too low. In typical models, momentum deposition in the winds is much larger than needed to drive the flow, indicating a fundamental discrepancy with the models. The discrepancy can be best illustrated by noting that a single strong resonance line can drive a mass-loss rate of $\sim L/c^2$ (Lucy & Solomon 1970; see also Lamers & Rogerson 1978). For 10 Lac, with $L = 10^5$ L$_\odot$, the predicted mass-loss rate is $\sim 7 \times 10^{-9}$ M$_\odot$ yr^{-1}, and since we do observe a strong saturated UV resonance line (the O VI λ1032 component is almost black in 10 Lac), this must be regarded as a lower limit to the mass-loss rate. Unfortunately, with this mass-loss rate, the theoretical C IV resonance line is much stronger than observed.

5. Rotation

While O stars are often rapid rotators, the rotation is generally ignored in stellar analysis, except that it provides a broadening mechanism which smears out line profiles. The influence of rotation on photospheric lines can be performed, to a good approximation, by a simple convolution. However, this procedure does not work for wind lines which form over a range of radii, and (because of conservation of angular momentum) over a range of rotation velocities. Thus for wind profiles, one must resort to 2D calculations (Perentz & Puls 1996, 2000; Busche & Hillier 2005). For simplicity, the work of Busche & Hillier (2005) assumes that the necessary opacities and emissivities can be obtained from 1D calculations using simple scaling relations, and accurate profiles are then computed using a 2D formal solution. While their effects on the derived mass-loss rate, for example, are not huge, their neglect can lead to erroneous results when trying to determine more subtle effects such as the variation of velocity or volume-filling factor with radius. This is highlighted in recent calculations undertaken by Bouret *et al.* (2008b, and this volume) for ζ Pup, which is a moderately fast rotator. By correctly allowing for rotation, considerably improved profile fits could be obtained, providing much greater confidence in the analyses. These calculations only allow for the effects of rotation on the radiative transfer — it remains to be seen whether departures in density from a spherical distribution and gravity darkening also have a significant influence on observed line profiles.

6. Conclusion

While considerable advances have been made in our understanding of stellar winds it is clear that major uncertainties remain. To make progress, it is essential that we allow for the correct boundary conditions at the base of the photosphere, and that the stochastic nature of the winds is taken into account. There are still uncertainties in the meaning and role of Abbott waves which need to be resolved. Further, the origin of X-rays and their influence on the ionization structure and wind dynamics needs to be considered. Detailed dynamical work, together with detailed quantitative spectral analysis using all spectral bands, should allow significant advances to be made over the next decade.

Acknowledgements

The author wishes to thank his collaborators Stan Owocki, Janos Zsargó, Jean-Claude Bouret, Thierry Lanz, Paco Najarro, Joachim Puls and others for many stimulating discussions on massive stars and their winds. Partial support for this work was provided by STScI grant HST-AR-10693.02-A, Chandra GO award TM6-7003X, and NASA ADP grant NNG04GC816(subaward Z602201).

References

Allen, C. W. 1973, *Astrophysical Quantities, 3rd. ed.* (London: University of London, Athlone Press)
Asplund, M., Nordlund, Å., Trampedach, R., *et al.* 2000, *A&A*, 359, 729
Berghöfer, T. W., Schmitt, J. H. M. M., & Cassinelli, J. P. 1996, *A&AS*, 118, 481
Bouret, J.-C., Lanz, T., Hillier, D. J., & Foellmi, C. 2008a, in: W.-R. Hamann, A. Feldmeier, L. M. Oskinova (eds.), *Clumping in Hot Star Winds* (Potsdam: Universitäts-Verlag), in press
Bouret, J.-C., Lanz, T., Hillier, D. J., & Foellmi, C. 2008b, in preparation
Bouret, J.-C., Lanz, T., Hillier, D. J., Heap, S. R., *et al.* 2003, *ApJ*, 595, 1182
Busche, J. R. & Hillier, D. J. 2005, *AJ*, 129, 454
Castor, J. I., Abbott, D. C., & Klein, R. I. 1975, *ApJ*, 195, 157
Cohen, D. H., de Messières, G. E., MacFarlane, J. J., *et al.* 2003, *ApJ*, 586, 495
Cox, A. N. 2000, in: A. N. Cox (ed.), *Allen's Astrophysical Quantities, 4th ed.* (New York: AIP Press; Springer)
Crowther, P. A., Hillier, D. J., Evans, C. J., *et al.* 2002, *ApJ*, 579, 774
Donati, J.-F., Howarth, I. D., Jardine M. M., *et al.* 2006, *MNRAS*, 370, 629
Eversberg, T., Lepine, S., & Moffat, A. F. J. 1998, *ApJ*, 494, 799
Feldmeier, A., Puls, J., & Pauldrach, A. W. A. 1997, *A&A*, 322, 878
Hillier, D. J. 1991, *A&A*, 247, 455
Hillier, D. J. 2003, in: K. van der Hucht, A. Herrero, & C. Esteban (eds.), *A Massive Star Odyssey: From Main Sequence to Supernova* (San Francisco: ASP), *Proc. IAU Symp 212*, 70
Hillier, D. J., Lanz, T., Heap, S. R., *et al.* 2003, *ApJ*, 588, 1039
Hillier, D. J. & Miller, D. L. 1998, *ApJ*, 496, 407
Hubeny, I. & Lanz, T. 1995, *ApJ*, 439, 875
Kaufer, A., Stahl, O., & Prinja, R. K. 2007, in: A. T. Okazaki, S. P. Owocki, & S. Stefl (eds.), *Active OB-Stars: Laboratories for Stellar and Circumstellar Physics* (San Francisco: ASP), *ASP Conf. Ser.*, 361, 179
Kaufer, A., Stahl, O., Wolf, B., *et al.* 1997, *A&A*, 320, 273
Kramer, R. H., Cohen, D. H. & Owocki, S. P. 2003, *ApJ*, 592, 532
Lamers, H. J. G. L. M. & Cassinelli, J. P. 1999, *Introduction to Stellar Winds* (Cambridge: Cambridge University Press)
Lamers, H. J. G. L. M. & Rogerson, Jr., J. B. 1978, *A&A*, 66, 417
Lanz, T., Hillier, D. J., Hubeny, I. J., *et al.* 2008, in preparation

Lépine, S. & Moffat, A. F. J. 1999, *ApJ*, 514, 909
Lucy, L. B. 2007a, *A&A*, 468, 649
Lucy, L. B. 2007b, *A&A*, 474, 701
Lucy, L. B. & Solomon, P. M. 1970, *ApJ*, 159, 879
Lucy, L. B. & White, R. L. 1980, *ApJ*, 241, 300
Maeder, A., Georgy, C., & Meynet, G. 2008, 479, L37
Markova, N., Puls, J., Repolust, T., & Markov, H. 2004, *A&A*, 413, 693
Martins, F., Schaerer, D., Hillier, D. J., & Heydari-Malayeri, M. 2004, *A&A*, 420, 1087
Martins, F., Schaerer, D., Hillier, D. J., et al. 2005, *A&A*, 441, 735
Mokiem, M. R., de Koter, A., Evans, C. J., et al. 2007, *A&A*, 465, 1003
Mokiem, M. R., de Koter, A., Puls, J., et al. 2005, *A&A*, 441, 711
Oskinova, L. M., Feldmeier, A., & Hamann, W.-R. 2006, *MNRAS*, 372, 313
Oskinova, L. M., Hamann, W.-R., & Feldmeier, A. 2007, *A&A*, 476, 1331
Owocki, S. P. 2008, in: W.-R. Hamann, A. Feldmeier, L. M. Oskinova (eds.), *Clumping in Hot Star Winds* (Potsdam: Universitäts-Verlag), in press
Owocki, S. P. & Cohen, D. H. 2006, *ApJ*, 648, 565
Owocki, S. P., & ud-Doula, A. 2004, *ApJ*, 600, 1004
Owocki, S. P. & Puls, J. 1999, *ApJ*, 510, 355
Owocki, S. P., Castor, J. I., & Rybicki, G. B. 1988, *ApJ*, 335, 914
Pauldrach, A. W. A., Kudritzki, R. P., Puls, J., & Butler, K. 1990, *A&A*, 228, 125
Petrenz, P. & Puls, J. 1996, *A&A*, 312, 195
Petrenz, P. & Puls, J. 2000, *A&A*, 358, 956
Poe, C. H., Owocki, S. P., & Castor, J. I. 1990, *ApJ*, 358, 199
Puls, J., Markova, N., Scuderi, S., et al. 2006, *A&A*, 454, 625
Sana, H., Rauw, G., Nazé, Y., et al. 2006, *MNRAS*, 372, 661
Shaviv, N. J. 2001, *ApJ*, 549, 1093
Simón-Díaz, S. & Herrero, A. 2007, *A&A*, 468, 1063
Stothers, R. B. 2003, *ApJ*, 589, 960
Vink, J. S., de Koter, A., & Lamers, H. J.G. L.M. 2000, *A&A*, 362, 295
Walborn, N. R., Lennon, D. J., Heap, S. R., et al. 2000, *PASP* 112, 1243

Discussion

OWOCKI: If turbulence in the transonic region "chokes" the wind, and leads to a lower mass-loss rate, you would expect the line force in the outer wind, which scales as $(1/\rho dv/dr)^\alpha$, to give a higher acceleration leading to a much higher terminal speed.

HILLIER: Agreed, however I think models fully consistent with the observations need to be analyzed to see if there are inconsistencies with the new lower mass-loss rates. For ζ Pup, the wind momentum deposition is reasonably consistent with that needed to drive the wind. It is a little too large in the outer region, and we struggle to drive the wind just above the sonic point but given the coarse treatment of clumping and the neglect of time dependent effects I don't think we have a major problem. For stars such as 10 Lac, however, we do have a major problem (as noted in the text) with the low mass-loss rates — the large force in the outer region would lead to terminal velocities much larger than those observed.

John Hillier.

Robert Stencel.

The Physical Properties of Red Supergiants: Comparing Theory and Observations

Philip Massey[1], Emily M. Levesque[2], Bertrand Plez[3] and Knut A. G. Olsen[4]

[1] Lowell Observatory, 1400 W. Mars Hill Rd., Flagstaff, AZ 86001, USA
email: Phil.Massey@lowell.edu

[2] Institute for Astronomy, University of Hawaii,
2680 Woodlawn Drive, Honolulu, HI 96822, USA
email: emsque@ifa.hawaii.edu

[3] GRAAL, Université Montpellier II, CNRS, 34095 Montpellier, France
email: Bertrand.Plez@graal.univ-montp.fr

[4] Gemini Science Center, NOAO, P.O. Box 26732, Tucson, AZ 85726-6732, USA
email: kolsen@noao.edu

Abstract. Red supergiants (RSGs) are an evolved stage in the life of intermediate massive stars ($\leqslant 25\,M_\odot$). For many years, their location in the H-R diagram was at variance with the evolutionary models. Using the MARCS stellar atmospheres, we have determined new effective temperatures and bolometric luminosities for RSGs in the Milky Way, LMC, and SMC, and our work has resulted in much better agreement with the evolutionary models. We have also found evidence of significant visual extinction due to circumstellar dust. Although in the Milky Way the RSGs contribute only a small fraction ($< 1\%$) of the dust to the interstellar medium (ISM), in starburst galaxies or galaxies at large look-back times, we expect that RSGs may be the main dust source. We are in the process of extending this work now to RSGs of higher and lower metallicities using the galaxies M31 and WLM.

Keywords. stars: atmospheres – circumstellar matter – stars: evolution – stars: late-type – stars: mass loss – supergiants

1. Introduction

Those of us here who have worked on massive stars for a while are probably all attracted by stellar physics at the extremes. For O-type stars, we are dealing with stars that are as massive and luminous as stars come, and as hot as main-sequence stars can get. To properly assess their physically properties through spectroscopic analysis has required not only the introduction of non-LTE atmosphere models (Mihalas & Auer 1970, Auer & Mihalas 1972) but an additional thirty years of developments, such as the inclusion of mass loss (Abbott & Hummer 1972, Kudritzki 1976), the inclusion of hydrodynamics of the stellar wind (Gabler *et al.* 1989; Kudritzki & Hummer 1990; Puls *et al.* 1996), and, most recently, the full inclusion of line blanketing (Hillier & Miller 1989; Hillier *et al.* 2003; Herrero, Puls, & Najarro 2002). (For a recent summary, see Massey *et al.* 2004 and Massey *et al.* 2005a). The study of LBVs and WRs is equally exciting, stars where radiation pressure dances with gravity to see who will lead (Lamers 1997, Smith & Owocki 2006), and where high mass loss rates are continuous rather than episodic due to high metal content in the stellar atmosphere (Crowther 2007 and references therein).

However, largely ignored until now are the red supergiants (RSGs). The physical conditions in these stars are, in their own way, equally extreme. They have the largest physical sizes of any stars (up to $1500\times$ the radius of the sun; see Levesque *et al.* 2005, here after Paper I). This large physical size invalidates the usual assumptions of plane parallel geometry. The velocities of the convective layers in these stars' atmospheres are supersonic, giving rise to shocks (Freytag, Steffen, & Dorch 2002), and making the stars' photospheres very asymmetric and invalidating mixing-length assumptions. Their extremely cool temperatures (3400 - 4300 K) lead to the the presence of molecules in their atmospheres, requiring the inclusion of extensive molecular opacity sources in any realistic model atmosphere. From an observational point of view, the large (negative) bolometric corrections and their sensitivity to the adopted temperature complicate the transformation from the observed color-magnitude diagram to the physical H-R diagram in much the same way as it does for the O-type stars.

Recent advances in stellar atmosphere models for cool stars (e.g., Plez 2003) have allowed the first reasonable determination of the physical properties of these stars, in much the same way that the non-LTE H and He models of Auer & Mihalas (1972) allowed the first reasonable determination of the physical properties of O-type stars by Conti (1973). And while we may still have a way to go, we believe our answers will hold up as well as those that Conti (1973) have, which is really pretty good (see Massey et al. 2005a).

I find it personally interesting that there has been a real aversion to looking at what happens to a massive star as it heads over to the far right side of the H-R diagram. I think this is cultural—for many years much of the "massive star community" really thought of itself as the "hot star community", with the exception of a few workers, most notably our good colleague Roberta Humphreys, whose early work on supergiants in the Milky Way and other Local Group galaxies (such as Humphreys 1978, 1979a, 1979b, 1980a, 1980b, 1980c, Humphreys & Davidson 1979, Humphreys & Sandage 1980) certainly spurred my own interest in the field, and whose presence at these symposia always reminds us that there's more to the life of a massive star than the O and WR stages.

Let us briefly review what the evolutionary tracks predict RSGs come from. In Fig. 1 we show the tracks covering a range of a factor of 10 in metallicity, from z=0.004 (SMC-like) to z=0.020 (solar) to z=0.040 (M31-like). I have drawn a vertical line at an effective temperature of 4300 K, which roughly corresponds to that of a K0 I. Stars to the right of that line we are calling RSGs. At solar metallicities we expect that stars with initial masses $\leqslant 25 M_\odot$ will become RSGs. At lower metallicities (SMC-like) the upper mass limit for RSGs is probably a bit higher—maybe $30 M_\odot$?—it's hard to tell because of the quantization of the tracks. The upper mass tracks go much further to the right at this low metallicity, but stop short of the RSG dividing lines—these $30\text{-}60 M_\odot$ stars become F- and G-type supergiants, but not K or M. At higher metallicity (M31) the limit is definitely lower, around $20 M_\odot$. In the case of solar metallicity the $25 M_\odot$ track turns back to the blue, and in fact such a star should become a WR after the RSG phase.

One more thing of note is that the tracks don't extend very far to the right of the K0 I (vertical) line in the SMC—the RSGs in the SMC shouldn't be very late, mostly K through M0, say. At higher metallicities they extend further to cooler effective temperatures. This is consistent with the change in the average RSG type observed in the SMC, LMC, and Milky Way (Elias, Frogel, & Humphreys 1985, Massey & Olsen 2003).

That said, when we began worrying about the issue of the physical properties of RSGs it was because if one placed the "observed" location of RSGs on the H-R diagram, they missed the tracks entirely: the alleged effective temperatures and luminosities were cooler and higher than those predicted by the evolutionary tracks. This was first noticed by Massey & Olsen (2003) for the SMC and LMC, but we quickly confirmed that the problem also existed for the Milky Way sample (Massey 2003a). When I mentioned this issue at the Lanzarote meeting (Massey 2003b) Daniel Schaerer came up to me afterwards and said

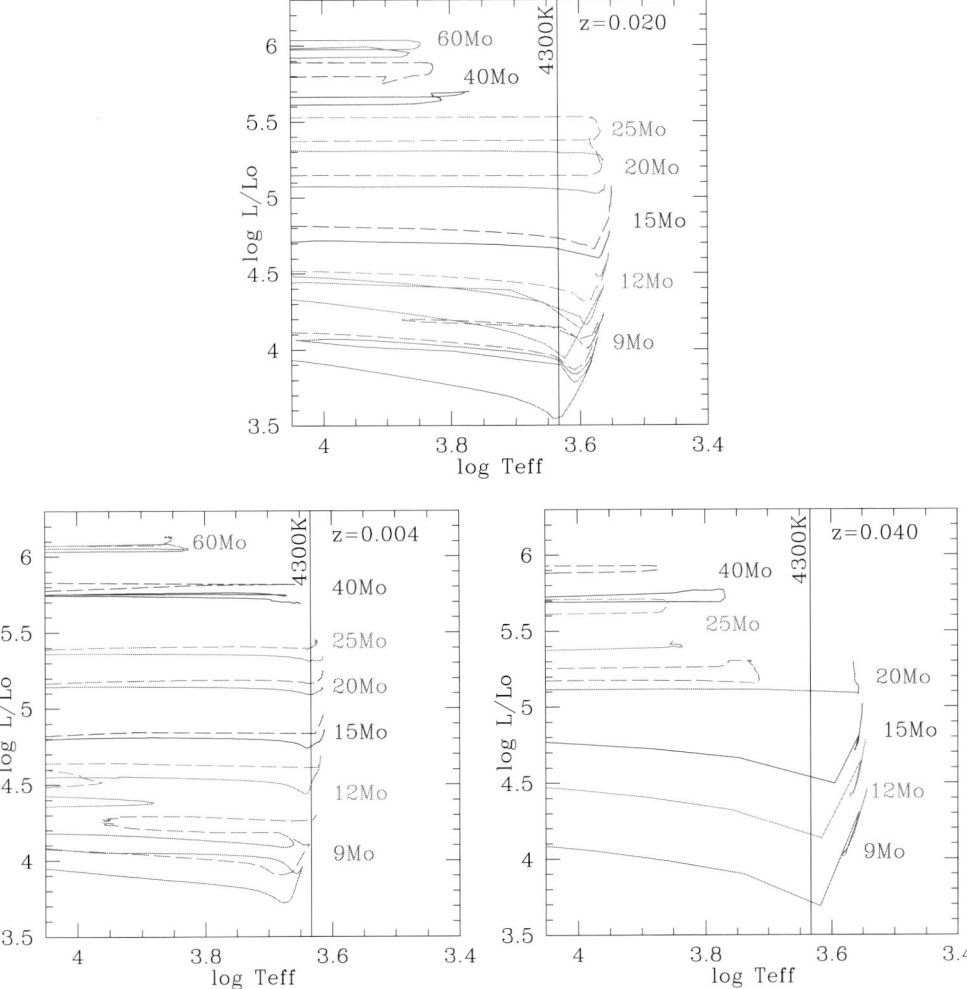

Figure 1. Effects of metallicity on the evolutionary tracks in the RSG region. The tracks for z=0.020 (solar) are from Meynet & Maeder (2003), for z=0.004 (SMC-like) are from Meynet & Maeder (2000), and for z=0.040 (M31-like) are from Meynet & Maeder (2005) and Meynet *et al.* 1994). Solid curves denote the tracks with no initial rotation, while the dashed lines correspond to initial rotations of 300 km s^{-1}. The black vertical line marks a temperature of 4300 K, roughly that of a K0 I at both solar and SMC metallicity (Papers I, II).

really, this was not a problem, since how far to the right the tracks went were heavily dependent upon such issues as how the mixing length was treated (see, for example, Maeder & Meynet 1987). However, this did not explain the issue of the luminosities being too high, and as an observer I was more concerned with what if the "observations" were wrong? Because, of course we don't "observe" effective temperatures and bolometric luminosities; instead, we obtain photometry and spectroscopy and use some relationship to convert these to physical properties. Indeed, further reading convinced me that there could be a serious problems, as much of what we "knew" about the effective temperature scale of RSGs were derived from lunar occultations of red *giants* (not supergiants); see discussion in Massey & Olsen (2003).

What would be involved in determining the effective temperatures of RSGs "right"? We really need models that have enough physics in them to correctly reproduce temperature-sensitive spectral features. The participants at this conference (mostly) understand what was involved in getting there for O-type stars. RSGs present their own challenges, as noted above. Fortunately, at the time I got intrigued by this problem, sophisticated models that were up to the task were becoming available. A modern version of these MARCS models was described by Plez, Brett, & Nordlund (1992), based upon the earlier work of Gustafsson et al. (1975). These are static, LTE, opacity-sampled models, and the current version (Plez 2003; Gustafsson et al. 2003) includes improved atomic and molecular opacities and sphericity.

2. RSGs in the Milky Way

In Paper I we obtained moderate-resolution spectrophotometry of 74 Galactic RSGs, which we then compared to the models. Our primary selection criterion was that the RSG had to be in a cluster or an association with a relatively well determined distance from the OB stars (Humphreys 1978, Garmany & Stencil 1992). We used a grid of models with effective temperatures 3000-4500 K in increments of 25 K, and $\log g = -1$ to $+1$ [cgs] in steps of 0.5 deg. We would typically begin by reddening the $\log g = 0.0$ model spectra of various effective temperatures by different amounts (using a Cardeli, Clayton, & Mathis 1989 reddening law with $R_V = 3.1$) until we got a good match to the depths of the molecular bands (principally TiO) *and* continuum shape of the spectra of the star. We would then see if the derived luminosity implied a surface gravity consistent with the $\log g = 0.0$; if not, we used a more appropriate model. In practice, the value we determined for the effective temperatures did not depend upon our 0.5 dex uncertainty in the adopted surface gravity, and the A_V was affected by < 0.3 mag.

When all was said and done, we had derived a new effective temperature scale which was significantly warmer than the older ones. It is shown by the points in Fig. 2. Their error bars reflect the standard deviation of the mean of our determinations; for the M stars (where we can use the TiO bands) our precision was $\sim$25 K, or about 0.7%—compare this to the typical 2000 K (5%) uncertainty we have when fitting O stars and their weak lines!

What did that do to the placement of stars in the H-R diagram? Just what we hoped! We show the situation (old and new) in Fig. 3. Now there is excellent agreement both in the effective temperatures, and in the upper luminosities, of RSGs in the Milky Way.

One of the cute things to come out of Paper I is the answer to "How large do normal stars get?" We see at the bottom of Fig. 3 a blowup of the upper right of our H-R diagram, now with lines of constant radii marked. The largest stars known in the Milky Way have radii of $1500\,R_\odot$, or about 7.2 AU. If you were to take one of these behemoths, and plunk it down where the sun is, its surface would extend to between the orbits of Jupiter and Saturn. Of course, "real" RSGs are known to have highly asymmetrical and messy "surfaces", as witness the high angular resolution images obtained of Betelgeuse by Young et al. (2000).

3. RSGs in the Magellanic Clouds: Weirder and Weirder

We naturally wanted to extend this work to RSGs in the Magellanic Clouds, where the metallicities are lower than in the Milky Way. Since the metallicity is lower, we expect that we will need cooler temperatures in order to form the same strength of TiO, the basis for the classification of mid-K through M stars. And, indeed that's just what we found (Paper II): M-type stars are 50 K and 150 K cooler in the LMC and SMC, respectively,

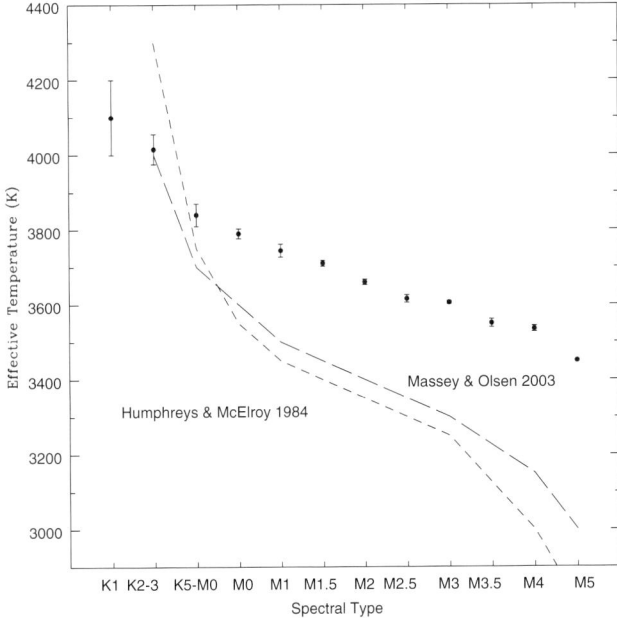

Figure 2. Effective temperature scale for Galactic RSGs. The new temperature scale for RSGs is shown by the points. For comparison, we include the much cooler effective temperatures of Humphreys & McElroy (1984) and Massey & Olsen (2003).

compared to their counterparts in the Milky Way. Just as we had for Galactic RSGs, we found great improvement between the observations and the models. For the LMC there is excellent agreement (not shown here; see Fig. 8 in Paper II). For the SMC the results were also improved (Fig. 4), but there were a substantial number of stars that were a bit cooler than the tracks allow.

This "no star zone" beyond the end of the tracks is known as the Hayashi forbidden region—stars in this region are no longer in hydrostatic equilibrium. They shouldn't exist. Even before we had these results, we were intrigued by the fact that there were *some* stars in the LMC and SMC that were classified as significantly later than the average type by Massey & Olsen (2003).

However, the real revelation came in our efforts to obtain a spectrum of HV 11423, one of the brightest RSGs in the SMC. It was on our observing list because it was classified as an M0 I by Elias, Frogel, & Humphreys (1985) (based upon photographic spectra obtained in 1978 and 1979), and we were tired of all of the K-types we had been observing. But, when we took spectra of it in early December 2004 it appeared to be of early K-type, probably K0-1 I. We honestly didn't think much about this at the time, but imagined that perhaps we had gotten the wrong star, although the two spectra we had obtained (on different nights) had matched. The next year (December 2005) we tried again, and took a couple of spectra. Much to our amazement the star was M4 I, much, *much* later than the spectral types seen for RSGs in the low metallicity SMC. We then went back and checked, and of course we *had* taken the spectrum of the correct star in 2004 as well—the coordinates left no doubt of that. We took another spectrum the following year, in September (2006), and the star was again an early K. Some digging in various data archives unearthed a VLT/UVES spectrum (apparently never published)

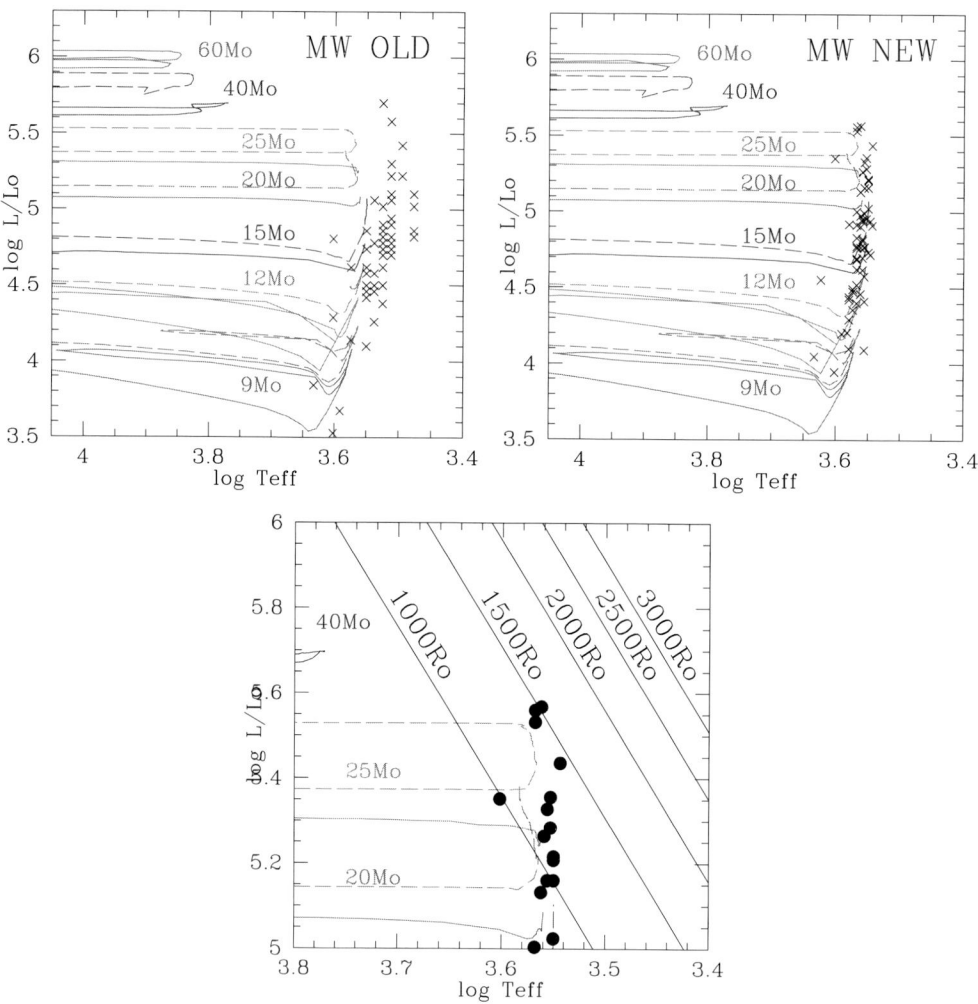

Figure 3. Agreement with evolutionary tracks for the Milky Way. On the left (top) we show the agreement (or rather lack thereof) between the evolutionary tracks and the "observed" location of RSGs using the old effective temperature scale and bolometric corrections given by Humphreys & McElroy (1984). On the right (top), we show the agreement using the results from Paper I. The evolutionary tracks are the same as those shown for $z = 0.020$ in Fig. 1 In the bottom figure we show an expansion of the upper-right part of the later figure, now with lines of constant radii indicated.

taken in December 2001; the star was clearly of even later type than our December 2005 M4 I type—more like an M4.5-5 I. So, here is one of the *brightest* RSGs in the SMC, and it is doing this funny little jig in the H-R diagram, changing effective temperatures from 3300 to 4300 K on the time scale of months and no one had noticed. Furthermore, the amount of visual extinction (A_V) changed by more than 1 mag, which we attribute to episodic dust formation (see below). Details can be found in Massey et al. (2007).

We concluded that this star is in an unstable period, maybe near the end of its life. Of course, one star is an oddity. The wonderful thing is that Leveque et al. (2007) found several more just like it! We think this underscores just exactly how lightly we've scratched the surface of stellar population studies of even the nearest galaxies.

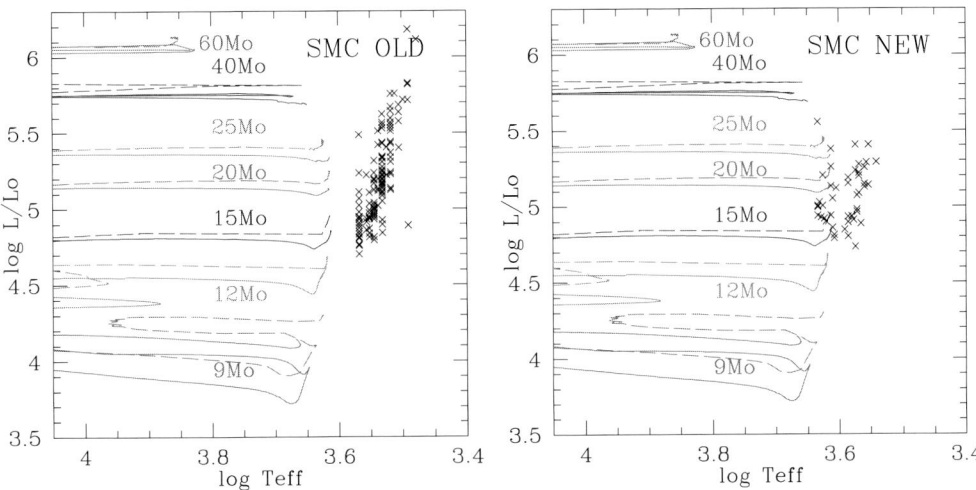

Figure 4. Agreement with evolutionary tracks for the SMC. On the left we show the lack there of) between the evolutionary tracks and the "observed" location of RSGs. On the right, we show the agreement using the results from Paper II. The evolutionary tracks are the same as those shown for $z = 0.004$ in Fig. 1.

4. Self-Consistency: Broad-band colors and VY CMa

If we only talked about our successes, we would be doing public relations and not science. One of the critical tests we performed was to see if our spectral fitting gave results that were consistent with what we would get from the models if we were to use the broad-band colors $(V-K)_0$ and $(V-R)_0$ instead. Such a test is not completely independent from our spectral fitting, as we must deredden the broad-band colors to make these comparisons, and for this we adopt the reddenings determined from the spectral fittings, but that is relatively minor. In Fig. 5 (left) we show the comparison between the derived effective temperatures from the spectral fittings, and that obtained from the $(V-K)_0$ colors. We see there is a systematic difference that is apparently metallicity-dependent: the median difference is 60 K for the Milky Way, 105 K for the LMC, and 170 K for the SMC, all in the sense the the effective temperatures derived from $(V-K)_0$ are hotter. To make the SMC data conform we would have to finagle the $(V-K)_0$ calibration by nearly 0.5 mag, so this is not due to some sort of subtle photometric transformation issue from K_s to K or something. Our first thought was that this was some sort of discrepancy having to do between the fluxes derived from the models and the strengths of the spectral features—the $(V-K)_0$ effective temperatures depend upon the former, while the spectral fitting effective temperatures depend upon the latter—but this notion was dispelled by looking at the results from $(V-R)_0$. Here we find very good agreement between the effectives temperatures derived from spectral fitting and those from photometry.

Instead, we now believe this is a discrepancy between the effective temperatures derived from the optical and those derived from the near-IR, and may just be due to the intrinsic limitations of static, 1-D models. We know that these stars likely contain cool and warm regions on their surfaces (Freytag, Steffen, & Dorch 2002), so it would not be unreasonable if the effective temperature one measures is wavelength dependent. This issue is discussed in greater depth in Paper II.

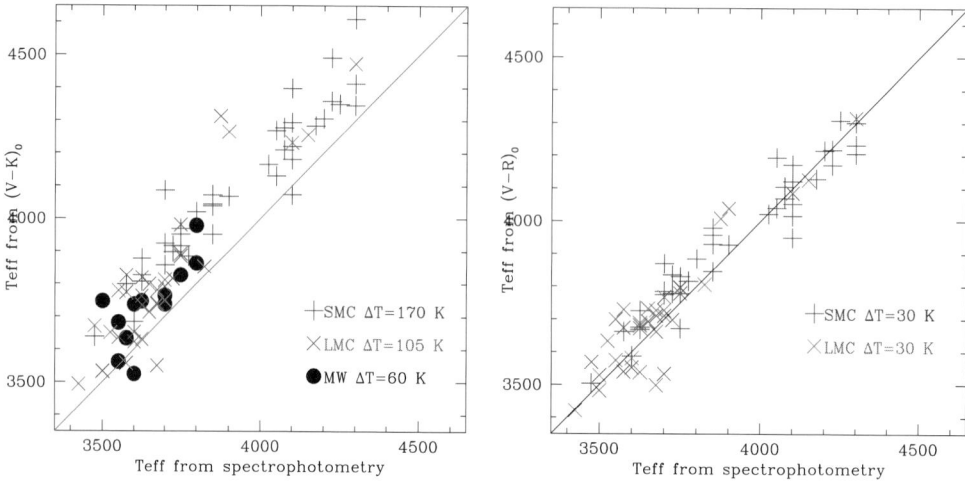

Figure 5. The effective temperatures derived from broad-band photometry are compared to those determined from spectral fitting.

Let us now turn briefly to an analysis we did of VY CMa (Massey, Levesque, & Plez 2006), where we got both the right and wrong answers. VY CMa is a Galactic RSG with some extreme properties claimed for it in the literature: a luminosity of 2 to 5×10^5 $L_\odot$, a mass-loss rate of $2 \times 10^{-4} M_\odot$ yr^{-1}, with an effective temperature usually quoted as 2800-3000 K. We were disturbed by these values, as if you plotted the star in the H-R diagram based on these, it would lie well into the Hayashi zone. Yet, the star is fairly stable—George Wallerstein has been observing it spectroscopically for many decades, and the only changes observed have to do with weak emission that originates in the extensive nebulosity around the star. We analyzed the star based upon new optical spectrophotometry and existing optical and *JHK* photometry, and concluded that the star had an effective temperature of 3650 K and a luminosity of 0.6×10^5 $L_\odot$.

There was only one itsy-bitsy problem with these results: they had to be wrong. Once our paper appeared, several colleagues called our attention to the fact that the luminosity we derived for the star was inconsistent with the total luminosity of the system (star plus dust). We should have realized there was a problem ourselves, as we had derived an effective temperature and radius for the surrounding dust shell. The corresponding luminosity of the dust (which we did not work out) is 2.3×10^5 $L_\odot$, about $4\times$ larger than what we got for the star itself. Since the dust is heated by the star, this is impossible.

About the only way we have found out of this would be if there was substantial extra grey extinction. We get A_V by reddening the models of appropriate temperature to match the shape of the stellar continuum using a $R_V = 3.1$ Cardelli, Clayton, & Mathis (1989) law. However, if the copious dust surrounding VY CMa has a distribution of grain sizes which is skewed towards larger sizes than usual, then we would underestimate A_V as the dust would be greyer than we assume. We would need about an additional 1.5-2.0 mag of such grey extinction. In any event, if we take our effective temperature, and a luminosity of 3 or 4×10^5 $L_\odot$ then the star sits in a very reasonable place in the H-R diagram, near the upper luminosity limit for RSGs.

Still, we don't think this reveals some fundamental flaw with what we're doing. The amount of dust around VY CMa is quite unusual (see Smith *et al.* 2001), and it will be of interest to determine the properties of this dust.

5. When Smoke Gets in Your Eyes

One of the things that worried us when we were doing our fits was that there were some stars for which there was very poor agreement in the near-UV (i.e., <4100Å, hereafter NUV), always in the sense that the star had more flux than the best-fitting model. We illustrate an example in Fig. 6 (left) for the star KY Cyg. Now, we considered a number of possibilities. Were these binaries, with the NUV being contributed by a hot companion? We didn't think so. We had indeed found some stars that clearly *were* binaries—but this was evident by having a composite spectrum, with Balmer lines clearly evident. The remaining stars that showed extra flux in the near-UV didn't exhibit any signs of a composite spectrum. So that didn't wash as an explanation.

We investigated this further in Massey *et al.* (2005b). Was this due to a problem with the models? We didn't think so, because there were lots of stars that didn't show this problem, and there didn't seem to be any correlation with effective temperature. What the NUV problem did show a correlation with was the amount of visual extinction—stars with the largest NUV problem also had the largest A_V. We looked into this a little more deeply, and indeed it turned out that the stars with the largest A_V actually had a considerable amount of *excess* extinction compared to OB stars in the same OB associations. This is illustrated in Fig. 6. The error bars show the typical sigma of A_V of the OB stars in a given association. For most RSGs there was good correlation between the A_V found for the OB stars, and the A_V found for the RSGs, but for some significant fraction of the RSGs there was significantly more extinction—up to several magnitudes†. We see the same thing for RSGs in the Magellanic Clouds: in both the SMC and LMC RSGs show greater extinction than the OB stars (Paper II).

If the extra extinction was due to circumstellar dust, then that could also explain the extra flux in the NUV—light near the star would be scattered by the dust, making the light more blue. (This is different than the effect of dust along the line of sight.) But, no one had suggested that the observed dust mass-loss rates would lead to significant circumstellar extinction around RSGs. We did the math, though (Massey *et al.* 2005b),

† The alert reader may realize that A_V determined by $E(B-V)$ from broad-band colors requires a different effective R'_V for intrinsically red stars than for intrinsically blue stars. We derive our A_V from spectrophotometry, so this issue doesn't apply. Still, if you are trying to do this for RSGs from broad-band photometry, then $R'_V = 4.1 + 0.1 E(B-V) - 0.2 \log g$; see discussion in Massey *et al.* (2005b).

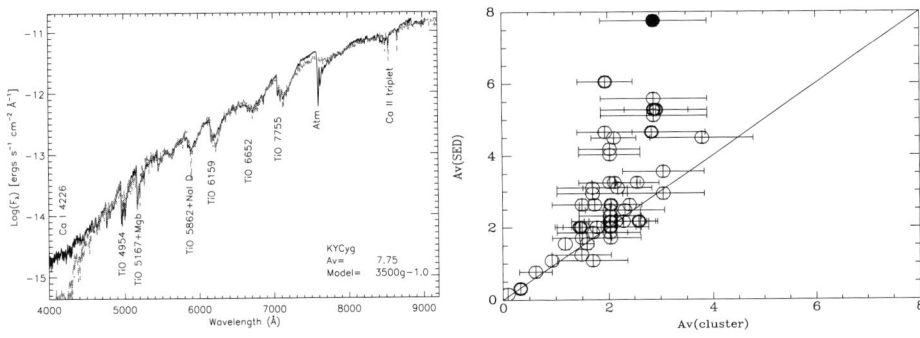

Figure 6. The effects of circumstellar dust? On the left we show the NUV problem for the star KY Cyg. Note that the star (solid line) has far more flux in the NUV than does the reddened model (dashed). On the right the RSG extinction is plotted against that of the OB stars in the same associations. The error bars denote the range of extinction of the OB stars. The filled circle at the top denotes KY Cyg, which has one of the worst NUV problems.

and this really was *exactly* what you would expect: a thin-shell approximation (10 yr of dust mass loss at a rate of 10^{-8} $M_\odot$ yr^{-1} condensing out at distance of ten stellar radii) should lead to > 1 mag of visual extinction.

Exactly how much dust do RSGs contribute to the ISM? In our work, we found a nice correlation between the bolometric luminosity of a star and its dust mass-loss rate. With this we were then able to estimate the amount of dust that RSGs contribute locally, about 3×10^{-8} $M_\odot$ yr^{-1} kpc^{-2}, in good agreement with the value estimated by Jura & Kleinmann (1990). This is probably 3× less than that contributed by late-type WCs in the solar neighborhood, and about 200× less than that contributed by asymptotic giant branch (AGB) stars. So, locally, they don't amount to much as dust producers. However, in a metal-poor starburst, or in galaxies at large look-back time, one would expect RSGs to *dominate* the production of dust, as late-type WCs are not found in metal-poor systems, and AGBs require several Gyr to form.

Before leaving this subject, I'd like to briefly address the subject of RSG mass-loss. We don't know what drives the mass-loss of RSGs: arguments have been presented both for pulsation and for having the dust drive the wind. But, I'd like to quote something my colleague Stan Owocki wrote in an email about all this, contrasting RSG mass-loss with O star mass-loss. The escape velocity from a star is just 620 km s^{-1} × $\sqrt{(M/R)}$. O stars have a M/R ratio that is of order unity, but not RSGs! There the ratio is much smaller, more like 0.02. So, the escape velocity is down by a factor of 7 or so, under 100 km s^{-1}. Stan argued that the mass-loss of a hot star is set by conditions outside the stellar interior, i.e., opacity in the atmosphere and wind, that results in the classic CAK mass loss (Castor, Abbott, & Klein 1975). A RSG, on the other hand, suffers mass loss because the "heavy lifting" has been done by the interior, as a significant fraction of the luminosity of the star has gone into making a bigger radius. So, Stan argues, it is kind of like walking with a nearly full glass of water (RSG) vs a glass that is only 1% full (O star)—even a small jiggle can lead to big changes in the mass loss for a RSG.

6. The Future

Where do we go from here? Our group is working on several projects. One of this is to extend these studies to other metal-poor galaxies, particularly WLM, and see if (for instance) we can find more wacky late-type RSGs like HV 11423 and its friends. Another is to extend this to M31, where the metallicity is 2× solar, at least according to studies of nebular abundances (for instance, Zaritsky, Kennicutt, & Huchra 1994). This brings us to one of our preliminary results. Abundance studies of several M31 A- and B-type supergiants by Venn *et al.* (2000) and Smartt *et al.* (2001) found abundances that were essentially solar, not 2× solar. This is of course confusing, as it flies in the face of everything we know (or thought we knew!) about one of our nearest neighboring galaxies.

We can comment on this briefly. The observed upper luminosity of the RSGs is consistent with that expected based on the 2× solar tracks but not the solar metallicity tracks. This is illustrated in Fig. 7. If the metallicity were truly solar, then where are of the high luminosity RSGs? The ones that would come from 25 $M_\odot$? (Compare also to Fig. 3, upper right.) But the 2× solar models work very well. I was gratified to learn at this conference that Norbert Przybilla finds similarly high abundances for these A-type supergiants in his reanalysis, using the improved photometry that our Local Group Galaxies Survey has provided.

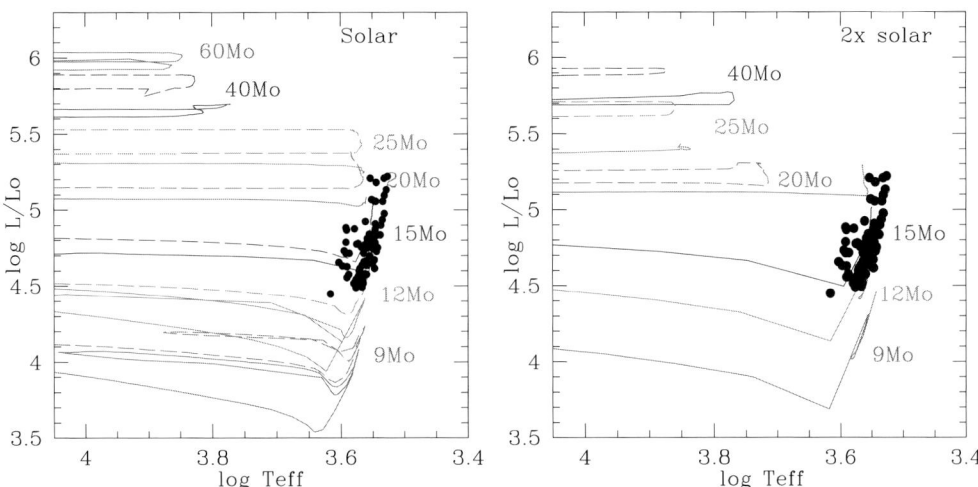

Figure 7. RSGs in M31 compared to solar (left) and 2× solar (right) tracks. The evolutionary tracks used are the same as in Fig. 1. The distribution works well for the 2× tracks, but not the solar tracks, which predict higher luminosity RSGs than what are observed.

Acknowledgements

We thank our colleagues Georges Meynet and Andre Maeder, who co-authored several of the papers we discussed here and who have always been very generous by making their work available to others. Geoff Clayton and David Silva are also working with us. This work is partially supported through the National Science Foundation (AST-0604569).

References

Abbott, D. C. & Hummer, D. G. 1985, *ApJ*, 294, 286
Auer, L. H. & Mihalas, D. 1972, *ApJS*, 24, 193
Cardelli, J. A., Clayton, G. C., & Mathis, J. S. 1989, *ApJ*, 345, 245
Castor, J. I., Abbott, D. C., & Klein, R. I. 1975, *ApJ*, 195, 157
Conti, P. S. 1973, *ApJ*, 179, 181
Crowther, P. A. 2007, *ARA&A*, 45, 177
Elias, J. H., Frogel, J. A., & Humphreys, R. M. 1985, *ApJS*, 57, 91
Freytag, B., Steffen, M., & Dorch, B. 2002, *Astron. Nach.*, 323, 213
Gabler, R., Gabler, A., Kudritzki, R. P., *et al.* 1989, *A&A*, 226, 162
Garmany, C. D. & Stencil, R. E. 1992, *A&AS*, 94, 211
Gustafsson, B., Bell, R. A., Eriksson, K., & Nordlund, Å. 1975, *A&A*, 42, 407
Gustafsson, B., Edvardsson, B., Eriksson, *et al.* 2003, in: I. Hubeny, D. Mihalas, & K. Werner (eds.), *Stellar Atmosphere Modeling* (San Francisco: ASP), *ASP Conf. Ser.*, 288, 331
Herrero, A., Puls, J., & Najarro, F. 2002, *A&A*, 396, 949
Hillier, D., Lanz, T., Heap, S. R., *et al.* 2003, *ApJ*, 588, 1039
Hillier, D. & Miller, D. L. 1989, *ApJ*, 496, 407
Humphreys, R. M. 1978, *ApJS*, 38, 309
Humphreys, R. M. 1979a, *ApS*, 39, 389
Humphreys, R. M. 1979b, *ApJ*, 345, 854
Humphreys, R. M. 1980a, *ApJ*, 238, 65
Humphreys, R. M. 1980b, *ApJ*, 241, 587
Humphreys, R. M. 1980c, *ApJ*, 241, 598
Humphreys, R. M. & Davidson, K. 1979, *ApJ*, 232, 409
Humphreys, R. M. & McElroy, D. B. 1984, *ApJ*, 284, 565

Humphreys, R. M. & Sandage, A. 1980, *ApJS*, 44, 319
Jura, M. & Kleinmann, S. G. 1990, *ApJS*, 73, 769
Kudritzki, R. P. 1976, *A&A*, 552, 11
Kudritzki, R. P. & Hummer, D. G. 1990, *ARA&A*, 28, 303
Lamers, H. J. G. L. M. 1997, in: A. Nota & H. J. G. L. M. Lamers (eds.), *Luminous Blue Variables: Massive Stars in Transition*, (San Francisco: ASP), *ASP Conf. Ser*, 120, 76
Levesque, E. M., Massey, P., Olsen, K. A. G., et al. 2005, *ApJ*, 628, 973
Levesque, E. M., Massey, P., Olsen, K. A. G., et al.. 2006, *ApJ*, 645, 1102
Levesque, E. M., Massey, P., Olsen, K. A. G., & Plez, B. 2007, *ApJ*, 667, 202
Maeder, A. & Meynet, G. 1987, *A&A*, 182, 243
Massey, P. 2003a, *ARA&A*, 41, 15
Massey, P. 2003b, in: K. A. van der Hucht, A. Herrero & C. Esteban (eds.), *A Massive Star Odyssey, from Main Sequence to Supernova* (San Francisco: ASP), *Proc. IAU Symp. 212*, 316
Massey, P., Bresolin, F., Kudritzki, R. P., et al. 2004, *ApJ*, 608, 1001
Massey, P., Levesque, E. M., Olsen, K. A. G., et al. 2007, *ApJ*, 660, 301
Massey, P., Levesque, E. M., & Plez, B. 2006, *ApJ*, 646, 1203
Massey, P., & Olsen, K. A. G. 2003, *AJ*, 126, 2867
Massey, P., Plez, B., Levesque, E. M., et al. 2005b, *ApJ*, 634, 1286
Massey, P., Puls, J., Pauldrach, A. W. A., et al. 2005a, *ApJ*, 627, 477
Meynet, G. & Maeder, A. 2000, *A&A*, 361, 101
Meynet, G. & Maeder, A. 2003, *A&A*, 404, 975
Meynet, G. & Maeder, A. 2005, *A&A*, 429, 581
Meynet, G., Maeder, A., Schaller, G., et al. 1994, *A&A*, 103, 97
Mihalas, D. & Auer, L. H. 1970, *ApJ*, 160, 1161
Plez, B. 2003, in: U. Munari (ed.), *GAIA Spectroscopy: Science and Technology* (San Francisco: ASP), *ASP Conf. Ser.*, 298, 189
Plez, B., Brett, J. M., & Nordlund, Å. 1992, *A&A*, 256, 551
Puls, J., Kudritzki, R. P., Herrero, A., et al. 1996, *A&A*, 305, 171
Smartt, S. J., Crowther, P. A., Dufton, P. L., et al. 2001, *MNRAS*, 325, 257
Smith, N., Humphreys, R. M., Davidson, K., et al. 2001, *AJ*, 121, 1111
Smith, N. & Owocki, S. P. 2006, *ApJ*, 645, 45
Venn, K. A., McCarthy, J. K., Lennon, D. J., et al. 2000, *ApJ*, 541, 610
Young, J. S., Baldwin, J. E., Boysen, R. C., et al. 2000, *MNRAS*, 315, 635
Zaritsky, D., Kennicutt, R. C., & Huchra, J. P. 1994, *ApJ*, 420 87

Discussion

KUDRITZKI: You use the TiO molecular lines as a T_{eff} diagnostic. This depends on an assumption about the oxygen abundances. Which did you use for the Milky Way? And then for LMC, SMC?

MASSEY: Relative to the Sun, we used log $Z=0.0$ for the Milky Way, log $Z=-0.3$ for the LMC, and log $Z=-0.75$ for the SMC, based primarily on the nebular oxygen abundances we know for the latter two. But, the oxygen abundance is actually irrelevant. The amount of TiO you get is controlled by the amount of Ti, not O, because Ti is so much rarer than O. There's good discussion of this point in Plez (2003).

PRZYBILLA: From the analysis of a couple of A-type supergiants in M31, we derive oversolar abundances, about 30-40% above the Grevesse & Sauval (1998, *Spa. Sci. Rev.* 85, 161) values.

MASSEY: Good! When I refer to '2 x solar' I mean log O/H+12 = 9.0 if I'm talking about the nebular abundances, or $Z=0.040$ if I'm talking about the available evolutionary models. If I understood you correctly, you've revised the abundances of the Venn *et al.* A-type supergiants to log O/H+12=9 due to improved estimates of the luminosities. So, maybe the agreement about the metallicity is better than I thought now.

MOFFAT: Exactly a year ago in Carilo celebrating Virpi's 70th birthday, I put you on the spot. Now it is a year later, so may I ask you the same question? Well, open cluster Westerlund 1 has a large number of RSGs yet its turn-off mass is close to $40\,M_\odot$, not $25\,M_\odot$. This would then be an exception to your findings that RSGs come from stars of mass $10\,M_\odot < M_i < 25\,M_\odot$. Do you have an explanation for this?

MASSEY: It would work if the cluster is not strictly coeval.

CROWTHER: One might reconcile the existence of RSGs in Westerlund 1 ($40\,M_\odot$ turn off) with Geneva rotating models at $Z_\odot$ by adopting reduced initial rotation velocities ($100\,\mathrm{km\,s^{-1}}$ cf. $300\,\mathrm{km\,s^{-1}}$ generally adopted). There is no evidence for age differences between RSGs and other massive stars in Westerlund 1.

LEITHERER: Did you include in your sample stars with good Hipparcos distances and/or diameter measurements like α Her, α Sco, or α Ori? That should give you an additional handle on T_{eff}.

MASSEY: We restricted our sample to stars in OB associations, both for the distances and because (naively) I thought we might need to know the typical reddenings from the OB stars. But we did include a few famous RSGs as spectral standards, such as α Her.

Phil Massey.

Norbert Przybilla, Maria Nieva and Florian Schiller.

Atsuo Okazaki.

The Evolutionary State of the Cool Hypergiants – Episodic Mass Loss, Convective Activity and Magnetic Fields

Roberta M. Humphreys

Astronomy Dept., University of Minnesota,
116 Church St. SE., Minneapolis, MN 55455, USA
email: roberta@umn.edu

Abstract. The evolved cool stars near the empirical upper luminosity boundary in the HR Diagram all show evidence for considerable instability perhaps due to their proximity to this stability limit and/or their evolutionary state. Recent high resolution imaging and spectroscopy of several of these stars have revealed a subset characterized by complex ejecta and evidence for episodic mass loss driven by convective activity and magnetic fields. This group includes famous stars such as the red supergiants VY CMa, NML Cyg and the post RSG IRC +10420. I will review the observational evidence and discuss the implications for the final stages of these evolved stars, their mass loss mechanism, and evolutionary state.

Keywords. stars: mass loss – stars: magnetic fields – stars: evolution

---

1. What is a Cool Hypergiant?

A few highly unstable, very massive stars lie on or near the empirical upper luminosity boundary in the HR diagram. These include the Luminous Blue Variables, the cool hypergiants, and even rarer objects, all related by high mass loss phenomena, sometimes violent, which may be responsible for the existence of the upper boundary (Humphreys & Davidson 1979, 1994). In this paper, I use the term 'cool hypergiant' for the stars that lie just below this upper luminosity envelope with spectral types ranging from late A to M. The cool hypergiants very likely represent a very short-lived evolutionary stage, and are distinguished by their high mass loss rates. Many of them also show photometric and spectroscopic variability, and some have large infrared excesses, and extensive circumstellar ejecta.

The evolutionary state of most of these stars is not known. They are all post main sequence stars, but the intermediate–type or "yellow" hypergiants could be either evolving to cooler temperatures or be post-red supergiant (RSG) stars in transition to warmer temperatures. de Jager (1998) has suggested that most if not all of the intermediate–type hypergiants are post-RSGs. In their post-RSG blueward evolution these very massive stars enter a temperature range (6000–9000 K) with increased dynamical instability, a semi-forbidden region in the HR diagram, that he called the *"yellow void"*, where high mass loss episodes occur.

To better understand the evolution of these cool, evolved stars near the upper luminosity boundary and the mass loss mechanisms that dominate the upper HR diagram, we have obtained high resolution multi-wavelength images with HST/WFPC2 of several of the most luminous known evolved cool stars - the M–type hypergiants, μ Cep (M2e Ia), S Per (M3-4e Ia), NML Cyg (M6 I), VX Sgr (M4e Ia–M9.5 I), and VY CMa (M4–M5 Ia) and the intermediate–type (F and G–type) hypergiants, IRC+10420 (A–F Ia) ρ Cas (F8p Ia), HR 8752 (G0-5 Ia) and HR 5171a (G8 Ia). The presence or lack

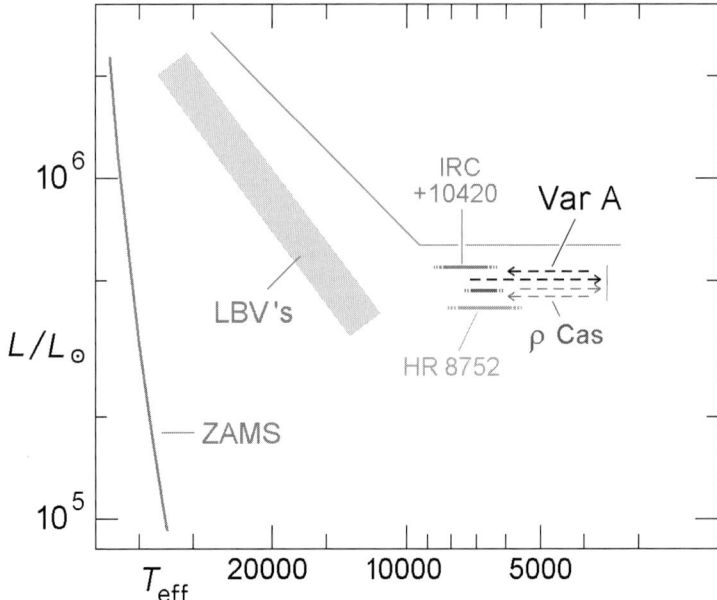

Figure 1. A schematic HR Diagram showing the positions of the intermediate – type hypergiants and their apparent shifts in temperature due to changes in their wind and formation of an optically thick cooler wind.

of fossil shells, bipolar or equatorial ejecta, and other structures in their ejecta will be a record of their current and prior mass loss episodes and provide clues to their evolutionary history. These stars were selected on the basis of their infrared emission, strong molecular emission, or peculiar spectroscopic variations to give us a snapshot of different steps in their evolution across the top of the HR Diagram.

Our results are quickly summarized: we found no detections of visible circumstellar material associated with ρ Cas, HR 8752, HR 5171a and μ Cep; VX Sgr and S Per, both OH/IR sources, were marginally resolved. NML Cyg's (OH/IR source) ejecta has been shaped by its environment and IRC+10420 and VY CMa (OH/IR sources) have extensive and complex circumstellar nebulae.

2. The Yellow or Intermediate–Type Hypergiants

Although ρ Cas is famous for its shell ejection episodes in 1945-47, 1985-87, and 2002, we found no evidence for visible CS ejecta in its HST images and the near-infrared adaptive optics images were also negative (Schuster *et al.* 2008). During these events, it develops TiO bands accompanied by a high but temporary mass loss rate ($\approx 10^{-2} M_\odot$ yr^{-1}) due to the formation of an optically thick wind. It quickly returns to its normal F supergiant type spectrum. Prior to this recent episode, its apparent temperature had been slowly increasing (see Figure 1), but after the event it returned to its former lower temperature and F-type spectrum. The star's photometric variability was indicative of pulsational instability (Lobel *et al.* 2003). Thus its apparent shift on the HR Diagram was not due to evolution but to changes its wind or envelope.

Variable A in M33, while not one of our imaging targets, experienced a similar high mass loss episode with the formation of a much cooler optically thick wind or false-photosphere like ρ Cas, but its event lasted $\approx$ 45 years! Var A was one of the original

Hubble- Sandage variables and one of the visually brightest stars in M33, when in 1950 it rapidly declined 3 mags. or more becoming very red. Its spectrum in 1985-86 was that of an M supergiant and it had a strong IR excess (Humphreys *et al.* 1987). In 2004-05 its spectrum had returned to its prior warmer F-type and it had gotten bluer (Humphreys *et al.* 2006). Like ρ Cas its apparent transits on the HRD are not due to interior evolution, but to changes its wind or photosphere.

2.1. *The Post–Red Supergiant IRC+10420*

IRC +10420 may be one of the most important stars in the HR diagram for understanding the final stages of massive star evolution. Jones *et al.* (1993) combined multi-wavelength spectroscopy, photometry, and polarimetry to confirm a large distance of 4–6 kpc and its resulting high luminosity. With its high mass loss rate and large infrared excess, they concluded that IRC +10420 is a post–red supergiant evolving back toward the blue side of the HR diagram, in an evolutionary phase perhaps analogous to the proto-planetary/post-AGB stage for lower mass stars. IRC+10420 is also the warmest maser source known and in the past 20 years or so its apparent spectral type has gone from late F-type to a mid-A (Oudmaijer *et al.* 1996, Oudmaijer 1998).

HST/WFPC2 images (Humphreys *et al.* 1997) revealed a complex circumstellar environment, with a variety of structures including condensations or knots, ray-like features, and several small, semi–circular arcs or loops within $2''$ of the star, plus one or more distant reflection shells. These features are all evidence for high mass loss episodes during the past few hundred years.

A few other intermediate-temperature hypergiants such as ρ Cas and HR 8752 occupy the same region in the HR diagram, but IRC +10420 is the only one with obvious circumstellar nebulosity, making it our best candidate for a star in transition from a red supergiant possibly to an S Dor–type variable (LBV), a Wolf-Rayet star, or a pre-supernova state. Moreover, its photometric history (Gottleib & Liller 1978, Jones *et al.* 1993) and apparent change in spectral type from late F to mid A indicate that it has changed significantly in the past century. Humphreys *et al.* (2002) obtained HST/STIS spatially resolved spectroscopy of IRC +10420 and its reflection nebula and demonstrated that at its temperature and with its high mass loss rate, the wind must be optically thick. Consequently, like ρ Cas and Var A, its observed variations in apparent spectral type and inferred temperature are due to changes in the wind, and not to interior evolution on such short timescales.

3. The Red Supergiants

Our WFPC2 images revealed visible CS ejecta around all of our M-type supergiant candidates except for the very luminous μ Cep, although near-IR adaptive optic imaging resolved its dust shell at the expected distance (Schuster *et al.* 2008). All of the stars with detectable ejecta are known supergiant OH/IR stars and are strong sources of maser emission.

3.1. *NML Cyg – Interacting with Its Environment*

The HST/WFPC2 images of the powerful OH/IR maser NML Cyg (M6 I) shows an optically obscured star embedded in a small asymmetric bean-shaped nebula (Schuster, Humphreys & Marengo 2006). This small CS nebula is remarkably similar in shape to the 21cm ionized hydrogen (H II) contours $\sim 30"$ from the star. The presence of ionized hydrogen surrounding an M supergiant like NML Cyg was somewhat of an enigma. Morris & Jura (1983) showed that the asymmetric "inverse" H II region was the result of the interaction of a spherically symmetric, expanding wind from NML Cyg and photo-ionization from plane parallel Lyman continuum photons from the luminous, hot stars in the nearby association Cyg OB2 (see Figures 1 and 2 in Morris & Jura). They suggested that the molecular material in the wind is photo-dissociated closer to the star so that it does not shield the atomic hydrogen from the ionizing photons.

Our images show circumstellar material much closer to NML Cyg than the surrounding H II region and coincident with the water masers and SiO masers, suggesting that we are likely imaging the molecular photo-dissociation boundaries. Schuster *et al.* (2006) show that the shape of the envelope seen in the WFPC2 images can be modeled as the result of the interaction between the molecular outflow from NML Cyg and the near–UV continuum flux from Cyg OB2, i.e. analogous to an "inverse Photo-Dissociation Region" (PDR). The water masers show a one-sided asymmetric distribution similar in extent to the reflection nebula, that also matches its convex shape (see Figure 6 in Schuster at al 2006). The dusty cocoon engulfing NML Cyg must be the consequence of high mass loss in the RSG stage, but its envelope has most likely been shaped by its interaction with and proximity to Cyg OB2. If the outflow from NML Cyg is bipolar (Richards *et al.* 1996), then it appears that the molecular material SE of the star is preferentially shielded from photo-dissociation. Even without assuming bipolarity, there is more maser emission to the ESE, consistent with our model for NML Cyg's circumstellar envelope.

3.2. *The Extreme Red Supergiant VY CMa*

The extreme red supergiant and powerful infrared source and OH maser VY CMa is one of the most luminous and largest evolved cool stars known. With its very visible asymmetric nebula, 10" across, combined with its high mass loss rate, VY CMa is a special case even among the cool hypergiants that define the upper luminosity boundary in the HR Diagram. VY CMa is ejecting large amounts of gas and dust at a prodigious rate, and is consequently one of our most important stars for understanding the high mass loss episodes near the end of massive star evolution.

Multi-wavelength HST/WFPC2 images of VY CMa (Smith *et al.* 2001) revealed a complex circumstellar environment dominated by the prominent nebulous arc to the northwest, which is also visible in ground-based data, two bright filamentary arcs to the southwest, plus relatively bright clumps of dusty knots near the star, all of which are evidence for multiple and asymmetric mass loss episodes (Figure 2). The apparent random orientations of the arcs suggested that they were produced by localized ejections, not necessarily aligned with an axis of symmetry or its equator. We therefore speculated that the arcs may be expanding loops caused by localized activity on the star's ill-defined surface.

We subsequently obtained obtained second epoch HST/WFPC2 images to measure the transverse motions which when combined with the Doppler velocities (Humphreys *et al.* 2005) provide a complete picture of the kinematics of the ejecta including the total space motions and directions of the outflows (Humphreys, Helton & Jones 2007). Our results show that the arcs and clumps of knots are moving at different velocities, in different directions, and at different angles relative to the plane of the sky and to the star, confirming their origin from eruptions at different times and from physically separate regions on the star. Independent polarimetry (Jones, Humphreys & Helton 2007) confirms the line of sight orientation of the primary features and together with the kinematics lets us determine the *three-dimensional morphology of the ejecta*.

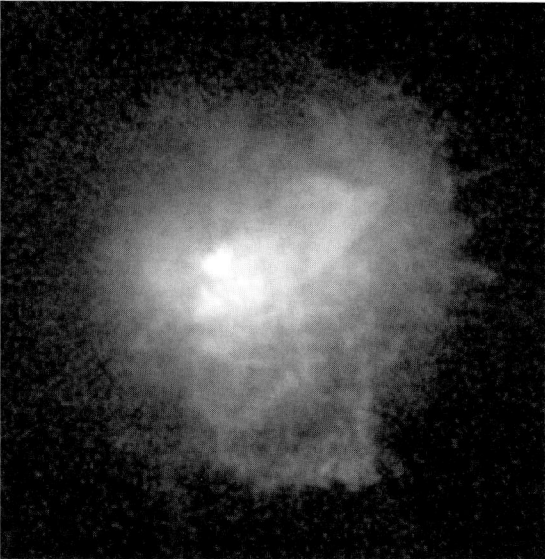

Figure 2. The HST/WFPC2 image of VY CMa showing the multiple arcs and knots in its circumstellar ejecta.

4. Episodic Mass Loss, Convective Activity and Magnetic Fields

The numerous arc-like structures, knots, and filaments in the circumstellar ejecta of VY CMa and IRC+10420 are evidence for multiple, asymmetric mass loss events at different times and apparently by localized processes from different regions on the star. This activity could be attributed to either non-radial pulsations or to magnetic/convective regions and events analogous to solar activity, that is large "starspots". The distinction may be vague for a red supergiant where the convective cells are expected to be comparable to the stellar radius in size (Schwarzschild 1975); although, nonradial pulsations would not be expected to produce the narrow arcs and loops observed in VY CMa. Starspots and large "asymmetries" have now been observed on several stars including red giants, AGB stars and supergiants.

Vlemmings et al. (2002, 2005) have measured the magnetic field strength from the circular polarization of H_2O masers in the ejecta of AGB stars and several evolved supergiants including the strong OH/IR sources VY CMa, VX Sgr, NML Cyg, and S Per. They show that the H_2O masers around VX Sgr can be fit by a dipole magnetic field (Vlemmings et al. 2005). Their analysis supports the Zeeman interpretation of the circular polarization of the SiO masers (Barvainis et al 1987; Kemball & Diamond 1997) only a few AU from the surface of these stars. Together with Zeeman splitting of the OH emission far out in the wind at a few thousand AU (Szymczak & Cohen 1997; Masheder et al 1999), these measurements confirm the presence of a magnetic field throughout the ejecta of these objects. The results for VY CMa imply magnetic fields of $\sim$ 8000G at the star's surface. Similarly, for IRC+10420, the circular polarization of the OH masers (Nedoluha & Bowers 1992) imply fields of $\sim$ 3000G at the surface. This may be high for a global field but perhaps not for large convective cells.

5. The Evolutionary State?

In published papers and at previous IAU symposia I have suggested that these RSGs with the visible CS ejecta represent a short-lived, high mass loss phase. That is very likely the case but we have to take a closer look at these stars and ask why. de Jager (1998) has suggested that the "yellow" hypergiants are all post-RSGs. IRC+10420 fits this model very well, but ρ Cas with its well documented mass loss episodes plus the similar high luminosity stars, HR5171a and HR8752†, show no visible CS ejecta. It is possible that they are not post-RSGs but are still evolving from the blue to the red supergiant region, or alternatively that they have only recently entered this region of dynamical instability.

It must be emphasized that all of these objects are evolved (post-MS) stars of high initial mass and high luminosity. Proximity to the Humphreys-Davidson limit however, does not appear to be the explanation for the presence of extensive CS ejecta; ρ Cas for example is quite close (Figure 1), while S Per is significantly fainter.

So what do these stars with the CS ejecta have in common – VY CMa, IRC+10420, NML Cyg, VX Sgr and S Per? They are all supergiant OH/IR stars with strong maser emission. Most OH/IR sources are AGB stars. Only a few are above the AGB limit and therefore recognized as supergiants or massive stars. I am therefore suggesting that these hypergiant-OH/IR stars have been on a blue loop in the HRD and are now in the RSG stage for the second time, analogous to the AGB stars at lower mass. IRC+10420 could indeed be a star evolving back to the blue for the second time similar to the post-AGB stars. If so, these stars must be very close to the end of their lives.

References

Barvainis, R., McIntosh, G., & Predmore, C. R. 1987, *Nature*, 329, 613
de Jager, C. 1998, *A&AR*, 8, 145
Gottlieb, E. W. & Liller, W. 1978, *ApJ*, 225, 488
Humphreys, R. M. & Davidson, K. 1979, *ApJ*, 232, 409
Humphreys, R. M., Jones, T. J., & Gehrz, R. D. 1987, *AJ*, 94, 315
Humphreys, R. M. & Davidson, K. 1994, *PASP*, 106, 1025
Humphreys, R. M., Smith, N., Davidson, K., et al 1997, *AJ*, 114, 2778
Humphreys, R. M., Davidson, K., & Smith, N. 2002, *AJ*, 124, 1026
Humphreys, R. M., Davidson, K., Ruch, G., & Wallerstein, G. 2005, *AJ*, 129, 492
Humphreys, R. M., Jones, T. J., Polomski, E., et al. 2006, *AJ*, 131, 2105
Humphreys, R. M., Helton, L. A., & Jones, T. J. 2007, *AJ*, 133, 2716
Jones, T. J., Humphreys, R. M., Gehrz, R. D., et al 1993, *ApJ*, 411, 323
Jones, T. J., Humphreys, R. M., Helton, L. A., et al. 2007, *AJ*, 133, 2730
Kemball, A. J. & Diamond, P. J. 1997, *ApJ*, 481, L111
Lobel, A., Dupree, A. K., Stefanik, R. P., et al. 2003, *ApJ*, 583, 923
Masheder, M. R.W., van Langevelde, H. J., Richards, A. M. S., et al. 1999, *New Astron.*, 43, 563
Morris, M. & Jura, M. 1983, *ApJ*, 267, 179
Nedoluha, G. E. & Bowers, P. F. 1992, *ApJ*, 392, 249
Oudmaijer, R. D. 1998, *A&AS*, 129, 541
Oudmaijer, R. D., Groenewegen, M. A. T., Matthews, H. E., et al. 1996, *MNRAS*, 280, 1062
Richards, A. M. S., Yates, J. A., & Cohen, R. J. 1996, *MNRAS*, 282, 665
Schuster, M. T., Humphreys, R. M., & Marengo, M. 2006, *AJ*, 131, 603
Schuster, M. T. 2008, in preparation
Schwarzschild, M. 1975, *ApJ*, 195, 137
Smith, N., Humphreys, R. M., Davidson, K., et al. 2001, *AJ*, 121, 1111
Szymczak, M. & Cohen, R. J. 1997, *MNRAS*, 288, 945
Vlemmings, W. H. T., Diamond, P. J., & van Langevelde, H. J. 2002, *A&A*, 394, 589
Vlemmings, W. H. T., van Langevelde, H. J., & Diamond, P. J. 2005, *A&A*, 434, 1029

† HR8752 has a nearby hot companion star which may prevent the formation of a dusty nebula.

Discussion

OWOCKI: How are the magnetic fields measured?

HUMPHREYS: The magnetic fields were measured from the circular polarization of the H_2O masers, the SiO masers and the Zeeman splitting of the OH masers. The references are given in the paper.

HILLIER: What is the argument that indicates ρ Cas had a shell ejection rather than a expansion of the outer envelope?

HUMPHREYS: "Shell episode" or "ejection" is the terminology used in the literature for the ρ Cas events. Modern observations show that these are really the formation of a cool optically thick wind or false photosphere similar to what is observed in the classical LBVs but for cooler stars.

MAEDER: You are associating these episodic ejections with convection. Could it also be due to pulsation?

HUMPHREYS: Regular pulsation cannot explain these spatially separate episodic ejections. The distinction between large convective cells and non-radial pulsation may be vague on these size scales. But non-radial pulsation would not produce the narrow arcs and loops seen in the ejecta of VY CMa. These visible structures, while resembling prominences, thousands of AU from the star, are not connected to the chromosphere, but are the giant loops or bubbles swept up by large gas outflows extending over several arc secs in the ejecta as revealed by the spectroscopy. They most closely resemble coronal mass ejections.

KUDRITZKI: Steffen and Freytag have done 3D pulsational simulations of α Orionis and show pronounced spot features and deviations from symmetry. Could this explain your observations?

HUMPHREYS: They may be able to simulate a pulsation that produces asymmetries. But remember these are cool stars with large convective envelopes and the magnetic fields are now observed in the ejecta of these stars and several AGBs. In the case of VX Sgr, a dipole magnetic field fits the observations. The magnetic field had to get there somehow.

Roberta Humphreys.

Nolan Walborn and friends.

Massive Binaries

Anthony F. J. Moffat

Département de physique, Université de Montréal,
C.P. 6128, Succursale Centre-Ville, Montréal, QC, H3C 3J7, Canada

Abstract. As with all binaries, those that contain massive stars reveal various degrees of interaction, depending mainly on orbital separation and age, although things happen much faster in massive binaries. Those massive binaries with initial periods exceeding ∼10 years generally only interact via wind-wind collisions, with little or no effect on their subsequent evolution (unless located in dense clusters). Shorter-period systems show even stronger wind-wind collisions as a rule, but also interact more directly via Roche Lobe Overflow or Common Envelope, with dramatic effects on their evolution. If we didn't have binaries among massive stars, we would be missing a whole host of interesting phenomena in the Universe, such as sources of enhanced stellar X-ray or non-thermal radio emission, WR dust-spirals, inverse mass-ratios, very rapid spin, rejuvenation and massive blue-stragglers, enhanced cluster dynamics, many runaways and possibly even SMBHs and GRBs! On the other hand, non(or little)-interacting massive binaries are also useful to provide information on Star-Formation processes and determination of stellar parameters (such as the mass) that would otherwise be difficult or impossible to obtain from single stars. In this review, I highlight some of the developments that have occurred during the past few years since the last IAU Symposium on Massive Stars in 2002.

Keywords. binaries: general – stars: early-type

1. Introduction

1.1. Defining "massive" in binaries

Most researchers at this meeting will agree that "massive" means having an initial mass on the Zero-Age Main Sequence (ZAMS) that leads a star to implode/explode as a core-collapse or pair-instability supernova. In the case of single stars, this means $M_i > 8 M_\odot$ (although this limit may be slightly higher, depending on which stars become white dwarfs (WDs) instead: note the possible impact of the recent discovery of pure-carbon WDs: Dufour *et al.* 2007). Furthermore, in the case of close binaries, the primary could prematurely lose mass via Roche-lobe overflow (RLOF) or common envelope (CE) processes, which could increase the required limit slightly (Vanbeveren *et al.* 1998). The reverse may be true for secondaries that might accrete mass from their primaries. In any case, at least one of the stars in the binary must be massive in the above sense to qualify the binary to be "massive".

1.2. Interacting vs. non-interacting binaries

A crucial distinction among massive binaries is whether the two stars interact or not during their evolution. Starting at the ZAMS, where we assume the binary is non-interacting (otherwise it would not be stable and exist as such for very long), there will be no subsequent interaction affecting either star's evolution for initial periods greater than $P_c \sim 10$ years. The exact critical period will depend on various factors, like eccentricity, masses, spin, etc. On the other hand, colliding winds (CW) will occur even for very large separations or long periods far in excess of 10 years, where X-ray and other emission from the

CW zone will go as $1/D$, where D is the orbital separation. Apart from CWs, the two stars in this case will behave almost as if each was single.

On the other hand, for $P < P_c$, not only will CWs become more intense (but not necessarily following a $1/D$ law for very small D, where the winds could collide at sub-terminal speeds, and other complications arise, e.g. non-adiabatic processes, heating effects), but (normally) the primary star can either fill its Roche lobe and start spilling over to the secondary, or the primary can engulf the secondary in a Common Envelope phase. The former leads (for the most extreme conservative case) to an increase in the separation and the period, while the latter invariably leads to a dramatic decrease in separation, often to a merger.

1.3. *RLOF or CE?*

Whether RLOF or CE ensues after the ZAMS depends on the mass-loss rate of the primary while it is expanding at its Roche lobe. Generally speaking for massive binaries, if the rate exceeds $\sim 10^{-3} M_\odot yr^{-1}$, then CE will be favoured (Vanbeveren *et al.* 1998).

As long as a single massive star can reach the red-supergiant (RSG) phase, in a binary such a star will eventually carry out RLOF if $P < P_c$. This was believed to be the case for stars up to $\sim 40\ M_\odot$; for stars above this limit, their LBV phase guarantees that they will not do RLOF. However, with the revision in the masses of RSGs in the Local Group down from 25-40 to 10-25 $M_\odot$ (Massey *et al.* this volume), implies that there will be no RLOF for most stars in binaries with primary masses above $\sim 25 M_\odot$, with the reasonable assumption that an LBV phase occurs instead of RSG for those initial masses. During the LBV phase, the *average* mass-loss rate is of order $10^{-4} M_\odot yr^{-1}$, depending probably mostly on the mass. However, what really counts is the episodic mass-loss rate, which can be orders of magnitude higher, when the star is also most puffed up. In those cases, huge quantities of mass can be lost from the system in such short-lived bloated stages, as witnessed by the (admittedly extreme) LBV binary η Car (P = 5.54 yrs; Corcoran *et al.* 2007). Unfortunately, little is known about the details of such general behaviour in LBVs, especially concerning the frequency and intensity of their eruptions.

Luckily, though, there are some empirical tests of whether RLOF or CE has taken place. One test, recently completed, is the evaluation of the binary frequency among WR stars in different metallicity environments. WR stars are like the canary in the mine, quite sensitive to metallicity-driven winds leading up to their formation. As one goes from the Solar neighbourhood in the Galaxy ($Z_\odot$) to the LMC ($Z_\odot/2$) and the SMC ($Z_\odot/5$), one expects, even allowing for increased rotation at low metallicity (Z), that the lower limit for the formation of single WR stars will increase systematically (Maeder & Meynet 2000). This means that, assuming a binary frequency that is independent of metallicity and assuming short-period binaries will remove the outer layers of massive stars down to some fixed mass limit independent of Z, one expects to find significantly increasing WR+O binary frequency as one goes from the Galaxy to the LMC and SMC. Schnurr *et al.* (2008) have just finished a comprehensive search for WR binaries among the WNL population of the LMC, completing the previous studies of Foellmi *et al.* (2003a,b) for Magellanic Cloud (MC) WNE stars and Bartzakos et al. (2001) for MC WC stars. (There are no WNL stars in the SMC.) Using the very laborious but necessary technique of searching for periodic RV varitions as signatures of WR + O binaries, we now have a complete, uniform study of all 144 WR stars in both MCs.

Table 1 summarizes the results for these stars, except for the 6 bright WNL stars in R136 (see below), compared to all the well-studied WR stars in the solar neighbourhood. This table shows *no compelling trends of binary frequency with metallicity* in any WR subgroup. (As an aside, it is curious to note that WN8 and WN9 stars appear to avoid

Table 1. Wolf-Rayet binary frequency (P<200d).

Group	Galaxy[1] ($Z_\odot$)	LMC ($Z_\odot/2$)	SMC ($Z_\odot/5$)
WN	9/30 = 0.30	16/102 = 0.15	4/11 = 0.36
WC/O	7/38 = 0.18	3/24 = 0.13	(1/1 = 1.00)
WR	16/68 = 0.24	19/126 = 0.15	5/12 = 0.42
WN8,9	0/4 = 0.00	0/8 = 0.00	not present

Notes:
[1] for d<4 kpc and 6.5<R(kpc)<9.5, assuming R_o = 8 kpc.

binaries, but they do tend to be runaways, both suggesting a peculiar formation mechanism.) Although the metallicity only varies over a factor ~5 here, the Z-dependence for minimum masses to become single WR stars is significant. We conclude that RLOF is not a significant factor in the formation of WR stars, contrary to what has been believed in the past (Vanbeveren *et al.* 1998). This may appear strange, especially given the reduction in mass-loss rates of O stars, the immediate progenitors of WR stars, by a factor ~3 (just like for WR stars: Moffat 2008). So how do WR stars form? One possibility is that the crucial phase to bring an O star to become WR, binary or single, is the LBV phase, where enormous quantities of mass are lost, possibly via continuum driving independently of metallicity (Smith & Owocki 2006).

Another prediction of RLOF is that after RLOF, all upper layers of the WR progenitor are removed, leaving zero H in the final WR star. Foellmi *et al.* (2006) have studied the eclipsing WN3(h)a + O5 LMC binary BAT99-129, with P = 2.77 d. The WR component here has H/He ~1 by number, making it highly unlikely that RLOF has occurred (however, see Langer & Petrovic 2007).

We take all of this to mean that, although RLOF can occur in some very tight massive systems, it is not generally common in systems that lead to most WR binaries. CE may be much more common, explaining the existence of numerous short-period WR + O systems. Of particular interest are the two WC4 + O systems Br 31 and Br32 in the LMC, with P = 3.0d and 1.9d, resp. (Moffat *et al.* 1990). RLOF is unlikely responsible for short periods like this in such highly evolved binaries; rather, CE evolution, probably during the LBV phase, is the most likely source.

2. Statistics

Still of considerable relevance is the O-star binary frequency study of Mason *et al.* (1998), who found an essentially constant binary frequency for short-period systems below $P \sim 1$ month and above $P \sim 10^4$ yrs. In between there is a wide, deep dip in numbers, where the usual techniques of spectroscopic RV-orbit and visual-binary determination are difficult. In the likely case that the gap will eventually be filled in, it thus appears very likely that Öpik's power-law also applies to O binaries for all separations, i.e. $f(s)ds \propto s^{-1}ds = d\log_e(s)$, and the same flat law in log P, since Kepler's 3rd law has power-law coupling between P and s. Furthermore, the overall initial binary frequency of O stars is probably close to 100%, at least in clusters where all O stars are believed to be formed, with the additional trend that short-period systems favour mass ratios close to order unity. Unfortunately, star-formation theory including binaries (e.g. Bonnell 2007) is unclear how to reproduce these binary trends for massive stars.

If we now adopt Öpik's law as being the same for all (high) masses, then we will have a binary frequency on the MS $N(P) = kd(\log P)$, with k = const, for periods lying in the range $P_{\min} < P < P_{\max}$. Then, assuming that all (massive) binaries with $P < P_{\rm sp}$ can be detected with current spectroscopic means, then the ratio of that number to the total will be $R = \log(P_{\rm sp}/P_{\min})/log(P_{\max}/P_{\min})$. Thus, taking reasonable values $P_{\min} \sim 1$ d, $P_{\rm sp} \sim 1$ yr and $P_{\max} \sim 10^7$ yr (corresponding to a separation of ~ 1 pc in a cluster where most massive stars are born), yields R ~ 0.3. This ratio is much like the typical spectroscopic binary frequency of massive stars observed in young clusters (Sana *et al.* this volume) and is thus compatible with $\sim 100\%$ binary frequency in the birth process of massive stars.

Although it will take some time, attempts are well underway to fill the Mason *et al.* gap (if real) for intermediate periods (e.g. Gamen et al. 2008; Maiz Appellaniz *et al.* 2008). Of particular interest is the overlap between spectroscopic *and* visual binaries (leading to precise distances and masses), which currently is essentially negligeable. However, there is one case that deserves special mention: the recent detection of a visual orbit for the closest known WR star, γ^2 Vel, that happens to be in a well-known, 78.53d-spectroscopic binary (North et al. 2007)-see Fig. 1. A revised and more precise distance of 336^{+8}_{-7} pc was obtained, compared to 258^{+41}_{-31} pc from Hipparcos (Schaerer *et al.* 1997, van der Hucht *et al.* 1997). The O-star radius $R_O = 17 \pm 2 R_\odot$ is now compatible with an O7.5II-III star, compared to O7.5III-V with Hipparcos. The masses are 28.5 ± 1.1 and 9.0 ± 0.6 M$_\odot$ for the O and WC8 components, respectively. *This is the first time ever that a complete visual orbit has been obtained for a WR+ O binary, finally opening the door to obtaining truly reliable parameters to constrain WR-star models.*

3. Life without massive binaries

Massive binaries are not just twice as interesting as single stars. Due to a whole host of interaction effects, some even at relatively large separations, massive binaries lead to a large variety of phenomena in the Universe that would simply be lacking if massive binaries did not exist. Inspired by a recent workshop devoted to Massive Stars in Interacting Binaries (MSIB, St-Louis & Moffat 2007), here are ten grand ways that life in the Universe would be less interesting without massive binaries:

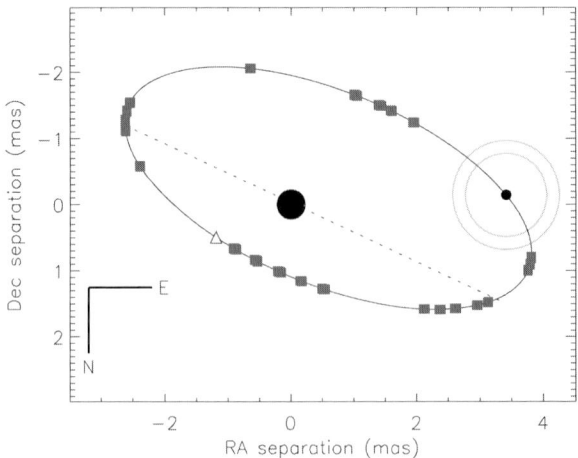

Figure 1. Visual orbit of γ^2 Vel relative to the O star (from North *et al.* 2007).

3.1. Enhanced stellar X-rays

In very young stellar clusters, the most luminous X-ray sources are often massive binaries. A good example of this is the Carina Nebula (Corcoran et al. 1995), in which the dominating X-ray sources are η Car, WR25, HD 93250, HD93205 and HD 93129, all relatively close binaries in which the X-ray flux enhancement is due to colliding winds (CW). The clusters in the Carina Nebula are too young to have already made X-ray binaries with accreting compact companions. The mutual interaction of all the hot-star winds in young clusters also leads to a diffuse X-ray background, best seen in the very dense, 1 Myr cluster NGC 3603 (Moffat et al. 2002).

CW effects have been found to be especially important in three key systems. Nazé et al. (2007) determined the first X-ray light curve in an extragalactic binary system, HD 5980 (LBV + WNE, P = 19.3d, e = 0.3, with deep optical eclipses): the CW X-ray flux reaches a maximum when the star with the faster, less-dense wind (the WNE companion) passes in front of the 1994 LBV-erupter, as expected. Corcoran et al. (priv. comm.) have compared the X-ray light curves of η Car (LBV + WNE, P = 5.54 yr, e $\sim$0.9) and WR 140 (WC7pd + O5I, P = 7.94 yr, e = 0.88): both systems show a general $1/D$ dependence (although asymmetric), broken by deep absorption due to eclipses when the WNE star (wide dip in η Car) or the O5 star (narrow dip) passes in front of the primary star and eclipses most of the CW X-ray emitting zone. **I call attention that both η Car and WR 140 will be going through their very active periastron passages, lasting only a few months, centred on January 2009, the International Year of Astronomy! During and around that interval, various observing campaigns using different techniques are being organized.**

In slightly older clusters, high-mass X-ray binaries (HMXRB) and X-ray emitting supernova remnants (SNR) begin to appear, as the more massive stars complete their evolution. In general, the most frequent type of HMXRB are those containing Be + NS (Negueruela 2007). Some recent X-ray accretion highlights are the following:

• FUSE observations have revealed the Hatchett-McCray ionization effect for the first time in the HMXRB with the highest-mass OB star known, HD 153919, O6Iaf + NS or BH (associated with 4U1700-37) (Iping & Sonneborn 2007).

• After the only reliably established WR + cc system so far, Cyg X-3 (P = 4.8h) in the Galaxy, two other similar cases have now been discovered, each with orbital periods just over 30d: IC 10 X-1 (Bauer & Brandt 2004) and NGC 300 X-1 (Carpano et al. 2007). Their rarity is reflected in the fact that one had to go well beyond the Galaxy to find other cases.

• A TeV gamma-ray light curve has been obtained for the first time for a HMXRB, that of the microquasar LS 5039, with a 3.9d orbital modulation as seen at other more conventional energies (Aharonian et al. 2006)-see Fig. 2. Phase-dependent inverse-Compton scattering is believed to be the source.

3.2. Non-thermal radio emission

The fraction of non-thermal (NT) radio emitters among OB and WR stars is quite high, of order $\sim$30%. Since single stars are believed to produce only thermal radio emission arising in the $\sim$100 km/s shocks associated with turbulent clumping, those systems that have an additional NT component are believed to be binaries in which the NT emission arises in the strong CW shocks between the winds (at $\sim$1000 km/s) of the two stars (Dougherty & Williams 2000). This has been borne out recently in many systems, both spatially resolved and unresolved, the latter by monitoring the radio flux and its spectrum around the orbit. Probably the most impressive result was obtained for WR140, whose multi-frequency radio emission has now been resolved using VLBA (Dougherty et al.

2005). The time-dependent behaviour of the bean-shaped NT emission coming from the CW bow head at phases when the radio emission is not attenuated, has been combined with the optical orbit (Marchenko *et al.* 2003) to deduce the complete orbit of WR140.

3.3. *Dust spirals*

Starting with WR104, WC9d, P = 241.5d (Tuthill et al. 1999, Tuthill *et al.* 2008), many dust-spirals have now been resolved in the NIR, always around pop I WC stars, either of type WC9 (and a few WC8; possibly always binary?) or other hotter WC types (always binaries). Recently, two more dust spirals have been imaged in the Quintuplet cluster (Tuthill *et al.* 2006); possibly all of the 5 extremely red stars in this cluster (from which it derives its name) are in fact dust spirals, some better resolved than others. In all cases, it is believed that the amorphous carbon-dust is formed by successive nucleation processes downstream in the CW shock cone of some massive binaries containing a WC star, where conditions are conducive to its formation.

Of particular interest in the case of WR140 from the VLBA data (see above) is that at periastron passage, the O star is located NW of the WR component on the sky, in a direction at odds with where one sees the bulk of the CW-produced dust emission to occur ($\sim$SE). This surprising result suggests that conditions around the stars may be too hot to produce significant dust around periastron passage; most of the dust is probably produced at phases intermediate between periastron and apastron (Williams *et al.*, in prep.).

Spectacular images have now been secured in the MIR for several dust spirals by Marchenko & Moffat (2007). Of special note is the system WR112, WC9d, surrounded by 5 distinct spirals, believed to be due to the dust formed and carried out with the general shock-cone flow, during successive orbits - see Fig. 3. However, both here and in other objects (e.g. WR140; WR48a, WC8ed, P >20 yrs), the spirals are broken up into several (4?) arcs, each homologously repeating faithfully as the dust spiral expansion proceeds. One mechanism to explain this is currently being studied by Hervé et al. (in prep.), involving axisymmetric winds (with stronger and faster winds at the poles: Owocki 2004) whose rotation axes are misaligned both with each other and with the orbital binary axis of revolution. This can in principal produce two pairs of directions of enhanced CW effects including dust production and emission, with arbitrary angle between the two pairs.

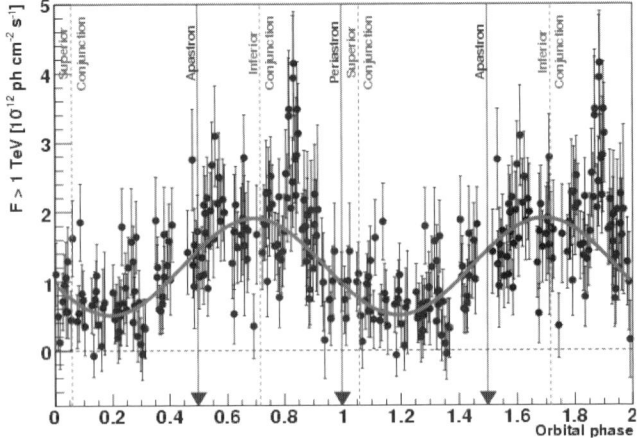

Figure 2. TeV light curve of microquasar LS 5039 (from Aharonian *et al.* 2006).

3.4. Inverse mass ratios

For the 7 next grand ways of what we would miss without massive binaries, I will necessarily be brief, partly because it is beyond my own expertise, but also because these are discussed in more detail in the MSIB workshop (St-Louis & Moffat 2007).

Besides the classic case of β Lyr, where the mass-losing star in the interacting binary is the less massive, a more recent and equally interesting example is RY Scuti, which appears to be going through CE evolution on the way to become a close WR + O binary (Smith et al. 2002; Grundstrom *et al.* 2007). Currently, the less massive star in RY Scuti is the more active mass-losing component.

3.5. Very rapid spin

Several types of stars may well owe their extreme spins to massive binary evolution: millisecond pulsars, Be stars and Gamma-Ray Bursts (see MSIB).

3.6. Rejuvenation, blue stragglers

The colour-magnitude diagrams (CMDs) of some young open clusters often reveal the presence of member stars that are located on younger isochrones to the blue side of the bulk of the cluster Main Sequence (e.g. Ahumada & Lapasset 2007). However, in extremely young clusters, like NGC 3603 (age 1 Myr), the upper MS is essentially vertical (especially in observed colour indices), so one should talk not of Blue Stragglers, but rather "Luminous Stragglers", i.e. very massive stars that might have been rejuvenated to higher luminosities, without becoming significantly bluer, by binary merger or "cannibalism". Another important question is whether rejuvenated ("second phase") WR stars exist in as large numbers as claimed by Vanbeveren *et al.* (1998) via binary separation. Presumably, most of them become single runaways, that would populate the field between their preferred formation places in young star clusters. However, the runaway nature of field WR stars remains to be systematically demonstrated.

3.7. Enhanced dense-cluster dynamics

Massive stars are known to have large cross sections when it comes to stellar collisions in dense systems. This is even more the case when the massive stars occur in binaries.

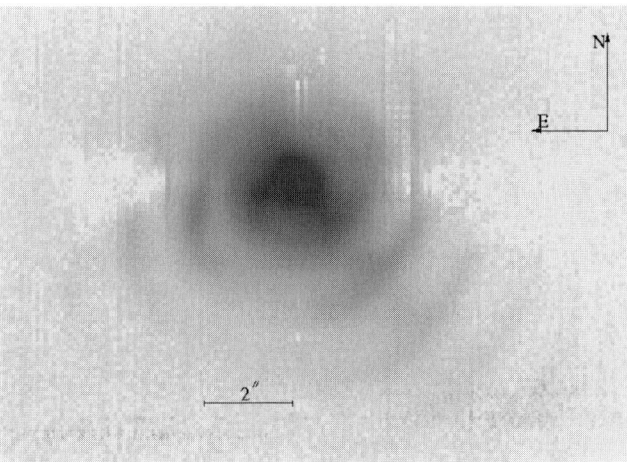

Figure 3. Mid-IR image of dust spirals around WC9d star WR112 (from Marchenko & Moffat 2007).

3.8. *Runaways*

In contrast to lower-mass stars, O stars have a relatively high frequency (∼25%) of being runaways, most likely leading to their presence in the field. Two independent processes probably operate to produce massive runaways, that are mostly single stars: SN in a close binary and the cluster "slingshot" effect (see MSIB). The fastest measured runaway WR star still appears to be WR124, with a peculiar radial velocity component of ∼200 km/s (Moffat *et al.* 1982).

3.9. *IMBHs and possibly even SMBHs*

Fabbiano (2006) defines ultra-luminous X-ray sources (ULX) as those having $L_X > 10^{39}$ erg s^{-1}. If this arises in a single source, then that object must be a black hole (BH) with mass > 100 M$_\odot$, i.e. an intermediate-mass BH, compared to stellar-mass BHs. Such IMBHs would bridge the gap between stellar (mass < 30 M$_\odot$) and supermassive BHs found in the centres of large galaxies, both in mass and in X-ray spectral hardness. Perhaps IMBHs are precursors to SMBHs. IMBHs could be either the collapsed remnant of a massive primordial star in the early Universe, or formed in the core-collapse of a young, dense star-cluster, ultimately facilitated by massive-binary effects. If these ULXs are not IMBHs, then they could represent the upper limit of a normal XRB population.

3.10. *Gamma Ray Bursts*

While the best candidates for the short, hard GRBs are merging NS + NS or NS + BH (not BH + BH, however) binaries, the more intense, long, soft bursts appear to involve the collapse via SN Ic of a rapidly spinning WR progenitor (core) most likely of type WC/WO. The short, hard GRBs involve massive-binary progenitors, while the long, soft GRBs might also, in order to sufficiently spin up the collapsing star. WC/O + O binaries with initial periods $P < $ 3-4d might meet the bill, given that no single WR star is known currently to be spinning fast (van den Heuvel 2007). As noted in Section 1, such binaries do exist, e.g. Br 31 and Br32 in the LMC are WC4 + O6 systems with periods of 3.0d and 1.9d (Moffat *et al.* 1990). Recent evidence of the alternative scenario of single WC/WO stars that remain spinning fast at low metallicity shows that this bias cannot be strong, if it exists at all (Fynbo *et al.* 2006).

4. Physical parameters of stars

Binaries are excellent tools for determining various stellar parameters because they act as yardsticks. I will concentrate here on the recent exciting example of the quest for the *masses* of the most massive stars in the local Universe. This is important for the following reasons:
- to fix the L(M) relation, which should flatten out at high M;
- to learn about the physics of MS stars near their Eddington limit;
- to provide links to population III stars, expected to be on average at least an order of magnitude more massive than today's average;
- to help decide whether the upper mass-limit for stars is due to the star-formation process or petering out of the IMF.

The bottom line appears now to be that *the most massive stars in the Local Universe are of type WN5-7ha*. These are not classical He-burning WR stars; they are rather often found to be the most luminous H-burning MS stars known. Luckily, such stars, although extremely rare, are intrinsically very luminous, allowing them to be found in a larger volume. Because of this and the huge winds they produce, their orbital inclinations can be obtained when in binaries, via four different time-dependent techniques: (1) (normally double per orbit) photospheric eclipses, (2) single-dip atmospheric (continuum) eclipses (Lamontagne et al. 1996), (3) broadband linear polarization (St-Louis et al. 1988), and (4) wind-wind collisions (Hill et al. 2002). This has advantages when it comes to determining stellar masses.

Where does one find such stars? According to some star-formation scenarios, the most massive stars tend to lie near the dense cores of rich clusters. However, the true situation may be more complicated, since one does find some of the earliest (thus massive) stars in much looser, young associations and outside the central dense parts of clusters/associations. A good example of this are the two bright WN6h stars in 30 Dor, R144 (believed to be the brightest WR star in the Local Group: Crowther, priv. comm.) and R145 (see below), that are located well away from the central R136 core. Another example is the rich but sparse association Cyg OB2, which contains some 150 OB and WR stars. It is also a good idea to look among the brightest MS stars of the youngest dense clusters, since evolved stars have their own complications.

4.1. NGC 3603/A1, C

The best example of a dense, young, rich cluster in the Galaxy is NGC 3603, with age ~1 Myr (Brandl et al. 1999; Melena et al. 2008) and the bulk of its stars within a radius of ~30" (1 pc for a distance of 7 kpc). Its HST-based RI CMD (Drissen 1999) reveals a tight upper MS, with 3 WN6ha stars (A1, B, C) located on average a full magnitude above the brightest cluster O stars, reaching the earliest known types of O2-3. Since these H-rich ($X_H \sim 0.6$: Crowther & Dessart 1998) WNL stars are now believed to have similar T_{eff} compared to O2-3 stars (Crowther, priv. comm.), this makes the WNL stars in NGC 3603 very luminous and hence potentially very massive. They also lie within 2" (0.06 pc) of the cluster centre.

Schnurr et al. (in prep.) have now obtained repeated (22 visits over 8 months during 2005) K-band spectra of the 3 WN6ha stars and two central bright O3 stars in NGC 3603, using IFU + AO in SINFONI at the VLT. Without such high spatial resolution, this would have been impossible in the tight core of NGC 3603. Schnurr et al.'s spectra confirm the binary nature of star A1 (with two unequally bright WN stars), for which deep double eclipses in a P = 3.7724d orbit were found previously from HST/NICMOS photometry (Moffat et al. 2004), leading to an orbital inclination $i = 71^o$. Combined with the RV orbits, this yields masses of 114±30 and 84±15 M$_\odot$ for A1. (Further observations are underway with the hope to improve on this estimate.) Then, while star B was found to have constant RV over a year, star C appears to be a short-period (9d) single-line spectroscopic binary with significant eccentricity (e = 0.3). With $L_X > 4 \times 10^{34}$ erg s^{-1}, star C has the highest known X-ray flux of any Galactic WR star, apart from the extremely eccentric-orbit system WR140 at X-ray maximum. Even A1 has $L_X > 10^{34}$ erg s^{-1}, while star B only has $L_X = 10^{33}$ erg s^{-1}. All of this is compatible with CW-enhanced X-rays, although it may appear surprising that WR binaries of such short period yield such high X-ray fluxes. The reason for this is probably that A1 and C contain MS WNLh stars, quite different from the more compact, classical dense-wind (although not necessarily greater mass-loss rate) He-burning WR stars.

4.2. Westerlund 2/WR20a

Another luminous star that has turned out to have high masses is WR20a, with identical, eclipsing components WN6ha + WN6ha in a P = 3.675d circular orbit (Rauw et al. 2004,

2005; Bonanos et al. 2004). It is a member of a moderately dense, young cluster, also containing early-O type stars, Westerlund 2, although WR20a is not located near its core. The masses are 83 ± 6 and 82 ± 6 M$_\odot$, which before NGC 3603/A1, made them the most massive stars ever "weighed" in a binary. The position on the HRD of these two stars also fits well the isochrones for 77 ± 4 M$_\odot$, although the models used for comparison do not include rotation.

4.3. Carina Nebula/WR25

The Carina Nebula and its several young clusters of varying ages also contains three WNLh stars (WR22, 24, 25), much like those in NGC 3603. However, none of these 3 stars is located at or near the core of any of the Carina Nebula's clusters. On the other hand, as in NGC 3603, two of the Carina Nebula's WNLh stars are binary, although of much longer period, while the third appears to be single (WR24). While WR22's WNLh component is of "only" moderately high mass (55 M$_\odot$ according to Schweickhardt et al. 1999; 72 M$_\odot$ from Rauw et al. 1996; P=80.3d), WR25 now finally has a long-awaited orbital solution (Gamen et al. 2006; Gamen et al. 2008): P = 207.7d, e = 0.56, M $\sin^3 i$ = 75 ± 7, 27 ± 3 M$_\odot$. However, the only source of orbital inclination so far is from WR25's X-ray light curve (Pollock 2008), with $i = 37 \pm 2^\circ$, which yields implausibly high absolute masses. For comparison, Hamann et al.'s (2006) atmospheric models for WR25 yield $L = 10^{6.8}$L$_\odot$, $T_\star = 50kK$, making it much closer to η Car in the HRD. The modeled mass is either 110 or 210 M$_\odot$, depending on the adopted distance [(m-M)$_o$ = 11.8 (à la Walborn) or 12.55 (à la Massey)].

4.4. 30 Dor/outside R136

In his detailed spectroscopic search for, and study of, all the WNL binary stars in the LMC, Schnurr (2007) (see also Schnurr et al., in prep.) finds that two stars stand out as being especially luminous and thus potentially very massive, both in the 30 Dor region (although not in the central dense cluster R136): R144 and R145. Both are WN6h but, as noted above, R144 is considered to be the most luminous WR star in the Local Group. With a likely long orbital period, it is still under study. R145 on the other hand, is about 0.5 mag fainter than R144 in intrinsic brightness, but now possesses a plausible RV orbit, with M $\sin^3 i$ = 140 ± 37, 59 ± 26 M$_\odot$ for its two components. However, with $i = 40 \pm 6^\circ$ from broadband linear polarimetry, one again obtains implausibly high masses. Furthermore, this system shows strong phase-dependent variation in CW excess emission in its HeII 4686 emission line, which appears to be compatible with this low-inclination value. The source of the problem of high masses (e.g. in the shift-and-add technique to extract the RV orbit of the faint companion?) is under investigation.

4.5. 30 Dor/R136

Schnurr et al. (in prep.) have also monitored for RV variations all of the 6 most luminous (all WNL-like) stars within R136 (i.e. R136a1, a2, a3, a5, b, c). Based on 9 spectra for each object spread over 22 days using VLT/SINFONI in NIR/AO spectroscopic mode, they find that none of these stars shows RV variations above the $\sigma \sim 20$ km/s level (worse for a5, which has weak lines). This is in contrast with the expectation of Moffat et al. (1985) that one component in R136 revealed (heavily diluted) binary motion with P = 4.377d. This result, although subject to small numbers, is in stark contrast with the central WNL stars in NGC 3603, where 2/3 are short-period binaries, and the Carina Nebula, where 2/3 are moderately long-period binaries. In addition, none of these 6 stars in R136 shows any evidence for runaway motion.

Table 2. Summary of luminous WN5-7h(a) stars (L>L$_\odot$) with Keplerian mass estimates.

Star	Spectrum	P(d)	e	i(^{o})	Msin^{3}i (M$_\odot$)	M(M$_\odot$)	L$_X$ (erg s^{-1})
Carina Nebula							
WR22	WN7h+O9V-III	80.3	0.60	90	55/21	55/21	3 10^{32}
WR25	WN6ha+O	207.7	0.56	37	75/27	(344/124)	8 10^{33}
NGC 3603							
A1	WN6ha+WN6:	3.8	0	71	96/71	114/84	>1 10^{34}
C	WN6ha+?	8.9	0.3	?	?	?	>4 10^{34}
Westerlund 2							
WR20a	WN6ha+WN6ha	3.7	0	74	74/73	83/82	8 10^{33}
30 Dor							
R144	WN6h	?(long)?	?	?	?	?	2 10^{33}
R145	WN6h+O	158.8	0.68	40	140/59	(519/218)	1 10^{33}

Notes:
Masses are for the primary/secondary, resp.

4.6. Summary

Table 2 shows a summary of the 7 most likely candidates for the highest mass among LG stars. All of them are of type WN5-7h or WN5-7ha. Among O stars, none has been found to exceed ~60 M$_\odot$ based on Keplerian orbits (Williams *et al.* 2008). Even the most luminous known O star, HD 93129A, O2If*, a visual binary, has masses *estimated* to be ~80 and 50 M$_\odot$ (Nelan et al. 2004). These WN5-7h/ha stars are not "classical" He-burning WR stars; rather they are a kind of super-Of star, whose optical emission lines are due to the extremely high stellar luminosity. Some are very young (~1 Myr), so they appear to represent an extension of the upper MS above that occupied by the hottest known O stars of types O2-3.

Why were these stars overlooked before as being the most massive? A likely reason is that with improved atmosphere models including blanketing effects, the hottest MS O stars have become ~10% cooler and thus fainter, while WNLh stars have become hotter and brighter (Crowther, priv. comm.)-see Fig. 4. Both seem to converge to $T_{\rm eff} \sim 45$ kK.

None of these stars gives a convincing Keplerian-based mass estimate that surpasses ~

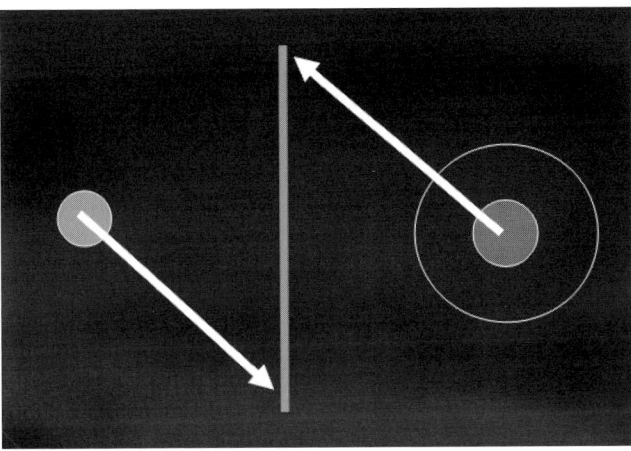

Figure 4. Schematic illustration in a CMD how models of luminous WNLh stars (right) have become hotter and brighter, while early-O stars (left) have become cooler and fainter, both converging to T$_{\rm eff} \sim 45$ kK.

150 $M_\odot$. This is entirely compatible with other independent estimates of the upper limit of star formation in the local Universe, based on number statistics of luminous stars in massive, young clusters (Figer 2005; Weidner & Kroupa 2004; Oey & Clark 2005; Koen 2006).

5. Final remark

As in most areas of Astronomy these days, the study of massive stars is currently enjoying a particularly active and exciting epoch of discovery and relevance to the Cosmos as a whole. This is especially true for massive binaries!

References

Aharonian, F., Akhperjanian, A. G., Bazer-Bachi, A. R., et al. 2006, *A&A*, 460, 743
Ahumada, J. & Lapasset, E. 2007, *A&A*, 463, 789
Bartzakos, P., Moffat, A. F. J., & Niemela, V. S. 2001, *MNRAS*, 324, 18
Bauer, F. E. & Brandt, W. M. 2004, *ApJ*, 601, 67
Bonanos, A. Z., Stanek, K. Z., Udalski, A., et al. 2004, *ApJ*, 611, L33
Bonnell, I. A. 2007, St-Louis, N. & Moffat, A. F. J. (eds.), *Massive Stars in Interacting Binaries*, (San Francisco: ASP) *ASP Conf Series*, 367, 303
Brandl, B., Brandner, W., Eisenhauer, F., et al. 1999, *A&A*, 352, 69
Carpano, S., Pollock, A. M. T., Wilms, J., et al. 2007, *A&A*, 461, L9
Corcoran, M. F., Swank, J., Rawley, G., et al. 1995, *Rev. Mex. A&A Conf. ser.*, 2, 97
Corcoran, M. F., Hamaguchi, K., Gull, et al. 2007, St-Louis, N. & Moffat, A. F.J. (eds.), *Massive Stars in Interacting Binaries*, (San Francisco: ASP) *ASP Conf Series*, 367, 249
Crowther, P. A. & Dessart, L. 1998, *MNRAS*, 296, 622
Dougherty, S. M. & Williams, P. M. 2000, *MNRAS*, 319, 1005
Dougherty, S. M., Beasley, A. J., Claussen, M. J., et al. 2005, *ApJ*, 623, 447
Drissen, L. 1999, in: K. A. van der Hucht, G. Koenigsberger & P. R. J. Eenens (eds.), *Wolf-Rayet Phenomena in Massive Stars and Starburst Galaxies*, (San Francisco: ASP), *Proc. IAU Symp*, 193, 741
Dufour, P., Liebert, J., Fontaine, G., & Behara, N. 2007, *Nature*, 450, 522
Fabbiano, G. 2006, *ARA&A*, 44, 323
Figer, D. F. 2005, *Nature*, 434, 192
Foellmi, C., Moffat, A. F. J., & Guerrero, M. A. 2003a, *MNRAS*, 338, 360
Foellmi, C., Moffat, A. F. J., & Guerrero, M. A. 2003b, *MNRAS*, 338, 1025
Foellmi, C., Moffat, A. F. J., & Marchenko, S. V. 2006, *A&A*, 447, 667
Fynbo, J. P. U., Starling, R. L. C., Ledoux, C., et al. 2006, *A&A*, 451, L47
Gamen, R., Gosset, E., Morrell, N., et al. 2006, *A&A*, 460, 777
Gamen, R., et al. 2008, in: P. Benaglia, G. Bosch & C. E. Cappa (eds.), *Massive Stars: Fundamental Parameters and Circumstellar Interactions*, *Rev. Mex. A&A Conf. ser.*, in press
Grundstrom, E. D., Gies, D. R., Hillwig, T. C., et al. 2007, *ApJ*, 667, 505
Hamann, W.-R., Gräfener, G., & Liermann, A. 2006, *A&A*, 457, 1015
Hill, G. M., Moffat, A. F. J., & St-Louis, N. 2002, *MNRAS*, 335, 1069
Iping, R. C., & Sonneborn, G. 2007, St-Louis, N., & Moffat, A. F. J. (eds.), *Massive Stars in Interacting Binaries*, (San Francisco: ASP) *ASP Conf Series*, 367, 459
Koen, C. 2006, *MNRAS*, 365, 590
Lamontagne, R., Moffat, A. F. J., Drissen, L., et al. 1996, *AJ*, 112, 2227
Langer, N., & Petrovic, J. 2007, St-Louis, N. & Moffat, A. F. J. (eds.), *Massive Stars in Interacting Binaries*, (San Francisco: ASP) *ASP Conf Series*, 367, 359
Maiz Apellaniz, J., Walborn, N. R., Morrell, N. I., et al. 2008, in: P. Benaglia, G. Bosch & C. E. Cappa (eds.), *Massive Stars: Fundamental Parameters and Circumstellar Interactions*, *Rev. Mex. A&A Conf. ser.*, in press
Maeder, A. & Meynet, G. 2000, *ARA&A*, 287, 803

Marchenko, S. V., Moffat, A. F. J., Ballereau, D., et al. 2003, *ApJ*, 596, 1295

Marchenko, S. V. & Moffat, A. F. J. 2007, St-Louis, N. & Moffat, A. F. J. (eds.), *Massive Stars in Interacting Binaries*, (San Francisco: ASP) *ASP Conf Series*, 367, 213

Mason, B. D., Gies, D. R., Hartkopf, W. I., et al. 1998, *AJ*, 115, 821

Melena, N. W., Massey, P., Morrell, N. I., & Zangari, A. M. 2008, *AJ*, 135, 878

Moffat, A. F. J., Lamontagne, R., & Seggewiss, W. 1982, *A&A*, 114, 135

Moffat, A. F. J., Seggewiss, W., & Shara, M. M. 1985, *ApJ*, 295, 109

Moffat, A. F. J., Niemela, V. S., & Marraco, H. G. 1990, *ApJ*, 348, 232

Moffat, A. F. J., Corcoran, M. F., Stevens, I. R., et al. 2002, *ApJ*, 573, 191

Moffat, A. F. J., Poitras, V., Marchenko, S. V., et al. 2004, *AJ*, 128, 2854

Moffat, A. F. J. 2008, in: Hamann, W.-R., Feldmeier, A., & Oskinova, L. M. (eds.), *International Workshop on Clumping in Hot-Star Winds*, (Potsdam: Universitätsverlag), in press

Nazé, Y., Corcoran, M. F., Koenigsberger, G., & Moffat, A. F. J. 2007, *ApJ*, 658, 25

Negueruela, I. 2007, St-Louis, N. & Moffat, A. F. J. (eds.), *Massive Stars in Interacting Binaries*, (San Francisco: ASP) *ASP Conf Series*, 367, 459

Nelan, E. P., Walborn, N. R., Wallace, D. J., et al. 2004, *AJ*, 128, 323

North, J. R., Tuthill, P. G., Tango, W. J., & Davis, J. 2007, *MNRAS*, 377, 415

Oey, M. S., & Clarke, C. J. 2005, *ApJ*, 620, 43

Owocki, S. P. 2004, in M. Heydari-Malayeri, P. Stee, & J.-P. Zahn (eds.), *Evolution of Massive Stars, Mass Loss and Winds*, EAS Publ. Ser., 13, 163

Pollock, A. M. T. 2008, in: P. Benaglia, G. Bosch & C. E. Cappa (eds.), *Massive Stars: Fundamental Parameters and Circumstellar Interactions*, Rev. Mex. A&A Conf. ser., in press

Rauw, G., Vreux, J.-M., Gosset, E., et al. 1996, *A&A*, 306, 771

Rauw, G., De Becker, M., Nazé, Y., et al. 2004 *A&A*, 420, L9

Rauw, G., Crowther, P. A., De Becker, M., et al. 2005, *A&A*, 432, 985

Schaerer, D., Schmutz, W., & Grenon, M. 1997, *ApJ*, 484, L153

Schnurr, O. 2007, Ph. D. thesis, Univ. de Montréal

Schnurr, O., Moffat, A. F. J., St-Louis, N., et al. 2008, *MNRAS*, submitted

Schweickhardt, J., Schmutz, W., Stahl, O., et al. 1999, *A&A*, 347, 127

Smith, N., Gehrz, R. D., Stahl, O., et al. 2002, *ApJ*, 578, 464

Smith, N., & Owocki, S. P. 2006, *ApJ*, 645, 45

St-Louis, N., Moffat, A. F. J., Drissen, L., et al. 1988, *ApJ*, 330, 286

St-Louis, N. & Moffat, A. F. J. 2007 (eds.), *Massive Stars in Interacting Binaries*, ASP Conf. Ser. 367 (MSIB)

Tuthill, P. G., Monnier, J. D., & Danchi, W. C. 1999, *Nature*, 398, 487

Tuthill, P. G., Monnier, J. D., Tanner, A., et al. 2006, *Science*, 313, 935

Tuthill, P. G., Monnier, J. D., Lawrance, N., et al. 2008, *ApJ*, 675, 698

Vanbeveren, D., Van Rensbergen, W., & De Loore, C. 1998, *The Brightest Binaries*, (Dordrecht: Kluwer Academic Publications)

van den Heuvel, E. P. J. 2007, St-Louis, N. & Moffat, A. F. J. (eds.), *Massive Stars in Interacting Binaries*, (San Francisco: ASP) *ASP Conf Series*, 367, 549

van der Hucht, K. A., Schrijver, H., Stenholm, B., et al. 1997, *New Astron.*, 2, 245

Weidner, C., & Kroupa, P. 2004, *MNRAS*, 348, 187

Williams, S. J., Gies, D. R., Henry, T. J., et al. 2008, *ApJ*, submitted arXiv:0802.4232

Discussion

ZINNECKER: As you have shown, the most massive binaries are not found in the centres of dense, young clusters, but outside. I wonder why? Both dynamical and primordial mass-segregation would suggest otherwise. Do you have any thoughts on this puzzling, counter-intuitive result?

MOFFAT: Well, *some* most massive stars *are* in cluster cores (e.g. A1 in NGC 3603). But yes, one has R145 in 30 Dor well outside the R136 core. Then there is the loose, but

O-star rich association Cyg OB2, where one finds very massive stars all over the place, albeit not WNLha.

CROWTHER: If R145 (WN6h + OB) has a minimum mass of 140 $M_\odot$, what would the expected mass of the much brighter R144 (WN6h) star be?

MOFFAT: R144 is $\sim$0.5 mag in M_v (or in M_{bol}, since the intrinsic colours are similar) brighter than R145, so R144 should be even more massive, indeed. The difference could be reduced, however, if R144s components are more nearly of equal brightness (like those in WR 20a) than those in R145.

MAÍZ-APELLÁNIZ: The Hipparcos distance to γ^2 Vel actually agrees with the recent VLTI results. First, the Lutz-Kelker correction has to be applied. Second, the recent reduction of the Hipparcos data by Floor van Leewen has pushed it farther away.

MOFFAT: Thanks, that's great to hear!

RAUW: Concerning these very high masses that make some sense in terms of an upper mass limit, there is also the core of HD 15558 (De Becker et al. 2006, A&A 456, 1121), where we find $M \sin^3 i$ well above 100 $M_\odot$. The primary star is O5 III and the period is 442 days. It seems quite strange that these systems with very large $M \sin^3 i$ are often found to have long orbital periods. This might indicate that there is something wrong in our interpretation of the radial velocities (RVs) of these stars.

MOFFAT: Yes, this is unusual for an O5 III star. But long-period systems should be easier and more reliable in RV. That is, as long as one is able to piece together lots of necessarily shorter snapshot observing runs, where there may also be systematic effects occurring between different runs. For R144 and R145 we are working to rectify this using more contiguous observing over long time scales at one (small) telescope.

Tony Moffat.

3-D SPH Simulations of Colliding Winds in η Carinae

Atsuo T. Okazaki[1], Stanley P. Owocki[2], Christopher M. P. Russell[3] and Michael F. Corcoran[4]

[1] Faculty of Engineering, Hokkai-Gakuen University, Toyohira-ku, Sapporo 062-8605, Japan
email: okazaki@elsa.hokkai-s-u.ac.jp

[2] Bartol Research Institute, University of Delaware, Newark, 19716 DE, USA
email: owocki@bartol.udel.edu

[3] Department of Physics and Astronomy, University of Delaware, Newark, 19716 DE, USA
email: crussell@udel.edu

[4] Universities Space Research Association, Goddard Space Flight Center,
Greenbelt, MD 20771, USA
email: corcoran@milkyway.gsfc.nasa.gov

Abstract. We study colliding winds in the superluminous binary η Carinae by performing three-dimensional, Smoothed Particle Hydrodynamics (SPH) simulations. For simplicity, we assume both winds to be isothermal. We also assume that wind particles coast without any net external forces. We find that the lower density, faster wind from the secondary carves out a spiral cavity in the higher density, slower wind from the primary. Because of the phase-dependent orbital motion, the cavity is very thin on the periastron side, whereas it occupies a large volume on the apastron side. The model X-ray light curve using the simulated density structure fits very well with the observed light curve for a viewing angle of $i = 54°$ and $\phi = 36°$, where i is the inclination angle and ϕ is the azimuth from apastron.

Keywords. stars: individual (η Car) – stars: winds, outflows – hydrodynamics – methods: numerical – binaries: general

1. Introduction

η Carinae is one of the most luminous and massive stars in the Galaxy. It has exhibited a series of mass ejection episodes, the most notable of which was the Great eruption in the 1840's when the star ejected mass of $\sim 10\,M_\odot$, from which the Homunculus nebula was formed. Its current mass and luminosity are $M \sim 10^2\,M_\odot$ and $L \sim 5 \times 10^6\,L_\odot$, respectively.

The spectrum of η Car is rich in emission lines and has no photospheric lines. Damineli (1996) first noticed a 5.5 yr periodicity in the variability of the He I 10830Å line. Later, other optical lines were also found to show variations with the same periodicity. In the X-ray band, the flux exhibits particularly interesting, periodic variability: After a gradual increase toward periastron, the X-ray flux suddenly drops to a minimum, which lasts for about three months. It then recovers to a level slightly higher than that at apastron (Ishibashi et al. 1999; Corcoran 2005; see also Fig. 3). All these variations are consistent with the system being a long-period ($P_\mathrm{orb} = 2{,}024$ days), highly eccentric ($e \sim 0.9$) binary. The X-ray emission is considered to arise from the wind collision region.

Although there is mounting evidence that η Car is a supermassive binary, it is very hard to directly observe the stars. They are buried deep inside dense winds, which are further engulfed by the optically thick, Homunculus nebula. As a result, even the viewing angle

Table 1. Stellar, wind, and orbital parameters

Parameters	η Car A	η Car B
Mass ($M_\odot$)	90	30
Radius ($R_\odot$)	90	30
Mass loss rate ($M_\odot$ yr^{-1})	2.5×10^{-4}	10^{-5}
Wind velocity (km s^{-1})	500	3,000
Wind temperature (K)	3.5×10^4	3.5×10^4
Orbital period $P_{\rm orb}$ (d)	2,024	
Orbital eccentricity e	0.9	
Semi-major axis a (cm)	2.3×10^{14}	

is not well constrained. It is therefore important to construct a 3-D dynamical model, on the basis of which the observed features are interpreted.

In this paper, we give a brief summary of the results from 3-D numerical simulations of colliding winds in η Car. Detailed results will be published elsewhere (Okazaki *et al.* 2008).

2. Numerical Model

Simulations presented here were performed with a 3-D Smoothed Particle Hydrodynamics (SPH) code. The code is based on a version originally developed by Benz (Benz 1990; Benz *et al.* 1990) and then by Bate and his collaborators (Bate, Bonnell & Price 1995). It uses the variable smoothing length, and the SPH equations with the standard cubic-spline kernel are integrated with individual time steps for each particle. In our code, the winds are modeled by an ensemble of gas particles, which are continuously ejected with a given outward velocity at a radius just outside each star. The artificial viscosity parameters adopted are $\alpha_{\rm SPH} = 1$ and $\beta_{\rm SPH} = 2$.

For simplicity, we take both winds to be isothermal and coasting without any net external forces, assuming in effect that gravitational forces are canceled by radiative driving terms. We set the binary orbit on the x-y plane and the major axis of the orbit along the x-axis (the apastron is in the $+x$-direction). The outer simulation boundary is set at either $r = 10.5a$ or $r = 105a$ from the centre of mass of the system, where a is the semi-major axis of the binary orbit. Particles crossing this boundary are removed from the simulation. In the following, $t = 0$ (Phase 0) corresponds to the periastron passage.

Table 1 summarizes the stellar, wind, and orbital parameters adopted in our simulations. With these parameters, the ratio η of the momentum fluxes of the winds from η Car A and B is $\eta \sim 4.2$. The parameters adopted here are consistent with those derived from observations (Corcoran *et al.* 2001; Hillier *et al.* 2001), except for the wind temperature of η Car A. As mentioned above, we take the same temperature for both winds for simplicity. Note that the effect of wind temperature on the dynamics of high-velocity wind collision is negligible.

3. Structure and evolution of colliding winds

Figure 1 shows the wind collision interface geometry at $t \sim 0$ (at periastron; top panel), $t \sim 50$ d (during the X-ray minimum; middle panel), and $t \sim 200$ d (after the X-ray minimum; bottom panel) in a simulation covering $r \leqslant 10.5a$. In each panel, the greyscale plot shows the density in the orbital plane (left panel) and the plane perpendicular to the

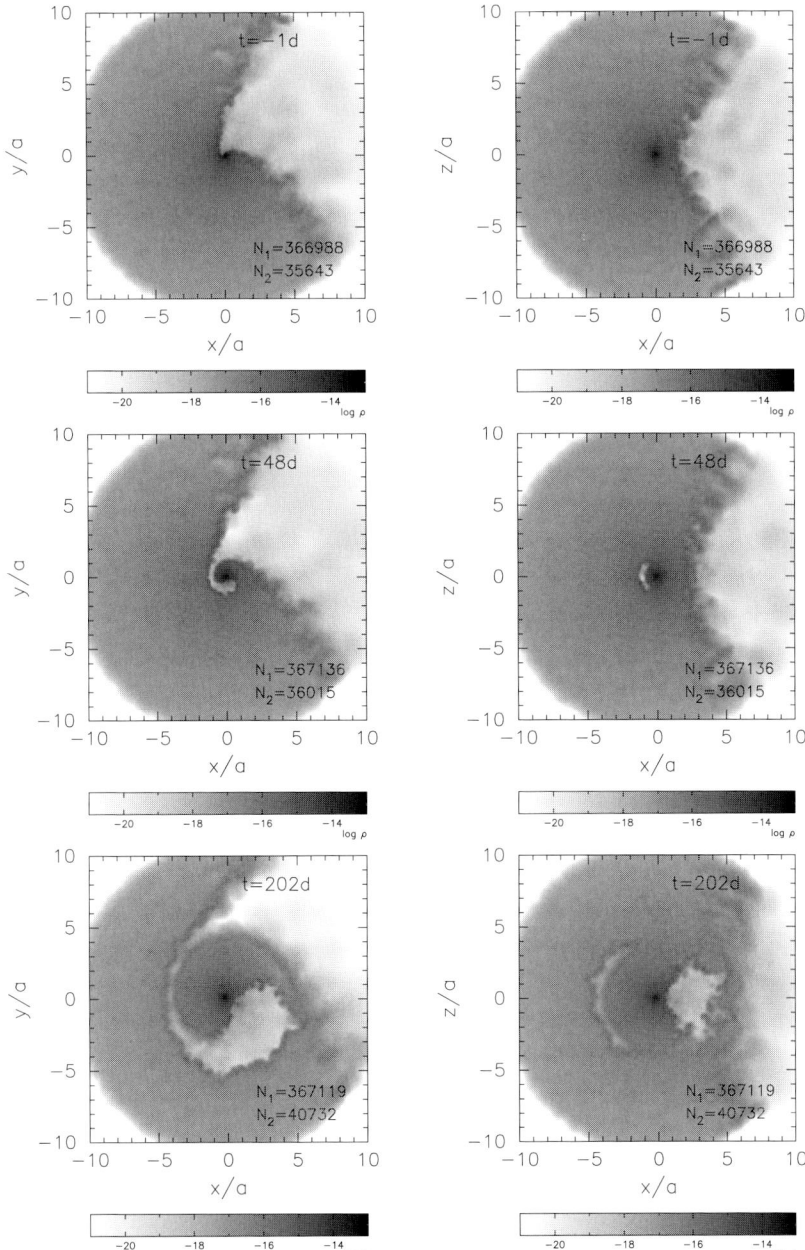

Figure 1. Wind collision interface geometry at $t \sim 0$ (top), $t \sim 50\,\mathrm{d}$ (middle), and $t \sim 200\,\mathrm{d}$ (bottom) for $r \leqslant 10a$. In each panel, the greyscale plot shows the density in the orbital plane (left) and the plane which is perpendicular to the orbital plane and through the major axis of the orbit (right), on a logarithmic scale with cgs units. The dark spot near the origin represents the primary, while the small dark spot close to the apex of the lower density wind represents the secondary. Annotations in each panel give the time (in days) from periastron passage and the numbers of particles in the primary wind, N_1, and in the secondary wind, N_2.

orbital plane and through the major axis of the orbit (right panel), on a logarithmic scale with cgs units. The dark spot near the origin represents the primary (η Car A), while

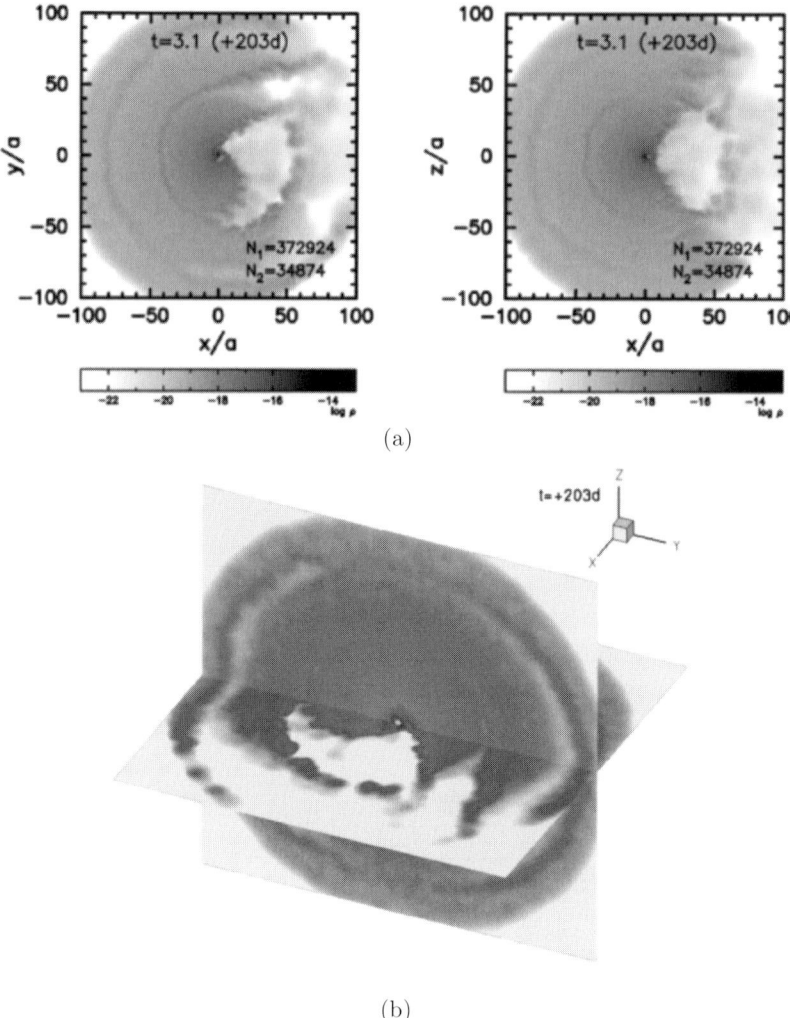

Figure 2. Wind collision interface geometry at $t \sim 200\,\mathrm{d}$ within $r = 100a$: (a) 2-D density maps in the orbital plane (left) and the plane perpendicular to the orbital plane and through the major axis of the orbit (right) and (b) the 3-D plot of the logarithmic density.

the small dark spot close to the apex of the lower density wind represents the secondary (η Car B).

Although the wind collision interface in the simulation exhibits variations from instabilities, its global shape is easily traced and illustrates how the lower density, faster wind from the secondary makes a cavity in the higher density, slower wind from the primary. As expected, the shape of the collision interface around apastron, where the orbital speed of the secondary is only $\sim 20\,\mathrm{km\,s^{-1}}$ with respect to the primary, is in agreement with the analytical one (e.g., Antokhin et al. 2004). As the secondary approaches the periastron, the interface begins to bend, and at phases around periastron, where the orbital speed of the secondary is $\sim 360\,\mathrm{km\,s^{-1}}$ with respect to the primary, the lower density wind from the secondary makes a thin layer of cavity along the orbit. Then, the thickness of the cavity increases as the secondary moves away from periastron.

In order to study the wind collision interface geometry on a larger scale, we have performed a simulation covering $r \leqslant 105a$. Figure 2(a) shows the 2-D density maps in

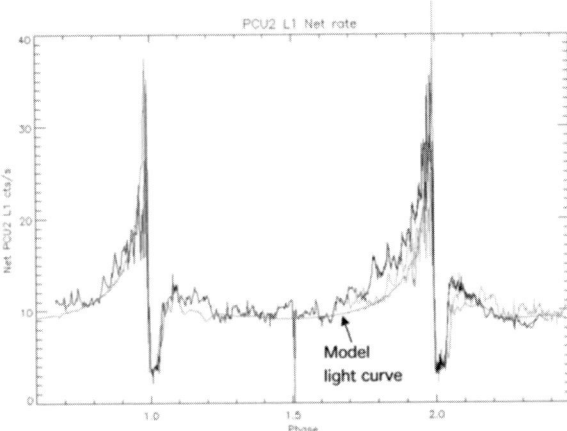

Figure 3. Comparison of RXTE data (black and gray lines; see Corcoran 2005) with the closest match model light curve (less-jaggy gray line). The light gray line shows the first cycle of RXTE data shifted by one period. Taken from Russell *et al.* (this volume).

the orbital plane and the plane perpendicular to it at the same phase (Phase 0.1) as that of the bottom panel of Fig. 1. The 3-D interface geometry is shown in Fig. 2(b) by the logarithmic density plot. From Fig. 2, we note that the lower density wind from the secondary carves out a large-scale, spiral cavity in the higher density wind from the primary. The shape of the cavity is very asymmetric. It is just a thin layer on the periastron side with respect to the primary, whereas it occupies a large volume on the apastron side.

It is interesting to study whether the model presented here can explain the variability in the RXTE X-ray light curve. Using the density distribution in our $r = 10a$ simulation, Russell *et al.* (this volume) modeled the X-ray light curve and compared it with the observed light curve. Assuming that the X-ray emission occurs at the head of the wind-wind interaction cone located at $D/(1 + \sqrt{\eta})$ from the secondary, where D is the binary separation, and varies in intensity with D^{-1} at any given orbital phase, they generated trial X-ray light curves by computing the phase variation of absorption to observers at various assumed lines of sight. They found that the RXTE X-ray light curve is very well fit with an optimal viewing angle of approximately 54 degrees of inclination and 36 degrees from apastron in the prograde direction (see Fig. 3 from Russell *et al.* (this volume) for the comparison between the model and observed X-ray light curves). The excellent fit seen in Fig. 3 confirms that the current model basically gives a correct picture of the wind-wind collision interaction in η Carinae.

4. Conclusions

We have studied the wind collision interaction in the supermassive binary η Carinae, carrying out 3-D SPH simulations. The results from simulations have clarified how the lower density, faster wind from the secondary (η Car B) carves out a cavity in the higher density, slower wind from the primary (η Car A). With an optimal viewing angle of $i = 54°$ and $\phi = 36°$, where i is the inclination angle and ϕ is the azimuth measured from apastron in the prograde direction, the model gives an excellent fit with the RXTE X-ray light curve (Russell *et al.* this volume).

Acknowledgements

A.T.O. thanks Japan Society for the Promotion of Science for the financial support via Grant-in-Aid for Scientific Research (16540218). SPH simulations were performed on HITACHI SR11000 at Hokkaido University Information Initiative Center.

References

Antokhin, I. I., Owocki, S. P., & Brown, J. C. 2004, *ApJ*, 611, 434
Bate, M. R., Bonnell, I. A., & Price, N. M. 1995, *MNRAS*, 285, 33
Benz, W. 1990, in: J. R. Buchler (ed.), *The Numerical Modelling of Nonlinear Stellar Pulsations* (Dordrecht: Kluwer), p. 269
Benz, W., Bowers, R. L., Cameron, A. G. W., & Press, W. H. 1990, *ApJ*, 348, 647
Corcoran, M. F., Ishibashi, K., Swank, J. H., & Petre, R. 2001, *ApJ*, 547, 1034
Corcoran, M. F. 2005, *AJ*, 129, 2018
Damineli, A. 1996, *ApJ*, 460, L49
Hillier, D. J., Davidson, K., Ishibashi, K., & Gull, T. 2001, *ApJ*, 553, 837
Ishibashi, K., Corcoran, M. F., Davidson, *et al.* 1999, *ApJ*, 524, 983
Okazaki, A. T., Bate, M. R., Ogilvie, G. I, & Pringle, J. E. 2002, *MNRAS*, 337, 967
Okazaki, A. T., Owocki, S. P., Russell, C. M. P., & Corcoran, M. F. 2008, *MNRAS*, in press

Discussion

DAVIDSON: Four essential points.

(*a*) One cannot derive useful orbit parameters for this object from Doppler velocities, because *every* available spectral feature evolves in a complex way. Eccentricity 0.9 is possible but not established.

(*b*) Several years ago, Kazunori Ishibashi proposed an orbit orientation based on the X-rays. It roughly matched the parameters you adopted.

(*c*) η Car's spectroscopic events are *not* primarily eclipses, not even eclipses by the wind. He II, He I, near IR, photometry, and X-ray flares all show that something *far* more interesting is involved. See Martin *et al.* (2006, *ApJ* 640, 474) and references therein.

(*d*) The wind-wind collision region is at low latitudes but nearly all of the primary wind is polar (except during an event!).

KUDRITZKI: The X-ray dip is not an eclipse in the classical stellar sense, but rather a "wind eclipse" or an interval when the X-ray emission from the wind-wind collision is embedded in the dense wind from the primary.

OKAZAKI: Studying the effect of such an asymmetry in the primary wind on the interaction geometry is interesting. It is easy to implement the wind asymmetry in my code.

The First Determination of the Rotation Rates of Wolf-Rayet Stars

André-Nicolas Chené[1] and Nicole St-Louis[2]

[1] Canadian Gemini Office, Herzberg Institute of Astrophysics,
5071, West Saanich Road, Victoria (BC), V9E 2E7, Canada
email: andre-nicolas.chene@nrc-cnrc.gc.ca

[2] Département de Physique & Observatoire du mont Mégantic, Université de Montréal,
C. P. 6128, succ. centre-ville, Montréal (Québec), H3C 3J7 , Canada
email: stlouis@astro.umontreal.ca

Abstract. The most recent stellar models have shown that the faster a massive star spins, the more its nuclear yields, mass-loss rate and lifetime are different from the standard model. One thus needs to know the rotation rate of massive stars to trace their evolutionary tracks adequately. In Wolf-Rayet (WR) stars, the direct measurement of the rotational velocity is impossible, since their continuum emission is formed in the dense wind that hides the hydrostatic, stellar surface. Here, we present a technique to derive the rotation rates of WR stars from a periodic wind phenomenon, the corotating interaction regions (CIR). For five WR stars, a first estimate of the rotation rates has been deduced from the CIR periods.

Keywords. stars: Wolf-Rayet – stars: rotation – stars: winds, outflows

1. Introduction

Rotation has a great impact on the structure and the evolution of massive stars, as shown by many articles in these proceedings. It is possible to measure the rotational velocity of an OB star directly from the Doppler broadening of its photospheric absorption lines. However, in the case of Wolf-Rayet (WR) stars, the hydrostatic surface is veiled by a dense wind. Hence, their spectra do not show any classical photospheric absorption lines. That is why no fruitful attempt to determine the rotation rates of WR stars has been made so far.

Nevertheless, it is still possible to derive the rotation rates of WR stars from a periodic wind phenomenon. Indeed, if a stellar spot is present at the surface of a hot and massive star (due to magnetic activity or pulsations), the mass-loss rate increases locally and a large-scale density structure in corotation with the stellar surface is formed in the wind (Cranmer & Owocki 1996). This structure, called Corotating Interaction Region (CIR), is observed in WR stars through flux variations in photometry and large-scale profile changes of broad WR emission lines. Hence, the period found in photometric and spectroscopic variations can be related to the period of rotation of the star. To get the rotational velocity, we multiply this period by the stellar radius (found in the literature).

This type of variability has already been observed in WR 6 (P=3.77 d) and WR 134 (P=2.34 d) (Morel *et al.* 1997, 1999). For each of the stars, a unique period has been found for both the photometric and spectroscopic variability. However, the variability pattern stays coherent only within a certain time range. Because of this epoch-dependency, the period determination is only possible if all the data have been taken contiguously and if this period is smaller than the coherence time.

A blind search for CIR-periods in WR spectra would be pricey. Moreover, not all the WR spectra show large-scale spectral variability. Hence a spectroscopic survey has been made by St-Louis et al. (2008, in prep) and Chené & St-Louis (2008, in prep) for all "single" galatic WR stars brighter than the 13th mag., in order to search for large-scale line-profile changes in 5 spectra randomly sparced in time. In the sample, all known binaries have been excluded, since large-scale line-profile variability is already expected from wind-wind collision. From this survey, a list of 10 new CIR-type variable candidates has been established.

Among the candidates are WR 1, WR 55, WR 58, WR 61, WR 67, WR 100, WR 115 and WR 120. A first light-curve has been obtained for all of these stars and spectroscopic

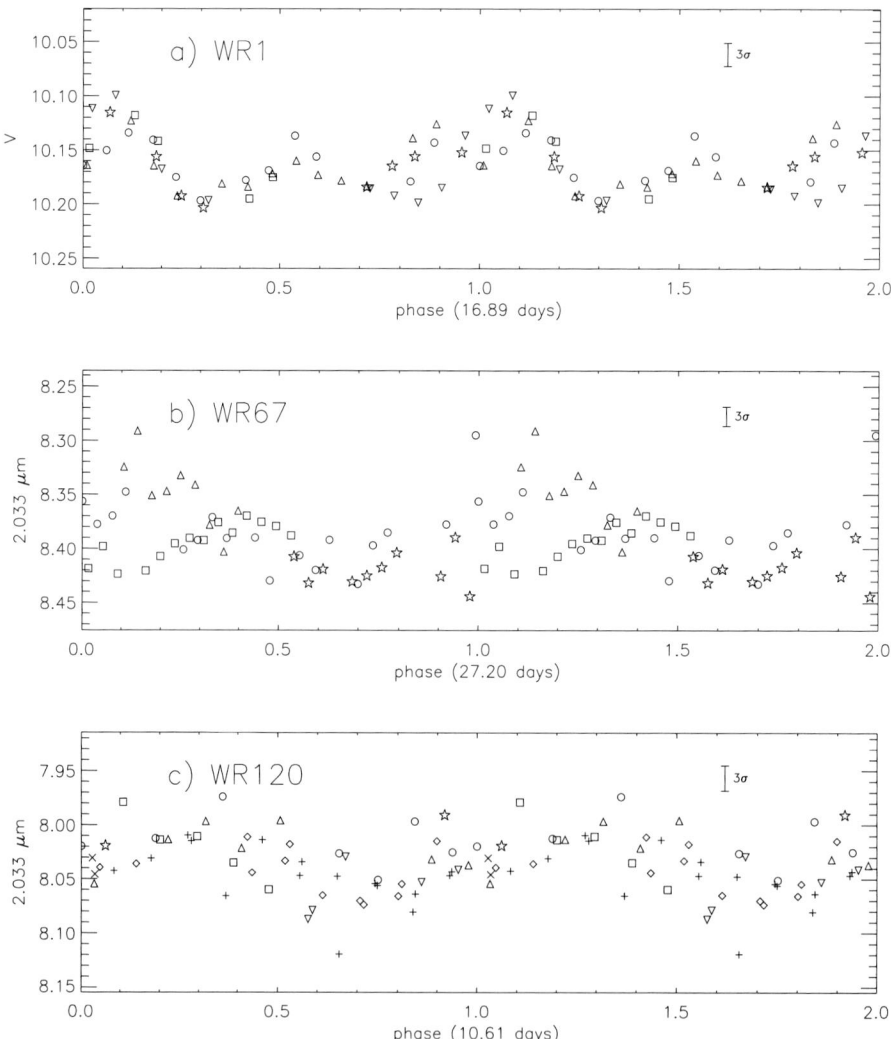

Figure 1. Light-curve of a) WR 1, b) WR 67 and c) WR 120 folded with a period of 16.89 d, 27.20 d and 10.61 d respectively. Each symbol represent a different cycle; circles for the first cycle, triangles for the second, stars for the third, squares for the fourth, upside down triangles for the fifth, diamonds for the sixth, plus signs for the seventh and x for the eighth. The vertical line indicates the 3σ error bar.

monitoring for some of them. Here, we present an up-to-date summary of our current results.

2. Results

2.1. WR 1

We monitored WR 1 in broadband V using CCD-imagery at the 0.81m Tenagra Observatory ltd. A search for periods in the light-curve has been done using both PDM and CLEAN techniques and have yielded a unique period of 16.89 d (see Fig. 1.a). Interestingly, the variability pattern does not stay the same during all the campaign. It is coherent during the first 4 cycles, when the light-curve shows three bumps having different amplitudes every cycle. However, the 5th cycle (the curve with upside down triangles) has a different shape. The bump centered near phase 0.8 and the dip centered near phase 1.0 have disappeared. Hence, during the time covered by observations in cycle 5, instead of two bumps, there is only one.

WR 1 has also been observed in spectroscopy at the 1.6m Observatoire du mont Mégantic and the 1.8m Dominion Astrophysical Observatory. In Fig. 3(*left* panel) is the spectroscopic variability of WR 1 shown folded with a 16.89 day period. In this montage of residual spectra, we see that the bulk line-profile variability changes according to the cycle, but still some patterns seem to be recurrent (see bumps and dips traced in dashed and dotted lines, respectively).

2.2. WR 67 and WR 120

WR 67 and WR 120 have been monitored in a narrowband centered at 2.033 μ m (a band where no emission lines are present) at the 1.5m CTIO telescope, using the near-infrared camera CPAPIR. The search for period in their light-curve has yielded a period of 27.20 d for WR 67 and 10.61 d for WR 120. In Fig. 1.b and 1.c the two folded light-curves are shown. The period for WR 67 is not completely convincing, since the variability patterns of the different cycles differ between phases 0.0 and 0.3. This could be explained by the epoch-dependency, but new data have to be obtained to confirm this period. WR 120's light-curve has a big scatter over the entire 10.61 d-period. This could either be explained by stochastic wind variability or pulsations. WR 120 has been observed also in spectroscopy at the 4m CTIO telescope, but over a time range smaller than the period found in photometry. Nevertheless, the spectroscopic data show variability on a time-scale of days (see the grayscale in the *right* panel of Fig. 3).

2.3. WR 58 and WR 61

No photometric period has been found so far for either WR 58 nor WR 61 (also observed with CPAPIR at the 1.5m CTIO). However, in Fig. 2.a and 2.b large-amplitude long-term photometric variability (more than 0.3 mag.) can clearly be seen. Longer time-series are needed to determine if the variability is repetitive.

2.4. *The high-scatter light-curves*

No period is found in the light-curve of WR 55, WR 63, WR 100 nor WR 115 (also observed with CPAPIR at the 1.5m CTIO), but the large scatter might indicate an important short time scale variability (from hours to one day). The actual time sampling frequency is inadequate for such short periods. WR115's light-curve is shown in Fig. 2.c and the grayscale of its spectral residuals in the middle panel of Fig. 3. The

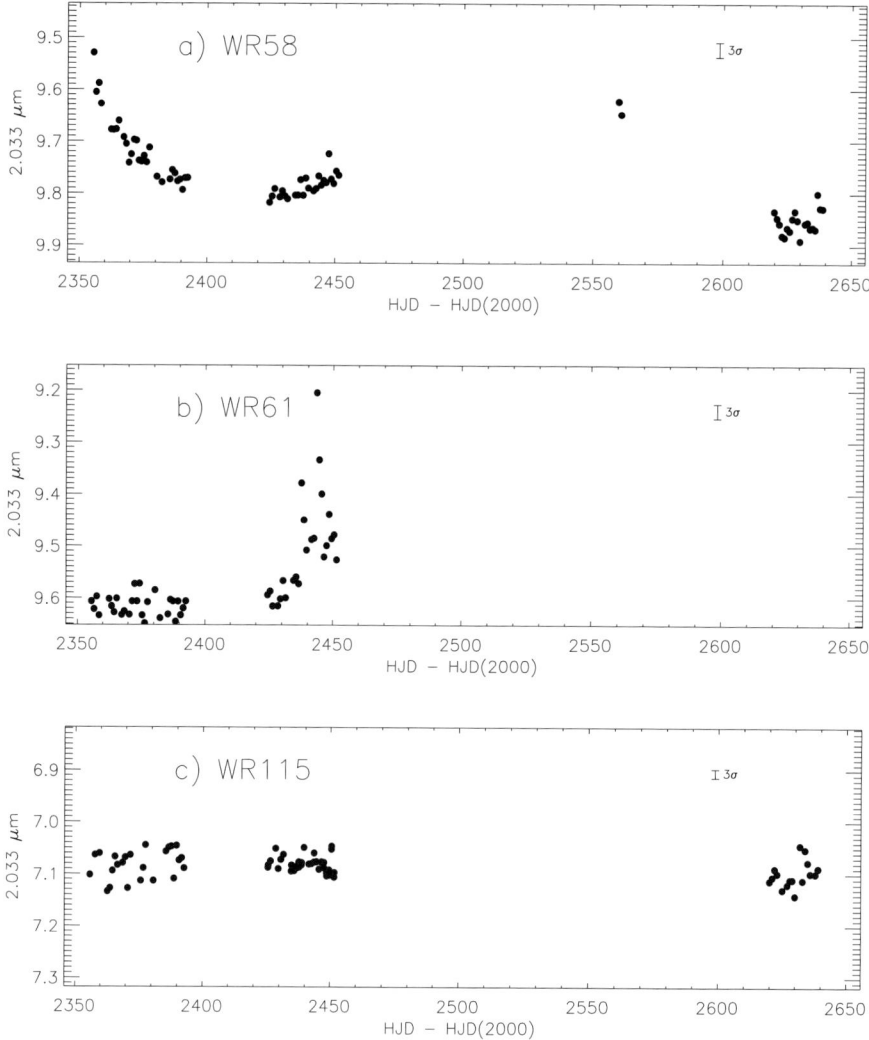

Figure 2. Light-curve of a) WR 58, b) WR 61 and c) WR 115. The vertical line indicates the 3σ error bar.

line-profile variability of WR115 seems to occur on a time-scale of days. Better time sampling is needed to determine the period of variability.

3. Conclusion

By taking from the literature the stellar radii of the WR stars for which a CIR periods have been determined (e.g. Hamann *et al.* 1995), a first estimate of the rotation rates of WR stars can be made. All the deduced rotational velocities are presented in Tab. 1. Also, for all the stars in our sample for which no period has been determined, the time-scale of the variability is indicated. All rotational speeds obtained so far are small. This corresponds to the predictions made by Meynet & Maeder. However, it remains to be shown that among our variable candidates, none have variability period. Indeed, all stars

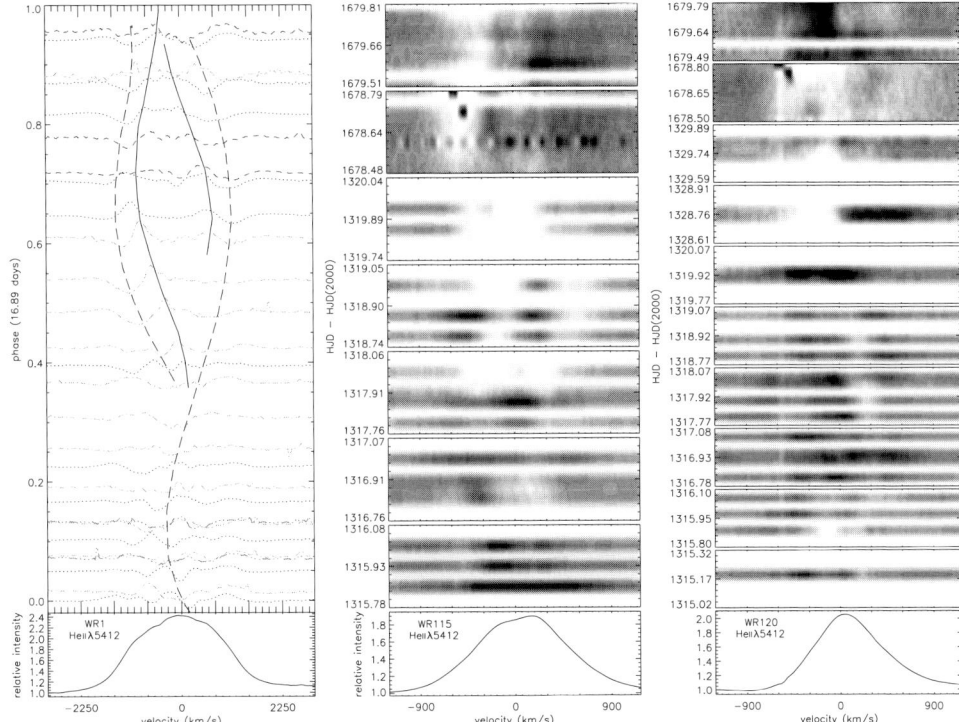

Figure 3. *Left* : Spectral residuals of the HeII λ5412 line of WR 1. The montage is folded with a period of 16.89 d. Each line-style corresponds to a cycle; dotted lines for the first cycle, short dashed line for the second, dash dotted line for the third and dash triple-dotted line for the forth. The solid and long dashed lines trace the motion of bumps and dips, respectively. *Middle* and *Right* : Grayscale of the spectral residual of the HeII λ5412 line of WR 115 and WR 120, respectively. In these grayscale, time increases from the bottom to the top of the y-axis. The brightest regions are bumps and the darkest are dips.

showing variability on time-scales of hours or days in photometry, and also WR 46 (not in the sample) which has periodic photometric and spectroscopic variability with a period of 7 to 8 hours (Veen *et al.* 2002), are all rapid rotator candidates, if we assume that their variability comes from CIRs.

This whole project is based on the assumption that CIRs are attached to the rotating surface of the star. However, in R. Blomme's poster (this volume), the possibility that CIR period could be associated with pulsation periods is discussed. Of course, a lot of theoretical work still has to be done to understand the origin and the characteristics of CIRs. Nevertheless, if all the periods found in photometric and spectroscopic variability of WR stars are purely pulsation periods, we expect them to be proportional to the stellar luminosity. If we assume that the distance to the WR stars is well known, by taking the bolometric magnitudes listed in van der Hucht (2001) (M_v=−3.51 for WR 1, M_v=−3.52 for WR 6, M_v=−3.41 for WR 67, M_v=−5.41 for WR 120 and M_v=−4.97 for WR 134), we see that it is not the case.

References

Cranmer, S. & Owocki, S. P. 1996, *ApJ*, 462, 469
Hamann, W.-R., Koesterke, L., & Wessolowski, U. 1995, *A&A*, 299, 151

Table 1. Rotation rates of WR stars

WR	sp. type	period	rot. vel. (km/s)[1]
1	WN4	16.89 d	6.5
6	WN4	3.77 d	40
55	WN7	few months?	–
58	WN4/WCE	few months?	–
61	WN5	few months?	–
67	WN4	27.20 d	2
100	WN7	few days?	–
115	WN6	few days?	–
120	WN7	10.61 d	70
134	WN6	2.34 d	60

Notes:
[1] Using stellar radii from Hamann *et al.* (1995).

van der Hucht, K. A. 2001, *New Astronomy Reviews*, 45, 135
Meynet, G. & Maeder, A. 2003, *A&A*, 404, 975
Morel, T., St-Louis, N., & Marchenko, S. V. 1997, *ApJ*, 482, 470
Morel, T., Marchenko, S. V., Eenens, P. R.J., *et al.* 1999, *ApJ*, 518, 528
Veen, P. M., van Genderen, A. M., van der Hucht, K. A. *et al.* 2002, *A&A*, 385, 585

Discussion

ZINNECKER: Which rotation rate would you need for a WR-star (surface) so that it would become a progenitor of a long Gamma-Ray Burst?

CHENÉ: According to the current models, a WR star should be rotating faster than 300 km s^{-1}. If we take a typical radius for a WR star, this would lead to a rotation period of a few hours.

André-Nicolas Chené.

Session II

Physics and Evolution of Massive Stars

Developments in Physics of Massive Stars

Georges Meynet[1], Sylvia Ekström[1], André Maeder[1],
Raphael Hirschi[2], Cyril Georgy[1] and Coralie Beffa[1]

[1]Observatory of Geneva University, Switzerland
email (G. Meynet): georges.meynet@obs.unige.ch

[2]EPSAM, University of Keele, UK
email: r.hirschi@epsam.keele.ac.uk

Abstract. New constraints on stellar models are provided by large surveys of massive stars, interferometric observations and asteroseismology. After a review of the main results so far obtained, we present new results from rotating models and discuss comparisons with observed features. We conclude that rotation is a key feature of massive star physics.

Keywords. stars: abundances – stars: early-type – stars: evolution – stars: mass loss – stars: emission-line, Be – stars: rotation – gamma rays: bursts

1. Large surveys of massive stars

Most of the developments in stellar physics arise from the necessity to better reproduce observed features. For instance the observation of strong nitrogen enrichments at the surface of main sequence OB stars indicates that some extra mixing process is at work. Since massive stars are fast rotators and since rotation triggers many instabilities able to drive the transport of chemical species, a lot of effort was put in order to properly modelize the effects of rotation. Results of these computations at their turn triggered large surveys of massive stars made possible thanks to the advent of powerful multispectrographs. Among the most recent surveys let us mention the following ones:

- Keller (2004) presents measurements of the projected rotational velocities of a sample of 100 early B-type main-sequence stars in the Large Magellanic Cloud (LMC). The sample is drawn from two sources: from the vicinity of the main-sequence turnoff of young clusters (ages $1-3 \cdot 10^7$ yr) and from the general field.
- Strom *et al.* (2005) measured the projected rotational velocities for 216 B0-B9 stars in the rich, dense h and χ Persei double cluster and compared with the distribution of rotational velocities for a sample of field stars having comparable ages (12-15 Myr) and masses (M 4-15 $M_\odot$).
- Huang & Gies (2006a) present projected rotational velocities for a total of 496 OB stars belonging to young galactic clusters. Surface helium abundances are given in Huang & Gies (2006b).
- Martayan *et al.* (2006; 2007b) have measurements of $v \sin i$ of B and Be stars in the SMC (202 B, 131 Be) and in the LMC (121 B, 47 Be).
- Wolff *et al.* (2007) have measured projected rotational velocities for about 270 B-type stars belonging to galactic clusters or associations.
- The VLT-Flames survey (still in progress) has for aim to analyze about 750 OB stars observed in 7 fields centered on young clusters, 3 in the Milky Way (NGC 3293, 4755, 6611), 2 in the LMC (NGC 2004, N11) and 2 in the SMC (NGC 330, NGC 346). The spectral classification and the radial velocities are discussed in Evans *et al.* (2005; 2006) together with some considerations on the populations of binaries and of Be stars.

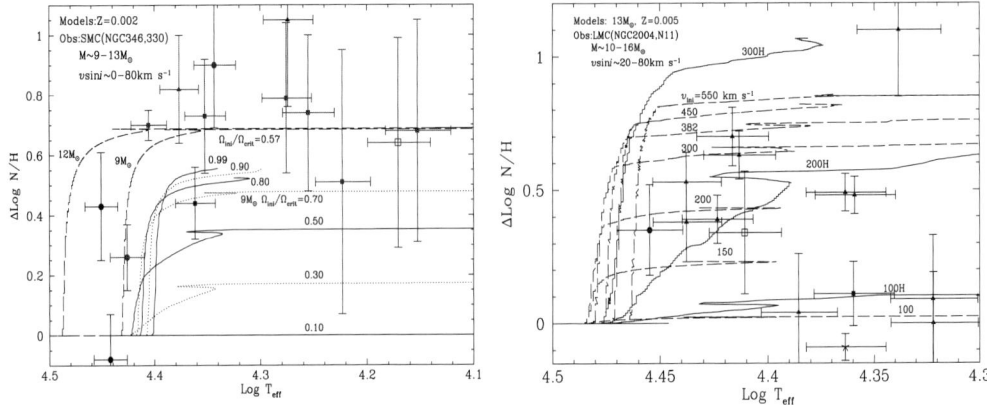

Figure 1. *Left panel*: Points with error bars are observations of SMC stars in the direction of NGC 346 by Hunter et al. (2007) and of NGC 330 by Trundle et al. (2007). Only stars with estimated masses between 9 and 13 $M_\odot$ have been considered. Continuous and dotted lines are the tracks computed by Ekström et al. (2008). Long-dashed lines are tracks computed by Ekström et al. (in preparation, see text). *Right panel*: Observed points are for stars with estimated masses between 10 and 16$M_\odot$ taken from the same references as in the left panel. Long-dashed lines are rotating tracks computed for the present work with initial abundances as given by Hunter et al. (2007) (see text). The continuous lines are models with rotation and magnetic fields (an H follows the initial equatorial velocity in that cases).

Rotation velocities are discussed in Mokiem et al. (2006; 2007), Dufton et al. (2006) and Hunter et al. (2008a) (609 measurements published so far). First results concerning the surface chemical compositions are presented in Hunter et al. (2007; 2008b), Trundle et al. (2007) (110 measurements of CNO and Mg Si published so far). Helium surface enrichments at the surface of O and early B-type stars are discussed in Mokiem et al. (2006; 2007) (57 measurements published so far).

Concerning the distribution of **rotational velocities**, the following results have been obtained:

• **O-type stars in the SMC**: Mokiem et al. (2006) deduce the underlying v distribution of the unevolved SMC O-type stars. They obtain a mean velocity v of about 150-180 km s^{-1} and an effective half width of roughly 100-150 km s^{-1}.

• **OB-type stars in the MCs**: Hunter et al. (2008a) present the atmospheric parameters and the projected rotational velocities for approximately 400 O- and early B-type stars in the Magellanic Clouds. The observed $v \sin i$ distributions can be modeled by Gaussians with a peak at respectively 100 and 175 km s^{-1} for the LMC and SMC and with a $1/e$ half width of 150 km s^{-1} in both cases.

• **B- and Be-type stars in the MCs**: Martayan et al. (2006; 2007a) obtain mean $v \sin i$ of 161±20 km s^{-1} and 155±20 km s^{-1} for SMC B-type stars of respectively 2-5 (111 stars) and 5-10 $M_\odot$ (81 stars). Analogous stars in the LMC have mean projected rotational velocities of 144±13 and 119±11 km s^{-1}. Let us note that the average velocities of Be stars are much greater (see the details in the above references). As an illustrative example, the mean $v \sin i$ for SMC Be stars in the mass ranges 2-5, 5-10, 10-12 and 12-18$M_\odot$ are respectively 277±34 (14 stars), 297±25 (81 stars), 335±20 (13 stars) and 336±40 (14 stars).

• **B-type stars in the Galaxy**: Huang & Gies (2006a) present projected rotational velocities for 496 OB stars belonging to 19 young galactic clusters with estimated ages between 6 and 73 Myr. Mean $v \sin i$ values of 139, 154 and 151 km s^{-1} have been obtained for groups of O9.5-B1.5, B1.5-B5.0 and B5.0-B9.0 type stars. These authors derived the underlying probability distribution for the equatorial velocities v and obtained a peak at 200 km s^{-1}. Dufton *et al.* (2006) obtain a peak of v at 250 km s^{-1} with a full-width-half-maximum of approximately 180 km s^{-1} for the unevolved targets in the galactic clusters NGC 3293 and 4755.

Some authors have discussed the variation of the velocity as a function of the age of the stars. The main results are the following:

- **O-type stars in the SMC:** According to Mokiem *et al.* (2006) who analyses O-type stars in the SMC, the observed distribution of $v \sin i$ for evolved stars (luminosity classes I-II) contains relatively fewer fast rotators and slow rotators compared to the distribution for unevolved stars (luminosity classes IV-V). A similar trend is obtained for galactic stars. These authors suggest that when the star evolves, it undergoes a spin down due to an increased radius, a loss of angular momentum through stellar winds. This may explain the smaller proportion of fast rotators. The smaller proportion of slow rotators might be due to excess of turbulent broadening among evolved stars.
- **OB-type stars in the MCs:** Hunter *et al.* (2008a) from a sample of 400 O- and early B-type stars in the Magellanic Clouds also find that supergiants are the slowest rotators in the sample, typically having rotational velocities less than 80 km s^{-1}.
- **OB-type stars in the Galaxy:** Huang & Gies (2006b) show that all OB stars of their sample experience a spin-down during the MS phase. A few relatively fast rotators are found near the TAMS. According to these authors, these stars may be spun up by a short contraction phase or by mass transfer in a close binary.
- Wolff *et al.* (2007) find that independent of environment, the rotation rates for stars in the mass range 6-12 M$_\odot$ do not change by more than 0.1 dex over ages between about 1 and 15 Myr.

Rotation may also depend on the mass and on the metallicity as is deduced from the results below:

- **Higher masses, lower velocities:** According to Hunter *et al.* (2008a) there is some evidences that the most massive objects rotate slower than their less massive counterparts. Dufton *et al.* (2006) find that the mean rotational velocity of stars which have strong winds is lower than that of the lower mass stars.
- **Lower Z, higher velocities:** Mokiem *et al.* (2006) find that among O-type stars, the distribution for unevolved SMC objects shows a relative excess of stars rotating with projected velocities between 120 and 190 km s^{-1} compared to analogous velocity distributions in the Galaxy. This can be interpreted as a decrease of angular momentum loss by stellar winds in lower metallicity environments. Hunter *et al.* (2008a) also obtain that SMC metallicity stars rotate on average faster than galactic ones (mainly field objects). No difference is found between galactic and LMC stars. Martayan *et al.* (2007a) find that, for B and Be stars, the lower the metallicity, the higher the rotational velocities.

Many authors find that rotational velocities of OB stars in clusters are greater than those of stars belonging to less dense systems like stellar associations or the field: for instance

- Keller (2004) obtains that the mean $v \sin i$ of early B-type stars in clusters with ages between 1-3·10^7 years of the LMC is 146 km s^{-1}, while it is 112 km s^{-1} for analogous field stars. A same trend has been found for galactic stars although with lower values for both the clusters (116 km s^{-1}) and the field (85 km s^{-1}).

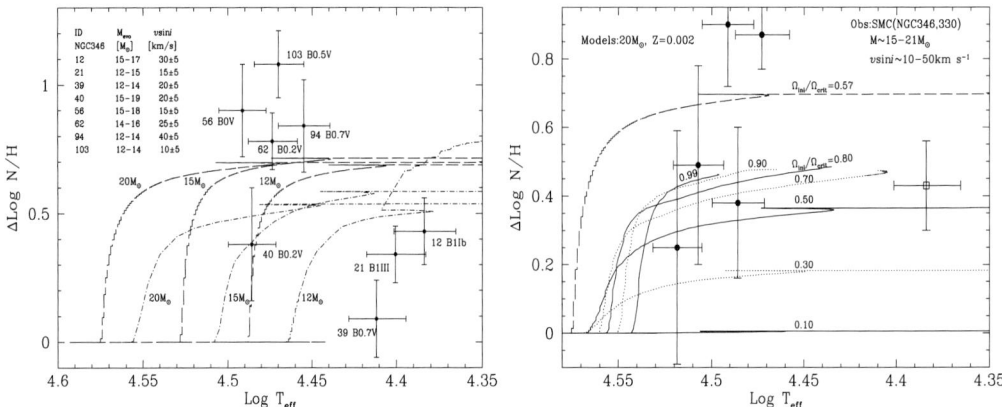

Figure 2. *Left panel*: Observed points are from Hunter *et al.* (2007). The long-dashed tracks are from the rotating models of Ekstöm *et al.* (in preparation, see text) with $v_{\rm ini}/v_{\rm crit} = 0.4$, the dot-short dashed tracks are rotating models with $v_{\rm ini} = 300$ km s^{-1} from Maeder & Meynet (2001). The models of Ekstöm *et al.* have initial values of $\Omega_{\rm ini}/\Omega_{\rm crit} = 0.57$, while the models of Maeder & Meynet (2001) have $\Omega_{\rm ini}/\Omega_{\rm crit} = 0.64$, 0.61 and 0.58 for respectively the 12, 15 and 20M$_\odot$. *Right panel*: Points with error bars are observations of SMC stars in the direction of NGC 346 by Hunter *et al.* (2007) and of NGC 330 by Trundle *et al.* (2007). Only stars with estimated masses between 15 and 21 M$_\odot$ have been considered. The star NGC330-003 has not been plotted being well outside the range of the other stars, it would lie at the position given by log T$_{\rm eff}$=4.235 and Δlog N/H=1.19. Continuous and dotted lines are the tracks computed by Ekstöm *et al.* (2008). Long-dashed lines are tracks computed by Ekstöm *et al.* (in preparation, see text).

- Strom *et al.* (2005) find that B-type stars members of h and χ Per (age between 12 and 15 Myr) have mean $v \sin i$ higher than analogous field stars. The difference between these two means depends on the evolutionary stage. For less evolved stars (4-5 M$_\odot$), the mean projected velocity is 183 km s^{-1} for cluster stars and 92 km s^{-1} for field stars, for somewhat more evolved stars (5-9 M$_\odot$), the cluster and field mean $v \sin i$ are 145 and 93 km s^{-1}, while for stars approaching the end of the Main-Sequence phase (9-15 M$_\odot$), one has respectively 104 and 83 km s^{-1}.
- Dufton *et al.* (2006) find that the projected velocities in the galactic clusters NGC 3293 and 4755 are systematically larger than those for the field. Huang & Gies (2006a) find from their study of galactic stars that there are more fast rotators among the B cluster stars than in the case of the field stars. The mean projected rotational velocities are 148±4 km s^{-1} and 113±3 km s^{-1} for the cluster and field stars respectively.
- Wolff *et al.* (2007) obtain that stars formed in high-density regions lack the cohort of slow rotators that dominate the low-density regions and young field stars.

The Be stars are stars surrounded by an expanding equatorial disks probably produced by the concomitant effects of both fast rotation and pulsation. These objects are wonderful laboratories to study the effects of extreme rotation. Many new results concerning them have been obtained:

- Objects with Be phenomena are the fastest rotators in the sample studied by Hunter *et al.* (2008a) (400 OB stars in the MCs). This trend is confirmed by Martayan *et al.* (2006; 2007a) who obtain that Be stars rotate faster than B stars whatever the metallicity is (see above).
- These last authors obtain that Be stars with masses below about 12M$_\odot$ are mainly observed in the second part of the MS whatever the metallicity. The more massive stars

are mainly in the first part of the MS in the MW, while in the Magellanic Clouds, they are all in the second part of the MS.
• Martayan *et al.* (2007b) find 13 Be stars among the sample of Be SMC stars with short-term periodicity and 9 of them are multi-periodic pulsators. The detected periods fall in the range of slowly pulsating B-type stars modes (from 0.4 to 1.60 days).
• Maeder *et al.* (1999) and Wisniewski & Bjorkman (2006) find that the fraction of Be stars with respect to the total number of B and Be stars in clusters with ages (in years) between 7.0 and 7.4 (in logarithm) increases when the metallicity decreases. This fraction passes from about 10% at solar metallicity to about 35% at the SMC metallicity. These significant proportions imply that Be stars may drastically affect the mean $v \sin i$ obtained for a given population of B-type stars.

Surface enrichments in helium and nitrogen have been observed with the following main trends (see also Maeder, this volume):

• **He in O-type stars in the SMC:** In the SMC, for 31 O-type stars, Mokiem *et al.* (2006) find values of $y = n_{He}/(n_H + n_{He})$ between 0.09 and 0.24, where n_{He} is the density number of helium and n_H of hydrogen. Note that $n_{He}/(n_H + n_{He}) = Y/(Y+4X)$, where Y and X are respectively the mass fraction of helium and of hydrogen. Setting $Y \sim 1-X$, one obtains values for $Y = 4y/(1+3y)$ between 0.28 and 0.56 (here 0.28 would correspond to y=0.09 *i.e.* to the initial helium mass fraction). These authors conclude that while rotation can qualitatively account for such enrichments, the observed enrichments are in many cases much stronger than those predicted by the models.

• **He in O-type stars in the LMC:** In the LMC, for 28 O-type stars, y values between about 0.09 and 0.28 are obtained by Mokiem *et al.* (2007), *i.e.* helium mass fractions between 0.28 and 0.61.

• **He in OB-type stars in the MW:** Huang & Gies (2006b) determine He abundances for OB galactic stars. In their high mass range ($8.5 M_\odot < M < 16 M_\odot$), the He enrichment progresses through the main sequence and is greater among the faster rotators†. On average He abundance increases of 23% ±13% between ZAMS and TAMS. These authors also obtain that He enrichments are higher for higher $v \sin i$ values. Lyubimkov *et al.* (2004) find a ZAMS to TAMS increase in He abundance of 26% for stars in the mass range 4-11$M_\odot$ and 67% for more massive stars in the range 12-19$M_\odot$.

• **N in O-type stars in the SMC:** Heap *et al.* (2006) study a sample of 18 O-type stars in the direction of the SMC cluster NGC 346. The surface of about 80% of the stars is moderately to strongly enriched in nitrogen, while showing the original helium, carbon and oxygen abundances.

• **N in OB-type stars in the MCs:** At the present time, the published values of nitrogen surface abundances from the VLT-Flames survey concern stars with relatively low $v \sin i$ (see Trundle *et al.* 2007 and Hunter *et al.* 2007‡). Some of these observations for SMC and LMC stars are shown in Figs. 1 and 2. A large spread of abundances is found spanning a range between 0 and 1.19 dex of N/H enhancements for masses between 9 and 21$M_\odot$.

• **N in B-type stars in the MW:** The galactic stars do not seem to present large spread of nitrogen abundances like that seen for Magellanic Cloud stars. Trundle *et al.* (2007) note that if the galactic stars underwent the same degree of enrichment as the LMC and SMC stars (in absolute value), this would amount to only a factor of two

† These authors also found many helium peculiar stars (He-weak and He-strong). These stars were not used to study the process of He-enrichment.
‡ In Hunter *et al.* (2008b) nitrogen surface abundances for stars with high $v \sin i$ are discussed but no detailed tables with the individual measurements are provided.

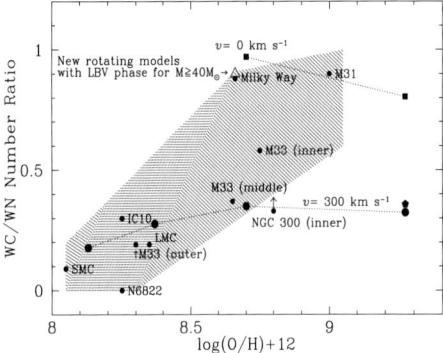 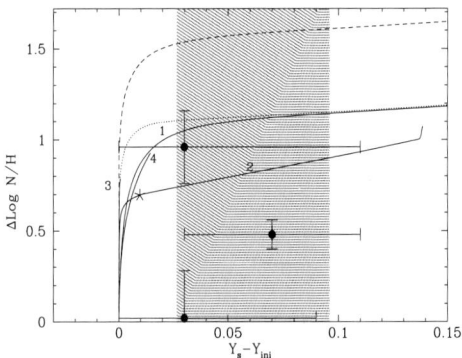

Figure 3. *Left panel:* Variation of the number ratios of WN to WC stars as a function of metallicity. The grey area encompasses the observed ratios. Individual measures are indicated by black circles labeled with the name of the Galaxy (see references in Meynet & Maeder 2005). Solar (O/H) value is taken from Asplund *et al.* (2005), the (O/H) values for the SMC and LMC are taken from Hunter *et al.* (2007). The dotted lines show the predictions of the rotating and non–rotating stellar models of Meynet & Maeder (2005). The black pentagon shows the ratio predicted by Z=0.040 models computed with the metallicity dependence of the mass loss rates during the WR phase. The open triangle shows the WC/WN ratio obtained from the new rotating models (see text). *Right panel:* Continuous lines show the evolution of the nitrogen surface enrichment as a function of the helium surface enrichment. Model 1 is the fast rotating 60M$_\odot$ model computed by Meynet & Maeder (2007) ($Z = 0.002$, v_{ini} =523 km s^{-1}. This model follows a nearly "homogeneous" evolution. Model 2 is a 20M$_\odot$ model for $Z = 0.002$ with an initial velocity on the ZAMS of 304 km s^{-1}, Models 3 and 4 are 13M$_\odot$ for $Z = 0.005$ with an initial velocity on the ZAMS of 382 km s^{-1} without and with a magnetic field. The dotted line shows the typical evolution at the centre for the initial composition used to compute the $Z = 0.002$ models, the dashed line shows the evolution at the centre when the initial composition of the LMC as given by Hunter *et al.* (2007) are adopted. For model 1 the whole track shown occurs during the MS phase, for model 2, the star indicates the end of the MS phase, for tracks 3 and 4 the track ends at the end of the MS phase. The grey area shows the He enrichment obtained by Huang & Gies (2006b) at the end of the MS. The three dots correspond to three stars observed in the direction of the LMC cluster N11 by Mokiem *et al.* (2007) and Hunter *et al.* (2007) for which He enrichments have been found From top to bottom, one has the stars N11-008 (B0.7 Ia, $\sim$ 31M$_\odot$), N11-072 (B0.2 III, $\sim$ 14M$_\odot$) and N11-042 (B0 III, $\sim$ 25M$_\odot$).

or 0.3 dex enhancement in the Galaxy. Such enhancements are similar to the degree of uncertainties in their measurements.

The problem of the mass discrepancy (see Herrero et al 1992), *i.e.* of the difference obtained between spectroscopic and evolutionary masses, has been significantly reduced thanks to improvements brought to stellar atmosphere models. However still some discrepancies are reported in general linked with strong helium surface enrichments:

• **In SMC:** Mokiem *et al.* (2006) find a mild mass discrepancy for stars with spectroscopic masses inferior to about 20M$_\odot$, which correlates with the surface helium abundance. These authors find that the discrepancies are consistent with the predictions of chemically homogeneous evolution. Most of the stars observed by Heap *et al.* (2006) exhibit the mass discrepancy problem although no surface He enrichment.

• **In LMC:** Mokiem *et al.* (2007) from the analysis of O-type stars in the LMC find that bright giants and supergiants do not show any mass discrepancy, regardless of the surface helium abundance. In contrast they find that the spectroscopically determined masses of the dwarfs and giants are systematically smaller than those derived from non-rotating evolutionary tracks. All dwarfs and giants having $y > 0.11$ ($Y > 0.33$) show this mass discrepancy.

Close binary stars have a very different evolution from single stars. Some clues pointing to these differences are indicated below:

• **Rotational velocities:** Hunter *et al.* (2008a) find that binaries tend to rotate slower than single objects. This results however may be somehow biased by the fact that binarity is easier to detect at low projected rotational velocities. Huang & Gies (2006a) on the contrary find that their binary candidates (*i.e.* those stars with a difference of radial velocity larger than 50 km s^{-1}), have a mean $v \sin i$ of 173±15 km s^{-1}, higher than 144±5 km s^{-1} the mean for the remaining constant radial-velocity stars. Huang & Gies (2006b) find that close binaries generally experience a significant spin-down around the stage where polar gravity is equal to 3.9 in logarithm. According to these authors, that is probably the result of tidal interaction and orbital synchronization.

• **Surface abundances:** According to Hunter *et al.* (2007) main-sequence binary objects have close to baseline nitrogen surface abundances. These systems thus do not present apparent signs of extra-mixing. In contrast several evolved binary objects have high nitrogen enhancements. These abundances are similar to those observed in apparently single stars. Thus it appears difficult to discriminate among the possible causes of these enrichments, *i.e.* between extra-mixing operating in single stars and mass-transfer events in close binary systems. The same result has been obtained by Trundle *et al.* (2007).

The above listed results are of course not exhaustive and many more might have been cited as the variation of the effective temperature scale for OB stars at different metallicities or the variation of the mass loss with the metallicity. Concerning this last point let us just mention that Mokiem *et al.* (2006) find that for stars with $\log L/L_\odot$ superior to about 5.4, the wind strengths are in excellent agreement with the theoretical predictions of Vink *et al.* (2001).

Obviously the evolution of rotation and of the surface enrichments depend on the mass, the age, the metallicity, the environment (field/clusters), binarity and probably on other factors as magnetic fields. Thus the task to disentangle observationally all these different effects is very challenging and requires in addition to large surveys detailed observations of a few systems for which many precise measurements can be performed.

2. Interferometry and asteroseismology

In addition to the above large surveys, there are at least two observational techniques which now begin to be applied to massive stars. The first one is interferometry. This technique allows to measure the shape of stars, the variation with the colatitude of the effective temperature, the inclination angle, the shape of the stellar wind as well as some characteristics of stellar disks. Among recent very interesting results let us mention:

• Meilland *et al.* (2007b) present the global geometry of the disk around the Be star α Arae. The global geometry of the disk is compatible with a thin keplerian disk and polar enhanced winds (see also Kervella & Domiciano de Souza 2006). They also obtain that α Arae is rotating very close to its critical rotation (for results on the Be star κ CMa see Meilland *et al.* 2007a).

• Domiciano de Souza *et al.* (2007) have performed the first high spatial and spectral resolution observations of the circumstellar envelope of a B[e] supergiant (CPD-572874).

• Millour *et al.* (2007), using AMBER/VLTI observations of the Wolf-Rayet and O (WR+O) star binary system $\gamma 2$ Velorum deduce that the binary system lies at a distance

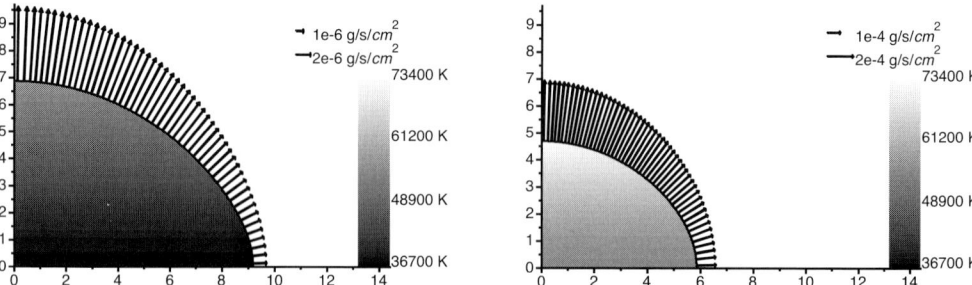

Figure 4. *Left panel*: Variation of the mass flux at the surface of an initial 35M$_\odot$ at a stage during the core H-burning phase when the mass fraction of hydrogen at the centre is $X_c = 0.42$. The axis are in units of solar radius. The velocity of the star on the ZAMS is 550 km s^{-1} corresponding to $\Omega/\Omega_{\rm crit} = 0.84$. The star follows a homogeneous evolution. At the stage represented $\Omega/\Omega_{\rm crit} \sim 1$. *Right panel*: Same as the left panel, but for a later stage with $X_c = 0.02$ and $\Omega/\Omega_{\rm crit} \sim 1$. Courtesy of C. Georgy.

of 368+38-13 pc, in agreement with recent spectrophotometric estimates, but significantly larger than the Hipparcos value of 258+41-31 pc.

- Weigelt *et al.* (2007) have made the first NIR spectro-interferometry of the LBV η Carinae. Their observations support theoretical models of anisotropic winds from fast-rotating, luminous hot stars with enhanced high-velocity mass loss near the polar regions.

The second technique, the asteroseismology, provides new insights on massive star interiors (see the paper by Aerts, this volume). At the moment, data for five B-type stars with masses between 8 and 14 M$_\odot$ have been obtained. The core overshoot parameter expressed in units of pressure scale height has been found to be of the order of 0.20 (two stars are compatible with that value, two with 0.10 and one with 0.44). For three stars the values of the ratio of core to envelope angular velocity have been obtained. The values are 5, 3.6 and 1 (solid body rotation).

In the next sections we discuss a few results recently obtained from massive star rotating models.

3. The WC/WN number ratios

The variation with metallicity of the number of WC to WN stars has been often discussed in this meeting. It is a well known fact that this ratio increases with the metallicity (see Fig. 3 left panel†). Many attempts have been performed to reproduce the observed trend: for instance the enhanced mass loss rate models of Meynet & Maeder (1994) provided a good agreement for solar and higher than solar metallicity but produced too few WN stars in metal-poor regions. The inclusion of rotation together with reduced mass loss rates accounting for the effects of clumping improved the situation in the metal poor region, but produced too many WN stars at solar and higher metallicities (Meynet & Maeder 2005). Eldridge & Vink (2006) show that models that include the mass-loss metallicity scaling during the WR phase closely reproduce the observed decrease of the relative population of WC over WN stars at low metallicities. However such models severely underestimate the fraction of WR to O-type stars. In that case, to improve the situation, a high proportion of Wolf-Rayet stars originating from mass transfer in close binaries have to be assumed at all metallicities. For instance at solar metallicity about

† we consider here regions having reached a stationary situation, *i.e.* regions where the star formation rate can be considered to have remained constant for the last twenty million years.

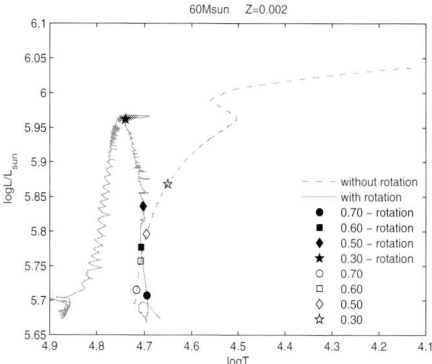 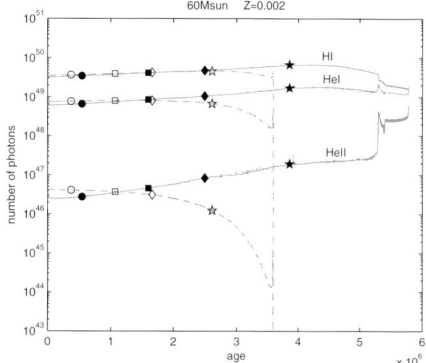

Figure 5. *Left panel*: Evolutionary tracks in the HR diagram for a rotating (continuous line) and a non-rotating 60M$_\odot$ stellar model (dashed line) at $Z = 0.002$. The rotating model has an initial velocity of 530 km s^{-1} or $\Omega/\Omega_{\rm crit} = 0.75$. Points indicate the position of the model when various values of the hydrogen mass fraction at the centre are reached. *Right panel*: Evolution as a function of the age of the ionising luminosity. Continuous lines are for the rotating model, dashed lines are for the non-rotating ones. Courtesy of C. Beffa.

75% of the WR stars should be produced in close binary systems (see Fig. 5 in Eldridge et al. 2008).

Recently we reexamined this question starting from our rotating stellar models (Meynet et al. in preparation). First let us recall that in Meynet & Maeder (2005, 2006), the most massive rotating stars enter into the WR regime already during the MS phase. This feature has good and bad effects. On one hand, it allows these models to well reproduce the variation of the number fraction of WR to O-type stars since it significantly increases the WR lifetimes. On the other hand, it produces very long WN phases since the star enters into the WR phase having still a huge H-rich envelope. As a consequence, too low values for the WC/WN ratio are obtained at solar and higher metallicities.

In the above computations, we made the hypothesis that when a star enters into the WR stage during the MS phase, it avoids the Luminous Blue Variable phase. This is probably not correct. A more realistic solution is to consider that a star which becomes a WR star during the MS phase, enters a LBV phase after the core H-burning phase, before evolving back into the WR regime. When this more realistic solution is applied, reasonable values for both the WR/O and the WC/WN ratios are obtained. Indeed the ratios of WR/O and of WC/WN given by these models at the solar metallicity are 0.06 and 0.9 which compare reasonably well with the observed values of 0.1 and 0.9 respectively. Both ratios are not reproduced by the non-rotating models to which a similar solution is applied. At the moment only the case at solar metallicity has been computed, but we are confident that such a scenario will also provide reasonable answers at other than solar metallicities. This discussion illustrates the possible key role that the LBV phase may play in shaping the WC/WN ratio.

4. The changes of the surface abundances

Models without rotation predict no surface enrichments before the red supergiant stage for stars less massive than about 40M$_\odot$. This is clearly in contradiction with the observations. In contrast, rotationally induced mixing produce changes of the surface abundances already during the MS phase. This is shown in Figs. 1 and 2. Model results depend on the physics included and on the choice of some parameters. In Figs. 1 left and

2 right, two series of models are plotted: the models of Ekström *et al.* (2008) have been computed with the same physics as in Meynet & Maeder (2005), *i.e.* the expression for the horizontal turbulence was taken as proposed in Zahn (1992), the value of α in the expression of the shear diffusion coefficient was taken equal to 2 (see Eq. 3 in Maeder & Meynet 2001), and an overshooting parameter of $0.1H_p$ was used (these models will be called models A hereafter), the models B by Ekström *et al.* (in preparation) differ in many aspects. We just mention the two most important ones: they use the expression for the horizontal turbulence by Maeder (2003) and the value of α in the expression of the shear diffusion coefficient was taken equal to 4. We can clearly see that, given an initial value of $\Omega_{ini}/\Omega_{crit}$, models B are more efficiently mixed than models A. One sees also that models A, even spanning the whole range of possible $\Omega_{ini}/\Omega_{crit}$ values cannot reproduce the highest enrichments. Models B in contrast do appear in better position to reproduce the observed range of values and are thus to be preferred.

In Fig. 1 right, rotating models computed with the same physics as models B have been computed for the LMC metallicity (long-dashed lines). Tracks computed with the effects of a magnetic field as in Meynet & Maeder (2007) are also shown for a few velocities (continuous lines). We see that models with magnetic fields are more efficiently mixed, they also show a more progressive surface enrichment. Both series of models can reproduce the observed range of values. One notes however that models with magnetic fields appear in a better position for explaining the most extreme cases. In Fig. 2 left, the models B (long-dashed lines) are plotted together with the models of Maeder & Meynet (2001). Again models B are more efficiently mixed and appear in better position for reproducing the observed enrichments.

On the whole, we see that the recent observations obtained in the large surveys described above support quite efficient mixing processes. The models of Ekström *et al.* (in preparation) can account for a great part of the observed range. More detailed comparisons will be presented elsewhere.

In Fig. 3 right, the continuous lines show the evolutions of the surface N-enrichment as a function of the He-enrichment at the surface of different models. The dotted line shows the evolution in the convective core of a $20M_\odot$ stellar model with $Z = 0.002$. The evolution in the convective core represents an upper envelope of what we can expect from a theoretical point of view. Note that this curve does not much depend on the initial mass, but depends on the initial CNO content as can be seen by comparing the dotted and the dashed line. The dashed line is obtained from models with initial abundances for the SMC as given by Hunter *et al.* (2007), while the dotted one is obtained starting with the relative abundances as given by Asplund *et al.* (2005).

Let us emphasize that progression along the "surface" (continuous lines) and "core" (dotted) lines goes at a different pace. This can be realized by noting that on the "surface" line the end of the MS phase in general occurs very early (typically the end of the MS is indicated by a star on the track 2, corresponding to the same model as the one used to draw the dotted line), while, on the "core" line, the end of the MS phase would occurred at $Y_s - Y_c = 0.75$ and $\Delta \log \mathrm{N/H}$ tending toward infinity since the hydrogen abundance is zero. The N-enrichment occurs very rapidly at the surface well before any surface helium enrichment. This is of course due to the fact that very rapidly (see the dotted line) the nitrogen abundance increases at the centre creating thus a strong chemical gradient between the core and the envelope. Since diffusive velocity is greater when the gradient of abundance is greater, the presence of such a strong gradient favors a rapid mixing. The gradient in helium is much more shallow and makes the diffusion of this element to occur on much longer timescales.

With these remarks in mind, let us now discuss each model: model 1 shows the behavior of a fast rotating 60 M$_\odot$ model at $Z = 0.002$ following a nearly homogeneous evolution. The evolution of its surface abundances approaches that of the central ones. The whole portion of the track shown in Fig. 3 right occurs during the MS phase. Such a model would easily account for the observed He enrichments during the MS phase. The model 2 (20 M$_\odot$ with $v_{\rm ini} = 304$ km s^{-1} and for $Z = 0.002$) shows at the end of the MS phase an increase in helium of 0.01 (in mass fraction) and an increase of 0.7 dex in N/H. The He-enrichment is lower than the observed one. Models 3 and 4 compare two 13 M$_\odot$ models for the LMC composition. Both models have very high initial rotation (382 km s^{-1}) but model 3 is computed without magnetic field while model 4 is computed with a magnetic field. We see that despite the high velocity, the model without magnetic field does not have any surface helium enrichment. Model with magnetic field reaches at the end of the MS phase an enrichment of nearly 0.015 with respect to its initial value. This enrichment is however too low to account for the He-enrichments obtained by Huang & Gies (2006b) at the end of the MS phase.

We conclude from these comparisons that the present-day estimates of He-enrichments at the surface of some stars require tracks following a nearly homogeneous evolution†.

5. Wind anisotropies and homogeneous evolution

Rotation has a deep impact on the way massive stars are losing mass (see the papers by Maeder and Hirschi, this volume). Here we focus on the anisotropies of the winds induced by fast rotation (Maeder 1999). When the surface velocity is near the critical one strong wind anisotropies appear. This is illustrated in Figs. 4 which shows the variation of the mass flux at the surface of a fast rotating 35 M$_\odot$ model following a homogeneous evolution. The lengths of the arrows are proportional to the mass flux. Figure 4 left shows a stage when the star is an O-type star, the right panel when it is a WR star. The scale for the mass flux was changed between left and right panels since during the WR phase the mass fluxes are about 2 orders of magnitudes higher. The grey scale varies as a function of the effective temperature (von Zeiple effect).

Accounting for wind anisotropies allows to keep in the star about 25-30% of the angular momentum which would be lost by an isotropic wind. This difference produces important differences in the way the angular momentum is distributed in the star at the pre-supernova stage. Typically, when wind anisotropies are considered, the specific angular momentum in the 3 inner solar masses are more than a factor 6 higher than when the isotropic winds are considered (see Meynet & Maeder 2007). Thus wind anisotropies are probably a key feature of homogeneous evolutionary tracks and of GRB progenitors.

Such homogeneous evolution will also produce higher outputs of ionising photons as can be seen in Fig. 5. The total number of photons released during the whole stellar lifetime with an energy sufficient to ionize H is about a factor 2 greater (passing from 1.46 10^{56} to 2.65 10^{56}) for the blueward track than for the normal one. Most of these photons (98%) are emitted during the core H-burning phase. The total number of HeI ionising photons passes from 2.4 10^{55} to 6.7 10^{55}. Most of these photons are released during the core H-burning phase (92%). The number of HeII ionising photons passes from 8.7 10^{52} to 184 10^{52}. In that case 2/3 are emitted during the core He burning phase. Thus homogeneous evolution increases the number of HeI and HeII photons by

---

† Note that many stars show N enrichments but no He enrichment, they would stand on the vertical line at the abscissa 0 in Fig. 3 right. These stars can in general be explained by models with normal rotation velocities.

about a factor 3 and 20 respectively. Therefore, if homogeneous evolution is a not too rare scenario for PopIII and very metal poor massive stars, then the budget of ionising photons in the early Universe should take into account such sources.

6. Conclusions

Considering a massive star of given initial mass and metallicity, one encounters for increasing angular momentum content the following effects of rotation: The first effect of rotation which already occurs for modest rotation rate is internal mixing. Of course when the velocity increases the mixing also increases. Then increasing the angular momentum content, the second series of effects induced by rotation concerns mass loss. Rotation may trigger mechanical mass loss, it may also at very low metallicity increase the "metallicity" of the outer layers (metals produces by the star itself) and thus increase the opacity and trigger radiative stellar winds (see the papers by Hirschi and Ekström in the present volume). Finally as an extreme case of the effects of rotation, there are the homogeneous evolution. Theory indicates that many of these effects vary as a function of Z, tending to be more important in metal poor regions. At the moment many indirect observational features seem to require high rotation rates at low Z for massive stars (see Meynet *et al.* 2008). Direct observations, limited to MC metallicities, seem also to support this view by indicating that rotation rates appear to be higher at lower Z. Therefore rotation is probably a key effect for understanding the first generations of massive stars.

References

Asplund, M., Grevesse, N., & Sauval, A. J. 2005, in: T. G. Barnes III & F. N. Bash (eds.), *Cosmic Abundances as Records of Stellar Evolution and Nucleosynthesis*, (San Francisco: ASP), *ASP Conf. Ser.*, 336 25
Domiciano de Souza, A., Driebe, T., & Chesneau, O. 2007, *AJ*, 464, 81
Dufton, P. L., Smartt, S. J., Lee, J. K., *et al.* 2006, *A&A*, 457, 265
Ekström, S., Meynet, G., Maeder, & A., Barblan, F. 2008, *A&A*, 478, 467
Eldridge, J. J. & Vink, J. S. 2006, *A&A*, 452, 295
Eldridge, J. J., Izzard, R. G., & Tout, C. A. 2008, *MNRAS*, 384, 1109
Evans, C. J., Smartt, S. J., Lee, J.-K., *et al.* 2005, *A&A*, 437, 467
Evans, C. J., Lennon, D. J., Smartt, S. J., & Trundle, C. 2006, *A&A*, 456, 623
Heap, S. R., Lanz, T., & Hubeny, I. 2006, *ApJ*, 638, 409
Herrero, A., Kudritzki, R. P., Vilchez, J. M., *et al.* 1992, *A&A*, 261, 209
Huang, W. & Gies, D. R. 2006a, *ApJ*, 648, 580
Huang, W. & Gies, D. R. 2006b, *ApJ*, 648, 591
Hunter, I., Dufton, P. L., Smartt, S. J., *et al.* 2007, *A&A*, 466, 277
Hunter, I., Lennon, D. J., Dufton, P. L., *et al.* 2008a, *A&A*, 479, 541
Hunter, I., Brott, I., Lennon, D. J., *et al.* 2008b, *ApJ*, 676, L29
Keller, S. 2004, *PASA*, 21, 310
Kervella, P. & Domiciano de Souza, A. 2006, *A&A*, 453, 1059
Kervella, P. & Domiciano de Souza, A. 2007, *A&A*, 474, 49
Lyubimkov, L. S., Rostopchin, S. I., & Lambert, D. L. 2004, *MNRAS*, 351, 745
Maeder, A. 1999, *A&A*, 347, 185
Maeder, A. & Meynet, G. 2001, *A&A*, 373, 555
Maeder, A., Grebel, E. K., & Mermilliod, J.-C. 1999, *A&A*, 346, 459
Martayan, C., Frémat, Y., Hubert, A.-M., *et al.* 2006, *A&A*, 452, 273
Martayan, C., Floquet, M., Hubert, A.-M., *et al.* 2007a, *A&A*, 472, 577
Martayan, C., Frémat, Y., Hubert, A.-M., *et al.* 2007b, *A&A*, 462, 683
Meilland, A., Millour, F., & Stee, P. 2007a, *A&A*, 464, 73

Meilland, A., Stee, P., & Vannier, M. 2007b, *A&A*, 464, 59
Meynet, G. & Maeder, A. 2005, *A&A*, 429, 581
Meynet, G. & Maeder, A. 2007, *A&A*, 464, L11
Meynet, G., Ekström, S., & Maeder, A. *et al.* 2008, in: T. Abel, A. Heger & B. O'Shea (eds.), *First Stars III*, (New York: AIP), in press (arXiv:0709.2275)
Millour, F., Petrov, R. G., & Chesneau, O. 2007, *A&A*, 464, 107
Mokiem, M. R., de Koter, A., Evans, C. J., *et al.* 2006, *A&A*, 456, 1131
Mokiem, M. R., de Koter, A., Evans, C. J., *et al.* 2007, *A&A*, 465, 1003
Strom, S. E., Wolff, S. C., & Dror, D. H. A. 2005, *AJ*, 129, 809
Trundle, C., Dufton, P. L., Hunter, I., *et al.* 2007, *A&A*, 471, 625
Vink, J. S., de Koter, A., & Lamers, H. J. G. L. M. 2001, *A&A*, 369, 574
Weigelt, G., Kraus, S., & Driebe, T. 2007, *A&A*, 464, 87
Wisniewski, J. P., & Bjorkman, K. S. 2006, *ApJ*, 652, 458
Wolff, S. C., Strom, S. E., Dror, D., & Venn, K. 2007, *AJ*, 133, 1092

Discussion

KOENIGSBERGER: Could you say a few words on how the gradient of mean molecular weight can affect the efficiency of rotational mixing?

MEYNET: When the molecular weight increases with depth as is the case in stars, mixing becomes more difficult, since greater energy is required to lift off heavy material and to mix down light one. This can be shown through the expression of the Richardson criterion (see Maeder & Meynet 1996, *A&A*, 313, 140). Meynet and Maeder (1997, *A&A*, 321, 165) have shown that the strict application of the Richardson criterion would prevent any mixing in regions with a molecular weight gradient. Only when the effects of the strong horizontal turbulence are accounted for in a proper way (Talon & Zahn 1997, *A&A*, 317, 749) can mixing be efficient enough.

LIMONGI: A comment on the anticorrelations observed in globular clusters. In principle AGB models can explain the anticorrelations maintaining the C+N+O=const. In the most massive AGB stars, HBB occurs and also if there are few dredge-up episodes the chemical composition observed in the turn off stars of globular clusters can be reproduced. the problem is that to quantitatively account for the observations a very peculiar IMF is required. Now a question concerning the WC/WN ratio: how much the WN/WC problem depends on the mass loss rate during the red supergiant phase?

MEYNET: In the scenario presented here we showed that we can reproduce the WR/O and the WC/WN ratios without modifying the mass loss rates in the red supergiant phase. Thus from these computations, the mass loss rate during the RSG does not appear as a critical point. Now a greater mass loss during the red supergiant phase might allow lower initial masses to enter the WN regime at the end of their life thus may have an impact on this ratio. At the moment however, except may be in the clusters near the galactic centre, there is no single-aged clusters showing both a significant red supergiant and Wolf-Rayet population. This indicates that the two populations originate from different initial mass ranges.

HUNTER: While fast rotation does change the surface gravities, it does not change the age of all the stars and hence the bulk of the "older" fast rotating unenriched stars remain. Secondly, given the rotational velocity distribution, the low $v \sin i$ stars are not due to $\sin i$ effects, but are indeed slowly rotating stars.

AERTS: 1) The study of Morel *et al.* (2006, *A&A* 457, 651) contains stars with N-enrichment and a seismic estimate of the equatorial rotational velocity, so here one is not bothered with $\sin i$ uncertainty. 2) The FLAMES survey contains two clusters with a lot of βCephei pulsators, so a line-profile analysis of these stars can also lead to $\sin i$. this could help to interpret the "Hunter-diagram".

Georges Meynet, with Mike Dopita in the foreground.

Carrie Trundle (left), with Rich Townsend (center) and Chris Evans (right).

Can Pulsational Instabilities Impact a Massive Star's Rotational Evolution?

Rich Townsend[1,2] and Jim MacDonald[2]

[1] Bartol Research Institute, University of Delaware, Newark, DE 19716, USA
email : rhdt@bartol.udel.edu

[2] Department of Physics & Astronomy, University of Delaware, Newark, DE 19716, USA
email : jimmacd@udel.edu

Abstract. We investigate whether angular momentum transport due to unstable pulsation modes can play a significant role in the rotational evolution of massive stars. We find that these modes can redistribute appreciable angular momentum, and moreover trigger shear-instability mixing in the molecular weight gradient zone adjacent to stellar cores, with significant evolutionary impact.

Keywords. stars: early-type – stars: rotation – stars: oscillations – stars: variables: other – instabilities – waves – methods: numerical

1. Background: Pulsation in Massive Stars

As the many, extensive variability surveys of the past couple of decades have revealed, pulsation in massive stars appears to be ubiquitous. Examples of these surveys include the *HIPPARCOS* astrometry mission, which photometrically discovered over a hundred new pulsating B stars (e.g., Waelkens *et al.* 1998); the study by Fullerton *et al.* (1996), revealing optical line-profile variations consistent with pulsation in 23 out of a sample of 30 O stars; and the *IUE* Mega campaign (Massa *et al.* 1995), which highlighted systematic variability in the wind of the early B supergiant HD 64760, subsequently attributed to co-rotating interaction regions rooted in photospheric pulsations (Fullerton *et al.* 1997; see also Kaufer *et al.* 2006).

Against this observational background, it seems reasonable to conjecture that those O and B stars already confirmed as pulsators could represent just the tip of the iceberg — that, in fact, a far greater proportion of massive stars are undergoing pulsations, albeit at amplitudes that fall below present-day detection thresholds. This expectation is lent considerable support by theoretical calculations (e.g., Pamyatnykh 1999, his Figs. 3 & 4) showing that any star with a mass $M_* \gtrsim 3\,M_\odot$ *must* pass through one or more pulsation instability strips as it evolves from ZAMS to TAMS.

These instability strips all arise from the operation of a thermodynamic engine within the star, which converts radiant heat into mechanical energy associated with periodic pulsation. As Eddington (1926) originally pointed out ('... *we require, in fact, something corresponding to the valve-mechanism of a heat engine...*'), a key component of this engine is a regulatory process that adds heat to the stellar material when at its hottest, and removes heat when at its coolest. In classical (δ) Cepheid pulsators, the regulatory process is the positive temperature dependence of the Rosseland mean opacity κ at temperatures $\log T \approx 4.5$ where second helium ionization occurs. For massive pulsators, a similar 'κ mechanism' operates on the opacity peak at $\log T \approx 5.3$ associated with bound-bound transitions of iron-group elements. This 'iron bump' leads to overstable p-mode pulsations in the β Cepheid stars ($M_* \gtrsim 7\,M_\odot$; Dziembowski & Pamyatnykh 1993), and to g-mode pulsations in the slowly pulsating B (SPB) stars ($3\,M_\odot \lesssim M_* \lesssim 7\,M_\odot$; Dziembwoski et al. 1993) and in supergiant B stars ($M_* \gtrsim 25\,M_\odot$; Pamyatnykh 1999). Here, the quoted mass ranges are for modes of harmonic degree $\ell = 0\ldots 2$; toward larger values of ℓ, the SPB and supergiant g-mode instability strips merge (see Balona & Dziembowski 1999).

2. Wave Transport of Angular Momentum

Traditionally, massive-star pulsation has been regarded simply as a dynamical phenomenon to be modeled: we see variations in the photospheric or wind diagnostics of a particular star, and we attempt to interpret these variations as arising from pulsation perturbations. More recently, the advent of specialized space observatories such as *MOST* (Walker et al. 2003) and *COROT* (Baglin et al. 2006) has opened the door to applying the techniques of asteroseismology to massive stars — using the oscillation spectrum of a pulsating star to place constraints on interior physics such as the incidence of convective overshoot, or the degree of differential rotation.

In both of these contexts, pulsation is seen as a passive player in a star's evolution. But what if, conversely, the star's evolutionary trajectory were determined to some extent by its pulsation? This idea has already been applied to low-mass stars; Talon & Charbonnel (2003, 2005), for instance, argue that internal gravity waves (IGWs — essentially, g-mode transients damped over a timescale commensurate with their period) play a role in braking the rotation in the inner regions of such stars.

To include the effects of IGWs on stellar evolution, an extra term is added to the equation governing angular momentum transport, so that

$$\rho \frac{\mathrm{d}}{\mathrm{d}t}\left[r^2 \Omega\right] = \frac{1}{5r^2}\frac{\partial}{\partial r}\left[\rho r^4 \Omega U\right] + \frac{1}{r^2}\frac{\partial}{\partial r}\left[\rho \nu r^4 \frac{\partial \Omega}{\partial r}\right] - \frac{3}{8\pi}\frac{1}{r^2}\frac{\partial}{\partial r}\mathcal{L}_J. \qquad (2.1)$$

This equation applies to shellular differential rotation (as argued by Zahn 1992, strong horizontal turbulence will tend to enforce uniform rotation across shells of constant radius r). Modulo geometrical factors of order unity, the term on the left-hand side represents the local rate of change of angular momentum per unit radius, with $\Omega(r)$ the local angular velocity. On the right-hand side, the first term represents angular momentum transport due to meridional circulation with a velocity $U(r)$. The second term represents diffusive processes with a transport coefficient ν; the major contribution to ν comes from convection and, in radiative zones, from secular shear instability (e.g., Maeder & Meynet 1996). Finally, the third term represents wave transport, as described by a luminosity function $\mathcal{L}_J(r)$ that quantifies the net amount of angular momentum carried per unit time through the shell at radius r. The main contribution to $\mathcal{L}_J$ comes from the Reynolds stress,

$$L_J = 4\pi r^2 \rho \langle r \sin\theta\, \mathbf{v}_r\, \mathbf{v}_\phi \rangle. \qquad (2.2)$$

Here, $\mathbf{v}_r$ and $\mathbf{v}_\phi$ are the radial and azimuthal velocity perturbations due to the wave, and $\langle\rangle$ denotes the average over all solid angles.

3. Application to Massive Stars

In low-mass stars, stochastic processes such as turbulent stresses or convective penetration are considered as the dominant wave excitation mechanism (e.g., Talon & Charbonnel 2003). Many authors have assumed that the same processes (albeit operating in different parts of the interior) are responsible for wave excitation in massive stars; for

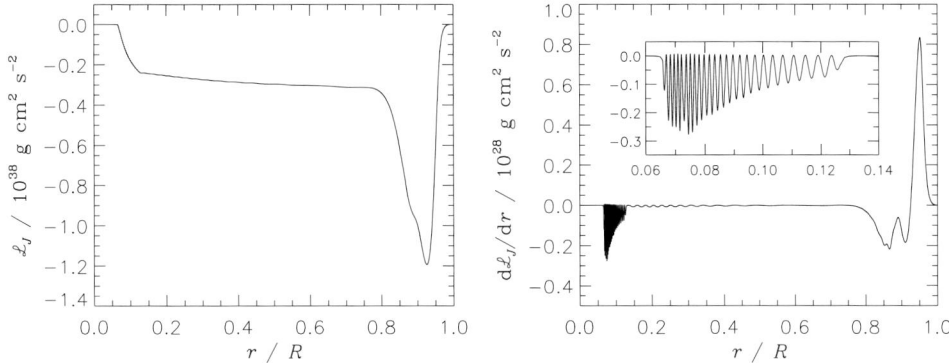

Figure 1. The angular momentum luminosity $\mathcal{L}_J$ (left), and its radial derivative $\mathrm{d}\mathcal{L}_J/\mathrm{d}r$ (right), plotted as a function of radius for the $\{n,\ell,m\} = \{40, 4, -4\}$ g mode of the $10\,\mathrm{M}_\odot$ model. The inset in the right-hand panel details the variation of the luminosity derivative in the μ-gradient zone adjacent to the core.

instance, Maeder & Meynet (2000) remark that '*...we could expect gravity waves to be generated by turbulent motions in the convective core.*'

However, as should be clear from §1, the waves observed in massive stars are not stochastic IGWs but unstable global standing oscillations, driven to large amplitudes by the iron-bump κ mechanism. Thus, the appropriate formalism for treating wave transport in these stars is a normal mode analysis with inclusion of excitation and damping processes — that is, nonradial, nonadiabatic pulsation theory. This was clearly recognized by Ando (1986, and self-references therein), who was the first to conjecture that nonadiabatic pulsation may play an important role in shaping the internal rotation profile of massive stars. Unfortunately, Ando's investigations predated the release (in the early 1990's) of updated opacity data that revealed the iron bump; thus, he was unable to reach any firm conclusions.

To build on this prior work, we examine angular momentum transport by high-order, intermediate-degree g modes in a $10\,\mathrm{M}_\odot$ stellar model near the end of its main-sequence evolution ($X_{\mathrm{core}} = 0.02$). (There is nothing particularly significant about this M_*; our results generalize to stars of both higher and lower masses. However, the late evolutionary stage is chosen to emphasize the deposition of angular momentum in the molecular weight gradient zone, discussed further below). We use the BOOJUM pulsation code (see Townsend 2005) to calculate the complex oscillation spectrum of the stellar model. Modes whose eigenfrequency ω has a negative imaginary part (i.e., $\Im(\omega) < 0$) are unstable; the eigenfunctions of these modes encapsulate all of the information necessary to evaluate the $\mathbf{v}_r$ and $\mathbf{v}_\phi$ terms in eqn. (2.2), with the exception of an arbitrary overall normalization.

Fig. 1 plots the angular momentum luminosity $\mathcal{L}_J$ as a function of fractional radius r/R_* for a single unstable g mode of the $10\,\mathrm{M}_\odot$ model, having indices $\{n,\ell,m\} = \{40, 4, -4\}$ and normalized so that the peak photospheric velocity perturbation is $1\,\mathrm{km\,s^{-1}}$ (this is a conservative choice; for reference, the typical photospheric velocities observed in pulsating massive stars are on the order of the sound speed, $\sim 10-20\,\mathrm{km\,s^{-1}}$). Also plotted is the luminosity derivative $\mathrm{d}\mathcal{L}_J/\mathrm{d}r$; as eqn. (2.1) indicates, this quantity is positive where angular momentum is extracted, and negative where it is deposited. The figure reveals angular momentum extraction from the surface layers, where the κ mechanism excites the g mode, and matching angular momentum deposition in the interior, primarily in two regions where the g mode is strongly damped. The outer damping region ($0.78 \lesssim r/R_* \lesssim 0.92$) arises from the κ mechanism operating in reverse: the opacity has

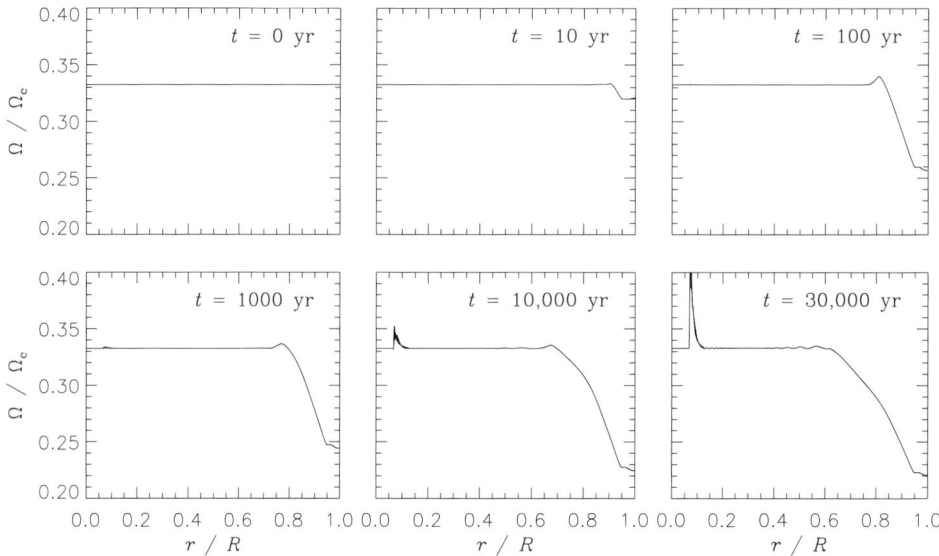

Figure 2. Snapshots of the angular velocity Ω of the $10\,\mathrm{M}_\odot$ model, plotted as a function of radius at six epochs during the HEIMDALL simulation.

a strongly negative temperature dependence, and so the thermodynamic engine converts mechanical energy into radiant heat. The inner damping region ($0.07 \lesssim r/R_* \lesssim 0.13$) is associated with the zone of varying molecular weight (μ) adjacent to the convective core. In this zone, the g mode has a very short wavelength due to the steep gravitational stratification; this leads to a spatially oscillatory pattern of angular momentum deposition, as can be seen from the inset in the right-hand panel of Fig. 1.

The angular momentum luminosity shown in Fig. 1 reaches a peak magnitude of $1.2 \times 10^{38}\,\mathrm{g\,cm^2\,s^{-2}}$; by way of comparison, Talon & Charbonnel (2003, their Fig. 4) find a net luminosity of $\sim 2 \times 10^{36}\,\mathrm{g\,cm^2\,s^{-2}}$ for IGWs in their model for a $1.2\,\mathrm{M}_\odot$ star. The two orders-of-magnitude difference between these values is simply a reflection of the far-higher amplitudes associated with the unstable modes found in massive stars, than the stochastically excited waves in low-mass stars. To give a rough estimate of the expected impact of the unstable modes, we note that over 1 Myr (a typical timescale for main-sequence evolution) the total angular momentum deposited in the μ-gradient zone by the $\{n, \ell, m\} = \{40, 4, -4\}$ mode would, *ceteris paribus*, be on the order of $8 \times 10^{50}\,\mathrm{g\,cm^2\,s^{-1}}$. This is approaching the total angular momentum $\sim 10^{51}\,\mathrm{g\,cm^2\,s^{-1}}$ stored in the core if the $10\,\mathrm{M}_\odot$ star were rotating uniformly at the critical rate. Thus, the angular momentum transport due to the pulsation can be expected to have an appreciable impact on the star's rotational evolution, and — at the most general level — the answer to the question posed in this paper's title is in the affirmative.

4. A Self-Consistent Simulation

Of course, the estimates given above neglect the fact that as the internal rotation profile evolves in response to the angular momentum transport, there will be a corresponding feedback effect on the pulsation. Clearly, some kind of self-consistent simulation is desirable, and to this end we have developed a prototype pulsation-transport code. The code, named HEIMDALL, solves the angular momentum transport equation (2.1) for an input

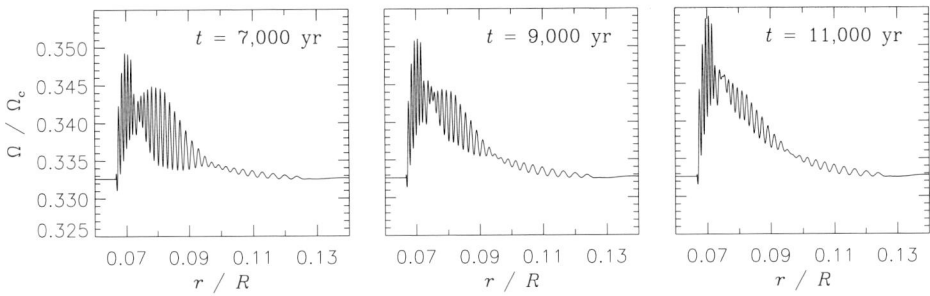

Figure 3. The evolution of the angular velocity Ω in the μ-gradient zone, for the same simulation shown in Fig. 2. Note how the steep shears in the left-hand panel have been mixed away by the shear instability in the center and right-hand panels.

stellar model. The meridional circulation term is neglected, because we are interested in transport occurring on timescales shorter than the circulation timescale R_*/U; however, the diffusion term is retained, to allow the rotation profile to relax from the steep angular velocity gradients created by the wave transport term. To evaluate the wave transport term, a modularized version of the BOOJUM code is used to calculate the complex oscillation spectrum of the stellar model at each simulation timestep. As in the preceding section, the angular momentum luminosity is obtained from mode eigenfunctions; however, rather than arbitrarily fixing mode amplitudes, HEIMDALL allows them to evolve over each timestep in accordance with individual linear growth/damping rates $-\Im(\omega)$.

Fig. 2 shows snapshots of the rotation profile $\Omega(r)$ for the $10\,M_\odot$ model, from a HEIMDALL simulation of transport by $\{\ell, m\} = \{4, -4\}$ g modes. The simulation begins in a state of uniform rotation at 33% of the critical rate Ω_c. Initially, a broad spectrum of g modes, with radial orders $n = 26\ldots 47$, are unstable toward the κ mechanism. As these g modes grow in amplitude, they transport angular momentum inwards from the surface layers. Because these layers contain little mass, they are braked quite rapidly; this established a broad shear region separating the interior from the surface, which acts to damp all but one of the initially unstable g modes. The radial order of the single remaining mode progressively increases from $n = 47$ to $n = 63$ in a sequence of mode-switching episodes; in between the switching, the mode hovers at the borderline of neutral stability, maintaining a surface amplitude of $\sim 1 - 2 \,\mathrm{km\,s^{-1}}$.

In the $10,000\,\mathrm{yr}$ panel of Fig. 2, the long-term impact of the single remaining mode begins to emerge: angular momentum is deposited in the μ-gradient zone, resulting in its gradual spin-up. For the reasons discussed previously the deposition is spatially oscillatory, and leads to the establishment of nested shear layers of very narrow extent ($\sim 10^{-3}\,R_*$). These shear layers are clearly revealed in the left-hand panel of Fig. 3; however, in the center and right-hand panels, the shear layers have been partly dissolved by diffusive transport associated with the secular shear instability, which tends to smooth out steep gradients in Ω.

This result represents perhaps the most exciting finding in our exploratory calculations. As it dissolves shear layers established by pulsation angular momentum transport, the shear instability will mix the chemical composition in the μ-gradient zone. Given that this zone plays a pivotal role in modulating angular momentum transport, in particular serving as an insulator which inhibits meridional circulation coupling between core and envelope, *we expect that the disruption of this zone by shear/pulsation-assisted mixing (SPAM) will have a profound impact on the rotational evolution of massive stars.* Building on the foundation established by our HEIMDALL simulations, we plan further calculations to examine the precise nature of this impact.

Acknowledgements

RHDT is supported by NASA grant *LTSA*/NNG05GC36G.

References

Ando, H. 1986, *A&A*, 163, 97
Baglin, A., Michel, E., Auvergne, M., & The COROT Team, 2006, in: Thompson, M. (ed.), *SOHO 18/GONG 2006/HELAS I: Beyond the spherical Sun*, p. 34
Balona, L. A. & Dziembowski, W. A. 1999, *MNRAS*, 309, 221
Dziembowski, W. A., Moskalik, P., & Pamyatnykh, A. A. 1993, *MNRAS*, 265, 588
Dziembowski, W. A. & Pamyatnykh, A. A. 1993, *MNRAS*, 262, 204
Eddington, A. S. 1926, *The Internal Constitution of the Stars* (Cambridge: CUP)
Fullerton, A. W., Gies, D. R., & Bolton, C. T. 1996, *ApJS*, 103, 475
Fullerton, A. W., Massa, D. L., Prinja, R. K., et al. 1997, *A&A*, 327, 699
Kaufer, A., Stahl, O., Prinja, R. K., & Witherick, D. 2006, *A&A*, 447, 325
Maeder, A. & Meynet, G. 1996 *A&A*, 313, 140
Maeder, A. & Meynet, G. 2000, *ARA&A*, 38, 143
Massa, D., Fullerton, A. W., Nichols, J. S., et al. 1995, *ApJ*, 452, L53
Pamyatnykh, A. A. 1999, *Acta Astronomica*, 49, 119
Talon, S. & Charbonnel, C. 2003, *A&A*, 405, 1025
Talon, S. & Charbonnel, C. 2005, *A&A*, 440, 981
Townsend, R. H. D. 2005, *MNRAS*, 360, 465
Waelkens, C., Aerts, C., Kestens, E., et al. 1998, *A&A*, 330, 215
Walker, G., Matthews, J., Kuschnig, R., et al. 2003, *PASP*, 115, 1023
Zahn, J.-P., 1992, *A&A*, 265, 115

Discussion

CRANMER: Are your plots showing the equatorial plane? If so, might the angular momentum be circulating back toward the surface at mid-latitudes, say?

TOWNSEND: No, since the rotation is shellular, the angular velocity is uniform over each spherical surface.

MAEDER: [In considering angular momentum transport by g modes], if we account for horizontal turbulence, with the coefficient by S. Mathys and myself, it introduces a strong damping factor which considerably reduces the efficiency of this transport process.

TOWNSEND: I think that the horizontal turbulence will be efficient at damping IGWs excited stochastically in the core; but the g modes I'm considering are excited by the iron-bump κ mechanism in the envelope, and are far more robust. Don't forget that we see direct observational evidence for these unstable g modes.

SKINNER: Could you comment on what observational data are now available or might be available in the near future to test the validity of the models/simulations?

TOWNSEND: Survey data will be most useful in looking for evidence (e.g., a correlation between pulsation and surface enrichment) that the μ-gradient zone has been disrupted by SPAM. In this respect, both the Large Synoptic Survey Telescope and the Kepler mission look to be promising developments.

Rotation and Massive Close Binary Evolution

Norbert Langer[1], Matteo Cantiello[1], Sung-Chul Yoon[2], Ian Hunter[3], Ines Brott[1], Danny Lennon[4], Selma de Mink[1] & Marcel Verheijdt[1]

[1] Astronomical Institute, Utrecht University, The Netherlands
[2] University of California, Santa Cruz, USA
[3] The Queen's University of Belfast, Northern Ireland, UK
[4] Isaac Newton Group, Santa Cruz de La Palma, Canary Islands, Spain

Abstract. We review the role of rotation in massive close binary systems. Rotation has been advocated as an essential ingredient in massive single star models. However, rotation clearly is most important in massive binaries where one star accretes matter from a close companion, as the resulting spin-up drives the accretor towards critical rotation. Here, we explore our understanding of this process, and its observable consequences. When accounting for these consequences, the question remains whether rotational effects in massive single stars are still needed to explain the observations.

Keywords. stars: early-type – stars: fundamental parameters – stars: mass loss – stars: rotation – stars: binaries: close

1. Why look at rotation?

Rotation has been identified as an important physics ingredient which needs to be considered to understand the evolution of massive star (e.g., Heger, Langer & Woosley 2000; Meynet & Maeder 2000). It is thought to gives rise to physical effects inside stars which cause observable quantities to change, and may even radically alter the evolutionary path of the stars. A drastic example is the occurrence of chemically homogeneous evolution, which may provide a progenitor path towards long gamma-ray bursts (Yoon & Langer 2005, Yoon et al. 2006, Woosley & Heger 2006).

One of the most relevant prediction of massive star models with rotation is that rotationally triggered internal transport processes are capable to bring nuclear processed material, most notably nitrogen, from the convective core of massive main sequence stars into their radiative envelope. For fast enough rotation, fresh nitrogen thus appears at the surface of the star, and becomes continuously more enriched as function of time during core hydrogen burning.

Numerous incidental evidences have been collected from observations which are in support of this picture (cf. Maeder & Meynet 2000, and references therein). However, while many observations refer to stars in their post-main sequence stages, abundance analyses of main sequence stars have mostly been restricted to apparent slow rotators and to rather small groups of stars. A major step forward in comparing massive star models and observations is provided by the FLAMES Survey of Massive Stars, which encompassed many hundred O and early B main sequence stars in the Galaxy and the Magellanic Clouds (Evans et al. 2005). The B star sample of this survey was analyzed in a way which allowed for the first time to obtain quantitative constrains on the nitrogen enhancement also in a large number of rapid rotators (Hunter et al. 2008, Fig. 1). Surprisingly, Hunter et al. could not unambiguously conclude that the effects of rotation

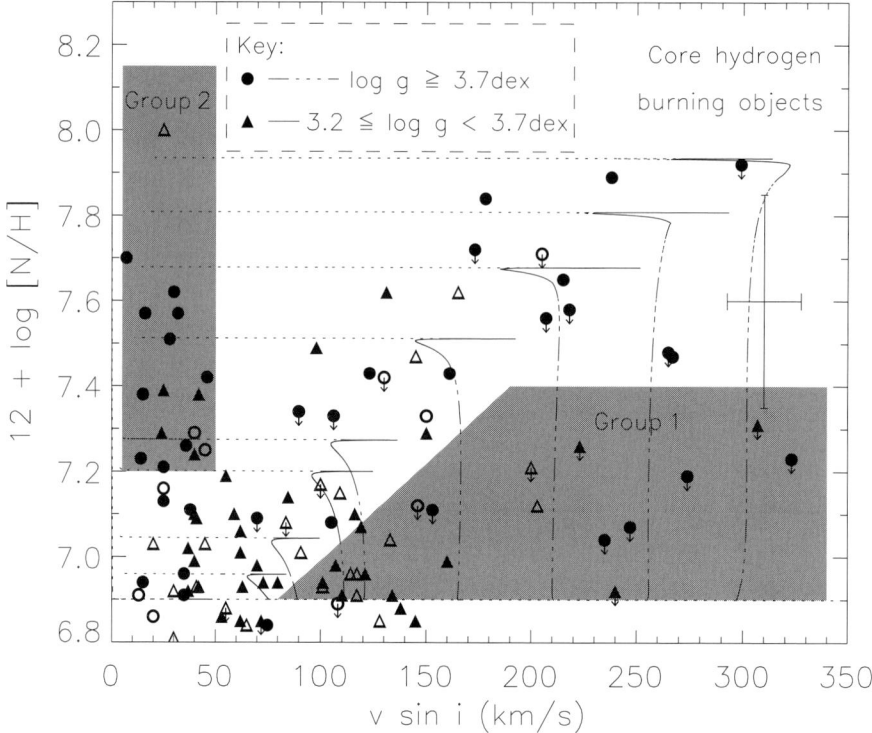

Figure 1. Nitrogen abundance (12 + log [N/H]) against the projected rotational velocity ($v \sin i$) for core hydrogen burning objects in two LMC fields centered on N11 and NGC2004, according to Hunter et al. (2008). Open symbols: radial velocity variables; downward arrows: abundance upper limits; lowest dotted line: LMC baseline nitrogen abundance. The mean uncertainty in the nitrogen abundance is 0.25 dex while that in $v \sin i$ is 10%. The bulk of the stars occupy a region at low $v \sin i$ and show little or modest nitrogen enrichment. The stellar evolution tracks are computed for an initial mass of 13 $M_\odot$ corresponding to the average mass of the sample stars, and their rotational velocity has been multiplied by $\pi/4$ to account for random inclinations. The surface gravity is used as indicator of the evolutionary status and the objects (see legend) and tracks have been split to indicate younger and older core hydrogen burning stars, respectively. Gray shading highlights two groups of stars which remain unexplained by the stellar evolution tracks.

are observed as expected. They rather found two groups of rapid rotators, one with significant enrichment and one without (the latter being dubbed Group 1 in Fig. 1). Both groups contain stars which are close to core hydrogen exhaustion, as indicated by their low surface gravity.

Hunter et al. (2008) gave two possible ways of interpretation. The one which saves the current picture of rotational mixing is that the enriched fast rotators in the FLAMES sample are indeed single stars, while the non-enriched fast rotators have a peculiar binary history. An observing campaign is underway to test the hypothesis that these latter stars are indeed all binaries — for which in the current FLAMES data there is no clear evidence. Alternatively, the results of Hunter et al. (2008) could imply that rotational mixing is not efficient, and that the enriched fast rotators are all spun-up accretion stars in binaries. The worry that the latter might be true is strengthened by the finding of yet another discrete group of massive main sequence stars by Hunter *et al.* (2008; named Group 2 in Fig. 1), namely intrinsically slowly rotating stars with a strong nitrogen enhancement. While this group of stars clearly needs an alternative explanation, it appears likely that previous reports of nitrogen enrichment in massive main sequence stars which served as support for rotational mixing picked up stars comparable to those in Group 2, as they were limited to low projected rotational velocities.

In the following, we discuss which possibilities are supported by current models of massive close binaries. As mentioned above, it appears impossible to understand the nitrogen pattern in fast rotators without invoking close binaries. One may actually wonder whether even *all* fast rotators could be produced by close binary effects.

2. Required physics

Massive close binary evolution is modeled by various groups (e.g., Podsiadlowski *et al.* 1992, Wellstein & Langer 1999, Belczynski *et al.* 2002, Vanbeveren *et al.* 2007, Vazquez *et al.* 2007, Eldridge *et al.* 2008). However, in order to predict the surface nitrogen abundances and the rotational velocities of binary components, a rather large amount of physical effects needs to be considered in binary evolution models.

It is desirable to include, in such binary models, the physics of rotation as it is currently used in models of rotating massive single stars. The reason is that in close binaries, even rather small amounts of matter transferred during Roche-lobe overflow spins up the mass gainer to extreme rotation rates (Packet 1981). Therefore, *if* rotational mixing is real, it might have the strongest effects in binary systems.

Angular momentum transport by internal magnetic fields is one ingredient in single star models which appears to be indispensable as well. Heger *et al.* (2005) showed that without this effect, young neutron stars are predicted to spin too rapidly. Suijs *et al.* (2008) showed that magnetic transport is also required to prevent too fast rotation in white dwarfs (Fig. 2).

For binary evolution models, there are two more pieces of physics which need to be included, which both relate to angular momentum exchange between the components of a binary system. Mass transfer within the Roche approximation is commonly applied in binary evolution calculations, but the corresponding angular momentum transfer is mostly neglected. The latter is crucial to model the spin-up of the accretion star, and thus to represent the most rapidly rotating stars at all. Spin-orbit coupling through tides is the other unmissable ingredient in close binary models, as it can lead to significant spin-down (mostly!) or spin-up (rarely) in massive binaries with periods below 10...20 d.

Wellstein (2001) and Petrovic *et al.* (2005a, 2005b; see also Detmers *et al.* 2008) have produced a binary code which includes all the required physics. While only few binary evolution models have been computed with this code so far, these models may help to answer a few of the questions raised above. Some of their properties are discussed below.

3. Luminosity and effective temperature

We want to briefly discuss two important effects of binarity on the distribution of stars in the HR diagram, in particular concerning the mass gainer, which will be the more prominent star of the two after a mass transfer event.

Fig. 3 shows that mass gainers become more luminous the more mass they gain. However, if the accreted amount of matter is large, and if it occurs late enough during the core hydrogen burning evolution of the mass gainer, its rejuvenation, i.e. in particular the growth of its convective core to adapt to the increased stellar mass, might be avoided

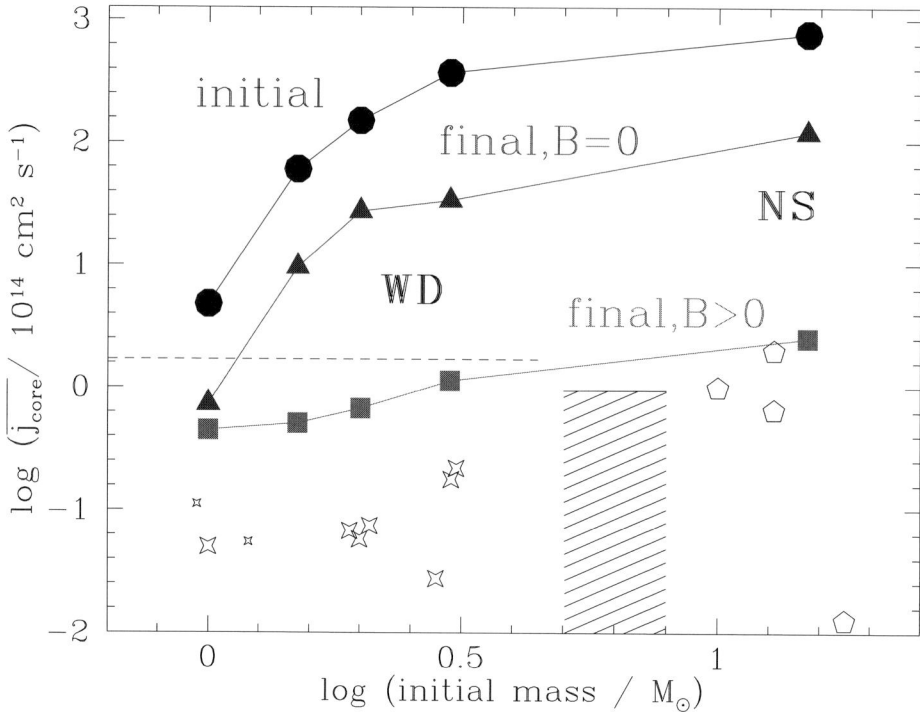

Figure 2. Average initial (upper line) and final core specific angular momentum of 1...3 $M_\odot$ stars according to Suijs et al. (2008), and 15 $M_\odot$ stars according to Heger et al. (2005). Filled triangles corresponds to the final models of non-magnetic sequences, and filled squares to the final models of magnetic sequences. The dashed horizontal line indicates the spectroscopic upper limit on the white dwarf spins obtained by Berger et al. (2005). Star symbols represent astroseismic measurements from ZZ Ceti stars (Bradley 1998, 2001; Dolez 2006; Handler 2001, Handler et al. 2002, Kepler et al. 1995, Kleinmann et al. 1998, Winget et al. 1994), where smaller symbols correspond to less certain measurements. The hatched area is populated by magnetic white dwarfs (Ferrario & Wickramasinghe 2005; Brinkworth et al. 2007). The three open pentagons correspond to the youngest Galactic neutron stars (Heger et al. 2005). The lowest pentagon is thought to roughly correspond to magnetars (Camilo et al. 2007).

(Braun & Langer 1995). In this case, core helium burning may take place very close to the main sequence, i.e., helium burning stars may be mistaken for main sequence stars. In this respect, we point out that Hunter et al. (2008) interpreted the sharp drop of the projected rotational as function of surface gravity as signaling the cool end of the main sequence band.

The increased luminosity of the mass gainers, which is also clearly visible in the evolutionary tracks shown in Fig. 4, may cause them to appear as blue stragglers in samples of stars with similar age (cf., Pols & Marinus 1994). If then a sample of stars is defined through a visual magnitude cut-off, which might favor evolved main sequence stars near the turn-off, it is conceivable that mass gainers constitute a significant fraction of the whole sample.

4. Distribution of rotational velocities

In order to model the distribution of stars in Fig. 1, one requires a distribution function for the initial rotational velocity of single stars (IRF). However, what can be measured is

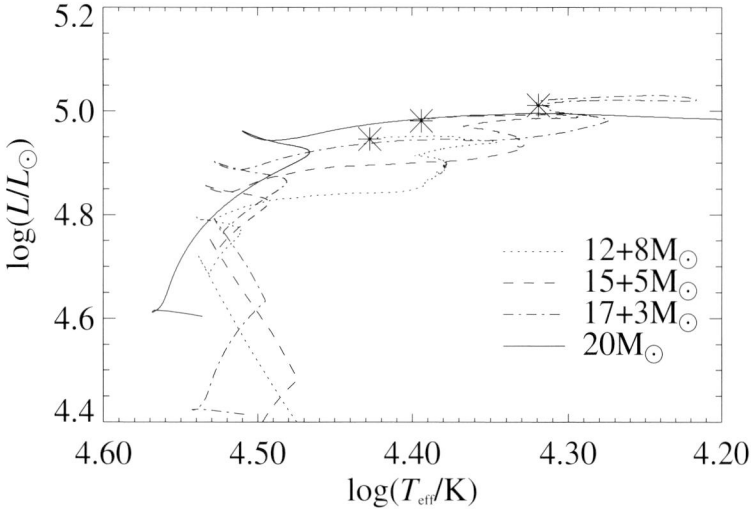

Figure 3. Tracks in the HR diagram for stars with a total mass of $20\,M_\odot$ after accretion at a central helium mass fraction of $Y = 0.7$, starting at $12\,M_\odot$, $15\,M_\odot$, and $17\,M_\odot$ (see legend), compared to the track of a $20\,M_\odot$ single star (Braun & Langer 1995). While the single star evolves to the red supergiant stage immediately after core hydrogen exhaustion, the accreting stars, which do not rejuvenate, remain blue supergiants throughout core helium burning. Their pre-supernova position is indicated be an asterisk.

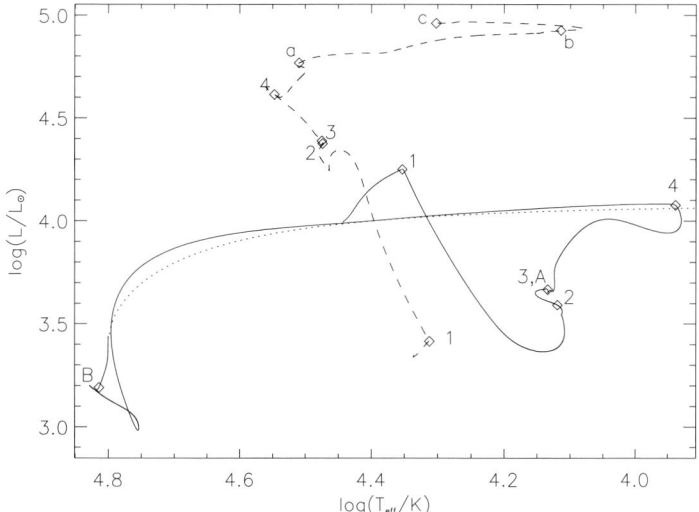

Figure 4. Evolutionary tracks of the primary (solid and dotted line) and secondary star (dashed line) of the case A binary system No. 31 of Wellstein *et al.* (2001; initial masses are $12\,M_\odot$ and $7.5\,M_\odot$, the initial period is $2.5\,\mathrm{d}$) in the HR diagram. Beginning and end of the mass transfer phases are marked with numbers; 1: begin of Case A, 2: end of Case A, 3: begin of Case AB, 4: end of Case AB. The labels A/a designate the end of central hydrogen burning of the primary/secondary, B/b the end of central helium burning of the primary/secondary, and c the point of the supernova explosion of the secondary. In this system, the secondary star ends its evolution first. The time of its supernova explosion marks the end of the solid line in the track of the primary. The further evolution of the primary is shown as dotted line. During this phase, it is treated as a single star since the system is likely broken up due to the secondary's explosion.

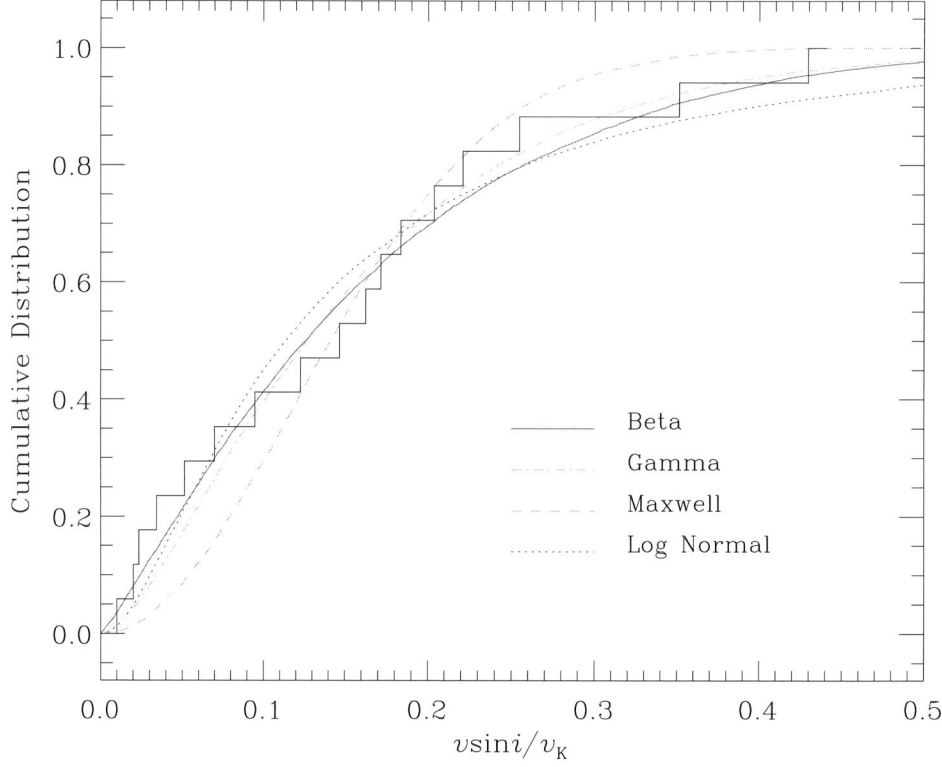

Figure 5. Cumulative distribution of the fraction of the Keplerian value of the observed rotational velocity (i.e., $v\sin i$) of unevolved young stars in NGC 346 in the Small Magellanic Cloud, according to Yoon et al. (2006). The data (step function) is from Mokiem et al. (2006). The dotted-dashed, solid, and dashed lines are the best fits of synthesized distribution functions using three different distribution laws: beta, gamma and Maxwellian, respectively. Here it is assumed that the stellar rotation axes are randomly oriented.

only the present-day distribution of rotational velocities (PRF). Fig. 5 gives an example derived within the FLAMES survey, which is the PRF of the O stars in an SMC field centered on NGC 346 (Mokiem et al. 2006). The Magellanic Clouds are good study grounds for this, since at least spin-down of single stars by radiation driven winds can be neglected for all except the very most massive main sequence stars. Worrisome about Fig. 5 is that all but the fastest three rotators have rotation rates of less than a quarter of critical rotation, while two of the three fast rotators appear to be runaway stars. We thus ask the question: could *all* rapid rotators be binary products?

The top panels of Figs. 6 and 7 show the evolution of the rotational velocity of mass gainers for two different binary evolution models (Wellstein 2001). The mass gainer will be the dominantly visible star after the first accretion event, and the mass loser may be hard to notice at all or even be ejected through its supernova explosion. These figures show that very rapid rotators are produced, which are long-lived main sequence stars. Due to the blue straggler effect, they may dominate certain parts of the HR-diagram (cf. Section 3).

Fig. 8 shows the range of contact-free evolution in the initial period versus initial mass ratio diagram for primary star masses of relevance here. Mass transfer and spin-up is expected everywhere within the contact-free regime (Wellstein et al. 2001). However, in

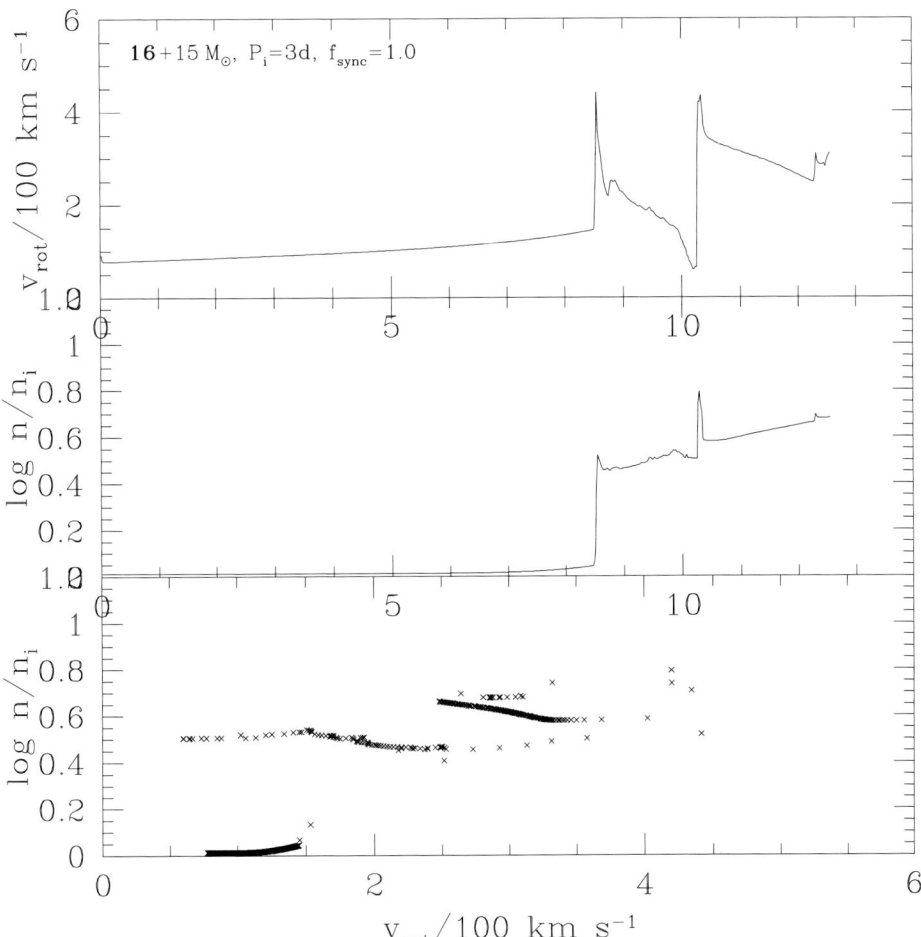

Figure 6. Equatorial rotational velocity (upper panel) and surface nitrogen mass fraction relative to the initial value (middle panel) as function of time, for a the mass gainer in a solar metallicity $16\,M_\odot + 15\,M_\odot$ binary with an initial period of 3 days. Time is given in Myr for the upper two panels. The computations include the physics of rotation for both components as in Heger et al. (2000), and Spin-Orbit coupling as in Detmers et al. (2008) with the nominal coupling parameter $f_{\rm sync} = 1$, and rotationally enhanced stellar wind mass loss (Langer 1998). Internal magnetic fields are not included. The bottom panel shows the evolution of the mass gainer in the nitrogen enhancement versus rotational velocity diagram, where each data point represents a duration of 20 000 yr. The spin-down of the star after the first accretion event ($t = 8.5...10\,{\rm Myr}$) is mostly due to tidal effects. This example shows that massive close binaries can produce rotating nitrogen rich stars which are rapidly rotating, but also such which are slowly rotating.

most regions outside of this, both stars in the binary are expected to merge as a result of their interaction (see also Podsiadlowski et al. 1992). While the details of the merger process are difficult to predict, the merger product will be an extreme rotator due to the enormous surplus of angular momentum. Merger stars will only be observed as single stars.

We see that due to mass transfer and merging, a stellar population with few or no rapid rotators initially may build up a certain number of rapidly rotating core hydrogen

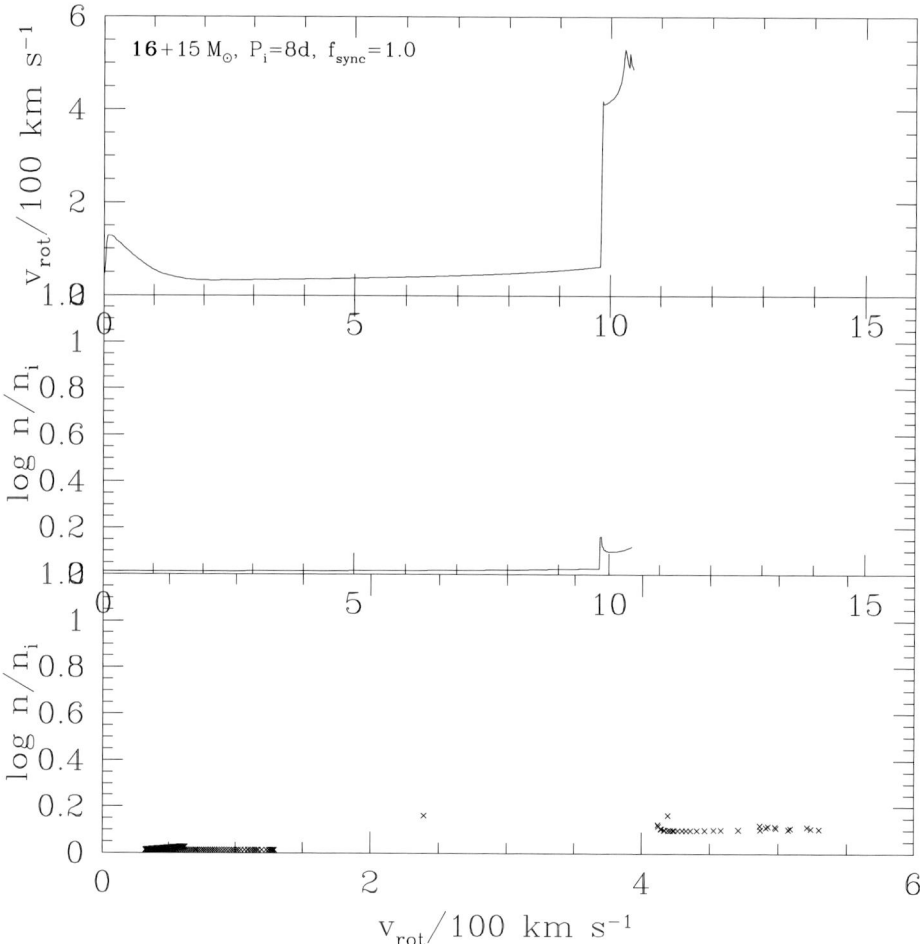

Figure 7. As Fig. 6, but for a $16\,M_\odot + 15\,M_\odot$ binary with an initial period of 8 days, which makes it a Case B system. Due to the larger initial period, the mass transfer is very non-conservative. Enough mass is accreted to spin-up the mass gainer, but not enough to create a large nitrogen surface enhancement. This example shows that massive close binaries can produce rapidly rotating, evolved main sequence stars which are *not* strongly nitrogen-enriched. Note that the calculations stops about 1 Myr after the mass transfer due to numerical difficulties. Possibly, the mass gainer would become nitrogen-enhanced later on due to rotational mixing.

burning stars. Some of them may slow down again due to their close companion (cf. Fig. 6, top panel), but many will not. It will be an important task for the near future to put quantitative limits on the relative number of those stars in stellar populations.

5. Nitrogen enrichment

The surface nitrogen abundance of mass gainers in close binary systems can be affected in two major ways. First of all, the matter which is accreted from the companion is often nitrogen-rich, as it stems from deep layers of the initial primary. While the transferred matter may be very nitrogen-rich, with values close to CNO-equilibrium being reached toward the end of the mass transfer, the enrichment on the mass gainer remains limited

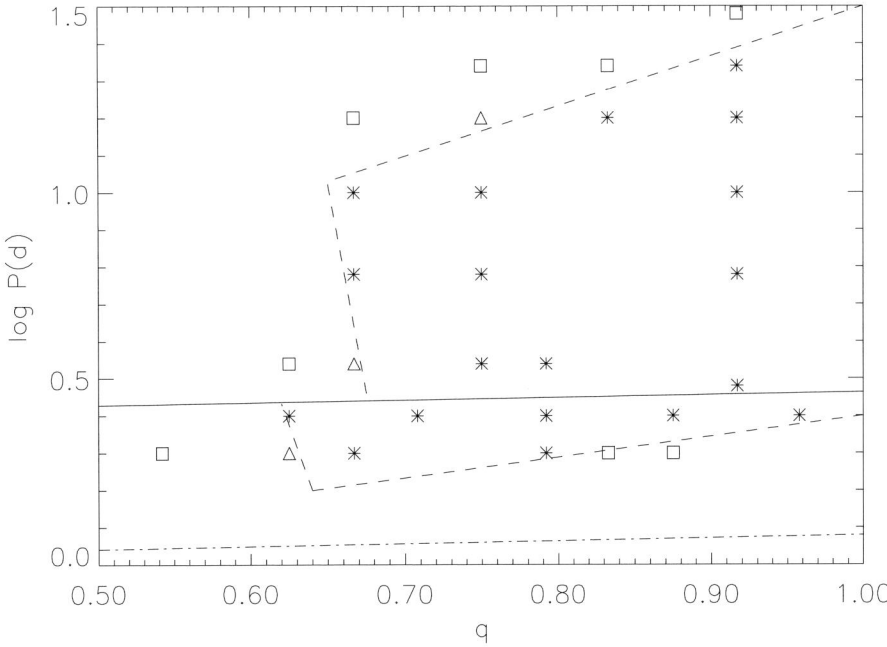

Figure 8. Distribution of all computed binaries with $12\,M_\odot$ primaries of Wellstein *et al.* (2001) in the initial period versus initial mass ratio diagram. Asterisks mark contact-free systems, while squares mark systems which evolve into contact. Systems marked with triangles are borderline cases, i.e., they evolve into a short contact phase but the secondary radius never exceeds its Roche radius by more than a factor 1.5. The solid line separates Case A (below) and Case B systems. All case A systems for this primary mass have a reverse supernova order. The dashed lines indicate the boundary between contact-free and contact evolution. The dashed-dotted line is defined by the condition that the primary fills its Roche lobe already on the zero age main sequence.

since thermohaline mixing dilutes the accreted matter with its whole envelope. Figure 9 gives an idea of the maximum obtainable enrichment, which, in the mass range considered here, amounts typically to 0.6 dex, where some Case A systems can produce as much as 0.8 dex. Non-conservative systems, where much of the overflowing matter is ejected from the binary system, are expected to obtain substantially smaller enrichments.

Secondly, nitrogen can be enhanced in mass gainers due to rotational mixing. Since these stars are amongst the most rapidly rotating main sequence stars, rotational mixing in these stars might be substantial. In fact the second panel in Fig. 6 shows that, while accretion has raised the nitrogen surface mass fraction by ~ 0.4 dex, it further-on slowly increases to a total enrichment of almost ~ 0.7 dex. Fig. 10 shows a more dramatic example: the same binary system as shown in Fig. 6, but with an initial period of 6 d instead of 3 d, accelerates its mass gainer to a rotational velocity of almost $500\,\mathrm{km\,s^{-1}}$, which gives an extra nitrogen enrichment of more than a factor two, to about 1 dex in total. At low metallicity, Cantiello *et al.* (2007) has found that the mass gainer of a $16\,M_\odot + 15\,M_\odot$ system with an initial period of 5 d evolves chemically homogeneously after the Case B mass transfer event.

So we see that, quite naturally, close binaries produce rapidly rotating nitrogen-rich stars. This remains true if rotational mixing is completely neglected, where the nitrogen

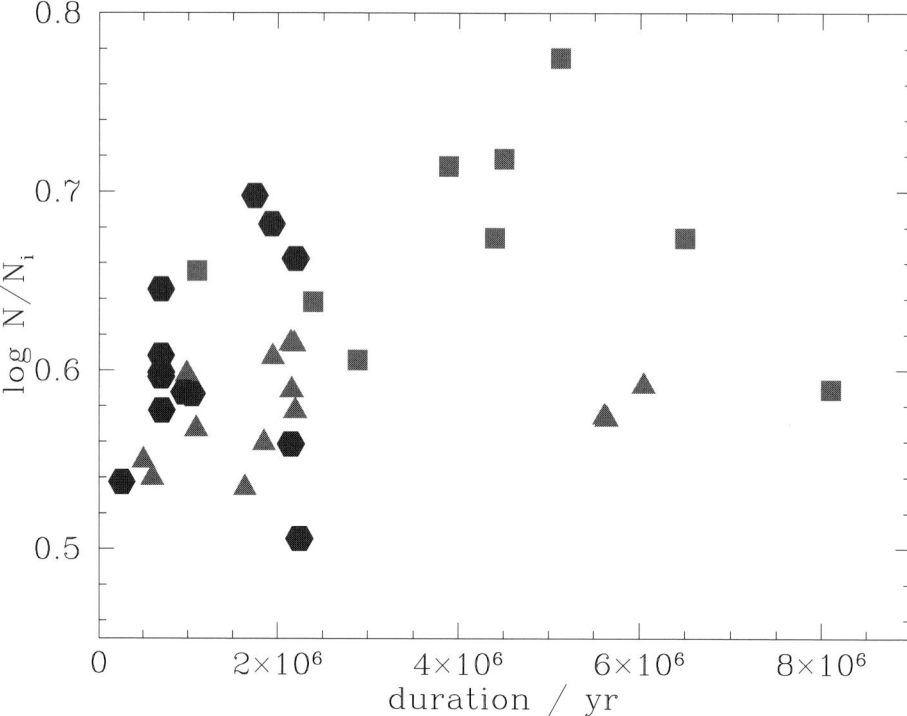

Figure 9. Surface abundance of nitrogen relative to its initial abundance for the solar metallicity binary models of Wellstein *et al.* (2001), as function of the duration of the post-mass transfer phases. The models are conservative, have $12\,M_\odot$ and $16\,M_\odot$ primaries, and initial mass ratios larger than 0.5. Case A mass gainers during slow Case A mass transfer (Algol-type systems) are marked by triangles. The same stars appear as squares after Case AB mass transfer. Post Case B mass gainers are marked by hexagons.

enhancement would just be a factor of 2...3. With rotational mixing included, a factor of 10 can be reached.

But Fig. 10 also shows that binaries can indeed produce rapid rotators with little enrichment (cf. lower two panels in Fig. 7). Note that the highly non-conservative evolution which is required to obtain this may be largely underrepresented in Fig. 10, as the corresponding binary evolution models are numerically difficult.

Finally, as the post-mass transfer period in very close systems can lead to rapid tidal spin-down (e.g., in the 2.15 d binary shown in Fig. 10), close binaries may also produce slowly rotating nitrogen-rich main sequence stars.

6. Conclusions

We have shown above that an effort is needed to compare the predictions of stellar evolution theory with the new results derived from the FLAMES Survey of Massive Stars (cf. Fig. 1). In particular, binary evolution models which include rotational mixing, mass and angular momentum transfer, and tidal interaction are required, of which only few exist today (cf., Fig. 10).

Fig. 10 can not be directly compared with Fig. 1. It does not contain a clean population study, and can therefore only indicate which parts of the diagram might be populated by binary systems, but not how many stars one might expect in those parts. Nevertheless,

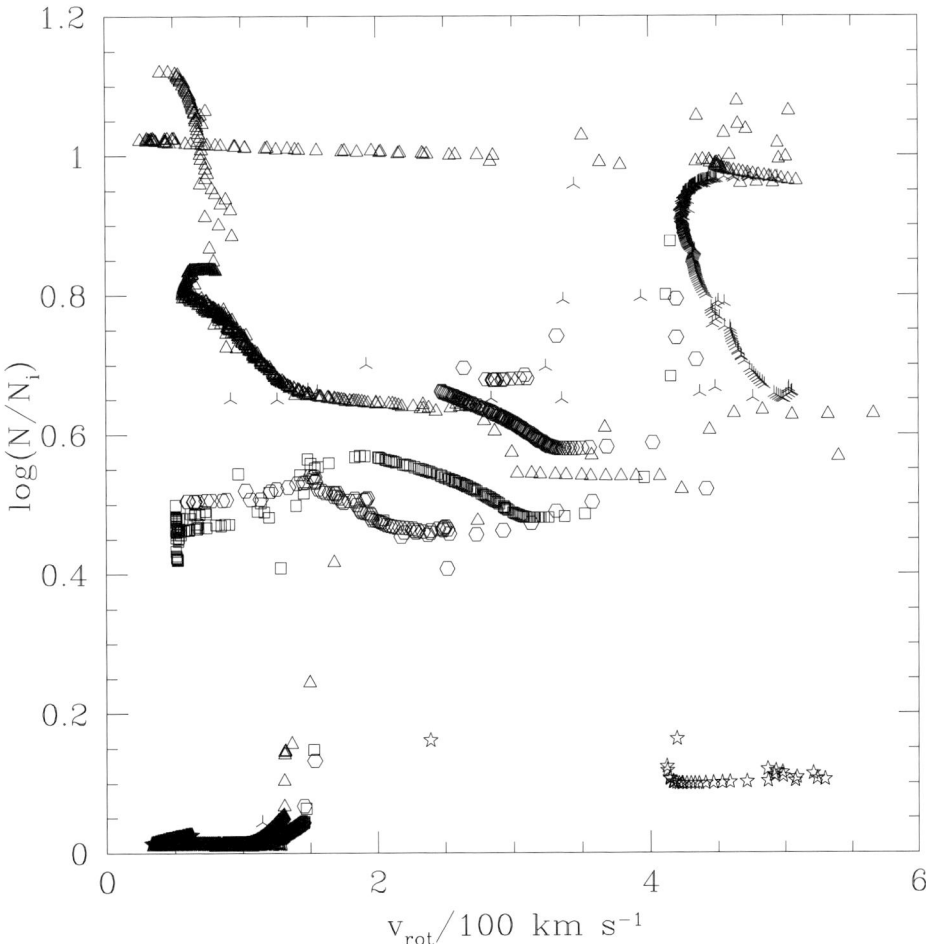

Figure 10. Nitrogen enhancement versus rotational velocity diagram, for the mass gainers of 5 computed binary systems, where each data point represents a duration of 20 000 yr. All but one system start out with $16\,M_\odot + 15\,M_\odot$. Open hexagons are used for the Case A system shown in Fig. 6, while open squares show the analogous system with a 10 time stronger tidal force parameter. Three-spiked stars designate a Case B system with an initial period of 6 d, while five-spiked stars show the 8 d system from Fig. 7. Triangles mark a $10\,M_\odot + 9\,M_\odot$ Case A system with an initial period of 2.15 d. Only the core-hydrogen burning stage of the mass gainers is shown. As in Fig. 9, it can be seen that the enrichment by mass transfer alone goes up to about $\log N/N_i \simeq 0.6$, with larger enrichments predominantly produced by rotational mixing.

it appears rather striking that the few models shown in Fig. 10 can populate each part of the diagram in which Fig. 1 shows a significant density of stars. In fact, one may ask the question whether single stars are needed at all to understand Fig. 1, except for the slowly rotating non-enriched stars.

This question ties in with the one asked in Sect. 4: Could all rapid rotators be spun-up mass gainers and merger products? While it seems difficult to answer this question currently, the following statement seems secure: The FLAMES result shows that it is impossible to understand the nitrogen enhancement in massive stars without considering binary evolution. Whether the same statement can be made for single stars is questioned by the results shown above.

Acknowledgements

SCY was supported by the DOE Program for Scientific Discovery through Advanced Computing and NASA.

References

Berger, L., Koester, D., Napiwotzki, R., et al. 2005, A&A, 444, 565
Belczynski, K., Kalogera, V., & Bulik, T. 2002, ApJ, 572, 407
Bradley, P. A. 1998, ApJS, 116, 307
Bradley, P. A. 2001, ApJ, 552, 326
Braun, H. & Langer, N. 1995, A&A, 297, 483
Brinkworth, C. S., Burleigh, M. R., & Marsh, T. R. 2007, in: R. Napiwotzki & M. R. Burleigh (eds.), *15th European Workshop on White Dwarfs* (San Francisco: ASP), ASP Conf. Ser., 372, 183
Camilo, F., Ransom, S. M., Halpern, J. P., & Reynolds, J. 2007, ApJ, 666, L93
Cantiello, M., Yoon, S.-C., Langer, N., & Livio, M. 2007, A&A, 465, L29
Detmers, R., Langer, N., Podsiadlowski, P., & Izzard, R. 2008, A&A, in press, astro-ph/0804.0014
Dolez, N., Vauclair, G., Kleinman, S. J., et al. 2006, A&A, 446, 237
Eldridge, J. J., Izzard, R. G., Tout, C. A. 2008, 384, 1109
Evans, C. J., Smartt, S. J., Lee, J.-K., et al. 2005, A&A, 437, 467
Ferrario, L. & Wickramasinghe, D. T. 2005, MNRAS, 356, 615
Handler, G. 2001, MNRAS, 323, L43
Handler, G., Romero-Colmenero, E., & Montgomery, M. H. 2002, MNRAS, 335, 399
Heger, A. & Langer, N. 2000, ApJ, 544, 1016
Heger, A., Langer, N., & Woosley, S. E. 2000, ApJ, 528, 368
Heger, A., Woosley, S. E., & Spruit, H. C. 2005, ApJ, 626, 350
Hunter, I., Brott, I., Lennon, D. J., et al., 2008, ApJ, 676, L29
Kepler, S. O., Giovannini, O., Wood, M. A., et al. 1995, ApJ, 447, 874
Kleinman, S. J., Nather, R. E., Winget, D. E., et al. 1998, ApJ, 495, 424
Langer, N. 1998, A&A, 329, 551
Maeder, A. & Meynet, G. 2000, ARAA, 38, 143
Maeder, A. & Meynet, G. 2005, A&A, 440, 1041
Meynet, G. & Maeder, A. 2000, A&A, 361, 101
Mokiem, M. R., de Koter, A., Evans, C. J., et al. 2006, A&A, 456, 1131
Packet, W. 1981, A&A, 102, 17
Petrovic, J., Langer, N., Yoon, S.-C., & Heger, A. 2005a, A&A, 435, 247
Petrovic, J., Langer, N., & van der Hucht, K. A. 2005b, A&A, 435, 1013
Podsiadlowski, P., Joss, P. C., & Hsu, J. J.L. 1992, ApJ, 391, 246
Pols, O. R. & Marinus, M. 1994, A&A, 288, 475
Spruit, H. C. 2002, A&A, 381, 923
Suijs, M. P. L., Langer, N., Poelarends, A.-J., et al. 2008, A&A, 481, L87
Vanbeveren, D., Van Bever, J., & Belkus, H. 2007, ApJ, 662, L107
Vazquez, G. A., Leitherer, C., Schaerer, D., et al. 2007, ApJ, 663, 995
Wellstein, S. 2001, PhD thesis, University Potsdam
Wellstein, S. & Langer, N. 1999, A&A, 350, 148
Wellstein, S., Langer, N., & Braun, H. 2001, A&A, 369, 939
Winget, D. E., Nather, R. E., & Clemens, J. C. 1994, ApJ, 430, 839
Woosley, S. E. & Heger, A. 2006, ApJ, 637, 914
Yoon, S.-C. & Langer, N. 2005, A&A, 443, 643
Yoon, S.-C., Langer, N., & Norman, C. 2006, A&A, 460, 199

The Effect of Massive Binaries on Stellar Populations and Supernova Progenitors

John J. Eldridge[1], Robert G. Izzard[2] and Christopher A. Tout[1]

[1] University of Cambridge, Institute of Astronomy, The Observatories, Madingley Road, Cambridge CB3 0HA, UK
e-mail (J.J. Eldridge): jje@ast.cam.ac.uk
[2] Sterrekundig Instituut Utrecht, Postbus 80000, 3508 TA Utrecht, The Netherlands

Abstract. We have calculated a large set of detailed binary models and used them to test the observed stellar population ratios that compare the relative populations of blue supergiants, red supergiants and Wolf-Rayet stars at different metallicities. We have also used our models to estimate the relative rate of type Ib/c to type II supernovae. We find, with an interacting binary fraction of about two thirds, that we obtain better agreement between our models and observations than with single stars. We discuss the use of models in determining the nature of supernova progenitors and show the surprising result that many type Ib/c supernova progenitors are less luminous and less massive in our models than the observed population of Wolf-Rayet stars.

Keywords. binaries: close – stars: evolution – supernovae: general – stars: Wolf-Rayet

---

1. Introduction

When we try to match the observed properties of massive stars with single star models we have always found a poor fit. For example, the ratio of the number of blue supergiants to red supergiants and its variation with metallicity cannot be reproduced (Langer & Maeder 1995; Massey & Olsen 2003). The sources of this disparity could be due to our limited models of convection or the lack of rotation in our stellar models (Maeder & Meynet 2001). Another possibility is that a large number of these stars have binary companions. It is well known that this can change stellar populations substantially (Vanbeveren, Van Rensbergen & De Loore 1998).

We have produced a large set of binary stellar models to predict their effect on the relative populations of blue supergiants (BSGs), red supergiants (RSGs) and Wolf-Rayet (WR) stars. The advantage of our study (Eldridge, Izzard & Tout 2008) is that we model the binary interactions in a detailed stellar evolution code rather than approximating the evolution as in rapid population synthesis that use tables or equations fitted to detailed models (Hurley, Tout & Pols 2002). These detailed models ensure that more uncertain phases of evolution, such as when the hydrogen envelope is close to being removed, are treated as accurately as possible. This is vital when we attempt to determine a stellar type for a model and its respective lifetime.

In this proceedings we first summarize our results as described in Eldridge, Izzard & Tout (2008), where we compare the relative stellar populations predicted by our code with observed populations. We then discuss how our results indicate that a large number of type Ib/c supernovae, those devoid of hydrogen, may not have WR stars as their progenitors.

2. Stellar population ratios

In Figures 1, 2 and 3 we compare observations of three stellar population ratios at various metallicities to ratios predicted from our models. The observed ratios are calculated by observing galaxies and counting the number of each type of star and dividing the number in one population by another (Massey & Olsen 2003). The model predictions are calculated by first determining how long each model spends as either a BSG, RSG or WR star. We use the definitions of BSGs and RSGs from Massey & Olsen (2003) and the definitions for WR stars from Maeder & Meynet (1994). In addition we also require that a WR star must have $\log_{10}(L/L_\odot) \geqslant 4.9$ to be consistent with the luminosity limit for BSGs and RSGs.

With the BSG, RSG and WR lifetimes we then use an initial mass function and assume a constant star formation rate to calculate the relative populations. For the binary population we take flat distributions in the ratio of the secondary to the primary mass and the logarithm of the initial separation. We find that about two thirds of our binaries interact while the remaining third evolve as single stars.

Figure 1 shows the relative number of BSGs, which are main sequence stars, to RSGs, post-main sequence stars that still have their hydrogen envelopes. Even though there are only two points on the figure, it demonstrates that binary models better match these observations. This is for two reasons. First more RSGs lose their hydrogen envelopes than for single stars by interaction with their companion and thus the number of RSGs decreases. Secondly, more massive stars are formed from low-mass companions accreting matter or by stellar mergers. Therefore the BSG population is increased.

The RSGs that lose their hydrogen envelopes become helium stars or WR stars. Figure 2 shows how the number of O stars (the hottest BSGs) against the number of WR stars, varies with metallicity. We see that the increased WR population does improve agreement with observations. Also the binary line agrees with predictions of the Geneva

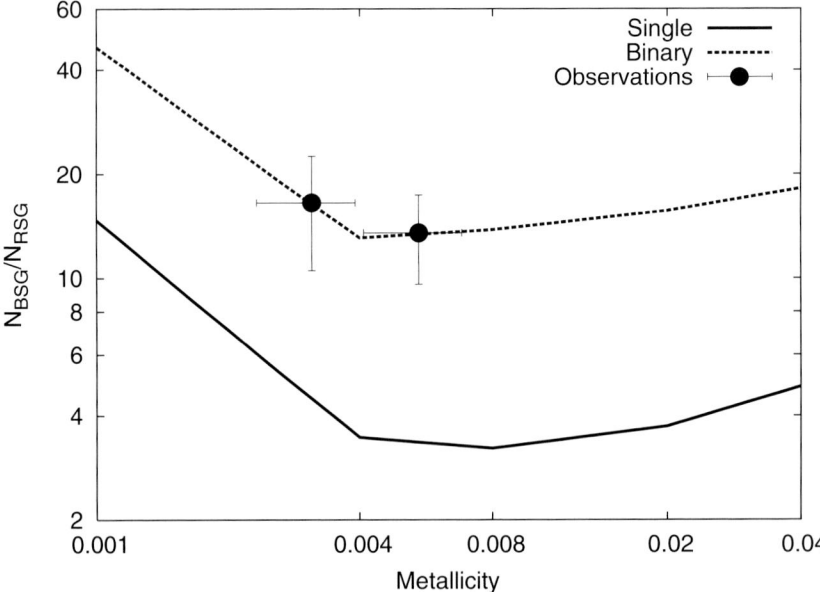

Figure 1. Ratio of the numbers of blue supergiants to red supergiants versus metallicity. Observations are taken from Massey & Olsen (2003). The solid line is from our single star models while the dashed line is from our binary models.

rotating models Meynet & Maeder (2005) so rotation and binaries appear to have the same effect on this stellar population ratio.

Binaries do not improve agreement for all the ratios. In Figure 3 we show the relative number of RSGs to WR stars. We see that neither the single nor binary star model predictions match the observed trend. This it at odds with the match in the previous two figures which show good agreement. Including missing details from our models (such as rotation) may improve agreement. However it is more likely that because the ratio is based on a small number of observed RSGs, especially when metallicity is greater than $Z = 0.008$ our assumptions in calculating the predicted ratios are not appropriate. For example with only a few stars we cannot be certain we are sampling the IMF fully nor looking at a sample with constant star formation.

In summary binary models improve agreement between predicted and observed stellar population ratios. The agreement is not always perfect and extra details still need to be added to our models.

3. Relative supernova rates

The final outcomes of massive stellar evolution, supernovae (SNe), can also be used to provide a constraint on stellar models because the SN type depends on the final stellar type. These events come in three broad types. Type Ia SNe are thought to be thermonuclear explosions of carbon-oxygen white dwarfs and are not of relevance here. The remaining two types are core-collapse SNe where either an oxygen-neon or iron core is formed which collapses when electron degeneracy pressure or nuclear burning can no longer provide support against gravitational collapse. This collapse releases a tremendous

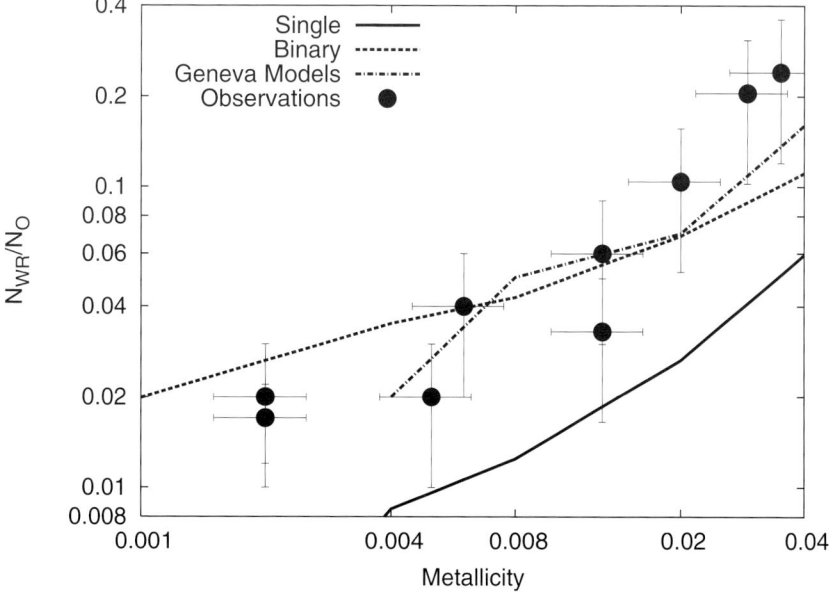

Figure 2. Ratio of the numbers of Wolf-Rayet stars to O-supergiants versus metallicity. Observations are taken from Maeder & Meynet (1994). The solid line is our single star models while the dashed line is our binary models. The dashed-dotted line is from the Geneva models (Meynet & Maeder 2005). The y-axis error bars are an assumed error of 50 percent of the values given by Maeder & Meynet (1994).

amount of energy that is transferred into the envelope causing the subsequent SN (Heger et al. 2003; Eldridge & Tout 2004).

While there are many subtypes of SN the broadest definitions are type II, when hydrogen is detected in the SN spectrum and type Ib/c where hydrogen is undetected. The relative rate of type Ib/c to type II SNe indicates the number of stars that experienced mass loss strong enough to remove their hydrogen envelope before core-collapse. Prantzos & Boissier (2003) first showed that this ratio decreases with metallicity. However, they only estimated the SN metallicities from the host galaxy magnitude. Recently Prieto, Stanek & Beacom (2008) presented a more detailed analysis, estimating the metallicity by spectroscopy. We compare both observed trends in Figure 4. They agree within the errors but Prieto, Stanek & Beacom (2008) tend to have slightly lower ratios around solar metallicity and predict a much shallower evolution with metallicity. The lowest metallicity bin is uncertain and is based on one type Ib/c SN.

Comparing to the model predictions we see that the binary models and Geneva rotating models provide the best agreement. A large fraction of the type Ib/c progenitors are not WR stars but are helium stars with $M < 5M_\odot$. In the stellar population ratios above for a star to be a WR star we required that $\log(L/L_\odot) \geqslant 4.9$ which is similar to the least luminous WR star that has been observed. However from our binary models there are many stars that lose their hydrogen envelopes and explode as type Ib/c SN but are not WR stars, this has been discussed by others (Vanbeveren, Van Rensbergen & De Loore 1998; Pols & Dewi 2002). The question then becomes where are these helium stars or low-mass WR stars. They remain unobserved. They would be similar to stars such as those described by Wood & Lockley (2000) and Oliveira, Steiner & Cieslinski (2003).

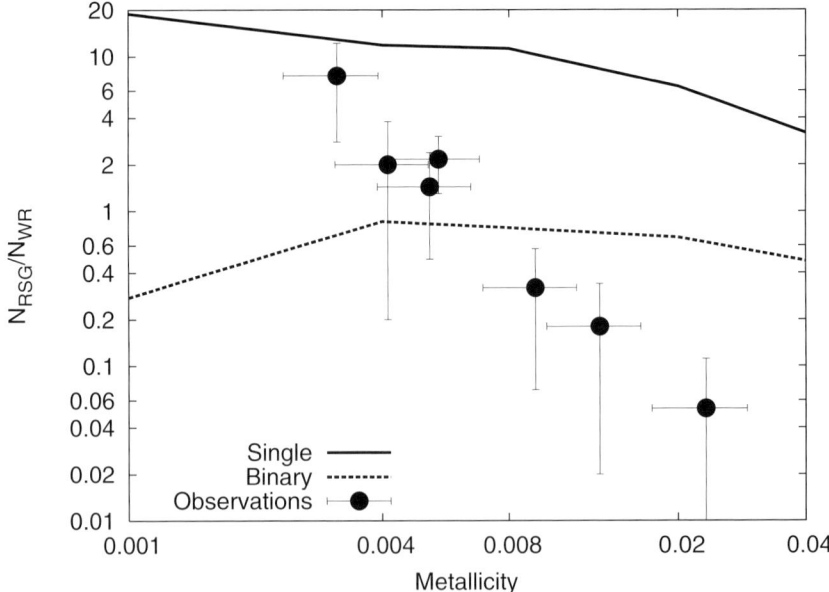

Figure 3. Ratio of the number of red supergiants to Wolf-Rayet stars versus metallicity. The observations are taken from Massey (2003). The solid line is from our single star models while the dashed line is from our binary models.

4. Progenitors of type Ib/c supernovae

To support the above conclusion that many type Ib/c SN progenitors must be low-mass helium stars we have compared our binary models to the detection limits placed on the progenitors by studies such as that of Crockett *et al.* (2007). If we compare the deepest B-band magnitude limit for a type Ib/c SN to date, that of SN 2002ap, to model B-band magnitudes calculated by combining our models and the WR atmosphere models of Hamann, Gräfener & Liermann (2006), then we find that helium/WR stars more massive than $3\,M_\odot$ would have been observed while only the less massive stars would remain undetected. This confirms the conclusion of Crockett *et al.* (2007) that if the progenitor was a normal massive WR star it would have been observed in the pre-explosion images. The number of such non-detections is growing and therefore a large number of type Ib/c SN may not have WR progenitors as previously thought.

5. For the future

The next step with this large set of binary models is to find further problems to apply it to. One extension is to model an instantaneous burst of star formation rather than continuous star formation.

Where are these helium stars or low-mass WR stars? We infer they exist but answering this question requires an observational solution. The reason why they have remained unobserved to date is that they may be in binaries and may be hidden by their companions, or they could be copious producers of dust and therefore obscured. Regions where they may exist must be identified and observed more extensively.

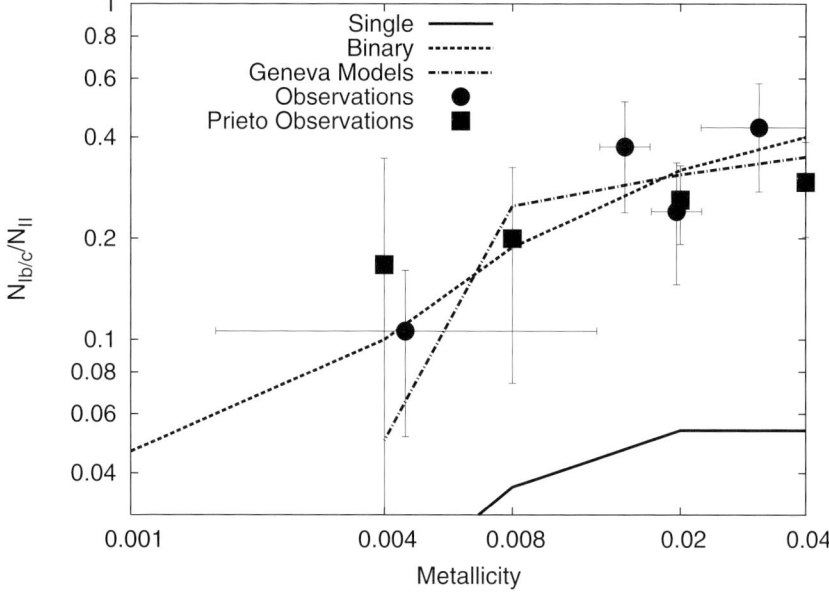

Figure 4. The observed and predicted ratios of the type Ib/c supernova rate to the type II supernova rate. Observations are taken from Prantzos & Boissier (2003) (*circles*) and Prieto, Stanek & Beacom (2008) (*squares*). The Geneva model predictions are taken from Meynet & Maeder (2005). The line is for their rotating models.

Acknowledgements

JJE carried out as part of the "Understanding the lives of massive stars from birth to supernovae" EURYI Award. CAT thanks Churchill College, Cambridge for his Fellowship. RGI thanks the NWO for his current fellowship in Utrecht.

References

Crockett R. M., Smartt S. J., Eldridge J. J., et al. 2007, *MNRAS*, 381, 835
Eldridge J. J. & Tout C. A. 2004, *MNRAS*, 353, 87
Eldridge J. J, Izzard R. G., & Tout C. A. 2008, *MNRAS*, 384, 1109
Hamann W.-R., Gräfener G., & Liermann A. 2006, *A&A*, 457, 1015
Heger A., Fryer C. L., Woosley S. E., et al., 2003 *ApJ*, 591, 288
Hurley J. R., Tout C. A., & Pols O. R. 2002, *MNRAS*, 329, 897
Langer N. & Maeder A. 1995, *A&A*, 295, 685
Maeder A. & Meynet G. 1994, *A&A*, 287, 803
Maeder A. & Meynet G. 2001, *A&A*, 373, 555
Massey P. 2003, *ARA&A*, 41, 15
Massey P. & Olsen K. A. G. 2003, *AJ*, 126, 2867
Meynet G. & Maeder A. 2005, *A&A*, 429, 581
Oliveira A. S., Steiner J. E., & Cieslinski D. 2003, *MNRAS*, 346, 963
Pols O. R. & Dewi J. D. M. 2002, *PASA*, 19, 233P
Prantzos N. & Boissier S. 2003, *A&A*, 406, 259
Prieto J. L., Stanek K. Z., & Beacom J. F. 2008, *ApJ*, 673, 999
Vanbeveren D., Van Rensbergen W., & De Loore C. D. 1998, *The Brightest Binaries* (Dordrecht: Springer) *Astrophys. Space Science Lib.*, v. 232,
Wood J. H. & Lockley J. J. 2000, *MNRAS*, 313, 789

Discussion

MAEDER: It would be interesting to combine the results of single and binary evolution with ratios between them based on the binary rates given yesterday by Tony Moffat. Now, as a side remark, I am surprised by the smallness of the blue loops of your tracks in the HR diagram.

ELDRIDGE: The ratio of the single and binary populations is fixed by the assumed initial binary parameter distribution. The number of interacting binaries we find is similar to the numbers given by Tony Moffat yesterday. We could consider this an independent check of to determine single/binary ratio. To answer the blue loops remark I should say myself and Richard Stancliffe are very interested in them and are looking into them. We are looking into them observationally and theoretically.

GAYLEY: The nomenclature of "single star" and "binary" is confusing because theorists tend to mean objects whose evolution is altered, and observers tend to mean objects that offer unique observational diagnostics. So I'd like to enter a plea that we routinely distinguish three types, rather than two: evolutionary binaries, observational binaries and stars of unknown or unobservable binarity. Note these overlap: interacting binaries are the overlap of the first two, runaways are the overlap of the first and the third.

ELDRIDGE: I think this is a great idea. One thing I intend to do is to calculate such details as how many binaries we would expect to observe, how many stars we might not observe as binaries and also details such as runaways. But the situation is confusing and you are right that we must be clear about how we apply the adjective "binary" to stars.

Thoughts on Core-Collapse Supernova Theory†

Adam Burrows[1,2], Luc Dessart[2], Christian D. Ott[2], Eli Livne[3] and Jeremiah Murphy[2]

[1]Department of Astrophysical Sciences, Princeton University, Princeton, NJ 08544
email: burrows@astro.princeton.edu
[2]Dept. of Astronomy, University of Arizona, Tucson, AZ 85721
[3]Racah Institute of Physics, The Hebrew University, Jerusalem, Israel

Abstract. An emerging conclusion of theoretical supernova research is that the breaking of spherical symmetry may be the key to the elusive mechanism of explosion. Such explorations require state-of-the-art multi-dimensional numerical tools and significant computational resources. Despite the thousands of man-years and thousands of CPU-years devoted to date to studying the supernova mystery, both require further evolution. There are many computationally-challenging instabilities in the core, before, during, and after the launch of the shock, and a variety of multi-dimensional mechanisms are now being actively explored. These include the neutrino heating mechanism, the MHD jet mechanism, and an acoustic mechanism. The latter is the most controversial, and, as with all the contenders, requires detailed testing and scrutiny. In this paper, we analyze recent attempts to do so, and suggests methods to improve them.

Keywords. supernovae: general – radiative transfer – stars: magnetic fields – stars: neutron – neutrinos – hydrodynamics – MHD

1. Introduction

After decades of theoretical exploration, it is now clear that the core collapse that leads to supernova explosions, and the supernovae themselves, are not spherical phenomena. During the violent dynamical sequence that gives birth to both a supernova and a neutron star (or black hole!), the core of a massive star runs a formidable gauntlet of hydrodynamic instabilities. First, the progenitor Chandrasekhar core experiences turbulent convection before collapse, ensuring density and entropy inhomogeneity at collapse. Then, the material behind the stalling bounce shock executes Rayleigh-Taylor overturn, aided later by neutrino heating from below that drives convection. The latter is boosted, if not overwhelmed, by the standing-accretion-shock instability (SASI; Blondin, Mezzacappa, & DeMarino 2003) that commences $\sim$100-250 milliseconds after bounce (Burrows et al. 2006), if there is no explosion before this, and has a frequency of from $\sim$20 to $\sim$80 Hz (Burrows et al. 2007a). The dominant SASI modes are $\ell = 1$ and $\ell = 2$ harmonics. Moreover, after bounce a shell below the neutrinospheres at $\sim$20-50 km in the inner core at a radius of $\sim$10-20 kilometers (km), executes convective overturn due to negative lepton (composition) gradients. It was hoped that such core convection, as well as doubly-diffusive instabilities ("neutron fingers"), could boost the neutrino luminosities that may ultimately be responsible for reenergizing the explosion. However, it has

† We acknowledge support for this work from the Scientific Discovery through Advanced Computing (SciDAC) program of the DOE, under grant numbers DE-FC02-01ER41184 and DE-FC02-06ER41452, and from the NSF under grant number AST-0504947.

been shown that the numbers are not encouraging (Dessart *et al.* 2006; Bruenn & Dineva 1996).

In addition, once the shock is launched, it progresses through a layer cake of outer zones, each bounded by density discontinuities and composition jumps. When the shock traverses these regions it trips further Rayleigh-Taylor-like and Richtmyer-Meshkov instabilities. Coupling all of the above mechanisms by which symmetry can be broken with rotation and magnetic stresses will only further enrich the multi-dimensional character of the hydrodynamics. Clearly, symmetry breaking is a key feature of core collapse supernova explosions.

Recently, Burrows *et al.* (2006) and Burrows *et al.* (2007abcd) have suggested that the violent turbulence around the inner core and the late-phase pounding of this core by accretion streams excite vigorous $\ell = 1$ and $\ell = 2$ core g-modes that damp by the emission of sound. If there had been no earlier explosion by other means, in their simulations Burrows *et al.* find that approximately ~ 1 second after bounce this sound can be sufficient to reenergize the shock. They suggest dumping acoustic power in the inner mantle at a rate of perhaps more than 10^{50} ergs s^{-1} for many seconds can in principle lead to a supernova. The core g-mode is very aspherical and leads to an anisotropic, oftimes unipolar, explosion. However, this g-mode/acoustic mechanism is quite controversial. Is the resolution (temporal as well as spatial) adequate? Can such simulations be trusted after $\sim$1,000,000 timesteps? Is the result code-dependent? Do other mechanisms, such as neutrino heating or MHD jets (Burrows *et al.* 2007e), trump the acoustic mechanism before it has a chance to operate?

Whatever the ultimate solution, a central theme for modern supernova theory has emerged – whatever the mechanism, be it neutrino heating, MHD jets, or acoustic (and it might be a mix of all three), the breaking of spherical symmetry is an organizing principle of the theoretical debate. This puts a premium on the development and testing of multi-physics, multi-dimensional radiation hydrodynamic and magnetohydrodynamic computational approaches and codes. This imperative, and the limitations of current computers and algorithms, are important reasons the supernova puzzle is as yet unsolved.

Recently, Marek (2007) and Marek & Janka (2008) have looked into the generation of core g-modes and have challenged some of the findings in our recent series of papers (Burrows *et al.* 2006,2007abcd) on the core-oscillation/acoustic power mechanism. While it is vitally important to scrutinize such novel and provocative ideas as the acoustic mechanism in detail and directly, we believe that these two studies fall somewhat short of adequately addressing the issues raised by our series on this topic. Below we explain our conclusions and make various observations we think are germane to the case at hand. Since the Marek (2007) work has more details on what is to be found in both works, we focus on it, though refer to each work where appropriate. Though our remarks and observations are critical of these works, they are meant merely to improve the science return of any group seeking to critically explore our provocative acoustic findings and we very much appreciate the effort expended by these researchers to seriously address, check, confirm, or refute our controversial findings.

Therefore, in this short paper, we map out some of the numerical challenges and methodologies of testing the core-oscillation hypothesis. We don't provide a systematic review of core-collapse supernova theory, such a task being beyond the scope of a short proceedings. For this, however, readers may profit from Burrows (2000) or Woosley & Janka (2005). Nevertheless, the enclosed discussion provides a snapshot of some of the current debate in supernova theory.

2. Summary Comments on a Recent Investigation into Core G-mode Oscillations

Here, we summarize some of our technical reservations about the Marek & Janka work, and then address more specific points in §3, §4, §5, and §6 below: 1) Their simulations do not extend to times after bounce when we say the core g-mode oscillations are large, but to times when we too say the excitation of such modes is minimal. Hence, they do not make the proper direct comparison, at the proper epoch; 2) They do their simulations in 1D in the central $1.6-1.7$ km, which we believe artificially dampens the growth of the $\ell = 1$ core g-mode; 3) The measure they use for the presence of the mode is $\Delta P/P$, where P is the pressure, at a given internal radius. In the convective region this is a measure of the strength of turbulence and of the potential for excitation of the interior g-mode by turbulence, but is not a good measure of the presence of the core g-mode itself; and 4) Marek & Janka (2008) claim that the onset of explosion might require a small nuclear incompressibility (K), and tout a value of $K = 180$ MeV. Their model with $K = 263$ MeV does not look as promising. However, the measured value of K is 240±20 MeV, much closer to the value employed in their non-exploding model, calling into question their central conclusion. We now proceed with a more specific discussion.

3. On $\Delta P/P$ as an Imperfect Measure of the G-mode

Marek (2007) and Marek & Janka (2008) use $\Delta P/P$ to discern the presence of the core oscillation itself, with Marek & Janka (2008) focussing only on the interior 10 and 20 kilometer (km) radii and Marek (2008) including a discussion of the outer radius at 35 km as well. However, $\Delta P/P$ (as plotted at 35 km in Fig. 7 of Burrows *et al.* 2006) is a direct signature not of the core g-mode, but of the pressure fluctuations in the region between the shock and the inner core. The primary origins of these "outer" pressure fluctuations are the SASI and neutrino-driven convection, not the core oscillation. The restoring force for core g-modes is buoyancy in the inner ∼10 km, not pressure, and by using $\Delta P/P$ as a g-mode index Marek & Janka focus unduly on a subdominant modal signature. A better measure of the presence and strength of an $\ell = 1$ core g-mode might be the overlap with the eigenfunction itself, particularly in the displacement or velocity, and the identification of the countervailing motion of the inner core with the shell around it; for the $\ell = 1$ core g-mode these regions oscillate out of phase.

Nowhere in Burrows *et al.* (2006, 2007abcd) do we propose $\Delta P/P$ as a measure for the strength of the core g-mode pulsations. Rather, we use it as a measure of the pressure fluctuations at the surface of the protoneutron star near $25-35$ km that can excite core g-mode oscillations in the first, weaker, phase of the excitation sequence we see (see also Goldreich & Keeley 1977; Goldreich & Kumar 1988,1990). The second phase is the more important, that of the "self-excited" oscillator during and after the onset of explosion, which in our calculations occurs more than ∼1 second after bounce. The calculations of Marek & Janka (2008) do not extend to more than ∼610 milliseconds after bounce, leaving a crucial gap of ∼400 milliseconds. At this earlier time, we too see small amplitudes for the core g-mode oscillations and they aren't yet having a significant dynamical influence (Burrows *et al.* 2007a).

Feedback ("extra") pressure waves from the resultant g-mode core motion are not so manifest early on at radii of ∼35–50 km, and certainly not at 10–20 km, even when the amplitude of the $\ell = 1$ core g-mode oscillation (better measured with the Lagrangian displacement in the inner 10 km) is modest, not small. g-modes are predominantly "gravity modes" and the p-mode character they have in the outer region is because of their *mixed* character and is sub-dominant. It is only when the core oscillation becomes *non-linear* that the p-mode character of the outer region of the core g-mode becomes interesting. Then, the outer pressure wave components of the *complex* g-mode can propagate out with vigor and steepen into shocks. All during this time the inner 10 km (with a node near 6−8 km) is oscillating in g-mode fashion, with gravitation/buoyancy as the restoring force. Note that this inner region is not convective and the inner g-mode is not, of course, in an evanescent region, though its outer tail is; it is only there in the evanescent region that the mode is predominantly p-mode in character. This is quite unexceptional (Unno *et al.* 1989).

In our simulations, supersonic accretion funnels that penetrate through the kinks in the outer shock structure created by the vigorous SASI are the ultimate agents of strong core g-mode excitation. It is the downflowing plumes and their *ram pressure* that excite the $\ell = 1$ core g-mode at very late times. Even when these accretion funnels are steadily impinging upon the inner core and do not have resonant frequency components, they can excite ∼300-Hz g-modes; witness the generation of gravity waves on a pond due to a steady jet of water. It is the width of the exciting stream, not its temporal fluctuation, that sets the "wavelength" of the g-modes that can be excited. The frequency spectrum of the excitation is a consequence of this wavelength and the dispersion relation of gravity waves on the inner core. This excitation frequency spectrum easily overlaps the core g-mode spectrum.

4. On the Amplitudes of $\Delta P/P$

$\Delta P/P$ and Mach number in post-shock regions grow with time in models for which the shock is stalled and it bounds the inner turbulent region. This includes almost all models published to date. Mach numbers approach a few tenths to ∼0.5 at late times (0.5-1.0 seconds). The Mach numbers provide a measure of the vigor of the motions in this region. In turbulent regions, high Mach numbers translate directly into high pressure fluctuations, with Mach numbers near one implying pressure fluctuations of order unity, i.e. large.

It is not clear why in the calculations of Marek (2007) $\Delta P/P$ in the turbulent regions is generically small, even at the latest times achieved (for most of his models, around ∼350 milliseconds after bounce). Both Burrows *et al.* (2006, 2007abc) and Yoshida, Ohnishi, & Yamada (2007) have published much larger values that seem more consistent with the character of the late post-shock turbulent flow. As noted, Marek (2007) generally does not simulate long enough after bounce. Also, his initial seed perturbations might be small. With small seeds it takes longer to erase the memory of initial conditions and to achieve a given amplitude. During ∼0.05 to ∼0.5 seconds after bounce, Yoshida, Ohnishi, & Yamada obtain values of $<\Delta P/P>$ for the $\ell = 1$ component near 35 km of ∼ 0.1, with excursions to 0.2. For this same quantity, Burrows *et al.* (2006) see a steady growth in its value from ∼0.05 at 0.3 seconds after bounce to ∼0.2 at ∼0.55 seconds after bounce. However, Marek (2007) obtains values of $\Delta P/P$ for the $\ell = 1$ component at $r = 35$ km and at ∼350 ms after bounce of only ∼0.005 to 0.02. Interestingly, the value Marek obtains depends upon the EOS employed (see Marek 2007, Figures 5.9 and 5.10), with ten times larger amplitudes for the softer EOS. Marek & Janka (2008) do calculate one model to ∼610 ms after bounce, but this model is rotating modestly and rotation is expected to partially suppress the SASI implicated in the turbulence generated in this region (Burrows *et al.* 2007a). Nevertheless, even in this model $<\Delta P/P>$ achieves a value of ∼0.05 and is still rising when it is halted. In fact, $<\Delta P/P>$ is rising at the end of all the Marek (2007) simulations. To directly compare the values of $\Delta P/P$ in the outer turbulent regions and the g-mode amplitudes in the inner core with those Burrows et al. (2007abc) obtain, it is important for Marek & Janka to continue their simulations for another ∼500 ms. In Burrows et al. (2007a), we needed to evolve for more than 0.9−1.0 seconds to see vigorous core g-mode oscillation. It takes a long time for the core oscillation to manifest itself and the simulation time needs to be commensurate.

5. On Simulating the G-mode

To show that the MPA hydro code can support and simulate $\ell = 1$ g-modes, Marek (2007) calculates a few test models in which he imposes such a mode in the inner core and follows it for 10−20 milliseconds. Ten to twenty milliseconds is only a few oscillation cycles. However, in all but one model, he constrains the inner 1.6 km to 1D motion. In Marek & Janka (2008), this constrained inner core has a radius of 1.7 km. The g-mode is, thereby, forced to flow *around* this inner 1.6/1.7 km. Since the node of this mode is near 6−8 km, we believe that 1.6 km is too large a region in which to inhibit the necessary multi-D flow. Importantly, the $\ell = 1$ g-mode has its greatest amplitude in this central region, where its eigenmode motion is straight through $r = 0$ (the modal velocities are the same for $+x$ and $-x$; spherical coordinates with a reflecting boundary at $r = 0$ unphysically flip the sign, by construction). Though in Marek (2007) and Marek & Janka's (2008) 2D/1D calculations the pressures and velocities are fluctuating around this 1.6/1.7-km region, the implied constraint force that keeps the very inner core from naturally responding will by its nature mute the expression of the $\ell = 1$ g-mode.

In the calculation reported by Marek & Janka (2008), they see the imposed g-mode oscillation decay within a few cycles, which is rather fast. They state in reference to their tests: "These demonstrate that our code is well able to track large-amplitude g-modes, also of dipole character, *if such modes are excited in the neutron star core* [our italics]." However, they do not in fact demonstrate that their code can track in a self-consistent fashion *long-term* excitation by anisotropic accretion and turbulence. This is the crucial question and they have not shown that their inner 1D region doesn't inhibit excitation. The decay they witness could easily be due in part or in large measure to the fact that their inner core is anchored. One way to address this would be to explore the dependence of the decay time on the size of the region that one does in 1D. For this test, the radius of the 1D region could be varied from, say, 0.5 km to 4 km.

Marek (2007) does indeed perform one test calculation all the way to the center ($r = 0$), and presumably imposes a reflecting boundary condition there. However, he calculates this model for a total of only ∼10 ms and starts the calculation only ∼30 ms after bounce. As seen in Burrows et al. (2006, 2007a), this is far too early to start and far too short a time to perform such a calculation if one wants to witness the excitation of core oscillations. Hence, by constraining the inner 1.6 km or employing a reflecting boundary at $r = 0$, the most important and largest amplitude region of the $\ell = 1$ mode is thereby neutered and its excitation inhibited. To clearly avoid this problem, we believe that simulations should be done with a quasi-Cartesian grid at the very center. Otherwise, the amplitude of the g-mode seen in a given hydrodynamic environment is artificially suppressed. This fact and our experience are the origins of our caveats in Burrows et al. (2006, 2007ab).

6. Additional Observations

Marek (2007) claims that the acoustic-driven explosions seen in the simulations of Burrows et al. (2006, 2007ab) arise rather abruptly. However, in fact, these explosions emerge in those simulations over a period of ∼100-200 ms. For comparison, when a neutrino-driven explosion is witnessed in the calculations of Buras *et al.* (2006), Kitaura *et al.* (2006), or Burrows et al. 2007c, it emerges on a timescale of ∼50-150 ms. We suggest that this is a characteristic timescale for any explosive instability in the "supernova" core and that the acoustic mechanism is not exceptional in this regard. Marek (2007) also claims that at the late times when Burrows *et al.* (2006, 2007ab) see vigorous core oscillations there is not sufficient accretion power to maintain it. As Figures 2 and 7 in Burrows *et al.* (2007a) clearly show, though the accretion rate has subsided by these times, there is still ample accretion power to maintain such oscillations. It is merely a matter of the efficiency of the conversion of accretion power into mechanical power, as opposed to neutrino luminosity. A ∼10% efficiency would be adequate for this purpose.

Marek (2007) suggests that when the shock radius achieves a value larger than ∼300 km this indicates the onset of explosion, that this is a "point of no return." The calculations of Marek (2007) and Marek & Janka (2008) are stopped near the time when the average shock radius achieves this value, though why is not clear. Marek & Janka (2008) add in the discussion of their results the timescale τ_{adv} versus τ_{heat} condition advocated by Thompson, Quataert, & Burrows (2005), but such arguments are no substitute for actual calculation. During the vigorous SASI phase, it is often the case that the shock radius substantially exceeds 300 km, only to recede again and continue non-linear and bounded SASI pulsation (Burrows *et al.* 2006, 2007a). Hence, we caution against using simple criteria for the onset of explosion. In fact, none of the SASI- and neutrino-aided "explosions" seen by the MPA group, neither the 11.2 $M_\odot$ model of Buras *et al.* (2006) nor the 15 $M_\odot$ model of Marek & Janka (2008), is actually followed for more than a few tens of *milliseconds* after explosion seems to ensue. It is crucially important that any claim of explosion be buttressed by calculations in which the shock actually achieves a radius of thousands of kilometers (preferably larger), and not just 300-400 km. The calculations of Burrows *et al.* (2006, 2007abcd) were carried out until the shock reached a radius of 4000-5000 km, and even this should not be considered far enough.

Importantly, we note that the incompressibility (K) of nuclear matter has been measured to be 240±20 MeV (Shlomo, Kolomietz, & Colò 2006; Lattimer & Prakash 2007). Marek & Janka (2008) calculate models using values of 180 MeV and 263 MeV, and conclude that only the 180-MeV model witnesses the onset of explosion. As a result, it is not clear, given the dependence on the EOS they identify in their calculations, that a more realistic value of K would lead to the same explosive behavior they claim for their model with $K = 180$ MeV.

In conclusion, we do not challenge the possibility that the neutrino mechanism might act on shorter timescales than the acoustic mechanism, and thereby abort it. However, we do hope that the community will redouble its efforts to test in a cogent and clear fashion its particulars and viability. This will require calculations to at least 1.2 seconds after bounce and the demonstration that these calculations do not suppress $\ell = 1$ and $\ell = 2$ core g-mode oscillations due to numerical exigencies. Conversely, we are redoubling our efforts to address afresh all the issues that attend core-collapse theory, including the neutrino, acoustic, and MHD mechanisms in all their particulars (see, for example, Dessart *et al.* 2007,2008; Hubeny & Burrows 2007; and Burrows *et al.* 2007e).

References

Blondin, J. M., Mezzacappa, A., & DeMarino, C. 2003, *ApJ*, 584, 971
Bruenn, S. W. & Dineva, T. 1996, *ApJ*, 458, L71
Buras, R., Janka, H.-T., Rampp, M., & Kifonidis, K. 2006, *A&A*, 457, 281

Burrows, A. 2000, *Nature*, 403, 727

Burrows, A., Dessart, L., & Livne, E. 2007c, in: R. McCray, K. Weiler, & S. Immler (eds.), *SUPERNOVA 1987A: 20 YEARS AFTER: Supernovae and Gamma-Ray Bursters* (New York: AIP) *AIP Conf. Proc.*, 937, 370

Burrows, A., Dessart, L., Livne, E., & Ott, C. D. 2007b, in: G. E. Brown (ed.), *Centennial Festschrift for Hans Bethe*, (Netherlands: Elsevier), *Phys. Rep.*, 442, 23

Burrows, A., Dessart, L., Livne, E., & Ott, C. D. 2007d, in: L. A. Antonelli, G. L. Israel, L. Piersanti, L., & A. Tornambe (eds.), *The Multicoloured Landscape of Compact Objects and their Explosive Origins* (New York: AIP), *AIP Conf. Proc.*, 924, 243

Burrows, A., Livne, E., Dessart, L., et al. 2006, *ApJ*, 640, 878

Burrows, A., Livne, E., Dessart, L., et al. 2007a, *ApJ*, 655, 416

Burrows, A., Dessart, L., Livne, E., et al. 2007e, *ApJ*, 664, 416

Dessart, L., Burrows, A., Livne, E., & Ott, C. D. 2006, *ApJ*, 645, 534

Dessart, L., Burrows, A., Livne, E., & Ott, C. D. 2007, *ApJ*, 669, 585

Dessart, L., Burrows, A., Livne, E., & Ott, C. D. 2008, *ApJ*, 673, L43

Goldreich, P. & Keeley, D. A. 1977, *ApJ*, 212, 243

Goldreich, P. & Kumar, P. 1988, *ApJ*, 326, 462

Goldreich, P. & Kumar, P. 1990, *ApJ*, 363, 694

Hubeny, I. & Burrows, A. 2007, *ApJ*, 659, 1458

Kitaura, F. S., Janka, H.-T., & Hillebrandt, W. 2006, *A&A*, 450, 345

Lattimer, J. M. & Prakash, M. 2007, *Phys. Rep.*, 442, 109

Marek, A. 2007, Ph. D. Thesis, Technical University of Munich (http://mediatum2.ub.tum.de/doc/604499/document.pdf)

Marek, A. & Janka, H.-T. 2008, *ApJ* submitted (arXiv:0708.3372)

Shlomo, S., Kolomietz, V. M., & Colò, G. 2006, *Eur. Phys. J.*, A30, 23

Thompson, T. A., Quataert, E., & Burrows, A. 2005, *ApJ*, 620, 861

Unno, W., Osaki, Y., Hiroyasu, A., et al. 1989, *Nonradial oscillations of stars, 2nd ed.* (Tokyo: University of Tokyo Press)

Woosley, S. E. & Janka, H.-T. 2005, *Nature Physics*, 1, 147

Yoshida, S., Ohnishi, N., & Yamada, S. 2007, *ApJ*, 665, 1268

Discussion

STANEK: Theorists fail to make core-collapse SNe. Does nature ever fail to make a SN from a massive star? Wht are the observational constraints?

BURROWS: I think that SN rates, OB star birth rates, pulsar birth rates, and nucleosynthetic constraints suggest that most massive stars must 'supernova', though a factor of two difference is still possible. However, I know of no compelling reason 'success' and 'failure' should alternate or vary wildly along the mass function. I don't think rotation is a key ingredient of explosion, except for hypernova and perhaps GRBs, rare subsets of the massive star family outcomes, so this parameter does not provide an acceptable out.

LIMONGI: One of the most important issues for masive stars is their final fate. Many people refer to the WHW or our picture, but in my opinion all the pictures must be taken very carefully because they have been obtained with simulated explosions that suffer many uncertainties. For how long can you perform your computations and can you say what is the real fate of a massive star at least in some cases?

BURROWS: We calculate for approximately 1-1.5 seconds of physical time after core bounce. While this is quite long by the standards of the field, it is not long enough to determine much of what we would like to know, such as final explosion energy, kick velocity, matter fallback, or the r-process. For the least-massive massive stars (8-9 $M_\odot$) we think we can conclude they are underenergetic (a few 10^{50} ergs), but much of what we want and need to know about the fate of a generic massive star is currently beyond credible theory, alas.

MODJAZ: Would an observational signature/prediction for your new acoustic SN mechanism be that the SN are unipolar, not bipolar? So, for example, polarization measurements of SN Ib/c could test that by constraining the geometry (bipolar vs. unipolar).

BURROWS: Yes, more often than not I would expect acoustic-driven explosions to be top-bottom asymmetric. However, late-time neutron-driven explosions, of all but the least-massive massive stars, should also be top-bottom asymmetric and slightly unipolar. So, 'unipolar' vs. 'bipolar' is a means of distinguishing MHD-driven jet models of explosion from both acoustic and neutron models. To distinguish acoustic and many neutrino driven models from one another via morphology or polarization requires more subtle discriminants.

DAVIDSON: This is a little off the track of your main concerns, but those lovely graphics make it irresistible. Of course you have shown some impeccable jets, but do cannonballs ever occur? The Crab Nebula includes a line of very well separated, dense, compact ovoidal things, which MacAlpine called 'argoknots' for reasons you can guess. Morphologically they are astounding.

BURROWS: We do not get sprays and clumps of material, but instabilities in the star and circumstellar material due to the passage and progress of the shock and later cooling are more likely culprits. However, the explosion asymmetry we see certainly set the stage for the variegated structures and condensations that emerge and are captured in SNR images.

Adam Burrows.

Episodic Mass Loss and Pre-SN Circumstellar Envelopes

Nathan Smith

Astronomy Department, University of California, Berkeley, CA 94720, USA
email: nathans@astro.berkeley.edu

Abstract. I discuss observational clues concerning episodic mass-loss properties of massive stars in the time before the final supernova explosion. In particular, I will focus on the mounting evidence that LBVs and related stars are candidates for supernova progenitors, even though current paradigms place them at the end of core-H burning. Namely, conditions in the immediate circumstellar environment within a few 10^2 AU of Type IIn supernovae require very high progenitor mass-loss rates. Those rates are so high that the only known stars that come close are LBVs during rare giant eruptions. I will highlight evidence from observations of some recent extraordinary supernovae suggesting that explosive or episodic mass loss (a.k.a. LBV eruptions like the 19th century eruption of Eta Car) occur in the 5-10 years immediately preceding the SN. Finally, I will discuss some implications for stellar evolution from these SNe, the most important of which is the observational fact that the most massive stars can indeed make it to the ends of their lives with substantial H envelopes intact, even at Solar metallicity.

Keywords. circumstellar matter – shock waves – stars: evolution – stars: mass loss – stars: winds, outflows – supernovae: general

1. Introduction

Supernovae are one of the most influential ways that massive stars act as cosmic engines, energizing and polluting the ISM (especially the early ISM). To evaluate the role of massive stars as cosmic engines observationally, we must first understand the relationship between massive star evolution (mass loss, rotation, metallicity) and the type of SN observed, since SNe and can be seen to large distances. The flip side to that coin is that if we can understand the relationship between different SN properties and the star's evolution locally, then SNe also become our best probe of mass loss, circumstellar structure, stellar evolution, the initial mass function, and star formation rates throughout cosmic time. In fact, one could argue that SNe are the *only* reliable way to directly probe mass loss of individual stars at large distances where spatially resolving individual stars and separating their light from their host environment is hopeless.

One of the most important questions concerns the connection between the type of SN observed and the star's initial mass. Whether a SN is a normal Type II with H (8–20 M$_\odot$ stars), a Type Ib/c that has shed its H envelope and probably died as a WR star, or a Type IIn with dense circumstellar material, is determined mainly by the star's mass-loss rate during evolution. A key prediction of most current stellar evolution models is that all stars above some mass, say $\sim$30 M$_\odot$, will fully shed their H envelopes and explode as WR stars, making SNe of Type Ib/c. Recent observations of SNe, on the other hand, are making it difficult to avoid the conclusion that this prediction is wrong.

Namely, SNe that have dense circumstellar material suggest to us that a small fraction of stars — probably the most massive stars — have violent mass-loss events that precede the final SN explosion. When H is present, these are seen as Type IIn SNe, making the most luminous SNe in the Universe.

I should preface the discussion below by saying that exploding LBVs or LBV-like stars represent a small fraction of observed core-collapse SN events. The vast majority are normal Type II's (see Smartt, this volume), while 15–20% are normal Type Ib/c SNe. That is perhaps not so surprising though, given that the most massive stars and LBVs themselves are quite rare.

2. LBVs as Supernova Progenitors

I'll start with a list of observational clues that some SNe occur when their progenitor star was in (or had recently been in) the LBV phase, contrary to expectations. Each of these on their own is compelling if not necessarily convincing, but taken together, they paint a consistent picture of LBVs as SN progenitors that is hard to dismiss.

- **SN 2006jc:** While this was not a Type IIn event (it was a peculiar Type Ib), it is unique so far in that it is the only SN that was actually *observed* to have an LBV-like outburst 2 yr before exploding as a SN. The eventual supernova showed an unusual spectrum and dust formation that was caused by very dense circumstellar material (see Foley *et al.* 2007; Pastorello *et al.* 2007; Smith et al. 2008).
- **SN 2005gl:** This Type IIn supernova had a progenitor star identified on pre-explosion images, which was a blue/yellow supergiant that had a luminosity and colors indicative of an LBV star (Gal-Yam et al. 2007).
- **Radio modulations:** Kotak & Vink (2006; and this volume) suggested that semi-periodic modulations in the radio lightcurves of a few SNe might be caused by the shock running into density variations caused by normal S Dor-type excursions of an LBV progenitor. The suggested cases were SNe 2001ig and 2003bg, both of which transitioned between Type Ib/c and Type II spectra (or the reverse) as they evolved, and possibly also SN 1979C and 1998bw. However, I should note that alternative interpretations of the radio modulations have been forwarded as well (see Ryder *et al.* 2004; Soderberg *et al.* 2006; Schwarz & Pringle 1996).
- **Circumstellar Nebulae:** There are three objects in our galaxy that are near twins of the unusual ring nebula around SN 1987A: HD 168625 is an LBV (Smith 2007), while SBW1 (Smith *et al.* 2007a) and Sher 25 (Smart *et al.* 2002) have abundances inconsistent with passage through a previous RSG phase. The nearly identical nebular ring structures suggest that the progenitor of SN 1987A had recently been in a similar evolutionary state when it exploded (see Smith 2007 and Smith *et al.* 2007a for further discussion).
- **Luminous Type IIn Supernovae:** The very high-luminosity of some Type IIn SNe can only be accounted for if the star had a huge mass-loss event in the decade before the SN. The clearest cases, which are also the three most luminous SNe ever observed, are SN 2006gy, SN 2005ap, and 2006tf (Smith *et al.* 2007b; Ofek *et al.* 2007; Smith & McCray 2007; Woosley *et al.* 2007; Quimby *et al.* 2007, and Smith *et al.*, 2008b); these require mass-loss rates on the order of 1 $M_\odot$ yr^{-1} in the few years before the SN. SN 1979C, 1988Z, and others, also suggest LBV-like mass loss based on their extended high-luminosity CSM interaction. In some cases, like SN2006gy and 2006tf, the progenitor's wind speed is *observed* in the narrow P Cyg Hα absorption, and its speed is $\sim$200 km s^{-1}. That's too fast for a RSG, but just right for an LBV.

This last point is probably the most interesting, in my view, because it highlights a relatively unfamiliar phenomenon that is truly remarkable, and that I think deserves more attention. Namely, the high luminosity of these Type IIn SNe require huge bursts of episodic mass loss *right before they explode*. In order to produce a Type IIn spectrum, the luminosity from CSM interaction must be comparable to or larger than the luminosity from recombination of the SN ejecta or radioactive decay, which are characteristically about 10^9 $L_\odot$. A convenient expression for the progenitor's mass-loss rate needed to produce an observed luminosity $L_9=L_{\rm SN}/(10^9 L_\odot)$ through CSM interaction, with an optimistic 100% efficiency of converting shock kinetic energy into visual light, is given by

$$\dot{M} = 0.04\, L_9 \frac{v_w}{200} \Big(\frac{v_{\rm SN}}{4000}\Big)^{-3} {\rm M_\odot yr^{-1}}$$

where v_w and $v_{\rm SN}$ are the progenitor's wind speed and the SN blast wave speed, respectively, in km s^{-1}. L can be measured from the light curve, while v_w and $v_{\rm SN}$ can usually be measured from the narrow and relatively broad components of the Hα line. For the main peak of SN 2006gy, the observed luminosity of L_9=50 at 70 d after explosion and the observed speeds of $v_w \simeq$200 km s^{-1} and $v_{\rm SN} \simeq$4,000 km s^{-1} required a mass-loss rate for the progenitor of $\sim$2 M$_\odot$ yr^{-1} for 5–10 yr. That's the most extreme example, but its easy to see that mass-loss rates more than about 10^{-2} M$_\odot$ yr^{-1} are needed in order for the CSM interaction luminosity to compete with the normal luminosity source of the SN.

If we look around us and ask "Among known stars in the Universe, which ones have the requisite mass loss to produce Type IIn SNe?", the only viable candidates with mass-loss rates above 10^{-2} M$_\odot$ yr^{-1} are *LBVs during a giant eruption*. If they are not bona-fide LBVs, then Type IIn progenitors are doing a darn good job of impersonating the H composition, wind speeds, and mass-loss rates of LBVs.

3. Synchronicity

The argument that the heavy mass loss occurs in the decade or so immediately preceding the SN is pretty straightforward. In the 100 days or so after explosion when the SN is bright and shows a Type IIn spectrum, it is sweeping through a radius of only a few 100 AU (typical observed blast wave speeds are only a few 1000 km s^{-1}, because the dense material decelerates the blast wave). In Type IIn SNe, the progenitor's wind speed can be observed in the narrow P Cygni absorption of Hα, usually indicating speeds of a couple 100 km s^{-1}. Thus, the radius out to which the blast wave reaches in the time it is being observed corresponds to the star's mass loss during the previous decade.

4. Deaths of the Most Massive Stars

Type IIn supernovae give us the most luminous SNe known in the Universe. A natural tendency is to associate them with the deaths of the most massive stars. Here are some reasons to favor that interpretation:

The classical LBVs that are expected to have giant eruptions *à la* Eta Carinae are the most luminous stars known, with initial masses of 60–150 M$_\odot$, although there are also some lower luminosity LBVs that are probably post-RSGs (Smith, Vink, & de Koter 2004) which may arise from stars with initial masses of perhaps 25–40 M$_\odot$. Since LBV eruptions provide our only observed precedent for the required mass-loss rates of Type IIn SNe, the simplest assumption is that Type IIn's do indeed represent the rare deaths of the most massive stars.

Aside from giant LBV eruptions, the "pulsational pair instability" is the best bet going (Woosley et al. 2007), as it is the only theoretically-proposed mechanism to produce mass loss similar to LBV eruptions, and it is expected to occur right before the SN. However, the instability only occurs for initial mass above $\sim$95 M$_\odot$ (Woosley et al. 2007). Thus, if this mechanism is responsible for the pre-SN mass loss of Type IIn's, then they *REALLY* must be the deaths of the most massive stars.

The brightest Type IIn SNe are energetic events, with combined luminous + kinetic energies well in excess of 10^{51} ergs (for example, 06gy emitted more than that in visible light alone). High-energy SN explosions are not something we associate with stars of moderate mass (i.e. 8-20 $M_\odot$).

(This is a bit of a tangent, but recent results show that ~30 $M_\odot$ black holes exist [Prestwich et al. 2007; Silverman & Filippenko 2008]. Those must come from the core-collapse deaths of very massive stars that did shed all their H, because the most massive H-free WR stars are less than that [Smith & Conti 2008].)

The luminous Type IIn's, like SN 2006gy, appear to eject 10's of $M_\odot$ in the decade or so before the SN. Very massive stars seem to be able to do this (Smith & Owocki 2006), but it is hard to believe that an 8–20 $M_\odot$ star could shed that much of its mass in a couple years. Furthermore, we would have no explanation for why only a small fraction of these stars have violent precursor events, whereas most die as normal Type II-P SNe. On the other hand, the most massive stars are rare compared to stars of 8–40 $M_\odot$, so its natural that their deaths would be a small fraction of all core-collapse SNe.

Lastly, wind speeds of progenitor stars can be gleaned from the narrow P Cygni Hα absorption in the spectra of Type IIn SNe, and they typically have fast winds of a few 10^2 km s^{-1}, characteristic of LBVs (see Smith et al. 2007b). Moderately massive stars (initial mass 8–20 $M_\odot$) should die as RSG's, with wind speeds of 10–20 km s^{-1}.

Conclusion: Type IIn progenitors are NOT moderately massive stars (initial masses of 8-20 $M_\odot$), but must be very massive stars, with initial masses that are probably above 50–60 $M_\odot$. Since they die with a lot of H, this is bad news for stellar evolution models.

5. LBVs, LBV Impostors, SN Impostors, and SNe

Since LBV-like eruptions are apparently responsible for the conditions that make the most luminous SNe in the Universe, our ignorance of their underlying physical mechanism is rather embarrassing.

1. One possibility is that the progenitor stars are in a regular LBV phase, that giant LBV eruptions occur repeatedly, and that some of these coincidentally occur shortly before the final SN explosion. The natural implication is that while some eruptions occur within a decade or so of the SN, many more probably will not. Statistically, this is a bit troubling, because Type IIn's represent 2–5% of core collapse SNe (Capallaro et al. 1997). If those correspond to the ones that have had giant eruption-like mass loss in the decades before the SN, then there must be at least 10 times more that are in a quiescent phase between giant eruptions. To achieve this from a normal Salpeter IMF, we would need to have *all* stars with initial masses above about 40 $M_\odot$ die as LBVs.

2. A second possibility is that these LBV-like eruptions really represent a *precursor* to the Type IIn SN, and their synchronization is not a coincidence. This could be the case if the SN-precursor mass ejections are in fact associated with an instability, perhaps the pulsational pair instability or some other instability leading to explosive mass loss in the very final nuclear burning stages. In that case, to get the observed rate of Type IIn's from a Salpeter IMF, we'd need all stars above 85–90 $M_\odot$ to explode in this way. Interestingly, this is also the range of masses that are supposed to be susceptible to the pulsational pair instability (Woosley et al. 2007).

In either case, how would we know? Can we tell the difference observationally between a classical LBV and a pulsational pair event? It is important to reiterate that regardless of the underlying physical mechanism (which is...what, again?), an observer who witnesses a brightening of several magnitudes accompanying an ejection of 0.1–10 $M_\odot$ from a massive H-rich star would classify it as a giant LBV eruption, because that's the definition of an

LBV eruption. So, whether one wishes to call them giant LBV eruptions, SN impostors, LBV impostors, pulsational pair instability ejections, failed SNe, explosive shell burning events, mergers, or some other name, the fact remains that if seen in an external galaxy, we'd probably call it a giant LBV eruption.

6. Massive Stars Can Die With Hydrogen

To me, one of the most interesting questions in massive star research is whether these Type IIn SN progenitors (1) are stars that really share the same evolutionary phase as local examples of LBVs but are exploding as SNe, or instead (2) are caused by some different underlying physical mechanism that causes LBV-like mass loss right before the SN explosion. There's also (3) the possibility that some of them are genuine pair instability SNe (Smith *et al.* 2007b). One or more of these is right, but no matter which one it is, a firm observational fact remains that is hard to avoid and which I want to emphasize:

At nearly Solar metallicity, observed supernovae tell us that very massive stars can make it to the ends of their lives and explode with massive H envelopes still intact.

This is a very important clue to understanding the evolution of massive stars, because it is in direct conflict with the predictions of stellar evolution models. One might conjecture that the predictions of stellar evolution models, which depend primarily on the adopted mass-loss rates, are wrong because they have assumed mass-loss rates that are too high...and we know they are too high. If this is true, then the current paradigm — that LBVs represent only a very brief transition phase between the end of core-H burning and the beginning of core-He burning lasting a few 10^4 yr — is probably wrong as well, and the idea that LBVs can explode as SNe becomes more compelling.

In order to make the normal LBV phase last until core collapse, the LBV lifetime must be longer than current estimates of a few 10^4 yr – more like a few 10^5 yr — because it must outlast core He burning. The main justification for a short LBV lifetime is that there are too few of them, so statistically, their lifetimes can't be too long.

Is it possible that the time during which an evolved massive star can potentially be an LBV is actually longer? Suppose, for the sake of argument, that the specific eruptive instability that leads us to call something an LBV actually represents an intermittent, possibly recurring active phase within a much longer blue supergiant/LBV phase. In other words, suppose LBV stars go through *dormant* phases, like volcanoes, which last much longer than their eruptive phases. What are the consequences of that? There should be many more H-rich evolved blue supergiants that are not caught at the right moment when they show wild variability, but may or may not have observable circumstellar material. This is, of course, known to be the case. There are many blue supergiants — often called LBV "candidates" because of their spectral similarity to LBVs — which do not exhibit the specific variability that earns them *bona-fide* LBV status. What are these stars, if not evolved massive stars that are potential dormant LBVs? Massey *et al.* (2007) argued a similar point, noting that there are several hundred LBV candidates in M31 and M33, compared to 8 known from their variability. **I would argue that this means the "greater LBV phase" is in fact much longer than the very rapid transition from core-H to core-He burning that we normally hear quoted.** If true, it would no longer be surprising to see LBVs exploding as SNe.

7. Addendum: Binaries, Binaries, Binaries....

At this point, especially at this meeting, an obvious proclamation comes to mind: "Binaries are the solution!" Namely, the requisite LBV precursor event could occur in a companion star in a massive binary system instead of the exploding star; the exploding star would be a more evolved WN/WC star, so the Type Ib/c SN then expands into its companion's dense H envelope, appearing as a Type IIn SN. This would nicely resurrect stellar evolution models, because then massive stars don't need to survive until core collapse with H envelopes intact. Phew!

The problem is that this actually makes things much worse. Remember that the SN and LBV-like eruption need to be synchronized to within about a decade to produce a Type IIn event. *What are the chances that the LBV star in a WR+LBV binary system would happen to have a giant LBV eruption within a decade before the WR star explodes?* At this meeting, J. Eldridge noted that in binary evolutionary models for massive stars, one expects that $\sim$2–5% of WR stars will have companions in the LBV phase. But here I'd emphasize that having a companion in the LBV phase is not enough for Type IIn SNe: we also need to have that companion suffer a giant eruption $\lesssim$10 yr before the other star's SN explosion. How likely is that? If the nominal time between recurring giant LBV eruptions is 1000 yr, then there's a 1% chance of an LBV eruption occuring within a decade of its companion's SN — but that occurs only in the $\sim$5% of massive binaries already in the WR+LBV phase — so now we are down to 0.05%. *This is not nearly enough.* Type Ib/c SNe make up about 15% of core-collapse SNe, while Type IIn's make up about 2–5% (Capallaro *et al.* 1997). Therefore, we would need about 10–25% of the Type Ib/c SNe to explode into a companion's LBV-eruption envelope to account for Type IIn SNe, compared to the 0.05% we might expect. What about mergers? Given the compact radii of WR stars, it seems a tall order to expect that $\sim$10% of massive binaries would merge fortuitously and eject a few solar masses of H only a decade before the SN.

References

Cappellaro, E., Turatto, M., Tsvetkov, D. Yu., *et al.* 1997, *A&A*, 322, 431
Foley, R. J., Smith, N., & Ganeshalingam, M., al. 2007, *ApJ*, 657, L105
Gal-Yam, A., Leonard, D. C., Fox, D. B., *et al.* 2007, *ApJ*, 656, 372
Kotak, R. & Vink, J. S. 2006, *A&A*, 460, L5
Massey, P., McNeill, R. T., Olsen, K. A. G., *et al.* 2007, *AJ*, 134, 2474
Ofek, E. O., Cameron, P. B., Kasliwal, M. M., *et al.* 2007, *ApJ*, 659, L13
Pastorello, A., Mazzali, P. A., Pignata, G., *et al.* 2007, *Nature*, 447, 829
Prestwich, A. H., Kilgard, R., Crowther, P. A., *et al.* 2007, *ApJ*, 669, L21
Ryder, S. D., Sadler, E. M., Subrahmanyan, R., *et al.* 2004, *MNRAS*, 349, 1093
Schwarz, D. H. & Pringle, J. E. 1996, *MNRAS*, 282, 1018
Silverman, J. M. & Filippenko, A. V. 2008, *ApJ*, 678, L17
Smith, N. 2007, *AJ*, 133, 1034
Smith, N. & Owocki, S. P. 2006, *ApJ*, 645, L45
Smith, N. & McCray R. 2007, *ApJ*, 671, L17
Smith, N. & Conti, P. S. 2008, *ApJ*, 679, 1467
Smith, N., Vink, J., & de Koter, A. 2004, *ApJ*, 615, 475
Smith, N., Bally, J., & Walawender, J. 2007a, *AJ*, 134, 846
Smith, N., Li, W., Foley, R. J., *et al.* 2007b, *ApJ*, 666, 1116
Smith, N., Foley, R. J., & Filippenko, A. V. 2008, *ApJ*, 680, 568
Smith, N. *et al.* 2008b, *ApJ*, in press (astro-ph/0804.0042)
Soderberg, A., Chevalier, R. A., Kulkarni, S. R., & Frail, D. A. 2006, *ApJ*, 651, 1005
Woosley, S. E., Blinnikov, S., & Heger, A. 2007, *Nature*, 450, 390

Discussion

DAVIDSON: Nathan, you mentioned the word "gonzo" and indicated that some astronomers imagine its negative! In science, *gonzo is good*. Astronomy has only a few dozen gonzo objects, typically with two attributes: (1) They're at extraordinarily revealing stages in their careers, and (2) are close enough to observe really well. Apart from Eta Car, you mentioned two or three of the others. *Each is worth hundreds of routine objects*, because they oft show where theory fails. These objects often go neglected by observers, sometimes at crucial stages.

Nathan Smith.

Kris Davidson.

Anilmis Nurdan and Mary Oksala.

André-Nicolas Chené and Olivier Schnurr.

The Progenitor Stars of Core-Collapse Supernovae

Stephen J. Smartt[1], R. Mark Crockett[1], John J. Eldridge[2] and Justyn R. Maund[3]

[1]Astrophysics Research Centre, School of Mathematics and Physics,
Queen's University Belfast, Belfast BT7 1NN

[2]Institute of Astronomy, The Observatories, University of Cambridge,
Madingley Road, Cambridge CB3

[3]Department of Astronomy and McDonald Observatory, University of Texas,
1 University Station C1402, Austin, TX 78712-0259

Abstract. Knowledge of the nature and mass of the progenitor stars of core-collapse supernovae are critical elements to test theoretical models of stellar evolution and stellar explosions. Here we describe the current limits and restrictions that can be placed on the progenitor stars of type II SNe and those of Ib/c. There are detections of some type II-P SN progenitors but the exploding stars that produce type Ib/c have eluded discovery. We discuss implications of these quantitative limits and the conclusions that we can now draw.

Keywords. supernovae: general – stars: evolution – supergiants

1. Introduction

There is huge diversity in the spectral and photometric evolution of supernovae (SNe) that arise from the death of massive stars. The diversity is much more pronounced than for the thermonuclear type Ia SNe. As discussed at this conference (see the contributions from Burrows, Dessart et al) the long favoured paradigm for energy release is the collapse of an iron core at the end of the nuclear burning life of a massive star. The observed core-collapse SN (CCSN) types are then thought to be related to the state of the stellar envelope when this explosion occurs deep in the heart of the star. The mass range that will produce CCSNe and the exact relation between CCSN type and progenitor star is now an area in which observers can contribute valuable and fundamental information. The core-collapse paradigm has even been challenged recently (for example, see Nathan Smith's contribution to this volume) with the exotic pair-instability or pair-creation process suggested to be responsible for some extremely luminous SNe. The process that produces gamma-ray bursts in some way also creates SNe (of types Ic and Ib)† hence it is important that we understand the nature of stars before they collapse to allow theoretical models of stellar structure and the core-collapse mechanism to be studied. With the availability of easy access HST and large ground-based archives searching for progenitor stars has become almost routine for nearby events. As will be shown in this paper, although the search mechanism is rather straightforward the detection of massive progenitors has not been easy. In fact it is somewhat surprising how few definite detections have been

† At the time of the conference in Dec. 2007 only type Ic SNe had been associated with GRBs or X-ray flashes. However in January 2008 a SN Ib (SN2008D) was found to be preceded by an Swift discovered XRF.

2. The relative rates of the supernovae types

In this paper we are presenting a summary of the results for the search for CCSN progenitors over a 10 year period. We have tended to apply a distance limit of ~25-30 Mpc when searching for SNe which are close enough that the massive stellar populations (at least the brightest stars) are resolved and detectable. Hence to make this search meaningfully complete and remove as much bias as possible we restrict this survey to galaxies within $V_{\rm vir} < 2000\,{\rm km\,s^{-1}}$, which corresponds to a distance of around 27 Mpc. It is interesting to also study the rates within this volume, as one would presume this reflects the initial mass function (IMF), stellar evolution, and in some way the influence of binarity. We have checked every SN reported to the IAU and found all those in galaxies which, after correction for the infall of the Local Group towards the Virgo cluster, lie within the velocity range. We have compiled the types from the most up to date literature sources, unpublished but publicly available spectra and photometry (including some amateur lightcurves) and assigned a most probable type. Out of the 135 SNe that make it into our catalogue only 2 have not been spectroscopically classified. The breakdown of the types is listed in Table 1. It is interesting to note that only one of these SNe has been reported to have shown spectral and photometric evolution that is similar to SN1987A : that is 1998a (Pastorello et al. 1995). Hence the rate of the explosions of blue supergiants does not appear to be large. At the time SN1987A was found, it was suggested by several authors that the faint magnitude during its early evolution ($M_V \sim -16$) could mean many SNe of this type have gone undiscovered (see for example Arnett et al. 1989). The last 10 years of SNe discoveries show that is unlikely and they are intrinsically rather rare.

3. Detection of progenitor stars

Apart from the now well known cases of SN1987A and SN1993J there are relatively few cases were a progenitor star has been identified. Our group, together with others around the globe (e.g. the UC Berkeley, Caltech and STScI groups) have been searching hard for the highest quality images of nearby galaxies in well maintained archives that allow interesting limits to be set on the nature of progenitor stars Although there were some attempts to do this with the early HST archive (e.g. Barth et al. 1996) it was not until the late 1990's that the HST archive became rich enough to allow the systematic searching for progenitor stars. Since around 1999, with the publication of interesting limits on the two very close SNe 1999em and 1999gi (Smartt et al. 2001, Smartt et al. 2002) there have been frantic attempts to identify progenitors soon after the explosion of nearby supernovae (e.g. Smartt et al. 2004, Li et al. 2006, Gal-Yam et al. 2005, Maíz-Apellániz et al. 2004). Astronomers interested in this field should beware, there are often many claims and counter-claims in the early days after an explosion in which the astrometric positioning of the supernova on the pre-explosion image (using ground-based astrometry) often leads to wrong results (e.g. Smartt et al. 2001, Richmond & Modjaz 2005, Li et al. 2004).

The most valuable source of archive imaging is of course HST due to its exquisite resolution across the optical and NIR, the well calibrated and easy to search archive, and the depth it achieves in exposures of even a few hundred seconds. However this resolution requires that the follow-up image that is used to astrometrically match the images is of similar resolution. Average seeing ground-based images are normally not good enough to unambiguously identify progenitor stars in HST images. The uncertainty in the alignment is usually of the order ~70-100 milliarcseconds, and often this is not

Table 1. The relative frequency of SNe types within a fixed volume and in a fixed time period of 10 years.

Type	No.	Relative Rates (per cent)	Core-Collapse Only Rates (per cent)
II-P	53	39.3	59.6
II-L	2.5	1.9	2.8
IIn	3.5	2.6	3.9
IIb	5	3.7	5.6
Ib	8	5.9	9.0
Ic	17	12.6	19.1
Ia	37	27.4	...
LBVs	7	5.2	...
Unclassified	2	1.5	...
Total	135	100	100
Total CCSNe	89	66	100

good enough to definitively associate a progenitor star with a SN (Crockett *et al.* 2007). Follow-up HST images or adaptive-optics assisted ground-based images are essential. An example is shown in Fig 1. In this case the astrometric accuracy of the alignment between the HST pre-explosion and the Gemini AO image is 20 milliarcseconds (Crockett *et al.* 2008).

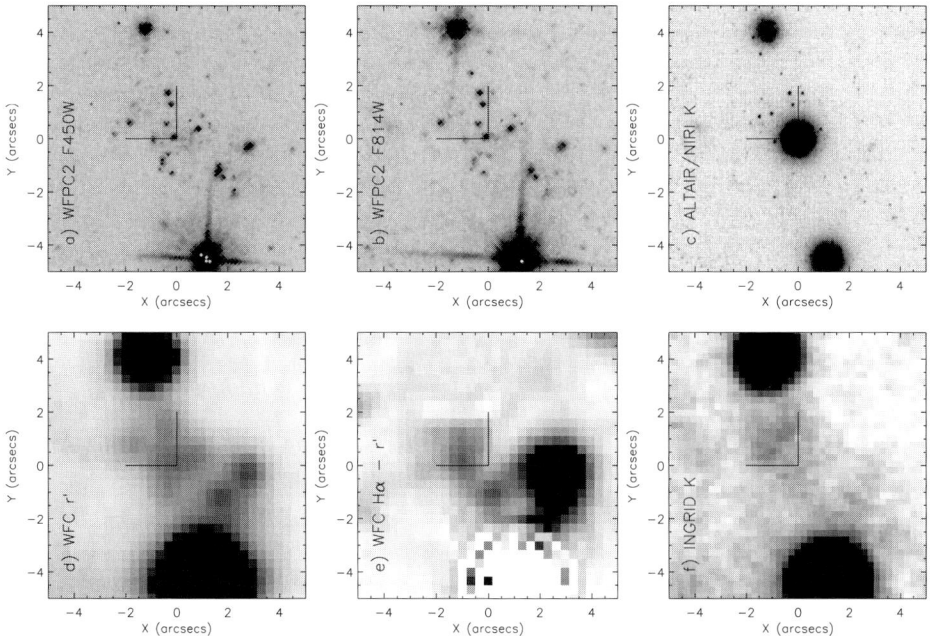

Figure 1. Figure taken from Crockett *et al.* (2008) which studied the pre-explosion environment of the type Ic SN2007gr. This series of images nicely illustrates the resolution of typical archive imaging from HST, ground-based imaging in average conditions (around 1 arcsecond quality) and diffraction limited K−band imaging from Gemini North and the Altair AO system. In this case the SN was shown to be very close, but not exactly coincident with, a bright point source of absolute magnitude of $M_{\rm F450W} = -9.0$. Crockett *et al.* (2008) suggest this may be a compact star cluster.

4. The Progenitor stars of Type II-P SNe

Smartt et al. (2008) will present a full list of all the nineteen Type II SNe which are within the volume limit as discussed above *and* which have deep enough pre-explosion images that they set interesting and restrictive limits on the progenitor stars. We have reanalyzed these with a consistent approach, homogeneous treatment of errors and application of one set of stellar evolutionary models. This removes systematic uncertainties when one compares results from the various published mass limits in the literature. The models we use are the Cambridge STARS tracks (Eldridge & Tout 2004). Seventeen of them are confirmed type II-P SNe from their lightcurves and spectra (the other two being of type II, but lack of photometric monitoring prevents a definitive subtype). Three of these have red supergiant progenitors detected (2003gd, 2004A and 2005cs) all with masses in the range $8^{+5}_{-2}M_\odot$, one is of unknown colour and its magnitude implies a likely mass of $16^{+6}_{-4}M_\odot$ (1999ev) and two are in compact stellar clusters with turn-off masses around 10-15$M_\odot$ (2004am and 2004dj ; Maíz-Apellániz et al. 2004 and Matilla et al. 2008). The rest all have upper limits as for example derived in Smartt et al. (2001, 2002). These upper limits assume that the progenitors were red supergiants at the time of their explosion, as one needs to assume a bolometric correction to apply to the observed magnitude limit to estimate a bolometric luminosity limit and hence a mass. This assumption is probably justified as the vast majority of the sample (17 out of 19) are confirmed II-P, and have fairly similar characteristics in terms of plateau lengths (Arnett 1980).

With these measurements of progenitor masses and upper limits it is now possible to estimate the parameters that describe the progenitor population. The three parameters that we are interested in are the minimum initial mass for a type IIP SN, the maximum initial mass and the IMF of the progenitors. We have employed an unbinned maximum likelihood method (e.g. as employed in Jegerlehner et al. 1996) and calculated the most likely values of minimum mass and upper mass that will produce the mass distribution that we see, assuming that the slope of the IMF is Salpeter ($\alpha = -1.35$). There appears to be no strong evidence in the Local Group that the IMF significantly varies†. The results can be seen in Figure 2. The parameters we estimate are therefore $m_{\rm min} = 7.5^{+1}_{-1.5}M_\odot$ and $m_{\rm max} = 15^{+3}_{-2}M_\odot$. We have also compared the mass - final luminosity relation of other models and find no major differences between those and the STARS code we employ here.

The lower mass limit is consistent with recent estimates for the upper mass limit that will produce a white dwarf. Dobbie et al.(2006) and Williams (2007) suggest that stars up to 6.8-8.6$M_\odot$ may produce white dwarfs. The upper mass limit however is problematic. We see red supergiants in the LMC and SMC with masses up to 20-25$M_\odot$ (see the contributions by Massey & Levesque at this conference), but why do we not see such high mass SN progenitors? The metallicity of the galaxies at the positions of the SNe are in the LMC-Solar range, thus one would not expect that all stars above $\sim$18$M_\odot$ would enter the WR phase. These more massive red supergiants would be rather easy to detect in the available pre-explosion images. We estimate that if the progenitors are from a massive stellar population with an Salpeter IMF, we would be missing around 3 massive progenitors in this sample. This is not a clear cut result yet, but is something that we will have to worry about if we continue not to detect massive progenitors.

† Outside perhaps the Galactic Center environment, see Don Figer's review, this volume

5. The Progenitor stars of Type Ib/Ic SNe

There have been no detections of a normal quiescent star at the position of any nearby type Ib/c SN despite around 10 attempts. The intriguing case of SN2006jc is the only event where one can claim that there was an object detected before explosion. In this case it appears that an LBV-like outburst was coincident with the SN, two years before explosion (Pastorello *et al.* 2007) The SN appeared to be Ic-like but enshrouded in a He rich envelope, which was likely ejected in the pre-collapse outburst, which has led to a "Ib-n" suggested classification.

Apart from this there is no detection of the suspected massive WR stars at the position of any Ib/c, and the deepest limits come from SN2002ap (Crockett *et al.* 2007) which has a pre-explosion detection limit of $M_B \simeq -4.3$. Other detection limits are in the absolute magnitude range (in B, V, R like filters) of -5 to -7. These limits would be sensitive to a large fraction of the typical WR population of the Magellanic Clouds, the Milky Way and M31 (see discussion in Crockett *et al.* 2007). However WR stars have a wide range in optical magnitudes due to their high and wide ranging temperatures of between -2 to -7. We are currently working to interpret these observational limits in terms of the WR populations we see locally. A very simple test we can do to check if WR stars are viable progenitors of all the Ib/c SNe is as follows. We hypothesize that the WR LMC population is indeed the sole progenitor population of type Ib/c SNe. Then we can estimate the probability that all of 6 events (for which we have deep pre-explosion

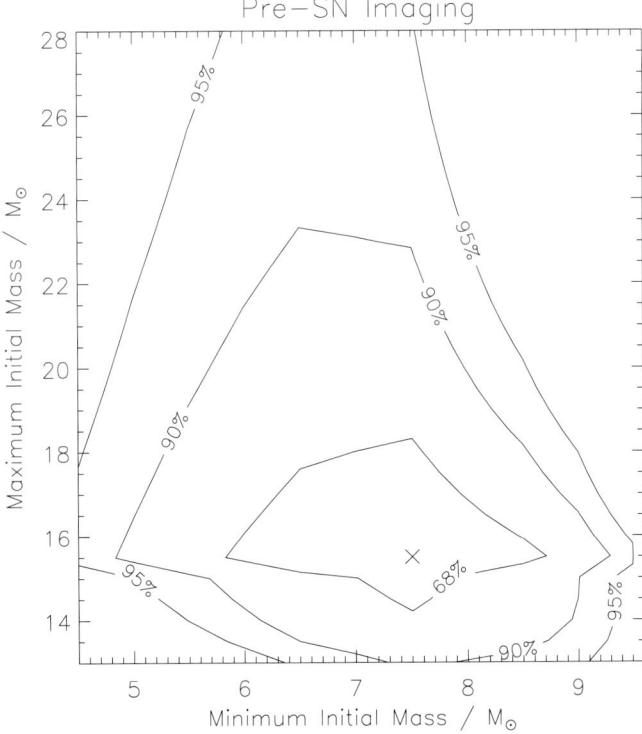

Figure 2. Figure taken from Smartt *et al.* (2008). This shows the likelihood contours for the minimum and maximum masses for type II-P SNe from the pre-explosion masses compiled from the literature and carefully re-analysed with one method and one stellar evolution code. This suggests a quite low mass for the minimum initial mass that can support a SN explosion.

limits) have not been detected due to statistical chance. This is simply the product of the individual probabilities and we estimate around 10 per cent. We still have to add in 4 other non-detections. Hence this suggests to us that the initial hypothesis is false, and that there is a fairly substantial population of progenitors that are not from initially very massive stars. These could be lower mass stars in interacting binaries, although there are not large numbers of these systems known in the Galaxy.

In summary then we have derived a most likely mass range for type II-P progenitors which includes a lower mass limit to produce a supernova explosion of around $m_{\min} = 7.5^{+1}_{-1.5} M_\odot$. The best estimate of the upper mass limit is $m_{\max} = 15^{+3}_{-2} M_\odot$. This is not consistent with the high mass red supergiant population in the Magellanic Clouds and it remains to be seen why this discrepancy exists. We postulate that higher mass stars may form black holes and faint explosions which are perhaps going undetected in current SNe surveys. Finding such a population would be a stunning discovery. It also appears that the observed Ib/c SN population cannot all be explained by massive Wolf-Rayet stars. Both the high rate of local Ib/c SNe and the lack of detections of any progenitors support this. We suggest that this is strong evidence that there must be an additional population of He stars and stripped CO stars within interacting binaries (e.g. see the models of Nomoto et al. 1995). Such a large population of progenitors have so far remained undiscovered in the Milky Way and the Local Group.

References

Arnett, W. D. 1980, *ApJS*, 237, 541
Arnett, W. D., Bahcall, J. N., Kirshner, R. P., & Woosley, S. E. 1989, *ARA&A*, 27, 629
Barth, A. J., van Dyk, S. D., Filippenko, A. V., et al. 1996, *AJ*, 111, 2047
Crockett, R. M., Smartt, S. J., Eldridge, J. J., et al. 2007, *MNRAS*, 381, 835
Crockett, R. M., Maund, J. R., Smartt, S. J., et al. 2008, *ApJ*, 672, L99
Dobbie, P. D., Napiwotzki, R., Burleigh, M. R., et al. 2006, *MNRAS*, 369, 383
Eldridge, J. J. & Tout, C. A. 2004, *MNRAS*, 353, 87
Gal-Yam, A., Fox, D. B., Kulkarni, S. R., et al. 2005, *ApJ*, 630, L29
Jegerlehner, B., Neubig, F., & Raffelt, G. 1996, *PRD*, 54, 1194
Li, W., Filippenko, A. V., & van Dyk, S. D. 2004, *IAUC*, 8388, 2
Li, W., Van Dyk, S. D., Filippenko, A. V., et al. 2006, *ApJ*, 641, 1060
Maíz-Apellániz, J., Bond, H. E., Siegel, M. H., et al. 2004, *ApJ*, 615, L113
Mattila, S., Smartt, S. J., et al. 2008, *MNRAS*, in prep.
Nomoto, K. I., Iwamoto, K., & Suzuki, T. 1995, *Phys. Rep.*, 256, 173
Pastorello, A., Baron, E., Branch, D., et al. 2005, *MNRAS*, 360, 950
Pastorello, A., Smartt, S. J., Mattila, S., et al. 2007, *Nature*, 447, 829
Richmond, M. W. & Modjaz, M. 2005, *IAUC*, 8555, 2
Smartt, S. J., Gilmore, G. F., Trentham, N., et al. 2001, *ApJ*, 556, L29
Smartt, S. J., Gilmore, G. F., Tout, C. A., & Hodgkin, S. T. 2002, *ApJ*, 565, 10
Smartt, S., Ramirez-Ruiz, E., & Vreeswijk, P. 2002, *IAUC*, 7816, 3
Smartt, S. J., Maund, J. R., Hendry, M. A., et al. 2004, *Science*, 303, 499
Smartt, S. J., et al. 2008, *MNRAS*, in prep.
Williams, K. A. 2007, in: R. Napiwotzki & M. R. Burleigh (eds.), *15th European Workshop on White Dwarfs* (San Francisco: ASP), *ASP Conf. Ser.*, 372, 85

Discussion

N. SMITH: I'm wondering about events like 1987A, the SN started out faint and the progenitor was a compact star. Are such SNe and progenitors under-represented because of selection effects or are they intrinsically rare ?

SMARTT: I think they are intrinsically rare. 87A was not so faint, with a peak during its plateau of -16, so such events would be detectable by the various surveys that have produced the 100 nearby SNe in the last 10 yrs. Only one of these events has a lightcurve and optical spectrum that one would confidently identify as 87A-like, so I think they are very rare. The progenitor itself was quite massive, around 20 solar masses, and such B-type supergiants would be detectable in many of the archive images. So again I don't think we are biassed against them.

CROWTHER: It seems clear that not all type Ib/c SNe result from very massive WN or WC stars since type Ic appear more common than type Ib SN (yet WN are more common than WC stars), plus these SN are often not associated with massive star forming regions in their host galaxies.

SMARTT: Yes, I agree but if all stars above 20 solar masses do actually explode one might expect that some fraction of our sample do include such massive stars.

LEITHERER: What limits your success rate, is it depth of the image or crowding of the field? A related question: could you miss some SN progenitors if they are heavily reddended or in a crowded field?

SMARTT: Surprisingly few SNe in my list were in very crowded regions, and only two are coincident with compact, spatially unresolved clusters (2004dj and 2004am). The major factor limiting the success rate of detecting progenitors is the depth of the archive images. I think we probably are missing some nearby SNe which are reddened by more than 3 mags in the visual, there are very few of these found and those that are tend to be very close (3–6 Mpc). It might be that the more massive progenitors produce these type of events, which are missing because the SNe are not discovered.

Nathan Smith (center left), Rolf Kudritzki (center right) and Steve Smartt (right).

Margarita Rosado.

Jean-Claude Bouret.

Can Very Massive Stars Avoid Pair-Instability Supernovae?

Sylvia Ekström, Georges Meynet and André Maeder

Geneva Observatory, University of Geneva, Maillettes 51 - CH 1290 Sauverny, Switzerland
email: sylvia.ekstrom@obs.unige.ch, georges.meynet@obs.unige.ch,
andre.maeder@obs.unige.ch

Abstract. Very massive primordial stars ($140\,M_\odot < M < 260\,M_\odot$) are supposed to end their lives as PISN. Such an event can be traced by a typical chemical signature in low metallicity stars, but at the present time, this signature is lacking in the extremely metal-poor stars we are able to observe. Does it mean that those very massive objects were not formed, contrarily to the primordial star formation scenarios ? Could it be possible that they avoided this tragic fate ?

We explore the effects of rotation, anisotropical mass loss and magnetic field on the core size of very massive Population III models. We find that magnetic fields provide the strong coupling that is lacking in standard evolution metal-free models and our $150\,M_\odot$ Population III model avoids indeed the pair-instability explosion.

Keywords. stars: evolution – stars: rotation – stars: chemically peculiar – stars: mass loss – stars: magnetic fields

1. Introduction

According to Heger *et al.* (2003), the fate of single stars depends on their He-core mass (M_α) at the end of the evolution. They have shown that at very low metallicity, the stars having $64\,M_\odot < M_\alpha < 133\,M_\odot$ will undergo pair-instability and be entirely disrupted by the subsequent supernova. This mass range in M_α has been related to the initial mass the star must have on the main sequence (MS) through standard evolution models: $140\,M_\odot < M_{\rm ini} < 260\,M_\odot$. Since we will present here a non-standard evolution, we will rather keep in mind the M_α range.

The typical mass of Population III (Pop III) stars is explored by early structure formation studies and chemistry considerations about cooling. Different studies (see Abel *et al.* 2002; Bromm *et al.* 2002, among others) give the same conclusion: Pop III stars are supposed to be massive or very massive, even when a bimodal mass distribution allows the formation of lower mass components (see Nakamura & Umemura 2001). Therefore we expect that many among them should die as pair-instability supernovae (PISN).

1.1. A typical chemical signature which remains unobserved

These PISN events are supposed to leave a typical chemical signature. According to Heger & Woosley (2002), the complete disruption of the star leads to a very strong odd-even effect: the absence of stable post-He burning stages deprives the star of the neutron excess needed to produce significant amounts of odd-Z nuclei. Also the lack of r- and s-process stops the nucleosynthesis around zinc. Even if one mixes these yields with the yields of zero-metallicity $12 - 40\,M_\odot$ models (which end up as standard Type II SNe), the PISN signature remains and should be observable.

However, using those yields, Tumlinson *et al.* (2004) have shown that it provides only a very poor fit to the abundances pattern observed in the metal-poor stars known today.

The odd-even effect is not observed, and the models significantly over-produce Cr and under-produce V, Co and Zn.

The most metal-deficient stars are supposed to be formed in a medium enriched by only one or a few SNe. The absence of the chemical signature of the PISN is a strong argument against their existence. But how could that be?

1.2. Simple solutions

The simplest solution to explain this absence is to suppose that the mass domain in question was not formed in the primordial clouds. Maybe the primordial IMF was not as top-heavy as we actually think, and the most massive stars formed then could very well be too small for such a fate. However, recent works on primordial stars formation seem to rule out this possibility (see Yoshida *et al.* 2006; O'Shea & Norman 2007).

Another possibility is that the signature was very quickly erased by the next generations of stars. Maybe the metal-poor stars we observe are enriched by more SNe than we actually think, and the later contributions are masking the primordial ones. Only the observation of more and more metal-deficient stars will provide an answer to that possibility.

One can also wonder whether there would be a way for those stars to avoid their fate. In this context, the simplest solution is to suppose that some mechanism could lead to such a high mass loss that the conditions for pair-instability would no more be met in the central regions. The aim of the present work is to explore this possibility.

2. Rotation at very low metallicity

Radiatively-driven winds are supposed to scale with metallicity as $Z^{(0.5-0.86)}$. Thus, very low or even no mass loss is expected when Z approaches to 0.

However, we have shown (Meynet *et al.* 2006) that rotation can change the mass loss history of low metallicity stars in a dramatic way. Two processes are involved:

(*a*) During MS, because of the low radiative winds, the star loses very little mass and thus very little angular momentum. As the evolution proceeds, the stellar core contracts and spins up. If a coupling exists between the core and the envelope (*i.e.* meridional currents or magnetic fields), the surface may be accelerated up to the critical velocity and the star may lose mass by a mechanical wind due to centrifugal acceleration. The matter is launched into a decretion disk, which may be dissipated later by the radiation field of the star.

(*b*) Rotation induces an internal mixing that enriches the surface in heavy elements. The effective surface metallicity is enhanced by a factor that has been shown to be very

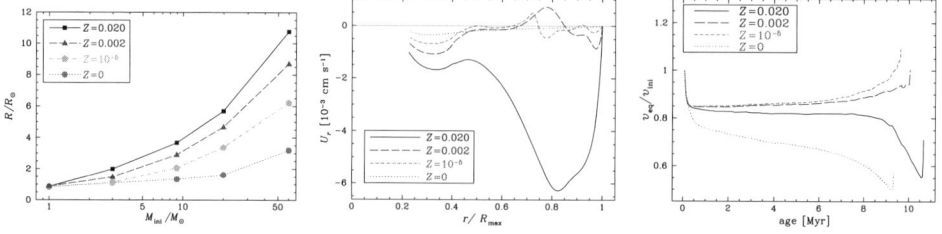

Figure 1. *Left:* Variations in the radius as a function of initial mass for various metallicities. *Centre:* Amplitude of the radial component of the meridional circulation inside $20\,\mathrm{M}_\odot$ models at the same evolutionary stage ($X_c = 0.4$) and having the same rotation rate ($v/v_{\mathrm{crit}} = 0.5$) but various metallicities. *Right:* Evolution of the equatorial velocity during the MS. The models are taken from Ekström *et al.* (2008).

large (up to 10^6 for a $60\,M_\odot$ at $Z_{\rm ini} = 10^{-8}$). Rotation also favours a redward evolution after the MS, allowing the star to spend more time in the cooler part of the HR diagram. The opacity of the envelope is increased, and the radiative winds may thus be drastically enhanced.

Both effects add up and lead to strong mass loss at very low metallicity.

2.1. Rotation at $Z = 0$

But what happens when the metallicity drops down to $Z = 0$ strictly? The absence of carbon prevents the star to start burning hydrogen through CNO-cycle, but pp−chains are not energetic enough to sustain the star, so it contracts longer during its formation. At metallicities $Z = 0.020$, 0.002, 10^{-5}, and 0 respectively, the radii on the ZAMS will be in a ratio $3.5 : 2.8 : 2 : 1$ for a $20\,M_\odot$ model (see Fig. 1, *left*), that is the $Z = 0$ star will be twice as compact as the $Z = 10^{-5}$ one. This has a direct influence on the amplitude of the outer cell of the meridional circulation, which depends on the compactness of the star as $U_r \propto \left[1 - \frac{\Omega^2}{2\pi G\bar\rho}\right]$. For the same metallicities, the amplitude of the outer cell will be in a ratio $100 : 17 : 4 : 1$ (Fig. 1, *centre*). This means that at $Z = 0$, the core-envelope coupling is almost null. Figure 1 (*right*) shows the evolution of the equatorial velocity during the MS for our $20\,M_\odot$ models. At standard metallicity, the meridional circulation is strong, but the mass loss is also strong, so the model spins down. At lower but non-zero metallicities, the mass loss drops, and though the core-envelope coupling is weak, the models spin up. At $Z = 0$, the mass loss is null, but the model evolves almost in a regime of local conservation of angular momentum. Thus, the natural inflation of the radius during the MS leads to a spin down of the model. The model may reach the critical limit, but later in its MS evolution (for a given initial mass and velocity), and the mechanical mass loss remains very small.

At central H exhaustion, the core of the $Z = 0$ model is already hot enough to burn some helium, so the transition between H-burning and He-burning is smooth, without any structural readjustment. The model remains in the blue part of the Hertzsprung-Russell diagram until very late in the central He-burning phase (Fig. 2). When it eventually evolves towards the red part, the outer convective zone is very thin, so the enrichment of the surface remains very low. Whereas at $Z = 10^{-8}$ we get a strong post-MS mass loss, at $Z = 0$ it is negligible.

Rotation alone seems thus to fail in providing a way to lose sufficient mass at $Z = 0$.

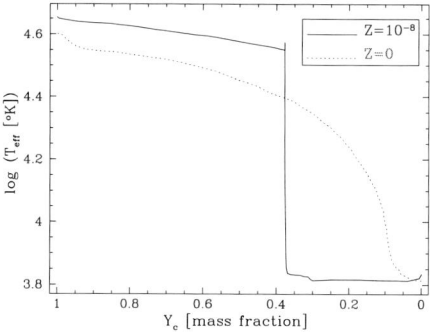

Figure 2. Evolution of $T_{\rm eff}$ during core He-burning for two models of $60\,M_\odot$. The $Z = 10^{-8}$ model is taken from Hirschi (2007).

2.2. Two natural effects of rotation

In the previous calculations, we have neglected two mechanisms that arise naturally with rotation and could change this picture.

The first one is the wind anisotropy (see Maeder 1999): when a star rotates fast, it becomes oblate. As shown by von Zeipel theorem, the poles become hotter than the equator, and the radiative flux is no more spherically symmetric: it gets much stronger in the polar direction than in the equatorial plane. Since the mass is lost preferentially near the poles, it removes much less angular momentum than in the spherical configuration, and thus the star may reach the critical limit much earlier.

The second mechanism is the magnetic fields. According to Spruit (2002) the differential rotation may amplify an existing magnetic field through the Tayler-Spruit dynamo mechanism, which provide a strong core-envelope coupling. This coupling will be able to accelerate the surface very early in the evolution, and maintain the star at critical limit throughout its entire evolution.

3. Evolution with anisotropical winds and magnetic fields

3.1. Ingredients

With the help of the two effects mentioned above, we have computed an exploratory model. Its mass has been chosen to be $150\,\mathrm{M_\odot}$ and the initial ratio between the equatorial velocity and the critical one to be $v_{\mathrm{ini}}/v_{\mathrm{crit}} = 0.56$. Let us mention that this ratio is higher than the one needed to account for the observed average velocities at solar metallicity, but it is still a "reasonable" value, not an extreme one.

The computation has been accomplished using the Geneva code with up to date nuclear reaction rates obtained with NETGEN (http://www-astro.ulb.ac.be/Netgen/). The opacity tables come from OPAL (http://www-phys.llnl.gov/Research/OPAL/opal.html) with the extension at low temperature by Ferguson et al. (2005). The initial composition is $X = 0.753$, $Y = 0.247$ and of course $Z = 0$.

The radiative mass loss prescription is an important ingredient for modelling massive stars. Here we have used Kudritzki (2002). Since this prescription is not aimed at the case $Z = 0$ strictly, we have used the same adaptations as in Marigo et al. (2003). The Wolf-Rayet (WR) mass loss rate is taken from Nugis & Lamers (2000) with the metallicity scaling from Eldridge & Vink (2006). For the calculation, we have taken the effective surface metallicity $Z_{\mathrm{eff}} = (1 - X - Y)_{\mathrm{surf}}$ so that the enrichment of the surface is accounted for. We must stress that this Z_{eff} is mainly composed by CNO elements but no iron. It is usually considered that WR winds are triggered by Fe lines, whereas the CNO lines determine only v_∞, so we expect that no WR winds can take place at $Z = 0$. But Vink & de Koter (2005) have shown that when the metallicity gets really low, the CNO lines take over the role of Fe lines in the line driving, and this is what we have assumed here.

When the model reaches the critical limit, the mechanical mass loss has been treated as described in Meynet et al. (2006).

The treatment of anisotropic winds has been implemented as in Maeder (2002) and the effect of magnetic fields as in Maeder & Meynet (2005).

3.2. Result

In Fig. 3, we present the evolution in the HR diagram (left panel) and the evolution of mass with time (right panel). The grey line shows a non-rotating model computed with the same physics for comparison.

During its whole evolution up to the end of core He-burning, the non-rotating model loses only 1.37 M$_\odot$. This illustrates the weakness of radiative winds at $Z = 0$.

The evolution of the rotating model (black line) can be described by four distinct stages:

(*a*) *(solid part*, lower left corner) The model starts its evolution on the MS with only radiative winds, losing only a little more than 0.002 M$_\odot$. During this stage, it accelerates quickly, mainly because of the strong coupling exerted by the magnetic fields.

(*b*) *(dashed part)* When its central content of hydrogen is still 0.58 in mass fraction, it reaches the critical velocity and starts losing mass by mechanical wind. It remains at the critical limit through the whole MS, but the mechanical wind removes only the most superficial layers that have become unbound, and less than 10% of the initial mass is lost at that stage (11.44 M$_\odot$). The model becomes also extremely luminous, and reaches the Eddington limit when 10% of hydrogen remains in the core. Precisely, it is the so-called $\Omega\Gamma$-limit that is reached here. Due to the fast rotation, the maximum Eddington factor allowed is reduced; at the same time, because of the high luminosity, the critical velocity is reduced in comparison with the one derived from the classical Ω-limit (see Maeder & Meynet 2000, for details).

(*c*) *(dotted part)* The combustion of helium begins as soon as the hydrogen is exhausted in the core, then the radiative H-burning shell undergoes a CNO flash, setting the model on its redward journey. The model remains at the $\Omega\Gamma$-limit and loses a huge amount of mass. The strong magnetic coupling keeps bringing angular momentum to the surface and even the heavy mass loss is not able to let the model evolve away from the critical limit. The mass lost during that stage amounts to 53.46 M$_\odot$. When the model starts a blue hook in the HR diagram, its surface conditions become those of a WR star ($X_{surf} < 0.4$ and $T_{\rm eff} > 10'000$ K). The luminosity drops and takes the model away from the Γ-limit, marking the end of that stage.

(*d*) *(solid part)* The rest of the core He-burning is spent in the WR conditions. The mass loss is strong but less than in the previous stage: another 26.34 M$_\odot$ are lost.

At the end of core He-burning, the final mass of the model is only $M_{\rm fin} = 58$ M$_\odot$, already below the minimum M_α needed for PISN ($M_\alpha \geqslant 64$ M$_\odot$). Note that the contraction of the core after helium exhaustion brings the model back to critical velocity, so this value for $M_{\rm fin}$ must be considered as an upper limit.

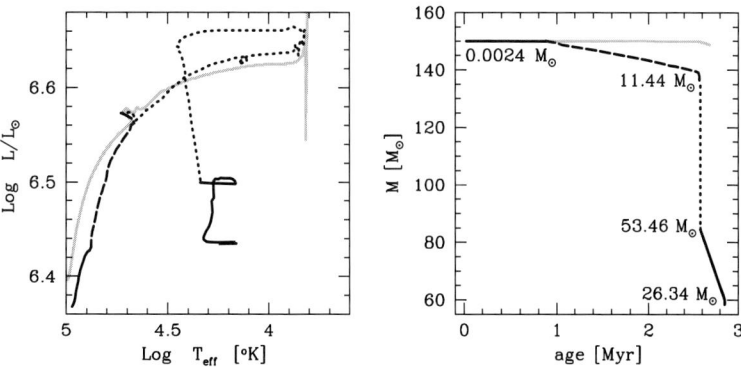

Figure 3. Black line: rotating model; *solid part*: beginning of MS ($X_c = 0.753$ down to 0.58; *dashed part*: rest of the MS; *dotted part*: beginning of core He-burning phase ($Y_c = 1.00$ down to 0.96); *solid part*: rest of the He-burning. Grey line: non-rotating model for comparison. **Left panel**: evolution in the Hertzsprung-Russell diagram; **Right panel**: evolution of the mass of the model. The mass indicated is the mass lost at each stage, not a summation.

4. Summary

The model we presented here is exploratory. We cannot draw general conclusions from it. But our model shows that heavy mass loss is possible even at $Z=0$, and the answer to our title's question is: *yes, under certain conditions, very massive stars can indeed avoid PISN*.

Some aspects need yet to be clarified. Is the WR mass loss rate we have used really valid? Can the CNO lines alone really drive a WR wind? In a more general perspective, we still lack a good mass loss recipe for the strict $Z=0$ case. A word of caution must also be cast on the inclusion of magnetic fields. The validity of the Tayler-Spruit dynamo is still under debate, and more work need to be done before we may confidently rely on results that have been obtained with the actual treatment. Moreover, there is not a clear consensus whether magnetic fields were present in the early Universe or not. In the Tayler-Spruit dynamo, the mechanism amplifies a pre-existing field, so if none were present at the start, we cannot use it.

Anyway, the physics used in the present model is today's "state of the art" and it is interesting to study what can be achieved with it. Our result is encouraging, because the computation has been accomplished with reasonable assumptions: the initial rotation rate used here was fast but not extreme, and the mechanisms called upon (anisotropy of the winds and magnetic fields) are not exotic ones, but two natural effects which arise when one treats properly the case of rotation.

After this first step, more work is needed. First, we have to refine the mass loss treatment at $\Omega\Gamma$-limit and check if the mass loss stays as strong as here. If it does so, we have to check if our results are valid in the whole PISN mass domain. Higher mass models should experience higher mass loss, but it is necessary to check if it would still be sufficient to help them avoid pair-instabilities.

References

Abel, T., Bryan, G. L., & Norman, M. L. 2002, *Science* 295, 93
Bromm, V., Coppi, P. S., & Larson, R. B. 2002, *ApJ* 564, 23
Ekström, S., Meynet, G., Maeder, A., & Barblan, F. 2008, *A&A* 478, 467
Eldridge, J. J. & Vink, J. S. 2006, *A&A* 452, 295
Ferguson, J. W., Alexander, D. R., Allard, F., et al. 2005, *ApJ* 623, 585
Heger, A., Fryer, C. L., Woosley, S. E., et al. 2003, *ApJ* 591, 288
Heger, A. & Woosley, S. E. 2002, *ApJ* 567, 532
Hirschi, R. 2007, *A&A* 461, 571
Kudritzki, R. P. 2002, *ApJ* 577, 389
Maeder, A. 1999, *A&A* 347, 185
Maeder, A. 2002, *A&A* 392, 575
Maeder, A. & Meynet, G. 2000, *A&A* 361, 159
Maeder, A. & Meynet, G. 2005, *A&A* 440, 1041
Marigo, P., Chiosi, C., & Kudritzki, R.-P. 2003, *A&A* 399, 617
Meynet, G., Ekström, S., & Maeder, A. 2006, *A&A* 447, 623
Nakamura, F. & Umemura, M. 2001, *ApJ* 548, 19
Nugis, T. & Lamers, H. J. G. L. M. 2000, *A&A* 360, 227
O'Shea, B. W. & Norman, M. L. 2007, *ApJ* 654, 66
Spruit, H. C. 2002, *A&A* 381, 923
Tumlinson, J., Venkatesan, A., & Shull, J. M. 2004, *ApJ* 612, 602
Vink, J. S. & de Koter, A. 2005, *A&A* 442, 587
Yoshida, N., Omukai, K., Hernquist, L., & Abel, T. 2006, *ApJ* 652, 6

Discussion

OWOCKI: Is the (second) large ($\geqslant 50\,\mathrm{M_\odot}$) mass loss event in your $M = 150\,\mathrm{M_\odot}$ model from the pole or equator?

EKSTRÖM: From both: there is a polar component because of the anisotropy of the winds and an equatorial component because of the critical velocity.

TOWNSEND: I worry that your mass-loss estimates (due to mechanical outflows at the equator) may be sensitive to the size of the decretion disk formed. For a very large disk, the star can shed a huge amount of angular momentum, and yet shed only a small amount of mass. So, modeling the disk physics correctly is important if we want to know the 'efficiency ratio' $\dot{M}/\dot{J}$ accurately.

EKSTRÖM: I fully agree with you and in fact it is a project we have with Stan Owocki to work on the accurate modeling of the decretion disk.

Sergio Simon-Diaz with real hula dancer.

Stellar Evolution at Low Metallicity

Raphael Hirschi[1], Cristina Chiappini[2,3], Georges Meynet[2], André Maeder[2] and Sylvia Ekström[2]

[1] Astrophysics group, Keele University, Lennard-Jones Lab., Keele, ST5 5BG, UK
email: r.hirschi@epsam.keele.ac.uk

[2] Observatoire Astronomique de l'Université de Genève, CH-1290, Sauverny, Switzerland

[3] Osservatorio Astronomico di Trieste, Via G. B. Tiepolo 11, I - 34131 Trieste, Italia

Abstract. Massive stars played a key role in the early evolution of the Universe. They formed with the first halos and started the re-ionisation. It is therefore very important to understand their evolution. In this review, we first recall the effect of metallicity (Z) on the evolution of massive stars. We then describe the strong impact of rotation induced mixing and mass loss at very low Z. The strong mixing leads to a significant production of primary ^{14}N, ^{13}C and ^{22}Ne. Mass loss during the red supergiant stage allows the production of Wolf-Rayet stars, type Ib,c supernovae and possibly gamma-ray bursts (GRBs) down to almost $Z = 0$ for stars more massive than 60 $M_\odot$. Galactic chemical evolution models calculated with models of rotating stars better reproduce the early evolution of N/O, C/O and ^{12}C/^{13}C. Finally, the impact of magnetic fields is discussed in the context of GRBs.

Keywords. stars: mass loss – stars: Population II – stars: rotation – supernovae: general – stars: Wolf-Rayet – Galaxy: evolution – gamma rays: bursts

1. Introduction

Massive stars started forming about 400 millions years after the Big Bang and ended the dark ages by re-ionising the Universe. They therefore played a key role in the early evolution of the Universe and it is important to understand the properties and the evolution of the first stellar generations to determine the feedback they had on the formation of the first cosmic structures. It is unfortunately not possible to observe the first massive stars because they died a long time ago but their chemical signature can be observed in low mass halo stars (called EMP stars), which are so old and metal poor that the interstellar medium out of which these halo stars formed are thought to have been enriched by one or a few massive stars. Since the re-ionisation, massive stars have continuously injected kinetic energy (via various types of supernovae) and newly produced chemical elements (by both hydrostatic and explosive burning and s and r processes) into the interstellar medium of their host galaxy. They are thus important players for the chemodynamical evolution of galaxies. Most massive stars leave a remnant at their death, either a neutron star or a black hole, which produce pulsar or X-ray binaries.

The evolution of stars is governed by three main parameters, which are the initial mass, metallicity (Z) and rotation rate. The evolution is also influenced by the presence of magnetic fields and of a close binary companion. For massive stars ($M \gtrsim 10\,M_\odot$) around solar metallicity mass loss plays a crucial role, in some cases removing more than half of the initial mass. Internal mixing, induced mainly by convection and rotation also significantly affect the evolution of stars. In this review, after a summary of the properties of low-Z stars, we discuss the possible impact of rotation induced mixing and mass loss at low Z. We then present the implication of strong mixing and mass loss for the nucleosynthesis and for galactic chemical evolution in the context of extremely metal poor stars. We also discuss the effects of magnetic fields. We end with conclusions and an outlook.

2. Properties of non-rotating low-Z stars

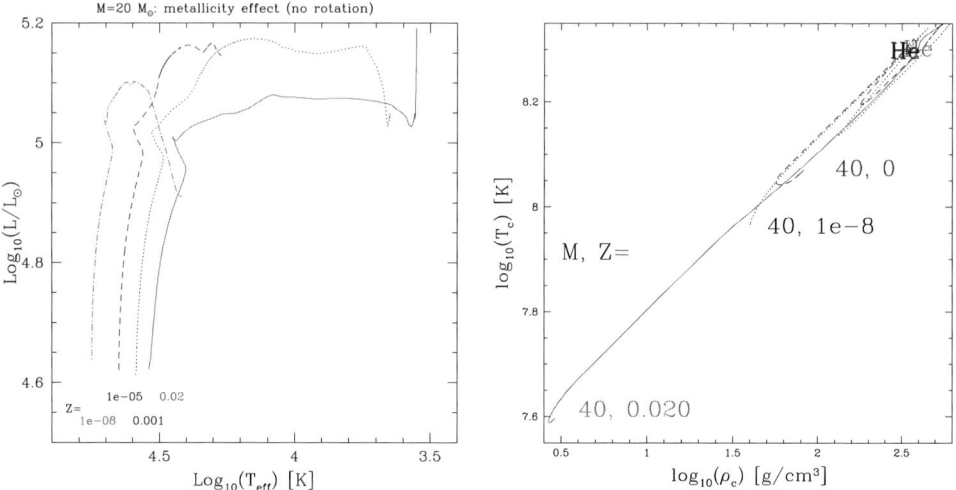

Figure 1. *Left*: H-R diagram for non-rotating 20 $M_\odot$ models with $Z = 10^{-8}, 10^{-5}, 0.001$ and 0.02, showing that more metal poor stars have more compact envelopes and are less likely to reach the red supergiant stage. *Right*: Central temperature versus central density diagram for 40 $M_\odot$ models with $Z = 0, 10^{-8}$ and 0.02. The evolutionary tracks start where hydrogen burning start and the He symbols are placed at the start of helium burning. The central conditions are much hotter and denser at very low Z. Note also the different initial H-burning conditions between $Z = 0$ and 10^{-8} (explained in the text).

The first stellar generations are different from solar metallicity stars due to their low metal content or absence of it. First, low-Z stars are more compact (see Fig. 1) due to lower opacity. Second, *metal free* stars burn hydrogen in a core, which is denser and hotter due to the lack of initial CNO elements (see start of dashed curve in Fig. 1 *right*). This implies that the transition between core hydrogen and helium burning is much shorter and smoother. Furthermore, hydrogen burns via the pp-chain in shell burning. These differences make the metal free stars different from the second or later generation stars (Ekström *et al.* 2008).

Third, mass loss is metallicity dependent and therefore is expected to become very weak at very low metallicity. The metallicity (Z) dependence of mass loss rates is usually described using the formula:

$$\dot{M}(Z) = \dot{M}(Z_\odot)(Z/Z_\odot)^\alpha \tag{2.1}$$

The exponent α varies between 0.5-0.6 (Kudritzki & Puls 2000, Kudritzki 2002) and 0.7-0.86 (Vink *et al.* 2001, Vink & de Koter 2005) for O-type and WR stars respectively (See Mokiem *et al.* 2007 for a recent comparison between mass loss prescriptions and observed mass loss rates). Until very recently, most models use at best the total metal content present at the surface of the star to determine the mass loss rate. However, the surface chemical composition becomes very different from the solar mixture, due either to mass loss in the WR stage or by internal mixing (convection and rotation) after the main sequence. It is therefore important to know the contribution from each chemical species to opacity and mass loss.

Recent studies (Vink *et al.* 2000, Vink & de Koter 2005) show that iron is the dominant element concerning radiation line-driven mass loss for O-type and WR stars. In the case of WR stars, there is however a plateau at low metallicity due to the contributions from light elements like carbon, nitrogen and oxygen (CNO). In between the hot and cool parts of the HR-diagram, mass loss is not well understood. Observations of the LBV stage indicate that several solar masses per year may be lost (Smith *et al.* 2003) and there is no indication of a metallicity dependence. In the red supergiant (RSG) stage, the rates generally used are still those of Nieuwenhuijzen & de Jager (1990). More recent observations indicate that there is a very weak dependence of dust-driven mass loss on metallicity and that CNO elements and especially nucleation seed components like silicon and titanium are dominant (van Loon 2000, van Loon 2006, Ferrarotti & Gail 2006). Van Loon *et al.* (2005) provide recent mass loss rate prescriptions in the RSG stage. In particular, the ratio of carbon to oxygen is important to determine which kind of molecules and dusts form. If the ratio of carbon to oxygen is larger than one, then carbon-rich dust would form, and more likely drive a wind since they are more opaque than oxygen-rich dust at low metallicity (Höfner & Andersen 2007).

Fourth, the binary interactions are probably changed by the greater compactness of low Z stars on the MS. Furthermore, the first generation stars below 40 $M_\odot$ do not evolve to the RSG stage. De Mink *et al.* (2008) show that as Z decreases, mass transfer is more likely to take place after the ignition of He burning (case C) and for the very low-Z stars, which do not reach the RSG stage, only the closest binaries would still interact.

Finally, the first stars are thought to be more massive than solar-metallicity stars (Bromm & Larson 2004, Schneider *et al.* 2006). Note that other studies suggest that both very massive and low-mass stars may form at $Z = 0$ (Nakamura & Umemura 2001). Since mass loss is expected to be very low at very low metallicities, the logical deduction from these two arguments is that a large fraction of the first stars were very massive at their death (> 100 $M_\odot$) and therefore lead to the production of pair-creation supernovae (PCSNe). Unfortunately, the first massive stars died a long time ago and will probably never be detected directly (see however Scannapieco *et al.* 2005, Tornatore *et al.* 2007). There are nevertheless indirect observational constraints on the first stars coming from observations of the most metal-poor halo stars (Beers & Christlieb 2005). These observations do not show the peculiar chemical signature of PCSNe (strong odd-even effects and low zinc, see Heger & Woosley 2002). This probably means that at most only a few of these very massive stars (>100 $M_\odot$) formed or that they lost a lot of mass even though their initial metal content was very low as discussed in the next section. Although there is no signature of PCSNe at very low Z, they might occur in our local Universe (Smith *et al.* 2007, Langer *et al.* 2007, Woosley *et al.* 2007).

The topic of low Z stellar evolution is not new (see for example Chiosi 1983, El Eid *et al.* 1983, Carr *et al.* 1984, Arnett 1996). The observations of extremely metal-poor stars (Beers & Christlieb 2005) have however greatly increased the interest in very metal-poor stars. There are many recent works studying the evolution of metal-free (or almost) massive (Heger & Woosley 2002, Limongi & Chieffi 2005, Umeda & Nomoto 2005, Meynet *et al.* 2006), intermediate mass (Siess *et al.* 2002, Herwig 2004, Suda *et al.* 2004, Gil-Pons *et al.* 2005) and low mass (Picardi *et al.* 2004, Weiss *et al.* 2004) stars in an attempt to explain the origin of the surface abundances observed. The fate of non-rotating massive single stars at low Z is summarised in Heger *et al.* (2003) and several groups have calculated the corresponding stellar yields (Heger & Woosley 2002, Chieffi & Limongi 2004, Tominaga *et al.* 2007).

3. Rotation, internal mixing and mass loss

Massive star models including the effects of both mass loss and especially rotation better reproduce many observables around solar Z (See contributions by Meynet and Maeder in this volume). For example, models with rotation allow chemical surface enrichments already on the main sequence (MS), whereas without the inclusion of rotation, self-enrichments are only possible during the RSG stage (Heger & Langer 2000, Meynet & Maeder 2000). Rotating star models also better reproduce the WR/O ratio and also the ratio of type Ib+Ic to type II supernova as a function of metallicity compared to non-rotating models, which underestimate these ratios (Meynet & Maeder 2005). The models at very low Z presented here use the same physical ingredients as the successful solar Z models. The value of 300 km s^{-1} used as the initial rotation velocity at solar metallicity corresponds to an average velocity of about 220 km s^{-1} on the main sequence (MS) which is close to the average observed value. See for instance Fukuda (1982) for one of the first surveys and the list in Meynet's contribution in this volume for the most recent surveys. It is unfortunately not possible to observe very low Z massive stars and measure their rotational velocity since they all died a long time ago. Higher observed ratio of Be to B stars in the Magellanic clouds compared to our Galaxy (Maeder *et al.* 1999) could point out to the fact the stars rotate faster at lower metallicities. Also a low-Z star having the same ratio of surface velocity to critical velocity, v/v_{crit} (where v_{crit} is the velocity for which the centrifugal force balances the gravitational force) as a solar-Z star has a higher surface rotation velocity due to its smaller radius (one quarter of $Z_\odot$ radius for a very low-Z 20 M$_\odot$ star). In the models presented below, the initial ratio v/v_{crit} is the same or slightly higher than for solar Z (see Hirschi 2007 for more details). This corresponds to initial surface velocities in the range of $600 - 800 \text{ km s}^{-1}$. These fast initial rotation velocities are supported by chemical evolution models of Chiappini *et al.* (2006b) discussed in the next section. The mass loss prescriptions used in the Geneva stellar evolution code are described in detail in Meynet & Maeder (2005). In particular, the mass loss rates depend on metallicity as $\dot{M} \sim (Z/Z_\odot)^{0.5}$, where Z is the mass fraction of heavy elements at the surface of the star.

How do rotation induced processes vary with metallicity? The surface layers of massive stars usually accelerate due to internal transport of angular momentum from the core to the envelope. Since at low Z, stellar winds are weak, this angular momentum dredged up by meridional circulation remains in the star, and the star more easily reaches critical rotation. At the critical limit, matter can easily be launched into a keplerian disk which probably dissipates under the action of the strong radiation pressure of the star.

The efficiency of meridional circulation (dominating the transport of angular momentum) decreases towards lower Z because the Gratton-Öpik term of the vertical velocity of the outer cell is proportional to $1/\rho$. On the other hand, shear mixing (dominating the mixing of chemical elements) is more efficient at low Z. Indeed, the star is more compact and therefore the gradients of angular velocity are larger and the mixing timescale (proportional to the square of the radius) is shorter. This leads to stronger internal mixing of chemical elements at low Z (Meynet & Maeder 2002).

Figure 2 shows the evolution of the convective zones in a rotating and a non-rotating 20 M$_\odot$ models at $Z = 10^{-8}$. The history of convective zones (in particular the convective zones associated with shell H burning and core He burning) is strongly affected by rotation induced mixing. The most important rotation induced mixing takes place while helium is burning inside a convective core. Primary carbon and oxygen are mixed outside of the convective core into the H-burning shell. Once the enrichment is strong enough, the H-burning shell is boosted (the CNO cycle depends strongly on the carbon and

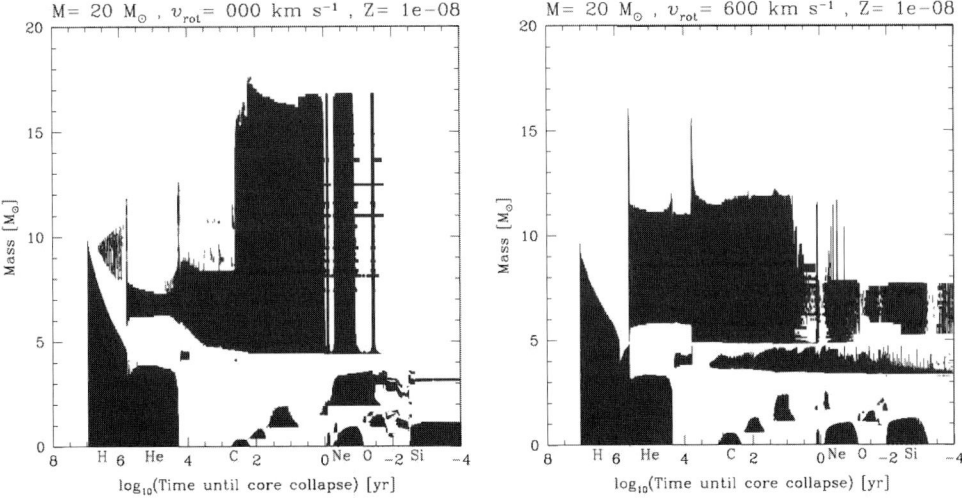

Figure 2. Structure evolution diagram for the non-rotating (*left*) and rotating (*right*) 20 $M_\odot$ models at $Z = 10^{-8}$. Black (coloured) areas correspond to convective zones along the Lagrangian mass coordinate as a function of the time left until the core collapse. The burning stage abbreviations are given below the time axis. Rotation strongly affects shell H burning and core He burning.

oxygen mixing at such low initial metallicities). The shell then becomes convective and leads to an important primary nitrogen production. In response to the shell boost, the core expands and the convective core mass decreases. At the end of He burning, the CO core is less massive than in the non-rotating model (see Fig. 2). Additional convective and rotational mixing brings the primary CNO to the surface of the star. This has consequences for the stellar yields. The yield of ^{16}O being closely correlated with the mass of the CO core, it is therefore reduced due to the strong mixing. At the same time the carbon yield is slightly increased. The relatively "low" oxygen yields and "high" carbon yields are produced over a large mass range at $Z = 10^{-8}$ (Hirschi 2007). This could be an explanation for the possible high [C/O] ratio observed in the most metal-poor halo stars (ratio between the surface abundances of carbon and oxygen relative to solar; see Fig. 14 in Spite *et al.* 2005).

Models of metal-free stars including the effect of rotation (Ekström *et al.* 2006) show that stars may lose up to 10 % of their initial mass due to the star rotating at its critical limit (also called break-up limit). The mass loss due to the star reaching the critical limit is non-negligible but not important enough to change drastically the fate of the metal-free stars. The situation is very different at very low but non-zero metallicity (Meynet *et al.* 2006, Hirschi 2007). The total mass of an 85 $M_\odot$ model at $Z = 10^{-8}$ is shown in Fig. 3 by the top solid line. This model, like metal-free models, loses around 5% of its initial mass when its surface reaches break-up velocities in the second part of the MS. At the end of core H burning, the core contracts and the envelope expands, thus decreasing the surface velocity and its ratio to the critical velocity. The mass loss rate becomes very low again until the star crosses the HR diagram and reaches the RSG stage. In the cooler part of the H-R diagram, the mass loss becomes very important. This is due to the dredge-up by the convective envelope of CNO elements to the surface increasing its overall metallicity. The total metallicity, Z, is used in this model (including CNO elements) for the metallicity dependence of the mass loss. Therefore depending on how much CNO is brought up to the surface, the mass loss becomes very large again. The CNO brought to the surface comes from primary C and O produced in the He-burning region and from primary N produced in the H-burning one.

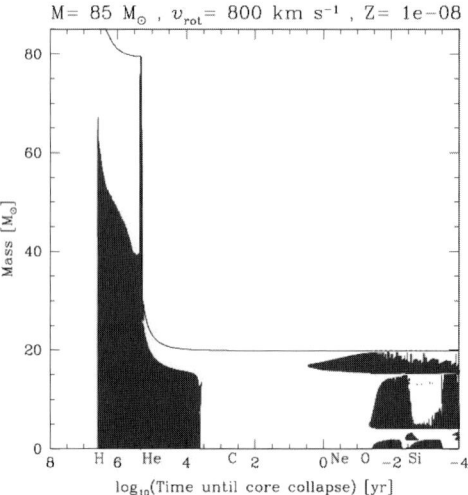

Figure 3. Structure evolution diagram (same as Fig. 2) for a 85 $M_\odot$ model with $v_{\rm ini} = 800$ km s^{-1} at $Z = 10^{-8}$. The top solid line shows the total mass of the star. A strong mass loss during the RSG stage removes a large fraction of total mass of the star.

Could such low-Z stars undergo dust-driven winds? For this to occur, the surface effective temperature needs to be low enough (usually $\log(T_{\rm eff}) < 3.6$) and carbon needs to be more abundant than oxygen. This last condition is fulfilled in our 85 $M_\odot$ model. However, it is presently unclear if: 1) Extremely low-Z stars reach such low effective temperatures. This depends on the opacity and the opacity tables used in our calculations did not account for the non-standard mixture of metals (high CNO and low iron abundance, see Marigo 2002 for possible effects). 2) At such low Z, enough metal is present to allow dust formation. Indeed, nucleation seeds (probably involving titanium) are necessary to form C-rich dust. There may also be other important types of wind, like chromospheric activity-driven, pulsation-driven, thermally-driven or continuum-driven winds.

The fate of rotating stars at very low Z is therefore probably the following:
• $M < 40\,M_\odot$: Mass loss is insignificant and matter is only ejected into the ISM during the SN explosion (see contributions by Nomoto et al, Limongi et al and Fröhlich et al in this volume), which could be very energetic if fast rotation is still present in the core at the core collapse.
• $40\,M_\odot < M < 60\,M_\odot$: Mass loss (at critical rotation and in the RSG stage) removes 10-20% of the initial mass of the star. The star probably dies as a black hole without a SN explosion and therefore the feedback into the ISM is only due to stellar winds, which are slow.
• $M > 60\,M_\odot$: A strong mass loss removes a significant amount of mass and the stars enter the WR phase. These stars therefore die as type Ib/c SNe and possibly as GRBs.

4. Nucleosynthesis and galactic chemical evolution

Rotation induced mixing leads to the production of primary nitrogen, ^{13}C and ^{22}Ne. In this section, we compare the chemical composition of our models with carbon-rich EMP stars and include our stellar yields in a galactic chemical evolution (GCE) model and compare the GCE model with observations of EMP stars.

4.1. The most metal-poor star known to date, HE1327-2326

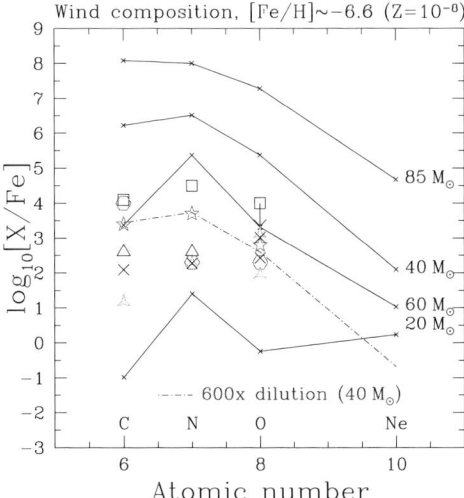

Figure 4. Composition in [X/Fe] of the stellar wind for the $Z = 10^{-8}$ models (solid lines). For HE1327-2326 (*stars*), the best fit for the CNO elements is obtained by diluting the composition of the wind of the 40 M$_\odot$ model by a factor 600 (see Hirschi 2007 for more details).

Significant mass loss in very low-Z massive stars offers an interesting explanation for the strong enrichment in CNO elements of the most metal-poor stars observed in the halo of the galaxy (see Meynet *et al.* 2006, Hirschi 2007). The most metal-poor star known to date, HE1327-2326 (Frebel *et al.* 2006) is characterised by very high N, C and O abundances, high Na, Mg and Al abundances, an s-process (sr) enrichment and depleted lithium. The star is not evolved so has not had time to bring self-produced CNO elements to its surface and is most likely a subgiant. By using one or a few SNe and using a very large mass cut, Limongi *et al.* (2003) and Iwamoto *et al.* (2005) are able to reproduce the abundance of most elements. However they are not able to reproduce the nitrogen surface abundance of HE1327-2326 without rotational mixing. The abundance pattern observed at the surface of that star present many similarities with the abundance pattern obtained in the winds of very metal poor fast rotating massive star models. HE1327-2326 may therefore have formed from gas, which was mainly enriched by stellar winds of rotating very low metallicity stars. In this scenario, a first generation of stars (PopIII) pollutes the interstellar medium to very low metallicities ([Fe/H]$\sim$-6). Then a PopII.5 star (Hirschi 2005) like the 40 M$_\odot$ model calculated here pollutes (mainly through its wind) the interstellar medium out of which HE1327-2326 forms. This would mean that HE1327-2326 is a third generation star. In this scenario, the CNO abundances are well reproduced, in particular that of nitrogen, which according to the latest values for a subgiant (see Frebel *et al.* 2006) is 0.9 dex higher in [X/Fe] than oxygen. This is shown in Fig. 4 where the abundances of HE1327-2326 are represented by the red stars and the best fit is obtained by diluting the composition of the wind of the 40 M$_\odot$ model by a factor 600. When the SN contribution is added, the [X/Fe] ratio is usually lower for nitrogen than for oxygen. It is interesting to note that the very high CNO yields of the

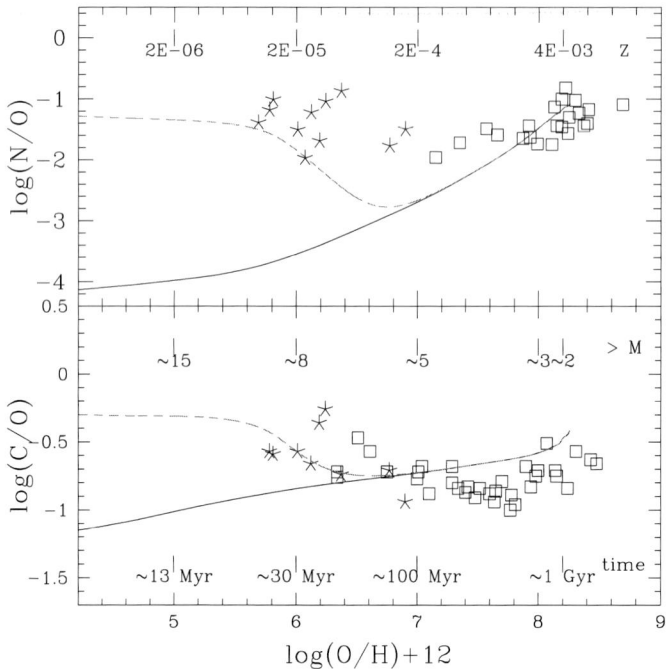

Figure 5. Chemical evolution model predictions of the N/O and C/O evolution, in the galactic halo, for different stellar evolution inputs. The solid curves show the predictions of a model without fast rotators at low metallicities. The dashed lines show the effect of including a population of fast rotators at low metallicities. For the data see Chiappini et al. (2006b) and references therein.

40 M$_\odot$ stars brings the total metallicity Z above the limit for low mass star formation obtained in Bromm & Loeb (2003).

4.2. Primary nitrogen and ^{13}C

The high N/O plateau values observed at the surface of very metal poor halo stars require very efficient sources of primary nitrogen. Rotating massive stars can inject in a short timescale large amount of primary N. They are therefore very good candidates to explain the N/O plateau observed at very low metallicity. According to the heuristic model of Chiappini et al. (2005), a primary nitrogen production of about 0.15 M$_\odot$ per star is necessary. Upon the inclusion of the stellar yields including the effects of fast rotation at $Z=10^{-8}$ in a chemical evolution model for the galactic halo with infall and outflow, both high N/O and C/O ratios are obtained in the very metal-poor metallicity range in agreement with observations (see details in Chiappini et al. 2006a). This model is shown in Fig. 5 (dashed magenta curve). In the same figure, a model computed without fast rotators (solid black curve) is also shown. Fast rotation enhances the nitrogen production by ∼3 orders of magnitude. These results also offer a natural explanation for the large scatter observed in the N/O abundance ratio of *normal* metal-poor halo stars: given the strong dependency of the nitrogen yields on the rotational velocity of the star, we expect a scatter in the N/O ratio which could be the consequence of the distribution of the stellar rotational velocities as a function of metallicity.

As explained above, the strong production of primary nitrogen is linked to a very active H-burning shell and therefore a smaller helium core. As a consequence, less carbon is

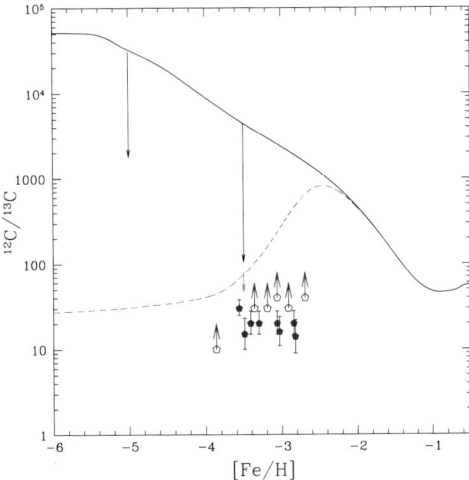

Figure 6. See a detailed description of this Fig. in Chiappini et al. (2008)

turned into oxygen, producing high C/O ratios. Although the abundance data for C/O is still very uncertain, a C/O upturn at low metallicities is suggested by observations (see Asplund 2005 and references therein). Note that this upturn is now also observed in very metal poor DLA systems (see the paper by M. Pettini in this volume).

In addition, stellar models of fast rotators have a great impact on the evolution of the ^{12}C/^{13}C ratio at very low metallicities (Chiappini et al. 2008). In this case, we predict that, if fast rotating massive stars were common phenomena in the early Universe, the primordial interstellar medium of galaxies with a star formation history similar to the one inferred for our galactic halo should have ^{12}C/^{13}C ratios between 30-300. Without fast rotators, the predicted ^{12}C/^{13}C ratios would be ~ 4500 at [Fe/H] = -3.5, increasing to ~ 31000 at around [Fe/H] = -5.0 (see Fig. 6).

Current data on EMP giant normal stars in the galactic halo (Spite *et al.* 2006) agree better with chemical evolution models including fast rotators. The expected difference in the ^{12}C/^{13}C ratios, after accounting for the effects of the first dredge-up (indicated by the arrows in Fig. 6), between our predictions with/without fast rotators is of the order of a factor of 2-3. However, larger differences (a factor of $\sim 60-90$) are expected for giants at [Fe/H]= -5 or turnoff stars already at [Fe/H]= -3.5. To test our predictions, challenging measurements of the ^{12}C/^{13}C in more extremely metal-poor giants and turnoff stars are required.

4.3. *Primary ^{22}Ne and s process at low Z*

Models at $Z = 10^{-8}$ show a production of primary ^{22}Ne during He burning. We also started calculating models at different Z to determine over which Z range the primary production of ^{22}Ne and also ^{14}N is important. In Fig. 7, we show the properties of a 20 M$_\odot$ model at $Z = 10^{-6}$ up to the end of He burning. Around 0.5% (in mass fraction) of ^{22}Ne is burnt during core He burning and therefore leads to a significant neutron release. Studies are underway to determine how much s process can be produced in these models.

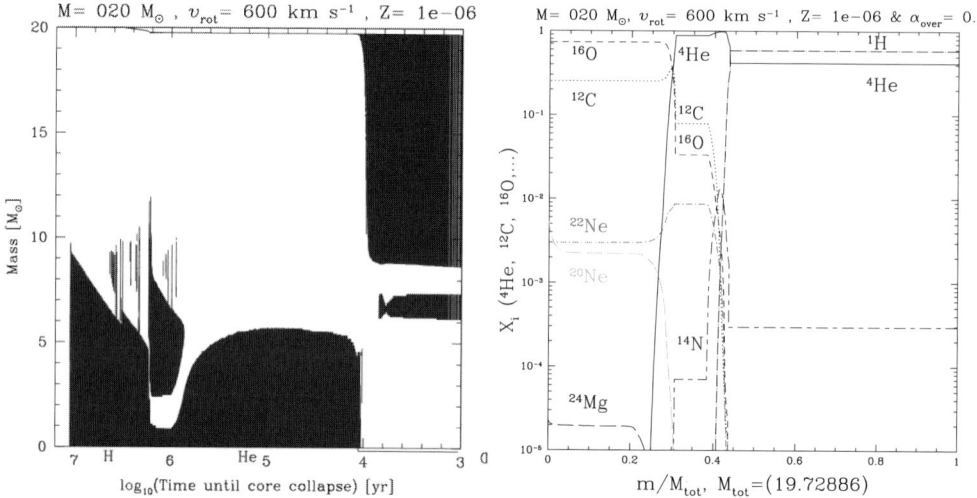

Figure 7. *Left*: Structure evolution diagram (see description of Fig. 2) for a rotating 20 M$_\odot$ models at $Z = 10^{-6}$ during H- and He-burning phases. *Right*: Chemical composition at the end of core He burning. Just above the core, one sees that the maximum abundance of ^{22}Ne is around 1% in mass fraction and at the end of core He burning, around 0.5% is burnt in the core, providing plenty of neutrons for s process.

5. Magnetic fields and GRBs

In this last section, we discuss the impact of magnetic fields. Models of rotating stars, which do not include the effect of magnetic fields predict gamma-ray bursts (GRBs) at almost all Z (Hirschi *et al.* 2005, Hirschi 2007). However, they also overestimate the initial rotation rate of pulsars. The inclusion of the effects of magnetic fields according to Spruit (2002) allows a better reproduction of the initial pulsar periods (Heger *et al.* 2005). Along with gravity waves, magnetic fields are also one possible cause for the flat rotation profile of the Sun (Eggenberger *et al.* 2005). Although it becomes much harder for the core to retain enough angular momentum until the core collapse, there is still an evolutionary scenario, the so-called chemically homogeneous evolution, leading to the production of fast rotating cores at the pre-SN stage and therefore enabling MHD explosions and GRBs (see Yoon et al. 2006, Woosley & Heger 2006 and also Yoon's contribution in this volume). The theoretical GRB event rates obtained by Yoon *et al.* (2006) are in good agreement with observations apart from the upper metallicity limit, which is lower than the observed one (see contribution by Stanek in this volume). Yoon *et al.* (2006) also predict that at $Z = 10^{-5}$, a large fraction of massive stars are GRB progenitors. We have calculated 40 M$_\odot$ models at $Z = 10^{-5}$ with $v_{\rm ini}/v_{\rm crit} = 0.59$ and at $Z = 10^{-8}$ with $v_{\rm ini}/v_{\rm crit} = 0.55$. The model at $Z = 10^{-5}$ confirms the possibility of producing GRBs down to very low Z. However, the model at $Z = 10^{-8}$ does not rotate fast enough to evolve chemically homogeneously. The difficulty of the very low Z models to evolve chemically homogeneously is due to the weakening of the meridional circulation. Indeed, in models including magnetic fields, meridional circulation becomes the dominant term for the mixing of chemical species (see Maeder & Meynet 2005). The meridional circulation becomes weaker at low Z because the meridional currents are less efficient in a denser medium, which is the case since low-Z stars are more compact. This means that not all stars in the first stellar generations will produce GRBs in this way. Finally, it is interesting to note that the presence of magnetic fields in metal-free stars may enhance mass loss significantly (see contribution by Ekström et al in this volume).

6. Conclusions and outlook

The inclusion of the effects of rotation changes significantly the simple picture in which stellar evolution at low Z is just stellar evolution without mass loss. A strong mixing is induced between the helium and hydrogen burning layers leading to a significant production of primary ^{14}N, ^{13}C and ^{22}Ne. Rotating stellar models also predict a strong mass loss during the RSG stage for stars more massive than 60 M$_\odot$. The chemical composition of the stellar winds is compatible with the CNO abundance observed in the most metal-poor star known to date, HE1327-2326. GCE models including the stellar yields of these rotating star models are able to better reproduce the early evolution of N/O, C/O and ^{12}C/^{13}C in our galaxy. These models predict a large neutron release during core He burning and thus a strong possibility of an s process at very low Z. These models predict the formation of WR and type Ib/c SNe down to almost Z=0. The inclusion of magnetic fields slows down the core of the stars and therefore reduces the probability of producing GRBs at metallicities around that of the Magellanic Clouds but GRBs are still predicted from single star models down to very low Z.

Large surveys of EMP stars (SEGUE, OZ surveys), of GRBs and SNe (Swift and GLAST satellites) and of massive stars (e. g. VLT FLAMES survey) are underway and will bring more information and constraints on the evolution of massive stars at low Z. On the theoretical side, more models are necessary to fully understand and study the complex interplay between rotation, magnetic fields, mass loss and binary interactions at different metallicities. Large grids of models at low Z will have many applications, for example to study the evolution of massive stars and their feedback in high redshift objects like Lyman-break galaxies and damped Ly-alpha systems.

Acknowledgements

R. Hirschi acknowledges financial support from EPSAM, from the organizers (IAU Grant), and from the Royal Society (Conference Grant round 2007/R3).

References

Arnett, D. 1996, in: H. L. Morrison & A. Sarajedini (eds.), *Formation of the Galactic Halo...Inside and Out*, (San Francisco: ASP) *ASP Conf. Ser.*, 92, 337
Asplund, M. 2005, *ARA&A*, 43, 481
Beers, T. C. & Christlieb, N. 2005, *ARA&A*, 43, 531
Bromm, V. & Larson, R. B. 2004, *ARA&A*, 42, 79
Bromm, V. & Loeb, A. 2003, *Nature*, 425, 812
Carr, B. J., Bond, J. R., & Arnett, W. D. 1984, *ApJ*, 277, 445
Chiappini, C., Matteucci, F., & Ballero, S. K. 2005, *A&A*, 437, 429
Chiappini, C., Hirschi, R., Matteucci, F., et al. 2006a, in: *Nuclei in the Cosmos IX*, CERN, PoS(NIC-IX)080
Chiappini, C., Hirschi, R., Meynet, G., et al. 2006b, *A&A*, 449, L27
Chiappini, C., Ekström, S., Meynet, G., et al. 2008, *A&A*, 479, L9
Chieffi, A. & Limongi, M. 2004, *ApJ*, 608, 405
Chiosi, C. 1983, *Memorie della Societa Astronomica Italiana*, 54, 251
de Mink, S. E., Pols, O. R., & Yoon, S. C. 2008, in: T. Abel, A. Heger & B. O'Shea (eds.), *First Stars III* (New York: AIP), *AIP Conf Proc.*, in press (arXiv0710.1010)
Eggenberger, P., Maeder, A., & Meynet, G. 2005, *A&A*, 440, L9
Ekström, S., Meynet, G., & Maeder, A. 2006, in: H. J.G. L.M. Lamers, N. Langer, T. Nugis, & K. Annuk (eds.), *Stellar Evolution at Low Metallicity: Mass Loss, Explosions, Cosmology* (San Francisco: ASP) *ASP Conf. Ser.*, 353, 141
Ekström, S., Meynet, G., & Maeder, A. 2008, in: T. Abel, A. Heger & B. O'Shea (eds.), *First Stars III*, (New York: AIP), *AIP Conf. Proc.*, in press (arXiv:0709.0202)

El Eid, M. F., Fricke, K. J., & Ober, W. W. 1983, *A&A*, 119, 54
Ferrarotti, A. S. & Gail, H.-P. 2006, *A&A*, 447, 553
Frebel, A., Christlieb, N., Norris, J. E., *et al.* 2006, *ApJ*, 638, L17
Fukuda, I. 1982, *PASP*, 94, 271
Gil-Pons, P., Suda, T., Fujimoto, M. Y., & García-Berro, E. 2005, *A&A*, 433, 1037
Heger, A. & Langer, N. 2000, *ApJ*, 544, 1016
Heger, A. & Woosley, S. E. 2002, *ApJ*, 567, 532
Heger, A., Fryer, C. L., Woosley, S. E., *et al.* 2003, *ApJ*, 591, 288
Heger, A., Woosley, S. E., & Spruit, H. C. 2005, *ApJ*, 626, 350
Herwig, F. 2004, *ApJS*, 155, 651
Hirschi, R. 2005, in: V. Hill, P. François & F. Primas (eds.), *From Lithium to Uranium: Elemental Tracers of Early Cosmic Evolution* (Cambridge: Cambridge University Press), *Proc. IAU Symp*, 228, 331
Hirschi, R. 2007, *A&A*, 461, 571
Hirschi, R., Meynet, G., & Maeder, A. 2005, *A&A*, 443, 581
Höfner, S. & Andersen, A. C. 2007, *A&A*, 465, L39
Iwamoto, N., Umeda, H., Tominaga, N., *et al.* 2005, *Science*, 309, 451
Kudritzki, R. P. 2002, *ApJ*, 577, 389
Kudritzki, R.-P. & Puls, J. 2000, *ARA&A*, 38, 613
Langer, N., Norman, C. A., de Koter, A., *et al.* 2007, *A&A*, 475, L19
Limongi, M. & Chieffi, A. 2005, in: V. Hill, P. François & F. Primas (eds.), *From Lithium to Uranium: Elemental Tracers of Early Cosmic Evolution* (Cambridge: Cambridge University Press), *Proc. IAU Symp*, 228, 303
Limongi, M., Chieffi, A., & Bonifacio, P. 2003, *ApJ*, 594, L123
Maeder, A., Grebel, E. K., & Mermilliod, J.-C. 1999, *A&A*, 346, 459
Maeder, A. & Meynet, G. 2005, *A&A*, 440, 1041
Marigo, P. 2002, *A&A*, 387, 507
Meynet, G. & Maeder, A. 2000, *A&A*, 361, 101
Meynet, G. & Maeder, A. 2002, *A&A*, 390, 561
Meynet, G. & Maeder, A. 2005, *A&A*, 429, 581
Meynet, G., Ekström, S., & Maeder, A. 2006, *A&A*, 447, 623
Mokiem, M. R., de Koter, A., Vink, J. S., *et al.* 2007, *A&A*, 473, 603
Nakamura, F. & Umemura, M. 2001, *ApJ*, 548, 19
Nieuwenhuijzen, H., & de Jager, C. 1990, *A&A*, 231, 134
Picardi, I., Chieffi, A., Limongi, M., *et al.* 2004, *ApJ*, 609, 1035
Scannapieco, E., Madau, P., Woosley, S., *et al.* 2005, *ApJ*, 633, 1031
Schneider, R., Omukai, K., Inoue, A. K., & Ferrara, A. 2006, *MNRAS*, 369, 1437
Siess, L., Livio, M., & Lattanzio, J. 2002, *ApJ*, 570, 329
Smith, N., Gehrz, R. D., Hinz, P. M., *et al.* 2003, *AJ*, 125, 1458
Smith, N., Li, W., Foley, R. J., *et al.* 2007, *ApJ*, 666, 1116
Spite, M., Cayrel, R., Plez, B., *et al.* 2005, *A&A*, 430, 655
Spite, M., Cayrel, R., Hill, V., *et al.* 2006, *A&A*, 455, 291
Spruit, H. C. 2002, *A&A*, 381, 923
Suda, T., Aikawa, M., Machida, M. N., *et al.* 2004, *ApJ*, 611, 476
Tominaga, N., Umeda, H., & Nomoto, K. 2007, *ApJ*, 660, 516
Tornatore, L., Ferrara, A., & Schneider, R. 2007, *MNRAS*, 382, 945
Umeda, H. & Nomoto, K. 2005, *ApJ*, 619, 427
van Loon, J. T. 2000, *A&A*, 354, 125
van Loon, J. T. 2006, in: H. J.G. L.M. Lamers, N. Langer, T. Nugis, & K. Annuk (eds.), *Stellar Evolution at Low Metallicity: Mass Loss, Explosions, Cosmology* (San Francisco: ASP), *ASP Conf. Ser.*, 353, 211
van Loon, J. T., Cioni, M.-R. L., Zijlstra, A. A., & Loup, C. 2005, *A&A*, 438, 273
Vink, J. S. & de Koter, A. 2005, *A&A*, 442, 587
Vink, J. S., de Koter, A., & Lamers, H. J.G. L.M. 2000, *A&A*, 362, 295
Vink, J. S., de Koter, A., & Lamers, H. J.G. L.M. 2001, *A&A*, 369, 574

Weiss, A., Schlattl, H., Salaris, M., & Cassisi, S. 2004, *A&A*, 422, 217
Woosley, S. E. & Heger, A. 2006, *ApJ*, 637, 914
Woosley, S. E., Blinnikov, S., & Heger, A. 2007, *Nature*, 450, 390
Yoon, S.-C., Langer, N., & Norman, C. 2006, *A&A*, 460, 199

Discussion

LEITHERER: Your models seem to have large mass loss, even at very low Z. Does this imply that the winds are dense enough to hide the hot core? Then we would not expect an extremely hard ionizing spectrum and little observable nebular He II λ1640, a possible telltale sign of the first generation of stars.

HIRSCHI: Mass loss takes place after the main sequence, so the ionizing flux can still be produced during the main sequence.

Raphael Hirschi.

Nolan Walborn (left) and Ian Hunter (right).

Geoffrey Burbidge.

Evolution of Progenitor Stars of Type Ibc Supernovae and Long Gamma-Ray Bursts

Sung-Chul Yoon[1], Norbert Langer[2], Matteo Cantiello[2], Stan E. Woosley[1] and Gary A. Glatzmaier[3]

[1]Department of Astronomy & Astrophysics, University of California, Santa Cruz
High Street, Santa Cruz, CA 95064, USA

[2]Astronomical Institute, Utrecht University, Utrecht, The Netherlands

3Department of Earth & Planetary Sciences, University of California, Santa Cruz
High Street, Santa Cruz, CA 95064, USA

Abstract. We discuss how rotation and binary interactions may be related to the diversity of type Ibc supernovae and long gamma-ray bursts. After presenting recent evolutionary models of massive single and binary stars including rotation, the Tayler-Spruit dynamo and binary interactions, we argue that the nature of SNe Ibc progenitors from binary systems may not significantly differ from that of single star progenitors in terms of rotation, and that most long GRB progenitors may be produced via the quasi-chemically homogeneous evolution at sub-solar metallicity. We also briefly discuss the possible role of magnetic fields generated in the convective core of a massive star for the transport of angular momentum, which is potentially important for future stellar evolution models of supernova and GRB progenitors.

Keywords. stars: evolution – binaries: close – stars: magnetic fields – stars: rotation – supernovae: general

1. Introduction

Rotation influences not only the evolution of massive stars, but also their supernova (SN) explosions (Maeder & Meynet 2000; Heger, Langer & Woosley 2000). In particular, recently many asymmetric supernovae with unusually large energy (broad-lined SNe or hypernovae) have been discovered (e.g. Mazzali $et\ al.$ 2007), which shows evidence for rapidly rotating progenitors. The most spectacular example may be long gamma-ray bursts (GRBs), which are generally believed to be produced by deaths of some massive stars retaining extremely large angular momenta in their cores ($j \geqslant \sim 10^{16}$ cm^2 s^{-1}; Woosley 1993; MacFadyen & Woosley 1999; see, however, Dessart $et\ al.$ 2008). Interestingly, such energetic core-collapse events seem to only occur in Wolf-Rayet (WR) stars: all of the broad-lined SNe/Hypernovae and the supernovae associated with long GRBs have been observationally identified as Type Ic (see Woosley & Bloom 2006 for a review). This raises the question which WR stars can produce broad-lined SNe Ic or long GRBs while most WR stars die as normal SNe Ibc. Here we present recent evolutionary models of massive stars that include binary interactions and the transport of chemical species and angular momentum via rotationally induced hydrodynamic instabilities and magnetic torques, and discuss how rotation and binary interactions may be related to the diversity of SNe Ibc and long GRBs.

2. Single star models

Redistribution of angular momentum and chemical species in a rotating star occurs by rotationally induced hydrodynamic instabilities and magnetic torques (See Talon 2007

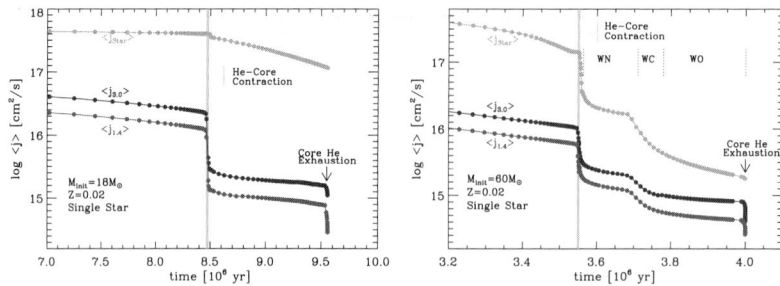

Figure 1. Mean specific angular momentum of the star and the innermost 1.4 M$_\odot$ and 3.0 M$_\odot$ as a function of the evolutionary time for 18 M$_\odot$ (left panel) and 60 M$_\odot$ (right panel) models. The time span for the helium core contraction is marked by the color shade as indicated by the label.

for a review). Eddington-Sweet circulations, shear instability and Goldreich-Schubert-Fricke instability among others have been considered in previous non-magnetic models (Maeder & Meynet 2000; Heger, Langer & Woosley 2000; Hirschi, Meynet & Maeder 2004), and the so-called Tayler-Spruit dynamo (Spruit 2002) has been implemented in recent magnetic models (e.g. Heger, Woosley & Spruit 2005; Maeder & Meynet 2005; Yoon & Langer 2005). In non-magnetic models, it is shown that the buoyancy due to the chemical gradient at the interface between the core and the envelope largely prohibits the considered rotationally induced hydrodynamic instabilities from transporting angular momentum. The amount of angular momentum retained in the core at the pre-supernova stage is thus close to its initial value even at solar metallicity (Heger, Langer & Woosley 2000; Hirschi, Meynet & Maeder 2004). Most massive stars are predicted to die with an enough amount of angular momentum in the cores to produce long GRBs via formation of millisecond magnetars or collapsar (i.e., $j \gtrsim \sim 10^{16}$ cm^2 s^{-1}), given that a large fraction of young massive stars in our Galaxy and Small/Large Magellanic Clouds are rapid rotators (e.g. Maeder & Meynet 2000; Mokiem et al. 2006; Hunter et al. 2008). On the other hand, in magnetic models adopting the Tayler-Spruit dynamo the core is effectively spun down by magnetic torques, and the predicted spin rates of white dwarfs and young neutron stars are smaller by two orders of magnitude than those from non-magnetic models, which can better explain observations (Heger, Woosley & Spruit 2005; Suijs et al. 2008).

Fig. 1 shows the evolution of the core angular momentum in the magnetic model sequences with $M_{\rm init}$ = 18 M$_\odot$ & $v_{\rm rot,init}$ = 144 km s^{-1}, and 60 M$_\odot$ & $v_{\rm rot,init}$ = 186 km s^{-1}. In the sequence with $M_{\rm init}$ = 18 M$_\odot$, the core loses a significant amount of angular momentum during the helium core contraction phase where a strong degree of differential rotation between the core and the envelope appears. A similar effect is also observed during the CO core contraction phase. In the sequence with $M_{\rm init}$ = 60 M$_\odot$, spinning-down of the core during He-core contraction becomes less significant as the star loses the hydrogen envelope, which leads to a smaller moment of inertia of the envelope. However, the core is further spun down by loss of mass due to LBV/WR winds during core He burning. At the neon burning stage, both stars retain a similar amount of angular momentum in their cores as shown in Fig. 1. In fact, the calculations by Yoon, Langer & Norman (2006) show that magnetic models give $<j_{1.4}> \simeq 2...3 \times 10^{14}$ cm^2 s^{-1} for different initial metallicities, masses and rotational velocities, implying that most massive stars including Type Ibc progenitors should die with a similar amount of core angular momentum. This conclusion remains the same even for binary stars, as discussed below.

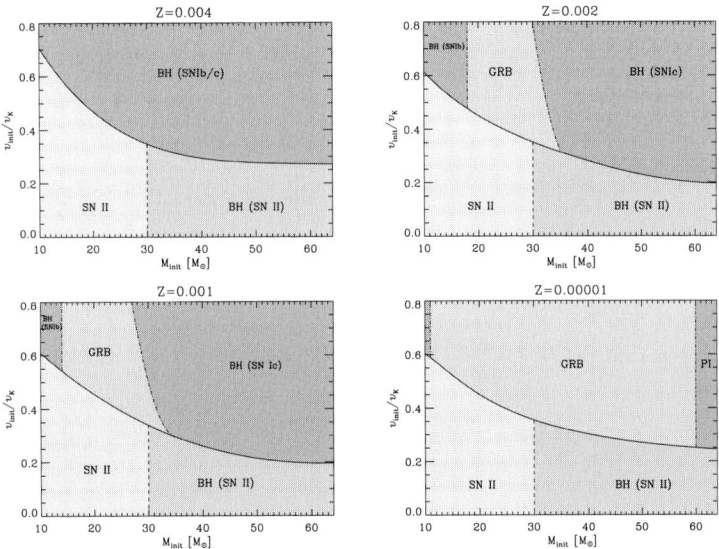

Figure 2. Final fate of our rotating massive star models at four different metallicities ($Z = 0.004, 0.002, 0.001$, & 0.00001), in the plane of initial mass and initial fraction of the Keplerian value of the equatorial rotational velocity. The solid line divides the plane into two parts, where stars evolve quasi-chemically homogeneous above the line, while they evolve into the classical core-envelope structure below the line. The dotted-dashed lines bracket the region of quasi-homogeneous evolution where the core mass, core spin and stellar radius are compatible with the collapsar model for GRB production (absent at Z=0.004). To both sides of the GRB production region for $Z = 0.002$ and 0.001, black holes are expected to form inside WR stars, but the core spin is insufficient to allow GRB production. For $Z = 0.00001$, the pair-instability might occur to the right side of the GRB production region, although the rapid rotation may shift the pair instability region to larger masses. The dashed line in the region of non-homogeneous evolution separates Type II supernovae (SN II; left) and black hole (BH; right) formation, where the minimum mass for BH formation is simply assumed to be 30 $M_\odot$. From Yoon, Langer & Norman (2006).

An exception is the case for the so-called chemically homogeneous evolution. If the initial rotational velocity is exceptionally high and if metallicity is sufficiently low, chemical mixing by Eddington-Sweet circulations may occur on a time scale even smaller than the nuclear time scale. Quasi-homogeneity of the chemical composition of a star is thus ensured on the main sequence and the star is gradually transformed into a massive WR star, avoiding the giant phase that would result in strong braking down of the core. All of the necessary conditions for producing long GRBs – massive core to make a black hole, removal of hydrogen envelope and retention of a large amount of angular momentum in the core – thus can be fulfilled by this type of evolution (Yoon & Langer 2005; Woosley & Heger 2006). This chemically homogeneous evolution scenario (CHES) favors low metallicity environment for producing long GRBs (Fig. 2) and predicts a higher ratio of GRB to SN rate at higher redshift, which should be tested by future observations (Yoon, Langer & Norman 2006; cf. Kistler *et al.* 2008).

3. Binary star models

A significant fraction of SNe Ibc may be produced in close binary systems (e.g. Podsiadlowski, Joss & Hsu 1992). An example is given in Fig. 3 that shows the evolution of of the primary star in a close binary system with $P_{\rm init} = 4$ days, $M_{\rm primary, init} = 18$ $M_\odot$, and

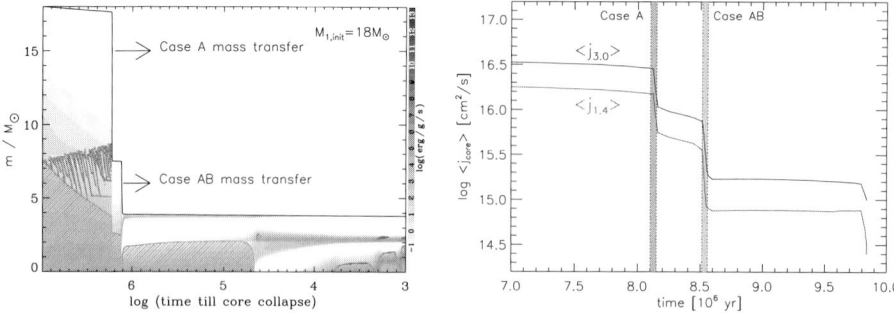

Figure 3. *Left panel* Evolution of the internal structure of the primary star in a binary system of $M_{\rm primary, init} = 18$ M$_\odot$, $M_{\rm secondary, init} = 17$ M$_\odot$ and $P_{\rm init} = 4$ days, from zero age main sequence to neon burning. *Right panel* Mean specific angular momentum of the innermost 1.4 M$_\odot$ and 3.0 M$_\odot$ of the primary star considered in the left panel as a function of time. The time span for Case A or Case AB mass transfer phase is marked by the color shades as indicated by the labels.

$M_{\rm secondary, init} = 17$ M$_\odot$. Both stars are tidally synchronized early on the main sequence. Once the primary star fills the Roche-lobe radius, it loses about 7 M$_\odot$ during the Case A mass transfer phase, and additional 3.5 M$_\odot$ later during the Case AB mass transfer phase, becoming a 4 M$_\odot$ WR star. The core loses angular momentum mostly during these Case A and AB mass transfer phases as shown in Fig. 3. The amount of angular momentum in the core at the neon burning stage turns out to be very similar to those in single star models (see Fig. 1). In fact, we find that mean specific angular momentum of the innermost 1.4 M$_\odot$ at the final evolutionary stage of a primary star in a binary system does not change much according to different initial parameters (primary mass, mass ratio and orbital separation): it remains within a narrow range of $2...3.5 \times 10^{14}$ cm^2 s^{-1} regardless of the detailed history of binary interactions, as long as the tidal synchronization is not unusually strong (Yoon, Woosley & Langer 2008, in prep.). This implies that the nature of most binary star progenitors of SNe Ibc may not much differ from that of single star progenitors, in terms of rotation.

The evolution of secondary stars has not yet been well understood. In particular, it sensitively depends on the uncertain efficiency of semi-convection whether the mass accreting star may be rejuvenated or not (Braun & Langer 1995). If a rather large semi-convection parameter is adopted, rejuvenation can significantly weaken the chemical gradient between the hydrogen burning core and the envelope, in favor of rotationally induced chemical mixing. As the secondary is spun up to the critical rotation by mass accretion, even the chemically homogeneous evolution can be occasionally induced if metallicity is sufficiently low and if the secondary is not strongly spun down by the tidal synchronization after the mass accretion phase (Cantiello *et al.* 2007). The secondary will eventually die as a GRB after traveling from a few to several hundreds PCs away, if the binary system is unbound due to the supernova kick as a result of the explosion of the primary. This scenario may explain the recent observational evidence that some GRBs are produced in runaway stars (Hammer *et al.* 2006).

Other types of binary interactions may also lead to formation of rapidly rotating WR stars to produce long GRBs. Tidal spinning-up of a WR star in a compact binary system with a neutron star or a black hole companion (Brown et. al. 2000; Izzard, Ramirez-Ruiz & Tout 2004; van Putten, M. H.P. M. 2004; van den Heuvel & Yoon 2007) and merger of two helium cores in a common evenlope (Fryer & Heger 2005) have been recently suggested among others. It remains uncertain, however, that such binary systems

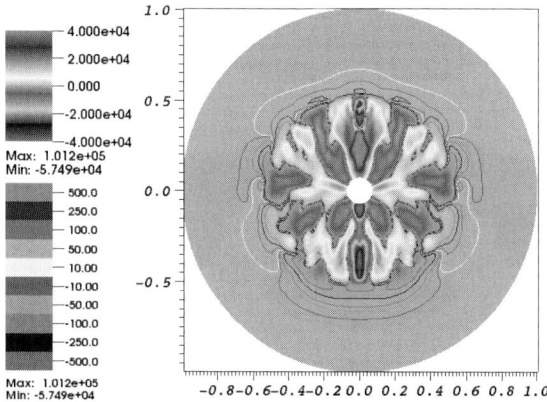

Figure 4. Mean radial fields $B_r(r,\theta)$ on the meridional plane in a 12 M$_\odot$ rotating star on the main sequence in a MHD simulation with a 3-D anelastic code (Glatzmaier 1984). The adopted angular velocity is 10^{-5} Rad s^{-1}. The inner region of $r \leqslant 7 \times 10^{10}$ cm ($r \leqslant 0.5$ in the code units) is the convective core.

could explain the observed GRB rate. For example, recent stellar evolution models by Detmers, Langer & Podsiadlowski (2008) show that a merger of the WR star with the compact object, which is not supposed to produce a classical long GRB, is the most likely outcome in the former case. Further detailed evolutionary models are certainly needed for observationally testing different evolutionary scenarios (e.g. van Marle et al. 2008). On the other hand, it is puzzling why no GRB-associated SNe Ib have been observed yet while most GRB progenitor scenarios predict their existence (e.g. Yoon, Langer & Norman 2006). Within the CHES, this puzzle might be solved if one considered anisotropic mass loss, as discussed in Meynet & Maeder (2007).

4. Concluding remarks

Our stellar evolution models including the Tayler-Spruit dynamo indicate that most SNe Ibc progenitors should explode with a similar amount of angular momentum in their cores, regardless of their single or binary star origin. This is due to the self-regulationary nature of the Tayler-Spruit dynamo. Loss of the hydrogen envelope due to stellar winds or mass transfer results in both removal of angular momentum from the core and weakening of the core braking by the extended hydrogen enveloped due to magnetic torques, and vice versa. Therefore, most different pre-supernova evolutionary paths may not contribute much to the diversity of Type Ibc supernovae in terms of core angular momentum, although different iron core masses, and thus different spin rates of young neutron stars may result (Heger, Woosley & Spruit 2005). GRB progenitors, which require unusually large angular momenta, may undergo the chemically homogeneous evolution, which may not be unusual at low metallicity. On the other hand, recent observations indicate that not all broad-lined SNe Ic are associated with long GRBs (e.g. Modjaz et al. 2008), which needs a theoretical explanation in future work.

Although the predicted spin rates of the stellar remnants from the magnetic models are consistent with observations, the validity of the Tayler-Spruit dynamo has been recently questioned by several authors (Denissenkov & Pinsonneault 2007; Zahn, Brun & Mathis 2008). Furthermore, we might have ignored potentially important physical ingredients in simulating the evolution of rotating massive stars. These include gravity waves (Townsend in this volume), and magnetic fields generated by the convective core. For instance, our recent 3-D simulations with an anelastic magnetohydrodynamics code (Glatzmaier 1984) show that the strength of poloidal fields generated in the convective core in a young massive star may amount to several thousand Gauss on average (Fig. 4), and its influence on the transport processes might be comparable to what the Tayler-Spruit dynamo predicts. This issue will be addressed in Yoon, Woosely & Glatzmaier (2008, in prep.).

Acknowledgements

This work was, in part, supported by the DOE Program for Scientific Discovery through Advanced Computing and NASA.

References

Braun, H. & Langer, N. 1995, *A&A*, 297, 483
Brown, G. E., Lee, C.-H., Wijers, R. A. M. J., *et al.* 2000, *NewA*, 5, 191
Cantiello, M., Yoon, S.-C., Langer, N., & Livio, M. 2007, *A&A*, 465, L29
Denissenkov, P. A. & Pinsonneault, M. 2007, *ApJ*, 655, 1157
Dessart, L., Burrows, A., Livne, E., & Ott, C. 2008, *ApJL*, 673, 43
Detmers, R., Langer, N., Podsiadlowski, Ph., & Izzard, R. G. 2008, *A&A*, submitted astro-ph/0804.0014
Fryer, C. L. & Heger, A. 2005, *ApJ*, 623, 302
Glatzmaier, G. A. 1984, *J. Comp. Phys.*, 55, 461
Hammer, F., Flores, H., Schaerer, D., *et al.* 2006, *A&A*, 454, 103
Heger, A., Langer, N., & Woosley, S. E. 2000, *ApJ*,
Heger, A., Woosley, S. E., & Spruit, H. C. 2005, *ApJ*, 626, 350
van den Heuvel, E. P. J & Yoon, S.-C. 2007, *Ap&SS*, 311, 177
Hirschi, R., Meynet, G., & Maeder, A. 2005, *A&A*, 425, 649
Hunter, I, Lennon, D. J., Dufton, P. I., *et al.* 2008, *A&A*, 479, 541
Izzard, R. G., Ramirez-Ruiz, E., & Tout, C. A. 2004, *MNRAS*, 348, 1215
Kistler, M. D., Yüksel, H., Beacom, J. F., & Stanek, K. Z. 2008, *ApJ*, 673, L119
MacFadyen, A. I. & Woosley, S. E. 1999, *ApJ*, 524, 262
Maeder, A. & Meynet, G. 2000, *ARA&A*, 38, 143
Maeder, A. & Meynet, G. 2005, *A&A*, 440, 104
Mazzali, P. A., Kawabata, K. S., Maeda, K., *et al.* 2007, *ApJ*, 670, 592
Meynet, G. & Maeder, A. 2007, *A&A*, 464, L11
van Marle, A. J., Langer, N., Yoon, S.-C., & Garcia-Segura, G. 2008, *A&A*, 478, 769
Modjaz, M, Kewley, L., Kirshner, R. P., *et al.* 2008, *AJ*, 135, 1136
Mokiem, M. R., de Koter, A., Evans, C. J., *et al.* 2006, *A&A*, 456, 1131
Podsiadlowski, Ph., Joss, P. C., & Hsu, J. J.L. 1992, *ApJ*, 391, 246
van Putten, M. H. P. M. 2004, ApJL, 611, 81
Suijs, M., Langer, N., Poelarends, A. J., *et al.* 2008, *A&A*, 481, L87
Spruit, H. C. 2002, *A&A*, 381, 923
Talon, S. 2007, astro-ph/0708.1499
Woosley, S. E. 1993, *ApJ*, 405, 273
Woosley, S. E. & Bloom, J. S. 2006, *ARA&A*, 44, 507
Woosley, S. E. & Heger, A. 2006, *ApJ*, 637, 914
Yoon, S.-C. & Langer, N. 2005, *A&A*,
Yoon, S.-C., Langer, N., & Norman, C. 2006, *A&A*, 460, 199
Zahn, J.-P., Brun, A. S., & Mathis, S. 2007, *A&A*, 474, 145

Core Overshoot and Nonrigid Internal Rotation of Massive Stars: Current Status from Asteroseismology†

Conny Aerts[1,2]

[1]Instituut voor Sterrenkunde, Katholieke Universiteit Leuven, Celestijnenlaan 200D, B-3001 Leuven, Belgium
[2]IMAPP, Radboud Universiteit Nijmegen, P.O. Box 9010, 6500 GL Nijmegen, the Netherlands
email: conny@ster.kuleuven.be

Abstract. Seismic estimates of core overshoot have been derived from extensive high-precision photometric and spectroscopic ground-based (multisite) campaigns for five main-sequence B-type stars. For three of these, the ratio of the near-core rotation frequency to the surface rotation frequency could be estimated as well, from the identified oscillation modes. We summarise these seismic results obtained for B stars. Now that the technique of asteroseismology was proven to work for probing the interior of massive stars, we expect a drastic increase in the precision of the structure parameters from the space missions CoRoT and Kepler, as well as from currently ongoing ground-based campaigns, in the coming years.

Keywords. stars: early-type – stars: oscillations – stars: evolution – stars: interiors – convection – line: profiles – techniques: spectroscopic – techniques: photometric – methods: data analysis – stars: fundamental parameters

---

1. The goals of asteroseismology of massive stars

Asteroseismology is a modern branch within stellar astrophysics whose goal is to improve stellar evolution models by exploiting the character of stellar oscillations. For an extensive recent review, with specific emphasis on the synergy between asteroseismology and interferometry, we refer to Cunha *et al.* (2007). With respect to massive stars, the largest unknowns in their interior structure are the amounts of rotational mixing and of core convective overshooting. These two phenomena are crucial for the evolution of O and B stars, while they are poorly understood and hardly calibrated from observational data. Parameterised formulae are used to describe these two physical processes, but most of the current observational data are not accurate enough to provide a stringent test of them. The goal is to improve this situation from asteroseismology by considering, besides classical high-precision spectroscopy and interferometry, extensive time series of (multi-colour) photometry and high-resolution spectroscopy to derive the details of the stellar oscillation properties. The latter are indeed an optimal probe of the interior physics of stars. Immense progress has been made in this field the last decade, see e.g. the summary by Kurtz (2006).

For a spherically symmetric star, each oscillation mode is characterised by its cyclic frequency $\nu_{n,\ell,m}$ and wavenumbers (n, ℓ, m), where n is the radial order of the mode and (ℓ, m) are the wavenumbers of the spherical harmonic describing the angular part of the eigenfunction (see, e.g., Cunha *et al.* 2007). The observational part of an asteroseismic

---

† A large part of this research was performed within the Belgian Asteroseismology Group (BAG): http://www.asteroseismology.be, in particular within an intensive Leuven–Liège collaboration.

study comes down to the derivation of $\nu_{n,\ell,m}$ and of the identification of (ℓ, m) from time series analysis. The theoretical part consists in the prediction of these quantities from a dense grid of stellar structure models.

Asteroseismology has the potential to derive high-precision (1–5%) estimates of the fundamental parameters of massive stars by fitting observed identified oscillation frequencies and to derive the core overshoot parameter as well as to estimate the nonrigidity of the internal rotation. We show here that this has been achieved for a few carefully selected slowly-rotating B stars in the past few years.

2. Procedure for core overshoot estimation

Several independent evolution and oscillation codes have been developed. In preparation of the CoRoT space mission, seven evolution codes have been compared in great detail. Only very minor differences due to the different numerical tools were found, provided that the same assumptions on the input physics of the models (equation of state, opacities, convection, etc.) and their physical parameters were made (Monteiro et al. 2006). This led to very similar seismic predictions for the stellar models and implies negligible theoretical uncertainty in the seismic interpretation of measured oscillation frequencies compared to the uncertainty induced by the present-day data, including those expected from CoRoT (Moya et al. 2008). Differences in the theoretical interpretation of measured oscillations thus only come from different adopted input physics of the models.

For O and B stars, standard stellar models typically cover a 5D parameter space, with input parameters the stellar mass M, the stellar age, the hydrogen fraction X, the metallicity Z and the core overshoot parameter $\alpha_{\rm ov}$ which expresses the thickness of the overshoot region around the stellar core. The overshoot parameter is usually expressed in terms of the local pressure scale height $H_{\rm p}$. Different descriptions of the temperature gradient and of the efficiency of the mixing in the overshoot region are in use and, thus, there is a theoretical uncertainty in the derived value of $\alpha_{\rm ov}$.

In order to estimate $\alpha_{\rm ov}$ from fitting of measured oscillation frequencies, one usually adopts a forward modelling method. This consists first of all of the computation of a very dense 5D grid of stellar models covering the (uncertain!) position of the star in the HR diagram, or, preferably in the less uncertain ($T_{\rm eff}, \log g$) diagram. Secondly, the oscillation frequencies $\nu_{n,\ell,0}$ and mode excitation for the grid models are computed. Subsequently, the quantum numbers (ℓ, m) of the detected frequencies must be derived from the data (e.g. Briquet & Aerts 2003, Zima 2006). Finally, a match between the observed frequencies of the $m = 0$ modes and those predicted by the models is made. For observed isolated frequencies with $m \neq 0$, this requires knowledge of the rotation frequency $\Omega(r)$ inside the star — see Eq.(3.1) discussed further on. As a good initial approximation, one assumes rigid internal rotation and derives $\overline{\Omega}$ from the measured $v \sin i$ and an estimate of the inclination angle from the mode identification (Mazumdar et al. 2006). If the inclination angle or the average rotation frequency cannot be estimated, then the $m \neq 0$ modes cannot be used in the seismic modelling.

A successful match is one for which the differences in observed and theoretical $m = 0$ frequencies is less than the accuracy of the measured identified mode frequencies. In that case, the models are sufficiently sophisticated in terms of their input physics to approximate the star, and one obtains a very precise seismic estimate of all the fundamental stellar parameters, including mass and age. If no satisfactory match is obtained, the conclusion must be that the input physics of the models is too different from reality. In the latter case, one can use the oscillation information to try and improve this input physics.

The situation may also occur that a good frequency match is achieved, but that (some of) the measured frequencies are predicted to be stable. That also allows to improve the input physics. Such a situation was encountered for the star HD 29248 whose detected oscillations pointed out a too low OPAL opacity value in the driving region (Pamyathnykh et al. 2004) or in the entire star (Ausseloos et al. 2004). This problem, as well as the occurrence of B-type pulsators in the Magellanic Clouds, have been solved meanwhile by considering recent OP opacity tables in combination with the new solar mixture (Miglio et al. 2007). Changes in the opacities typically have a large effect on the mode excitation as the driving is due to the κ mechanism, but not on the frequency values as these are mainly determined by the density structure within the star.

3. Nonrigidity of the internal rotation

The rotation of a star implies that the oscillation frequencies are split into multiplets of the same ℓ but with different m, compared to the non-rotating case where all modes of the same ℓ and different m have the same frequency value (e.g. Cunha et al. 2007). Thus, once the fundamental stellar parameters are fixed from a frequency match of the $m = 0$ modes, one is able to exploit such frequency splittings, if multiplets or parts thereof are detected. The frequency spectra of the three B stars where this is the case are shown in Fig. 1. It can be derived from this figure that the radial mode, which is the radial fundamental for all three stars, occurs at lowest, middle and highest frequency for HD 29248, HD 129929 and HD 157056, respectively. Since the three stars have similar mass, this thus provides the order of age of these objects, which is compatible with their luminosity class. The multiplet structures seen in HD 129929 and HD 29248 are such that rotational splitting does not mix up frequencies of different radial order. HD 157056 (equatorial rotation velocity of about $30\,\mathrm{km\,s^{-1}}$, see Table 1), on the other hand, is a case where the rotation frequency is such that the multiplets are at the limit of being well separated. In particular, the $\ell = 2, m = 0$ mode is not observed for this star, but its frequency can be estimated from the three other frequencies of the $\ell = 2$ multiplet, as they are identified as $m = -1, 1, 2$ (Briquet et al. 2007). Larger splitting would have implied difficulties to interpret the observed frequencies in terms of their (ℓ, m) values.

To test the rigidity of the rotation, so-called kernels $K_{n,\ell,m}(r)$ are computed (see, e.g., Cunha et al. 2007). These are functions over the radial distance inside the star which capture the relative sensitivity of the different mode frequencies to the internal properties of the star, such as the rotation, the kinetic energy density, etc. The frequency splittings are dependent on rotational kernels through

$$\nu_m = \nu_0 + m \int_0^R K_{n,\ell,m}(r)\, \frac{\Omega(r)}{2\pi}\, \frac{\mathrm{d}r}{R}, \qquad (3.1)$$

where ν_0 is the frequency in the absence of rotation and where we assume Ω to depend only on the radial distance r to the stellar centre. Thus, one can exploit the observed splittings within multiplets by comparing them with those predicted from Eq. (3.1). Usually, it is first tested if one can fulfil Eq.(3.1) with Ω constant throughout the star by using all the observed multiplets. If the data are in conflict with this assumption, the rotation cannot be rigid. Ideally, one would then want to invert the integral in Eq.(3.1) to derive $\Omega(r)$. This can only be done if many well-identified multiplets whose kernels have different probing power in the stellar interior are detected. This is the case for the Sun, but for no other main-sequence star so far. For the three B stars shown in Fig. 1, we only have splitting values from two different multiplets. As a result, we can only test how much $\Omega(r)$ near the core differs from the one in the envelope, e.g. by assuming a linear slope. This is how the values listed in Table 1 were derived.

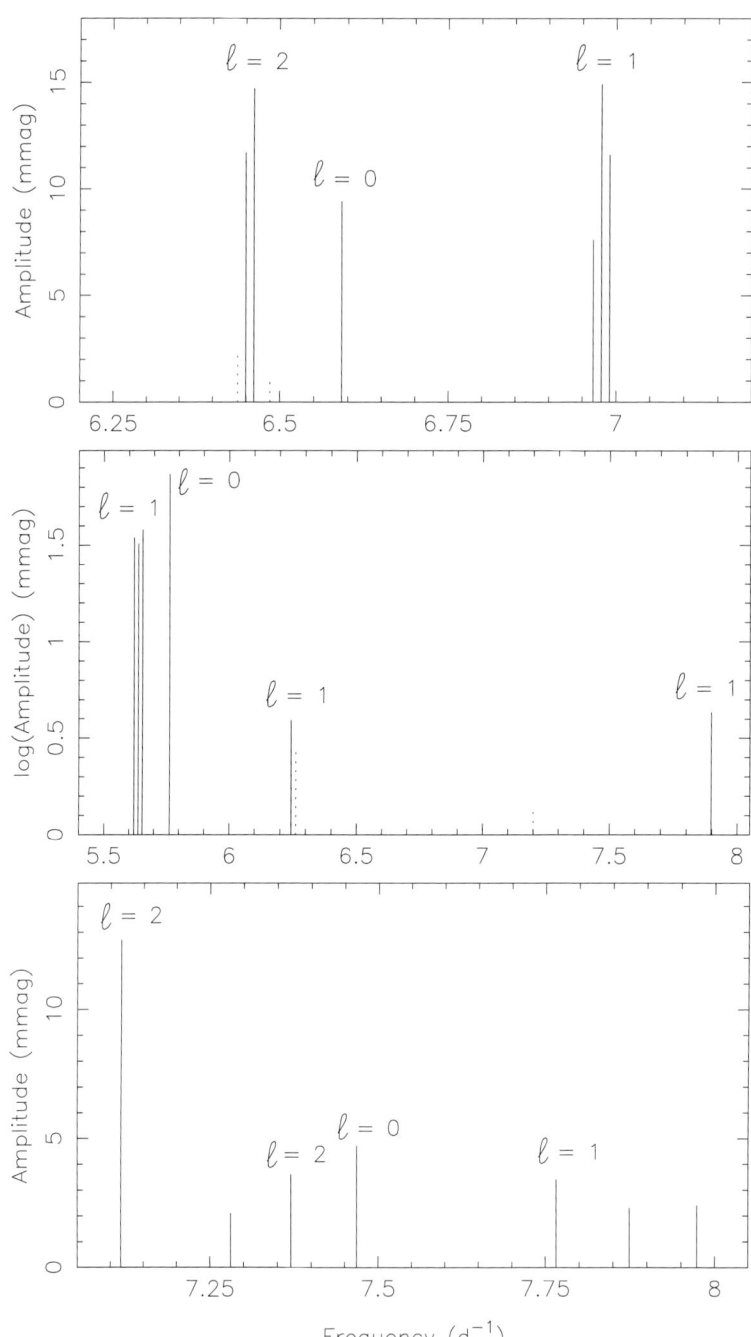

Figure 1. Schematic frequency spectra derived from photometric data of the β Cep stars HD 129929 (top), HD 29248 (middle) and HD 157056 (bottom), reproduced from Aerts et al. (2004), De Ridder et al. (2004) and Handler et al. (2005), respectively. The lower and higher frequency ranges of HD 29248 do not contain identified modes and have been omitted for visibility reasons. The identified degrees are indicated. Full lines indicate frequencies above 4σ, while dotted lines represent frequencies present in the data but with significance below 4σ. Note the different axis scales.

Table 1. Summary of the seismic results obtained so far. Reference numbers refer to (1): Aerts et al. (2006), (2): Pamyatnykh et al. (2004) and Ausseloos et al. (2004), (3): Mazumdar et al. (2006), (4): Aerts et al. (2003, 2004) and Dupret et al. (2004), (5): Briquet et al. (2007).

Ref.	Star	Mass (M$_\odot$)	SpT	$\alpha_{\rm ov}$ (H$_{\rm p}$)	ΩR (km s^{-1})	$\Omega_{\rm core}/\Omega_{\rm env}$
(1)	HD 16582	10.2±0.2	B2 IV	0.20±0.10	28(14?)	
(2)	HD 29248	9.2±0.6	B2 III	0.10±0.05	6±2	∼ 5
(3)	HD 44743	13.5±0.5	B1 III	0.20±0.05	31±5	
(4)	HD 129929	9.4±0.1	B3 V	0.10±0.05	2±1	3.6
(5)	HD 157056	8.2±0.3	B2 IV	0.44±0.07	29±7	∼ 1

It is evident that the seismic tuning of the internal rotation is very dependent on the frequency accuracy, besides the nature of the probing power of the detected mode kernels. The frequency accuracy is mainly a function of the total time base of the data and also of the number of data points (i.e. the duty cycle). It is of order 10^{-6} d^{-1} for HD 129929 (21 years of low-duty single-site data), 10^{-5} d^{-1} for HD 29248 (150 days of high-duty 11-site data), and 10^{-4} d^{-1} for HD 157056 (5 months of moderate-duty 3-site data). Long-term monitoring is in any case required to probe the internal rotation of massive stars.

4. Summary of the seismic results so far

The procedures outlined in the previous two sections were successfully applied to five B-type pulsators so far. Many other, often extensive but failed attempts to derive $\alpha_{\rm ov}$ are omitted here, in particular all those of rapidly rotating B and Be stars. Lack of secure mode identification is almost always the reason for the failure, besides frequency alias confusion. The results for the stars where $\alpha_{\rm ov}$ could be tuned are summarised in Table 1.

We point out that the quoted errors for the mass and core overshoot parameter do not include the theoretical uncertainties due to the changes of input physics of the models. The adopted opacities and metal mixtures have changed over the past few years, between the applications to HD 129929 and HD 157056, as a consequence of better opacity computations and a change in the solar mixture (Asplund et al. 2005). The effect of a change in the metal mixture was investigated for the case of HD 129929 by Thoul et al. (2004) and it was found that the new solar mixture generally leads to somewhat higher $\alpha_{\rm ov}$ compared to the one obtained from the old solar mixture (Grevesse & Noels 1993). The results of $\alpha_{\rm ov}$ for the first four stars in Table 1 were done with the old solar mixture and are therefore probably lower limits, but the modelling would have to be redone to make a quantitative assessment of this effect.

The results mentioned in Table 1 are of very high quality compared to what can be achieved from classical methods not using oscillations, even when considering the yet unquantified theoretical uncertainties due to the metal mixture. This becomes obvious when comparing the seismic results with those obtained from eclipsing binaries or from cluster isochrone fitting. Claret (2007) recently compiled results for eclipsing binaries and obtained $\alpha_{\rm ov}$-values between 0.2 and 0.4, with typical uncertainties of 0.1 to 0.3, for five systems with OB-type components. Earlier on, Mermilliod & Maeder (1986) made an extensive study of $\alpha_{\rm ov}$ needed to explain the HR diagrams of 25 young open clusters and they concluded that $\alpha_{\rm ov} \sim 0.3$ is needed to explain the observational data. This value was later revised to $\alpha_{\rm ov} \sim 0.2$ due to the use of other opacity computations (Meynet et al. 1993). These estimates from two independent classical methods are fully compatible with, but less precise than the seismic results. We note that they suffer from the same theoretical uncertainties as the seismology.

5. Future plans

It is now well established seismically that core convective overshooting occurs in massive main sequence stars. The results summarised here were based on the fitting of only two to four $m = 0$ oscillation modes. It is evident that a reduction in the uncertainty of the overshooting value will occur whenever several more well identified modes can be used to derive $\alpha_{\rm ov}$. This is one of the major goals for massive stars to be studied in the framework of the seismology programmes of the space missions CoRoT (Michel et al. 2006) launched on 27/12/2006 and Kepler (Christensen-Dalsgaard et al. 2007) to be launched in 2009.

Large activities are continued to get more results from ground-based multi-site campaigns, in order to increase the sample with more evolved stars and stars of gradually faster rotation and higher mass-loss rate. The goal is to reach a sample that covers the evolution from the main sequence until the pre-supernova stage. In this way, stellar evolution models in the upper HR diagram can be calibrated seismically to a level of precision that has never been achieved before.

References

Aerts, C., Marchenko, S. V., Matthews, J. M., et al. 2006, ApJ, 642, 470
Aerts, C., Thoul, A., Daszyńska, J., et al. 2003, Science, 300, 1926
Aerts, C., Waelkens, C., Daszyńska-Daszkiewicz, J., et al. 2004, A&A, 415, 241
Asplund, M., Grevesse, N., Sauval, A. J., et al. 2005, A&A, 431, 693
Ausseloos, M., Scuflaire, R., Thoul, A., & Aerts, C. 2004, MNRAS, 355, 352
Briquet, M. & Aerts, C. 2003, A&A, 398, 687
Briquet, M., Morel, T., Thoul, A., et al. 2007, MNRAS, 381, 1482
Christensen-Dalsgaard, J., Arentoft, T., Brown, T. M., et al. 2007, CoAst, 150, 350
Claret, A. 2007, A&A, 475, 1019
Cunha, M. S., Aerts, C., Christensen-Dalsgaard, J., et al. 2007, A&AR, 14, 217
De Ridder, J., Telting, J. H., Balona, L. A., et al. 2004, MNRAS, 351, 324
Dupret, M.-A., Thoul, A., Scuflaire, R., et al. 2004, A&A, 415, 251
Grevesse, N. & Noels, A. 1993, in: Prantzos, N., Vangioni-Flam, E., & Casse, M. (eds.), Origin and evolution of the elements, (Cambridge: Cambridge University Press), 14
Handler, G., Shobbrook, R. R., & Mokgwetsi, T. 2005, MNRAS, 362, 612
Kurtz, D. W. 2006, in Sterken, C., & Aerts, C. (Eds.), Astrophysics of Variable Stars, (San Francisco: ASP) ASP Conf Ser, 349, 101
Mazumdar, A., Briquet, M., Desmet, M., & Aerts, C. 2006, A&A, 459, 589
Mermilliod, J.-C. & Maeder, A. 1986, A&A, 158, 45
Meynet, G., Mermilliod, J.-C., & Maeder, A. 1993, A&AS, 98, 477
Michel, E., Baglin, A., Auvergne, M., et al. 2006, in Fridlund, M., Baglin, A., Lochard, J., & Conroy, L. (Eds.), The CoRoT Mission, ESA-SP, 1306, 39
Miglio, A., Montalbán, J., & Dupret, M.-A. 2007, MNRAS, 375, L21
Monteiro, M. J. P. F. G., Lebreton, Y., Montalban, J., et al. 2006, in Fridlund, M., Baglin, A., Lochard, J., & Conroy, L. (Eds.), The CoRoT Mission, ESA-SP, 1306, 363
Moya, A., Christensen-Dalsgaard, J., Charpinet, S., et al. 2008, ApSS, in press (arXiv:0711.2587)
Pamyatnykh, A. A., Handler, G., & Dziembowski, W. A. 2004, MNRAS, 350, 1022
Thoul, A., Scuflaire, R., Ausseloos, M., et al. 2004, CoAst, 144, 35
Zima, W. 2006, A&A, 455, 227

Discussion

PRZYBILLA: The pulsators often show chemical peculiarities. How would the differences relative to the solar mixture (which the asteroseismic models are based upon) influence the interpretation?

AERTS: So far, we have too few frequencies to tune the individual abundances seismically. Rather, we derive the abundances from classical spectroscopy, determine the opacity, and check the mode excitation to see if we have consistency with the observed oscillations. If we have many well identified frequencies (several tens to hundreds), we will be able to tune the metal mixture seismically. We are still far from that in massive stars, but CoRoT and Kepler will hopefully help improve the situation in the coming decade

MAEDER: Can you tell us why you are also happy with the new solar abundances?

AERTS: 1) The seismic model of the Sun, even though it degraded somewhat in quality with the Asplund *et al.* (2005) abundances, is still very good! If I compare the error bars in the interior structure parameters of the Sun with those for massive stars, then I'd say there is not need for worries in this community. 2) As discussed in Miglio *et al.* (2007), some outstanding issues about mode excitation in B-type pulsators have now been solved, by using the OP opacities together with the new solar abundances. 3) For studies of massive stars, one may first as well use the B stars in the solar neighbourhood to compare with.

MEYNET: Do you have plans to do seismic studies of more rapid rotators?

AERTS: Yes, we have several multisite campaigns (finished and ongoing) focused on stars of higher $v\sin i$, but still only up to $v\sin i \simeq 60$ km s^{-1}. On the other hand, campaigns are also ongoing on young open clusters with B type pulsators (one such cluster is also in the FLAMES survey). Those have typically $v\sin i \sim 100-150$ km s^{-1}. Note, however, that the mode identification becomes very difficult for more rapid rotators, because the oscillation frequency multiplets are merged due to the rotational splitting, and they are therefore hard to disentangle.

Conny Aerts.

Norbert Langer.

Ben Davies.

Session III

Massive Star Populations in the Nearby Universe

# Young Massive Clusters

Donald F. Figer

Chester F. Carlson Center for Imaging Science
Rochester Institute of Technology
Rochester, NY 14623-5604
email: `figer@cis.rit.edu`

Abstract. Over the past ten years, there has been a revolution in our understanding of massive young stellar clusters in the Galaxy. Initially, there were no known examples having masses $>10^4$, yet we now know that there are at least a half dozen such clusters in the Galaxy. In all but one case, the masses have been determined through infrared observations. Several had been identified as clusters long ago, but their massive natures were only recently determined. Presumably, we are just scratching the surface, and we might look forward to having statistically significant samples of coeval massive stars at all important stages of stellar evolution in the near future. I review the efforts that have led to this dramatic turn of events and the growing sample of young massive clusters in the Galaxy.

Keywords. open clusters and associations: general – stars: early-type

---

1. Introduction

Massive stellar clusters are the birthplaces of massive stars. They are also the places where many massive stars reside all of their lives, having little time to wander before exploding as supernovae. The astrophysical importance of massive clusters largely derives from their content of massive stars, objects that have extraordinary effects on their surroundings. Indeed, massive stars are key ingredients and probes of astrophysical phenomena on all size and distance scales, from individual star formation sites, such as Orion, to the early Universe during the age of reionization when the first stars were born. As ingredients, they control the dynamical and chemical evolution of their local environs and individual galaxies through their influence on the energetics and composition of the interstellar medium. They likely play an important role in the early evolution of the first galaxies, and there is evidence that they are the progenitors of the most energetic explosions in the Universe, seen as GRBs. As probes, they define the upper limits of the star formation process and their presence may end further formation of nearby lower mass stars and planets.

Despite the importance of massive stars, and the clusters in which they reside, no truly massive clusters were known to exist in the Galaxy before about ten years ago, when the Arches and Quintuplet clusters were identified as being at least that massive (Figer *et al.* 1999b). Since then, a number of efforts have led to the identification of about a half dozen more such massive clusters in the Galaxy.

The evolution of massive stars is difficult to study because many of the most important phases are short. To date, one of the most effective techniques to overcome this problem is to identify massive clusters at ages when its most massive stars are largely in a single phase of evolution., We see this in the Arches cluster (the most massive stars, i.e. hydrogen burning Wolf-Rayet stars on the main sequence), rhe Central Cluster (Ofpe/WN9 stars), Westerlund 1 (Wolf Rayet stars), and the Scutum red supergiant clusters (RSGs).

Table 1. Properties of massive clusters in the Galaxy[a]

Cluster	Log(M) M$_\odot$	Radius pc	Log(ρ) M$_\odot$ pc^{-3}	Age Myr	Log(L) L$_\odot$	Log(Q) s^{-1}	OB	YSG	RSG	LBV	WN	WC
Westerlund 1[b]	4.7	1.0	4.1	4–6	...	...	...	6	4	2	16	8
RSGC2[c]	4.6	2.7	2.7	14–21	...	...	0	0	26	0	0	0
RSGC1[d]	4.5	1.3	3.5	10–14	...	...	1	1	14	0	0	0
Quintuplet[e]	4.3	1.0	3.2	4–6	7.5	50.9	100	1	2	6	13	
Arches[f]	4.3	0.19	5.6	2–2.5	8.0	51.0	160	0	0	0	6	0
Center[g]	4.3	0.23	5.6	4–7	7.3	50.5	100	0	4	1	18	12
NGC 3603[h]	4.1	0.3	5.0	2–2.5	...	...	60	0	0	0	3	0
Trumpler 14[i]	4.0	0.5	4.3	<2	...	...	31	...	...	...	...	...
Westerlund 2[j]	4.0	0.8	3.7	1.5–2.5	...	...	...	...	...	...	2	...
Cl 1806-20[k]	3.8	0.8	3.5	4–6	...	...	5	0	...	1	2	2

[a] An ellipsis has been entered in cases where data are not reliable or available. "M" is the total cluster mass in all stars extrapolated down to a lower-mass cutoff of 1 M$_\odot$, assuming a Salpeter IMF slope and an upper mass cutoff of 120 M$_\odot$ (unless otherwise noted) "Radius" gives the average projected separation from the centroid position. "ρ" is M divided by the volume. This is probably closer to the central density than the average density because the mass is for the whole cluster while the radius is the average projected radius. "Age" is the assumed age for the cluster. "Luminosity" gives the total measured luminosity for observed stars. "Q" is the estimated Lyman continuum flux emitted by the cluster. [b] Figer et al. (2006). [c] Davies et al. (2007). [d] Figer et al. (2006). [e] Figer et al. (1999b). [f] Mass estimates have been made based upon the number of stars having $M_{\mathrm{initial}} > 20$ M$_\odot$ given in Figer et al. (1999b) and the mass function slope in Kim et al. (2006). The age, luminosity and ionizing flux are from Figer et al. (2002). [g] Krabbe et al. (1995). The mass, M has been estimated by assuming that a total $10^{3.5}$ stars have been formed. The age spans a range covering an initial starburst, followed by an exponential decay in the star formation rate. [h] Harayama (2007). [i] Harayama (2007). [j] Harayama (2007). [k] Figer et al. (2005).

We are on the cusp of a revolution in massive stellar cluster research, as the identified sample is likely the "tip of the iceberg." In the next ten years, we can expect to identify perhaps a factor of ten more massive clusters in the Galaxy than are currently known. With this sample, we can expect to routinely address many of the long-pursued questions in massive star research to determine, for example: 1) the most massive star that can form, 2) the binary frequency of massive stars, 3) the properties of massive stellar winds, e.g. clumping, 4) the evolutionary sequence of massive stars, and 5) the end states of massive stars. Indeed, Barbosa & Figer (2004) summarize many interesting questions that might be addressed with future studies of massive stars.

In this paper, I review the Galactic sample of massive stellar clusters (M>10^4 M$_\odot$), the efforts that led to their discovery, and the identification of a stellar upper mass limit. Finally, I speculate on the hidden population of massive stellar clusters in the Galaxy that will likely be revealed in the near future.

2. The sample of massive young clusters

There are approximately ten known Galactic clusters with masses $\gtrsim 10^4$ M$_\odot$, with three being located within the central 50 pc of the Galactic center. Table 1 gives the known properties of these clusters in a list that is ordered with decreasing mass. In some cases, the properties are poorly determined, in which case the entries are filled with an ellipsis. The most massive, Westerlund 1, has a mass of $\approx$50,000 M$_\odot$. The densest are in the Galactic center, with the Arches cluster having a density of $\approx$300,000 M$_\odot$ pc^{-3}. While most are quite young, the RSG (red supergiant) clusters are distinctly older. Presumably, the table reflects an observational bias in that younger clusters are generally more concentrated, and thus easier to identify as clusters.

In the following, I review some of the clusters with more well-established characteristics.

Figure 1. GLIMPSE image of RSGC1. W42, to the upper left, is an unrelated star formation region along the line of sight.

2.1. Westerlund 1

Westerlund 1 is the most massive young cluster known in the Galaxy (Clark *et al.* 2005; Negueruela & Clark 2005; Skinner *et al.* 2006; Groh *et al.* 2006; Crowther *et al.* 2006; Brandner *et al.* 2008). Given its age of ≈4 Myr, it contains more evolved massive stars than any other cluster in the Galaxy, including half the known population of yellow supergiants and the most WR stars. The cluster contains a magnetar, a highly magnetized neutron star that may be descended from a particularly massive progenitor (Muno *et al.* 2006). Oddly, the massive stellar content in Westerlund 1 has only been recently revealed, more than 40 years after the cluster's discovery.

2.2. Red Supergiant clusters

Figer *et al.* (2006), Davies *et al.* (2007), and Davies *et al.* (2008), identified two massive clusters in the red supergiant (RSG) phase, the first containing 14 (see Figure 1) and the second containing 26 RSGs, or collectively about 20% of all such stars known in the Galaxy. The inferred cluster masses are $\approx 3 \times 10^4$ $M_\odot$ for the former, and $\approx 4 \times 10^4$ $M_\odot$ for the latter. Interestingly, the two clusters are near each other, located within 1 degree on the sky and within 1 kpc along the line of sight at the base of the Scutum-Crux arm. The first cluster was first identified as a candidate cluster in the Bica *et al.* (2003) catalog, and some of the stars in the second cluster had already been identified as RSGs in Stephenson (1990) and Nakaya *et al.* (2001). Just like in the case of Westerlund 1, we see that a cluster once thought to be of relatively low mass can turn out to be quite massive on further inspection. Perhaps there are more massive clusters amongst the many clusters that have already been identified.

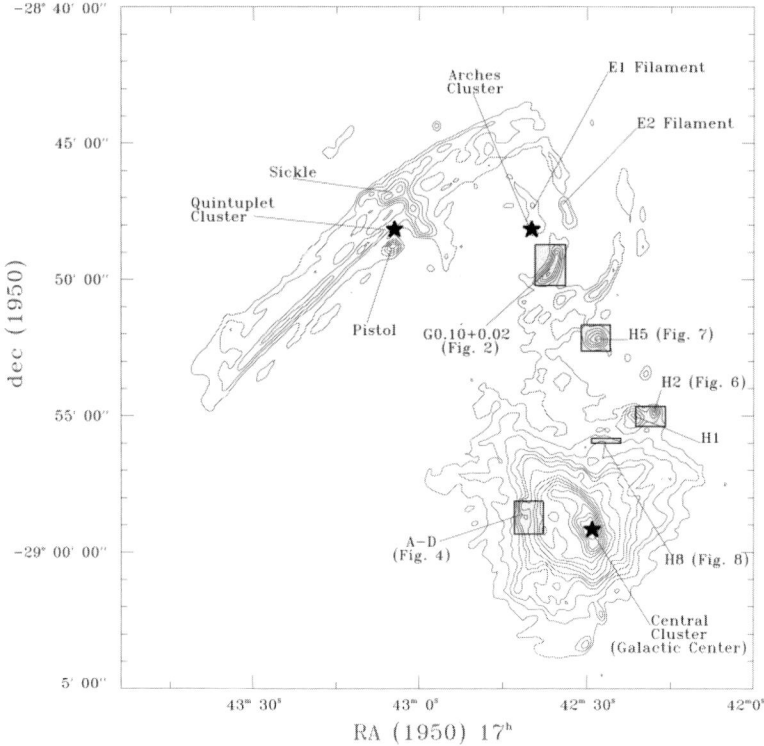

Figure 2. Radio emission from the GC region at 6 cm, adapted by Cotera *et al.* (1999) from Yusef-Zadeh & Morris (1987). The star symbols represent the three massive clusters.

2.3. *Central cluster*

The Central cluster resides in the central parsec of the Galaxy and contains many massive stars formed in the past 10 Myr (Becklin *et al.* 1978; Rieke, Telesco, & Harper 1978; Lebofsky, Rieke, & Tokunaga 1982; Forrest *et al.* 1987; Allen, Hyland, & Hillier 1990; Krabbe *et al.* 1991; Najarro *et al.* 1994; Krabbe *et al.* 1995; Najarro 1995; Blum, Sellgren, & Depoy 1995b; Genzel *et al.* 1996; Najarro *et al.* 1997). There are at least 80 massive stars in the Central cluster (Eisenhauer *et al.* 2005), including ≈50 OB stars on the main sequence and 30 more evolved massive stars (see Figure 3). These young stars appear to be confined to two disks (Genzel *et al.* 2003; Tanner *et al.* 2006). There is also a tight collection of a dozen or so B stars (the "s" stars) in the central arcsecond, highlighted in the small box in the figure. The formation of so many massive stars in the central parsec remains as much a mystery now as it was at the time of the first infrared observations of the region. Most recently, this topic has largely been supplanted by the even more improbable notion that star formation can occur within a few thousand AU of the supermassive black hole. See Figer (2008), and references therein, for a review of massive star formation and the "s" stars in the Galactic center.

Figure 3. K-band image of the Central cluster obtained with NAOS/CONICA from Schödel et al. (2007). The 100 or so brightest stars in the image are evolved descendants from main sequence O-stars. The central box highlights the "s" stars that are presumably young and massive ($M_{\text{initial}} \approx 20$ M$_\odot$).

2.4. Arches cluster

The Arches cluster is the densest young cluster in the Galaxy, and it has a relatively young age (Figer et al. 2002). Being so young and massive, it contains the richest collection of O-stars and WNL stars in any cluster in the Galaxy (Harris et al. 1994; Nagata et al. 1995; Figer 1995; Cotera 1995; Cotera et al. 1996; Serabyn, Shupe, & Figer 1998; Figer et al. 1999b; Blum et al. 2001; Figer et al. 2002; Figer 2005). The WNL stars are particularly interesting, as they represent the largest collection of the most massive stars (M>100 M$_\odot$) in the Galaxy. As seen elsewhere, e.g. R136 and NGC 3603, these stars are still burning hydrogen on the main sequence. Given its unique combination of characteristics, the cluster is ideal as a testbed for measuring the upper mass cutoff to the IMF (see Section 3). The strong stellar winds from the most massive stars in the cluster are detected at radio wavelengths (Lang, Goss, & Rodríguez 2001; Yusef-Zadeh et al. 2003; Lang et al. 2005; Figer et al. 2002), and x-ray wavelengths (Yusef-Zadeh et al. 2002; Rockefeller et al. 2005; Wang, Dong, & Lang 2006).

2.5. Quintuplet cluster

The Quintuplet cluster was originally noted for its five very bright stars (Glass, Moneti, & Moorwood 1990; Okuda et al. 1990; Nagata et al. 1990), but is now known to contain many massive stars (Geballe et al. 1994; Figer, McLean, & Morris 1995; Timmermann et al. 1996; Figer et al. 1999a). It is ≈ 4 Myr old and had an initial mass of $>10^4$ M$_\odot$ (Figer et al. 1999a). The hot stars in the cluster ionize the nearby "Sickle" HII region (see Figure 2). The Quintuplet is most similar to Westerlund 1 in mass, age, and spectral content. Some of the stars in the cluster have been detected at x-ray wavelengths (Law &

Yusef-Zadeh 2004), and at radio wavelengths (Lang et al. 1999, 2005). Recently, Tuthill et al. (2006) convincingly show that the five red stars in the cluster are dusty WC stars, characteristic of binary systems containing WCL plus an OB star (Tuthill, Monnier, & Danchi 1999). This may indicate that either the binary fraction for massive stars is extremely high (Nelan et al. 2004), or only binary massive stars evolve through the WCL phase (van der Hucht 2001). The Quintuplet cluster also contains two Luminous Blue Variables, the Pistol star (Harris et al. 1994; Figer et al. 1998, 1999c), and FMM362 (Figer et al. 1999a; Geballe, Najarro, & Figer 2000). Both stars are extraordinarily luminous ($L > 10^6$ $L_\odot$), yet relatively cool ($T \approx 10^4$ K), placing them in the "forbidden zone" of the Hertzsprung-Russell Diagram, above the Humphreys-Davidson limit (Humphreys & Davidson 1994). They are also both confirmed photometric and spectroscopic variables (Figer et al. 1999a). The Pistol star is particularly intriguing, in that it is surrounded by one of the most massive (10 $M_\odot$) circumstellar ejecta in the Galaxy (Figer et al. 1999c; Smith 2008). Both stars are spectroscopically (Figer et al. 1999a) and photometrically variable (Glass et al. 2001), as expected for LBVs.

3. An upper limit to the masses of stars

Massive star clusters can be useful testing grounds for a variety of theoretical predictions, e.g. the IMF, binary fraction, n-body interactions, etc. Although only recently discovered, the currently known set of Galactic massive clusters have already yielded an interesting result regarding the upper limit to the masses of stars.

Theoretically, one might expect stellar mass to be limited by pulsational instabilities (Schwarzschild & Härm 1959) or radiation pressure (Wolfire & Cassinelli 1987). Although stellar evolution models have been computed for massive stars up masses of 1000 $M_\odot$, no such stars have ever been observed. Indeed, some of the most famous "massive stars" have turned out to be multiple systems.

Observationally, the problem is difficult because massive stars are rare. They are formed in small numbers with respect to lower mass stars and they only live a few million years. A star formation event must produce about 10^4 $M_\odot$ in stars to have a statistically meaningful expectation of stars with masses greater than 150 $M_\odot$. For example, such a cluster should have about three such stars, assuming a Salpeter initial mass function (Salpeter 1955) extrapolated to zero probability. Unfortunately, the cluster must satisfy a number of other criteria in order to use it for identifying an upper mass cutoff. For instance, it must be young enough so that its most massive members have not yet exploded as supernovae. Yet, it must be old enough to be free of natal molecular material. In order to identify its individual stars, it must be close enough to us. To relate apparent magnitude to absolute magnitude, it must be at known distance. The stars must be coeval enough so that the star formation episode that produced the cluster can be considered to constitute a single event. Finally, the age must be known relatively accurately.

It is difficult to satisfy all these criteria, and, indeed, only the Arches cluster does. Being in the Galactic center happens to be useful in this regard, as its distance is then relatively well known as compared to other clusters. The result for the Arches cluster is shown in Figure 4. In this plot, it is apparent that there is an absence of stars with masses above about 150 $M_\odot$, where many are expected, i.e. there is an upper mass cutoff. See Weidner & Kroupa (2004), Weidner & Kroupa (2006) and Oey & Clarke (2005) for other arguments for such a cutoff.

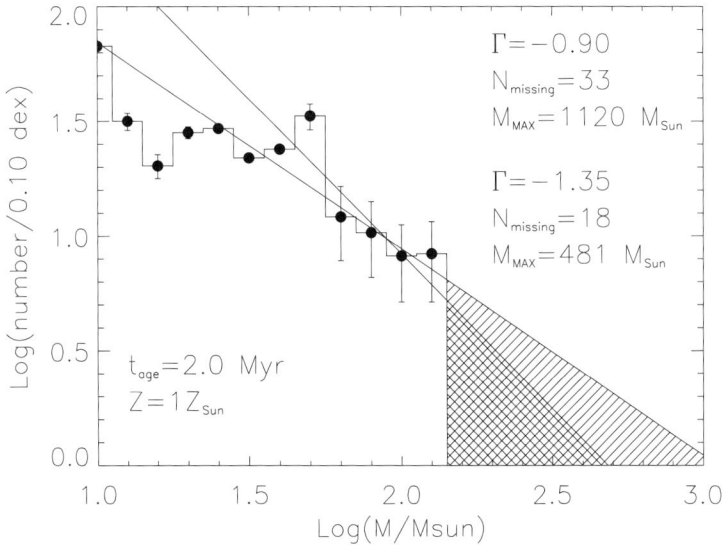

Figure 4. Number versus mass for stars in the Arches cluster from Figer (2005). There is a clear deficit of stars with initial masses greater than 150 $M_\odot$, as seen in the hatched regions, for a reasonable range of IMF slopes.

4. The hidden population of massive stellar clusters

2MASS (Skrutskie et al. 1997) and GLIMPSE (Benjamin et al. 2003) have heralded a new era in massive star cluster research. With these surveys, we will be able to probe much further into the Galactic plane than ever before. We can expect an order of magnitude increase in the number of known young clusters in the Galaxy as these surveys are further investigated. Figure 5 shows the Galactic distribution of known young clusters from WEBDA (dots), candidate clusters from Bica *et al.* (2003) (triangles), and the verified massive clusters discussed in this paper, with the addition of several very likely massive clusters (hexagons). The Galactic center is at (0,0) and the Sun is at (0,8). The visual clusters are mostly within 3 kpc from the Sun. The infrared clusters are a bit further away; however, it is clear that the number is highly incomplete and that the far side of the Galaxy as well as the central regions must contain many new clusters yet to be discovered. We estimate that the total number of Galactic stellar clusters should exceed 20,000 (poster by Messineo, this volume).

Acknowledgement

I thank the following individuals for discussions related to this work: Ben Davies, Paco Najarro, Rolf Kudritzki, Maria Messineo, Lucy Hadfield, Qingfeng Zhu, and Sungsoo Kim. The material in this paper is based upon work supported by NASA under award No. NNG05-GC37G, through the *Long Term Space Astrophysics* program. This research has made use of the SIMBAD database, Aladin and IDL software packages, and the GSFC IDL library. This research was performed in the Rochester Imaging Detector Laboratory with support from a NYSTAR Faculty Development Program grant. Some of the data presented herein were obtained at the W. M. Keck Observatory, which is operated as a scientific partnership among the California Institute of Technology, the University of

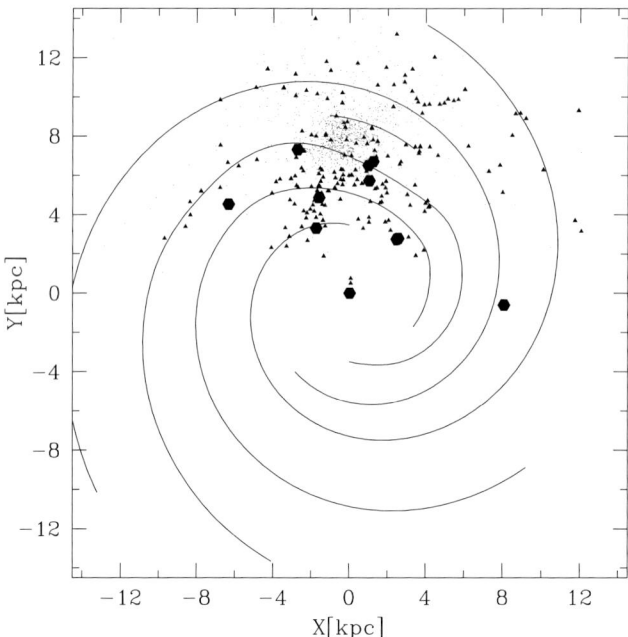

Figure 5. Galactic distribution of known young clusters (dots), candidate clusters (triangles), and verified massive clusters (hexagons). The Galactic center is at (0,0) and the Sun is at (0,8). Distance for the known young clusters are from Dias *et al.* (2002). Distances for the candidate clusters were determined by fitting the line-of-sight velocities of nearby HII regions (Kuchar & Bania 1994) to the Galactic rotation curve, assuming that these regions are associated with the clusters. This figure is courtesy of Maria Messineo.

California, and the National Aeronautics and Space Administration. The Observatory was made possible by the generous financial support of the W. M. Keck Foundation.

References

Allen, D. A., Hyland, A. R., & Hillier, D. J. 1990, *MNRAS*, 244, 706
Barbosa, C. & Figer, D. 2004, (arXiv:astro-ph/0408491)
Becklin, E. E., Matthews, K., Neugebauer, G., & Willner, S. P. 1978, *ApJ*, 219, 121
Benjamin, R. A., Churchwell, E., Babler, B. L. *et al.* 2003, *PASP*, 115, 953
Bica, E., Dutra, C. M., Soares, J., & Barbuy, B. 2003, *A&A*, 404, 223
Blum, R. D., Schaerer, D., Pasquali A. *et al.* 2001, *AJ*, 122, 1875
Blum, R. D., Sellgren, K., & Depoy, D. L. 1995b, *ApJ*, 440, L17
Blum, R. D., Sellgren, K., & Depoy, D. L. 1996a, *AJ*, 112, 1988
Brandner, W., Clark, J. S., Stolte, A. *et al.* 2008, *A&A*, 478, 137
Clark, J. S., Negueruela, I., Crowther, P. A., & Goodwin, S. P. 2005, *A&A*, 434, 949
Cotera, A. S. 1995, *Ph.D. Thesis*
Cotera, A. S., Erickson E. F., Colgan, S. W.J. *et al.* 1996, *ApJ*, 461, 750
Cotera, A. S., Simpson, J. P., Erickson E. F. *et al.* 1999, *ApJ*, 510, 747
Crowther, P. A., Hadfield, L. J., Clark, J. S. *et al.* 2006, *MNRAS*, 372, 1407
Davies, B., Figer D. F., Kudritzki, R.-P. *et al.* 2007, *ApJ*, 671, 781
Davies, B., Figer D. F., Law, C. J. *et al.* 2008, *ApJ*, 676, 1016
Dias, W. S., Alessi, B. S., Moitinho, A., & Lépine, J. R. D. 2002, *A&A*, 389, 871
Eisenhauer, F., Genzel, R., Alexander, T. *et al.* 2005, *ApJ*, 628, 246

Figer, D. F. 1995, *Ph.D. Thesis*
Figer, D. F. 2005, *Nature*, 434, 192
Figer, D. F. 2008, in press (arXiv:0803.1619)
Figer, D. F., Najarro, F., Gilmore, D. et al. 2002, *ApJ*, 581, 258
Figer, D. F., Kim, S. S., Morris, M. et al. 1999b, *ApJ*, 525, 750.
Figer, D. F., MacKenty, J. W., Robberto, M. et al. 2006, *ApJ*, 643, 1166
Figer, D. F., McLean, I. S., & Morris, M. 1995, *ApJ*, 447, L29
Figer, D. F., McLean, I. S., & Morris, M. 1999a, *ApJ*, 514, 202
Figer, D. F., Morris, M., Geballe, T. R. et al. 1999c, *ApJ*, 525, 759
Figer, D. F., Najrro, F., Geballe, T. R. et al. 2005, *ApJ*, 622, L49
Figer, D. F., Najarro, F., Morris, M. et al. 1998, *ApJ*, 506, 384
Forrest, W. J., Shure, M. A., Pipher, J. L., & Woodward, C. E. 1987, in: D. C. Backer (ed.), The Galactic Center (New York: AIP), *AIP Conf Proc* 155, 153
Geballe, T. R., Genzel, R., Krabbe, A. et al. 1994, in: I. S. McLean (ed.), Infrared Astronomy with Arrays, The Next Generation, *Astrophys Space. Sci. Lib.* 190, 73
Geballe, T. R., Najarro, F., & Figer, D. F. 2000, *ApJ*, 530, L97
Genzel, R., Schödel, R., Ott, T. et al. 2003, *ApJ*, 594, 812
Genzel, R., Thatte, N., Krabbe A. et al. 1996, *ApJ*, 472, 153
Glass, I. S., Matsumoto, S., Carter, B. S., & Sekiguchi, K. 2001, *MNRAS*, 321, 77
Glass, I. S., Moneti, A., & Moorwood, A. F. M. 1990, *MNRAS*, 242, 55P
Groh, J. H., Damineli, A., Teodoro, M., & Barbosa, C. L. 2006, *A&A*, 457, 591
Harayama, Y. 2007, *PhD Thesis*, LMU München
Harris, A. I., Krenz, T., Genzel, R. et al. 1994, in: R. Genzel & A. I. Harries (eds.), The Nuclei of Normal Galaxies: Lessons from the Galactic Center, (Dodrecht: Kluwer), *NATO Advanced Science Inst. Series C*, 445, 223
Humphreys, R. M., & Davidson, K. 1994, *PASP*, 106, 1025
Kim, S. S., Figer, D. F., Kudritzki, R. P., & Najarro, F. 2006, *ApJ*, 653, L113
Krabbe, A., et al. 1995, *ApJ*, 447, L95
Krabbe, A., Genzel, R., Drapatz, S., & Rotaciuc, V. 1991, *ApJ*, 382, L19
Kuchar, T. A., & Bania, T. M. 1994, *ApJ*, 436, 117
Lang, C. C., Figer, D. F., Goss, W. M., & Morris, M. 1999, *AJ*, 118, 2327
Lang, C. C., Goss, W. M., & Rodríguez, L. F. 2001, *ApJ*, 551, L143
Lang, C. C., Johnson, K. E., Goss, W. M., & Rodríguez, L. F. 2005, *AJ*, 130, 2185
Law, C. & Yusef-Zadeh, F. 2004, *ApJ*, 611, 858
Lebofsky, M. J., Rieke, G. H., & Tokunaga, A. T. 1982, *ApJ*, 263, 736
Muno, M. P., Clark, J. S., Crowther P. A. et al. 2006, *ApJ*, 636, L41
Nagata, T., Woodward, C. E., Shure, M., & Kobayashi, N. 1995, *AJ*, 109, 1676
Nagata, T., Woodward, C. E., Shure, M. et al. 1990, *ApJ*, 351, 83
Najarro, F. 1995, *Ph.D. Thesis*
Najarro, F., Hillier, D. J., Kudritzki, R.-P. et al. 1994, *A&A*, 285, 573
Najarro, F., Krabbe, A. Genzel, R. et al. 1997, *A&A*, 325, 700
Nakaya, H., Watanabe, M., Ando, M. et al. 2001, *AJ*, 122, 876
Negueruela, I. & Clark, J. S. 2005, *A&A*, 436, 541
Nelan, E. P., Walborn, N. R., Wallace, D. J. et al. 2004, *AJ*, 128, 323
Oey, M. S. & Clarke, C. J. 2005, *ApJ*, 620, L43
Okuda, H., Shibai, H., Nakagawa, T. et al. 1990, *ApJ*, 351, 89
Rieke, G. H., Telesco, C. M., & Harper, D. A. 1978, *ApJ*, 220, 556
Rockefeller, G., Fryer, C. L., Melia, F., & Wang, Q. D. 2005, *ApJ*, 623, 171
Salpeter, E. E. 1955, *ApJ*, 121, 161
Schwarzschild, M. & Härm, R. 1959, *ApJ*, 129, 637
Schödel, R., Eckart, A., Alexander, T. et al. 2007, *A&A*, 469, 125
Serabyn, E., Shupe, D., & Figer, D. F. 1998, *Nature*, 394, 448
Skinner, S. L., Simmons, A. E., Zhekov, S. A. et al. 2006, *ApJ*, 639, L35

Skrutskie, M. F., Schneider, S. E., Stiening, R. *et al.* 1997, in: F. Garzon (ed.), The Impact of Large Scale Near-IR Sky Surveys (Dordrecht: Kluwer), *Astrophys Space Sci Lib*, 210, 25
Smith, N. 2008, in: M. Livio (ed.), Massive Stars: From Pop III and GRBs to the Milky Way, (Cambridge: CUP) in press (arXiv:astro-ph/0607457)
Stephenson, C. B. 1990, *AJ*, 99, 1867
Tanner, A., Figer, D. F., Najarro, F. *et al.* 2006, *ApJ*, 641, 891
Timmermann, R., Genzel, R., Poglitsch, A. *et al.* 1996, *ApJ*, 466, 242
Tuthill, P. G., Monnier, J. D., & Danchi, W. C. 1999, *Nature*, 398, 487
Tuthill, P., Monnier, J., Tanner, A. *et al.* 2006, *Science*, 313, 935.
van der Hucht, K. A. 2001, *New Astronomy Review*, 45, 135
Wang, Q. D., Dong, H., & Lang, C. 2006, *MNRAS*, 371, 38
Weidner, C. & Kroupa, P. 2004, *MNRAS*, 348, 187
Weidner, C. & Kroupa, P. 2006, *MNRAS*, 365, 1333
Wolfire, M. G. & Cassinelli, J. P. 1987, *ApJ*, 319, 850
Yusef-Zadeh, F., Law, C., Wardle M. *et al.* 2002, *ApJ*, 570, 665
Yusef-Zadeh, F. & Morris, M. 1987, *ApJ*, 320, 545
Yusef-Zadeh, F., Nord, M., Wardle, M. *et al.* 2003, *ApJ*, 590, L103

Discussion

VANBEVEREN: In 1982, I published a study where it was shown (from a theoretical point of view) that the maximum mass of the stars in a cluster depends on the total mass of the cluster and, from the cluster (total) mass function, I proposed a mass function for the maximum mass. This idea has been picked up by Weidner and Kroupa in 2004. Is there any observational evidence that he maximum mass of the stars in a cluster depends on the cluster mass?

FIGER: Some work to answer that question has been done by Oey & Clarke (2005); however, the clusters in that study have very uncertain ages and it is thus not clear if their most massive members have already exploded as supernovae.

DAMINELI: Your open cluster candidates do not fit the "radii" portion of the arms, specifically in the Carina arm. How did you assign distances to the clusters?

FIGER: We considered clusters associated with HII regions for which the line-of-velocities are available (Kuchar & Bania 1994). We therefore obtained distances by using the equations in Burton (1988), assuming near-distances.

LANGER: You use the argument that no WNE or WC stars are found in the Arches cluster to prove that it is young enough so that the most massive stars have not exploded off yet. Is this challenged by the evidence put forward by Nathan Smith that the most massive stars die as hydrogen-rich LBVs?

FIGER: Nathan Smith's speculation suggests that the most massive stars die as WNh. If this is true, then it is possible that we can no longer see stars with initial masses above 150 $M_\odot$ in the Arches cluster because they have already exploded as supernovae, NOT because they did not form. This would also be true for R136 in 30 Dor. One interesting observational test of such a claim is to search for coeval clusters in which there are simultaneously WNh stars and more evolved subtypes (presumably from the highest initial masses in the cluster). Perhaps NGC 3603 is promising for such a test where there are WNh stars and a more evolved subtype, Sher 25; however, one would have to clearly show that these stars are coeval.

Massive Stars in the Galactic Center

F. Martins[1][†], D. J. Hillier[2], R. Genzel[1,3], F. Eisenhauer[1], T. Ott[1], S. Gillessen[1] and S. Trippe[1]

[1]MPE,
Postfach 1312, D-85741, Garching, Germany
email: martins@graal.univ-montp2.fr

[2]Dept. of Physics and Astronomy, University of Pittsburgh,
PA-15260 Pittsburgh, USA

[3]Dept. of Physics, University of California,
CA-94720 Berkeley, USA

Abstract. We present results of two studies aiming at better understanding the properties of massive stars in the Galactic Center. We focus on the youngest and oldest of the three massive clusters harboring this region, namely the Arches and central cluster. We show that the development of powerful observational techniques in the near infrared spectral range (mainly 3D spectroscopy) allows to uncover the entire massive star population in these clusters. Using CMFGEN models, we derive the classical stellar and wind properties of 46 stars, as well as their surface abundances. The latter allow us to investigate in detail their evolutionary status and to identify evolutionary sequences between different types of stars. We thus constrain stellar evolution in the upper part of the HR diagram.

Keywords. stars: early-type – stars: Wolf-Rayet – stars: fundamental parameters – Galaxy: center

1. Introduction

The Galactic Center hosts three massive clusters: the Quintuplet, the Arches and the central cluster. They are all young ($2 <$ age < 8 Myr), massive ($M > 10^4$ M$_\odot$) and consequently host a large number of massive stars: several hundreds of O stars, as well as about one third of the total number of Wolf-Rayet stars known in the Galaxy. A very interesting feature of the clusters is that they have different ages: the Arches cluster is the youngest (2-3 Myr), followed by the Quintuplet cluster (4 Myr) and the central cluster (6 Myr). Hence, the contain different populations of massive stars and represent a unique opportunity to study massive star evolution. Here, we report on the detailed analysis of the massive star content of the youngest (Arches) and the oldest (central) cluster and provide improved evolutionary sequences for massive stars.

2. The central cluster

The central parsec of the Galaxy is a fascinating environment hosting the radio source and supermassive black hole Sgr A*. Its stellar content was first uncovered in the late eighties by Forrest *et al.* (1987) and Allen *et al.* (1990). They revealed the presence of bright infrared sources with strong emission lines in the K-band, especially the HeI 2.058 μm line. At that time, such objects were observed only in the central parsec and they were thought to be peculiar to this region. This population of about 10 stars was subsequently

[†] Present address: GRAAL - CNRS - Université Montpellier II, Place Eugène Bataillon, 34095 Montpellier, France

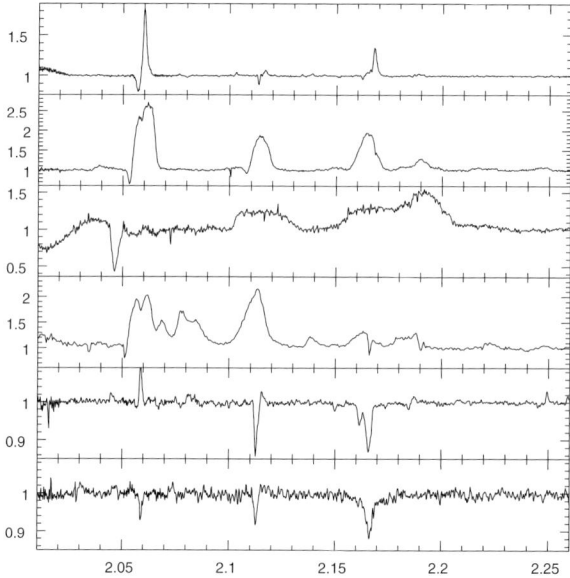

Figure 1. Mosaic of massive stars spectra observed in the central parsec with SINFONI. From top to bottom: IRS16C (Ofpe/WN9), IRS9W (WN8), IRS16SE2 (WN5-6), IRS7SE (WC9), average of the brightest O supergiants, average of O dwarfs.

analyzed by Najarro et al. (1994) and Najarro et al. (1997). They showed that the HeI emission line stars were luminous but rather cool ($T_{\rm eff} \sim 25{,}000$ K) massive stars in a post-main sequence evolutionary state. This population was somewhat puzzling since it was not producing enough ionizing photons to account for the nebular HeI emission. Besides, it was inconsistent with a 6-7 Myr starburst model, an age inferred from the very presence of these stars. Indeed, such a model produced many Wolf-Rayet stars not observed in the central cluster. This lead people to speculate that stellar evolution (the building block of starburst models) was failing to reproduce the central cluster population and that stellar evolution was proceeding differently in the Galactic Center. The alternative to this explanation was that the predicted population of hotter massive stars was not yet observed because it was fainter.

The solution to this puzzle came in the last few years. It benefited from two major improvements: new observational techniques/instruments and improved atmosphere models for hot stars. In terms of observational capabilities, the advent of adaptive optics assisted integral field spectrographs such as SINFONI on the ESO/VLT has allowed a detailed and deep (complete down to mK$\sim$14) spectroscopic mapping of the central parsec of the Galaxy. It resulted in the discovery of more than 100 new massive stars (a sample is shown in Fig. 1; see also Paumard et al. (2006)). Among them, one finds evolved late type O and early B supergiants, late type WC stars (mainly WC9), WN stars (WN5 to WN9) and even main sequence B stars such as the famous "S stars" (Ghez et al. (2003), Eisenhauer et al. (2005)). All these new stars are hotter than the HeI emission line stars. Consequently, there is a population of hot evolved stars as expected from a 7 Myr old starburst. Qualitatively, this tells us that there is probably no special evolution in this region.

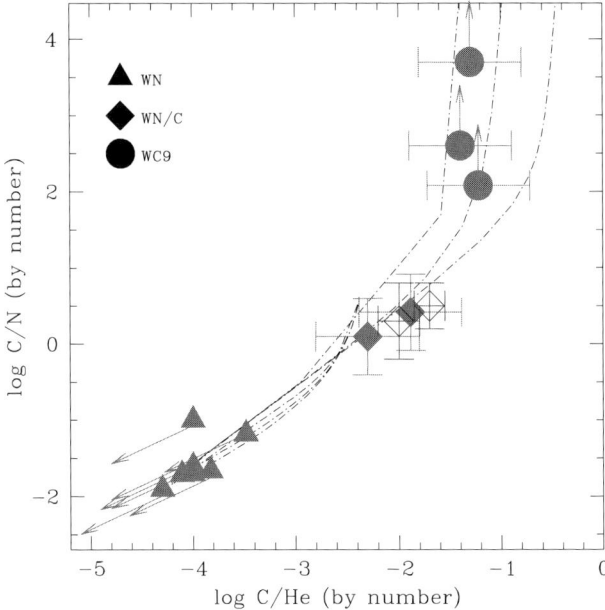

Figure 2. C/He as a function of N/He for WN8 (triangles), WN8/WC9 (diamonds) and WC9 stars (circles). The Geneva evolutionary tracks with rotation are shown by the dot-dashed lines. The agreement between model predictions and derived properties is very good, indicating a normal stellar evolution in the central cluster.

To test this idea quantitatively, we used new generation atmosphere models to derive the stellar and wind properties of this new population (Martins *et al.* (2007)). The models were computed with the code CMFGEN (Hillier & Miller(1998)) which includes a non-LTE treatment, spherical extension and winds, as well as line-blanketing. From the fit of observed K-band spectra, we derived the classical parameters: effective temperature, luminosity, mass loss rate, wind terminal velocity, ionizing flux. After summing up the contribution of all stars to obtain the total ionizing budget (using the calibrations and spectra of Martins *et al.* (2005) for stars of known spectral type but not analyzed directly in this study), we computed a photoionization model of the region with CLOUDY. We showed that the population of massive stars is now fully able to account for the nebular properties of the region. This is an indirect proof that this population is not peculiar, and that stellar evolution is most likely "normal".

A more direct evidence for a normal evolution came from the determination and analysis of surface abundances in post-main sequence. Using the amount of H, He, C and N observed at the surface of those stars, we were able to derive convincing evolutionary sequences between various types of stars. Fig. 2 illustrates this point. We show the C/He ratio as a function of the N/He ratio for WN8, transition WN8/WC9 and WC9 stars. The carbon enrichment and nitrogen depletion from WN8 to WC9 stars is obvious. Most importantly, the Geneva evolutionary tracks including rotation shown by dot-dashed lines reproduce perfectly the observed sequence, confirming that standard stellar evolution accounts for the Galactic Center massive star population.

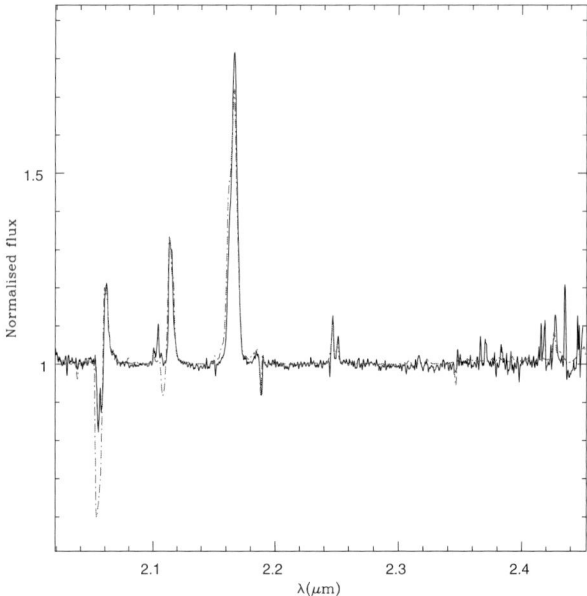

Figure 3. Best fit model (red dashed line) of the observed spectrum (black solid line) of a WN8-9h star in the Arches cluster.

3. The Arches cluster

The Arches cluster is the youngest of the three massive clusters in the Galactic Center (2-3 Myr, see Blum *et al.* (2001), Figer *et al.* (2002)). As such, it contains a different population of massive stars: early O stars and WN stars. Since they are rather young, they also have larger initial masses. Hence, studying the properties of these objects can bring new constraints on stellar evolution at very high luminosity and mass.

We have thus obtained SINFONI observations of the central part of the cluster as well as of bright stars located at larger distance. We have confirmed the presence of a large number of O4-6 supergiants, and we have improved the classification of the WN stars: they all appear to be WN7-9h stars, i.e. late type WN stars still showing a significant amount of hydrogen in their atmosphere (Martins *et al.* 2008).

As for the central cluster, we have derived the stellar and wind properties of the 28 brightest stars (Martins *et al.* 2008). A typical fit is shown in Fig. 3. They turned out to be all luminous objects, some of them reaching $10^{6.3}$ $L_{\odot}$. In particular, all the WNLh stars are very luminous stars. Translated into initial masses, this means that they have progenitors with initial masses larger then 70 $M_{\odot}$, up to 120 $M_{\odot}$. These interesting objects are barely evolved in terms of He enrichment: some of them have the initial composition and are similar to O supergiants (see Fig. 5). However, all the WNLh stars show nitrogen enrichment and carbon depletion compared to O supergiants (see Martins *et al.* 2008). This is a strong indication that they are in an early state of evolution, already showing products of the CNO cycle at their surface, but young enough not to show a large hydrogen depletion (and helium enrichment). As a consequence, we identify them as the slightly advanced versions of very massive objects with spectral types earlier than O4-6 supergiants and still in the hydrogen burning phase.

Another interesting property of these WN7-9h stars concerns their winds. As shown in Fig. 4, they follow a very well defined modified wind momentum - luminosity relation.

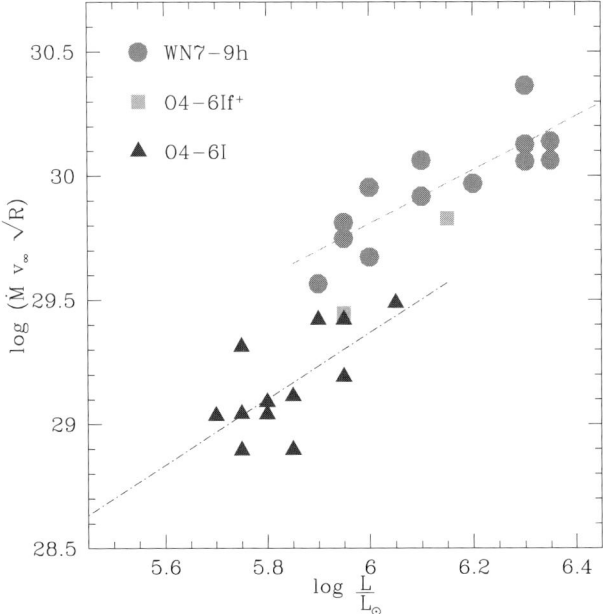

Figure 4. Modified wind momentum - Luminosity relation for the Arches stars. Blue triangles are O supergiants, green squares are OIf$^+$ stars, and magenta circles are WN7-9h stars. The latter stars follow a well defined relation, indicating that their winds are most likely radiatively driven.

This relation is shifted and has a different slope compared to that followed by O supergiants. This is however a strong indication that their winds are radiatively driven. This is another indirect evidence that these extreme objects are quite similar to O stars.

4. Summary

We have derived the stellar and wind properties of 46 stars in the central and Arches cluster in the Galactic Center. Since these clusters have different ages, they host different populations. We have thus analyzed different types of massive stars: very massive ($M \sim$60-120 $M_\odot$) and luminous objects in the Arches cluster, older (6 Myr) objects with initial masses in the range 30–60 $M_\odot$ in the central cluster. From the detailed study of their abundances, we have been able to define evolutionary sequences between different class of objects. Fig. 5 illustrates this point. The hydrogen mass fraction of various types of objects is shown as a function of luminosity. The Geneva evolutionary tracks with rotation are overplotted (solid lines). From such a diagram, and using the predicted sequences, one can define the following sequences:

$M \sim$ 30-60$M_\odot$:   Ofpe/WN9 → WN8 → WN8/WC9 → WC9
$M >$ 70$M_\odot$:   <O4-6I → WN7-9h

These sequences are in qualitative agreement with those summarized by Crowther (2007) in his recent review on Wolf-Rayet stars. In the particular cases studied here, we have been able to refine the sequences and to show the relation between different spectral subtypes. This illustrates the power of analysis of large sample of stars in clusters to

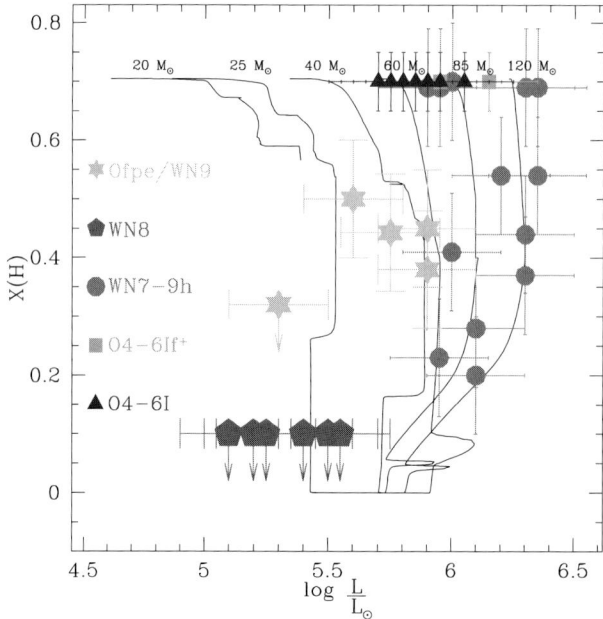

Figure 5. Hydrogen mass fraction as a function of luminosity for most of the stars studied in the Arches and central cluster. Solid lines are the recent Geneva evolutionary tracks including rotation (Meynet & Maeder (2005)). Studying clusters of different ages allows to populate different parts of this diagram and to pinpoint the evolutionary link between various types of stars.

constrain stellar evolution. One of the most important conclusion is that in the central cluster, there is no peculiar stellar evolution as believed until recently.

References

Allen, D. A. Hyland, A. R., & Hillier, D. J. 1990, *MNRAS*, 244, 706
Blum, R. D. Schaerer, D., Pasquali A. *et al.* 2001, *AJ*, 122, 1875
Crowther, P. A. 2007, *ARA&A*, 45, 177
Eisenhauer, F. Genzel, R., Alexander, T. *et al.* 2005, *ApJ*, 628, 246
Figer, D. F. Najarro, F., Gilmore, D. *et al.* 2002, *ApJ*, 581, 258
Forrest, W. J., Shure, M. A., Pipher, J. L., & Woodward, C. E. 1987, in: D. C. Backer (ed.), The Galactic Center (New York: AIP), *AIP Conf Proc* 155, 153
Ghez, A., Duchêne, G., Matthews, K. *et al.* 2003, *ApJ*, 586, L127
Hillier, D. J. & Miller, D. L. 1998, *ApJ*, 496, 407
Martins, F., Schaerer, D., & Hillier, D. J. 2005, *A&A*, 436, 1049
Martins, F., Genzel, R., Hillier, D. J. *et al.* 2007, *A&A*, 468, 233
Martins, F., Hillier, D. J., Paumard, T. *et al.* 2008, *A&A*, 478, 219
Meynet, G. & Maeder, A. 2005, *A&A*, 429, 581
Najarro, F. Hillier, D. J., Kudritzki, R.-P. *et al.* 1994, *A&A*, 285, 573
Najarro, F., Krabbe, A. Genzel, R. *et al.* 1997, *A&A*, 325, 700
Paumard, T. Genzel, R., Martins, F. *et al.* 2006, *ApJ*, 643, 1011

Discussion

HANSON: You did not mention the very luminous red supergiant IRS7, in the Galactic central cluster. For the mass range and ages you list, it seems consistent with being coeval, do you agree?

MARTINS: Yes, I agree. The central cluster is 6 ± 2 Myr old. This corresponds to an age at which the first red supergiants appear. At the same time, there is still a number of Wolf-Rayet objects coming from stars with initial masses of 40-50 $M_\odot$.

HANSON: Now that you have a deeper study of the central cluster, is the very high number of HeI emission line stars, seen initially because of observational selection effect, no longer anomalous?

MARTINS: The so-called HeI emission line stars were initially discovered because they the most luminous stars in the near-IR (due to their low effective temperature). We have uncovered 10-15 times more hotter massive stars, including WN, WC and O stars. These stars are fainter in the near IR and were not seen initially. Consequently, the HeI stars are outnumbered by other types of stars and do not represent a peculiar population as thought a decade ago. Note also that the HeI stars have been observed in other environments (especially the Magellanic clouds) since their discovery in the Galactic Center.

MOFFAT: You indicated that the WN7-9h stars are preceded by O4-6I stars. But is this necessary? Could the WNLh stars not have started as WNLh with strong emission lines because of their high luminosity even on the ZAMS? E.g. NGC 3603 has an age of 1 Myr, barely enough to evolve an O star to WN.

MARTINS: Given that we see both WN7-9h and O4-6I stars in the Arches, I argued that the former probably came from stars earlier than O4-6I stars, maybe O2-3 stars. One cannot exclude that very luminous stars such as the one observed in the Arches cluster or NGC 3603 were born as WNLh stars, but the evidence I showed that WNLh stars are already evolved in terms of CNO processing (i.e. they are N rich/ C poor) indicate that there must exist a class of objects with almost the same luminosity, but different surface abundance patterns. Such objects could be very early O stars. Note also that the timescale to see significant N enrichment is of the order of 1 Myr. This could explain the presence of WNLh stars in NGC3603.

From left to right: Amparo Marco, Ana Ursúa and Alfredo Sota.

Ronny Blomme.

Metallicity Studies in the IR: Unveiling Obscured Clusters of Our Galaxy

Francisco Najarro

Instituto de Estructura de la Materia, CSIC, Serrano 121, 28006 Madrid, Spain
email:najarro@damir.iem.csic.es

Abstract. We review direct and indirect methods to derive metallicity through infrared spectroscopy of massive stars. The choice of different spectral types to obtain abundances allows to trace metallicity for a wide range of ages of the cluster hosting the massive stars. These methods have a great potential to understand the evolution of large amount of heavily obscured galactic clusters, which are currently being discovered through infrared surveys.

Keywords. stars: abundances – stars: winds, outflows – stars: atmospheres – infrared: stars – Galaxy: abundances

1. Introduction

With the advent of new detectors, the field of quantitative spectroscopy of massive stars in the infrared (IR) has undergone a major revolution during the last decade. Most of the achieved improvements in this field were triggered by the detection of a He I emission line cluster (Krabbe et al. 1991) in the central parsec of our Galaxy (GC), and our curiosity to unveil the star formation processes and energetics taking place there. The NLTE atmospheric models accounting only for H and He (Najarro et al. 1994,1997) used a decade ago to study the IR spectra of massive stars have been recently upgraded (Hillier & Miller 1998) to account for blanketing and clumping, allowing us to obtain a detailed picture of the region (Figer et al. 2002; Najarro et al. 2004, 2006; Martins et al. 2007a, 2007b). Since the discovery of the massive cluster at the central parsec, and the Quintuplet and Arches Clusters within 30 pc of the GC, systematic IR searches have been undertaken in the Milky Way (e.g., Dutra & Bica 2000, Herrero et al., these proceedings). These are starting to reveal an important presence of obscured massive clusters in our Galaxy (see review by Figer, these proceedings). The presence of these clusters all over the Galaxy offers an unique opportunity to estimate metallicity and hence obtain the current two dimensional abundance pattern of our Galaxy placing, this way, crucial constraints on models of galactic chemical evolution. Interestingly, metallicity estimates of massive stars in the IR have been also triggered through observations of the three dense and massive star clusters in the GC and subsequent quantitative analysis of both their cool (Carr et al. 2000; Cunha et al. 2007) and hot (Najarro et al. 2004, 2006; Martins et al. 2007a, 2007b) stellar population. Given that the three clusters display different ages, they present an ideal laboratory to perform metallicity studies of stars at different evolutionary phases. Hence, they provide reliable templates to analyse the Galactic obscured massive clusters which are currently being discovered. In this paper, we present different methods to obtain both direct and indirect estimates metallicity of massive stars as a function of their spectral type.

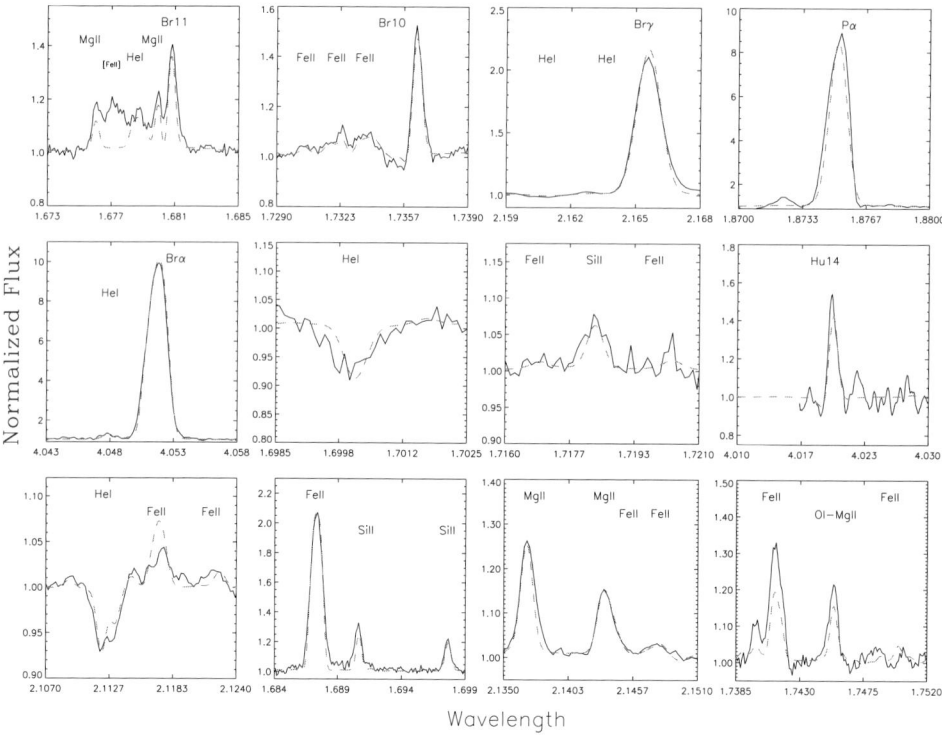

Figure 1. Model fits (*dashed lines*) to the observed infrared diagnostic lines (*solid lines*) of the Pistol Star (see Najarro *et al.* 2008).

2. IR metallicity studies. What solar is solar?

Ironically, one of the main problems arising when discussing metal abundances is **which solar abundance values should be linked to solar metallicity**. This controversy arises from the abundances adopted as solar in the evolutionary models, which are "delayed" with respect to the solar abundance values as determined from detailed spectroscopic modelling of the sun's photosphere. Here, we adopt the solar composition of Grevesse & Noels (1993, hereafter GN93). Although their abundances have been recently revised Asplund *et al.* (2005, hereafter AS05) (but see also Pinsonneault & Delahaye (2008)), they are the ones used by Iglesias & Rogers(1996) to compute stellar interior opacities and adopted in the most recent evolutionary models for massive stars with rotation (Meynet & Maeder 2003, 2005). Previously published evolutionary tracks for massive stars (Schaller *et al.* 1992, Meynet *et al.* 1994) used opacity tables calculated with solar composition from Anders & Grevesse, (1989, hereafter AN89), which differ significantly from GN93 only in Fe (A(Fe/H)=7.67 vs 7.50 in GN93)† and very slightly in the CNO ratios (A(C/H)=8.56, A(N/H)=8.05, A(O/H)=8.93 in AN89 vs A(C/H)=8.55, A(N/H)=7.97, A(O/H)=8.87 in GN93). Si and Mg are the same in all evolutionary models, and have been only slightly revised downward ($\sim$0.05 dex) by AS05. Thus, the reader should be warned about discussions of derived α-elements vs. Fe ratio in the literature, which may depend critically on the assumed Fe solar abundance.

† A(X/Y)=log[n(X)/n(Y)]+12

3. Direct Methods

Due to the lack of major features in their IR H- and K-Band spectra, hot stars are usually employed as standards when reducing IR spectra of other objects. In fact, in massive stars hot enough to allow accurate temperature determination through fitting of their He I and He II lines, their H and K spectra only display weak CNO features and some of them are blended. Thus, for these hotter objects, only CNO estimates may be obtained, while no direct information on α-elements other than O (Si, Mg, Ca) or iron group elements is available. Since some of these objects will already show some N enrichment and C and O depletion from C/N and O/N equilibrium processes, the derived CNO abundances have to be interpreted with great care. For very late O and early B stars the situation is even worse as not only they run out of metal lines in their H- and K- spectra but also the He II lines disappear disabling the possibility to obtain an accurate temperature determination.

However, for mid and late B stars with strong winds (LBVs and supergiants), the situation is drastically improved. Apart from the H and He I lines, the stars display several lines of Si II, Mg II and Fe II. Thus, a direct estimate of metallicity and α-elements vs Fe ratio can be obtained. Note that in order to host LBVs (or late B-supergiants) the cluster has to be at least 4 Myr old. This is case of the Quintuplet Cluster (Glas et al. 1987) which hosts two LBVs: the "Pistol Star" (Figer et al. 1998) and FMM362 (Geballe, Najarro & Figer 2000). Najarro et al. (2008) have recently presented a detailed quantitative spectroscopic analysis of both stars and have been able to obtain direct estimates of Fe and Mg and Si abundances in these objects. Figure. 1 shows the excellent quality of the model fits to the relevant lines in the "Pistol Star". Below we discuss the main metal diagnostic lines (see Najarro et al. 2008 for further discussion).

Iron. Two types of Fe II lines are found in the spectra. The first are the strong semi-forbidden lines that form in the outer wind and have small oscillator strengths (gf$\sim 10^{-5}$). The second are the weak permitted (gf~ 1) lines connecting higher lying levels that form much closer to the photosphere. The permitted lines are more robust iron abundance indicators, having only weak dependences on other parameters, such us turbulent velocity. The strengths of the semi-forbidden lines depend on the mass loss rate and the run of the iron ionization structure in the outer wind, which is sensitive to the hydrogen ionization structure due to the strong coupling to the Fe/H charge-exchange reactions. We obtain roughly solar iron abundances for both LBVs, with ± 0.15dex as plausible uncertainties.

Magnesium. The strongest Mg II lines observed in the H and K bands share the $5p^2P$ level. Those lines with it as the upper level, the 2.13/14 μm and 2.40/41 μm doublets (see Fig. 1) are much stronger than those with it as the lower level (H band lines), revealing that pumping through the resonance $3s^2S$-$5p^2P$ line must be a significant populator of the $5p^2P$ levels due to Lyβ fluorescence. We estimate about twice solar Mg abundance and an associated uncertainty of about ± 0.25dex (due to uncertainties related to the fluorescence contribution).

Silicon. The Si II doublet $5s^2S_{1/2}$-$5p^2P_{3/2}$ 1.691 μm and $5s^2S_{1/2}$-$5p^2P_{1/2}$ 1.698 μm constitutes a powerful diagnostic tool, as it appears in emission for only a very narrow range of stellar temperatures and wind density structures, indicating the presence of amplified NLTE effects. However, since it forms at the base of the wind, its strong dependence on the details of the velocity field there hinders a precise silicon abundance determination. Instead, we use the well-behaved recombination line Si II $3s^26g^2$G-$3s^25f^2$F at 1.718 μm which shows a stronger dependence on the silicon abundance. From our model fits (see Fig. 1) we derive roughly twice solar abundance (± 0.20 dex) for silicon in each LBV, similar to magnesium.

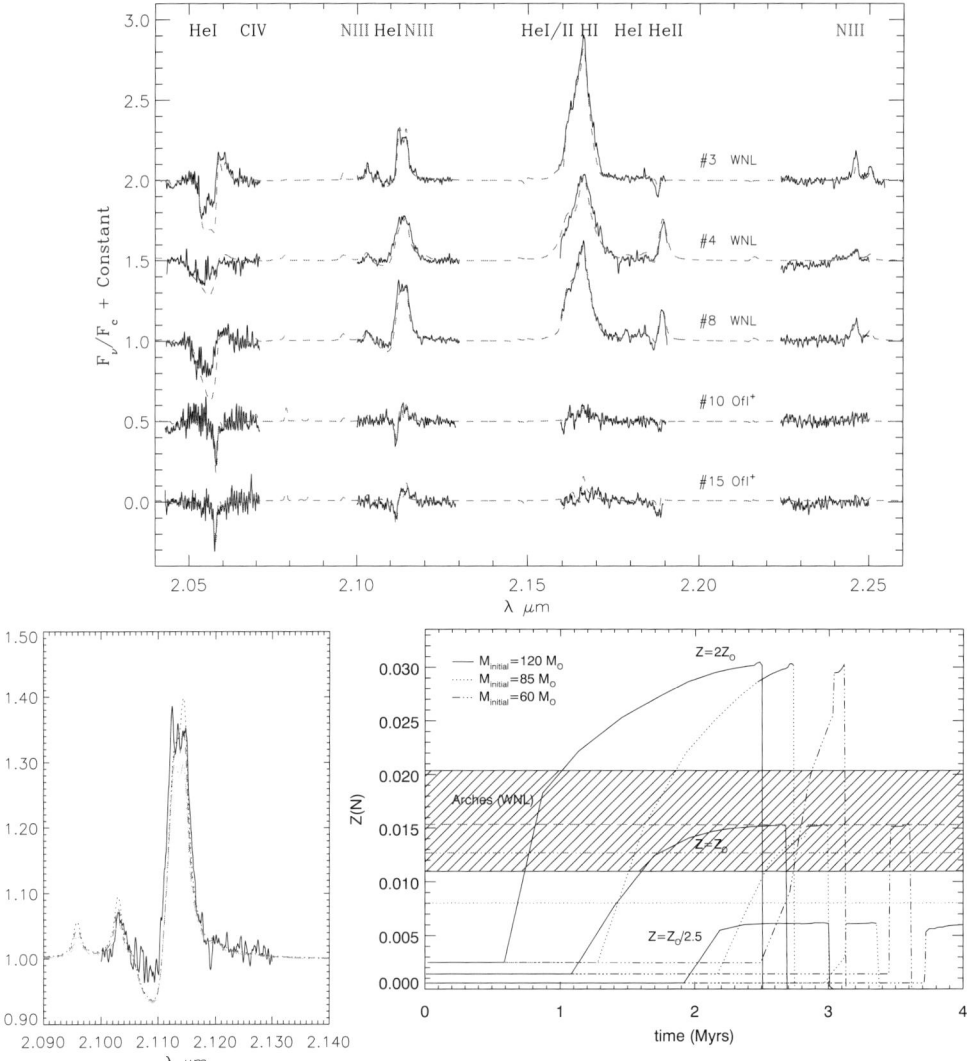

Figure 2. Top Observed spectra (solid) and model fits (dashed) for three WNL and two OfI$^+$ Arches stars. **Bottom-Left**. Leverage of error estimates on N abundance. Best fit (dashed) and 30% enhanced (dotted) and 30% depleted (dashed-dotted) nitrogen mass fractions are displayed. **Bottom-Right**. Nitrogen mass abundance versus time using Geneva models. Maximum nitrogen surface mass fractions for solar composition are also displayed as horizontal lines for different assumed initial CNO patterns (dashed for AN89, dashed-dotted for GN93 and dotted for AS05).

4. Indirect Methods: the WNL phase

Najarro *et al.* (2004) presented a method to determine the initial oxygen abundance and, to a lesser degree, carbon abundance, of the natal cloud where the massive stars are formed. The method assumed that the star is undergoing its WNL evolutionary phase, which implies that nitrogen has reached its maximum surface abundance value while star still shows hydrogen in its spectra. Evolutionary models indicate that 95% of that value is already attained by the time that H/He<2 (by number). Such scenario is currently met by the most massive stars in the Arches cluster and some WN stars in the Central parsec

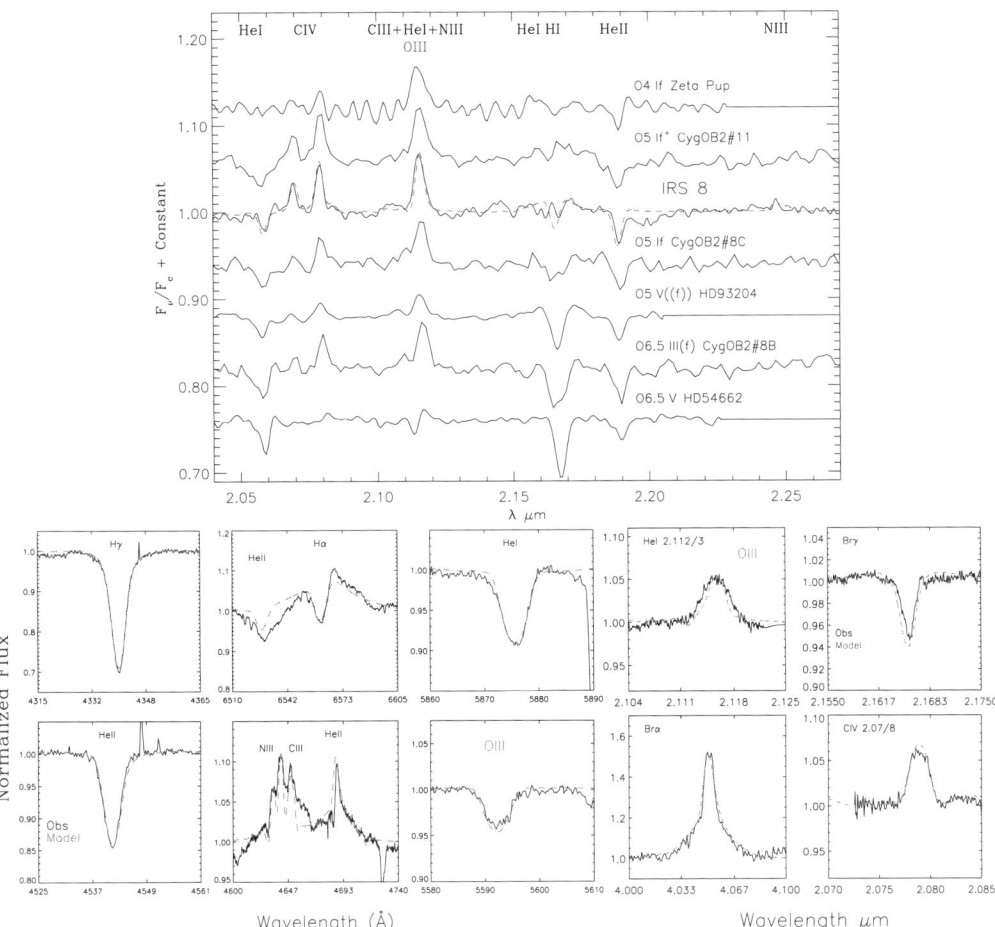

Figure 3. Top Spectral type determination of IRS 8* and identification of the O III 2.116μm feature. Model fit (dashed) with stellar parameters corresponding to an O5.5If star. **Bottom** Consistency check of the method through simultaneous fitting of the optical and IR lines, including oxygen, for CygOB2 #8C. (Najarro *et al.* in prep.)

cluster. Being very young ($\leqslant$2.5 Myr), the Arches cluster hosts WNLs and OIf$^+$ having infrared spectra dominated by H, He I, He II, N III lines. Some have weak C III/IV lines (see Fig. 2). Figer, McLean, & Najarro(1997) showed that these N III lines appear only for a narrow range of temperatures and wind densities, which occur in the WN9h (WNL) stage. The fairly distinct nature and energies of the multiplets involved in each of both N III line sets provide strong constraints for the determination of the nitrogen abundance. Najarro *et al.* (2004) followed the metallicity patterns from the Geneva evolutionary models and assumed no selective enrichment of CNO or α-elements vs Fe, in concluding that the stars in the Arches Cluster have solar α-element abundances (see Fig. 2). However, estimates of solar abundances have varied considerably over the past 15 years. Thus, depending on the assumed solar CNO composition, the derived nitrogen abundance by Najarro *et al.* (2004) could imply solar (AN89), 1.2 × solar (GN93) or 2.0 × solar (AS05) CNO composition (see Fig. 2). Using the same approach, Martins *et al.* (2007a, 2007b) have recently analyzed a larger sample of hot stars in the Arches and Central Parsec clusters and find similar results. Interestingly, if one considers only the objects in Martins *et al.*

(2008) with He/H> 0.1 and those with Z(C)< 0.05, i.e., fulfilling the condition to be close enough to $Z(N)_{max}$, their average value of Z(N) shows excellent agreement with the one obtained by Najarro *et al.* (2004).

5. New direct method: Oxygen abundance in early and mid O stars

Again, the GC has played a starring role, as the identification of the central source of IRS 8 in the cluster at the central parsec as a mid O supergiant (Geballe *et al.* 2006) enabled the development of a new method to estimate metallicity in these objects. When addressing the spectral type classification of IRS 8* by comparing its K-band spectrum with on line-available K-band spectra (Hanson *et al.* 1996, see Fig.3, top), Geballe *et al.* (2006) raised the question about the nature of the emission feature at 2.116μm. This had been attributed in the past to C III and N III n=8–7 transitions and is present over a very wide range of O spectral types and luminosities (Hanson *et al.* 1996). Our investigation (Geballe *et al.* 2006) indicated that the 2.116 μm feature in IRS 8* is dominated by O III n=8–7 transitions. Further, the O III component of the 2.116 μm feature largely depends on the oxygen abundance and only slightly on gravity, effective temperature, wind density, and velocity field. Hence, this feature constitutes a powerful diagnostic of oxygen abundance, and therefore an important metal abundance determiner, over a wide range of O spectral types (Najarro *et al.* in preparation). Given the strong spectral similarities of IRS 8* with the O5-6 supergiants in Cyg OB2 (see Fig. 3), we calibrated our method and checked for consistency by fitting simultaneously our available UV, optical and IR spectra of stars in Cyg OB2. Figure 3-*bottom* demonstrates the consistency of our fits to the optical and IR spectra for Cyg OB2#8C (O5If), and the excellent agreement of the models with both O III lines. The oxygen abundance of A(O/H)=8.91 obtained IRS 8* is fully consistent with the values obtained for the Arches WNLs and the α-element enrichment derived for the Quintuplet LBVs, providing a consistent picture of the metallicity in the GC.

Acknowledgement

I would like to thank Don Figer, John Hillier, Rolf Kudritzki and Tom Geballe for invaluable discussions. F. N. acknowledges AYA2004-08271-C02-02 grant. Special thanks to Miguel & Fabio for the great late sessions organized at Stevenson's Library.

References

Anders, E. & Grevesse, N., 1989, *Geochimica et Cosmochimica Acta*, Vol. 53, 197
Asplund, ., Grevesse, N., & Sauval, A. J. 2005, in: T. G. Barnes III & F. N. Bash (eds.), *Cosmic Abundances as Records of Stellar Evolution and Nucleosynthesis* (San Francisco: ASP), *ASP Conf Ser*, 336, 25
Dutra, C. M. & Bica, E., 2000, *A&A*, 359, 9
Carr, J. S., Sellgren, K. & Balachandran, S. C. 2000, *ApJ*, 530, 307
Cunha, K., Sellgren, K., Smith, V. V. *et al.* 2007, *ApJ*, 669, 1011
Figer, D. F., Najarro, F., Morris, M., *et al.* 1998, *ApJ*, 506, 384
Figer, D. F., McLean, I. S., & Najarro, F. 1997, *ApJ*, 486, 420
Figer, D. F., Najarro, F., Gilmore, D. *et al.* 2002, *ApJ*, 581, 258
Geballe, T. R., Najarro, F., & Figer, D. F. 2000, *ApJ*, 530, L97
Geballe, T. R., Najarro, F., Rigaut, F., & Roy, J.-R. 2006, *ApJ*, 652, 370
Glass, I. S., Catchpole, R. M., & Whitelock, P. A. 1987, *MNRAS*, 227, 373
Grevesse, N. & Noels, A. 1993, N. Prantzos, E. Vangioni-Flam & M. Casse (eds.), *Origin and evolution of the elements* (Cambridge: CUP), 14

Hanson, M. M., Conti, P. S., & Rieke, M. J. 1996, *ApJS*, 107, 281
Hillier, D. J. & Miller, D. L. 1998, *ApJ*, 496, 407
Iglesias, C. A. & Rogers, F. J. 1996, *ApJ*, 464, 943
Krabbe, A., Genzel, R., Drapatz, S., & Rotaciuc, V., 1991, *ApJ*, 382, L19
Martins, F., Genzel, R., Hillier, D. J. et al. 2007, *A&A*, 468, 233
Martins, F., Hillier, D. J., Paumard, T. et al., 2008, *A&A*, 478, 219
Meynet, G., Maeder, A., Schaller, G., Schaerer, D., & Charbonnel, C. 1994, *A&AS*, 103, 97
Meynet, G. & Maeder, A. 2003, *A&A*, 404, 975
Meynet, G. & Maeder, A. 2005, *A&A*, 429, 581
Moneti, A., Glass, I. S. & Moorwood, A. F. M. 1994, *MNRAS*, 268, 194
Nagata, T., Woodward, C. E., Shure, M. et al. 1990, *ApJ*, 351, 83
Najarro, F., Hillier, D. J., Kudritzki, R.-P. et al. 1994, *A&A*, 285, 573
Najarro, F., Krabbe, A., Genzel, R. et al. 1997, *A&A*, 325, 700
Najarro, F., Hillier, D. J., Figer, D. F., & Geballe, T. R. 1999, in: H. Falcke, A. Cotera, W. J. Duschl et al. (eds.), *The Central Parsecs of the Galaxy*, (San Francisco: ASP), *ASP Conf Ser*, 186, 340
Najarro, F. 2001, in: M. de Groot & C. Sterken (eds.), *P Cygni 2000: 400 Years of Progress*, (San Francisco: ASP), *ASP Conf Ser*, 233, 133
Najarro, F., Figer, D. F., Hillier, D. J., & Kudritzki, R. P. 2004, *ApJ*, 611, L105
Najarro, F. 2006, *Journal of Physics: Conf. Ser.*, 54, 224
Najarro, F., Figer, D. F., Hillier, D. J., Geballe, T. R., & Kudritzki, R. P. 2008, *ApJ*, submitted
Pinsonneault, M. H., & Delahaye, F. 2008, *ApJ*, submitted (arXiv:astro-ph/0606077)
Schaller, G., Schaerer, D., Meynet, G., & Maeder, A. 1992, *A&AS*, 96, 269

Discussion

STOLOVY: Why, of all of the massive stars that you have discussed, is the Pistol Star the *only* one with a nebula around it? Why doesn't the LBV candidate FMM362 have a nebula?

NAJARRO: It could very well be, as in the case of P Cygni, that for FMM362 the nebula may be there but very faint to be detected.

LEITHERER: What do you know about the abundances in nearby giants and supergiants? The comparison can give insight into the chemical evolution and also confidence in the modeling.

NAJARRO: As I showed, we are currently investigating these stars in Galactic regions, like the Cyg OB2 cluster where observations are available from the UV to the optical. First results seem to confirm consistency between optical and IR studies.

Paco Najarro.

Margaret Hanson.

Massive Stars in the Nuclei and Arms of Spirals

Fabio Bresolin

Institute for Astronomy, University of Hawai'i
2680 Woodlawn Drive, 96822 Honolulu, Hawai'i, USA
email: bresolin@ifa.hawaii.edu

Abstract. Many of the properties of massive stars in external galaxies, such as chemical compositions, mass functions, and ionizing fluxes, can be derived from the study of the associated clouds of ionized gas. Moreover, the signatures of Wolf-Rayet stars are often detected in the spectra of extragalactic H II regions. This paper reviews some aspects of the recent work on the massive star content of nearby spiral galaxies, as inferred from the analysis of giant H II regions. Particular attention is given to regions of high metallicity, including nuclear hot spots, and to the chemical abundance comparison between supergiant stars and ionized gas.

Keywords. galaxies: spiral – galaxies: ISM – galaxies: stellar content – galaxies: abundances – stars: early-type

1. Introduction

Hot and massive stars are among the most conspicuous objects visible in the arms and the nuclei of spiral galaxies. However, there is more than meets the eye. Young massive stars in external galaxies can be detected well beyond the optical galactic boundary defined by the isophotal radius. In the nearby NGC 2403, Davidge (2007) has found young main sequence stars out to 8 disk scale lengths (16 kpc) from the galactic center. The GALEX mission has revealed the presence of UV-bright complexes, i.e. clusters of young massive stars, in the outskirts of about 1/4 of the nearby spiral galaxy sample it has observed. For example, in M83 Thilker *et al.* (2005) have detected UV-bright knots out to 4 disk scale lengths. The star formation rate in these external regions is fairly low, as also indicated by the fact that only 5-10% of the UV knots are detected in Hα (Zaritsky & Christlein 2007). The oxygen abundance in the outer H II regions in M83 has been determined by Gil de Paz (2007) to be around 10% of the solar value, while the center of the galaxy reaches approximately twice the solar value (Bresolin *et al.* 2005; Pellerin & Robert 2007). Finally, what appear to be intergalactic H II regions have been found out to 30 kpc from their nearest galaxy (Ryan-Weber *et al.* 2004).

A large amount of additional work needs to be done on the properties of massive stars in the outer regions of spiral galaxies, in order to better understand the mechanisms that lead to their formation, and in general to learn more about the evolution of galaxies, for example about the relationship between neutral gas surface density and star formation rate, or the processes that lead to the build-up of the chemical elements. This review focuses on more 'normal' sites of massive star formation, represented by giant H II regions located in the optically bright portions of spiral disks. Since the line emission from these objects is directly linked to the ionization provided by their massive star content, the study of giant H II regions in nearby galaxies, as well as in distant star-forming galaxies, provides a wealth of information on the properties of the embedded clusters of ionizing stars (mass functions, hierarchical structuring), of the individual massive stars (initial mass functions, ionizing properties) and of the ionized gas (chemical abundances). For brevity's sake, only classical nebulae, photoionized by clusters of hot O and Wolf-Rayet stars, will be considered here, as space does not allow to cover other sources of line emission found in spiral galaxies, such as Seyfert nuclear activity, LINERS or supernova remnants. Also of particular interest, and the concluding topic of this article, are the rings of circumnuclear star formation that are present in a relatively large fraction of spiral galaxies in the nearby universe.

Some of the most interesting developments in our understanding of massive star formation and galaxy evolution derive from a multi-wavelength observational approach in the study of galaxy properties. In particular, recent space missions, such as GALEX in the ultraviolet and Spitzer in the mid- and far-infrared, have provided high-sensitivity access to star formation rate estimators, such as the UV stellar continuum and the monochromatic IR emission, that complement more traditional optical means, such as the optical recombination lines (Hα in particular). The cross-correlation between these different star formation rate estimators, obtained locally in galaxies within a few tens of Mpc (Schmitt et al. 2006), is crucial, for example, to gain confidence about the results obtained for high-redshift star-forming galaxies. Recent results from the Spitzer Infrared Nearby Galaxies Survey (SINGS) are particularly interesting in this context. Calzetti et al. (2007) and Kennicutt et al. (2007) have investigated the use of the 24 μm emission as a reliable star formation rate indicator in galaxies. Looking at both the optical and infrared line emission, Prescott et al. (2007) concluded that the fraction of highly obscured large star-forming regions in spiral disks is small, less than 4%, implying that optical studies that rely on the Hα emission to trace massive stars do not miss the bulk of massive star formation. This conclusion 'justifies' the remaining part of this review, in which we look at massive star properties in connection with optically bright H II regions. It is also worth mentioning the existence of an additional project, the Survey for Ionization in Neutral Gas Galaxies (SINGG), in which the massive star-forming regions, detected from their Hα emission, are studied in a radio-selected galaxy sample, providing a bias-free view of star-forming galaxies in the nearby universe (Meurer et al. 2006; Hanish et al. 2006).

In the following sections I will briefly discuss some recent findings concerning: (1) luminosity and mass functions of H II regions and star clusters; (2) Wolf-Rayet star content, chemical abundance studies and ionizing fluxes in metal-rich H II regions; (3) chemical abundance comparisons between blue supergiants and ionized gas; (4) circumnuclear regions of massive star formation in spiral galaxies.

2. The power of luminosity functions

One of the simplest techniques used for the analysis of extragalactic H II regions and star clusters, the construction of luminosity functions, can provide powerful conclusions on the properties of the evolving populations of massive stars in other galaxies.

H II region luminosity functions have been studied in detail for about two decades. Recently, Bradley et al. (2006) have published a composite luminosity function, modeled, as is customary, with a power law of the form $N(L)dL \propto L^{-\alpha}dL$, comprising approximately 18,000 H II regions in 56 spiral galaxies, and offering a clear view of the fact that the slope, $\alpha \simeq 2$, becomes shallower at lower Hα luminosities, as already observed by Kennicutt, Edgar & Hodge (1989). This can be due to a transition from ionization-bounded to density-bounded nebulae at a well-defined Hα luminosity (Beckman et al. 2000), or to the presence of stochastic variations in the number of ionizing stars at low H II region luminosity (Oey & Clarke 1998). Alternatively, it can be interpreted as the convolution of the time-dependent Hα luminosity due to an evolving massive star population with a lognormal cluster mass function (Dopita *et al.* 2006a).

The luminosity functions of young clusters in spiral galaxies can also be modeled using a power law, with a slope similar to that found for the H II region luminosity function (Larsen 2002). A break in the slope observed in M51 by Gieles *et al.* (2006) has been interpreted as due to a truncation of the cluster mass function above 10^5 M$_\odot$.

It has been pointed out by Elmegreen (2006) that the observed power-law form of the young cluster mass function in spiral galaxies, $N(M)dM \propto M^{-2}dM$, derives from the fact that the composite, system-wide stellar Initial Mass Function (IMF) of galaxies appears to have a slope that is close to the Salpeter value, as is found in the constituent individual stellar clusters and OB associations. This is the expectation from models of scale-free, hierarchical distributions of young star structures and star-forming sites within galaxies, from kpc-scale down to the size of individual stellar clusters. Recent observational confirmations of this hierarchical structuring have been found by Bastian *et al.* (2007) in the galaxy M33, and by Elmegreen *et al.* (2006) in NGC 628.

3. The utility of nebular spectra

The optical and near-IR spectra of H II regions are dominated by emission lines due to the recombination of H and He atoms, and to forbidden transitions, due to collisional excitation, of several metal ions, some of the strongest lines being [O II] $\lambda 3727$, [O III] $\lambda\lambda 4959,5007$, [N II] $\lambda\lambda 6548,6583$, [S II] $\lambda\lambda 6717,6731$, [S III] $\lambda\lambda 9069,9532$, [Ar III] $\lambda 7135$ and [Ne III] $\lambda 3868$. The study of line ratios allows to measure chemical compositions and the excitation state of the gas. The latter depends on the energy input from the massive stars that ionize the gas. The chemical abundances of the gas are expected to be virtually the same as for the young stars (modulo some effects that include depletion on dust grains and mixing at the stellar surfaces). Most of our current knowledge on the chemical composition of spiral galaxies, and of star-forming galaxies in general, including those at high-redshift, comes from emission-line studies of the ionized gas. Abundances from stellar spectra in external galaxies are more difficult to obtain. Some stellar results are discussed in connection with nebular abundances in section 5.

The spectroscopic study of extragalactic H II regions is being used to constrain massive star parameters in different ways. Some examples are only briefly mentioned here:

• the H II region chemical abundances provide key observational constraints for galactic evolution models, and therefore on the input parameters that characterize these models, such as the stellar IMF, the star formation rate and efficiency, the stellar yields, and the importance of gas inflows and outflows. As a recent example, Magrini *et al.* (2007a) have used the chemical abundances of H II regions, blue supergiants, planetary nebulae and RGB stars measured across the disk of M33, to model the temporal evolution of the chemical abundance gradient in this galaxy.

• as shown by Morisset *et al.* (2004), the ionizing flux output predicted for massive stars by different stellar atmosphere codes (CMFGEN, WM-basic, CoStar, TLUSTY, FAST-WIND) can differ by orders of magnitude at high photon energy (above 40 eV). Modeling nebulae with the prescriptions for the spectral energy distribution obtained from the different codes can therefore provide useful indications on the actual ionizing continuum flux distribution of O stars (Simón-Díaz & Stasińska 2008).

• stellar masses and effective temperatures of the ionizing stars can be estimated from optical and IR diagnostic diagrams, as recently shown by Dopita et al. (2006b) for Galactic ultracompact H II regions, using Spitzer-accessible line ratios, such as [Ne III]15.5μm/[Ne II]12.8μm and [S IV]10.5μm/[Ar III]9.0μm.

• the upper limit of the stellar initial mass function can be estimated using measurements of the equivalent width of the Hβ emission line in extragalactic H II regions that contain Wolf-Rayet star signatures (Pindao et al. 2002). Together with reliable chemical abundance measurements, this method indicates that the cut-off mass of the stellar IMF is not peculiar at high metallicity (Bresolin 2005), against several suggestions of the contrary present in the literature.

4. Metal-rich H II regions

4.1. Wolf-Rayet stars

Emission features from Wolf-Rayet (W-R) stars, e.g. the WN bump at 4660 Å, are frequently observed in high-metallicity H II regions (Bresolin, Kennicutt & Garnett 2004). This is well understood in terms of stellar evolution models, that show that the duration of the W-R phase and the W-R/O number ratio in an evolving ensemble of massive stars increase with metallicity (see the discussion in the recent paper on the Geneva models with rotation and Starburst99 by Vazquez et al. 2007). The growing number of deep spectra of extragalactic H II regions obtained in recent years shows that the C III 5696 Å line of late WC (WCL) stars is only found in the central regions of spiral galaxies, which, in virtue of the radial abundance gradients, are the most metal-rich. For example, WCL features are detected in H II regions in M101 only where the oxygen abundance is close to the solar value (Bresolin 2007). This agrees with early results from Phillips & Conti (1992) that WCL stars are preferentially found in metal-rich environments. Crowther et al. (2002) showed that the C III 5696 Å line strength increases with mass-loss rate, and therefore with metallicity.

A clear example of the effect of metallicity on the W-R star content is given by Hadfield et al. (2005), who detected W-R features in a large fraction of H II regions in M83, a nearby metal-rich galaxy (Bresolin & Kennicutt 2002), with WC8-9 stars being the dominant types. The WC/WN number ratio has been shown by many authors to increase with gas-phase metallicity (see Crowther 2007 for a recent review), in rough agreement with predictions by Eldridge & Vink (2006). It is worth mentioning here that in the study of this and similar trends with metallicity, most of the information on metallicities is derived from nebular studies. When comparing quantitative results with model predictions it is therefore important to remember that the reliability of the nebular abundances is still somewhat in question, especially at high metallicity (solar and above).

4.2. Chemical abundances

There are many additional situations in which an accurate knowledge of nebular abundances is crucial for a comparison with the model predictions, for example in the study of the metallicity dependence of the number ratio of type Ibc to type II supernovae (Prieto et al. 2008), and the determination of the mass-metallicity relation for high-redshift galaxies (Erb et al. 2006). Somewhat surprisingly, however, despite the fact that the study of nebular abundances is quite a mature field, the abundance scale is still somewhat uncertain, by at least a factor of two at the high-metallicity end (approximately solar metallicity and above). Skipping the details (see reviews by Bresolin 2008 and Stasińska 2008), we can focus here on two major observational projects related to the determination of nebular abundances that have been developed since the previous massive star symposium in Lanzarote: the determination of 'direct' abundances in high-metallicity nebulae, and the measurement of metal recombination lines in Galactic and extragalactic H II regions.

1. The classical method of chemical abundance analysis in H II regions is based on the detection of faint auroral lines, such as [O III] λ4363 and [N II] λ5755, that correspond to collisionally excited transitions that take place from the second lowest excited level of metal ions to the lowest excited level. In combination with nebular lines of the same ions originating from transitions from the lowest excited level to the ground level (e.g. [O III] λ5007 and [N II] λ6583) one can obtain a direct measurement of the electron temperature T_e, and therefore fix the value of the line emissivity, which is highly sensitive to T_e.

In recent years this technique has been extended towards the central, metal-rich zones of spiral galaxies, where the increased cooling from line emission leads to a strong decrease in the auroral-to-nebular line ratios (Bresolin et al. 2005; Bresolin 2007). There are some theoretical expectations that the method could generate erroneous results (Stasińska 2005), although observationally there is no indication yet that abundance biases are significant for H II regions with metallicity up to the solar value. If the auroral line method is reliable at high metallicity, then the oxygen abundances measured in the central regions of the most metal-rich nearby spiral galaxies lie in the range 12+log(O/H)=8.60–8.75 (Pilyugin, Thuan & Vílchez 2006), i.e. close to the solar value [adopting 12+log(O/H)$_\odot$=8.66 after Asplund et al. 2004], and 2-3 times smaller than found earlier, based on indirect methods that rely on semi-empirical calibrations of 'strong-line' diagnostics rather than on the direct measurement of the nebular electron temperature. An important consequence of the lower abundances is that the effective yields used in closed box galactic evolution models are reduced, again by factors of 2-3 (Bresolin, Kennicutt & Garnett 2004; Pilyugin, Thuan & Vílchez 2007).

2. Metal recombination lines, in particular O II λ4651 and C II λ4267, have been measured in a number of Galactic and extragalactic H II regions (Peimbert et al. 2007). The abundances derived from metal recombination lines have the advantage of being only mildly dependent on the value of the electron temperature (while collisionally excited lines depend exponentially on it). However, these lines are very faint, and this has insofar limited their use in extragalactic work to only a handful of targets. Garcia-Rojas & Esteban (2007) summarize the work done on nebular metal recombination lines, and show that oxygen abundances derived from recombination lines are, on average, a factor of two larger than those derived from collisionally excited lines (i.e. using the auroral line method). The origin of this abundance discrepancy is still object of debate. While Peimbert and collaborators invoke the presence of temperature fluctuations, alternative explanations have been proposed. In particular, Stasińska et al. (2007) have recently proposed an explanation based on the existence of abundance inhomogeneities, in the form of metal-rich droplets, rained down upon the interstellar medium of spiral disks following the explosion of supernovae. However, Lopez-Sanchez et al. (2007) question this interpretation, claiming that it is not really supported by the observational data.

In summary, the values one derives for the oxygen abundances of H II regions still depend on the method used. Choosing the 'direct' method (based on collisionally excited lines) produces a 'low abundance scale' (e.g. Pilyugin & Thuan 2005), that differs by a factor of 2-3 relative to the 'high abundance scale' obtained from the use of metal recombination lines or photoionization models (e.g. Kewley & Dopita 2002). This obviously has an immediate impact on the calibration of the so-called 'strong line methods', used throughout the literature to estimate nebular chemical abundances in star-forming galaxies from the strength of the most prominent optical forbidden emission lines, such as in the widely used parameter $R_{23} = ([\text{O{\sc ii}}] + [\text{O{\sc iii}}])/\text{H}\beta$.

4.3. Radiation field

The softening of the ionizing radiation and the decrease of the effective temperature of the ionizing stars of extragalactic H {\sc ii} regions with increasing metallicity have been investigated since the early extragalactic nebular studies in the 1970's. Optical diagnostics of effective temperature, such as He {\sc i} $\lambda 5876/\text{H}\beta$ and [O {\sc iii}] $\lambda 5007/\text{H}\beta$, show an apparent softening of the radiation field at high metallicity (Dors & Copetti 2003; Bresolin, Kennicutt & Garnett 2004). Infrared diagnostics (e.g. [Ne {\sc iii}]15.5μm/[Ne {\sc ii}]12.8μm) confirm this finding (Rigby & Rieke 2004). This has often given support to the idea that the upper mass cut-off of the stellar mass function is lowered, or that the IMF slope is steepened (Zhang *et al.* 2007), in metal-rich environments. The softening, however, can be explained without invoking changes in the mass function when hot star models in non-LTE and including line blanketing are used in population synthesis models to predict the evolution of the emission line ratios (Smith, Norris & Crowther 2002). The consideration of extended bursts of star formation instead of single-burst episodes also helps to discard anomalous IMFs in metal-rich starbursts (Fernandes *et al.* 2004).

Martin-Hernandez *et al.* (2002) attributed the decrease in the degree of ionization of neon with metallicity, observed in Galactic and extragalactic H {\sc ii} regions and in starbursts with the ISO satellite, to a metallicity effect on the spectral energy distribution of the ionizing stars. A similar conclusion has been drawn by Morisset *et al.* (2004) and Mokiem *et al.* (2004).

Rubin *et al.* (2007) have published one of the first tests between theoretical expectations and mid-IR observations obtained with the IRS spectrograph aboard the Spitzer space telescope. Since high-metallicity galaxies provide significant testbeds for theoretical stellar atmospheres, these authors have observed a number of H {\sc ii} regions across the disk of the metal-rich galaxy M83. When using different stellar atmosphere codes to produce nebular models, the best match with the observed emission line ratios [S {\sc iv}]10.5μm/[S {\sc iii}]18.7μm and [Ne {\sc iii}]15.6μm/[S {\sc iii}]18.7μm is obtained with WM-basic supergiant models. Interestingly, these authors, together with Simpson *et al.* (2004), point out that modern stellar atmosphere codes are still generally unable to reproduce the observed high values of the $\text{Ne}^{2+}/\text{O}^{2+}$ ratio at low $\text{O}^{2+}/\text{S}^{2+}$, known as the [Ne {\sc iii}] problem, and that was apparently considered solved in the mid-1990's with the first generation of non-LTE stellar atmosphere codes (Sellmaier *et al.* 1996). Further tests, in particular involving spectroscopic observations of both the ionizing source and the line-emitting gas of single-star H {\sc ii} regions compared to model nebulae, will help shed some light on the reliability of currently available model atmospheres of hot stars (Simón-Díaz *et al.* 2007).

5. Nebular vs. stellar abundances

For a long time measuring the chemical abundances of the young populations in spiral galaxies has been done mainly via the analysis of optical and IR emission lines of the ionized regions surrounding hot stars. The more direct approach of analyzing single stars in other spirals has been mostly limited to bright supergiants in nearby galaxies of the Local Group, namely M33 and M31. In the last few years high signal-to-noise spectra of single supergiant stars have been obtained with 8m-class telescopes at distances of up to

7 Mpc (Bresolin *et al.* 2001). At larger distances one can rely on the integrated spectrum of the ionizing clusters, for example in the UV, and deduce the chemical composition from the comparison with population synthesis models. One recent example of this 'integrated' approach is the IR spectroscopic study of the super-star cluster in NGC 6946 by Larsen *et al.* (2006). Using synthetic stellar spectra to reproduce the H- and K-band spectral features, which are dominated by red supergiants ($\sim$15 Myr of age), these authors have derived an oxygen abundance 12+log(O/H)=8.66 (i.e. solar), with an indication of an α/Fe element ratio enhanced by about 0.2 dex relative to solar. From the known nebular abundance gradient, I have obtained for the surrounding H II regions 12+log(O/H)=8.95, using strong-line methods calibrated in the high abundance scale, and 12+log(O/H)=8.55 using the low abundance scale (P method of Pilyugin). In NGC 1569 Larsen *et al.* (2008) have obtained a 0.25 dex discrepancy in oxygen content between stars and ionized gas.

Since both hot, massive stars and ionized gas sample the present-day chemical composition of galactic disks, we expect a good match between the results obtained from the two kinds of objects. The comparison can be complicated by a number of factors, including dust depletion in the gas, mixing at the stellar surface, and in general by the details of the chemical analysis that is done with different techniques and sets of spectral lines. The comparison is commonly restricted to the oxygen abundances, because this element can be measured relatively easily both in stars (B supergiants) and H II regions (typically the strongest metal emission lines). Thus far, comparing single-star chemical abundances with those in ionized nebulae has been carried out in a very restricted number of spiral galaxies. The best example in the northern hemisphere is arguably M33, while in the southern sky the best case is offered by NGC 300, a galaxy that is very similar to M33 in appearance, but located at approximately 2.5 times the distance ($\sim$2 Mpc).

5.1. *NGC 300*

A variety of massive star indicators have been studied in NGC 300, including the blue supergiants (Bresolin *et al.* 2002; Kudritzki *et al.* 2008), Cepheids (Gieren *et al.* 2005), W-R stars (Schild *et al.* 2003; Crowther *et al.* 2007) and SN remnants (Pannuti *et al.* 2000; Payne *et al.* 2004). The oxygen abundances of 6 B-type supergiants have been determined by Urbaneja *et al.* (2005a). The only nebular line emission information available until recently was based on spectra that were obtained with 4m-class telescopes (e.g. Deharveng *et al.* 1988), and in which the temperature-sensitive auroral lines remained undetected. The comparison between stellar and nebular abundances has therefore been complicated by the assumptions made about the calibration of the strong-line methods used to derive oxygen abundances for the H II regions. The situation has been recently improved by the author with the acquisition of deep optical spectra of H II regions in NGC 300 with the ESO Very Large Telescope. The faint [O III] λ4363 line has been measured in some 20 regions. The preliminary results obtained for the oxygen abundance distribution across the disk of NGC 300 agrees very well with the stellar abundances of Urbaneja *et al.* (2005a) and with the A-type supergiant metallicities measured by Kudritzki *et al.* (2008; see also Kudritzki's contribution to this volume), as shown in Fig. 1.

5.2. *M33*

Considerably more observational work is available for M33 than for NGC 300, especially concerning the ionized gas. Recent work has been carried out on the H II regions by Crockett *et al.* (2006) and Magrini *et al.* (2007b). The slope of the nebular oxygen abundance gradient agrees with the B supergiant value by Urbaneja *et al.* (2005b), but in this case a $\sim$0.2 dex offset exists (stellar value is higher). It is also worth noticing that (a) the beat Cepheids studied by Beaulieu *et al.* (2006) provide oxygen abundances that are somewhat intermediate between those found for the H II regions and the B supergiants; (b) the slope from the recent publication by Rosolowsky & Simon (2008) is about half the value from the works mentioned above; (c) the oxygen abundance determined by Esteban et al. (2002) from metal recombination lines for the giant H II region NGC 604 is about 0.2 dex higher than the value determined from collisionally excited lines, and is more in agreement with the B supergiant abundances. In this regard, it is important to note that in the Orion nebula, the three B dwarfs analyzed by Simón-Díaz et al. (2006) provide an oxygen abundance that agrees with the nebular value based on metal recombination lines by Esteban et al. (2004), if the depletion by dust, estimated at 0.08 dex by Esteban et al., is neglected.

6. Hot spots

In the nearby universe about one spiral galaxy out of five possesses circumnuclear rings of star-formation (Knapen 2005). These sites of massive star formation, known as 'hot spots', are important to understand the processes that lead to inflows of gas into the central regions of barred spiral galaxies, and the connection between bars and circumnuclear rings. Hot spots are ideal places to study massive star formation in

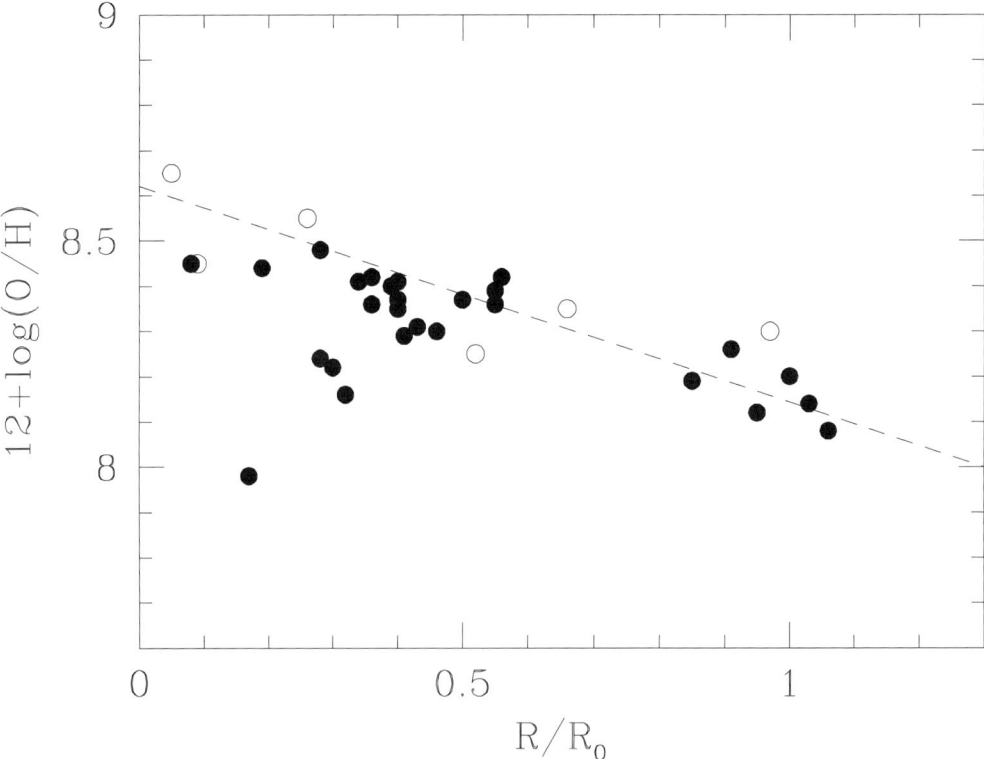

Figure 1. The radial oxygen abundance gradient in NGC 300 measured from the six B supergiants analyzed by Urbaneja et al. (2005a: open circles), the A-type supergiants studied by Kudritzki et al. (2008: only the fit to their data is shown here by the dashed line), and the H II regions observed by Bresolin et al. (2008, in prep., and this work: full circles). The radial coordinate is in units of the isophotal radius. The nebular abundances have all been determined from the strength of the [O III] $\lambda 4363$ auroral line.

high-metallicity environments, and to understand the effects of bars, merging events and interactions between galaxies on the star formation process.

In their study of NGC 7742 Mazzuca et al. (2006) have provided a remarkable example of the power of integral field spectroscopy in the detailed investigation of the physical conditions in the nuclear regions of spiral galaxies. Their maps of emission line ratios have shown that the main source of ionization in the ring of hot spots is photoionization, while shocks dominate just outside the ring. The metallicity deduced from the [N II]/[S II] diagnostic is approximately solar. A similar conclusion on the chemical abundance has been drawn by Sarzi et al. (2007) for a sample of eight spirals with circumnuclear star forming activity. Using instead a strong-line method, calibrated with a sample of extragalactic H II regions with auroral line detections, Diaz et al. (2007) have measured central abundances for three hot spot galaxies in the range 12+log(O/H)=8.60–8.85, i.e. between solar and 1.5× solar. These values are therefore not very high, and do not reach the 2-3× solar values that were quoted just a few years ago. N/O abundance ratios are found to lie between 2 and 4 times the solar value.

The study of the gas kinematics in the circumnuclear regions of spirals is also of particular interest. Evidence for non-circular motions of the ionized gas has been found in NGC 3351 by Hagele et al. (2007), consistent with an infall of gas at 25 km s^{-1}. A similar result has been obtained by Allard et al. (2006) in M100. These authors have also found evidence for the presence of azimuthal age gradients along the circumnuclear ring of this galaxy. The youngest massive stars are located at the contact points (resonances) between the ring and the dust lanes, where the gas inflow from the disk, under the action of the bar, intersects the circumnuclear ring.

The star formation history of the circumnuclear rings in eight different hot spot galaxies has been studied by Sarzi et al. (2007), based on the comparison of absorption line indices with population synthesis models. Their conclusion is that the star formation in the rings occurs in episodes extending over hundreds of Myr, thus excluding continuous star formation as well as single burst events. The circumnuclear rings are therefore rather stable features. The models by Sarzi et al. also allow for the build-up of the metallicity to approximately the solar value, as well as of the stellar mass in the central regions of spiral galaxies.

No hot spots with spectroscopic signatures from W-R stars have been detected by Sarzi et al. (2007) in their sample, perhaps as a consequence of the fact that the bulk of the massive stars are formed over extended periods, rather than in short bursts. However, Hattori et al. (poster, this volume) have presented evidence for a large number of W-R stars in the circumnuclear hot spots in the starburst galaxy NGC 7469. W-R spectroscopic features have been found by Bresolin & Kennicutt (2002) in a nuclear hot spot of M83. The metallicity measured from auroral lines by Bresolin et al. (2005) is approximately 1.6× solar, or 12+log(O/H)=8.94. The central region of M83 is interesting from other points of view. From the analysis of the velocity field Diaz et al. (2006) have found evidence for the presence of a hidden mass concentration ($\sim 1.6 \times 10^7$ M$_\odot$) near, but not coincident with, the optical nucleus of the galaxy. This mass lies at the end of a nuclear star-forming arc, whose size is compatible with the dynamical crossing time of the nuclear region. Along this arc the ages of the young clusters decrease towards the mass concentration. Diaz et al. propose that this hidden mass represent either bar-driven material funneled into the central regions of the galaxy, or the remnant of a recent merger event. We could therefore be witnessing an ongoing process that can eventually lead to the formation of a collapsed object and perhaps to the onset of AGN activity in a spiral galaxy at our galactic doorstep.

References

Allard, E. L., Knapen, J. H., Peletier, R. F., & Sarzi, M. 2006, *MNRAS*, 371, 1087
Bastian, N., Ercolano, B., Gieles, M., et al. 2007, *MNRAS*, 379, 1302
Beaulieu, J.-P., Buchler, J. R., Marquette, J.-B., et al. 2006, *ApJ*, 635, L101
Bresolin, F. 2005, in: E. Corbelli, F. Palla & H. Zinnecker (eds.), *The Initial Mass Function 50 years later*, (Dordrecht:Springer), 209
Bresolin, F. 2007, *ApJ*, 656, 186
Bresolin, F. 2008, in: G. Israelian & G. Meynet (eds.), *The Metal-Rich Universe*, (Cambridge: CUP), in press (astro-ph/0608410)
Bresolin, F. & Kennicutt, R. C. 2002, *ApJ*, 572, 838
Bresolin, F., Gieren, W., Kudritzki, R. P., et al. 2002, *ApJ*, 567, 277
Bresolin, F., Kennicutt, R. C., & Garnett, D. 2004, *ApJ*, 615, 228
Bresolin, F., Kudritzki, R. P., Mendez, R., & Przybilla, N. 2001, *ApJ*, 548, L159
Bresolin, F., Schaerer, D., Gonzalez Delgado, R. M., & Stasińska, G. 2005, *A&A*, 441, 981
Beckman, J. E., Rozas, M., Zurita, A., et al. 2000, *AJ*, 119, 2728
Bradley, T. R., Knapen, J. H., Beckman, J. E., & Folkes, S. L. 2006, *A&A*, 459, 13
Calzetti, D., Kennicutt, R. C., Engelbracht, C. W, et al. 2007, *ApJ*, 666, 870
Crockett, N. R., Garnett, D. R., Massey, P., & Jacoby, G. 2006, *ApJ*, 637, 741
Crowther, P. A. 2007, *ARA&A*, 45, 177
Crowther, P. A., Carpano, S., Hadfield, L. J., & Pollock, A. M. T. 2007, *A&A*, 469, L31
Crowther, P. A., Dessart, L., Hillier, D. J., et al. 2002, *A&A*, 392, 653
Davidge, T. J. 2007, *ApJ*, 664, 820
Deharveng, L., Caplan, J., Lequeux, J., et al. 1988, *A&AS*, 73, 407
Diaz, A. I., Terlevich, E., Castellanos, M., & Hagele, G. F. 2007, *MNRAS*, 382, 251
Diaz, R. J, Dottori, H., Aguero, M. P., et al. 2006, *ApJ*, 652, 1122
Dopita, M. A., Fischera, J., Sutherland, R. S., et al. 2006a, *ApJ*, 647, 244
Dopita, M. A., Fischera, J., Crowley, O., et al. 2006b, *ApJ*, 639, 788
Dors, O. L., & Copetti, M. V. F. 2003, *A&A*, 404, 969
Eldridge, J. J., & Vink, J. S. 2006, *A&A*, 452, 295
Elmegreen, B. G. 2006, *ApJ*, 648, 572
Elmegreen, B. G., Elmegreen, D. M., Chandar, R., et al. 2006, *ApJ*, 644, 897
Erb, D. K., Shapley, A. E., Pettini, M., et al. 2006, *ApJ*, 644, 813
Esteban, C., Peimbert, M., Garcia-Rojas, J., et al. 2004, *MNRAS*, 355, 229
Esteban, C., Peimbert, M., Torres-Peimbert, S., & Rodriguez, M. 2002, *ApJ*, 581, 241
Fernandes, I. F., de Carvalho, R., Contini, T., & Gal, R. R. 2004, *MNRAS*, 355, 728
Gieles, M., Larsen, S. S., Scheepmaker, R. A., et al. 2006, *A&A*, 446, 9
Gieren, W., Pietrzynski, G., Soszynski, I., et al. 2005, *ApJ*, 628, 695
Garcia-Rojas, J., & Esteban, C. 2007, *ApJ*, 670, 457
Gil de Paz, A., Madore, B. F., Boissier, S., et al. 2007, *ApJ*, 661, 115
Hadfield, L. J., Crowther, P. A., Schild, H., & Schmutz, W. 2005, *A&A*, 439, 265
Hagele, G. G., Diaz, A.I, Cardaci, M. V., et al. 2007, *MNRAS*, 378, 163
Hanish, D. J., Meurer, G. R., Ferguson, H. C., et al. 2006, *ApJ*, 649, 150
Kennicutt, R. C., Calzetti, D., Walter, F., et al. 2007, *ApJ*, 671, 333
Kennicutt, R. C., Edgar, B. K., & Hodge, P. W. 1989, *ApJ*, 337, 761
Kewley, L. J., & Dopita, M. 2002, *ApJS*, 142, 35
Knapen, J. H. 2005, *A&A*, 429, 141
Kudritzki, R. P., Urbaneja, M. A., Bresolin, F., et al. 2008, *ApJ*, in press
Larsen, S. S., Origlia, L., Brodie, J. P., & Gallagher, J. S. 2008, *MNRAS*, 383, 263
Larsen, S. S., Origlia, L., Brodie, J. P., & Gallagher, J. S. 2006, *MNRAS*, 368, 10
Larsen, S. S. 2002, *AJ*, 124, 1393
Lopez-Sanchez, A. R., Esteban, C., Garcia-Rojas, J., et al. 2007, *ApJ*, 656, 168
Magrini, L., Corbelli, E., & Galli, D. 2007a, *A&A*, 470, 843
Magrini, L., Vílchez, J. M., Mampaso, A., et al. 2007b, *A&A*, 470, 865
Martin-Hernandez, N. L., Vermeij, R., Tielens, A. G.G. M., et al. 2002, *A&A*, 389, 286

Mazzuca, L. M., Sarzi, M., Knapen, J. H., et al. 2006, ApJSS, 649, L79
Meurer, G. R., Hanish, D. J., Ferguson, H. C., et al. 2006, ApJSS, 165, 307
Mokiem, M. R., Martin-Hernandez, N. L., Lenorzer, A., et al. 2004, A&A, 419, 319
Morisset, C., Schaerer, D., Bouret, J.-C., & Martins, F. 2004, A&A, 415, 577
Oey, M. S., & Clarke, C. J. 1998, AJ, 115, 1543
Pannuti, T. G., Duric, N., Lacey, C. K., et al. 2000, ApJ, 544, 780
Payne, J. L., Filipović, M. D., Pannuti, T. G., et al. 2004, A&A, 425, 443
Peimbert, M., Peimbert, A., Esteban, C., et al. 2007, RMxAA, 29, 72
Pellerin, A. & Robert, C. 2007, MNRAS, 381, 228
Phillips, A. C. & Conti, P. S. 1992, ApJ, 395, 91
Pilyugin, L. S., Thuan, T. X., & Vílchez, J. M. 2007, MNRAS, 376, 353
Pilyugin, L. S., Thuan, T. X., & Vílchez, J. M. 2006, MNRAS, 367, 1139
Pilyugin, L. S. & Thuan, T. X. 2005, ApJ, 631, 231
Pindao, M., Schaerer, D., Gonzalez Delgado, R. M., & Stasińska, G. 2002, A&A, 394, 443
Prescott, M. K. M., Kennicutt, R. C., Bendo, G. J., et al. 2007, ApJ, 668, 182
Prieto, J. L., Stanek, K. Z., & Beacom, J. F. 2008, ApJ, 673, 999
Rigby, J. R. & Rieke, G. H. 2004, ApJ, 606, 237
Rosolowsky, E. & Simon, J. D. 2008, ApJ, 675, 1213
Rubin, R. H., Simpson, J. P., Colgan, S. W. J., et al. 2007, MNRAS, 377, 1407
Sarzi, M., Allard, E. L., Knapen, J. H., & Mazzuca, L. M. 2007, MNRAS, 380, 949
Schild, H., Crowther, P. A., Abbott, J. B., & Schmutz, W. 2003, A&A, 397, 859
Schmitt, H. R., Calzetti, D., Armus, L., et al. 2006, ApJ, 643, 173
Sellmeier, F. H., Yamamoto, T., Pauldrach, A. W. A., & Rubin, R. H. 1996, A&A, 305, 37
Simón-Díaz, S., & Stasińska, G. 2008, MNRAS, in press (arXiv:0805.1362)
Simón-Díaz, S., Stasińska, G., García-Rojas, J., et al. 2007, in: *Massive Stars: Fundamental Parameters and Circumstellar Interactions*, (arXiv:astro-ph/0702363)
Simón-Díaz, S., Herrero, A., Esteban, C., & Najarro, F. 2006, A&A, 448, 351
Simpson, J. P., Rubin, R. H., Colgan, S. W. J., et al. 2004, ApJ, 611, 338
Smith, L. J., Norris, R. P. F., & Crowther, P. A. 2002, MNRAS, 337, 1309
Stasińska, G. 2008, in: J., Cepa & F., Sanchez (eds.), *The emission line Universe*, (Cambridge: CUP), in press (arXiv:0704.0348)
Stasińska, G., Tenorio-Tagle, G., Rodriguez, M., & Henney, W. J. 2007, A&A, 471, 193
Stasińska, G. 2005, A&A, 434, 507
Thilker, D. A., Bianchi, L., Boissier, S., et al. 2005, ApJ, 619, L79
Urbaneja, M. A., Herrero, A., Bresolin, F., et al. 2005a, ApJ, 622, 862
Urbaneja, M. A., Herrero, A., Kudritzki, R.-P., et al. 2005b, ApJ, 635, 311
Vazquez, G. A., Leitherer, C., Schaerer, D., et al. 2007, ApJ, 663, 995
Zaritsky, D., & Christlein, D. 2007, AJ, 134, 135
Zhang, W., Kong, X., Li, C., et al. 2007, ApJ, 655, 851

Discussion

DOPITA: Given the disagreement between the various ways of determining the metallicity from H II regions and their comparison with stars, it might be timely to return to the method I used with D'Odorico and Benvenuti many years ago - to use evolved SNR. The new shock models are now much better, and the observation of individual SNR in nearby spiral galaxies is much easier than 20 years ago.

BRESOLIN: I agree, and once again NGC 300, where many SNR are known from recent observational work, can offer us a good starting point for this kind of investigation.

HERRERO: Fabio, I have a comment and a question. Comment: Esteban et al. do not derive the O content of the dust in Orion, but derive it by comparing the results by Cunha & Lambert (1994, ApJ, 426, 170) for stars in Orion with their own results for the gas phase. Our results for B stars in Orion agree with those of Esteban *et al.* for the gas phase. There is no need to assume any O depletion into dust. Question: You have shown some preliminary results from your recent work about abundances in H II regions in NGC 300. One of the H II regions is very close to the center of the galaxy and yet has an oxygen abundance that is extremely low. How accurate is that value? Is there something peculiar about that H II region or is it spectrum remarkable?

BRESOLIN: First, I need to remind you that the results I presented for NGC 300 are still very preliminary. In the coming months I will work to produce more definitive values for the chemical abundances in the H II regions I studied. I can expect some variations in some of the individual values once the proper analysis is completed, although I do not think that the general trend and the good agreement with the stars will change. Second, I think that it is perhaps a little naive to expect that the chemical composition of a single H II region or star always compares very well with that expected from its galactocentric position and the radial abundance gradient. The intrinsic scatter in abundance can also be rather large, and some peculiarites of a few individual H II regions and/or stars can also be expected.

Fabio Bresolin (left), Artemio Herrero (center) and Sergio Simón-Díaz (far right).

Massive Stars as Cosmic Engines
Proceedings IAU Symposium No. 250, 2007
F. Bresolin, P.A. Crowther & J. Puls, eds.

© 2008 International Astronomical Union
doi:10.1017/S1743921308020607

UCHII Regions and Newly Born O-type Stars

Peter S. Conti[1], Jeonghee Rho[2], James Furness[3] and Paul A. Crowther[3]

[1]JILA, University of Colorado, Boulder CO 80309-0440
e-mail:pconti@jila.colorado.edu

[2]Spitzer Science Center, CalTech, Pasadena CA 91125
email: rho@ipac.caltech.edu

[3]Department of Physics and Astronomy, University of Sheffield, S3 7RH,UK
email: J.Furness@sheffield.ac.uk, Paul.Crowther@sheffield.ac.uk

Abstract. We have obtained *Spitzer* IRAC and MIPS mid-IR images of a sample of 43 radio selected UCHII region sources to ascertain (*a*) whether the newly born O stars within are found with other stars in their birthplaces and (*b*) the nature of the surroundings. 37 of the sources appear to be in small clusters, and 33 are found in connection with other hot star formation activity. Thus, for the most part, O stars are not born in isolation. Here we give examples of the mid-IR images of the various types of UCHII regions.

Keywords. stars: formation – ISM: HII regions – infrared: ISM

1. Introduction

UCHII regions are dense, highly compact (< 0.1 pc) radio sources of line and continuum hydrogen emission surrounding newly born O-type stars. All newly born O stars go through this phase, which results from the ionization of the hydrogen in the dense natal cocoon remaining immediately after the birth processes. The inner cocoon is composed of ionized hydrogen, while further out neutral and molecular compounds, along with extensive dust, are present. This dust cocoon is roughly ten times the size of the UCHII region and its extensive extinction renders direct detection of the central exciting star(s) difficult at wavelengths shorter than the mid-IR (MIR).

Eventually, the cocoon material is dissipated by the central star's very high luminosity, its winds, and prodigious Lyman continuum radiation. Then the central object is revealed to our view. It might be appropriate here to note that the nearby and well known Orion Trapezium Cluster is being observed *long after* its UCHII region phase.

While our eventual objective is to determine the nature of the central exciting star(s) and relate them to the evolution and interaction with the natal material, here we wish to consider the following simpler questions:

Are the central exciting sources single or multiple?

Is the formation of O-type stars isolated in space or is there evidence of other nearby activity, or other hot luminous stars (O-type or early-B)?

2. Target acquisition

We have obtained *Spitzer* MIR images of 43 relatively isolated UCHII regions in our Galaxy, extracted from the catalogs of Wood & Churchwell (1989), Kurtz, *et al.* (1994),

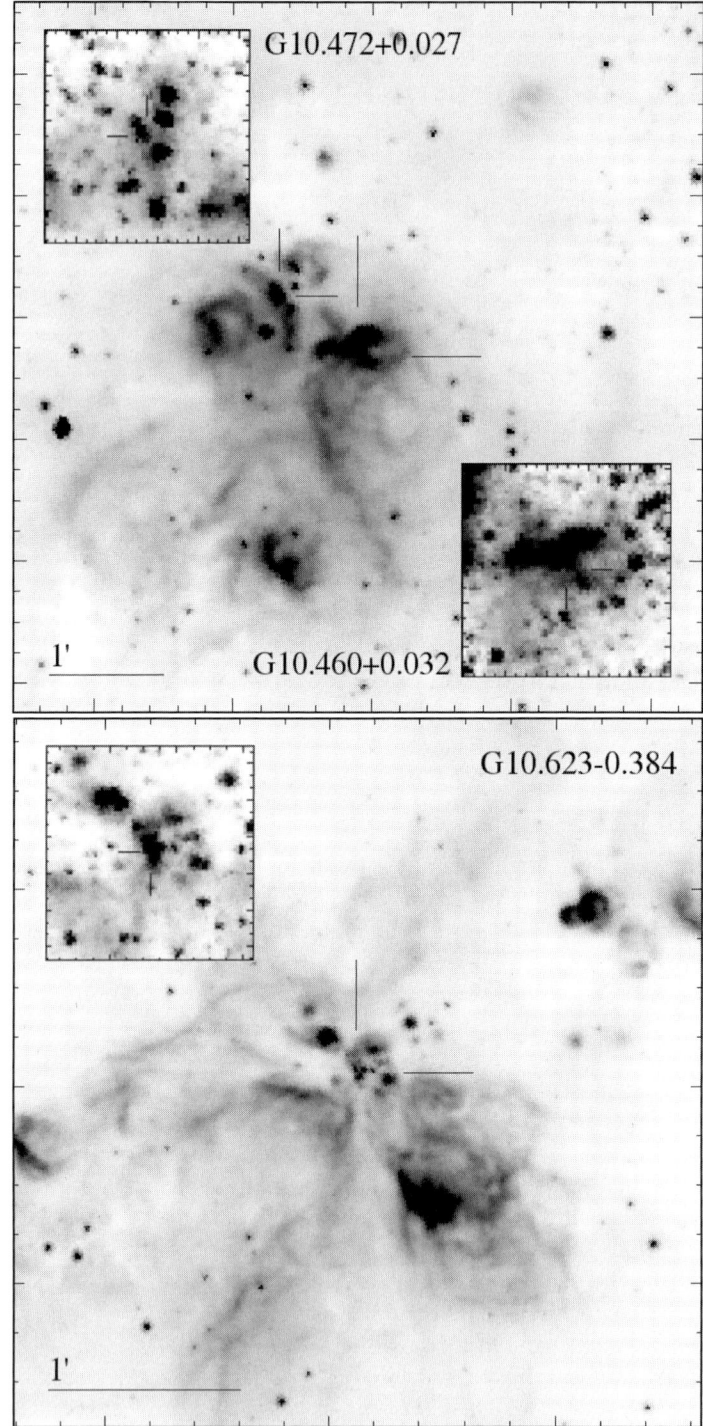

Figure 1. *Spitzer*/IRAC 8μm images of G10.460+0.032 and G10.472+0.027 (top) and G10.623-0.384 (bottom), with a scale of 10×10 pc for their adopted distance. Inset within these are 4.5μm images of the central 2×2 pc region presented at a higher spatial scale.

and Crowther & Conti (2003). The first two of these papers were radio surveys along the Galactic plane specifically searching for UCHII regions. Some 200 were found, nearly all of them not associated with giant HII regions which harbor many O-type stars. The latter paper presented MSX satellite MIR imaging data of 53 of these UCHII region sources. Here we consider 43 of the MSX list, utilizing the higher spatial resolution now available with the *Spitzer* instrument.

Generally speaking, UCHII regions are readily detected by radio surveys in uncrowded fields. They do exist in HII and GHII regions and clusters but they are difficult to identify there unless specific detailed searches have been carried out. For example, dozens are known in the GHII region W49A (de Pree *et al.* 1997). In our sample, only G49.490-0.370 is found in a GHII region, namely within IRS2 of W51 (see also poster by Barbosa *et al.* in this volume).

The *Spitzer* IRAC data we present here is at wavelengths of $4.5\mu m$ and $8\mu m$, obtained at a spatial resolution of ~ 2 arcsec. We also have $24\mu m$ MIPS images of the same sources but they are not considered here given the limited discussion we can carry out in a short presentation such as this one. A more detailed and lengthy paper is being generated for publication which will consider all our data.

3. Examples of *Spitzer* imaging

We will present representative data for four UCHII regions, illustrating the wide divergence of MIR spatial morphologies. These figures are uniformly scaled to 10×10 pc at $8\mu m$ with an insert of 2×2 pc at $4.5\mu m$. The former can be used to identify star formation activity near to and associated with the UCHII region as indicated by dust emission. The shorter wavelength samples the stellar content and is used to investigate possible multiplicity of the exciting object.

The distances of the UCHII regions are derived from the radio kinematic method whereby the radial velocities are used to infer that number from a Galactic rotation model and an assumption of purely circular galacto-centric motion. Our distance determination is described in Furness *et al.* (2008) It needs to be kept in mind that while these values are, so far, the only ones available, there could be problems with the near side/far side ambiguity for objects within the Solar circle. Additionally, if random motions are present, the derived distances could well be in error. Extensive studies of GHII regions using their detected and classified O-type stars, and spectroscopic parallaxes to obtain distance estimates, have frequently led to numbers which are discrepant with the published radio kinematic distances (e.g., Damineli *et al.* 2005).

3.1. *G10.460+0.032 and G10.472+0.027*

In the upper panel of Figure 1, we show a close pair of UCHII regions, each one within a somewhat larger region of hot star activity as evidenced by the extensive dust emission. They have similar kinematic distances. These two active regions are separated by 2 pc (in projection) and appear to be spatially connected. Another active region is found to the south at 5 pc, which may also be associated with the pair.

The 2×2 pc insert clearly shows the multiplicity of stellar images surrounding each UCHII region strongly suggesting that each resides in a small cluster of size about a parsec. We infer from these data that both UCHII regions are in clusters and other star birth activity is found nearby.

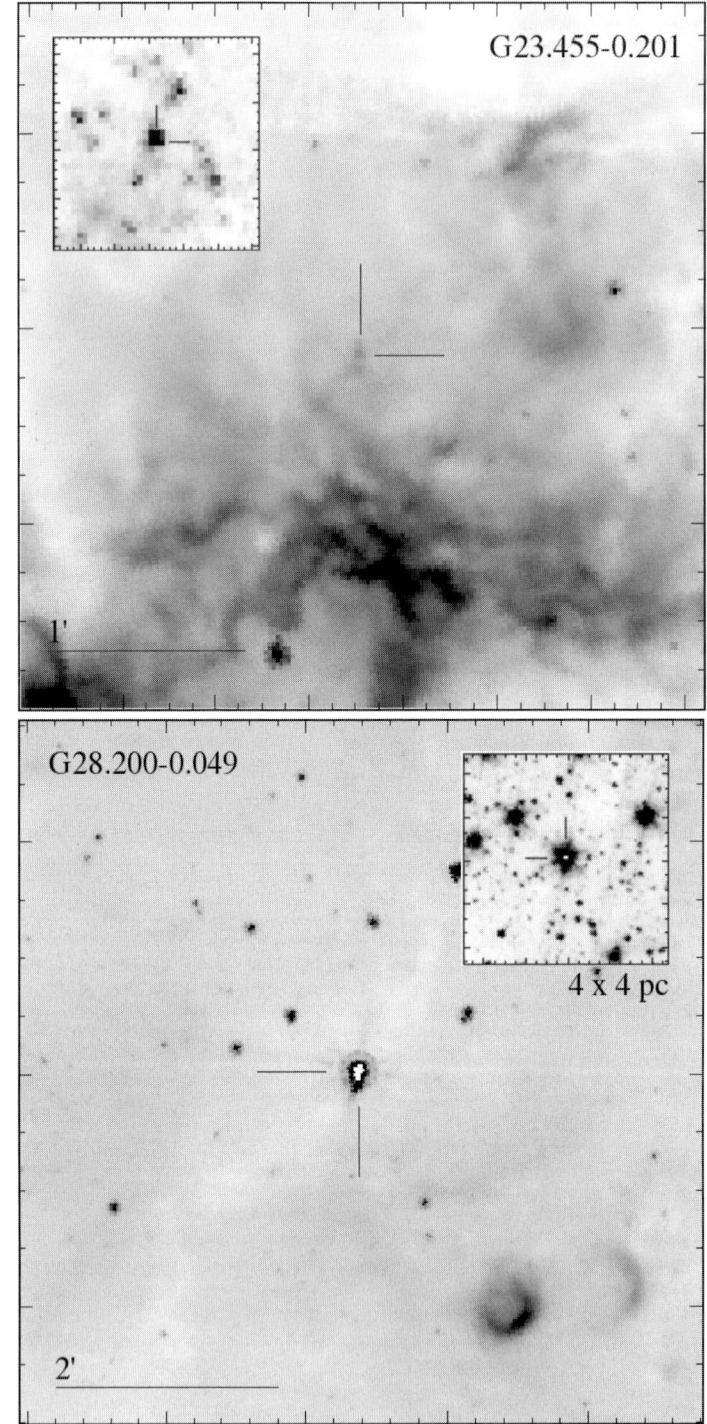

Figure 2. *Spitzer*/IRAC 8µm images of G23.455-0.201 (top) and G28.200-0.049 (bottom), with a scale of 10×10 pc for their adopted distance. Inset within these are 4.5µm images of the central 2×2 pc region (4×4 pc for G28.200-0.049) presented at a higher spatial scale.

3.2. G10.623-0.384

In the lower panel of Figure 1, we present an image of an UCHII region in the midst of a cluster of stars, along with a much stronger active region (G10.598-0.383) a few pc to the SW. While we do not have kinematic information concerning this object it appears to be enveloped in the same dusty material surrounding the UCHII region. We thus have evidence for a cluster and for nearby activity.

3.3. G23.455-0.201

As indicated by the upper panel of Figure 2, this UCHII region sits on the very outskirts of a very strong IR source, G23.437-0.209, which dominates the dust emission. Kim & Koo (2001) provide low resolution 21-cm radio data suggesting this dust source is also a strong HII emitter. The UCHII region appears unresolved at 4.5μm and there is no evidence for a cluster surrounding this object. Without kinematic information for the strong dust source we cannot say if they are connected. This object might be considered to be isolated.

3.4. G28.200-0.049

The lower panel of Figure 2 shows very strong IR source with a fainter very red companion less than a parsec to the south, but there is no evidence for any cluster nearby. The UCHII region also appears isolated, one of the very few in our survey.

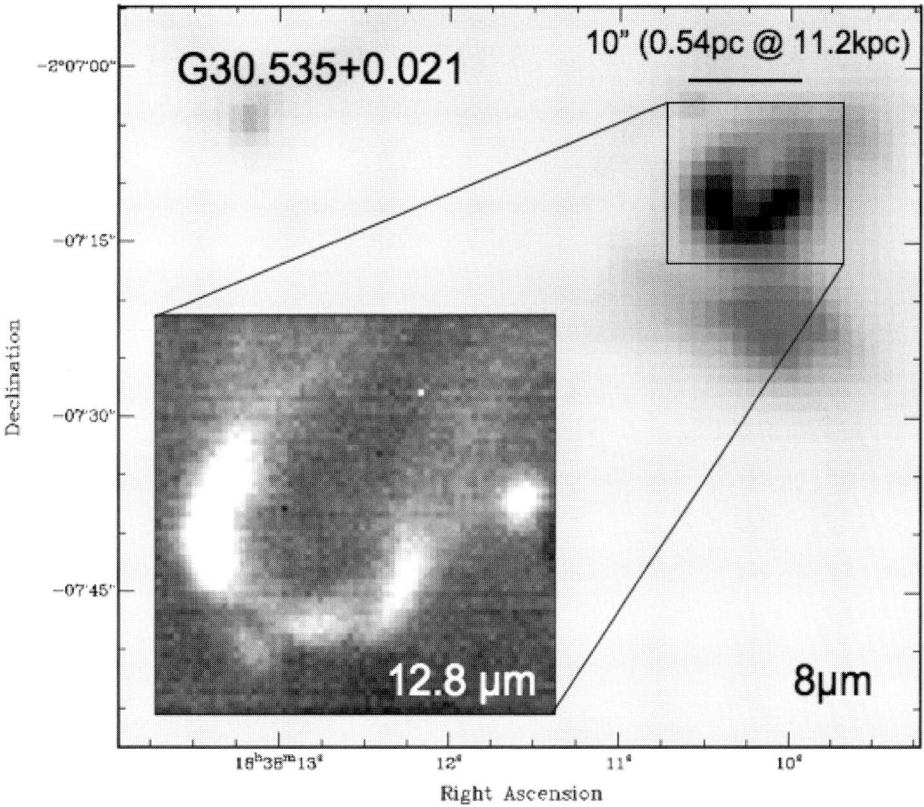

Figure 3. *Spitzer*/IRAC 8μm image of G30.535+0.021 (3.6 × 3.6 pc at 11.2 kpc) together with a higher spatial resolution VLT/VISIR image at 12.8μm (inset).

Source	d(kpc)	Cluster?	Environment
G10.460+0.032	5.8	yes	similar nearby
G10.472+0.027	5.8	yes	similar nearby
G10.623-0.384	17.5	yes	others nearby
G23.455-0.201	9.0	no	stronger nearby
G28.200-0.049	3.4	no	isolated

Table 1. Summary of properties for UCHII regions discussed here

4. Results

In Table 1 we compare the properties of the UCHII regions considered here. Turning now to the entire sample of 43 sources, of which we have shown only four, we find that 38 of them (86%) have evidence of being connected to a small cluster of stars. This conclusion comes from careful examination of the 4.5μm wavelength images similar to those shown in our figures here. Of the other 5 sources, a couple of them appear to be at large distances such that duplicity might remain undetected. Examination of the environments of the 43 sources, using the 8μm wavelengths, indicates that 32 (72%) appear to be connected to other nearby star formation activity. Star formation activity thus appears not to be isolated in most cases.

The very small fraction of UCHII region O-type stars that appear to be truly single is roughly consistent with recent extensive star counts of isolated (4%) field stars (e.g., Parker & Goodwin (2007) and references therein). Also, from the examinations of the surroundings of UCHII regions, it would appear that most of them have nearby, although not extensive, star formation activity.

Of course, the $\sim$2 arcsec spatial resolution of IRAC at 8μm corresponds to a physical scale of 0.1 pc at a distance of 10 kpc, which is comparable to the size of UCHII regions, so one would need to observe these regions at much greater spatial resolution to directly probe within the ionized region. Fortunately, efficient MIR imagers are now available at current 8–10m ground-based telescopes, such as MICHELLE at Gemini-North and VISIR at the ESO Very Large Telescope. One such example is presented in Fig. 3, showing a 12.8μm VLT/VISIR image of G30.535+0.021 obtained in July 2007 at a spatial resolution of $\sim$0.4 arcsec, corresponding to 0.02 pc at the 11.2 kpc distance of G30.535+0.021, sampling the warm dust within the UCHII region. The dust spatial morphology is reminiscent of 2–6cm radio continuum contours from Wood & Churchwell (1989) obtained at a similar resolution.

PSC would like to thank JPL and the NSF for support in the investigations reported here. VLT/ISAAC observations were obtained at ESO Paranal Observatory under programme 079.C-0093(A).

References

Crowther, P. A. & Conti, P. S. 2003, *MNRAS*, 343, 143
Damineli, A., Blum, R. D., Figueredo, E., & Conti, P. S. 2005, in: R. Cesaroni, M. Felli, E. Churchwell, & M. Walmsley, (eds.), *Massive Star Birth: A Crossroads of Astrophysics* (Cambridge: CUP), *Proc IAU Symp. 227*, 407
de Pree, C. G., Mehringer, D. M. & Goss, W. M. 1997, *ApJ*, 482, 307
Furness, J., Crowther, P. A., Conti, P. S., Rho, J. & Goodwin, S. P. 2008, *MNRAS*, submitted
Kim, K.-T. & Koo, B.-C. 2001, *ApJ*, 549, 979
Kurtz, S., Churchwell, E., & Wood, D. O. S. 1994 *ApJS*, 91, 651
Parker, R. J. & Goodwin, S. P. 2007, *MNRAS*, 380, 1271
Wood, J. J. & Churchwell, E. 1989, *ApJS*, 69, 831

Discussion

ZINNECKER: Your last conclusion that massive stars are not born in isolation is important, but probably premature in view of your limited sample. If we find one truly isolated massive star (not a runaway) it may be more significant for the understanding of massive star formation than discovering another 10 embedded clusters of massive stars. Larger unbiased surveys, perhaps in the LMC, are needed to settle this issue (cf. Zinnecker & Yorke 2007 *ARA&A* 45, 481, their section 2.3).

CONTI: The sample of 43 objects is nearly a fifth of the known radio selected objects outside of GHII regions.

CROWTHER: In addition to MIR Spitzer data, we can also use the ratio of the FIR to radio emission to enable the stellar content of UCHII regions to be determined. In general, most sources are consistent with star clusters (Kroupa) of which some appear isolated. This does not necessarily mean they are isolated but perhaps associated only with a low mass star cluster (Furness *et al.* 2008; Parker & Goodwin 2007).

OEY: To clarify: you said that 38 of the 43 UCHII regions show small clusters. Does that mean that 5 are isolated?

CONTI: The number of non-isolated UCHII was 32 out of 43, thus the word "many". Those in detected clusters are 37 of 43. Some resident clusters might have too small a diameter to be detected with the Spitzer instruments. These might show up in the improved ground based MIR data I presented at the very end.

OEY: Also, to follow Zinnecker's point, my student, Joel Lamb and I are indeed studying the field of O stars in the SMC to constrain the numbers of O stars found in isolation.

SKINNER: In the high resolution 12μm VLT images that you showed, can you speculate on what the faint emission extending away from the bright central object might be? Stellar or non-stellar?

CONTI: I would imagine it is locally heated dust. The scale size shown was something like 0.01 pc, much more spatially detailed than the Spitzer data.

Peter Conti.

Takashi Hattori discusses with Mike Dopita.

Lindsey Smith.

Binary Populations and Stellar Dynamics in Young Clusters

D. Vanbeveren[1,2], H. Belkus[1], J. Van Bever[3] and N. Mennekens[1]

[1]Astrophysical Institute, Vrije Universiteit Brussel, Brussels, Belgium
email: dvbevere@vub.ac.be, hbelkus@vub.ac.be

[2]GroupT, Association K.U. Leuven, Louvain, Belgium
email: dany.vanbeveren@groept.be

[3]Institute of Computational Astrophysics, St.-Mary's University, Halifax, Canada
email: vanbever@penguin.stmarys.ca

Abstract. We first summarize work that has been done on the effects of binaries on theoretical population synthesis of stars and stellar phenomena. Next, we highlight the influence of stellar dynamics in young clusters by discussing a few candidate UFOs (unconventionally formed objects) like intermediate mass black holes, η Car, ζ Pup, γ^2 Velorum and WR 140.

Keywords. binaries: close – Galaxy: kinematics and dynamics

1. Introduction

Many (most?) of the massive stars are formed in clusters and the observations reveal that many cluster members are binary components (to illustrate, see Sana *et al.*, 2008). The effect of binaries on population studies has been discussed many times by different research groups (the Brussels group was one of them) but the following points may be worth repeating:

• binaries with initial primary mass larger than 40 $M_\odot$ and with a period that is large enough so that LBV mass loss happens before the Roche lobe overflow (RLOF) would start, avoid RLOF (the LBV scenario as it was introduced in Vanbeveren, 1991); this may not be true for case A binaries

• most of the primaries of binaries with orbital period as large as 10 yrs and with initial mass smaller than 40 $M_\odot$ lose most of their hydrogen rich layers by RLOF; the evolution of these stars is therefore quite different from the evolution of single stars with the same mass

• the RLOF in late case B and case C binaries may be accompanied by the common envelope process that may result in the merger of the two components; it can be expected that this merger will be a rapid rotator with peculiar chemical abundances

• the RLOF in Case A and early case B binaries implies mass and angular momentum transfer towards the gainer; together with the merger process discussed above, this process is a natural way to make rapid rotators and therefore massive close binaries may be natural progenitors of long gamma ray bursts (Cantiello *et al.*, 2007)

• the favored model for short gamma ray bursts is the binary merger of two relativistic compact stars and it cannot be excluded that such mergers are important sites of galactic r-process enrichment (De Donder & Vanbeveren 2003a).

• the observed massive binary frequency is not necessarily the binary frequency at birth. This is due to the fact that a significant fraction of all massive binaries become single stars during their evolution (many binaries with small mass ratio merge, late case

B and case C binaries evolve through a common envelope phase and may merge as well, the supernova explosion of one of the components disrupts the binary), e.g. many single stars may have a binary past with an evolution which is distinctly different from stars that are born as single stars.

The proceedings of the conference on *Massive Stars in Interacting Binaries* (Moffat & St-Louis 2007) were published a few months ago. They can be considered as an update of the monograph *The Brightest Binaries* (Vanbeveren *et al.* 1998c) and we recommend careful reading to anyone with an open mind willing to admit that many theoretical population studies, where the effects of binaries are ignored, have mainly an academic value. In section 2 we list a few extra references that may be interesting in order to learn more about the effects of massive binaries on population studies.

Many massive objects (single or binary) are born in dense (embedded) clusters, containing a few 10s up to a few 1000s massive stars and even more (e.g., the clusters in the Carina association, the Arches and Quintuplet clusters in the galactic bulge, the Orion nebula cluster, R136 in the LMC, MGG-11 in M82, etc.). In these clusters the evolution of each object may be affected by the presence of all the others, or at least by the closest neighbors. In sections 3 and 4 we discuss a few peculiar massive objects that may be the result of the dynamical evolution of stars in dense clusters, we like to call them Unconventionally Formed Objects (UFOs).

2. The effects of binaries on theoretical massive star population studies

- *The WR population*: in the sixties and seventies, before it was realized that LBV and/or RSG stellar wind mass loss rates are large enough to remove most of the hydrogen rich layers of a massive star, it was generally accepted that most of the WR stars are formed through binary mass loss processes. It may sound as a surprise, but the study of Vanbeveren & Conti (1980) was probably among the first where convincing arguments were presented that no more than 40-50% of all galactic WR stars are binary members, indicating that massive single stars have to evolve into WR stars as well. This binary percentage still holds today. The effect of binaries on the WR population was described in detail in Vanbeveren *et al.* (1998a, b, c), Vanbeveren *et al.* (2007) (see also Eldridge in the present volume). Since all primaries of interacting binaries become hydrogen deficient stars at the beginning of core helium burning, all these primaries become WR-like stars. Therefore, a crutial parameter that affects theoretical WR-binary population studies is the minimum mass a binary component should have so that the WR-like star will be observed as a real WR star.
- *The supernova rates*: the effect of binaries on supernova rates (types Ia and Ib/c) has been investigated in much detail by Tutukov *et al.* (1992), Belczynski *et al.* (2002), also in the framework of a galactic evolutionary model (De Donder & Vanbeveren 2003b, 2004). More recent studies confirm this earlier work, although proper referencing is sometimes missing.
- *The Be stars*: many Be stars are the optical component of Be-X ray binaries, which may indicate that they became Be stars via the binary evolutionary channel (the binary mass transfer process). Accounting for the fact that the supernova explosion of a binary component disrupts most of the binaries, the large number of Be-X ray binaries may indicate that there are a much larger number of Be single stars which have a similar evolutionary past as the Be stars in the X-ray binaries. More information on how binaries affect the Be-star population is given in Pols *et al.* (1991), Pols & Marinus (1994), Van

Bever & Vanbeveren (1997). In the latter paper it was argued that the rich population of Be stars in some clusters can not be explained by binary evolution alone.

- *Binaries and the spectral synthesis of massive starbursts*: Belkus *et al.* (2003) and Van Bever and Vanbeveren (2000, 2003) studied the effect of binaries on the spectral synthesis of massive starbursts. Note that binaries play a very important role when the starburst is at least 4-5 Myr old.

- *Binaries and galactic chemical evolution*: many stars are formed in binaries and it therefore looks quite strange that most of the galactic chemical evolutionary studies account for single star chemical yields only. In Brussels we have a project where the effect of binaries on galactic evolution is studied in detail. Results have been published in an extended review by De Donder & Vanbeveren (2004) and in references quoted therein. Notice that as far as the C, N, O, ... Fe elements and r-process elements are concerned, binary yields are very different from single star yields and the effect on a galactic model is at least as large as the effect of fast rotation. But including binaries in a galactic code is not something as simple as replacing one set of yields by another.

3. Very massive stars and intermediate mass black holes: UFOs of the first kind

The X-ray luminosity in ultra-luminous X-ray sources (ULXs) may be as high as 10^{42} erg s^{-1} (Ptak & Colbert, 2004). Many (most) of these ULXs are found in young dense star clusters (e.g., the ULX in the cluster MGG-11 in M82), preferentially outside the cluster core. To understand these high luminosities by means of sub-Eddington mass accretion onto a relativistic object, one has to accept that a black hole (BH) component is present with a mass as high as 1000 $M_\odot$ (the term intermediate mass BH or IMBH is used). However, note that there exists an alternative model where the high luminosities are explained by supra-Eddington accretion onto a BH with a much smaller mass of 50-100 $M_\odot$ (Soria, 2007 and references therein).

The dynamical evolution of dense massive clusters (like MGG-11) obviously depends on the initial star density, but, most interestingly, is characterized by the processes of mass segregation and core collapse, which happen on a timescale similar to the evolutionary timescale of the most massive stars in the cluster. Core collapse is accompanied by real physical stellar collisions between the most massive stars and the possible formation of an intermediate mass object with a mass of 1000 $M_\odot$ or larger. It is therefore tempting to link the core collapse process and the existence of ULXs. This has been investigated in many papers by the team working with the direct N-body STARLAB software (e.g., Portegies Zwart *et al.*, 2007, and references therein) and working with the statistical MONTE-CARLO software (e.g., Fregeau & Rasio, 2007, and references therein).

The link with ULXs has two major uncertainties: core collapse is a fact and the successive collision of the most massive stars in the core (the term runaway merger is used) is a fact as well, but, when this merger process stops, it is at present unclear what happens with that merger object? Does it form a star with a mass similar to the mass of the merger object, and, when a very massive star is formed, how does it evolve? It is clear that when it forms, its evolution will be critically affected by stellar wind mass loss during core hydrogen burning (CHB) and during core helium burning (CHeB).

Belkus *et al.* (2007) investigated the evolution of very massive stars with an initial mass up to 1000 $M_\odot$, with a metallicity Z between 0.001 and 0.04, using the CHB stellar wind mass loss rate formalism of Kudritzki (2002) and the CHeB wind formalism proposed by Vanbeveren *et al.* (1998a, b, c), which is very similar to the one proposed by Nugis &

Lamers (2000). Belkus *et al.* concluded that very massive stars with solar or super solar metallicity end their life as a BH with a mass not larger than 70-80 $M_\odot$. Yungelson *et al.* (2008) recently published their study on the evolution of very massive stars with solar metallicity. Although they use 'ad hoc' (the term used by the authors) stellar wind mass loss rate formalisms, they essentially arrive at the same conclusion as Belkus *et al.*

Belkus *et al.* presented an easy evolutionary recipe for very massive stars. The latter is combined with our own direct N-body code and with our massive star evolution handler to simulate the early evolution of MGG-11. We adopt a starburst model with 3000 massive stars with solar type metallicity and a King density distribution with a half mass radius of 0.5 pc (more details are given in Vanbeveren *et al.*, 2008 and in Belkus *et al.*, 2008). The main conclusion is that stellar wind mass loss during CHB does not prevent the formation of a very massive object, but when the merger process stops and when this object becomes a very massive star, stellar wind mass loss determines its further evolution and an IMBH is not formed. This argues against the sub-Eddington model of the ULX in MGG-11.

4. UFOs of the second kind: why not η Car, γ^2 Velorum, ζ Pup, WR 140?

When a (small or large) number of massive objects (single stars or binaries) form together in a cluster with small radius, the evolution of each member may be affected by the presence of the others. The dynamical evolution of such clusters implies close encounters leading to physical collisions and stellar mergers. The collision process of two massive stars (two 88 $M_\odot$ stars and a 88 $M_\odot$ star with a 28 $M_\odot$ star) has been investigated by Suzuki *et al.* (2007) by using SPH and the following conclusions are striking:

- the merging lasts a few days and during the merging process 10-12 $M_\odot$ are lost
- the merger product is a mixed star where the degree of homogenization is larger in case both stars have similar masses.

η Car: it is tempting to link the collision process discussed above and the loss of about 10 $M_\odot$ during the 19th century event of this object. If the η Car progenitor experienced a dynamical encounter with a binary or with a single star that resulted in a stellar merger, the sudden very large mass loss and the very large eccentricity of the binary at present can be explained in a natural way.

γ^2 Velorum: The formation of WR+OB binaries in young dense stellar systems may be quite different from the conventional binary evolutionary scenario. Mass segregation in dense clusters happens on a timescale of one or a few million years which is comparable to the evolutionary timescale of a massive star. Within the lifetime of a massive star, close encounters may therefore happen very frequently. When we observe a WR+OB binary in a dense cluster of stars, its progenitor evolution may be very hard to predict. Our simulations predict the following UFO-scenario of WR+OB binaries. After 4 million years the first WR stars are formed, either single or binary. Due to mass segregation, this happens most likely when the star is in the starburst core. Dynamical interaction with another massive object becomes probable, especially when the other object is a binary. We encountered a situation where a WR star (a single WC-type with a mass = 10 Mo) interacts with a 16 $M_\odot$ + 14 $M_\odot$ circularized binary with a period P = 6 days. We explored the result of such an encounter by using the FEWBODY software of Fregeau *et al.* (2004). The details of all these simulations will be discussed elsewhere (Belkus *et al.*, 2008). One of the results is the following: the two binary components merge and

the 30 M$_\odot$ merger forms a binary with the WC star with a period of 80 days and an eccentricity e = 0.3. This binary very well resembles the WR+OB binary γ^2-Velorum but it is clear that conventional binary evolution has not played any role in its formation.

ζ Pup: ζ Pup, λ Cep and BD+43°3654 are 3 massive runaways with a runaway velocity between 40 km s^{-1} and 70 km s^{-1}. Their location in the HR diagram suggests that they belong to the most massive star sample of the solar neighborhood (Vanbeveren *et al.*, 1998b, c; Hoogerwerf *et al.*, 2001; Comeron & Pasquali, 2007). Runaways can be formed by the binary-SN scenario (Blaauw, 1961), where the original massive primary (the mass loser when the Roche lobe overflow process happens) explodes and eventually disrupts the binary, leaving a neutron star remnant and a runaway secondary (the mass gainer when the Roche lobe overflow happens). Such a scenario for ζ Pup was presented by Vanbeveren *et al.* (1998b, c). To explain the significant surface helium enrichment of the star, its rapid rotation and its runaway velocity (= 70 km s^{-1}), the mass transfer phase and the accretion process must be accompanied by spinning-up and quasi-homogenization of the mass gainer (the full mixing model as it was introduced by Vanbeveren and De Loore, 1994) whereas the overall evolution should have resulted in a pre-SN binary with a period of the order of 4 days. The latter requires some fine-tuning.

To illustrate that the dynamical ejection mechanism is a very valuable alternative, the FEWBODY software of Fregeau *et al.* (2004) was used to reproduce the observed properties of ζ Pup. We performed over 1 million single star-binary and binary-binary scattering experiments. The details of these experiments will also be given in Belkus *et al.* (2008). We explored the effects of different masses and different binary periods and eccentricities and, obviously, many experiments reproduce ζ Pup, but to obtain a runaway velocity as observed, the binaries participating in the scattering process always have to be very close (periods smaller than 100 days). Most interestingly, in all our experiments, *ζ Pup turns out to be a merger of 2 or 3 stars*.

WR 140: this WC5 + O4-5 binary has a period of 7.9 yrs, an eccentricity e = 0.85 and minimum component masses = 23 + 62 M$_\odot$. The O4-5 star is much younger than the WC5 star. Accounting for the very large period of the binary and the very large masses of both components, a typical binary rejuvenation process during RLOF seems unlikely. The only alternative then is a dynamical process, e.g. single-binary or binary-binary. Interestingly, a dynamical process explains the eccentricity in a natural way and it is tempting to link WR 140 and η Car.

5. An experiment

About 10% of all the O-type stars in the solar neighborhood may be runaway stars with a peculiar space velocity > 30-40 km s^{-1} (Gies & Bolton, 1986). Runaways can be formed by the binary-SN scenario (Blaauw, 1961) or by two-body dynamical interactions in star clusters where at least one body is a binary (Leonard & Duncan, 1990). It is clear that the number of O-type runaways predicted by both processes depends on the adopted binary frequency. We performed the following experiment:

calculate the O-type runaway frequency as function of the primordial binary cluster frequency accounting for cluster stellar dynamics and for the SN explosion in binaries, using typical initial parameters of solar neighborhood type clusters (like the Orion Nebula Cluster)

Preliminary results indicate that a 10% O-type runaway frequency can be obtained only by a model where the primordial (interacting) massive binary frequency in the solar neighborhood is larger than 60%.

References

Belczynski, K., Kalogera, & V., Bulik, T., 2002, *ApJ*, 572, 407
Belkus, H., Van Bever, J., & Vanbeveren, D., 2007, *ApJ*, 659, 1576
Belkus, H., Van Bever, J., & Vanbeveren, D., 2008, in preparation
Belkus, H., Van Bever, J., Vanbeveren, D., & Van Rensbergen, W., 2003, *A&A*, 400, 429
Blaauw, A., 1961, *Bull. Astron. Inst. Netherlands*, 15, 265
Cantiello, M., Yoon, S.-C., Langer, N., & Livio, M. 2007, *A&A*, 465, 29
Comeron, F. & Pasquali, A., 2007, *A&A*, 467, L23
De Donder, E. & Vanbeveren, D., 2003a, *NewA*, 9, 1
De Donder, E. & Vanbeveren, D., 2003b, *NewA*, 8, 817
De Donder, E. & Vanbeveren, D., 2004, *NewA Rev*, 48, 864
Fregeau, J. M., Cheung, P., Portegies Zwart, & S. F., Rasio, F. A., 2004, *MNRAS*, 352, 1
Fregeau, J. M. & Rasio, F. A., 2007, *ApJ*, 658, 1047
Gies, D. R. & Bolton, C. T., 1986, *ApJS*, 61, 419
Hoogerwerf, R., de Bruijne, J. H.J., & de Zeeuw, P. T., 2001, *A&A*, 365, 49
Kudritzki, R. P., 2002, *ApJ*, 577, 389
Leonard, P. J. T. & Duncan, M. J., 1990, *AJ*, 99, 608
Nugis, T. & Lamers, H. J. G. L. M., 2000, *A&A*, 360, 227
Pols, O. R., Cote, J., Waters, L. B.F. M., & Heise, J., 1991, *A&A*, 241, 419
Pols, O. R. & Marinus, M., 1994, *A&A*, 288, 475
Portegies Zwart, S. F., McMillan, S. L.W., & Makino, J., 2007, *MNRAS*, 374, 95
Ptak, A. & Colbert, E., 2004, *ApJ*, 606, 291
Sana, H., Gosset, E., Naze, Y., et al., 2008, *MNRAS*, 386, 447
Soria, R., 2007, *Astrophys. Space Sci*, 311, 213
Suzuki, T. K., Nakasato, N., Baumgardt, H. et al., 2007, *ApJ*, 668, 435
Tutukov, A., Yungelson, L. R., & Iben, I.Jr, 1992, *ApJ*, 386, 197
Van Bever, J. & Vanbeveren, D., 1997, *A&A*, 332, 116
Van Bever, J. & Vanbeveren, D., 2000, *A&A*, 358, 462
Van Bever, J. & Vanbeveren, D., 2003, *A&A*, 400, 63
Vanbeveren, D., 1991, *A&A*, 252, 159
Vanbeveren, D., Belkus, H., Van Bever, J., & Mennekens, N., 2008, *Astrophys. Space Sci.* in press (ArXiv:0712.3343)
Vanbeveren, D. & Conti, P. S., 1980, *A&A*, 88, 230
Vanbeveren, D. & De Loore, C., 1994, *A&A*, 290, 129
Vanbeveren, D., De Loore, C., & Van Rensbergen, W. 1998b, *A&A Review*, 9, 63
Vanbeveren, D., de Donder, E., Van Bever, J. et al. 1998a, *NewA*, 3, 443
Vanbeveren, D., Van Bever, J., & Belkus, H., 2007, *ApJ*, 662, 107
Vanbeveren, D., Van Rensbergen, W., & De Loore, C., 1998c, *The Brightest Binaries* (Dordrecht: Kluwer)
Yungelson, L. R., van den Heuvel, E. P.J., Vink, J. S. et al., 2008, *A&A*, 477, 223

Discussion

MAÍZ APELLÁNIZ: Would you comment on the fact that there is no observational indication for the existence of very massive stars (with masses = several 100 to 1000 $M_\odot$) in dense massive clusters?

VANBEVEREN: Well, first of all, when very massive stars form due to runaway collision, they form very fast. Then, after the main merger event, with our preferred mass loss rate formalisms the timescale where they remain a very massive star is very short as well. So, the lack of observations of such very massive objects is predicted by our simulations which account for large but realistic mass loss rates of very massive objects.

HERRERO: Young O-type stars may still be hidden in their parent cloud and therefore difficult to observe. This means that the 10% OB-type runaway frequency may be an upper limit. What would a smaller percentage imply for the initial binary frequency in your calculations?

VANBEVEREN: The predicted OB-type runaway frequency depends in a nearly linear way on the adopted initial binary frequency, so a lower OB-type runaway frequency would imply that in the population model we can start with a lower initial binary frequency in order to explain the runaway frequency.

OWOCKI: In the merging of two massive stars, radiation forces are likely to be important in driving mass loss at a rate much greater than can be achieved by the standard line-driving process. In the immediate post- merger the star is likely to be overluminous and above the Eddington limit, and during this time continuum driving could eject material up to the energy ("photon tiring") limit.

VANBEVEREN: Our main thesis is that already due to 'normal' stellar wind mass loss, it is unlikely that intermediate mass blank holes form in clusters with solar or super-solar metallicity. Any extra mass loss process (extra with respect to the one that we apply in our calculations) will strengthen this thesis.

Dany Vanbeveren.

Different water-sliding techniques: Artemio Herrero (top), Ana Ursúa (middle) and Paco Najarro (bottom).

Westerlund 1 as a Template for Massive Star Evolution

Ignacio Negueruela[1], J. Simon Clark[2], Lucy J. Hadfield[3]† and Paul A. Crowther[3]

[1] Departamento de Física, Ingeniería de Sistemas y Teoría de la Señal,
Universidad de Alicante, Apdo. 99, E03080 Alicante, Spain
email: ignacio@dfists.ua.es

[2] Department of Physics and Astronomy, The Open University, Walton Hall,
Milton Keynes MK7 6AA, United Kingdom

[3] Department of Physics and Astronomy, University of Sheffield,
Sheffield, S3 7RH, United Kingdom

Abstract. With a dynamical mass $M_{\rm dyn} \sim 1.3 \times 10^5\ M_\odot$ and a lower limit $M_{\rm cl} > 5 \times 10^4\ M_\odot$ from star counts, Westerlund 1 is the most massive young open cluster known in the Galaxy and thus the perfect laboratory to study massive star evolution. We have developed a comprehensive spectral classification scheme for supergiants based on features in the 6000–9000Å range, which allows us to identify >30 very luminous supergiants in Westerlund 1 and ~ 100 other less evolved massive stars, which join the large population of Wolf-Rayet stars already known. Though detailed studies of these stars are still pending, preliminary rough estimates suggest that the stars we see are evolving to the red part of the HR diagram at approximately constant luminosity.

Keywords. stars: early-type – stars: evolution – supergiants – stars: Wolf-Rayet – open clusters and associations: individual (Westerlund 1)

1. Introduction

As they evolve towards the Wolf-Rayet (WR) phase, massive stars must shed most of their outer layers. Models predict that massive stars will become supergiants (SGs) and evolve redwards at approximately constant $L_{\rm bol}$, but observations reveal a complex zoo of transitional objects, comprising Blue SGs, Red SGs, Yellow Hypergiants (YHGs), Luminous Blue Variables (LBVs) and OBfpe/WNVL stars, whose identification with any particular evolutionary phase is difficult. Understanding this evolution is, however, crucial because the mass loss during this phase completely determines the contribution that the star will make to the chemistry of the ISM and even the sort of post-supernova remnant it will leave.

Unfortunately, massive stars are scarce and, as this phase is very short on evolutionary terms, examples of massive stars in transition are rare. For most of them, distances are unknown and so luminosities are known at best to order-of-magnitude accuracy, resulting in huge uncertainties in other parameters (M_*, R_*). The difficulty to place these objects in the evolutionary sequence is obvious.

The open cluster Westerlund 1 (Wd 1) offers an unprecedented opportunity to improve this situation. It contains a large population of evolved stars which have formed at the same time, are at the same distance and have the same chemical composition. With an age of 4–5 Myr, Wd 1 contains a rich population of stars in transitional states that can be used to constrain evolutionary models.

† Present address: Chester F. Carlson Center for Imaging Science, Rochester Institute of Technology, 54 Lomb Memorial Drive, Rochester NY, 14623, USA

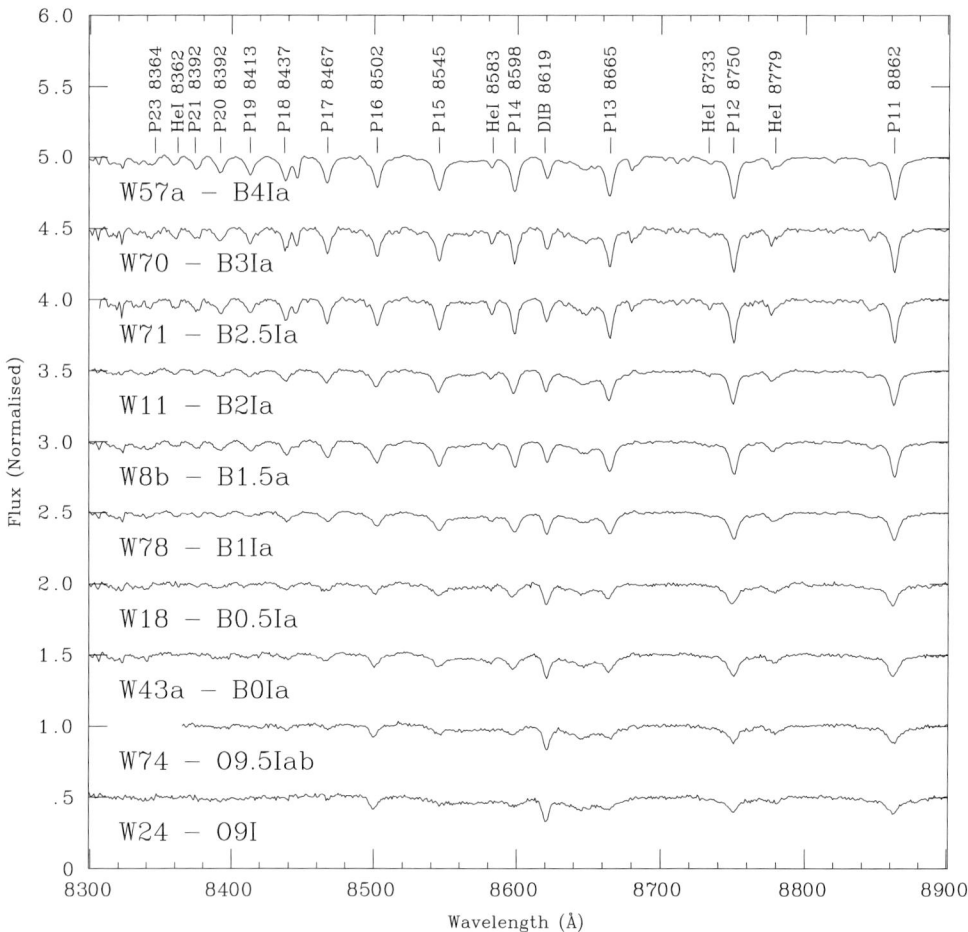

Figure 1. Sequence of *I*-band spectra of blue supergiants in Wd 1, showing the evolution of the main features. Note the disappearance of N I features between Pa 12 and Pa 13 and O I 8446Å around B2 and the development of Pa 16, which shows the presence of a strong C III line for stars B0 and earlier.

2. Cluster parameters

The parameters of Wd 1 are still poorly determined, though significant progress has been made in recent years. Two estimates of its mass have been recently made. Mengel & Tacconi-Garman (2008) have used the radial velocity dispersion of the ten stars brightest in the infrared ($\sigma = 8.4\,{\rm km\,s^{-1}}$) to estimate a mass of $\sim 1.3 \times 10^5\,{\rm M}_\odot$. This estimate suffers from a large uncertainty due to the low number of stars used and the spectral peculiarities of some. Brandner *et al.* (2008) have used star counts in the infrared to set a lower limit on the cluster mass $M_{\rm cl} > 5 \times 10^4\,{\rm M}_\odot$. Again, there are important uncertainties coming into this estimation. Large discrepancies have been found in the determinations of distance and (hence) masses based on infrared pre-main-sequence tracks and optical post-main-sequence tracks. In any case, the integrated initial mass function appears consistent with Salpeter's and then the observed population of massive stars would imply a mass $\sim 10^5\,{\rm M}_\odot$, consistent with both determination.

An important source of uncertainty in both measurements is the role of binarity, which affects both radial velocities and transformation of magnitudes into masses. Recent results suggest that the fraction of binaries with similar mass components is very high amongst cluster members. Most Wolf-Rayet stars show indirect indications of binarity, presence of dust in WC stars (Crowther *et al.* 2006) and hard X-ray emission in WN stars (Clark *et al.* 2008). Some of them also show photometric variability and at least one is an eclipsing binary (Bonanos 2007). The identification of a large number of X-ray sources detected by *Chandra* with a population of evolved late-O stars (Clark *et al.* 2008) suggests that the high binary fraction extends to lower masses.

The distance to the cluster is not very well constrained either. Photometry is affected by the very strong reddening ($A_V \approx 12$ mag). Analysis of the $E(B-V)$ colours suggests that the reddening deviates from the standard law (Clark *et al.* 2005). A recent determination, making use of atomic hydrogen in the direction to the cluster, gives $d = 3.9 \pm 0.7$ (Kothes & Dougherty 2007), compatible with, though slightly shorter than, estimates based on the stellar population (e.g., Crowther *et al.* 2006).

The age of the cluster can be constrained from the observed population. The ratio of WR stars to red and yellow hypergiants favours an age of 4.5 or 5.0 Myr, with the progenitors of the Wolf-Rayet stars having initial masses in the $40-55\,M_\odot$ range (Crowther *et al.* 2006). Such age is fully compatible with the observed population of blue supergiants (see below), which should be descended from stars with initial masses of $\sim 35\,M_\odot$. This again would imply masses of $\leqslant 30\,M_\odot$ for the stars at the top of the main sequence, in good agreement with the O7–8 V spectral type, again appropriate for the age.

3. Observations and analysis

Observations of stars in Wd 1 were carried out on the nights of 2003 June 12th and 13th using the spectro-imager FORS2 on Unit 1 of the VLT (Antu) using three different modes: longslit, multi-object spectroscopy with masks (MXU) and multi-object spectroscopy with movable slitlets (MOS). We used grisms G1200R and G1028z to obtain intermediate resolution spectroscopy. With this setup, we obtained almost continuous coverage over the 5800–9500Å range at intermediate resolution.

Within the central $5' \times 5'$ field of view, we selected our targets from the list of likely members of Clark *et al.* (2005). For the external regions, targets were selected at random amongst relatively bright stars. In total, we took three MXU and one MOS mask with both G1200R and G1028z grisms, and two further MXU masks with only the G1200R (these were aimed at relatively faint objects, which were expected to be OB stars near the MS and so not to have strong features in the range covered by G1028z). This resulted in ~ 100 stars observed with G1200R and ~ 70 stars observed with G1028z. More than 90% of the spectra turned out to correspond to OB stars, and hence cluster members.

We have used these observations to derive spectral types for cluster members. The use of *I*-band spectra to classify OB stars has been explored by Caron *et al.* (2003) and is further discussed in Appendix A of Clark *et al.* (2005). We have studied further spectral type and luminosity indicators in the *R* and *I* bands and derived spectral types from a combination of features. Features that contribute to our analysis are: the shape and strength of emission (P-Cygni profiles) in Hα, the strength of the C II 6578,6582Å doublet (in absorption), the presence and strength of the C II 7231, 7236Å doublet (in emission, a wind feature), the strength of the O I 7774Å triplet and the strength of the N II 6482Å line. In the *I* band, apart from the shape and strength of the Paschen lines (see Clark *et al.* 2005), a main indicator is the presence of the C III 8502Å line, which appears, blended into Pa 16, for stars B0 and earlier (see Fig. 1).

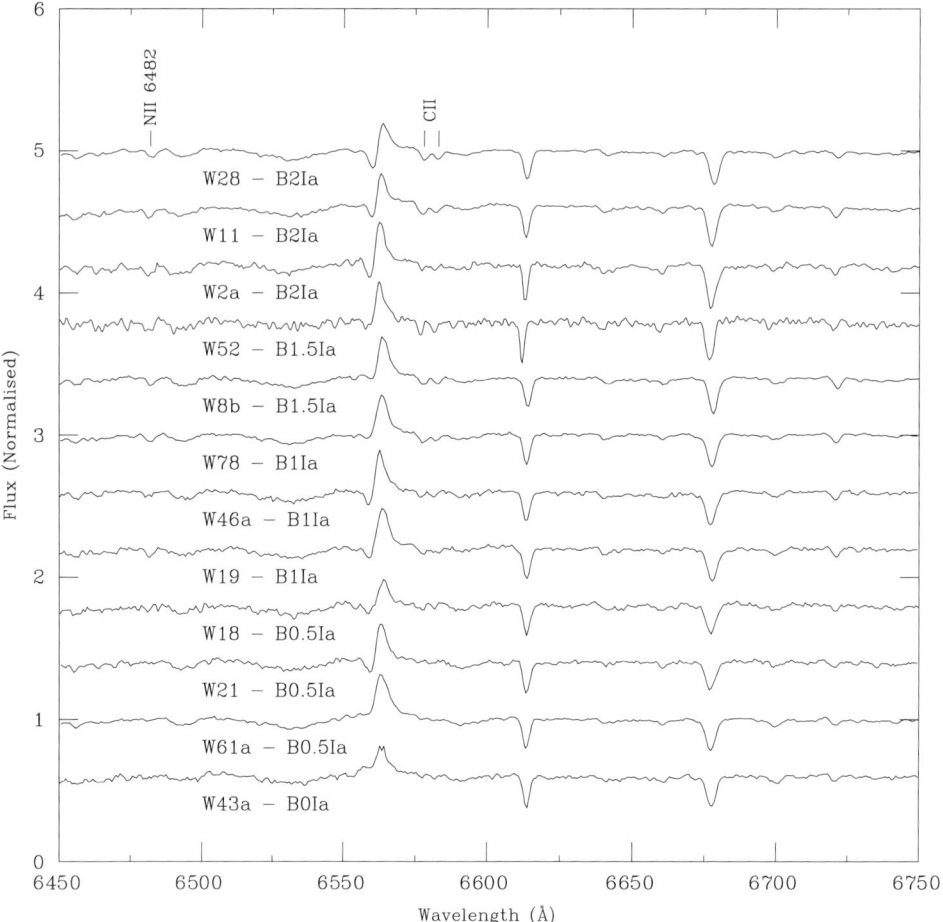

Figure 2. Sequence of red spectra of blue supergiants in Wd 1. The shape of Hα changes from a P-Cygni profile to pure emission around spectral type B0.5. The N II 6482Å line is only prominent in Ia supergiants and its strength peaks sharply at B2.

4. Implications

The upper HR diagram. The stellar content revealed by these observations is as follows. There is an A supergiant showing spectral variability (W243; Clark & Negueruela 2004) and six other A/F hypergiants of very high luminosity, with $M_V \approx -9$ to -10 (Clark *et al.* 2005). We call these objects hypergiants because of their very high intrinsic magnitudes, without making any direct inference of their evolutionary status.

The same can be said of at least three late-B luminous stars: W42a (B8 Ia$^+$), W33 (B5 Ia$^+$) and W7 (B5 Ia$^+$). In addition, there are ~ 20 supergiants that can be unambiguously classified as Ia, covering the B0 Ia–B4 Ia range. The earliest very luminous star that we find is W74, with spectral type O9.5 and a luminosity class that could be Ia or Iab.

Even though the exact luminosity class of O-type supergiants is more difficult to determine using red spectra, we do not find evidence for any other luminous O-type supergiant, though there are > 50 and (taking into account the incompleteness of our dataset) likely

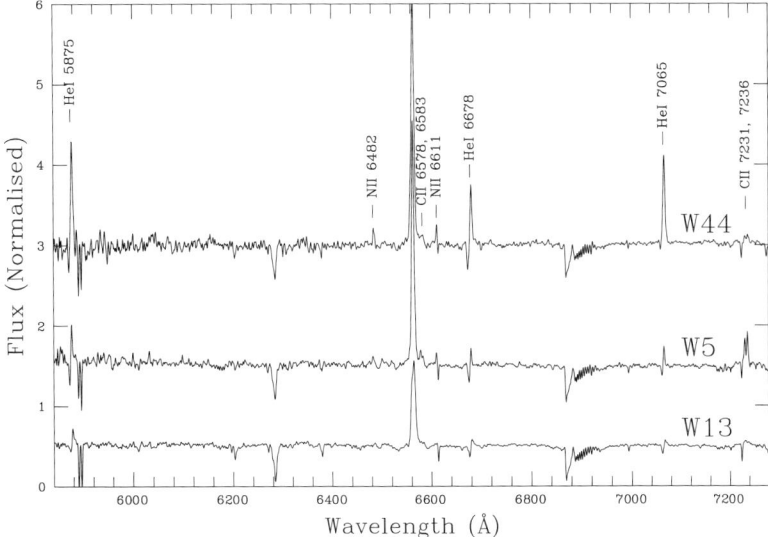

Figure 3. Spectra of three transitional objects in Westerlund 1, which can be classified as very late WN stars or early-B hypergiants. These objects might be examples of stars looping back to the blue region from the red supergiant phase.

∼ 100 stars which we generically classify as O9 I, meaning late-O stars well above the main sequence.

Evolution. We have made a rough attempt at calculating the intrinsic luminosities of all the supergiants, by using their V and I magnitudes and assuming that the relation between $E(V - I)$ and M_I is standard. We use the intrinsic $(V - I)$ colours of Ducati et al. (2001) and bolometric corrections from Martins et al. (2005) and Humphreys & McElroy (1984). Even though there is a large scatter, we find no clear trend. Stars of all spectral types seem to have similar bolometric magnitudes, a result consistent with the idea of redwards evolution at constant bolometric luminosity. Note that we have not compared the bolometric magnitudes of the red supergiants, as their spectral types are uncertain, and the bolometric corrections of such bright red stars are unknown.

This result does not imply that there are no stars looping back towards the blue in Wd 1. Obviously, the WR stars must have come back to the blue before reaching their present stage. There are three emission-line stars in the cluster which can be classified as very late WN stars or extreme B supergiants (see Fig. 3). These objects might be evolving into Wolf-Rayet stars. Detailed analysis of their spectra, in order to derive chemical abundances, will be necessary before their evolutionary status may be ascertained. Similar analyses should be conducted for the A and F hypergiants and W243.

The next steps in our work involve an investigation of the binary fraction in the cluster and detailed analysis of the spectra with state-of-the art models in order to obtain a better constraint on their evolutionary status.

References

Bonanos, A. Z. 2007, *AJ*, 133, 2696
Brandner, W., Clark, J. S., Stolte, A., et al. 2008, *A&A*, 478, 137
Caron G., Moffat A. F. J., St-Louis N., et al. 2003, *AJ*, 126, 1415
Clark J. S., Negueruela I., 2004, *A&A*, 413, L15

Clark, J. S., Negueruela, I., Crowther, P. A., & Goodwin, S. P. 2005, *A&A*, 434, 949
Clark, J. S., Muno, M. P., Negueruela, I., *et al.* 2008, *A&A*, 477, 147
Crowther, P. A., Hadfield, L. J., Clark, J. S., *et al.* 2006, *MNRAS*, 372, 1407
Ducati, J. R., Bevilacqua, C. M., Rembold, S. B., & Ribeiro, D. 2001, *ApJ*, 558, 309
Humphreys, R. M. & McElroy, D. B. 1984, *ApJ*, 284, 565
Kothes, R. & Dougherty, S. M. 2007, *A&A*, 468, 993
Martins, F., Schaerer, D., & Hillier, D. J. 2005, *A&A*, 436, 1049
Mengel, S. & Tacconi-Garman, L. E. 2008, in: Vesperini, M. Giersz, & A. Sills, (eds.), *Dynamical Evolution of Dense Stellar Systems* (Cambridge: CUP), *Proc IAU Symp 246* in press (arXiv:0711.1779)

Discussion

LANG: What does the distribution of radio emission reveal in this cluster – both the extended emission and the point-like sources?

NEGUERUELA: There are point sources associated with most of the cool and cold supergiants, which, we believe, are due to photoionisation of their extended atmospheres or winds by the hot stars. There is no extended emission obviously associated with the cluster, suggesting that supernova explosions have blown away any diffuse material. A more detailed summary may be found in the contribution by Dougherty & Clark to *Massive Stars: Fundamental Parameters and Circumstellar Interactions*, (arXiv:0705.0971)

CROWTHER: As your title suggests, the presence of large numbers of massive stars in Wd 1 allows robust tests of evolutionary models. By way of example, N(WN with H) < N(WC) < N(WN without H), yet both single (e.g., Geneva) and binary models (e.g., Eldridge et al. this volume) predict the completely wrong subtype distributions, namely N(WN without H) < N(WN with H) < N(WC) for an instantaneous burst of $4-5$ Myr.

NEGUERUELA: Indeed. The observed population is most likely compatible with a single burst of star formation and hence offers an observational test of evolutionary models. Also, as I mentioned, the fact that we can observe a large population of massive ($M_* > 30\,M_\odot$) stars evolving away from the main sequence and a very large population of low mass stars evolving towards the main sequence gives us an excellent opportunity to try to set post and pre-main-sequence isochrones in the same reference time and compare their predictions.

Ignacio Negueruela (center) with Amparo Marco (right) and Artemio Herrero (left).

One Hundred 30 Dors?

M. M. Hanson and B. Popescu

Department of Physics, University of Cincinnati, Cincinnati, OH, USA
email: margaret.hanson@uc.edu, popescb@email.uc.edu

Abstract. There are a few ways to estimate the number of massive open clusters expected in the disk of the Milky Way, such as the total star formation rate of the Galaxy, or the open cluster mass function extrapolated to include the entire Galaxy. Surprisingly, they give similar predictions: the Milky Way should contain about 100 clusters as massive as 30 Doradus. Are we seeing them? We look closely at these predictions and compare them to what has been found so far in our Galaxy. We present sophisticated image simulations our group is developing to estimate the selection biases faced by current infrared searches for these massive clusters.

Keywords. Galaxy: stellar content – open clusters and associations: general – stars: early-type

1. Does the Milky Way contain, 'Super Star Clusters'?

Paul Hodge (1961) was among the first to study very young massive clusters; these are extragalactic clusters of very high mass, but significantly younger than the known globular clusters. The popular term, 'super star cluster', was coined by Sydney van den Bergh (1971). He used it to describe a dozen enormously bright knots of emission seen in the nucleus of M82. However, this new term did not come into popular use until the *Hubble Space Telescope* era of high-resolution ultra-violet imaging of merging and starbursting galaxies (Holtzman *et al.* 1992; Whitmore *et al.* 1993, O'Connell, *et al.* 1994). Perhaps among the first to use the phrase super star cluster to describe a *Milky Way* cluster was Serabyn *et al.* (1999) and Knodlseder (2000), in their studies of the Arches and Cyg OB2 clusters, respectively. Since then, the term has been used to describe other Milky Way clusters, most recently, Westerlund 1. At the time of this contribution, Westerlund 1, with a mass of nearly 10^5 M$_\odot$, is the most massive young cluster known in the entire Local Group (Clark & Negueruela 2002, Brandner *et al.* 2008, Negueruela *et al.* this volume).

Should we have expected to find such massive, young stellar clusters lurking in the inner Milky Way? As surprising as it may at first sound, it is consistent with most observations (and theory!) that our Galaxy should contain numerous young star clusters similar in scale to R136 in 30 Doradus ($> 10^4$ M$_\odot$). It is important to recognize that a mass of just a few $\times 10^4$ M$_\odot$ is clearly at the low-mass tail of the 'super star cluster' mass distributions typically studied by extragalactic astronomers (see for example, Whitmore & Schweizer 1995; Zhang & Fall 1999; who report on young clusters with $M > 10^6$ M$_\odot$). Nonetheless, the idea of super star clusters being found within the Milky Way is a highly novel and still heavily debated notion.

2. Predicting the Super Star Cluster Population in the Milky Way

There are several compelling arguments which provide reasonable and consistent estimates of the number of massive clusters our Milky Way galaxy should be expected to harbor.

2.1. The Global Star Formation Rate of Normal Spiral Galaxies

A Galactic astronomer can become jealous with the full view of the grand design spiral arms and near complete census of star clusters allowed extragalactic astronomers studying face-on galaxies. It is of great interest to know how our Milky Way galaxy would appear among its spiral galaxy brethren. But direct comparisons are hard since our knowledge of the Milky Way's exact properties are so poorly known. We do however have a reasonable estimate of its global star formation rate, 2-5 $M_\odot \text{yr}^{-1}$ (Prantzos & Aubert 1995; Naab & Ostriker 2006;). Larsen & Richtler (2000) have studied young massive clusters found in nearby, normal spiral galaxies. They have shown that among these normal disk starforming spiral galaxies, a useful correlation is found between the galaxies global star formation rate and it's most massive, young cluster. In Figure 1, we show an image taken directly from Weidner *et al.* (2004, their Figure 3) that graphs this relationship. The rectangles represent the Milky Way star forming 'regions' of Taurus and Orion, while the rectangle at higher mass represents 30 Doradus in the LMC. Based on its global star formation rate, Weidner *et al.* (2004) argue that the most massive young cluster in the Milky Way should have a mass of 10^5 $M_\odot$ (several times the mass of R 136).

2.2. Extrapolating the Locally derived Cluster Mass Function

The mass function of stellar clusters has been measured within the Galaxy, in other galaxies, and in extreme starbursting systems. Extreme starbursting systems, such as the Antennae Galaxies, show clusters of extremely high masses, many over 10^6 $M_\odot$ or even 10^7 $M_\odot$ (Zhang & Fall 1999) yet, the mass function follows a simple power law, $\delta N/\delta M \propto M^{-2}$. Among normal, non-merging or bursting galaxies, a power-law of slope -2 is also found (Larsen 2002). Within our Galaxy, it was long ago shown by van den Bergh & Lafontaine (1984, vdBL), that the nearby open clusters follow a power law mass distribution, with slope -2. Even embedded young clusters measured in the Milky Way

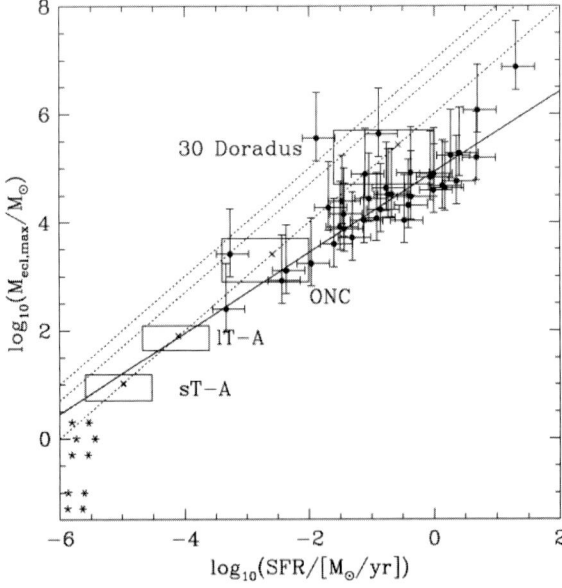

Figure 1. The global star formation rate of normal, disk-star-forming, spiral galaxies verses the most massive young cluster observed in that galaxy. The global star formation rate of the Milky Way is 2-5 $M_\odot \text{yr}^{-1}$, suggesting the most massive cluster present in the Milky Way should have a mass of about 10^5 $M_\odot$. *Reproduced from Weidner et al. (2004) with author permission.*

Table 1. Current Census of Massive Young Clusters in the Milky Way[1]

Cluster	l(deg)	Dist (kpc)	Age(myr)	Mass ($M_\odot$)	Mass Reference
Westerlund 1	339	3.5	4-5	50,000	Brandner et al. 2007
RSGC2	26	5.6	15-20	40,000	Davies et al. 2007
W49a	43	11.8	<2?	...	Homeier & Alves 2005
RSGC1	25	5.8	10-14	30,000	Davies et al. 2008
GC Central	0	8	6	20,000	Paumard et al. 2006
Arches	0	8	2-4	20,000	Figer et al. 2002
Quintuplet	0	8	3-5	20,000	Figer et al. 1999
NGC 3603	291	7	<2.5	12,500	Harayama et al. 2008
Westerlund 2	284	2.8	2	10,000	Ascenso et al. 2007
DBS2003 179	348	8.8	3-5	8,000	Borissova et al. 2008
CL 1806-20	10	9	3-4?	3,000	Bibby et al. 2008
Glimpse 30	298	7.2	4-5	3,000	Kurtev et al. 2007

[1] Taken mostly from Messineo et al. (this volume)

galaxy show the same power-law slope (Lada & Lada 2003). It has also been argued on theoretical grounds (Elmegreen & Efremov 1997, Weidner & Kroupa 2006) that a power-law mass distribution should be near universal.

Calibrating to the known clusters of the time, vdBL extrapolated the observed luminosity function to include the entire disk of the Galaxy. Such an exercise suggested 100 clusters should exist within the Galaxy with $M_V = -11$, about the luminosity of R136. Such a notion seemed 'hard to believe' at the time, and vdBL concluded that the cluster function must 'turn over' above $M_V = -8$. This indeed brings up a significant point: is there an upper mass limit for clusters and will this occur globally at the same mass or is it dependent on the galactic system? This question might already have been answered by studies of the luminosity function of extragalactic H II regions. These observations support the idea that there is a turn over in the luminosity function of H II regions in some galaxies, and it may be a function of the galaxy's Hubble Type (Kennicutt et al. 1989; Oey & Clarke 1998). McKee & Williams (1997) claim to see just such a truncation of young massive clusters in our own Galaxy. They estimate the truncation point for making high mass star clusters in the Galaxy is around 2.4×10^5 $M_\odot$ and predict there may be as many as 10 young stellar clusters with mass 10^5 $M_\odot$ presently in the Milky Way (C. McKee, private communication).

3. What is the census thus far of the Milky Way's super star clusters?

Beginning with the 2MASS infrared survey (Skrutskie et al. 2006) and the presently ongoing infrared surveys using the Spitzer Space Telescope, numerous groups have invested enormous energy into searching for extinguished young star clusters in the inner Galaxy. Messineo et al. (this volume) provide a current table we will reproduce here (Table 1) with a few additions and updates (Borissova et al. 2008, Bibby et al. 2008). Ages and particularly masses for all of the clusters listed here can differ in the literature, sometimes greatly, based on the method of study. But, we expect the masses to be secure to perhaps a factor of 2. Eight clusters are listed to have reasonably secure masses of at least 10^4 $M_\odot$. That represents barely 10% of the number we predicted based on the Milky Way's observed properties. Interestingly, of those most massive eight clusters, none are seen on the far side of the Galaxy. Assuming the Milky Way to be well balanced in star formation properties, it safe to assume we must be missing at least half the clusters (those 'eight' on the other side of the Galaxy). But could we be missing still more?

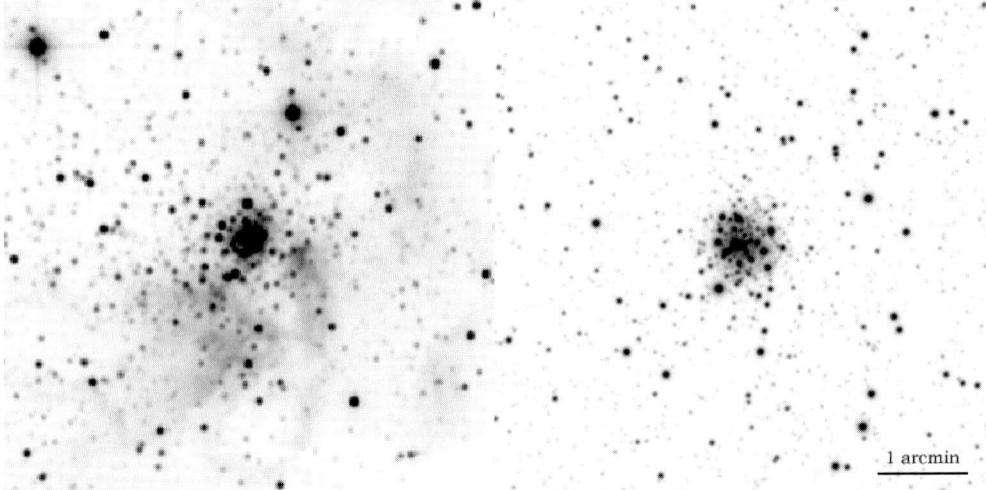

Figure 2. J-Band image of NGC 3603. On the left is the image taken from 2MASS[2] Skrutskie (Skrutskie *et al.* 2006). On the right is our MASSCLEAN simulation, assuming: $\log(age) = 6.00$, $M_{total} = 10^4 M_\odot$ and solar metallicity, $A_V = 4.5$ mag, $R_V = 3.1$, distance $d = 6$ kpc (distance modulus 13.9 mag); for the spatial distribution we used $r_t = 4.4'$ and $r_c = 0.4'$ and small segregation rate; Kroupa IMF with $\alpha_1 = 0.3$, $\alpha_2 = 1.3$ and $\alpha_3 = 2.4$.

4. Deriving the selection effects of current infrared surveys

Its difficult to make a quantitative estimate of the clusters that current infrared searches may be systematically missing. Certainly, it is a function of cluster properties (mass, density, age) but also its location in the Galaxy and line of sight extinction. All these characteristics need to be considered in estimating the selection effects of current imaging searches. In order to answer this question, we are developing a sophisticated and rigorous program which accurately simulates images of very massive clusters with a variety of characteristics (age, stellar density, distance, extinction) in the infrared. These will be used with current infrared search algorithms to directly determine the selection effects as a function of cluster characteristics and location in the Galaxy.

We have completed the first stage in our simulation program, called MASSCLEAN (Massive Cluster Evolution and Analysis Package, Popescu & Hanson 2008, in prep.). With a surprisingly small number of parameters we are able to accurately construct the entire evolution of a cluster. The cluster model is built in a modular way. For the mass distribution of the stars in the cluster we use a Kroupa-Salpeter initial mass function [IMF] (Kroupa 2002; Salpeter 1955). The versatility of the package allows us to use a one-, two- or three-power law function (Kroupa-Salpeter type) as the IMF. A truncated IMF is also supported (Oey & Clarke 2005). The stellar evolution is given by the Geneva Database (Lejeune & Schaerer 2001). This provides absolute magnitudes for all stars in the cluster in all the needed photometric bands over an enormous range of evolutionary stages (ages). We apply the CCM Model (Cardelli, Clayton & Mathis 1989) for extinction to each stars photometric band and a distance modulus to produce apparent magnitudes. The spatial distribution of stars within the cluster on the sky is given by the King Model (King 1962). Combining all this, MASSCLEAN can then generate simulated FITS images using SKYMAKER (Bertin 2001; Bertin and Fouqué 2007) for a range of masses, ages, distances, extinctions, and stellar density.

Our MASSCLEAN package has been tested against several real clusters and over a mass range of $10^3 - 10^6 M_\odot$. One thing not included in the simulation is nebulosity (see Fig. 2). However, such morphologies are only significant in the first 1 - 2 million years of the cluster's life. Although the package can simulate low mass clusters, we emphasize its use for massive clusters due to the difficulty of such a low mass simulation (lower number of stars to statistically populate the cluster properties). Once we are confident our simulations accurately reproduce massive clusters of known characteristics, we will begin development of completeness tests of current search algorithms to find our simulated clusters within a model of the Milky Way's disk.

5. Summary

While infrared searches have greatly increased our ability to study the massive star population of our inner Galaxy, it is not clear just how many massive clusters are still missing from these searches. When considering the Milky Way as like other normal spiral galaxies with similar star formation rates and other such general properties, we may be missing nearly 90% of these clusters to date. To address this question, we are developing a sophisticated cluster imaging simulation program. The first phase, to develop such a code and test our simulations against known massive clusters, is nearly complete. The next phase is to determine the detectability of such clusters once placed in various locations through out the Galactic disk. It is our hope that we can provide estimates of the completeness of current surveys as a function of cluster properties (distance, extinction, age, core density, etc). This should allow us to derive the biases inherent in current cluster search methods that rely on infrared imaging surveys. We suspect clusters that currently exist in certain age ranges (such as when the red supergiant phase is just starting, 4–5 million years old) and clusters with rather low stellar densities are being missed by the current searches, even if they have very high cluster mass. This will be particularly true if a cluster is behind very high extinction or is at a great distance. Finally, our study of current search algorithms will not only help us better understand the search biases, it may allow us to design a search algorithm which is better tailored to find a higher percent of the massive young clusters presently existing in the inner Milky Way.

References

Ascenso, J., Alves, J., Vicente, S., & Lago, M. T. V. T. 2007, *A&A*, 476, 199
Brandner, W., Clark, J. S., Stolte, A. et al. 2008, *A&A*, 478, 137
Bibby, J. L., Crowther P. A., Furness, J. P., & Clark, J. S. 2008, *MNRAS*, 386, 23
Bertin, E. 2001, SKYMAKER, http://terapix.iap.fr/cplt/oldSite/soft/skymaker/
Bertin, E. & Fouqué, P. 2007, SKYMAKER,
    http://terapix.iap.fr/rubrique.php?id_rubrique=221
Borissova, J., Ivonov, V., & Hanson, M. M. 2008, *A&A*, accepted
Cardelli, J. A., Clayton, G. C., & Mathis, J. S. 1989, *ApJ*, 345, 245
Clark, J. S. & Negueruela, I. 2002, *A&A*, 396, L25
Davies, B. Figer D. F., Kudritzki R.-P. et al. 2007, *ApJ*, 671, 781
Davies, B., Figer, D. F., Law C. J. et al. 2008, *ApJ* 676, 1016
Elmegreen, B. G. & Efremov, Y. N. 1997, *ApJ*, 480, 235
Figer, D. F., McLean, I. S. & Morris, M. 1999, *ApJ*, 514, 202
Figer, D. F. Najarro, F., Gilmore, D. et al. 2002, *ApJ*, 581, 258
Harayama, Y., Eisenhauer, F., Martins, F. 2008, *ApJ*, 675, 1319
Hodge, P. W. 1961, *ApJ*, 133, 413
Holtzman, J. A., Faber S. M., Shaya, E. J. et al. 1992, *AJ*, 103, 691
Homeier, N. L. & Alves, J. 2005, *A&A*, 430, 481
Kennicutt, R. C., Edgar, B. K., Hodge, P. 1989, *ApJ*, 337, 791

King, I. 1962, *AJ*, 67, 471
Kroupa, P. 2002, *Sci*, 295, 82
Kurtev, R., Borissova, J., Georgiev, L., et al. 2007, *A&A*, 475, 209
Lada, C. J. Lada, E. A. 2003, *ARA&A* 41, 57
Larsen, S. S. & Richtler, T. 2000, *A&A*, 354, 836
Larsen, S. S. 2002, *AJ*, 354, 836
Lejeune, T. & Schaerer, D. 2001, *A&A*, 366, 538
McKee, C. F. & Williams, J. P. 1997, *ApJ*, 476, 144
Naab, T. & Ostriker, J. P. 2006, *MNRAS*, 366, 899
O'Connell, R. W., Gallagher, J. S., Hunter, D. A. 1994, *ApJ*, 433, 65
Oey, M. S. & Clarke, C. J. 1998, *AJ*, 115, 1543
Oey, M. S. & Clarke, C. J. 2005, *ApJ*, 620, L43
Paumard, T. Genzel, R., Martins, F. et al. 2006, *ApJ*, 643, 1011
Prantzos, N. & Aubert, O. 1995, *A&A*, 302, 69
Salpeter, E. E. 1955, *ApJ*, 121, 161S
Serabyn, E., Shupe, D. & Figer, D. F. 1998, *Nature*, 394, 448
Skrutskie, M. F., Cutri, R. M., Stiening, R., et al., 2006, *AJ*, 131, 1163
van den Bergh, S 1971, *A&A*, 12, 474
van den Bergh, S, Lafontaine, A. 1984, *AJ*, 89, 1822
Weidner, C., Kroupa, P. & Larsen, S. S. 2004, *MNRAS* 350, 1503
Weidner, C., Kroupa, P. 2006, *MNRAS* 365, 1333
Whitmore, B. C. Schweizer, F., Leitherer, C. et al. 1993, *AJ*, 106, 1354
Whitmore, B. C. & Schweizer, F. 1995, *AJ*, 109, 960
Zhang, Q, Fall, M. 1999, *ApJL*, 527, 81

Discussion

OEY: Just a comment that unlike the local stellar mass function, which seems to show a universal upper mass cutoff, clusters do not have a universal upper mass cutoff. Oey & Clarke 1998 showed that the form of the H II region liminosity function demonstrates that the cluster upper-mass cutoff varies with Hubble type and the Milky Way is an Sb Galaxy. So you would not necessarily expect to extrapolate the power low mass function to infinity. The same has been shown by Kennicutt in the 1980's.

LEITHERER: The plot showing the location of the Galaxy relative to other objects has a misleading abscissa. One should not plot star-formation rate but the specific star formation rate, i.e. normalized to unit surface area. In your units, the Galaxy and the starburst prototype M82 would have nearly the same x-location because their total star formation rates are similar. Yet the specific rates differ by a factor of 100 because M82 is a dwarf galaxy. Would you use M82 as a guide for the Milky Way because there are hundreds of luminous clusters in M82?

HANSON: The plot I showed (Figure 1 in this proceedings) includes only normal, non-interacting spiral galaxies with normal disk star formation similar to what we would expect for the Milky Way.

Extragalactic Stellar Astronomy with the Brightest Stars in the Universe

Rolf Kudritzki[1], Miguel A. Urbaneja[1], Fabio Bresolin[1] and Norbert Przybilla[2]

[1] Institute for Astronomy, University of Hawaii
2680 Woodlawn Drive, Honolulu, HI 96822, USA
email: kud@ifa.hawaii.edu, urbaneja@ifa.hawaii.edu, bresolin@ifa.hawaii.edu

[2] Dr. Remeis-Sternwarte Bamberg, Erlangen University
Sternwartstr. 7, D-96049 Bamberg, Germany
email: przybilla@sternwarte.uni-erlangen.de

Abstract. A supergiants are objects in transition from the blue to the red (and vice versa) in the uppermost HRD. They are the intrinsically brightest "normal" stars at visual light with absolute visual magnitudes up to -9. They are ideal to study young stellar populations in galaxies beyond the Local Group to determine chemical composition and evolution, interstellar extinction, reddening laws and distances. We discuss most recent results on the quantitative spectral analysis of such objects in galaxies beyond the Local Group based on medium and low resolution spectra obtained with the ESO VLT and Keck. We describe the analysis method including the determination of metallicity and metallicity gradients. A new method to measure accurate extragalactic distances based on the stellar gravities and effective temperatures is presented, the flux weighted gravity – luminosity relationship (FGLR). The FGLR is a purely spectroscopic method, which overcomes the uncertainties introduced by interstellar extinction and variations of metallicity, which plague all photometric stellar distance determination methods. We discuss the perspectives of future work using the giant ground-based telescopes of the next generation such as the TMT, the GMT and the E-ELT.

Keywords. galaxies: distances and redshifts – galaxies: individual (NGC 300) – galaxies: abundances – galaxies: stellar content – stars: early-type – stars: abundances – stars: distances

---

1. Introduction

It has long been the dream of stellar astronomers to study individual stellar objects in distant galaxies to obtain detailed spectroscopic information about the star formation history and chemodynamical evolution of galaxies and to determine accurate distances based on the determination of stellar parameters and interstellar reddening and extinction. At the first glance, one might think that the most massive and, therefore, most luminous stars with masses higher than 50 $M_\odot$ are ideal for this purpose. However, because of their very strong stellar winds and mass-loss these objects keep very hot atmospheric temperatures throughout their life and, thus, waste most of their precious photons in the extreme ultraviolet. As we all know, most of these UV photons are killed by dust absorption in the star forming regions, where these stars are born, and the few which make it to the earth can only be observed with tiny UV telescopes in space such as the HST or FUSE and are not accessible to the giant telescopes on the ground.

Thus, one learns quickly that the most promising objects for such studies are massive stars in a mass range between 15 to 40 $M_\odot$ in the short-lived evolutionary phase, when they leave the hydrogen main-sequence and cross the HRD in a few thousand years as blue supergiants of late B and early A spectral type. Because of the strongly reduced absolute value of bolometric correction when evolving towards smaller temperature these objects increase their brightness in visual light and become the optically brightest "normal" stars in the universe with absolute visual magnitudes up to $M_V \cong -9.5$ rivaling with the integrated light brightness of globular clusters and dwarf spheroidal galaxies. These are the ideal stellar objects to obtain accurate quantitative information about galaxies.

2. Studies in the Milky Way and Local Group

There has been a long history of quantitative spectroscopic studies of these extreme objects. In a pioneering and comprehensive paper on Deneb Groth (1961) was the first to obtain stellar parameters and detailed chemical composition. This work was continued by Wolf (1971, 1972, 1973) in studies of A supergiants in the Milky Way and the Magellanic Clouds. Kudritzki (1973) using newly developed NLTE model atmospheres found that at the low gravities and the correspondingly low electron densities of these objects effects of departures from LTE can become extremely important. With strongly improved model atmospheres Venn (1995a), Venn (1995b) and Aufdenberg et al. (2002) continued these studies in the Milky Way. Most recently, Przybilla et al. (2006) and Schiller & Przybilla (2008) used very detailed NLTE line formation calculations including ten thousands of lines in NLTE (see also Przybilla et al. 2000, Przybilla et al. 2001a, Przybilla et al. 2001b, Przybilla & Butler 2001, Przybilla 2002) to determine stellar parameters and abundances with hitherto unknown precision (T_{eff} to $\leqslant 2\%$, $\log g$ to ~ 0.05 dex, individual metal abundances to ~ 0.05 dex). At the same time, utilizing the power of the new 8m to 10m class telescopes, high resolution studies of A supergiants in many Local Group galaxies were carried out by (Venn 1999, SMC), (McCarthy et al. 1995, M33), (McCarthy et al. 1997, M31), (Venn et al. 2000, M31), (Venn et al. 2001, NGC 6822), (Venn et al. 2003, WLM), and (Kaufer et al. 2004, Sextans A), yielding invaluable information about the stellar chemical composition in these galaxies. In the research field of massive stars, these studies have so far provided the most accurate and most comprehensive information about chemical composition and have been used to constrain stellar evolution and the chemical evolution of their host galaxies.

3. The Challenging Step beyond the Local Group

The concept to go beyond the Local Group and to study A supergiants by means of quantitative spectroscopy in galaxies out to the Virgo cluster has been first presented by Kudritzki, Lennon & Puls (1995) and Kudritzki (1998). Following-up on this idea, Bresolin et al. (2001) and Bresolin et al. (2002) used the VLT and FORS at 5 Å resolution for a first investigation of blue supergiants in NGC 3621 (6.7 Mpc) and NGC 300 (1.9 Mpc). They were able to demonstrate that for these distances and at this resolution spectra of sufficient S/N can be obtained allowing for the quantitative determination of stellar parameters and metallicities. Kudritzki, Bresolin & Przybilla (2003) extended this work and showed that stellar gravities and temperatures determined from the spectral analysis can be used to determine distances to galaxies by using the correlation between absolute bolometric magnitude and flux weighted gravity $g_F = g/T_{\text{eff}}^4$ (FGLR). However, while these were encouraging steps towards the use of A supergiants as quantitative diagnostic tools of galaxies beyond the Local Group, the work presented in these papers had still a fundamental deficiency. At the low resolution of 5 Å it is not possible to use ionization equilibria for the determination of T_{eff} in the same way as in the high resolution work mentioned in the previous paragraph. Instead, spectral types were determined and a simple spectral type – temperature relationship as obtained for the Milky Way was used to determine effective temperatures and then gravities and metallicities. Since the spectral type – $T_{\rm eff}$ relationship must depend on metallicity (and also gravities), the method becomes inconsistent as soon as the metallicity is significantly different from solar (or the gravities are larger than for luminosity class Ia) and may lead to inaccurate stellar parameters. As shown by Evans & Howarth (2003), the uncertainties introduced in this way could be significant and would make it impossible to use the FGLR for distance determinations. In addition, the metallicities derived might be unreliable. This posed a serious problem for the the low resolution study of A supergiants in distant galaxies.

This problem was overcome only very recently by Kudritzki *et al.* (2008), hereafter KUBGP), who provided the first self-consistent determination of stellar parameters and metallicities for A supergiants in galaxies beyond the Local Group based on the detailed quantitative model atmosphere analysis of low resolution spectra. They applied their new method on 24 supergiants of spectral type B8 to A5 in the Sculptor Group spiral galaxy NGC 300 (at 1.9 Mpc distance) and obtained temperatures, gravities, metallicities, radii, luminosities and masses. The spectroscopic observations were obtained with FORS1 at the ESO VLT in multiobject spectroscopy mode. In addition, ESO/MPI 2.2m WFI and HST/ACS photometry was used. The observations were carried out within the framework of the Araucaria Project (Gieren *et al.* 2005b). In the following we discuss the analysis method and the results of this pilot study.

4. A Pilot Study in NGC 300 – Analysis Method

For the quantitative analysis of the spectra KUBGP use the same combination of line blanketed model atmospheres and very detailed NLTE line formation calculations as Przybilla *et al.* (2006) in their high signal-to-noise and high spectral resolution study of galactic A-supergiants, which reproduce the observed normalized spectra and the spectral energy distribution, including the Balmer jump, extremely well. They calculate an extensive, comprehensive and dense grid of model atmospheres and NLTE line formation covering the potential full parameter range of all the objects in gravity (log g = 0.8 to 2.5), effective temperature ($T_{\rm eff}$ = 8300 to 15000K) and metallicity ([Z] = log $Z/Z_\odot$ = −1.30 to 0.3). The total grid comprises more than 6000 models.

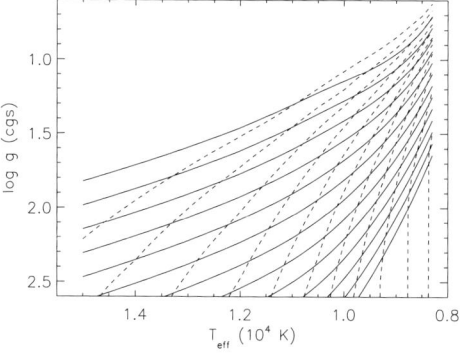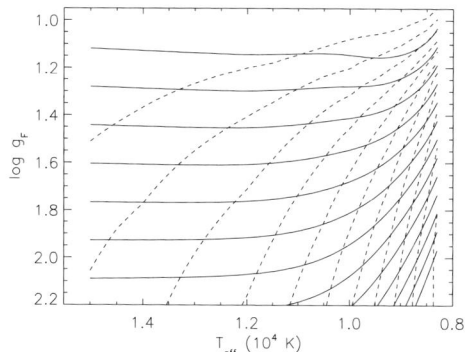

Figure 1. Left: Isocontours of Hδ equivalent widths (solid) and Balmer jump D_B (dashed) in the (log g, log $T_{\rm eff}$) plane. Hδ isocontours start with 1 Å equivalent width and increase in steps of 0.5 Å. D_B isocontours start with 0.1 dex and increase by 0.1 dex. **Right:** Same as left but for the flux weighted gravity log g_F instead of gravity log g. Note that this diagram is independent of metallicity, since both the strengths of Balmer lines and the Balmer jump depend only very weakly on metallicity.

The analysis of each of the 24 targets in NGC 300 proceeds in three steps. First, the stellar parameters (T_{eff} and log g) are determined together with interstellar reddening and extinction, then the metallicity is determined and finally, assuming a distance to NGC 300, stellar radii, luminosities and masses are obtained. For the first step, a well established method to obtain the stellar parameters of supergiants of late B to early A spectral type is to use ionization equilibria of weak metal lines (OI/II; MgI/II; NI/II etc.) for the determination of effective temperature T_{eff} and the Balmer lines for the gravities log g. However, at the low resolution of 5 Å the weak spectral lines of the neutral species disappear in the noise of the spectra and an alternative technique is required to obtain temperature information. KUBGP confirm the result by Evans & Howarth (2003) that a simple application of a spectral type - effective temperature relationship does not work because of the degeneracy of such a relationship with metallicity. Fortunately, a way out of this dilemma is the use of the spectral energy distributions (SEDs) and here, in particular of the Balmer jump D_B. While the observed photometry from B-band to I-band is used to constrain the interstellar reddening, D_B turns out to be a reliable temperature diagnostic, as is indicated by Fig. 1. A simultaneous fit of the Balmer lines and the Balmer jump allows to constrain effective temperature and gravity independent of assumptions on metallicity. Fig. 2 demonstrates the sensitivity of the Balmer lines and the Balmer jump to gravity and effective temperature, respectively. At a fixed temperature the Balmer lines allow for a determination of log g within 0.05 dex uncertainty, whereas the Balmer jump at a fixed gravity yields a temperature uncertainty of 2 percent. However, since the isocontours in Fig. 1 are not orthogonal, the maximum errors of T_{eff} and log g are 5 percent and 0.2 dex, respectively. These errors are significantly larger than for the analysis of high resolution spectra, but they still allow for an accurate determination of metallicity and distances.

The accurate determination of T_{eff} and log g is crucial for the use of A supergiants as distance indicators using the relationship between absolute bolometric magnitude M_{bol} and flux weighted gravity log g_F defined as

$$\log g_F = \log g - 4 \log T_{\text{eff},4} \tag{4.1}$$

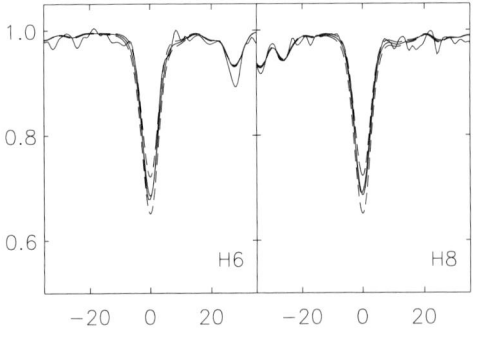
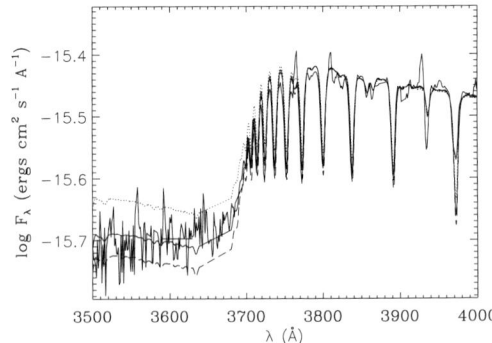

Figure 2. Left: Model atmosphere fit of two observed Balmer lines of NGC 300 target No. 21 of KUBGP for $T_{\text{eff}} = 10000$ K and log $g = 1.55$ (solid). Two additional models with same T_{eff} but log $g = 1.45$ and 1.65, respectively, are also shown (dashed). **Right:** Model atmosphere fit of the observed Balmer jump of the same target for $T_{\text{eff}} = 10000$ K and log $g = 1.55$ (solid). Two additional models with the same log g but $T_{\text{eff}} = 9750$ K (dashed) and 10500 K (dotted) are also shown. The horizontal bar at 3600 Å represents the average of the flux logarithm over this wavelength interval, which is used to measure D_B.

where $T_{eff,4} = T_{eff}/10^4$ K (see Kudritzki, Bresolin & Przybilla 2003). The relatively large uncertainties obtained with this fit method may cast doubts whether $\log g_F$ can be obtained accurately enough. Fortunately, the non-orthogonal behaviour of the fit curves in the left part of Fig. 1 leads to errors in T_{eff} and $\log g$, which are correlated in a way that reduces the uncertainties of $\log g_F$. This is demonstrated in the right part of Fig. 1, which shows the corresponding fit curves of the Balmer lines and D_B in the ($\log g_F$, $\log T_{eff}$) plane. As a consequence, much smaller uncertainties are obtained for $\log g_F$, namely ± 0.10 dex. KUBGP give a detailed discussion of physical reason behind this.

Knowing the stellar atmospheric parameters T_{eff} and $\log g$ KUBGP are able to determine stellar metallicities by fitting the metal lines with their comprehensive grid of line formation calculations. The fit procedure proceeds in several steps. First, spectral windows are defined, for which a good definition of the continuum is possible and which are relatively undisturbed by flaws in the spectrum (for instance caused by cosmic events) or interstellar emission and absorption. A typical spectral window used for all targets is the wavelength interval 4497 Å $\leqslant \lambda \leqslant$ 4607 Å. Fig. 3 shows the synthetic spectrum calculated for the atmospheric parameters of target No. 21 (the previous example) and for all the metallicities of the grid ranging from $-1.30 \leqslant [Z] \leqslant 0.30$. It is very obvious that the strengths of the metal line features are a strong function of metallicity. In Fig. 4 the observed spectrum of target No. 21 in this spectral region is shown overplotted by the synthetic spectrum for each metallicity. Separate plots are used for each metallicity, because the optimal relative normalization of the observed and calculated spectra is obviously metallicity dependent. This problem is addressed by renormalizing the observed spectrum for each metallicity so that the synthetic spectrum always intersects the observations at the same value at the two edges of the spectral window (indicated by the dashed vertical lines). The next step is a pixel-by-pixel comparison of calculated and normalized observed fluxes for each metallicity and a calculation of a χ^2-value. The minimum $\chi([Z])$ as a function of $[Z]$ is then used to determine the metallicity. This is also shown in Fig. 4. Application of the same method on different spectral windows provides additional independent information on metallicity and allows to determine the average metallicity obtained from all windows. A value of is $[Z] = -0.39$ with a very small dispersion of only 0.02 dex. However, one also need to consider the effects of the stellar

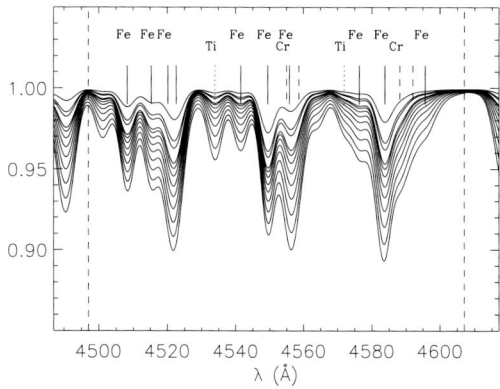

Figure 3. Synthetic metal line spectra calculated for the stellar parameters of target No.21 as a function of metallicity in the spectral window from 4497 Å to 4607 Å. Metallicities range from $[Z] = -1.30$ to 0.30, as described in the text. The dashed vertical lines give the edges of the spectral window as used for a determination of metallicity.

parameter uncertainties on the metallicity determination. This is done by applying the same correlation method for [Z] for models at the extremes of the error box for T_{eff} and $\log g$. This increases the uncertainty of [Z] to ± 0.15 dex, still a very reasonable accuracy of the abundance determination.

The fit of the observed photometric fluxes with the model atmosphere fluxes was used to determine interstellar reddening $E(B-V)$ and extinction $A_V = 3.1\, E(B-V)$. Simultaneously, the fit also yields the stellar angular diameter, which provides the stellar radius, if a distance is adopted. Alternatively, for the stellar parameters (T_{eff}, $\log g$, [Z]) determined through the spectral analysis the model atmospheres also yield bolometric corrections BC, which we use to determine bolometric magnitudes. These bolometric magnitudes then also allow us to compute radii. The radii determined with these two different methods agree within a few percent. Gieren et al. (2005a) in their multi-wavelength study of a large sample of Cepheids in NGC 300 including the near-IR have determined a new distance modulus m-M = 26.37 mag, which corresponds to a distance of 1.88 Mpc. KUBGP have adopted these values to obtain the radii and absolute magnitudes.

5. A Pilot Study in NGC 300 – Results

As a first result, the quantitative spectroscopic method yields interstellar reddening and extinction as a by-product of the analysis process. For objects embedded in the dusty disk of a star forming spiral galaxy one expects a wide range of interstellar reddening $E(B-V)$ and, indeed, a range from $E(B-V) = 0.07$ mag up to 0.24 mag was found. The individual reddening values are significantly larger than the value of 0.03 mag adopted in the HST distance scale key project study of Cepheids by Freedman et al. (2001) and demonstrates the need for a reliable reddening determination for stellar distance indicators, at least as long the study is restricted to optical wavelengths. The average over the observed sample is $\langle E(B-V)\rangle = 0.12$ mag in close agreement with the value of 0.1 mag found by Gieren et al. (2005a) in their optical to near-IR study of Cepheids in NGC 300. While Cepheids have somewhat lower masses than the A supergiants of our study and are consequently somewhat older, they nonetheless belong to the same population and are found at similar sites. Thus, one expects them to be affected by interstellar reddening in the same way as A supergiants.

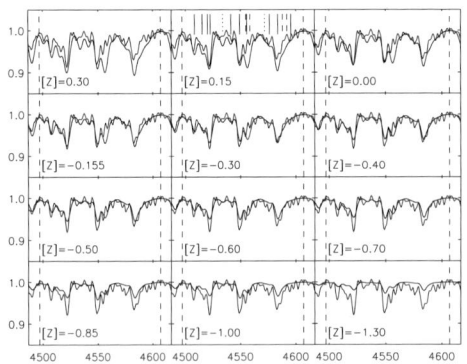
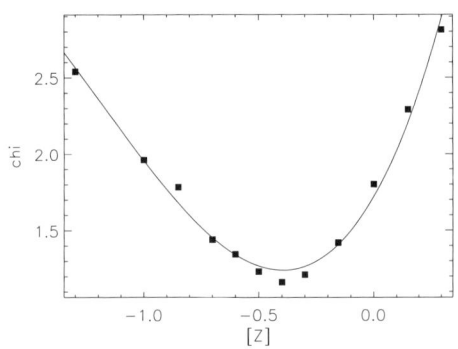

Figure 4. Left: Observed spectrum of target No. 21 for the same spectral window as Fig. 3 overplotted by the same synthetic spectra for each metallicity separately. **Right:** $\chi([Z])$ as obtained from the comparison of observed and calculated spectra. The solid curve is a third order polynomial fit.

Fig. 5 shows the location of all the observed targets in the (log g, log $T_{\rm eff}$) plane and in the HRD compared to the early B-supergiants studied by Urbaneja et al. (2005). The comparison with evolutionary tracks gives a first indication of the stellar masses in a range from 10 $M_\odot$ to 40 $M_\odot$. Three targets have obviously higher masses than the rest of the sample and seem to be on a similar evolutionary track as the objects studied by Urbaneja et al. (2005). The evolutionary information obtained from the two diagrams appears to be consistent. The B-supergiants seem to be more massive than most of the A supergiants. The same three A supergiants apparently more massive than the rest because of their lower gravities are also the most luminous objects. This confirms that quantitative spectroscopy is – at least qualitatively – capable to retrieve the information about absolute luminosities. Note that the fact that all the B supergiants studied by Urbaneja et al. (2005) are more massive is simply a selection effect of the V magnitude limited spectroscopic survey by Bresolin et al. (2002). At similar V magnitude as the A supergiants those objects have higher bolometric corrections because of their higher effective temperatures and are, therefore, more luminous and massive.

Fig. 6 shows the stellar metallicities and the metallicity gradient as a function of angular galactocentric distance, expressed in terms of the isophotal radius, ρ/ρ_0 (ρ_0 corresponds to 5.33 kpc). Despite the scatter caused by the metallicity uncertainties of the individual stars the metallicity gradient of the young disk population in NGC 300 is very clearly visible. A linear regression for the combined A- and B-supergiant sample yields (d in kpc)

$$[Z] = -0.06 \pm 0.09 - (0.083 \pm 0.022)\, d. \qquad (5.1)$$

Note that the metallicities of the B supergiants refer to oxygen only with a value of log N(O)/N(H) = −3.31 adopted for the Sun (Allende Prieto et al. 2001). On the other hand, the A supergiant metallicities reflect the abundances of a variety of heavy elements such as Ti, Fe, Cr, Si, S, and Mg.

With these results KUBGP extended the discussion started by Urbaneja et al. (2005) to compare with oxygen abundances obtained from HII-region emission lines. Urbaneja

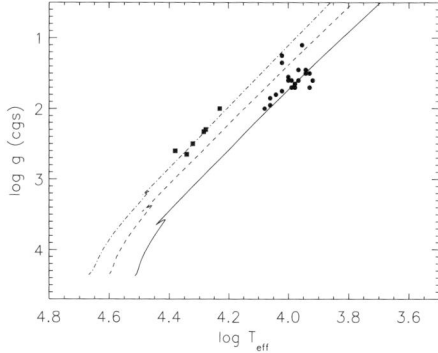 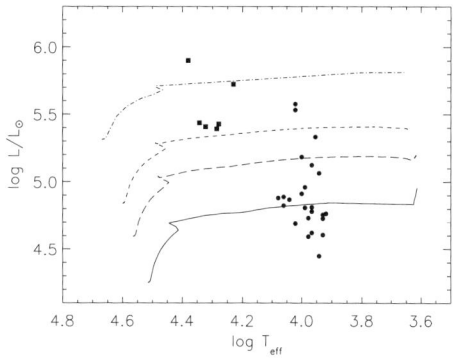

Figure 5. Left: NGC 300 A supergiants (filled circles) and early B supergiants (filled squares) in the (*log g, log $T_{\rm eff}$*) plane compared with evolutionary tracks by Meynet & Maeder (2005) of stars with 15 $M_\odot$ (solid), 25 $M_\odot$ (dashed), and 40 $M_\odot$ (dashed-dotted), respectively. **Right:** NGC 300 A and early B supergiants in the HRD compared with evolutionary tracks for stars with 15 $M_\odot$ (solid), 20 $M_\odot$ (long-dashed), 25 $M_\odot$ (short-dashed), and 40 $M_\odot$ (dashed-dotted), respectively. The tracks include the effects of rotation and are calculated for SMC metallicity (see Meynet & Maeder 2005).

320   R. Kudritzki *et al.*

et al. (2005) used line fluxes published by Deharveng *et al.* (1988) and applied various different published strong line method calibrations to determine nebular oxygen abundances, which could then be used to obtain similar regressions as above.

The different strong line method calibrations lead to significant differences in the central metallicity as well as in the abundance gradient. The calibrations by Dopita & Evans (1986) and Zaritsky *et al.* (1994) predict a metallicity significantly supersolar in the center of NGC 300 contrary to the other calibrations. On the other hand, the work by KUBGP yields a central metallicity slightly smaller than solar in good agreement with Denicolo *et al.* (2002) and marginally agreeing with Kobulnicky *et al.* (1999), Pilyugin (2001), and Pettini & Pagel (2004). At the isophotal radius, 5.3 kpc away from the center of NGC 300, they obtain an average metallicity significantly smaller than solar [Z] = −0.50, close to the average metallicity in the SMC. The calibrations by Dopita & Evans (1986), Zaritsky *et al.* (1994), Kobulnicky *et al.* (1999) do not reach these small values for oxygen in the HII regions either because their central metallicity values are too high or the metallicity are gradients too shallow.

In light of the substantial range of metallicies obtained from HII region emission lines using different strong line method calibrations it seems to be extremely valuable to have an independent method using young stars. It will be very important to compare our results with advanced work on HII regions, which will avoid strong line methods, and will use direct information about nebular electron temperatures and densities (Bresolin *et al.* 2008, in preparation, see also this volume).

KUBGP discuss the few outliers in Fig. 6 and claim that these metallicities seem to be real. Their argument is that the expectation of homogeneous azimuthal metallicity in patchy star forming galaxies seems to be naive. Future work on other galaxies will show whether cases like this are common or not.

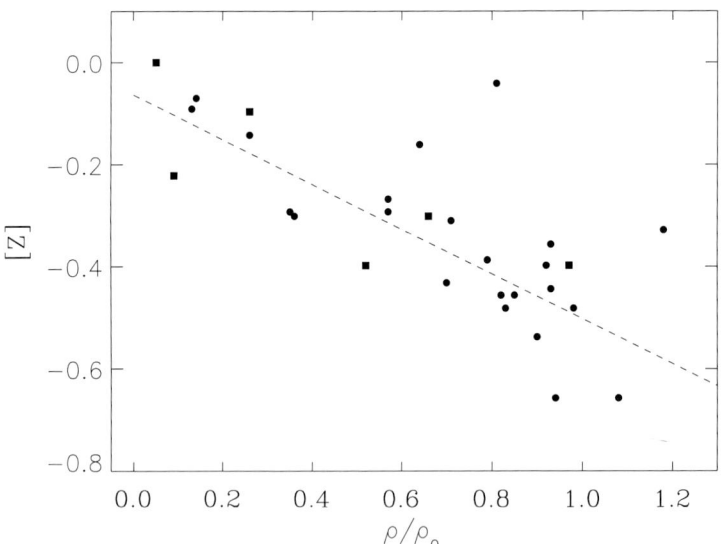

Figure 6. Metallicity [Z] as a function of angular galacto-centric distance ρ/ρ_0 for the A supergiants (filled circles) and the early B-supergiants studied by Urbaneja *et al.* (2005) (filled squares). Note that for the latter metallicity refers to oxygen only. The dashed curve represents the regression discussed in the text.

6. Flux Weighted Gravity – Luminosity Relationship (FGLR)

Massive stars with masses in the range from 12 $M_\odot$ to 40 $M_\odot$ evolve through the B and A supergiant stage at roughly constant luminosity (see Fig. 5). In addition, since the evolutionary timescale is very short when crossing through the B and A supergiant domain, the amount of mass lost in this stage is small. This means that the evolution proceeds at constant mass and constant luminosity. This has a very simple, but very important consequence for the relationship of gravity and effective temperature along each evolutionary track. From

$$L \propto R^2 T_{\rm eff}^4 = {\rm const.}; \qquad M = {\rm const.} \tag{6.1}$$

follows immediately that

$$M \propto g\, R^2 \propto L\, (g/T_{\rm eff}^4) = L\, g_F = {\rm const.} \tag{6.2}$$

Thus, along the evolution through the B and A supergiant domain the *"flux-weighted gravity"* $g_F = g/T_{\rm eff}^4$ should remain constant. This means that each evolutionary track of different luminosity in this domain is characterized by a specific value of g_F. This value is determined by the relationship between stellar mass and luminosity, which in a first approximation is a power law

$$L \propto M^x \tag{6.3}$$

and leads to a relationship between luminosity and flux-weighted gravity

$$L^{1-x} \propto (g/T_{\rm eff}^4)^x \,. \tag{6.4}$$

With the definition of bolometric magnitude $M_{\rm bol} \propto -2.5 \log L$ one then derives

$$-M_{\rm bol} = a_{\rm FGLR} (\log\, g_F - 1.5) + b_{\rm FGLR} \,. \tag{6.5}$$

This is the *"flux-weighted gravity – luminosity relationship"* (FGLR) of blue supergiants. Note that the proportionality constant $a_{\rm FGLR}$ is given by the exponent of the mass – luminosity power law through

$$a_{\rm FGLR} = 2.5x/(1-x) \,, \tag{6.6}$$

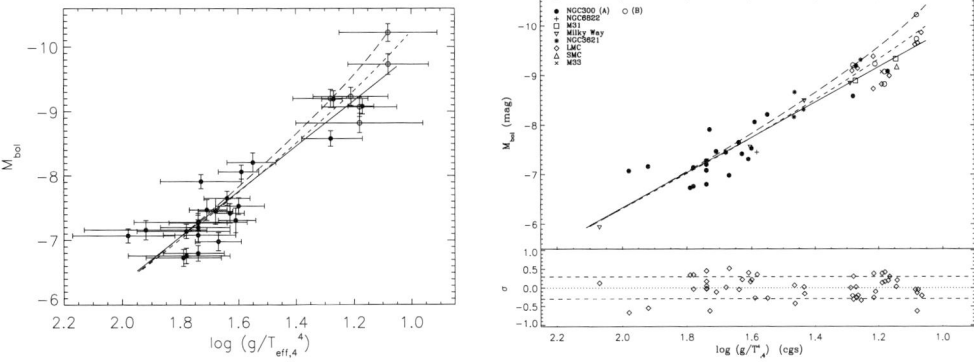

Figure 7. Left: The FGLR of A (solid circles) and B (open circles) supergiants in NGC 300 and the linear regression (solid). The stellar evolution FGLRs for models with rotation are also overplotted (dashed: Milky Way metallicity, long-dashed: SMC metallicity). **Right:** Same as left, but with additional objects from 7 other galaxies (see text).

for instance, for $x = 3$, one obtains $a_{\rm FGLR} = -3.75$. Note that the zero point of the relationship is chosen at a flux weighted gravity of 1.5, which is in the middle of the range encountered for blue supergiant stars.

KUBGP use the mass-luminosity relationships of different evolutionary tracks (with and without rotation, for Milky Way and SMC metallicity) to calculate the different FGLRs predicted by stellar evolution. Very interestingly, while different evolutionary model types yield somewhat different FGLRs, the differences are rather small.

Kudritzki, Bresolin & Przybilla (2003) were the first to realize that the FGLR has a very interesting potential as a purely spectroscopic distance indicator, as it relates two spectroscopically well defined quantities, effective temperature and gravity, to the absolute magnitude. Compiling a large data set of spectroscopic high resolution studies of A supergiants in the Local Group and with an approximate analysis of low resolution data of a few targets in galaxies beyond the Local Group (see discussion in previous chapters) they were able to prove the existence of an observational FGLR rather similar to the theoretically predicted one.

With the improved analysis technique of low resolution spectra of A supergiants and with the much larger sample studied for NGC 300, KUBGP resumed the investigation of the FGLR.

The result is shown in Fig. 7, which for NGC 300 (left diagram) reveals a clear and rather tight relationship of flux weighted gravity $\log g_F$ with bolometric magnitude $M_{\rm bol}$. A simple linear regression yields $b_{\rm FGLR} = 8.11$ for the zero point and $a_{\rm FGLR} = -3.52$ for the slope. The standard deviation from this relationship is $\sigma = 0.34$ mag. Within the uncertainties the observed FGLR appears to be in agreement with the theory.

In their first investigation of the empirical FGLR Kudritzki, Bresolin & Przybilla (2003) have added A supergiants from six Local Group galaxies with stellar parameters obtained from quantitative studies of high resolution spectra (Milky Way, LMC, SMC, M31, M33, NGC 6822) to their results for NGC 300 to obtain a larger sample. They also added 4 objects from the spiral galaxy NGC 3621 (at 6.7 Mpc) which were studied at low resolution. KUBGP added exactly the same data set to their new enlarged NGC 300 sample, however, with a few minor modifications. For the Milky Way they included the latest results from Przybilla *et al.* (2006) and Schiller & Przybilla (2008) and for the two objects in M31 we use the new stellar parameters obtained by Przybilla *et al.* (2008). For the objects in NGC 3621 they applied new HST photometry. They also re-analyzed the LMC objects using ionization equilibria for the temperature determination.

Fig. 7 (right diagram) shows bolometric magnitudes and flux-weighted gravities for this full sample of eight galaxies revealing a tight relationship over one order of magnitude in flux-weighted gravity. The linear regression coefficients are $a_{\rm FGLR} = -3.41 \pm 0.16$ and $b_{\rm FGLR} = 8.02 \pm 0.04$, very similar to the NGC 300 sample alone. The standard deviation is $\sigma = 0.32$ mag. The stellar evolution FGLR for Milky Way metallicity provides a fit of almost similar quality with a standard deviation of $\sigma = 0.31$ mag.

7. Conclusions and Future Work

The astrophysical potential of low resolution spectroscopy of A supergiant stars for studying galaxies beyond the Local Group is quite remarkable. By introducing a novel method for the quantitative spectral analysis one is able to determine accurate stellar parameters, which allow for a detailed test of stellar evolution models. Through the spectroscopic determination of stellar parameters one can also constrain interstellar reddening and extinction by comparing the calculated SED with broad band photometry. The study of NGC 300 finds a very patchy extinction pattern as to be expected for a star forming spiral galaxy. The average extinction is in agreement with multi-wavelength studies of Cepheids including K-band photometry.

The method also allows to determine stellar metallicities and to study stellar metallicity gradients. Solar metallicity is found in the center of NGC 300 and a gradient of -0.08 dex/kpc. To our knowledge this is the first systematic stellar metallicity study in galaxies beyond the Local Group focussing on iron group elements. In the future the method can be extended to not only determine metallicity but also the ratio of α- to iron group elements as a function of galactocentric distance. The stellar metallicities obtained can be compared with oxygen abundance studies of HII regions using the strong line method. This allows to discuss the various calibrations of the strong line method, which usually yield very different results.

The improved spectral diagnostic method makes it possible to very accurately determine stellar flux weighted gravities $\log g_F = \log g/T_{\rm eff}^4$ and bolometric magnitudes. Above a certain threshold in effective temperature a simple measurement of the strengths of the Balmer lines can be used to determine accurate values of $\log g_F$.

Absolute bolometric magnitudes $M_{\rm bol}$ and flux-weighted gravities $\log g_F$ are tightly correlated. It is shown that such a correlation is expected for stars, which evolve at constant luminosity and mass. This "flux-weighted gravity - luminosity relationship" (FGLR) agrees with stellar evolution theory.

With a relatively small residual scatter of $\sigma = 0.3$ mag the observed FGLR is an excellent tool to determine accurate spectroscopic distance to galaxies. It requires multicolor photometry and low resolution (5 Å) spectroscopy to determine effective temperature and gravity and, thus, flux-weighed gravity directly from the spectrum. With effective temperature, gravity and metallicity determined one also knows the bolometric correction, which is small for A supergiants, which means that errors in the stellar parameters do not largely affect the determination of bolometric magnitudes. Moreover, one knows the intrinsic stellar SED and, therefore, can determine interstellar reddening and extinction from the multicolor photometry, which then allows for the accurate determination of the reddening-free apparent bolometric magnitude. The application of the FGLR then yields absolute magnitudes and, thus, the distance modulus. With the intrinsic scatter of $\sigma = 0.3$ mag and 30 targets per galaxy one can estimate an accuracy of 0.05 mag in distance modulus (0.1 mag for 10 target stars). The results for WLM by Miguel Urbaneja et al. and for M33 by Vivian U et al. presented in these proceedings are a first demonstration of the power of the method.

The advantage of the FGLR method for distance determinations is its spectroscopic nature, which provides significantly more information about the physical status of the objects used for the distance determination than simple photometry methods. Most importantly, metallicity and interstellar extinction can be determined directly. The latter is crucial for spiral and irregular galaxies because of the intrinsic patchiness of reddening and extinction.

Since supergiant stars are known to show intrinsic photometric variability, the question arises whether the FGLR method is affected by such variability. For the targets in NGC 300 this issue has been carefully investigated by Bresolin et al. (2004), who studied CCD photometry lightcurves obtained over many epochs in the parallel search for Cepheids in NGC 300. They concluded that amplitudes of photometric variability are very small and do not affect distance determinations using the FGLR method.

The effects of crowding and stellar multiplicity are also important. However, in this regard A supergiants offer tremendous advantages relative to other stellar distance indicators. First of all, they are significantly brighter. Bresolin et al. (2005) using HST ACS photometry compared to ground-based photometry have studied the effects of crowding on the Cepheid distance to NGC 300 and concluded that they are negligible. With A supergiants being 3 to 6 magnitudes brighter than Cepheids it is clear that even with ground-based photometry only crowding is generally not an issue for these objects at the distance of NGC 300 and, of course, with HST photometry (and in the future JWST) one can reach much larger distances before crowding becomes important. In addition, any significant contribution by additional objects to the light of an A supergiant will become apparent in the spectrum, if the contaminators are not of a very similar spectral type, which is very unlikely because of the short evolutionary lifetime in the A supergiant stage. It is also important to note that A supergiants have evolutionary ages larger than 10 million years, which means that they have time to migrate into the field or that they are found in older clusters, which are usually less concentrated than the very young OB associations.

It is evident that the type of work described in this paper can be in a straightforward way extended to the many spiral galaxies in the local volume at distances in the 4 to 7 Mpc range. Pushing the method we estimate that with present day 8m to 10m class telescopes and the existing very efficient multi-object spectrographs one can reach down with sufficient S/N to V = 22.5 mag in two nights of observing time under very good conditions. For objects brighter than $M_V = -8$ mag this means metallicities and distances can be determined out to distances of 12 Mpc ($m - M = 30.5$ mag). This opens up a substantial volume of the local universe for metallicity and galactic evolution studies and independent distance determinations complementary to the existing methods. With the next generation of extremely large telescopes such as the TMT, GMT or the E-ELT the limiting magnitude can be pushed to V = 24.5 equivalent to distances of 30 Mpc ($m - M = 32.5$ mag).

References

Allende Prieto, C., Lambert, D. L., & Asplund, M. 2001, *ApJ*, 556, L63
Aufdenberg, J. P., Hauschildt, P. H., Baron, E. *et al.* 2002, *ApJ*, 570, 344
Bresolin, F., Kudritzki, R.-P., Méndez, R. H., & Przybilla, N. 2001, *ApJ*, 548, L159
Bresolin, F., Gieren, W., Kudritzki, R.-P., *et al.* 2002, *ApJ*, 567, 277
Bresolin, F., Pietrzyński, G., Gieren, W., *et al.* 2004, *ApJ*, 600, 182
Bresolin, F., Pietrzyński, G., Gieren, W. & Kudritzki, R. P. 2005, *ApJ*, 634, 1020
Deharveng, L., Caplan, J., Lequeux, J., *et al.* 1988, *A&AS*, 73, 407
Denicolo, G., Terlevich, R.& Terlevich, E. 2002, *MNRAS*, 330, 69
Dopita, M. A., Evans, I. N. 1986, *ApJ*, 307, 431
Evans, C. J. & Howarth, I. D. 2003, *MNRAS*, 345, 1223
Freedman *et al.* 2001, *ApJ*, 553, 47
Gieren, W., Pietrzyński, G., Soszynski, I., *et al.* 2005, *ApJ*, 628, 695
Gieren, W. *et al.* 2005, ESO Messenger, 121, 23
Groth, H. G. 1961, *ZAp*, 51, 231
Kaufer, A., Venn, K. A., Tolstoy, E., *et al.* 2004, *AJ*, 127, 2723
Kobulnicky, H. A., Kennicutt, R. C. & Pizagno, J. L. 1999, *ApJ*, 514, 544
Kudritzki, R.-P. 1973, *A&A*, 28, 103
Kudritzki, R.-P., 1998, in: A. Aparicio, A. Herrero & F. Sanchez (eds.), *Stellar Physics for the Local Group* (Cambridge: CUP) 149
Kudritzki, R.-P., Bresolin, F., & Przybilla, N. 2003, *ApJ*, 582, L83
Kudritzki, R.-P., Lennon, D. J. & Puls, J. 1995, in: J. R. Welsh & I. J. Danziger (eds.), *Science with VLT*, (Berlin: Springer-Verlag), 246
Kudritzki, R.-P., Puls, J., Lennon, D. J. *et al.* 1999, *A&A*, 350, 970
Kudritzki, R.-P., Urbaneja, M. A., Bresolin, F., *et al.* 2008, *ApJ*, in press (astro-ph/0803.3654)
McCarthy, J. K., Lennon, D. J., Venn, K. A., *et al.* 1995, *ApJ*, 455, L135

McCarthy, J. K., Kudritzki, R. P., Lennon, D. J., *et al.* 1997, *ApJ*, 482, 757
Meynet, & Maeder, A. 2005, *A&A*, 429, 581
Pettini, M. & Pagel, B. E.J. 2004, *MNRAS*, 348, L59
Pilyugin, L. S. 2001, *A&A*, 369, 594
Przybilla, N., Butler, K., Becker, S. R., *et al.* 2000, *A&A*, 359, 1085
Przybilla, N., Butler, K., Becker, S. R., & Kudritzki, R. P. 2001a, *A&A*, 369, 1009
Przybilla, N., Butler, K. & Kudritzki, R. P. 2001b, *A&A*, 379, 936
Przybilla, N. & Butler, K. 2001, *A&A*, 379, 955
Przybilla, N. 2002, thesis, Fakultaet fuer Physik, Ludwig-Maximilian University, Munich
Przybilla, N., Butler, K., Becker, S. R., & Kudritzki, R. P. 2006, *A&A*, 445, 1099
Przybilla, N., Butler, K., & Kudritzki, R. P. 2008, in: G. Israelian & G. Meynet (eds.), *The Metal-Rich Universe*, (Cambridge: CUP), in press (astro-ph/0611044)
Schiller, F. & Przybilla, N. 2008, *A&A*, 479, 849
Urbaneja, M. A., Herrero, A. J., Bresolin, F., *et al.* 2005, *ApJ*, 622, 877
Venn, K. A. 1995a, *ApJS*, 99, 659
Venn, K. A. 1995a, *ApJ*, 449, 839
Venn, K. A. 1999, *ApJ*, 518, 405
Venn, K. A., Lennon, D. J., Kaufer, A., *et al.* 2001, , 547, 765
Venn, K. A., McCarthy, J. K., Lennon, D. J., *et al.* 2000, *ApJ*, 541, 610
Venn, K. A., Tolstoy, E. Kaufer, A. *et al.* 2003, *AJ*, 126, 1326
Wolf, B. 1971, *A&A*, 10, 383
Wolf, B. 1972, *A&A*, 20, 275
Wolf, B. 1973, *A&A*, 28, 335
Zaritski, D., Kennicutt, R. C. & Huchra, J. P. 1994, *ApJ*, 420, 87

Discussion

MASSEY: This is absolutely beautiful work! Let me ask one question, though! You have described the precision with which you can measure effective temperatures, surface gravities, abundances, etc. Do you have a sense of their systematic accuracy? Your models must be the best possible now, but what might affect the results of the model that you are unsure of?

KUDRITZKI: There is always the possibility of systematic effects caused by the imperfectness of the models and the NLTE radiative transfer. There are several ways how one can try to address those. For the effective temperatures one can use different methods, ionization equilibria and Balmer jumps, for instance. When the results agree, this is a good indication. With our IfA student Ben Granett we have also tried the alternative approach to determine $T_{\rm eff}$ from the full integral over the complete UV, optical, IR energy distribution and we found good agreement again. You can test gravities with detached eclipsing binaries. This has been done in the most recent paper by Bonanos *et al.* (2006, *ApJ*, 652, 313). For abundances, the rule is to use as many lines as possible, also in different spectral windows, and to look at the scatter. This work has been done very carefully by Przybilla and collaborators (see his poster papers, this volume).

BURBIDGE: I wish to make three points. (1) This is marvellous work. The spectroscopic method of distance determination is very powerful. (2) Don't be convinced by those who claim that the Hubble Constant is well determined. The value obtained by Sandage and Tammann of about 55 km sec^{-1} Mpc^{-1} is much more likely correct than the value of 71 km sec^{-1} Mpc^{-1} being claimed. (3) How long will it take to extend the method to the Virgo Cluster?

KUDRITZKI: Many thanks. I do not have a bias towards any value of the Hubble constant. I think that it is important to have as many independent reliable methods as possible. That is the reason why I am convinced that the FGLR method is really important. It is an independent way to check existing results. This is also the philosphy of the Araucaria-Project, where we compare Cepheid distances with TRGB distances and with the FGLR method. As for the Virgo cluster, this is a real challenge with Keck, but it will be easy with the TMT.

D'ODORICO: With ELTs like the TMT you expect to achieve the maximum angular resolution at IR wavelengths, where you can use AO at best. Do you think it would be possible to do this type of studies of the most luminous supergiant stars based on IR spectra?

KUDRITZKI: A lot of work has been done already on the IR quantitative spectroscopy of supergiants, as for instance presented by the papers by Paco Najarro and Fabrice Martins (this volume). Norbert Przybilla has done a lot already on A supergiants. It will certainly be worthwhile to extent this work to the IR, although the B and A supergiants emit less flux in this spectral window. At the same time, I think, it will also be important to push the AO limit as much towards the visual as by any means possible. While this will certainly not be possible for the ELT first light instruments, it is not excluded in the long term.

Rolf Kudritzki (left) and Hans Zinnecker (right).

VLT/FORS Surveys of Wolf-Rayet Stars in the Nearby Universe

Lucy J. Hadfield[1] and Paul A. Crowther[2]

[1] Center for Imaging Science, Rochester Institute of Technology,
54 Lomb Memorial Drive, Rochester, NY 14623, USA
email: `hadfield@cis.rit.edu`

[2] Dept. of Physics & Astronomy, The University of Sheffield,
The Hicks Building, Hounsfield Road, Sheffield, S3 7RH, UK
email: `Paul.Crowther@sheffield.ac.uk`

Abstract. We present results from a series of VLT/FORS narrow-band imaging and spectroscopic surveys of Wolf-Rayet (WR) stars in nearby spiral galaxies and compare observed populations in high- and low metallicity environments. The metal-rich galaxy M 83 is seen to host an exceptional WR content, with over 1000 WR stars being detected. N(WC)/N(WN) $\sim$ 1.2 and late-type WC subtypes dominate the WC population. At low metallicity, $\sim$100 stars has been identified within NGC 1313, with N(WC)/N(WN) $\sim$ 0.5. In contrast to M83, the WC population of NGC 1313 comprises solely early subtypes plus a WO star (the first WO star to be identified beyond the Local Group). Consequently, the dominant WC subtype may serve as a crude metallicity diagnostic for WR galaxies.

In addition, the WR content of the blue compact dwarf galaxy NGC 3125 is examined. Previous UV and optical spectroscopic studies of knot A in NGC 3125 derive WR populations which differ by more than an order of magnitude. New VLT observations and archival HST spectroscopy reconcile this discrepancy via the use of LMC WR spectral templates and a reduced nebular-derived interstellar extinction. Empirical N(WR)/N(O) ratios for clusters within NGC 3125 are a factor of two higher than evolutionary synthesis predictions but are consistent with those observed for other young massive clusters.

Keywords. stars: Wolf-Rayet – galaxies: individual (NGC 1313, M 83, NGC 3125) – galaxies: stellar content

---

1. Introduction

Wolf-Rayet (WR) stars represent the penultimate evolutionary stage in the evolution of the most massive stars. Believed to represent the bare cores of their O star precursors, their spectra are characterised by broad emission lines of nitrogen (WN), carbon (WC) or oxygen (WO). Their unique spectral appearance allows them to be readily identified in external galaxies such that WR stars can be been detected as individual stars in nearby galaxies (e.g. Massey 1998) and in the integrated starlight of more distant galaxies (Schaerer *et al.* 1999a).

Metallicity, Z, is a key factor in determining the absolute number and subtype distribution of a WR population. Prior to the WR phase, O star winds has been empirically established to depend on metallicity, with the latest results revealing $\dot{M} \propto Z^{\sim 0.8}$ for SMC, LMC and Milky Way O stars (Mokeim *et al.* 2007). Enhanced mass-loss rates in metal-rich environments reduces the minimum mass required for WR formation such that the WR mass cut-off is expected to decrease from ~ 30 M$_\odot$ in the SMC to ~ 25 M$_\odot$ for the Milky Way (Meynet *et al.* 2004). Consequently, observed stellar populations provide a vital test of stellar evolution models.

The advent of 8m class telescopes and efficient multi-object spectrographs has allowed WR surveys to move beyond the Local Group, permitting a vital probe into massive stellar evolution under conditions which are not accessible on a local scale. Using a combination of narrow-band imaging filters tuned to WR emission features, Schild et al. (2003) identified a significant numbers of WR stars in the nearby (D∼2 Mpc) spiral galaxy NGC 300. Therefore, to increase current WR statistics across a broad range of metallicities our team has undertaken a series of VLT/FORS surveys of the WR populations in several, nearby galaxies.

2. The Wolf-Rayet Population of Nearby Star-forming Spiral Galaxies

2.1. The WR Population of NGC 1313

NGC 1313 is an isolated, face-on SB(s)d spiral situated at a distance of 4.1 Mpc (Mendez et al. 2002). Oxygen abundance studies reveal that NGC 1313 is intermediate in metallicity between the irregular Magellanic Cloud galaxies, whilst morphologically NGC 1313 is reminiscent of late-type spirals such as NGC 300 and M 33.

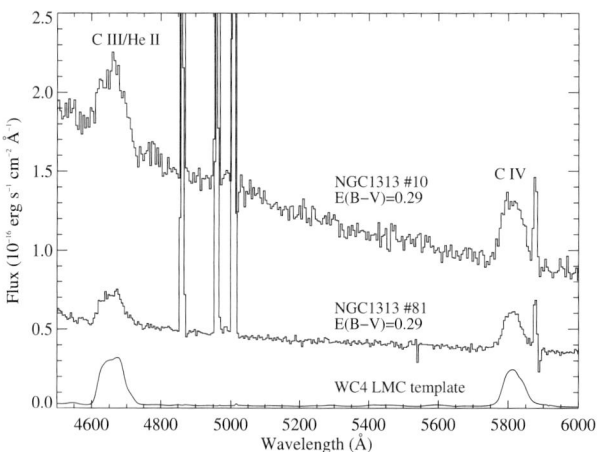

Figure 1. Dereddened spectral comparison between WC regions within NGC 1313 and template LMC WC4 spectra. Individual spectra have been offset by $0.1 \times 10^{-16} \mathrm{erg\,s^{-1}\,cm^{-2}\,\mathring{A}^{-1}}$.

Our VLT/FORS photometric survey of NGC 1313 has identified 94 potential WR regions in NGC 1313. Follow-up spectroscopy of 82 candidates confirms 85% host WR stars. For the majority of confirmed sources, observed line luminosities are consistent with single WR stars, as illustrated in Fig. 1. Ground-based imaging reveals that most sources appear relatively isolated and photometry suggests that they belong to binary systems or small clusters/associations.

Of the 12 cases where WR emission was not detected, two displayed nebular He II λ4686 whilst four spectra started longward of the He II λ4686 feature required for WN classification. Only six candidates showed no evidence of WR emission, of which three were consistent with foreground late-type stars.

Using template LMC WR stars, spectroscopy reveals N(WR)=84, with N(WC)/N(WN) = 0.6 (Hadfield & Crowther 2007). Accounting for the remaining candidates we estimate that the true WR content of NGC 1313 is N(WR)∼115, with N(WC)/(WN)∼0.4 (assuming photometric classifications). The WN stars are evenly distributed amongst early and

Table 1. Summary of nearby spiral galaxies which have been surveyed with FORS1/2. O star numbers are derived from Hα imaging, assuming an O7V Lyman continuum flux of 10^{49} ph/s.

Galaxy	D (Mpc)	log(O/H)+12	N(O7V)	N(WC)	N(WN)	Reference
NGC 300	1.9	8.6:	800	⩾16	∼15	Schild et al. 2003
						Crowther et al. 2007
NGC 1313	4.1	8.23	6 500	⩾51	∼33	Hadfield & Crowther 2007
M 83	4.5	<9.2	40 000	⩾470	∼560	Hadfield et al. 2005

late subtypes and include a rare WN/C4 transition star. The WC population consists exclusively of early-type stars, to one of which we assign a WO classification. This represents the first WO star to be identified beyond the Local Group. The WR population of NGC 1313 is presented in Table 1.

2.2. The WR Population of M 83

The study of WR stars in metal-rich environments has so far been restricted to M 31 (which has an unfavourable inclination) within the Local Group and the integrated light of star forming regions within starburst galaxies. Therefore, we have also investigated the WR population of the nearby (D=4.5 Mpc), metal-rich ($Z \sim 2Z_\odot$) environment of the grand design spiral galaxy M 83.

Our VLT/FORS imaging survey encompassed the entire galaxy, but our statistics excludes the starburst nucleus since the central $15''$ appears saturated on all FORS images. In excess of 280 WR candidate regions have been identified within the disc of M 83, of which 198 have been spectroscopically observed. The presence of WR stars has been confirmed in 131 regions i.e. a success rate of 66%.

Using Galactic WR stars as templates, we infer a WR population of ∼1100 stars (Hadfield et al 2005), ten times that estimated for NGC 1313. Observed line luminosities suggests that some sources host a single WR star, whilst others contain larger WR populations (N∼10). Both the WC and WN populations of M 83 are dominated by late subtypes, with WO subtypes absent. In contrast to NGC 1313, the majority of WR stars in M 83 are located within bright star forming regions and given that both galaxies are located at comparable distances the lower success rate of our M 83 survey reflects resolution issues of ground-based surveys.

2.3. Comparison with Evolutionary Predictions

Surveys for WR stars in Local Group galaxies over the past three decades have revealed a strong correlation between the relative number of WC to WN stars and oxygen content of the host galaxy (Massey & Johnson 1998). Undoubtedly, completeness should be kept in mind given that WC stars are more readily identified due to their intrinsically stronger lines. Nevertheless, our imaging surveys are optimised for net emission at λ4686 and in M 83 we achieved 4σ spectroscopic WNL detections with W_λ(He II λ4686) ∼ 1Å.

Extrapolating from previous observations, one would expect N(WC)/N(WN) ⩾ 1 for a galaxy with twice the Solar oxygen content (Massey & Johnson 1998) whereas based on the low metallicity of NGC 1313 one would expect N(WC)/N(WN)∼0.1. M 83 continues the observed trend with N(WC)/N(WN)∼1.2. For NGC 1313, we estimate a significantly higher subtype ratio of ∼0.6, or ∼0.4 if we include outstanding candidates. This is intermediate between that observed in the outer, sub-solar regions of M 33 [N(WC)/N(WN)∼0.35; Massey & Johnson 1998] and the inner region of NGC 300 [N(WC)/N(WN)∼0.7; Crowther et al. 2007].

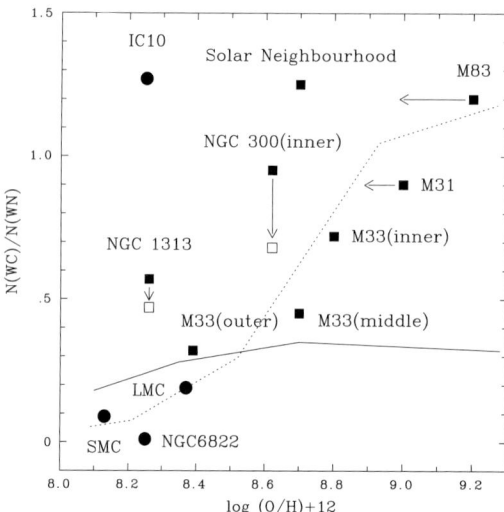

Figure 2. Observed N(WC)/N(WN) ratios for nearby spiral (squares) and irregular (circles) galaxies. Open symbols adopt photometric classifications for remaining WR candidates. Also shown are single star evolutionary model predictions for the rotating Geneva models (solid line; Meynet & Maeder 2005) and those of Eldridge & Vink (2006) (dotted). The latter models include a metallicity wind scaling for WR stars.

At low-metallicities, predictions from rotating evolutionary models are in good agreement with observed WC to WN ratios (see Fig. 2). However, at higher metallicities the Geneva models underestimate the number of WC stars e.g., predicting WC/WN=0.36 at $Z=0.04$. In contrast, the models which neglect rotational mixing, but include metallicity dependent WR winds, provide a better match to observations across the full metallicity range.

2.4. Using the WC subtype as a metallicity indicator

It has long been recognised that late-type WC stars are preferentially associated with metal-rich environments. For example, in the Milky Way WC9 stars are universally located within the inner, metal-rich regions (Conti & Vacca 1990), while metal-poor WC stars such as those observed in the Magellanic Clouds are exclusively early-type WC/WO stars. Results from our VLT/FORS surveys are fully consistent with such conclusions, since we identify an overwhelming WC8–9 population in M 83 and late-type WC stars are notably absent in NGC 1313.

To illustrate the difference in observed WC populations, Fig. 3 compares the ratio of WC7–9 (WCL) to WC4–6 (WCE) stars for a wide range of environments. At high metallicities ($\log{(O/H)} + 12 \geqslant 8.8$), late WC subtypes dominate the population (e.g., M 83). For intermediate metallicities ($\log{(O/H)} + 12 \sim 8.5 - 8.8$), the WC population is composed of a mixture of early and late WC subtypes, as observed for the Solar neighbourhood, Finally, in metal-poor galaxies such as NGC 1313 ($\log{(O/H)} + 12 \lesssim 8.5$) the WC population comprises solely of early-type WC and WO stars.

Exceptions to this general trend do occur. In the metal-poor galaxy IC 10 ($\log{(O/H)} + 12 = 8.26$) Crowther et al. (2003) identify one WC star as a WC7 subtype. Nevertheless, the dominant WC subtype may serve as a crude metallicity diagnostic for integrated stellar populations (e.g., SDSS WR galaxies). Indeed, the high metal content of NGC 1365 ($\log{(O/H)} + 12 = 9.3 - 9.5$) is confirmed by the presence a dominant WCL population (Phillips & Conti 1992).

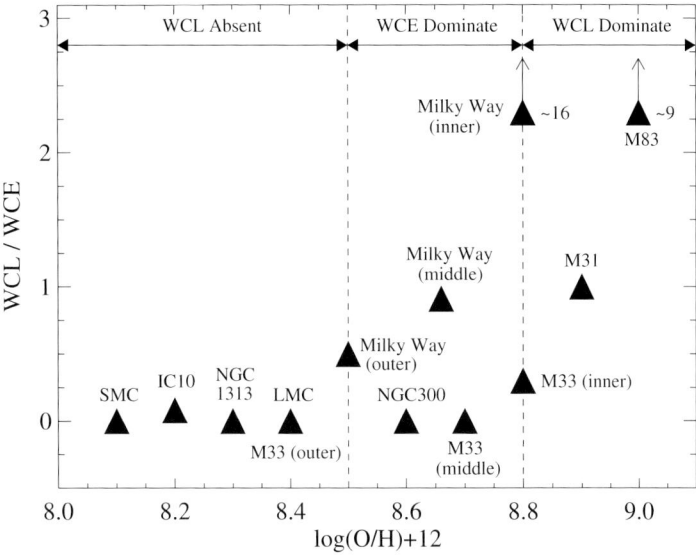

Figure 3. The distribution of WC4–6 to WC7–9 stars in well studied galaxies versus oxygen content (Hadfield & Crowther 2007; their Fig 10).

3. The Starburst Galaxy NGC 3125

NGC 3125 (Tol 3) is a LMC-metallicity irregular dwarf galaxy which is dominated by a central starburst region containing two main knots of star formation (3125-A and -B). From UV spectroscopy, Chandar *et al.* (2004) estimate a WR population of ~ 5000 and N(WR)/N(O)$\geqslant 1$ for knot A; whilst optical studies infer a WR population of only ~ 500 and N(WR)/N(O)~ 0.1 (Schaerer *et al.* 1999b). To resolve discrepancies between UV and optically derived WR populations we have re-investigated the massive stellar content of NGC 3125 using new VLT/FORS1 imaging and spectroscopy, plus archival *HST* imaging and spectroscopy.

Table 2. The WR population of NGC 3125.

	Diagnostic	A1	A2	B
N(WN5–6)	Optical	105	105	40
N(WC4)	Optical	20	–	20
N(WN)	UV	110	–	–
N(O)	UV (SB99)	550	750	350
N(WR)/N(O)	Optical/UV	0.23	0.14	0.14

New FORS1 narrow-band imaging confirms that 3125-A and -B represent the primary sites of WR stars, whilst the superior spatial resolution of *HST* resolves both regions into two dominant clusters. Both clusters within region A (A1 and A2) host WR stars, but the optically fainter cluster A2 appears to be heavily reddened. The resolution of our ground-based narrow-band images is insufficient to identify which cluster within region B hosts WR stars.

In contrast to other studies of unresolved WR populations, the WR content of 3125-A1 and -B has been estimated by matching LMC template WR spectra to the observed WR emission features. For A1, we find that the composite spectrum of 105 WN and 20 WC stars (Hadfield & Crowther 2006) reproduces the blue and red WR bumps exceptionally

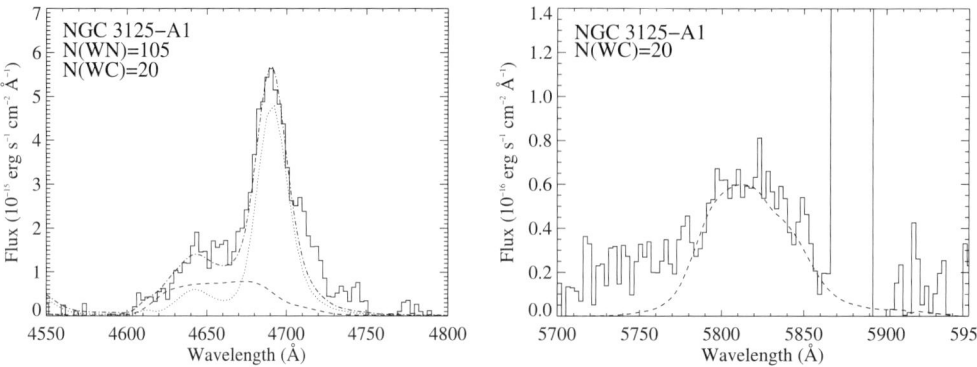

Figure 4. Spectral comparison between the observed (solid) and generic (dashed-dotted) WR emission features for cluster 3125–A1. Generic WC4 (dashed) and WN5–6 (dotted) features are marked. Spectra have been dereddened (SMC law) and continuum subtracted.

well (see Fig 4). This is factor of ∼3 lower than previous optical studies as a result of a reduced Hα/Hβ derived interstellar reddening. Applying this reduced reddening to archival UV STIS spectroscopy, together with an SMC extinction law (Bouchet *et al.* 1985), reveals that 110 generic LMC WN5-6 stars are required to reproduce the observed λ1640 emission, in excellent agreement with our VLT/FORS1 optically derived WN population.

From UV spectroscopy, we derive an O star content of ∼550 for A1 assuming a burst age of 4Myr. Similar results are obtained for region B based upon archival *HST* UV and optical photometry. We estimate N(WR)/N(O)∼0.1–0.2 for clusters within 3125-A and B, significantly larger than single star evolutionary models at LMC metallicities predict. However, our results are consistent with WR populations derived for other young massive clusters in the literature (Moll *et al.* 2007, Sidoli *et al.* 2006).

References

Bouchet, P., Lequeux, J., Maurice, E., *et al.* 1985, *A&A*, 149, 330
Chandar, R., Leitherer, C., Tremonti, C. A. 2004, *ApJ*, 604, 153
Conti, P. S. & Vacca, W. D. 1990, *AJ*, 100, 2
Crowther, P. A. 2007, *ARA&A*, 45, 177
Crowther, P. A., Carpano, S., Hadfield, L. J. & Pollock, A. M. T. 2007, *469*, L31
Crowther, P. A., Drissen, L., Abbott, J. B., *et al.* 2003, *A&A*, 404, 483
Eldridge, J. J. & Vink, J. S., 2006, *A&A*, 452, 295
Hadfield, L. J. & Crowther, P. A., 2006, *MNRAS*, 368, 1822
Hadfield, L. J. & Crowther, P. A., 2007, *MNRAS*, 381, 418
Hadfield, L. J. Crowther, P. A., Schild, H. & Schmutz, W. 2005, *A&A*, 439, 265
Massey, P. & Johnson, O., 1998, *ApJ*, 505, 793
Méndez, B., Davis, M., Moustakas, J., *et al.* 2002, *AJ*, 124, 213
Meynet, G. & Maeder, A. 2005, *A&A*, 429, 58
Mokiem, M. R., de Koter, A., Vink, J. S. *et al.* 2007, *A&A*, 473, 603
Moll, S. L., Mendel, S., de Gris, R., *et al.* 2007, *MNRAS*, 382, 1877
Phillips, A. & Conti, P. S., 1992, *ApJ*, 385, L91
Schaerer, D., Contini, T. & Pindao, M. 1999a, *A&AS*, 136, 35
Schaerer, D., Contini, T., & Kunth, D. 1999b, *A&A*, 341, 399
Schaerer D., & Vacca W. D. 1998, *ApJ*, 497, 618
Schild, H., Crowther, P. A., Abbott, J. B., & Schmutz, W. 2003, *A&A*, 397, 859
Sidoli, F., Smith, L. J., & Crowther, P. A., 2006, *MNRAS*, 370, 799

LBT Discovery of a Yellow Supergiant Eclipsing Binary in the Dwarf Galaxy Holmberg IX

J. L. Prieto[1], K. Z. Stanek[1], C. S. Kochanek[1] and D. R. Weisz[2]

[1]Department of Astronomy, Ohio State University, Columbus, OH 43210
email: prieto, kstanek, ckochanek@astronomy.ohio-state.edu

[2]Department of Astronomy, University of Minnesota, Minneapolis, MN 55455
email: dweisz@astro.umn.edu

Abstract. In a variability survey of M81 using the Large Binocular Telescope we have discovered a peculiar eclipsing binary ($M_V \simeq -7.1$) in the field of the dwarf galaxy Holmberg IX. It has a period of 271 days and the light curve is well-fit by an overcontact model in which both stars are overflowing their Roche lobes. It is composed of two yellow supergiants ($V - I \simeq 1$ mag, $T_{\rm eff} = 4800$ K), rather than the far more common red or blue supergiants. Such systems must be rare. While we failed to find any similar systems in the literature, we did, however note a second example. The SMC F0 supergiant R47 is a bright ($M_V \simeq -7.5$) periodic variable whose All Sky Automated Survey (ASAS) light curve is well-fit as a contact binary with a 181 day period. We propose that these massive systems are the progenitors of supernovae like SN 2004et and SN 2006ov, which appeared to have yellow progenitors. The binary interactions (mass transfer, mass loss) limit the size of the supergiant to give it a higher surface temperature than an isolated star at the same core evolutionary stage.

Keywords. galaxies: individual (Holmberg IX) – binaries: eclipsing

1. Introduction

Although small in number, massive stars are critical to the formation and evolution of galaxies. They shape the ISM of galaxies through their strong winds and high UV fluxes, and are a major source of the heavy elements enriching the ISM (Massey 2003; Zinnecker & Yorke 2007). A large fraction of massive stars are found in binaries (e.g., Kiminki *et al.* 2007). Eclipsing binaries are of particular use because they allow us to determine the masses and radii of the components and the distance to the system. Many young, massive eclipsing binaries have been found and studied in our Galaxy, the LMC, and the SMC, primarily in OB associations and young star clusters (e.g., Bonanos *et al.* 2004; Peeples *et al.* 2007; Gonzalez *et al.* 2005; Hilditch *et al.* 2005). The study of massive eclipsing binaries beyond the Magellanic clouds has been limited until very recently, when variability searches using medium-sized telescopes with wide-field CCD cameras, coupled with spectroscopy using 8-meter class telescopes, have yielded the first systems with accurately measured masses in M31 (Ribas *et al.* 2005) and M33 (Bonanos *et al.* 2006).

We conducted a deep variability survey of M81 (distance ~ 3.6 Mpc; Freedman *et al.* 2001) and its dwarf irregular companion, Holmberg IX, using the Large Binocular Camera (LBC) mounted on the Large Binocular Telescope (LBT). Holmberg IX is a young dwarf galaxy (age $\leqslant 200$ Myr), with a stellar population dominated by blue and red supergiants with no signs of old stars in the red giant branch (Makarova *et al.* 2002). The dwarf may have formed during a recent tidal interaction between M81 and NGC 2976 (e.g.,

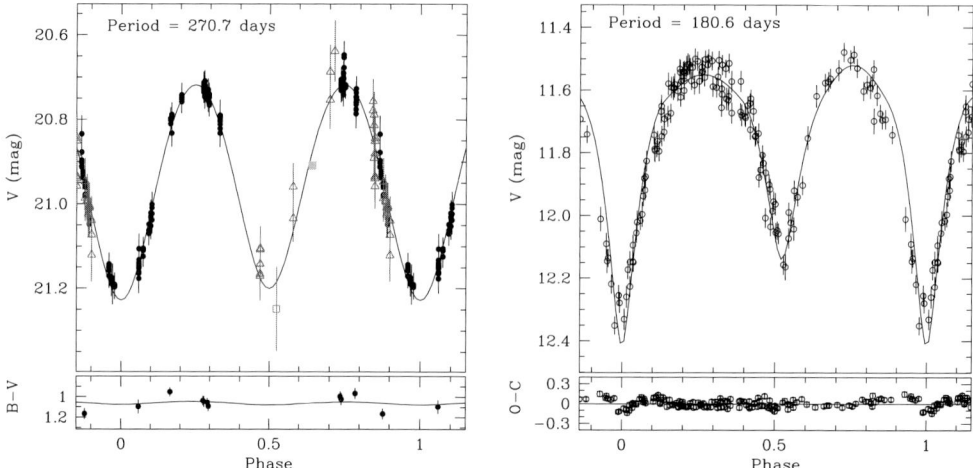

Figure 1. *Left:* V-band light curve (*top*) and $B-V$ color (*bottom*) of Holmberg IX V1. The data are from LBT (*filled bcircles*), MDM (*triangles*), SDSS (*square*) and HST ACS (*filled square*). The solid line is best-fit eclipsing binary model. *Right:* ASAS V-band light curve of the eclipsing binary SMC R47. The solid line shows the best-fit eclipsing binary model.

Boyce *et al.* 2001). The gas-phase metal abundance of Holmberg IX of between 1/8 and 1/3 solar (e.g., Miller 1995; Makarova *et al.* 2002) is consistent with this hypothesis (e.g., Weilbacher *et al.* 2003). A normal, isolated dwarf on the luminosity-metallicity relationship would have a metallicity of $\sim 1/20$ solar (Lee *et al.* 2006).

Here we report on the discovery of a 271 day period, evolved, massive eclipsing binary in Holmberg IX using data from the LBT. The overcontact system is the brightest periodic variable discovered in our LBT variability survey. It has an out-of-eclipse magnitude of $V_{max} = 20.7$ mag and is located at $\alpha = 09^{\rm h}57^{\rm m}37.14$, $\delta = +69°02'11''$ (J2000.0).

2. Observations

Holmberg IX was observed as part of a variability survey of the entire M81 galaxy conducted between January and October 2007 with the LBT 8.4-meter telescope, using the LBC-Blue CCD camera during Science Demonstration Time. The survey cadence and depth were optimized to detect and follow-up Cepheid variables with periods between 10-100 days (10% photometry at $V = 24$ mag). We discovered ~ 20 periodic variables in the dwarf, most of them Cepheids with periods between $10-60$ days. The light curve of the brightest periodic variable (hereafter V1) is well-fit by an eclipsing binary model with both stars overflowing their Roche lobes. Figure 1 shows the phased V-band light curve and $B - V$ color curve of the eclipsing binary system V1. We include LBT photometry, as well as contemporaneous photometry obtained with the MDM 2.4-meter telescope, and archival data obtained from SDSS and HST.

3. Massive Yellow Supergiant Binaries

The discovery of a 271 day period, luminous ($M_V \sim -7.1$), overcontact eclipsing binary system in our LBT survey of the dwarf galaxy Holmberg IX was unexpected. The colors of the binary V1 are consistent with an effective temperature of $T_{\rm eff} = 4800$ K. Given

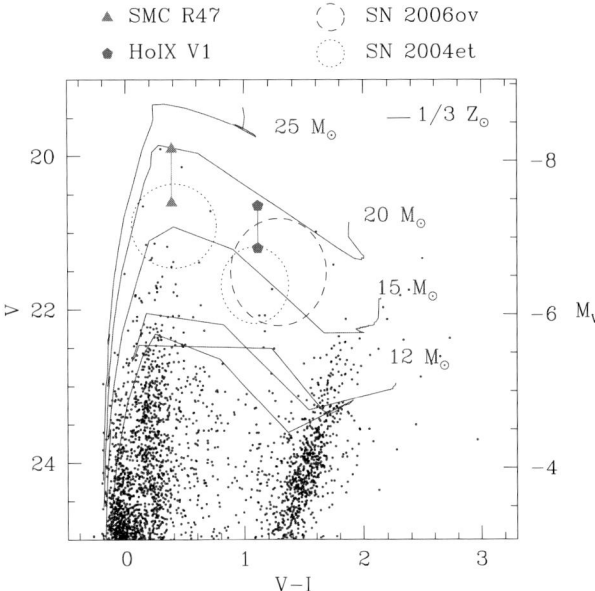

Figure 2. CMD with HST ACS photometry of Holmberg IX (*dots*) and the position at maximum and minimum of the yellow supergiant binaries: Holmberg IX V1 (*pentagons*) and SMC R47 (*triangles*). The blue lines are Geneva evolutionary tracks for single stars (Lejeune & Schaerer 2001). The ellipses show error circles with the location of the progenitors of the type IIP supernovae 2004et (*dotted*) and 2006ov (*dashed*).

Table 1. Best-fit Binary Model Parameters.

Parameter	Holmberg IX V1	SMC R47
Period, P	270.7 ± 2.3 days	181.58 ± 0.16 days
Time of primary eclipse, T_{prim}	2454186.0 ± 0.6	2452073.1 ± 0.2
Inclination, i	$55.7° \pm 0.6°$	$82.2° \pm 0.2°$
Primary temperature, T_1	4800 ± 150 K	7500 ± 100 K
Temperature ratio, T_2/T_1	1.05 ± 0.05	1.17 ± 0.02
Eccentricity, e	0.00	0.039 ± 0.002
Roche Lobe Filling factors	1.23 ± 0.02	1.02 ± 0.02

the absence of color variations and equal depths of the eclipses, both stars in the binary system are yellow supergiants.

We expected that such systems were rare†, but were surprised to find none in the literature. However, we found in the ASAS catalog of periodic variables (Pojmanski 2002) a luminous ($M_V \sim -7.5$ mag), 181 day period, contact eclipsing binary in the SMC (see Fig. 1). The star, SMC R47, had been spectroscopically classified as an F0 supergiant (Humphreys 1983). Table 1 gives the main parameters of the two long-period, yellow supergiant binaries, Holmberg IX V1 and SMC R47. From their position in the CMD (see Fig. 2), we estimate that at least one of the stars in each binary is $15 - 20 \, \mathrm{M}_\odot$ (main sequence age $\sim 10 - 15$ Myr).

† While the relative numbers of eclipsing binaries is a much more complicated problem, we note that the relative abundances of red, blue and yellow supergiants is 4:13:1 for the Geneva evolutionary track of a single, non-rotating star with $M = 15 \, \mathrm{M}_\odot$ and $Z = 0.004$.

4. Supernova Progenitors?

The stellar evolutionary path of stars of a given mass in binary systems can differ significantly from their evolution in isolation (Paczynski 1971). In particular, binary interactions through mass loss, mass accretion, or common-envelope evolution, play a very important role in the pre-supernova evolution (e.g., Podsiadlowski *et al.* 1992). Most of the massive stars with masses $30\,M_\odot \geqslant M \geqslant 8\,M_\odot$ are expected to explode as supernova when they are in the red supergiant phase, with a small contribution from blue supergiants (e.g., 1987A; West *et al.* 1987). Surprisingly, Li *et al.* (2005) identified the progenitor of the Type IIP supernova 2004et in pre-explosion archival images and determined that it was a yellow supergiant with a main-sequence mass of $\sim 15\,M_\odot$. Also, the position in the CMD of the likely progenitor of the Type IIP supernova 2006ov (Li *et al.* 2007) is remarkably similar to the position of the eclipsing binary in Holmberg IX.

We propose that the binaries we discovered in Holmberg IX and the binary found in the SMC are possible progenitors for these supernovae. A close binary provides a natural means of slowing the transition from blue to red, allowing the star to evolve and then explode as a yellow supergiant. As the more massive star evolves and expands, the Roche lobe limits the size of the star, forcing it to have a surface temperature set by the uncoupled core luminosity and the size of the Roche lobe. It can expand further and have a cooler envelope only by becoming a common envelope system, which should only occur as the secondary evolves to fill its Roche lobe. This delayed temperature evolution allows the core to reach SN II conditions without a red envelope.

Acknowledgements

J.L.P. and K.Z.S. acknowledge support from NSF through grant AST-0707982.

References

Bonanos, A. Z., Stanek, K. Z., Udalski, A. *et al.* 2004, *ApJ*, 611, L33
Bonanos, A. Z., Stanek, K. Z., Kudritzki, R.-P. *et al.* 2006, *ApJ*, 652, 313
Boyce, P. J., Minchin R. F., Kilborn, V. A. *et al.* 2001, *ApJ*, 560, L127
Freedman, W. L., Madore, B. F., Gibson, B. K. *et al.* 2001, *ApJ*, 553, 47
González, J. F., Ostriov, P., Morrell, N., & Minnitim, D. 2005, *ApJ*, 624, 946
Hilditch, R. W., Howarth, I. D., & Harries, T. J. 2005, *MNRAS*, 357, 304
Humphreys, R. M. 1983, *ApJ*, 265, 176
Kiminki, D. C., Kobulnicky, H. A., Kinemuchi, K. *et al.* 2007, *ApJ*, 664, 1102
Lee, H., Skillman, E. D., Cannon, J. M. *et al.* 2006, *ApJ*, 647, 970
Lejeune, T., & Schaerer, D. 2001, *A&A*, 366, 538
Li, W., Van Dyk, S. D., Filippenko, A. V. & Cuillandre, J.-C., 2005, *PASP*, 117, 121
Li, W., Wang, X., Van Dyk, S. D. *et al.* 2007, *ApJ*, 661, 1013
Makarova, L. N., Grebel, E. K., Karachentsev, I. D. *et al.* 2002, *A&A*, 396, 473
Massey, P. 2003, *ARA&A*, 41, 15
Miller, B. W. 1995, *ApJ*, 446, L75
Paczyński, B. 1971, *ARA&A*, 9, 183
Peeples, M. S., Bonanos, A. Z., DePoy, D. L. *et al.* 2007, *ApJ*, 654, L61
Podsiadlowski, P., Joss, P. C., & Hsu, J. J. L. 1992, *ApJ*, 391, 246
Pojmanski, G. 2002, *AcA*, 52, 397
Ribas, I., Jordi, C., Vilardell, F. *et al.* 2005, *ApJ*, 635, L37
Weilbacher, P. M., Duc, P.-A. & Fritze-v. Alvensleben, U. 2003, *A&A*, 397, 545
West, R. M., Lauberts, A., Schuster, H.-E. & Jorgensen, H. E. 1987, *A&A*, 177, L1
Zinnecker, H. & Yorke, H. W. 2007, *ARA&A*, 45, 481

Jose Prieto (right) with Kris Stanek (left).

Lucy Hadfield.

From left to right: Miguel Urbaneja, Norbert Przybilla, Artemio Herrero, Ignacio Negueruela and Cesar Esteban.

Session IV

Hydrodynamics and Feedback from Massive Stars in Galaxy Evolution

Bubbles and Superbubbles: Observations and Theory

You-Hua Chu

Astronomy Department, University of Illinois, 1002 W. Green Street, Urbana, IL 61801, USA
email: chu@astro.uiuc.edu

Abstract. Massive stars inject energy into the surrounding medium and form shell structures. Bubbles are blown by fast stellar winds from individual massive stars, while superbubbles are blown by fast stellar winds and supernova explosions from groups of massive stars. Bubbles and superbubbles share a similar overall structure: a swept-up dense shell with an interior filled by low-density hot gas. Physical properties of a bubble/superbubble can be affected by magnetic field, thermal conduction, turbulent mixing, inhomogeneous ambient medium, etc. I will review recent progresses on observations and compare them to theoretical expectations for (1) swept-up dense shells, (2) hot interiors, and (3) interface between a dense shell and its interior hot gas

Keywords. ISM: bubbles – ISM: structure – circumstellar matter – stars: mass loss – stars: Wolf-Rayet – stars: winds, outflows – supernova remnants

1. Introduction - Bubbles, Superbubbles, and Supergiant Shells

Massive stars inject energy into the interstellar medium (ISM) via UV radiation, fast stellar wind, and ultimately supernova explosion. Depending on the concentration and evolutionary status of the massive stars, these energetic interactions produce different shell structures with sizes ranging from fractions of 1 pc to greater than 1,000 pc. These shells are commonly detected in the Hα line, but the large shells without ionizing fluxes are detected only in the H I 21-cm line.

The hierarchy of interstellar shells is best illustrated by the supergiant shell (SGS) LMC-4. SGSs with diameters of $\sim$1000 pc are the largest interstellar structures in a galaxy. Nine SGSs in the Large Magellanic Cloud (LMC) have been identified using Hα images (Goudis & Meaburn 1978; Meaburn 1980), but they do not have a one-to-one correspondence with the SGSs identified from H I 21-cm maps (Kim *et al.* 1999) because these SGSs have different star formation histories and available ionizing fluxes (Book *et al.* 2008). LMC-4 is the largest SGS in the LMC, and as shown in Figure 1 its overall Hα emission originates from the ionized inner wall of its H I shell. The periphery of LMC-4 is dotted with prominent H II regions, superbubbles, supernova remnants (SNRs), and bubbles. While the SGS LMC-4 stretches over 1,000–1,500 pc, the superbubbles N51D and N57A on the south rim of LMC-4 are 100–200 pc across, and the bubble N57C is only $\sim$25 pc in size.

SGSs, superbubbles, and bubbles are three distinct classes of objects powered by massive stars. SGSs have sizes $\sim 10^3$ pc, dynamic ages of $\sim 10^7$ yr, and require multiple generations or episodes of star formation; for example, LMC-4 contains older OB associations in its interior and younger associations along its periphery (Book *et al.* 2008). Superbubbles have sizes $\sim 10^2$ pc, dynamic ages of $\sim 10^6$ yr, and require only one episode of star formation; for example, the OB associations LH 54 and LH 76 are responsible for creating the superbubbles N51D and N57A, respectively (Braunsfurth & Feizinger 1983; Oey & Smedley 1998). Bubbles have sizes up to a few $\times 10$ pc, and are powered by the

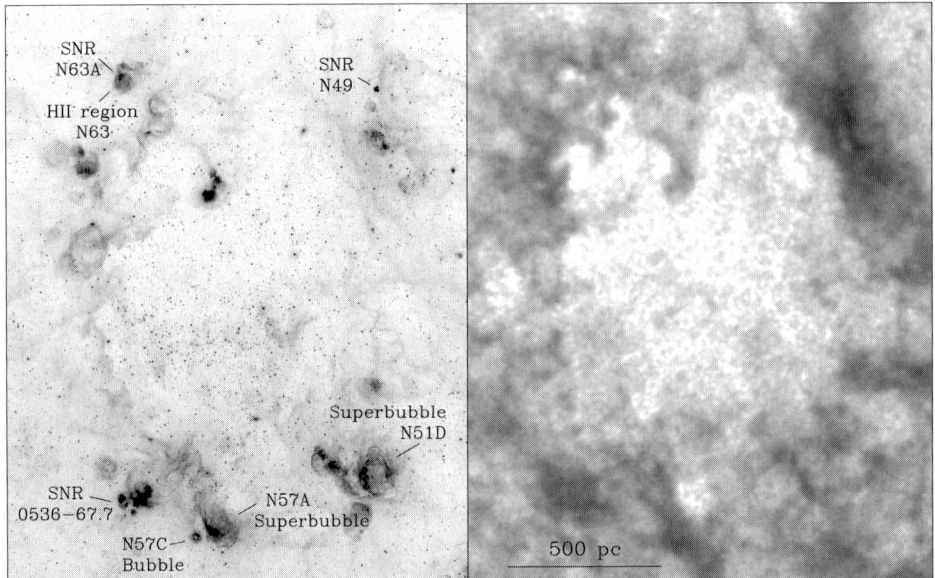

Figure 1. Hα (left) and H I 21-cm (right) images of the supergiant shell LMC-4. The Hα image of LMC-4 shows recent star formation activity along its rim, and the stellar energy feedback has produced superbubbles and bubbles, as well as prominent H II regions and supernova remnants. This image nicely illustrates the relative sizes of supergiant shell, superbubble, and bubble.

stellar wind of individual massive stars; for example, the bubble N57C is blown by the Wolf-Rayet (WR) star HDE 269748 (Chu & Lasker 1980).

Bubbles blown by individual massive stars evolve along with the central stars. During the main sequence phase, a massive star is surrounded by the ISM and its fast stellar wind blows an *interstellar* bubble. As the massive star evolves into the red supergiant (RSG) or luminous blue variable (LBV) phase, the copious mass loss forms a small *circumstellar* nebula inside the central cavity of the main sequence bubble. As the massive star evolves further into a WR phase, the fast stellar wind sweeps up the circumstellar material and forms a *circumstellar* bubble. The evolution of nebulae around massive stars has been modeled by, e.g., García-Segura *et al.* (1996a,b) and Freyer *et al.* (2003, 2006), and observations of such nebulae throughout the HR diagram were reviewed by Chu (2003).

2. Theories of Bubbles and Superbubbles

Theories of wind-blown bubbles were initiated in the late 60's (e.g., Mathews 1966; Pikel'ner 1968; Pilkel'ner & Shcheglov 1969), but the term "bubble" was not coined until the mid 70's (Castor *et al.* 1975). Bubbles were modeled assuming "momentum conservation" in which case fast stellar wind impinges on the bubble shell directly and imparts the out-going momentum (Steigman *et al.* 1975), or "energy conservation" in which case fast stellar wind is adiabatically shocked to high temperature and the thermal pressure of the hot gas drives the expansion of of the bubble shell (Dyson & de Vries 1972; Castor *et al.* 1975).

The first and most comprehensive model of interstellar bubbles that aimed at explaining UV and X-ray observations was presented by Weaver *et al.* (1977). This model assumed a homogeneous ambient medium that might not be realistic, but it included essential physical processes and became a seminal paper, a true classic. The physical

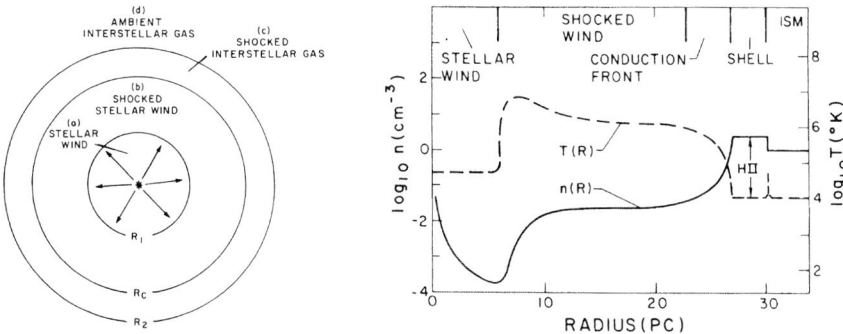

Figure 2. Left - A schematic drawing of the structure of an interstellar bubble. Right - temperature and density profiles of a pressure-driven interstellar bubble. Both figures are adopted from Weaver *et al.* (1977).

structure of a bubble is schematically shown in Figure 1 of Weaver *et al.* (1977) and reproduced in the left panel of Figure 2. At the center of a bubble, zone (a), the fast stellar wind expands freely from a massive star. It encounters an adiabatic stagnation shock at radius R_1, and the shocked stellar wind accumulates in zone (b). In the outer parts of a bubble, the ambient interstellar gas in zone (d) is shocked by the expanding bubble at radius R_2, and the shocked interstellar gas accumulates in zone (c). The shocked stellar wind and the shocked interstellar gas are separated by a contact discontinuity at radius R_c. The temperature and density profiles of Weaver *et al.*'s bubble are reproduced in the right panel of Figure 2. These profiles qualitatively manifest three physical actions. First, the stellar wind shock at R_1 is adiabatic, because the wind has high velocity and low density and the cooling time scale of the post-shock gas is long. Second, the interstellar shock at R_2 is isothermal because the interstellar gas is dense and the post-shock gas cools rapidly. Third, thermal conduction takes place at R_c, the interface between the shocked stellar wind and the shocked interstellar gas.

Weaver *et al.* (1977) model has formed the basis of numerous bubble models with more complex conditions in the ambient medium and time-dependent stellar winds. García-Segura *et al.* (1996a,b) considered mass-loss rates and stellar wind velocities that varied along stellar evolution, and modeled the development of interstellar bubble during the main sequence and circumstellar bubble during the WR phase. Circumstellar bubbles are notably different from Weaver *et al.*'s interstellar bubble because of the $\propto r^{-2}$ density profile in the circumstellar medium. Instabilities in the dense swept-up circumstellar bubble shells cause fragmentation and clumpy morphologies. Freyer *et al.* (2003, 2006) furthered García-Segura *et al.* models by adding radiation effect; their radiative hydrodynamic simulations show that photoionization significantly influence the morphological evolution of interstellar bubbles formed during the main sequence stage. Arthur *et al.* (1993, 1996) and Pittard *et al.* (2001a,b) added another important physical process to bubble models – mass-loading by conductive evaporation or hydrodynamic ablation of cold clumpy material embedded in the hot bubble interior. The net effect of mass-loading is a faster cooling of the bubble interiors.

Superbubbles are powered initially by fast stellar winds and later by additional supernova explosions. For supernovae exploding in the low-density interior of a superbubble, the SNR shocks heat the already-hot surroundings; the shocks may be decelerated and become thermalized without ever reaching the dense swept-up shell. The effects of supernova explosions can thus be approximated by a stellar wind with a mechanical luminosity equal to the average supernova energy injection rate. Therefore, Weaver *et al.* model for interstellar bubbles has also formed the basis of superbubble models (e.g., Mac Low & McCray 1988). Superbubbles modeled with the consideration of interstellar density gradient out of the galactic plane can produce blowouts into the galactic halo (Mac Low et al. 1989). The expansion and blowout of a superbubble can be impeded in the direction perpendicular to the magnetic field of the ISM (Tomisaka 1992).

3. Observations Confronting Theories in the 80's and 90's

Bubble models were inspired by optical observations of nebular dynamics and *Copernicus* observations of interstellar O VI absorption (Jenkins & Meloy 1974). Advances in optical high-dispersion spectroscopic observations and NASA's UV and X-ray missions in the 80's and 90's made it possible to compare observations with models.

Optical observations can be used to determine the density, radius, and expansion velocity of the swept-up bubble shells. UV spectroscopic observations of massive stars can be used to determine stellar wind velocities and mass loss rates. If all of these observables are available for a bubble, it will be possible to critically test bubble models. It is found that circumstellar bubbles, most notably the well-observed NGC 6888, expand too slowly for their observed stellar wind velocity and mass loss rate, and nebular size and density (García-Segura et al. 1996a). The stellar wind mass loss rate can be revised down by a factor of a few because of the clumping of wind (Moffat & Robert 1994), but the discrepancy between observations and model expectations is still significant.

Einstein X-ray observations detected diffuse emission from bubbles and superbubbles, but the data quality did not allow spectral analysis to determine plasma temperature and foreground absorption; assumptions of temperature and absorption had to be made in order to estimate X-ray luminosity. X-ray emission from the circumstellar bubble NGC 6888 was an order of magnitude lower than expected from models (Bochkarev 1988). A number of superbubbles in the LMC were also detected by *Einstein*, but their X-ray luminosities were higher than expected; it was suggested that off-center supernova explosions were responsible for the X-ray emission (Chu & Mac Low 1990).

ROSAT X-ray observations had a higher angular resolution and higher sensitivity for soft X-rays than *Einstein*. *ROSAT* PSPC observations detected diffuse X-ray emission from the circumstellar bubbles NGC 6888 and S 308. The X-ray emission from NGC 6888 shows a plasma temperature of $\sim 2 \times 10^6$ K and a luminosity of $\sim 1.6 \times 10^{34}$ erg s^{-1}, confirming the *Einstein* estimate of luminosity (Wrigge et al. 1994). The *ROSAT* PSPC observation of S 308 was centered on the WR star HD 50896. The PSPC entrance window's support structure happened to contain a ring similar in size to S 308, and thus occulted a significant fraction of diffuse X-ray emission from the shell rim. Diffuse X-ray emission was convincingly detected in S 308, but the low X-ray surface brightness and low S/N made the spectral analysis difficult and unreliable (Wrigge 1999). *ROSAT* observations of a number of other bubbles did not detect diffuse X-ray emission.

ROSAT PSPC observations detected diffuse X-ray emission from the Galactic superbubble in the Omega Nebula (M17). Spectral analysis shows the existence of plasma with temperatures as high as $\sim 8.5 \times 10^6$ K and an X-ray luminosity, 2.5×10^{33} erg s^{-1}, two orders of magnitude lower than expected, possibly caused by the strong magnetic field that hinders thermal conduction (Dunne et al. 2003). The OB association in M17 is so young that no supernova has occurred; the hot gas in M17 is likely powered by stellar winds solely. *ROSAT* PSPC observations have detected diffuse X-ray emission from the Eridanus superbubble, which is much larger and older than the M17 superbubble. The enhanced diffuse X-ray emission from the Eridanus superbubble, with a luminosity of

10^{35}–10^{36} erg s^{-1} and a plasma temperature of $\sim 2\times 10^6$ K, is likely caused by additional heating from a supernova explosion (Burrows *et al.* 1993).

ROSAT PSPC observations of 13 X-ray-bright superbubbles in the LMC show that their diffuse X-ray emission are more luminous than expected and are most likely energized by recent supernovae (Chu *et al.* 1993; Dunne *et al.* 2001). Deep *ROSAT* PSPC observations of three X-ray-faint LMC superbubbles yielded non-detections, and the 3σ upper limits of their X-ray luminosities are similar to the expectations of models (Chu *et al.* 1995). Examinations of bubble dynamics indicate that X-ray-bright superbubbles expand faster than expected and X-ray-faint superbubbles roughly expand as expected from models, confirming the role played by supernovae (Oey 1996).

IUE and *HST* observations of UV absorption from conduction layers have been obtained in Si IV $\lambda\lambda$1393,1402, C IV $\lambda\lambda$1548,1550, and N V $\lambda\lambda$1238,1242 lines. With the ionization potentials of Si III, C III, and N IV being 33.5, 47.0, and 77.5 eV, respectively, Si IV and C IV can be easily produced through photoionization by hot massive stars, but not N V, which needs to be produced through collisional ionization in a 10^5 K medium, such as the conduction layer. As bubbles and superbubbles do encompass hot massive stars, the only reliable conduction layer indicator is the N V absorption, which is expected to be weak as the product of its abundance and oscillator strength is the lowest among the three species. *IUE* observations of interstellar absorption lines toward the central stars of NGC 6888 (Nichols-Bohlin & Fesen 1993) and S 308 (Howarth & Phillips 1986) show complex velocity structures in the C IV absorption lines with components corresponding to both the approaching side of the bubble shell and the foreground ISM. *IUE* observations of C IV and Si IV absorption from LMC superbubbles show mainly photoionized components from the superbubble shells, although some X-ray-bright superbubbles show velocity offsets between C IV absorption and Hα emission, possibly an influence of SNR shocks (Chu *et al.* 1994). The only convincing detection of conduction layer was provided by the *HST* GHRS observations of HD 50896 in S 308, based on the velocity structures of the C IV and N V lines (Boroson *et al.* 1997). Constrained by the *Copernicus* measurements of the O VI absorption, the observed N V absorption strength can be explained by a conduction front of Weaver *et al.* (1977) if the nitrogen abundance is enhanced by a factor of $\sim$10, which is likely, as HD 50896 is a WN star.

In summary, observations of bubbles and superbubbles in the 80's and 90's have shown discrepancies from model expectations in bubble dynamics and X-ray luminosities. There is observational evidence of thermal conduction, but details are still uncertain.

4. Astrophysical Observations of Bubbles and Superbubbles

In 1999, the launch of *Far UV Spectroscopic Explorer* (*FUSE*), *Chandra X-ray Observatory*, and *XMM-Newton X-ray Satellite* made it possible to observe the hot interiors and conduction layers in bubbles and superbubbles with unprecedented resolution and accuracy so that bubble models can be critically examined. The launch of *Spitzer Space Telescope* in 2003 made it easy to identify massive young stellar objects so that their interstellar environments can be examined and used as realistic initial conditions for the formation of interstellar bubbles and superbubbles. In this section, I will review advances made in the 2000's for the three distinct layers of a bubble model: the swept-up shell, the hot interior, and the interface between them.

4.1. *Dense Swept-Up Shells*

It has been puzzling that Weaver *et al.* (1977) model of an interstellar bubble produces a dense swept-up shell, which should be ionized and visible in Hα images, but hardly any known main sequence O stars are surrounded by shell nebulae. The Bubble Nebula around the O star BD+60 2522 is an exception rather than the rule. In contrast, H I 21-cm line observations frequently find expanding shells around evolved stars and the shell properties are consistent with those expected from interstellar bubbles (Cappa *et al.* 2003). Why aren't interstellar bubbles around main sequence O stars visible?

This puzzle is solved by the kinematically detected bubble shells in the LMC H II regions N11B and N180B (Nazé *et al.* 2001). Expanding shells exist around the O stars in these young H II regions, but their expansion velocities are only 15–20 km s^{-1}. For a photoionized medium at 10^4 K, the isothermal sound velocity is 10 km s^{-1} and the observed shell expansion velocity drives only a weak shock into the ambient medium. Without a strong compression to produce sharp density contrast, these bubble shells cannot be identified morphologically over a complex background. If such a shell expands into a neutral medium that has an isothermal sound velocity of $\sim$1 km s^{-1}, strong shocks and compression are generated, and the resultant large density contrast between the swept-up shell and the ambient medium makes the H I shell easily detectable.

The swept-up bubble shells in N11B and N180B have been compared with Weaver *et al.* (1977) interstellar bubble models. For the observed shell size, expansion velocity, and density, the required stellar wind mechanical luminosity is 1–2 orders of magnitude lower than that implied by the observed stellar content (Nazé *et al.* 2001). This problem is similar to that for the circumstellar bubble NGC 6888. While it is possible that stellar wind luminosity has been over-estimated, it is also possible that the interstellar environment is much more complex than a homogeneous medium assumed by Weaver *et al.* (1977). For example, the bubble structure in N44F ("Celestial Geode") shows dust pillars along the shell rim (Nazé *et al.* 2002); furthermore, massive young stellar objects have been identified in some dust pillars, indicating on-going star formation (Chen *et al.* 2008). For a bubble in a clumpy medium, the evaporation and ablation of dense clumps in the hot interior of a bubble raises the density and cooling rate, and thus may significantly change the bubble dynamics. It therefore may be over-simplistic to compare a bubble formed in a complex environment to a Weaver *et al.* bubble in a homogeneous medium.

H I surveys of the Milky Way and nearby galaxies have found large numbers of shells, e.g., 300 shells in the outer Milky Way (Ehlerová & Palouš 2005), 500 shells in the Small Magellanic Cloud (Staveley-Smith *et al.* 1997), 23 supergiant shells and 103 giant shells in the LMC (Kim *et al.* 1999). These shells have a range of physical conditions in their ambient ISM, history of star formation, and evolutionary stages. Some of the "shells" may not be physical, especially those with sizes and expansion velocities comparable to the resolution limit of the H I observations. It is not clear whether physical insights can be unambiguously gained from the observed distributions of shell sizes or expansion velocities, although the size distribution of superbubbles in galaxies has been modeled by Oey & Clarke (1997), assuming all shells have the same lifetime. A non-negligible fraction of the shells do not show underlying population of massive stars that are responsible for the formation of the shells (Hatzidimitriou *et al.* 2005). Deep observations of the stellar population within these shells to search for the massive stars' surviving lower-mass siblings are needed before resorting to exotic origins, such as gamma-ray bursts.

4.2. *Hot Interiors*

Chandra observations of NGC 6888 and *XMM-Newton* observations of S 308 have yielded the highest-quality X-ray image and spectra of the hot interior gas on single-star bubbles. Figure 3 shows images of S 308 in [O III] λ5007 and in X-ray. The X-ray image shows a strong limb-brightening, indicating that the emitting layer is thin. A close-up comparison between optical and X-ray images (Figure 4) shows a gap between the fronts of [O III]

Bubbles and Superbubbles   347

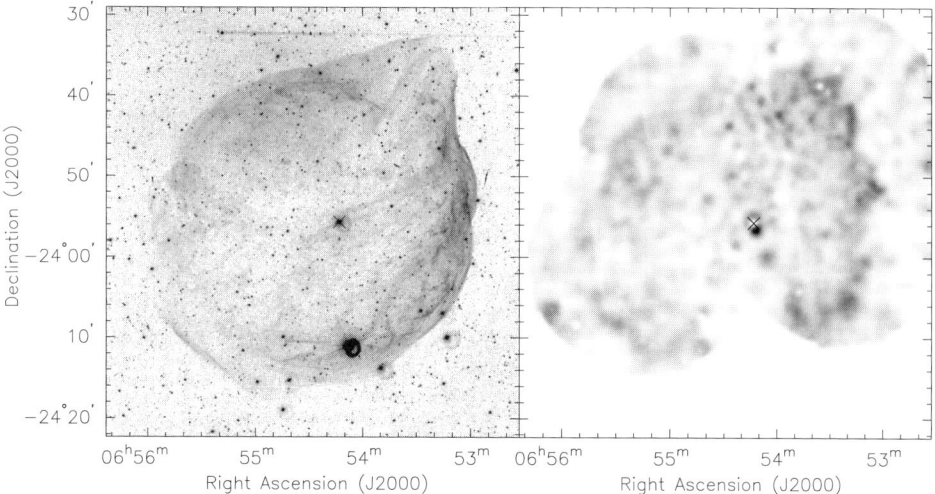

Figure 3. Left - [O III] λ5007 image of S 308. Right - *XMM-Newton* EPIC-pn X-ray image of S 308 in 0.3-1 keV band. Obvious point sources have been excised and adaptive smoothing has been applied. Diffuse X-ray emission inside S 308 shows a limb-brightened morphology. The central WN star HD 50896 is marked with a "×".

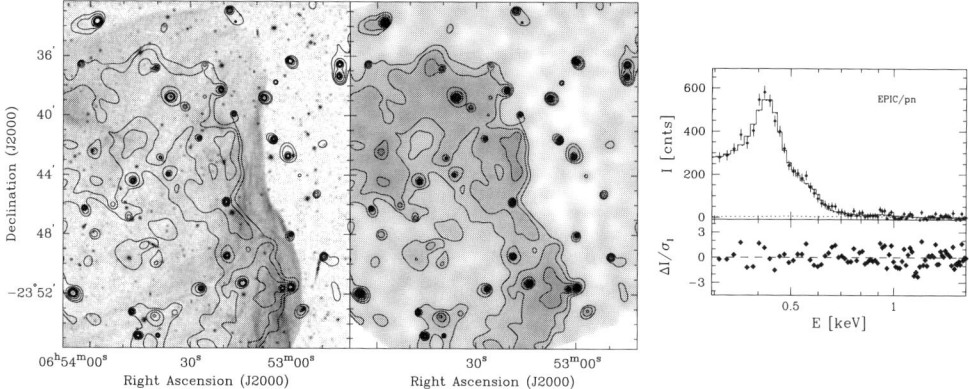

Figure 4. Left - [O III] and *XMM-Newton* EPIC/pn images of the NE quadrant of S 308 overlaid with X-ray contours. Point X-ray sources are not removed in this image; these point sources must have contributed to the X-ray emission above 1 keV detected in the *ROSAT* PSPC data (Wrigge 1999). Right - *XMM-Newton* EPIC-pn spectrum of the diffuse emission in the NE quadrant of S 308. These figures are from Chu *et al.* (2003).

and X-ray emission, and the X-ray spectrum shows that the X-ray emission is very soft, with hardly any emission above 1 keV (Chu *et al.* 2003). Such soft X-ray emission is most susceptible to interstellar absorption. Indeed, S 308 is detectable because it is nearby, at 1.5±0.3 kpc, and has a small foreground absorption column density, $N_{\rm H} \sim 1 \times 10^{21}$ cm^{-2}. The plasma temperature derived from spectral fits is $\sim 1.1 \times 10^6$ K, and the rms electron density of the hot gas is 0.3–0.6 cm^{-3} (Chu *et al.* 2003).

Chu *et al.* (2003) compared X-ray and optical observations of S 308 with the analytical model of circumstellar bubbles by García-Segura & Mac Low (1995), assuming that the optical shell expanding at 65 km s^{-1} consisted of swept-up RSG wind and that the expansion velocity of the RSG wind was ⩽30 km s^{-1}. They speculated that the gap between the X-ray emission edge and the optical emission edge might correspond to the

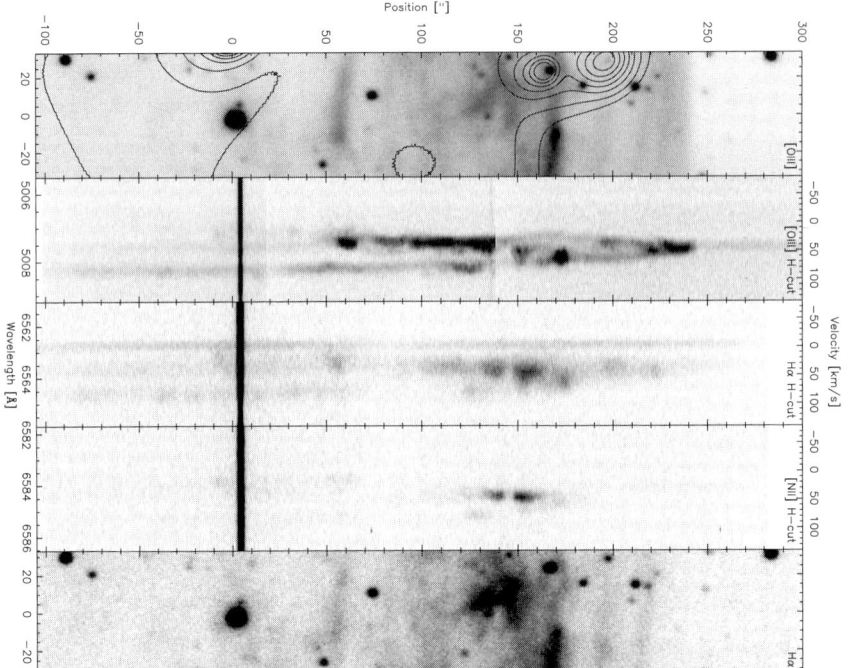

Figure 5. A narrow slice of the western shell rim of S 308. Top to bottom – [O III] image with X-ray contours; long-slit [O III] echellogram EW-oriented and passing through the star at the origin of the image above; Hα echellogram at the same position; [N II] echellogram at the same position; Hα+[N II] image of the same field shown in the [O III] image on top.

conduction layer. However, our recent high-dispersion long-slit echelle observations of the nebular shell of S 308 indicate a different interpretation. As shown in Figure 5, the "gap" between the [O III] front and the X-ray front has a "filled" line morphology in the [O III] echellogram. The continuous presence of emission at the systemic velocity across the gap indicates that material exists continuously along the radial direction perpendicular to the line of sight. Therefore, the "gap" between the [O III] front and the X-ray front is filled with nebular material. This filled layer most likely consists of RSG wind expanding at a uniform velocity of 65 km s^{-1}. The bubble dynamics presented by Chu et al. (2003) therefore needs revision. Freyer et al. (2005) suggested that the diffuse X-ray emission originates from the swept-up RSG wind shell. The limb-brightening in both the direct images and the echelle line images in Hα and [O III] indicates the presence of a denser region that may correspond to the swept-up RSG wind shell. The low temperature and large mass of the X-ray-emitting material suggest that significant mixing of nebular material into the shocked fast wind has occurred.

Chandra observations of NGC 6888 reveal diffuse X-ray emission extending toward the optical shell rim, and the spectral fits confirm the plasma temperature of $\sim 2 \times 10^6$ K previously determined from *ROSAT* PSPC observations (Gruendl et al. 2003). Since a range of plasma temperatures is expected from the hot interior to the conduction front, the observed X-ray spectrum needs to be fitted by those calculated using bubble models (e.g., Strickland & Stevens 1997).

Diffuse X-ray emission from Galactic H II regions, such as the Orion Nebula and the Rosette Nebula, were reported using *Einstein* observations, but *ROSAT* PSPC observations with higher resolution have resolved the diffuse emission into point sources. The high angular resolution and high sensitivity of *Chandra* and *XMM-Newton* make it possible to

Table 1. Physical Properties of Hot Gas in Bubble Interiors

Bubble Type	$T_{\rm e}$ [10^6 K]	$N_{\rm e}$ [cm^{-3}]	$L_{\rm X}$ [erg s^{-1}]
Orion Bubble	2	0.2–0.5	5×10^{31}
WR Bubble	1–2	1	$10^{33} - 10^{34}$
M17 Superbubble	1.5, 7	0.3	3.4×10^{33}
Planetary Nebula	2–3	100	$10^{31} - 10^{32}$

separate point sources from faint diffuse emission. Diffuse X-ray emission from the Orion Nebula has been convincingly detected by *XMM-Newton*, within a bubble-like structure ~ 3.5 pc in diameter (Güdel *et al.* 2008). The fast stellar wind from θ^1 Ori is likely responsible for the bubble structure. Compared with NGC 6888, the bubble of θ^1 Ori has a similar plasma temperature, $\sim 2 \times 10^6$ K, but its X-ray luminosity, $\sim 5.5 \times 10^{31}$ erg s^{-1}, is almost 3 orders magnitude lower than that of NGC 6888. This lower X-ray luminosity is caused by a weaker stellar wind and a lower ambient gas density.

The OB associations in the Rosette Nebula and the Omega Nebula (M17) are so young that no supernovae have occurred, and thus the hot gas in these two superbubbles has been powered solely by fast stellar winds. *Chandra* observations of the Rosette Nebula did reveal diffuse X-ray emission (Townsley *et al.* 2003), but a deep subsequent observation resolved more point sources from the faint diffuse emission (Townsley, personal communication). *Chandra* observations of the Omega Nebula show a dominant component at 7×10^6 K and a secondary component at 1.5×10^6 K. This high-temperature component is not commonly seen in diffuse X-ray emission powered by fast stellar winds, as shown in Table 1. The presence of 7×10^6 K hot gas in the Omega Nebula may be attributed to colliding stellar winds among the O stars in the cluster and/or the strong magnetic field that hinders thermal conduction at its interface with the cool dense nebular material (Dunne *et al.* 2003).

XMM-Newton observations of the superbubble N51D (Figure 6, left panel) in the LMC H II complex N51 have been used in conjunction with optical observations of the ionized swept-up shell, radio H I 21-cm line observations of the neutral swept-up shell, and spectroscopic and photometric data of the underlying massive star population to examine the energy budget. It is found that the total energy retained in the thermal and kinetic

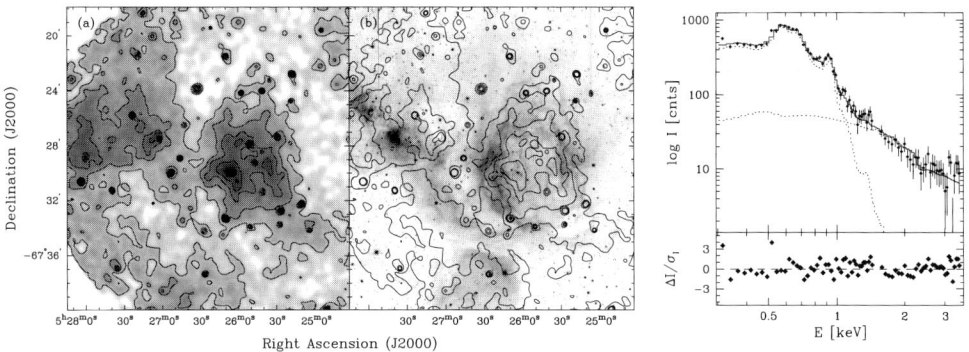

Figure 6. Left - *XMM-Newton* EPIC/pn X-ray and optical Hα images of the H II complex N51. The X-ray contours are over-plotted on the Hα image. Right - *XMM-Newton* EPIC/pn spectrum overlaid with the spectral components of the best-fit model. The residuals are plotted underneath the spectrum. These figures are taken from Cooper *et al.* (2004).

energies in the superbubble N51D is only about 1/3 of total stellar mechanical energy injected into the ambient medium (Cooper *et al.* 2004). It is possible that some energy has been used to accelerate particles that generate nonthermal X-ray emission (Figure 6, right panel). Similar energy budget problem is observed in the LMC superbubble 30 Dor C, which also exhibits nonthermal X-ray emission (Smith & Wang 2004).

Nonthermal X-ray emission is perhaps the most unexpected discovery from *Chandra* and *XMM-Newton* observations of superbubbles: RCW 38 (Wolk *et al.* 2002), 30 Dor C (Bamba *et al.* 2004; Smith & Wang 2004), and N51D (Cooper *et al.* 2004). Recent *Suzaku* observations of the superbubble in the LMC H II complex N11 also detected nonthermal X-ray emission (Maddox *et al.* 2008). Nonthermal emission at X-ray wavelengths requires very energetic particles. Smith & Wang (2004) considered synchrotron radiation and inverse Compton scattering of optical photons, but concluded that the origin of the nonthermal X-ray emission was uncertain. It is possible that energetic particles in superbubbles are generated by repeated shock acceleration due to the high supernova explosion rate (Parizot *et al.* 2004).

4.3. *Interface Layers*

At the interface between a cold swept-up shell and its hot interior, thermal conduction takes place through diffusion, resulting in mass evaporation from the cold dense shell into the hot interior. Evaporation or ablation of dense clumps left behind in a bubble interior also injects mass into the hot gas. Interface layers play a very important role in the thermal evolution of a bubble because its lower temperatures (10^5 K) and higher densities lead to higher cooling rates, which regulate the temperature of the hot interior.

Interface layers are traditionally observed through interstellar absorption lines of highly ionized species, e.g., C IV, N V, and O VI. Such observations, however, are made at the mercy of the availability of probe stars whose continua provide the backdrop of the absorption lines. Chu *et al.* (2004) compared *FUSE* observations of O VI $\lambda\lambda$1031,1037

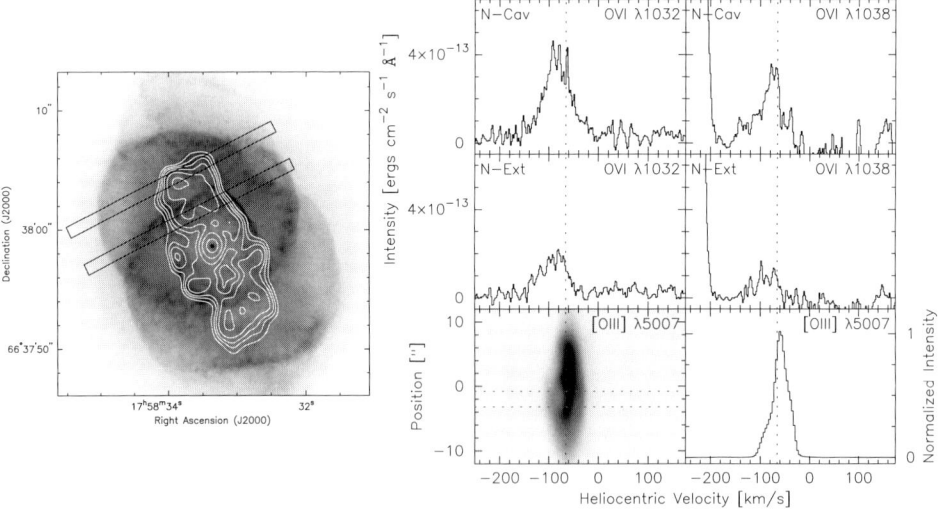

Figure 7. Left - *HST* Hα image of NGC 6543 overlaid with X-ray contours extracted from a smoothed *Chandra* ACIS-S image. Locations of the *FUSE* HIRS apertures (N-Ext and N-Cav) are marked. Right - The O VI line profiles in the top four panels. The bottom left panel shows a EW-oriented long-slit echellogram centered at 4.5″ north of the central star. The bottom right panel shows the [O III] velocity profile near the *FUSE* apertures. These figures are taken from Gruendl *et al.* (2004).

lines in absorption and in emission for planetary nebulae, and suggested that conduction layers were easier to observe in emission rather than absorption. Figure 7 shows *FUSE* observations of two slit positions in the planetary nebula NGC 6543, clearly detecting the O VI emission from the interface layer (Gruendl *et al.* 2004).

NGC 6888 and S 308 are the only two circumstellar bubbles detected in X-rays, but their central stars are too bright for *FUSE* observations of O VI absorption. Spatially resolved *FUSE* observations of O VI emission and *HST* STIS observations of the N V emission from S 308's interface layer would shed light on the conduction front or mixing layer; these observations were awarded but neither was carried out because of the premature demise of the instruments. Planetary nebulae are smaller versions of circumstellar bubbles. Diffuse X-ray emission from NGC 6543 has been detected; the limb-brightened morphology of the X-ray emission is expected from thermal conduction, but the X-ray luminosity is an order of magnitude lower than circumstellar bubble models with conduction (Chu *et al.* 2001). This raised doubt about thermal conduction. *FUSE* observations of the O VI emission from the conduction layer of NGC 6543 (Figure 7) were analyzed. Compared with circumstellar bubble models with thermal conduction, the O VI emission is an order of magnitude lower than expected; however, for the observed X-ray luminosity (an order or magnitude lower than expected) the O VI emission strength is roughly consistent with that expected from a conduction layer (Gruendl *et al.* 2004).

FUSE observations of O VI absorption from LMC superbubbles do not show obvious enhanced column density over the overwhelming LMC halo component (Howk *et al.* 2002). O VI emission from LMC superbubbles was serendipitously detected from archival *FUSE* observations, but it is not clear whether the emission arises from conduction interface, SNR shocks, or turbulent mixing layers (Sankrit & Dixon 2007).

Besides *FUSE*, the *SPEAR* imaging spectrometer also provides useful O VI observation of superbubble. Kregenow *et al.* (2006) have presented *SPEAR* observations of the Eridanus superbubble and used the spatial distribution of O VI and C IV emission along the shell rim to illustrate that the emission originates from the thermal conduction layer.

The current lack of functioning high-dispersion UV spectroscopic observing facilities is worrisome; without such, it is difficult to advance our understandings of the interface layer. Bubble models tending details of thermal conduction at the interface (Arthur, in this volume) are needed to interpret UV and X-ray observations of bubbles.

References

Arthur, S. J., Dyson, J. E., & Hartquist, T. W. 1993, *MNRAS*, 261, 425
Arthur, S. J., Henney, W. J., & Dyson, J. E. 1996, *A&A*, 313, 897
Bamba, A., Ueno, M., Nakajima, H., & Koyama, K. 2004, *ApJ*, 602, 257
Bochkarev, N. G. 1988, *Nature*, 332, 518
Book, L. G., Chu, Y.-H., & Gruendl, R. A. 2008, *ApJS*, 175, 165
Boroson, B., Blair, W. P., Davidsen A. F. *et al.* 1997, *ApJ*, 478, 638
Braunsfurth, E., & Feitzinger, J. V. 1983, *A&A*, 127, 113
Burrows, D. N., Singh, K. P., Nousek, J. A., Garmire, G. P., & Good, J. 1993, *ApJ*, 406, 97
Cappa, C. E., Arnal, E. M., Cichowolski, S., *et al.* 2003, in: K. A. van der Hucht, A. Herrero, & C. Esteban (eds.), A Massive Star Odyssey: From Main Sequence to Supernova, (San Francisco: ASP), *Proc. IAU Symp.* 212, 596
Castor, J., McCray, R., & Weaver, R. 1975, *ApJ*, 200, L107
Chen, C.-H. R., *et al.* 2008, submitted to *ApJ*
Chu, Y.-H. 2003, in: K. A. van der Hucht, A. Herrero, & C. Esteban (eds.), *A Massive Star Odyssey: From Main Sequence to Supernova*, (San Francisco: ASP), *Proc. IAU Symp.* 212, 585

Chu, Y.-H., Chang, H.-W., Su, Y.-L., & Mac Low, M.-M. 1995, *ApJ*, 450, 157
Chu, Y.-H., Gruendl, R. A., & Guerrero, M. A. 2004, in: M. Meixner, J. H. Kastner, B. Balick & N. Soker (eds.), *Asymmetrical Planetary Nebulae III: Winds, Structure and the Thunderbird*, (San Francisco: ASP), *ASP Conf. Series*, 313, 254
Chu, Y.-H., Guerrero, M. A., Gruendl, R. A., 2001, *ApJ*, 553, L69
Chu, Y.-H., Guerrero, M. A., Gruendl, R. A., et al. 2003, *ApJ*, 599, 1189
Chu, Y. H., & Lasker, B. M. 1980, *PASP*, 92, 730
Chu, Y.-H., & Mac Low, M.-M. 1990, *ApJ*, 365, 510
Chu, Y.-H., Mac Low, M.-M., García-Segura, G. et al. 1993, *ApJ*, 414, 213
Chu, Y.-H., Wakker, B., Mac Low, M.-M. & García-Segura, G. 1994, *AJ*, 108, 1696
Cooper, R. L., Guerrero, M. A., Chu, Y.-H., et al. 2004, *ApJ*, 605, 751
Dunne, B. C., Chu, Y.-H., Chen, C.-H. R. et al. 2003, *ApJ*, 590, 306
Dunne, B. C., Points, S. D., & Chu, Y.-H. 2001, *ApJS*, 136, 119
Dyson, J. E., & de Vries, J. 1972, *A&A*, 20, 223
Ehlerová, S., & Palouš, J. 2005, *Ap&A*, 437, 101
Freyer, T., Hensler, G., & Yorke, H. W. 2003, *ApJ*, 594, 888
Freyer, T., Hensler, G., & Yorke, H. W. 2006, *ApJ*, 638, 262
García-Segura, G., & Mac Low, M.-M. 1995, *ApJ*, 455, 145
García-Segura, G., Langer, N., & Mac Low, M.-M. 1996a, *A&A*, 316, 133
García-Segura, G., Mac Low, M.-M., & Langer, N. 1996b, *A&A*, 305, 229
Goudis, C., & Meaburn, J. 1978, *A&A*, 68, 189
Gruendl, R. A., Chu, Y.-H., & Guerrero, M. A. 2004, *ApJ*, 617, L127
Gruendl, R. A., Guerrero, M. A., & Chu, Y.-H. 2003, *BAAS*, 35, 746
Güdel, M., Briggs, K. R., Montmerle, T. et al. 2008, *Science*, 319, 309
Hatzidimitriou, D., Stanimirovic, S., Maragoudaki, F. et al. 2005, *MNRAS*, 360, 1171
Howarth, I. D. & Phillips, A. P. 1986, *MNRAS*, 222, 809
Howk, J. C., Sembach, K. R., Savage, B. D. et al. 2002, *ApJ*, 569, 214
Jenkins, E. B. & Meloy, D. A. 1974, *ApJ*, 193, L121
Kim, S., Dopita, M. A., Staveley-Smith, L., & Bessell, M. S. 1999, *AJ*, 118, 2797
Kregenow, J., Edelstein, J., Korpela, E. J. et al. 2006, *ApJ*, 644, L167
Mac Low, M.-M. & McCray, R. 1988, *ApJ*, 324, 776
Mac Low, M.-M., McCray, R., & Norman, M. L. 1989, *ApJ*, 337, 141
Maddox, L. A., et al. 2008, *ApJ* submitted
Mathews, W. G. 1966, *ApJ*, 144, 206
Meaburn, J. 1980, *MNRAS*, 192, 365
Moffat, A. F. J. & Robert, C. 1994, *ApJ*, 421, 310
Nazé, Y., Chu, Y.-H., Points, S. D. et al. 2001, *AJ*, 122, 921
Nazé, Y., Chu, Y.-H., Guerrero, M. A. et al. 2002, *AJ*, 124, 3325
Nichols-Bohlin, J. & Fesen, R. A. 1993, *AJ*, 105, 672
Oey, M. S. 1996, *ApJ*, 467, 666
Oey, M. S. & Clarke, C. J. 1997, *MNRAS*, 289, 570
Oey, M. S. & García-Segura, G. 2004, *ApJ*, 613, 302
Oey, M. S. & Smedley, S. A. 1998, *AJ*, 116, 1263
Parizot, E., Marcowith, A., van der Swaluw, E., et al. 2004, *Ap&A*, 424, 747
Pikel'Ner, S. B. 1968, *ApL*, 2, 97
Pikel'Ner, S. B. & Shcheglov, P. V. 1969, *Soviet Astronomy*, 12, 757
Pittard, J. M., Dyson, J. E., & Hartquist, T. W. 2001a, *A&A*, 367, 1000
Pittard, J. M., Hartquist, T. W., & Dyson, J. E. 2001b, *A&A*, 373, 1043
Sankrit, R. & Dixon, W. V. D. 2007, *PASP*, 119, 284
Smith, D. A. & Wang, Q. D. 2004, *ApJ*, 611, 881
Staveley-Smith, L., Sault, R. J., Hatzidimitriou, D. et al. 1997, *MNRAS*, 289, 225
Steigman, G., Strittmatter, P. A., & Williams, R. E. 1975, *ApJ*, 198, 575
Strickland, D. K. & Stevens, I. R. 1998, *MNRAS*, 297, 747
Tomisaka, K. 1992, *PASJ*, 44, 177

Townsley, L. K., Feigelson, E. D., Montmerle, T. et al. 2003, ApJ, 593, 874
Weaver, R., McCray, R., Castor, J. et al. 1977, ApJ, 218, 377
Wolk, S. J., Bourke, T. L., Smith, R. K. et al. 2002, ApJ, 580, L161
Wrigge, M. 1999, Ap&A, 343, 599
Wrigge, M., Wendker, H. J., & Wisotzki, L. 1994, Ap&A, 286, 219

Discussion

MORRIS: We've managed to obtain Spitzer/IRS spectra over some regions of the S 308 nebula, including what you identify as the thermal conduction zone. We don't see [Ne V] however. Is this a serious problem to constrain the conditions or even the existence of the thermal conduction zone?

CHU: The "conduction zone" speculated by Chu et al. (2003) is really the unperturbed RSG wind layer, as we have found from follow-up echelle observations of the shell rim (see Figure 5). The conduction zone would be immediately adjacent to the X-ray emission edge. Boroson et al. (1997) detected C IV and N V absorption toward the central star, and demonstrated convincingly the existence of conduction layer in S 308. Previously, Shull (1974, ApJ, 212, 102) detected O VI absorption, which also unambiguously demonstrates the existence of conduction layer. Recently, Freyer et al. (2006) suggest that the soft X-ray emission originates from the swept-up RSG wind, then the conduction zone would be closer in, but the nebular gas does not show any material expanding at 400 km s^{-1} expected from their model. More observations of the region near the X-ray emission front are needed to search for and locate the conduction zone.

OEY: Earlier in this meeting we heard of the importance of mass loss from near break-up rotation velocities. Can you comment on the effect of this kind of mass loss on the circumstellar media?

CHU: There are no observations that indicate a link between the stellar rotation velocities and the physical properties of circumstellar nebulae. The aspherical geometry of a circumstellar nebula may be caused by rapid stellar rotation, but the effect of a close binary companion may also be very important.

VANBEVEREN: I noticed that the table where you list the WR stars and their bubbles only contains WR single stars. Are there WR+OB binaries with bubbles and, if not, is there a reason why only WR single stars have these bubbles?

CHU: As far as I know, the majority of WR stars in bubbles are single WN stars. One bubble (DEM L39) has a WC central star and another (G2.4+1.4) has a WO central star. The central star of DEM L231 (N57C shown in Figure 1) has a spectral type of WN4+B, the only WR binary in a nice bubble. More single than binary WR stars have bubbles. I would guess that colliding winds in a close binary system breaks the spherical symmetry and makes it difficult to form bubbles.

You-Hua Chu.

Sally Oey, Maria Nieva and Sergio Simón-Díaz.

The Evolution of the Circumstellar and Interstellar Medium Around Massive Stars

S. Jane Arthur

Centro de Radioastronomía y Astrofísica, Universidad Nacional Autónoma de México,
Apartado Postal 3-72, 58090 Morelia, México
email: j.arthur@astrosmo.unam.mx

Abstract. Throughout their lives massive stars modify their environment through their ionizing photons and strong stellar winds. Here, I present coupled radiation-hydrodynamic calculations of the evolution of the bubbles and nebulae surrounding massive stars. The evolution is followed from the main sequence through the Wolf-Rayet stage and shows that structures are formed in the ISM out to some tens of parsecs radius. Closer to the star, instabilities lead to the breakup of swept-up wind shells. The photoevaporated flows from the resulting clumps interact with the stellar wind from the central star, which leads to the production of soft X-rays. I examine the consequences for the different observable structures at all time and size scales and evaluate the impact that the massive star has on its environment.

Keywords. stars: mass loss – ISM: bubbles – HII regions

1. Introduction

Massive stars interact with their environment both through their ionizing photons and their strong stellar winds. At early stages in their lives they produce bright emission-line nebulae such as the Trifid nebula and the Orion nebula. on size scales of up to a few parsecs. At the end of their lives, faint, photoionized Hα shells of tens of parsecs radius surround Wolf-Rayet (WR) stars.

Evidence for stellar wind bubbles around main sequence stars is not easy to come by because of the low density and high temperature of the shocked gas in the stellar wind bubble and confusion with the HII region. However, stellar wind shells have been detected kinematically (Nazé et al. 2001). The recent discovery of soft (2 million degrees), diffuse X-rays in the Orion nebula (Güdel et al. 2008) presents a challenge to theorists. Diffuse X-ray emission from star-forming regions had previously been detected at $\sim 10^7$ K in the Omega and Rosette nebulae, both of which surround clusters of massive stars (Dunne et al. 2003, Townsley et al. 2003). Stellar wind bubbles also manifest themselves as shells and holes of diameters of tens of parsecs in the distribution of HI gas around WR stars (e.g., Vasquez et al. 2005). Stellar wind bubbles of a few parsecs diameter are observed in optical emission lines around some WR stars and are distinguished from photoionized regions by their short dynamical timescales.

Traditional modeling of HII regions and stellar wind bubbles assumes a constant source of ionizing photons or stellar wind and a uniform ambient medium (Strömgren 1939, Kahn 1954, Dyson & de Vries 1972, Weaver et al. 1977). The variation of stellar wind parameters as a function of stellar evolution was taken into account by García-Segura et al. (1996a,b) and the ionizing photon flux of the central star was included by Freyer et al. (2003, 2006). However, little attempt has been made to take into the account the effects of a non-uniform ambient medium during the initial stages. Another topic often mentioned but never modeled self-consistently is thermal conduction. Weaver et al. (1977) included

355

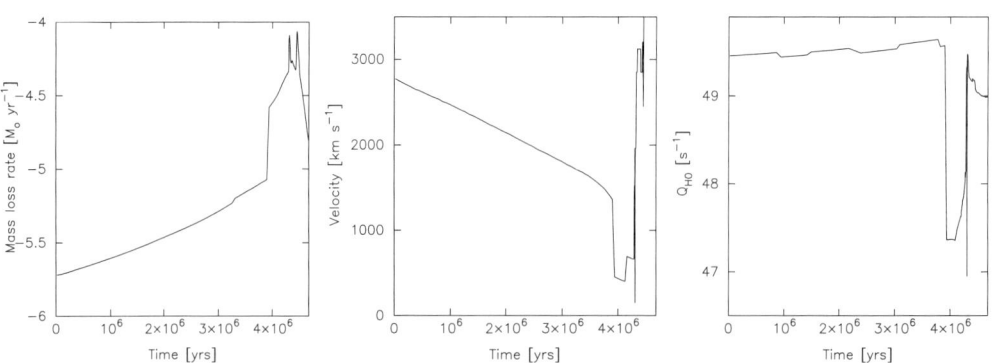

Figure 1. Mass-loss rate, stellar wind velocity and ionizing photon rate for a 60 $M_\odot$ star at solar metallicity and with an initial rotation velocity of 300 km s^{-1}.

it in order to explain (mistakenly) the diffuse interstellar OVI emission detected with the *Copernicus* satellite. Theorists argue that the magnetic field in the swept-up shell will inhibit thermal conduction by electrons. In this paper we address both of these aspects.

2. Modeling

The variation of the stellar wind parameters with time can be found from stellar evolution models such as those of Meynet & Maeder (2003). The most recent models, incorporating stellar rotation, suggest that such stars have double the WR lifetime of non-rotating models. The ionizing photon rate can be found from following the procedure outlined by Smith *et al.* (2002) (and included in the Starburst 99 code, Leitherer *et al.* 1999) using the appropriate values for effective temperature and stellar radius from the new stellar evolution models. In Figure 1 we show the mass-loss rate, stellar wind velocity and ionizing photon rate as a function of time for a 60 $M_\odot$ star of solar metallicity and initial rotation rate of 300 km s^{-1}. It is clear that the star experiences periods of enhanced mass loss, when the stellar wind velocity and ionizing photon rate both fall dramatically.

It is also important to take into account the metallicity and ionization state of the gas when calculating the radiative cooling rate. Failure to do so will result in incorrect expansion speeds for the shells and bubble.

The ambient medium into which the HII region and stellar wind bubble expand is more difficult to define. In the initial stages, the star will still be embedded in the dense molecular cloud from which it formed. At some point, however, the photoionized region will break out of the cloud into the intercloud medium. Furthermore, many well-known WR stars have large proper motions, which indicates that they have moved a long way from their birthplace during their lifetime.

3. HII regions in turbulent molecular clouds

Massive stars are born in the dense cores of molecular clouds and so their initial evolution will take place in such an ambient medium. Molecular clouds are now thought to be clumpy as a result of supersonic turbulence due to colliding streams of gas. We have studied the structure and evolution of HII regions in such an environment using a state-of-the-art radiation hydrodynamics code (Mellema *et al.* 2006). As initial conditions we took the output of a 3D turbulence calculation (Vázquez-Semadeni et al. 2005). This 512^3 grid represents a volume 4 pc on a side. The average number density within the cube is 1000 cm^{-3} but the peak density in the clumps is $> 10^6$ cm^{-3}. The ionizing source is

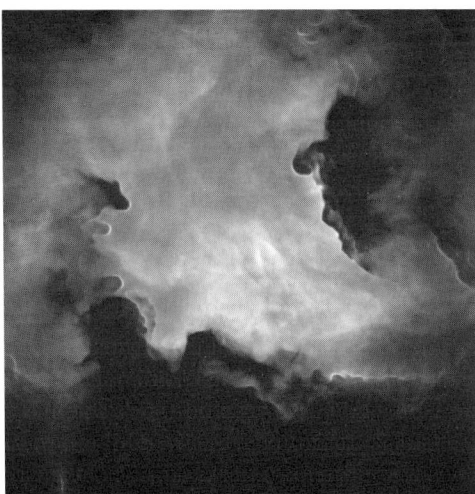 

Figure 2. Synthetic optical emission-line image of a simulated HII region in a turbulent molecular cloud. Two views through the computational cube. Colors are the same as the standard *HST* filters—*blue:* OIII, *green:* Hα, *red:* NII. Black regions represent foreground dust absorption.

located in the densest clump, which can be easily placed at the center of the grid because of the periodic boundary conditions of the turbulence calculation. The ionizing photon rate is set at $10^{48.5}$ s^{-1} for a star with effective temperature 37,000 K.

Figure 2 shows a synthetic emission-line image after 250,000 yrs of evolution. The numerical simulation bears a remarkable resemblance to real HII regions like the Trifid nebula. Bright-rimmed fingers and columns show where the ionization front is eroding the dense neutral material. Photoevaporated flows from the tips of the fingers shock against each other in the inner part of the nebula leading to density and temperature enhancements there. Such simulations strongly suggest that the structure of young HII regions is mainly due to structure in the underlying molecular cloud rather than instabilities in the ionization front, as has been previously suggested (García-Segura & Franco 1996).

The simulations described here do not yet include the stellar wind from the central star. This is principally due to the huge increase in the (already long) computational time required once the stellar wind is included because of the decrease in the timestep that this entails (Arthur & Hoare 2006). It is interesting that the main features of real HII regions can be produced with our simulations without stellar winds for these early stages in the life of a massive star.

4. Stellar wind bubbles and HII regions

At some point the stellar wind will become important and the swept-up shell of material driven by the hot shocked stellar wind bubble will trap the ionization front. The trapped HII region expands with time since its temperature is fixed at $\sim 10^4$ K and the pressure is set by the hot shocked bubble. Moreover, the ionization front requires a shock in the neutral gas ahead of it and so the stellar wind bubble is surrounded by a thick shell of shocked neutral material.

In Figure 3 we show the variation of density and temperature of the gas with radius in the stellar wind bubble towards the end of the main-sequence stage of the 60 M$_\odot$ star for cases with and without time-dependent thermal conduction. The stellar parameters are those depicted in Figure 1. This is a spherically symmetric calculation in order to

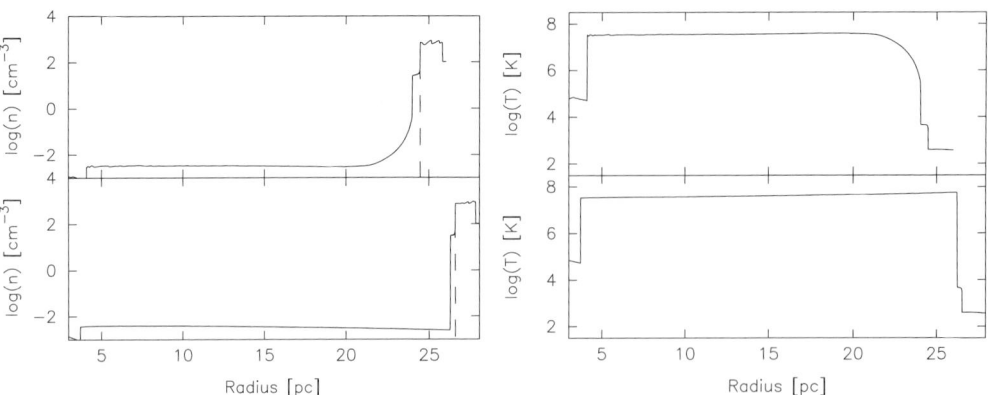

Figure 3. Density and temperature distributions of a main sequence stellar wind bubble. *Top:* With time-dependent thermal conduction. *Bottom:* Without thermal conduction. *Solid line:* total number density. *Dashed line:* Ionized number density, showing limit of HII region.

be able to follow several million years of stellar evolution and we have assumed that the ambient medium is uniform and neutral with a number density $n_0 = 100$ cm^{-3} and temperature ~ 500 K representing the photodissociated region ahead of the ionization front. The stellar wind is modeled as a region of uniform mass and energy deposition (e.g., Chevalier & Clegg 1985) since this treatment makes it easier to follow changes in the stellar wind. The resulting thermal wind becomes supersonic well before the inner stellar wind shock, hence the details of the wind treatment have no effect on the formation and expansion of the stellar wind bubble.

We see that thermal conduction enhances the density in the hot shocked bubble close to its edge. The increased cooling rate here lowers the temperature at the edge of the bubble and the average pressure in the bubble decreases. This slows the expansion rate and also allows the stellar wind shock to move outwards further from the star than in the corresponding case without thermal conduction.

When the stellar wind velocity drops during a period of enhanced mass loss, such as a red supergiant or LBV stage, the ram pressure of the stellar wind is much less than the thermal pressure in the hot shocked bubble and so the bubble begins to back fill. Moreover, when the ionizing photon rate becomes so low that it cannot sustain the HII region at the edge of the stellar wind bubble, the gas here recombines. This has the consequence that the pressure here drops to half its previous value, and so the shocked neutral material expands into this region. When the ionizing photon rate increases again in a Wolf-Rayet phase, the HII region reforms at the edge of the stellar wind bubble.

5. Instabilities

The slow, dense wind material of a red supergiant or LBV stage is squeezed into a thin shell by the high pressure of the surrounding hot shocked bubble. This thin, dense, neutral shell is subject to a variety of instabilities such as the thin shell instability and Rayleigh-Taylor instabilities, which break it up into dense neutral clumps, as shown in Figure 4. Once the ionizing photon rate increases again, these clumps acquire photoionized skins and the resulting transonic, dense, photoevaporated flows shock against the stellar wind, leading to densities and temperatures that should be observable in X-rays. In this scenario, the optical emission would come from the shocked neutral shell material, while the X-ray emission is produced by the shocked photoevaporated flows.

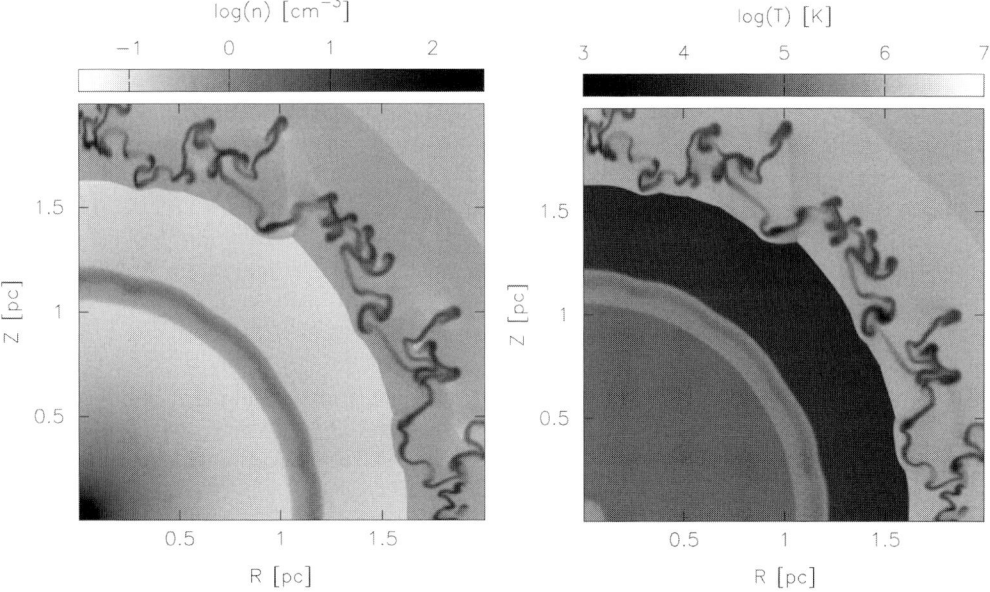

Figure 4. Interaction of a fast Wolf-Rayet wind with a dense red supergiant wind shell. *Left:* Ionized number density. *Right:* Temperature.

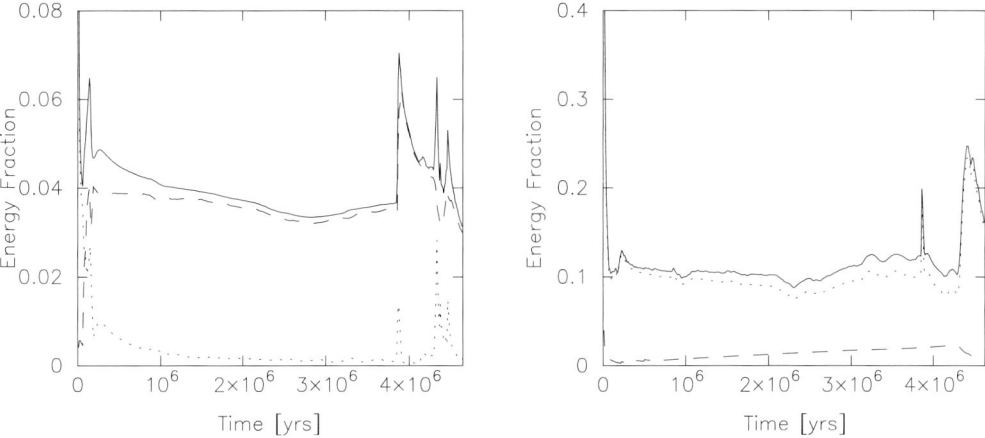

Figure 5. *Left:* Gas kinetic energy as a fraction of total wind mechanical energy for the model with time-dependent thermal conduction. *Right:* Gas thermal energy. *Solid line:* Total gas kinetic (thermal) energy. *Dashed line:* Neutral gas kinetic (thermal) energy. *Dotted line:* Ionized gas kinetic (thermal) energy. Note the different scales on the two graphs.

6. Energy budget

Observational studies of the HI shells around some Galactic WR stars (Vasquez *et al.* 2005) and also of the superbubble N51D in the LMC (Cooper *et al.* 2004) reveal that only a small fraction of the total stellar wind (and supernova) mechanical energy is imparted to the interstellar medium and argue that this constitutes an energy crisis for these bubbles. However, the present calculations show that even though the gas thermal and kinetic energies are a small fraction of the total stellar wind energy over the star's lifetime (see Fig. 5), the explanation for the 'missing' energy is simply that it was radiated away.

7. Summary

Radiation-hydrodynamic modeling shows how different structures form and evolve in the circumstellar and interstellar medium around massive stars. Structure in optical HII regions results from differences in densities and hence opacities in an underlying clumpy medium. Main-sequence stellar wind bubbles are surrounded by HII regions and thick neutral shells, making their direct detection difficult. Thermal conduction, if present, leads to enhanced radiative energy losses at the edge of the hot, shocked wind bubble, reducing its pressure and slowing the expansion. The interaction of the fast wind from the WR stage with a dense neutral shell from a previous red supergiant or LBV stage leads to instabilities, which break up the neutral shell into clumps.

The amount of energy imparted to the interstellar medium as kinetic or thermal energy of the gas is only a small fraction of the total stellar wind energy released over the star's lifetime, with the remainder of the energy simply being radiated away. A single massive star can affect the structure and dynamics of the interstellar medium at distances of up to tens of parsecs and the compression and subsequent fragmentation of dense shells can lead to new episodes of star formation.

References

Arthur, S. J., & Hoare, M. G. 2006, *ApJS*, 165, 283
Chevalier, R. A., & Clegg, A. W. 1985, *Nature*, 317, 44
Cooper, R. L., Guerrero, M. A., Chu, Y.-H., *et al.* 2004, *ApJ*, 605, 751
Dunne, B. C., Chu, Y.-H., Chen, C.-H. R., *et al.* 2003, *ApJ*, 590, 306
Dyson, J. E., & de Vries, J. 1972, *A&A*, 20, 223
Freyer, T., Hensler, G., & Yorke, H. W. 2003, *ApJ*, 594, 888
Freyer, T., Hensler, G., & Yorke, H. W. 2006, *ApJ*, 638, 262
García-Segura, G. & Franco, J. 1996, *ApJ*, 469, 171
García-Segura, G., Mac Low, M.-M., & Langer, N. 1996a, *A&A*, 305, 229
García-Segura, G., Langer, N., & Mac Low, M.-M. 1996b, *A&A*, 316, 133
Güdel, M., Briggs, K. R., Montmerle, T., *et al.* 2008, *Sci*, 319, 309
Kahn, F. D. 1954, *Bull. Astron. Inst. Netherlands*, 12, 187
Leitherer, C., et al. 1999, *ApJS*, 123, 3
Mellema, G., Arthur, S. J., Henney, W. J., Iliev, I. T., & Shapiro, P. R. 2006, *ApJ*, 647, 397
Meynet, G., & Maeder, A. 2003, *A&A*, 404, 975
Nazé, Y., Chu, Y.-H., Points, S. D., *et al.* 2001, *AJ*, 122, 921
Smith, L. J., Norris, R. P. F., & Crowther, P. A. 2002, *MNRAS*, 337, 1309
Strömgren, B. 1939, *ApJ*, 89, 526
Townsley, L. K., Feigelson, E. D., Montmerle, *et al.* 2003, *ApJ*, 593, 874
Vasquez, J., Cappa, C., & McClure-Griffiths, N. M. 2005, *MNRAS*, 362, 681
Vázquez-Semadeni, E., Kim, J., Shadmehri, M., & Ballesteros-Paredes, J. 2005, *ApJ*, 618, 344
Weaver, R., McCray, R., Castor, J., *et al.* 1977, *ApJ*, 218, 377

Discussion

WALBORN: It's very interesting that you get the pillars with just the radiation and no winds. Do you have any indications of the effects of the winds on the formation (or disruption?) of these structures?

ARTHUR: The pillars are the result of the interaction of the ionizing photons with the dense clumps in the molecular cloud and show the position of the ionization front. A stellar wind bubble would be internal to the ionization front and, in regions where there is a density gradient, would tend to blow out in the direction away from the dense cloud. I don't think a wind would make any difference to the formation of these structures.

Massive Stars as Cosmic Engines
Proceedings IAU Symposium No. 250, 2007
F. Bresolin, P.A. Crowther & J. Puls, eds.

© 2008 International Astronomical Union
doi:10.1017/S174392130802070X

Infrared Tracers of Mass-Loss Histories and Wind-ISM Interactions in Hot Star Nebulae

Patrick Morris[1] and the *Spitzer* WRRINGS team

[1]NASA Herschel Science Center, IPAC, Caltech, M/C 100-22, Pasadena, CA 91125
email: `pmorris@ipac.caltech.edu`

Abstract. Infrared observations of hot massive stars and their environments provide a detailed picture of mass loss histories, dust formation, and dynamical interactions with the local stellar medium that can be unique to the thermal regime. We have acquired new infrared spectroscopy and imaging with the sensitive instruments onboard the *Spitzer* Space Telescope in guaranteed and open time programs comprised of some of the best known examples of hot stars with circumstellar nebulae, supplementing with unpublished Infrared Space Observatory spectroscopy. Here we present highlights of our work on the environment around the extreme P Cygni-type star HDE316285, providing some defining characteristics of the star's evolution and interactions with the ISM at unprecented detail in the infrared.

Keywords. stars: mass loss – circumstellar matter – stars: winds, outflows – infrared: stars

1. Introduction

Observations of the circumstellar environments of Luminous Blue Variables (LBVs) with ESA's Infrared Space Observatory (ISO) and modern ground-based imaging devices have demonstrated how sensitive mid-infrared imaging and spectroscopy can substantially improve our knowledge of the distribution of material, the physical and chemical properties of nebular dust and gas, and therefore on the history of mass loss from the surface of the massive central star during its evolution from the Main Sequence. Other than the bright and compact environments of most LBV nebulae, none of the lower surface brightness nebulae around OB or Wolf-Rayet stars could be observed with ISO due mainly to instrumental sensitivity limitations. The instruments on the *Spitzer* Space Telescope have since offered significant gains in sensitivity and sky coverage; for comparison, the *Spitzer* Infrared Spectrometer (IRS; 5.3 – 38 μm) is factor of $\sim$50-100 more sensitive than the ISO Short Wavelength Spectrograph (SWS; 2.4-45.2 μm) at 5 μm, depending on resolution modes of the two instruments. We have exploited the *Spitzer* IRS, the Infrared Array Camera (IRAC) with imaging bands at 3.4, 4.6, 5.8, and 8.0 μm, and the Multiband Imaging Photometer for *Spitzer* (MIPS) with imaging capabilities at 24, 70, and 160 μm in Guaranteed and Open Time programs to observe the environments around a number of hot, massive stars. Most of these stars are surrounded by ring nebulae and are well known for their optical and/or radio properties; NGC2359 (WR), M1-67 (WR), G79.29+00.46 (BIe), and HD148937 (Ofp?e) are examples.

In this paper we summarize our study of the massive B supergiant and candidate LBV HDE316285 from our program. This highly luminous P Cyg-type star has been suspected of being surrounded by an extended, cold nebula (McGregor, Hyland & Hillier 1998); we have obtained conclusive observations of the environment around this star, which harbor telling characteristics of the star's mass loss history, wind-wind and wind-ISM interactions, the molecular content of the gas and thermal history of the dust.

2. The ring nebula around the extreme P Cygni star HDE316285

The combined 1.4-38 μm spectrum of HDE316285 is shown in Fig. 1. This star has been quantitatively characterized by Hiller *et al.* (1998) using line-blanketed non-LTE wind models as an "extreme" P Cygni-type star, with an optical and near-IR stellar wind spectrum that is quite similar to that of P Cyg. Spectral variability, the stellar properties and chemical content of HDE 316285 are similar to known LBVs, but is more extreme than P Cyg because of its high wind performance number (= ratio of wind momentum to radiative momentum), some 30 times greater. This number, however, is based on the assumption of a smooth, homogeneous wind and is subject to downward revision in the case of a clumped wind and associated reduction to the mass loss rate (cf. Hillier & Miller 1999). Somewhat unexpectedly, however, the infrared continuum, which can be used to constrain the volume filling factor (e.g., Morris *et al.* 2000), is contaminated by a strong thermal excess which becomes evident at around 16μm (see Fig. 1). In fact McGregor, Hyland, & Hillier (1988) had pointed out that HDE316285 is a moderate IRAS source, exhibiting cold dust ($\sim$60 K) removed from the central star. Since this star is projected $\sim 1°$ from the Galactic Center and is now recognized to be in close (apparent) proximity to the infrared star forming region (SFR) Sgr D (discussed below), a significant thermal component arising from surround molecular material heated by nearby young stars would provide the most likely explanation for the excess emission in the IRAS 25 and 60 μm passbands.

Infrared imaging. We now confirm that HDE316285 is definitely surrounded by its own circumstellar nebula. Following careful inspection of archival Midcourse Space Experiment 21.3 μm data in which a bipolar nebula can be seen just barely above the detection threshold, our MIPS 24μm imaging (see Fig. 2) reveals a limb-brightened, $3'.4 \times 4'.1$ ring nebula with the major axis oriented $\sim 40°$ from the north. The nebula is *not* discernable overall in the IRAC bands, though one bright knot of 8.0 μm emission is detected $\sim 1'$ NE of the central star. This knot and the brightened southern limb (respectively labelled "2" and "1" in Fig. 1) exhibit peak nebular fluxes of 624 and 450 MJy sr^{-1}, and were targeted by us for IRS spectroscopy, shown in Fig. 2 and discussed below. The nebula at 24 μm also clearly exhibits a clumpy structure, and filaments directed radially outwards

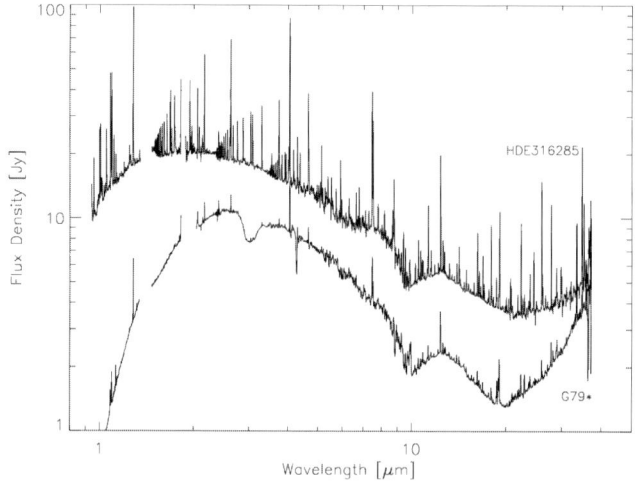

Figure 1. Combined near-IR (1.4-2.4μm), ISO/SWS (2.4-10μm), and *Spitzer*/IRS (10.0-38μm) SEDs of HDE316285 (upper) and G79.29+00.46 (lower). From Morris *et al.* (2008) submitted.

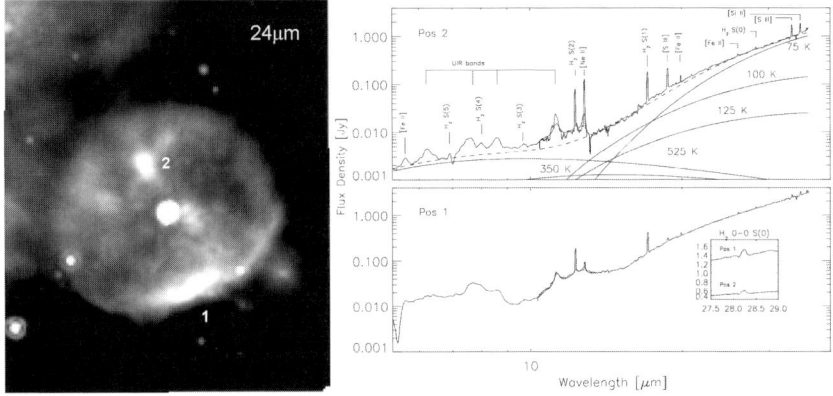

Figure 2. *Left:* HDE316285 at 24 μm with Spitzer/MIPS. The image is 5'.5 × 8, north is up, east is left. Two regions targeted for spectroscopy are indicated. *Right:* IRS 5.3-38μm spectra at two positions in the nebula (background-subtracted), with main features identified.

particularly evident in the NE half. The SW rim of the nebula appears to be interacting with material emitting at 8.0 μm in the surrounding ISM. (see also Fig. 3).

Spectral content: PAHs and rotational H_2. Spectra extracted at the two positions indicated in Fig. 2 and fully corrected for background emission with a dedicated off-position observation reveal thermal spectra with two principle components: a dominant cool component in the 75-125 K range, and a warm component at $T_{\rm dust} > 350$ K. Atop the warm component we see the family of polyaromatic hydrocarbon (PAH) bands at 6.22, 7.63, 8.63, 11.22, and 12.75 μm, and possibly a number of secondary peaks. These bands are common in reflection nebulae and the shells of AGB and post-AGB stars with mixed C and O chemistries, but they have also been detected mixed with the crystalline silicate dust in the nebulae around LBVs AG Car and R71 (Voors *et al.* 1999, 2000). We return to these features at the end of this paper.

The HDE316285 nebular spectra also exhibit pure rotational lines of H_2 arising from optically thin quadrupole transitions, from S(0) at 28.3 μm through S(6) at 6.1 μm, though the S(4) 8.0 μm and S(6) lines are blended with PAH emission. *This is the first known detection of rotational transitions of H_2 in the nebula of a hot massive star.* St. Louis *et al.* (1998) have previously detected the ro-vibrational 1-0 S(1) 2.112 μm line in the NGC2359 nebula around WR7, in a region where the nebula may be interacting with a surrounding bow shock. From this single line, however, St. Louis *et al.* could not deduce the excitation mechanism. This is relevant to the debate on the contribution of hot stars to the H_2 luminosity in starburst galaxies and ULIRGs, where fluorescent excitation in the UV radiation fields of OB and WR stars has been generally preferred over collisionally-induced emission for lack of cases of the latter in local settings.

The H_2 rotational lines are readily thermalized at moderate volume densities, allowing us to estimate H_2 column densities and excitation temperatures under the assumption of LTE in order to gain insight into the excitation conditions of the line forming regions. We follow excitation diagram methods developed by Burton (1992) and Gredel (1994), and applied to comparable environments (as in, e.g., NGC7129 by Morris *et al.* 2004 and references therein). The column density for transition j is $N_j = 4\pi F_j / \Omega E_j g_j A_j$, where E_j is the energy of the upper level, g_j is the statistical weight of ortho- and para-transitions, and A_j is the transition probability. The line column density is related to the total column density and excitation temperature $T_{\rm ex}$ by $\ln(N_j/g_j) = \ln(N_{\rm tot}/Q) - E_j/(kT_{\rm ex})$, and may be solved for with a Boltzmann excitation diagram. The partition function Q was

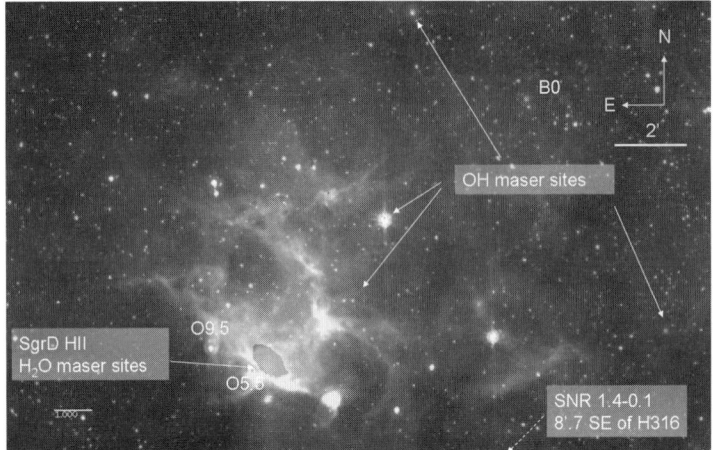

Figure 3. Mosaic of the Sgr D region, with radio sources identified by Mehringer *et al.* (1998) indicated. Blue corresponds to 4.6 μm, green to 8.0 μm, and red to 24 μm.

determined with an ortho-para ratio of 3 under the assumption of thermal equilibrium. For the observed line fluxes per unit solid angle F_j/Ω, we we have extracted over regions of 40 arcsec^2 and 62 arcsec^2 in the 5.3 – 14 μm and 14 – 35 μm ranges, respectively. In these areas we must view our measurements as spatially integrated. The measured line intensities were corrected for reddening $A_V = 6.0$ mag (Hillier *et al.* 1998). Measurement uncertainties of the S(3) – S(5) lines arise predominantly from the blending with the PAH features and the flux calibration. The S(6)/S(5) or S(4)/S(5) line ratios could be also affected by an ortho-para ratio different than the canonical (thermal equilibrium) value of 3, as in the presence of shocks, or by the superposition of multiple layers along the line of sight with different physical conditions. From our computed values of $\ln(N_j/g_j)$ and $E_{\rm up}/K$ we obtain estimates of the total H$_2$ column density $N_{\rm tot} \simeq 1.5 \times 10^{20}$ cm^{-2} and excitation temperature $T_{\rm ex} \simeq 240$ K, averaged over the two positions.

The relatively low value of $T_{\rm ex}$ at the high H$_2$ column density (compared to typical photodissociation regions) seem quite reasonable for line emission in the outflow arising from collisional (dissociative J-shock) excitation. We favor the J-shock mechanism over C-shocks in the outflow, based on the S(0) line intensity and strong [Si II] 34.815 μm emission. From J-shock models (Hollanbach & McKee 1989) we see also by the presence of [Ne II] 12.81 μm that the velocity is probably greater than 50 km sec^{-1}.

Is HDE316285 a hyper-luminous counterpart of the Pistol Star? In order to estimate the physical size and age of the nebula around HDE316285 we must know the distance to the star. Hillier *et al.* (1998) adopted 2 kpc, as a rough average of literature values between 1 and 3.4 kpc, and conceded that there is no real constraint on the distance. At that time it was not recognized that HDE316285 lies in close (apparent) proximity to the infrared SFR Sgr D (see Fig. 3). Assuming HDE316285 to be equidistant with Sgr D has the appeal of placing this B supergiant in the same dynamic ecosystem as the SFR and neighboring YSOs and maser sources, as well as a supernova remnant SNR 1.4-0.1 a few arcmin to the SE. Mehringer *et al.* (1998) have estimated the distance of Sgr D as 8 kpc based on H$_2$CO and CS line kinematics measured at the VLA. At this distance, the stellar luminosity of HDE316285 must be corrected from $\log(L_\star/L_\odot) = 5.5$ to 6.7! This takes into account a small reduction to $T_{\rm eff}$ ($\simeq$14kK) in the most recent line-blanketed models, but does very little to moderate *a stellar luminosity matched only by the Pistol Star in the GC and η Carina*. Assuming a distance of 8 kpc, then the diameter of the

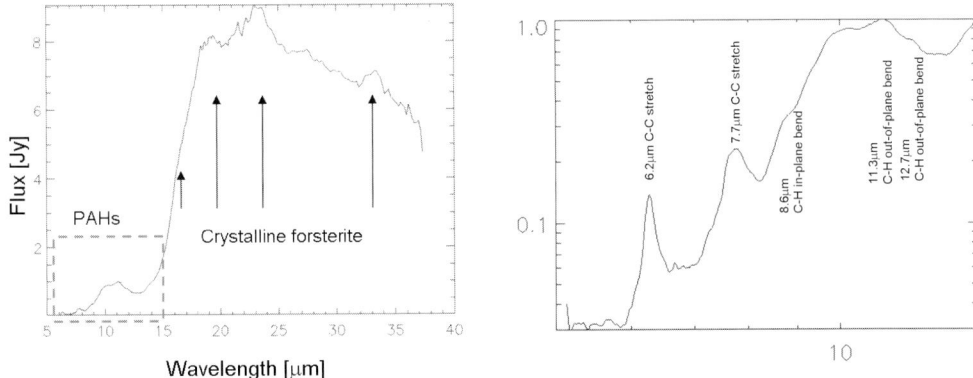

Figure 4. *Left: Spitzer*/IRS spectrum of the LMC LBV R71, with principle crystalline silicate bands and PAH features indicated. Nebular dust properties have been studied by Voors *et al.* (1999) using ISO/SWS data to 25 μm. *Right:* Zoom on the PAH bands in the IRS spectrum. Was the material comprised of crystalline dust and hydrocarbons condensed in the C-rich envelope of an RSG phase?

nebula is 8.6 pc, and must have a dynamical age of at least 9300 years using an upper constraint of 450 km/s for the time-averaged expansion velocity set by the unresolved IRS lines ($\lambda/\Delta\lambda \simeq 650$).

We cannot settle comfortably with this new distance to HDE316285, unfortunately. Blum & Damineli (1999) have presented K-band imaging of Sgr D, from which they argue that the SFR is located between 4-7.9 kpc (for $d_{GC} = 8$ kpc) and suggest $A_K = 1.9$ mag from the nebular Br-γ/radio flux ratio. No point sources are detected in the K-band image within the core of Sgr D, but focusing on sources within a few arcmin of the SFR, 18 of 34 sources have $A_K \leqslant 1.5$ mag (versus $A_K = 3$ for the GC itself). They devised a model with a uniform $A_K = 2.2$ mag screen of extinction at 4 kpc from the Sun, matching the actual star counts. HDE316285 has $A_V = 8$ mag ($A_K \sim 0.8$ mag), which is substantially lower, suggesting that the star is in the foreground. Indeed, University College London Echelle Spectrograph data taken in May 1995 show NaI D velocities of around +5 km/s, in agreement with a low distance to HDE316285 from comparison with the Brand & Blitz (1993) rotation curve using $d_{GC} = 8$ kpc.

In summary, we are confronted with conflicting evidence on the distance to Sgr D and HDE316285: on the one hand H_2CO and CS radio line kinematics place Sgr D at 8 kpc, while near-IR modeling of the reddened star population close to SFR can satisfy observed star counts at half the distance. Admittedly the latter method involves many more approximations and ensuing uncertainties, but the optical line (Na D) kinematics and comparatively low reddening of HDE316285 taken at their face values do not favor a large distance. If the B supergiant star is in the foreground, between 2 and 4 kpc, then either a substantial amount of foreground material is present also in the immediate surrounding ISM to interact with the stellar outflow, or else the H_2 lines are being formed in collisions between the present-day stellar wind and the outflow detected in our 24μm observations.

Implications and Outlook. The fact that HDE316285 is surrounded by a bipolar nebula supports the notion suggested by Hillier *et al.* (1998) that this luminous, strong-winded B supergiant has recently passed through an LBV phase. This is more plausible at a lower distance to the star with a commensurately lower limit age of the nebula, (i.e., $\geqslant 2300$). We have also noted the lack of specific dust bands that are the signatures of crystalline material which normally condenses in the thick, slowly-expanding envelopes of RSGs, suggesting that the material forming the present-day nebula was rapidly ejected and could not crystallize under conditions of rapid cooling and low monomer densities. The explosive outflows from η Carina also lack crystalline silicates, though there are a number of dust bands in this N-rich, C-depleted environment which are yet to be properly identified (Morris *et al.* 1999). In contrast, ISO spectroscopy of the LBVs AG Car, WRA 751, and R71 (Voors *et al.* 1999, 2000) reveal crystalline properties of the O-rich dust and the presence of C-rich hydrocarbons. See Fig. 4 for the IRS spectrum of R71. These properties are more consistent with formation during a RSG phase, supplying appropriate temperature and density condensation conditions as well as surface chemistries in which $N(C)/N(O) > 1$, and fits with the rather low outflow velocities and time-averaged mass-loss rates which yield the nebular masses estimated by Voors *et al.*

We must be careful to consider certain potential attenuating effects on dust crystallinity when interpreting the mass-loss history. Specifically, silicates initially condensed amorphously in an eruption might be annealed under electron bombardment (Carrez *et al.* 2002b). Forsterite, an Mg-rich crystalline silicate detected in the aforementioned LBVs (see Fig. 4), can be formed at fluences of $\sim 10^{17}$ e$^-$ cm^{-2} where (in the laboratory) particles are accelerated to several 10^6 eV. This may occur in YSOs with magnetic fields. OB supergiants are not magnetically active (except in the rare Ofp?e cases) and therefore the mechanism of electron bombardment would require acceleration by other means. Conversely, amorphization of initially crystalline dust may occur under heavy proton bombardment, which may explain the lack of crystalline silicates in the ISM (Carrez *et al.* 2002a, Brucato *et al.* 2004). Both effects of electron (re-)annealing and proton amorphization require study applied to BSG/RSG environments in order to improve the use of dust grain properties as tracers of the evolutionary state of the underlying star during condensation.

References

Blum, R. D. & Damineli, A. 1999, *ApJ*, 512, 237
Brand, J. & Blitz, L. 1993, *A&A*, 275, 67
Brucato, J. R., Strazzulla, G., Baratta, G., & Colangeli, L. 2004, *A&A*, 413, 395
Burton, M. G. 1992, *AuJPh*, 45, 463
Carrez, P., Demyk, K., Cordier P. *et al.* 2002a, *Meteoritics & Plan. Sci.*, 37, 1599
Carrez, P., Demyk, K., Leroux H. *et al.* 2002b, *Meteoritics & Plan. Sci.*, 37, 1615
Gredel, R. 1994, *A&A*, 292, 580
Hillier, D. J. & Miller, D. L. 1999, *ApJ*, 519, 354
Hillier, D. J., Crowther, P. A., Najarro, F., & Fullerton, A. W. 1998, *A&A*, 340, 483
McGregor, P. J., Hyland, A. R., & Hillier, D. J. 1988, *ApJ*, 324, 1071
Mehringer, D. M., Goss, W. M., Lis, D. C., *et al.* 1998, *ApJ*, 493, 274
Morris, P. W., Waters, L. B. F. M., Barlow, M. J. *et al.* 1999, *Nature*, 402, 502
Morris, P. W., van der Hucht, K. A., Crowther P. A. *et al.* 2000, *A&A*, 353, 624
Morris, P. W., Noriega-Crespo, A., Marleau, F. R. *et al.* 2004, *ApJS*, 154, 339
St-Louis, N., Doyon, R., Chagnon, F., & Nadeau, D. 1998, *AJ*, 115, 2475
Voors, R. H. M., Waters, L. B. F. M., Morris, P. W. *et al.* 1999, *A&A*, 341, L67
Voors, R. H. M., Waters, L. B. F. M., de Koter, A. *et al.* 2000, *A&A*, 356, 501

Stellar Feedback Through Cosmic Time: Starbursts & Superwinds

Michael A. Dopita

Research School of Astronomy & Astrophysics, The Australian National University
Cotter Road, Weston Creek, ACT 2611, Australia
email: Michael.Dopita@anu.edu.au

Abstract. Throughout cosmic time, the feedback of massive star winds and supernova explosions has been instrumental in determining the phase structure of the interstellar medium, controlling important aspects of both the formation and evolution of galaxies, producing galactic winds and enriching the intergalactic medium with heavy elements. In this paper, I review progress made in our theoretical understanding of how these feedback processes have operated throughout cosmic time from the epoch of the first stars through to the present day.

Keywords. galaxies: evolution – galaxies: high-redshift – intergalactic medium – ISM: evolution – ISM: jets and outflows – stars: winds, outflows

1. Introduction

Feedback from massive stars exercises an important control on a wide number of aspects of the evolution of our Universe. Specifically, the formation of massive stars controls the re-ionization of the Universe at the termination of the "dark age". The ionization of the intergalactic gas radically alters the development of structure in the Universe and the subsequent collapse of galaxies. Once galaxies have been formed, the energy and ionization feedback of massive stars determines the phase structure of the interstellar medium in galaxies, and controls structural parameters of disk galaxies such as the scale height of the gaseous component. In starburst regions, the over-pressure can encourage collapse of molecular clouds, triggering new generations of stars to form. This over-pressure can in turn drive mass-loaded galactic winds and outflows to enrich the inter-galactic medium. In dwarf galaxies, the loss of material into the halo may be sufficient to reduce the apparent yield of the heavy elements. The formation of massive stars in the nuclear regions of galaxies may also assist the growth of the central Black Holes, by providing a source of viscosity to assist the circum-nuclear gas to shed its angular momentum. In turn, the jets from the resulting AGN can themselves trigger star formation during their passage through the galactic medium. Thus, massive stars truly act as "cosmic engines" exercising an important control on both the structural evolution of the Universe and on the structure and evolution of the galaxies which it contains.

There have been a number of excellent recent reviews covering particular aspects of the feedback process. The issue of massive star feedback on controlling the structure and porosity of the interstellar medium (ISM) through the formation of bubbles and super-bubbles has been reviewed by Oey & Clarke (2008), see also Chu (this volume). A useful earlier review concerning the role of massive stars in controlling the morphology and physical structure of the ISM was published by Chu (1999). The physics of feedback processes such as accretion luminosity, ionizing radiation, and stellar winds in controlling the formation and growth of massive stars has been recently dealt with by Mac Low (2008). Various other aspects of feedback are covered in this conference. In particular I refer the reader to the paper by Sally Oey dealing with radiative feedback processes, Jay Gallager's report on feedback in starbursts, and both Yoshiaki Taniguchi's and Volker Bromm's reviews on the high redshift Universe and the first stars.

In this review, I will concentrate on the rapid progress which has been made in understanding feedback and its effects from a theoretical perspective. In particular, I will concentrate on our increasing understanding of the importance of feedback from the first stars in controlling the structural evolution of the Universe, and on the role of feedback in starburst environments.

2. The Early Universe

2.1. *The First Stars: An Overview*

For the issues associated with the formation of the first stars, we refer the reader to the extensive and fairly recent review by Bromm & Larson (2004). The details of the problems associated with feedback from the first stars and its influence on the development of structure in the early universe has recently be reviewed by Ciardi (2008). She distinguishes between three types of feedback process - mechanical feedback which alters the phase structure and the distribution of gas following the formation of the first stars, chemical feedback which occurs once there are sufficient heavy elements to allow the cooling and fragmentation of dense clouds into stars with an initial mass function (IMF) similar to today's, and radiative feedback processes driven by the ionizing radiation of the massive stars. Since this review is so comprehensive and recent, we refer the interested reader to it, and here I will emphasize only a few very recent results.

The first Population III stars are probably described by a top-heavy IMF, a consequence of H_2 cooling. Although it has been generally assumed that these formed at redshift $z \sim 20$ *e.g.* Yoshida *et al.* (2003), they may well have been formed as little as ~ 30 Myr after the Big Bang corresponding to $z \sim 65$; Naoz, Noter & Barkana (2006). At this time the overdense regions contained as little as $10^5 M_\odot$. However, it is not until ~ 500 Myr after the Big Bang ($z \sim 12$) that systems as large as the Milky Way start to assemble.

The formation of the Population III stars is strongly controlled by the formation and destruction of the H_2 molecule, since this is the principle cooling agent. This physics and its consequences has been recently been discussed in detail by Mesinger, Bryan & Haiman (2006). The strength of the UV radiation field is critical in controlling the H_2 fraction, since molecular hydrogen is readily destroyed by photoabsorbtion into auto-ionizing states in the 11.2-13.6 eV Lyman-Werner Bands by the Solomon Process (Stecher & Williams (1967), Abel *et al.* (1997)):

$$H_2 + \gamma \rightarrow H_2^* \rightarrow 2H \quad (2.1)$$

Such a process provides a radiative negative feedback for star formation by suppressing cooling (Ciardi & Ferrara (2005), Mesinger, Bryan & Haiman (2006)). However, H_2 formation is facilitated both by the availability of free electrons and by elevated temperature:

$$H + H_2^+ \rightarrow H^+ + H_2 \quad (2.2)$$
$$H + H^- \rightarrow e^- + H_2. \quad (2.3)$$

Such conditions are found in regions shock-heated by colliding and merging filaments in the cosmic web, which therefore serve to facilitate the formation of the earliest stars. They are also to be found in cooling and recombination zones associated with fossil

HII regions, in regions of X-ray heating by regions shock-excited by supernovae, *i.e.* regions surrounding these first stars during the time following the lifetime of the star as a powerful sources of UV radiation. In particular, Johnson, Greif & Bromm (2007) find that the build up of H_2 in the relic HII regions surrounding the first population of stars makes a large difference to the Lyman-Werner band radiative transfer. While the IGM is optically thick to Lyman-Werner photons over physical distances of ~ 30kpc at redshifts $z \sim 20$, the high molecule fraction that is built up in relict H II regions as well as their increasing volume-filling factor renders their local IGM optically thick to Lyman-Werner photons over path lengths as short as a kiloparsec or so. For the halo population of stars, Mesinger, Bryan & Haiman (2006) find that a Lyman-Werner background of as little as $J_{\rm UV} \sim 10^{-21}$ erg s^{-1} cm^{-2} Hz^{-1} sr^{-1} may be sufficient to strongly inhibit star formation in the early universe. Similar conclusions are reached by Johnson, Greif & Bromm (2007)

In the absence of a strong radiation field, the H_2 fraction is correlated only with the virial temperature of the gas in any forming mass condensation. Yoshida *et al.* (2003b) find that the H_2 fraction can be approximated by a power-law in terms of the virial temperature; $f_{H_2} \propto T_{\rm vir}^{1.52}$. In order to form cold clouds which can subsequently form the first Population III stars, the cooling timescale has to be shorter than the free-fall timescale. This then implies a mass boundary ($M \sim 7 \times 10^5 M_\odot$) below which dark matter halos are unable to form cold clouds. These will subsequently lose their baryonic gas at the time of re-ionization, and will remain as dark satellites until they are incorporated into larger structures by merging.

2.2. *The Epoch of Re-ionization*

Re-ionization is an important phase transition in the history of the Universe, since it terminates the formation of the first stars and clusters in dwarf mini-haloes. Subsequently, star formation has to occur in much larger mass agglomerations as the first recognizable galaxies collapse and become self-shielding.

The process of re-ionization is driven by the first massive stars which formed in the Universe. The process of evolution of the first HII regions was treated in a 1-D fashion followed by a ray tracing methodology to trace the ionization of the surrounding medium by Alvarez, Bromm & Shapiro (2006). They find, in contradiction to O'Shea *et al.* (2005) that that neighboring minihalos are not fully ionized before the initial star dies. The ionization rapidly moving ionization front gets trapped and converted into a D-type front outside the core region. Since the photoevaporation time for the minihalo exceeds the lifetime of the ionizing star, the core region remains shielded.

Recently, sophisticated 3-D radiative hydrodynamical models have been developed to attack this problem, first by Yoshida (2006) and Yoshida *et al.* (2007) and more recently by Wise & Abel (2008a). These authors find that even single stars input sufficient energy to blow away and ionize much of the baryonic matter in their mini-halos. A consequence of this is that fairly large mini-halos with masses as large as $2 \times 10^7 M_\odot$ can be expected to be dark, *see* Read, Pontzen & Viel (2006). The star initially ionizes a large region which first starts to expand from the local over-pressure before slowly recombining. The subsequent supernova evolves in the fossil HII region. Wise & Abel (2008a) find that it is these SNR which exercise the greatest control on the re-ionization of the intergalactic medium, by increasing the porosity of this medium, and heating an appreciable fraction of the total volume to coronal temperatures. Subsequent generations of stars are then able to ionize a much larger region, since in many directions the absorption of the UV photons is low. Wise & Abel (2008b) show that the voids also become highly metal enriched by this process, with the lowest density phases reaching close to solar metallicity at $z \sim 15$, while most of the denser phases are enriched with pair-instability supernova nucleosynthetic products only up to $\log Z/Z_\odot \sim -3$.

The metallicity history of the Universe throughout cosmic time has been modeled by Kobayashi, Springel & White (2007) using SPH methods of computation. This simulation is available on the web as a movie†. They also found that the enrichment history depends strongly on the environment. In large galaxies, enrichment takes place so quickly that [O/H] ~ -1 at $z \sim 7$, consistent with observations of Lyman break galaxies. The average metallicity of the universe reaches [O/H] ~ -2 and [Fe/H] ~ -2.5 at $z \sim 4$ ($t = 1.5$ Gyr), but takes until $z \sim 3$ ($t = 2.1$ Gyr) to reach this average value in the IGM. These models predict that the "metallicity floor" in the local universe would be of order $\log Z/Z_\odot \sim -2$.

To summarize the salient points of this section:
- The first stars form in dark matter halos having masses $\sim 10^6 M_\odot$ at $z \geqslant 18$.
- Weak UV fluxes stimulate both the formation of H_2 and the rate of star formation in the Population III stars, but feedback becomes negative once the UV flux density exceeds a critical value.
- Radiation feedback expels the gas from the shallow potential wells of their dark matter halos. Below $\sim 2 \times 10^7 M_\odot$, galaxies will be dark, and in halos as large as $10^8 M_\odot$, as much as 75% of the baryons can be driven out into the intergalactic medium.
- The combination of radiation feedback and mechanical energy feedback from SNe creates a spatially complex ISM phase structure, with hot voids and a photoionized low-density phase which together occupy most of the volume. The lower density phases are rapidly enriched up to metallicities between 10^{-2} and 10^{-3} solar.

3. Feedback at the Epoch of Massive Galaxy Formation

The star formation history of the Universe is usually summarized in the form of a Madau - Lilly (or Lilly - Madau) plot, the star formation rate per unit co-moving volume of the Universe plotted against redshift (Lilly et al. 1996; Madau et al. 1996). Apart from the uncertainty introduced by the variety of heterogeneous samples and techniques, there is considerable uncertainty attached to the intrinsic reddening corrections, especially at high redshift. This said, it is remarkable that a fair degree of convergence is reached, with various methods agreeing within a factor of three or so. For a recent compilation from the literature, see Fardal et al. (2007), and for $z < 0.5$, see Hanish et al. (2006).

From this plot, we can infer that there was a broad peak of star formation when the Universe was between 1 and 6 Gyr old. This was the epoch of massive galaxy assembly which gave rise to the passively-evolving "red-and-dead" Elliptical galaxies we see in the local volume. When we look back to this epoch of galaxy formation, we see these galaxies as infrared-bright, dusty galaxies detected either in the sub-mm region of the spectrum, the sub-mm galaxies (SMGs), or in the form of their radio-loud equivalents, the high-redshift radio galaxies (HizRGs)

For the SMGs Blain et al. (2004) has shown that, at a given dust temperature, the SMGs are typically 30 times as luminous as their ULIRG counterparts. Takagi et al (2003a, 2003b) had previously found that most ULIRGS have a constant surface brightness of order 10^{12} $L_\odot$kpc^{-2}. Our own results, Dopita et al. (2005), indicate that this corresponds to an ISM with a pressure of order $P/k \sim 10^7$ cm^{-3}K. These parameters probably characterise "maximal" star formation, above which gas is blown out into the halo of the galaxy and star formation quenched. Only mergers, which provide an additional ram-pressure confinement of the star formation activity may exceed this surface

† *see:* http://th.nao.ac.jp/ chiaki/research-e.html

brightness. Thus, in order to scale the star formation up to the rates inferred for SMGs ($\sim 1000 - 5000$ $M_\odot yr^{-1}$), we must involve a greater area of the galaxy in star formation, rather than trying to cram more star formation into the same volume. For a typical value of 10^{13} $L_\odot$ kpc^{-2}, we require "maximal" star formation over an area of ~ 10 kpc^2, and the most luminous SMGs require star formation to be extended over an area of at order ~ 100 kpc^2. We can conclude that SMGs are starbursts extended on a galaxy-wide scale, unlike the starburst regions seen in the nearby universe.

This violent galaxy-wide star formation epoch is terminated by the feeding and growth of the nuclear black hole which, when it grows beyond a critical mass is capable of expelling the gas from the host galaxies through the action of its relativistic jets. Evidence for this process comes from the study of the high redshift radio galaxies (Hi-zRGs) at $z > 3$. These rank among the most luminous, largest, and most massive galaxies known in the early Universe (De Breuck *et al.* 2002) and which are observed early in the Epoch of Galaxy Formation at a time when their super-massive black holes (SMBHs) are highly active, and while their relativistic jets are interacting most strongly on their host galaxies. Spectroscopy by Reuland *et al.* (2007) has shown that the Lyman$-\alpha$ halos exhibit disturbed kinematics, with broad lines, large velocity shears, and expanding shells associated with the radio lobes. In addition, these halos halos are both chemically enriched by star formation and ionized throughout the majority of their volume. The relativistic jets drive strong shocks into the galactic medium and trigger shock-induced star formation on their peripheries.

These observations confirm the general physical idea of Silk & Rees (1998) that it is the feedback of the black hole on its host galaxy which eventually limits the growth of the black hole. However, the observed feedback is not supplied by radiation pressure but rather by the mechanical energy input delivered by the relativistic jets. In the Hi-zRGs we observe the moment where galaxy collapse gives way to mass ejection, and a newly born galaxy is revealed. We may further speculate that this is the defining moment where the such galaxies enjoy one last violent burst of shock-induced star formation before beginning their evolution to become the "red and dead" massive Elliptical galaxies we see in our local universe. This shock-induced star formation proceeds at several thousands of solar masses a year for at least a dynamical timescale, and is therefore capable of producing as much as $10^{11} - 10^{12} M_\odot$ of new stars. Stars formed by such shock interactions should have orbital characteristics quite unlike those of the earlier stellar generations. Traces of this epoch may therefore be found by a detailed study of the orbital dynamics of the massive elliptical galaxies in the local Universe.

4. Starbursts & Superwinds

The basic physics of starburst-powered winds in galaxies is rather simple, and has been described by Chevalier & Oegerle (1979). To power a wind, we require (at least) that the energy injection per unit area into a fraction, f, of the disk gas will exceed its binding energy. As Chevalier & Oegerle (1979) has shown, this means that the gas escaping into the halo has an initial temperature greater than some critical value related to the binding energy. The physical parameters control the production of a galactic wind can be easily derived, The input energy $E_{\rm in}$ per unit area is proportional to the surface rate of star formation $\dot{\Sigma}_*$ integrated over the lifetime of the starburst, $\tau_{\rm SB}$;

$$E_{\rm in} = \alpha \dot{\Sigma}_* \tau_{\rm SB}. \tag{4.1}$$

The escape energy of the gas is

$$E_{\rm esc} = 0.5 f \Sigma_{\rm gas} v_{\rm esc}^2, \qquad (4.2)$$

where $\Sigma_{\rm gas}$ is the surface density of gas in the disk. The star formation rate is related to the gas surface density through the Kennicutt (1998) star formation law:

$$\dot{\Sigma}_* = \beta \Sigma_{\rm gas}^{1.5} \qquad (4.3)$$

Thus, a galactic wind becomes possible when:

$$\frac{\tau_{\rm SB} \Sigma_{\rm gas}^{0.5}}{v_{\rm esc}^2} \geqslant \frac{f}{2\alpha\beta} \sim {\rm const.} \qquad (4.4)$$

Therefore, a wind is favoured when the escape velocity is low (such as will be the case in small galaxies), when the starburst is long continued (which is not surprising, since long lived star formation events can put in more energy to remove the gas), and (somewhat more surprisingly) when the gas surface density is high.

4.1. Galactic Chimneys, Blowouts & Fountains

Our own galaxy is known to host a number of superbubbles, blowouts and chimneys. This had been demonstrated many years ago Heiles (1979a) and Heiles (1979b), but has been confirmed in exquisite detail by the high-resolution HI survey of McClure-Griffiths *et al.* (2002) and by her subsequent studies.

Two massive blow-outs are known in the Galaxy. McClure-Griffiths *et al.* (2003) have studied the Galactic chimney GSH 277+00+36 in HI. This chimney is more than 600 pc in diameter, and extends at least 1 kpc above and below the Galactic midplane. The shell and chimney walls exhibit a great deal of small-scale cold gas structures in HI in the form of loops or 'drips'. These exhibit narrow line widths in the range $1.5 - 2.5$ km s^{-1}. More recently, McClure-Griffiths *et al.* (2006) found a second example of a super bubble which has blown out of both sides of the galactic plane; GSH 242-03+37. This has a radius ~ 600 pc and would require an input energy of 3×10^{53} ergs to have imparted the kinetic energy of the expansion. Their HI images reveal a complicated shell with multiple chimney structures, capped by narrow filaments about 1.6 kpc above and below the Galactic mid-plane.

The formation of these fine-scale HI structures remains a bit of a mystery, but a good deal of insight can be derived from the new 3-D radiative hydrodynamical computations made by Ralph Sutherland (in preparation), using his new code Fyris Alpha. He finds that breakout occurs in his simulations about 7 Myr after the commencement of the starburst. The filamentary structure can be caused simply by the initial density fluctuations in the disk HI gas. Cooled HI droplets similar to those seen in the observations occur soon after breakout, and are visible up to the termination of the simulation at 25 Myr. Recombination of the denser material in the plane occurs after about 15 Myr, when the supernova energy input decreases. However, these later phases depend rather critically upon the lower mass limit for core-collapse supernovae, since the disappearance of supernova energy input terminates the phase in which the ejecta are being re-heated and driven into outflow.

The physics which drives these localized outflows is undoubtedly very similar to that which drives the high velocity low-intensity components seen in the ionized gas associated with supergiant HII regions in external galaxies. Relaño & Beckman (2005) and Rozas *et al.* (2006) find that these are found in a substantial fraction of supergiant HII regions in nearby disk galaxies, and indicate expansion velocities of 100-160 km s^{-1} in the high velocity components of the Hα-emitting gas. In Figure 1 we show the total kinetic energy

(Hα line core plus the broad wings) we have derived for the Rozas *et al.* (2006) sample of supergiant HII regions, plotted against the mass of the exciting stellar cluster. This mass is derived from the Hα luminosity, assuming a Salpeter stellar Initial Mass Function (*see* Dopita *et al.* (2006)). As can be seen, at high cluster masses, the he kinetic energy content is proportional to the mass of the central stars (indicated by the solid line). This is in accordance with equation 4.1, above.

On a larger scale, Booth, Theuns & Okamoto (2007) have used SPH simulations to provide the temporal evolution of disk galaxies based on a molecular cloud regulated model of star formation in the disk. For very young galaxies, a starburst-driven nuclear wind appears, but at later phases, a more generalized galactic wind is blown, with local hot-spots in the halo, and distinct dynamical structures in the wind. However, the resolution of these simulations in the halo regions is too low to give one full confidence in the results.

4.2. *Superwinds*

The poster child of Galactic winds is M82. However, it should always be recalled that M82 provides only a modest-sized starburst and associated galactic wind, hosted in a low-mass galactic system. McCarthy, van Breugel & Heckman (1987) were the first to suggest that the starburst in this object was driving a wind, and this has since been abundantly confirmed by the exquisite HST, Spitzer and Chandra images e.g. Mutchler *et al.* (2007) which show a complex bipolar system of dusty Hα filaments being driven

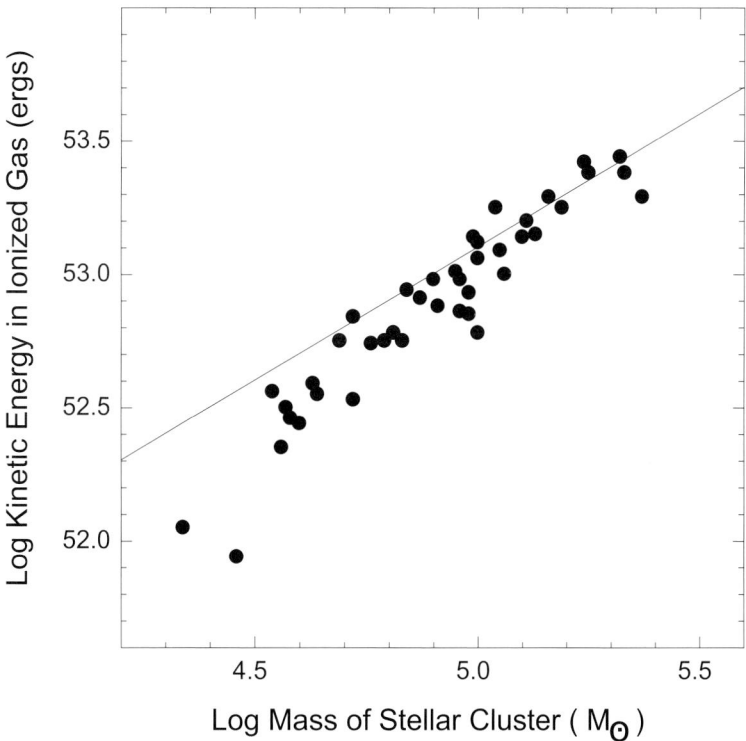

Figure 1. The kinetic energy content of the Rozas *et al.* (2006) HII regions as a function of the mass of the exciting stars. For the larger HII regions, the kinetic energy content is proportional to the mass of the central stars, in accordance with the expectation that the kinetic energy output of the stars drives both the turbulent and organized expansion velocities.

out by a hot X-ray emitting plasma originating in the starburst core of this galaxy. Such bipolar winds are commonly found in galaxy interactions, e.g. in the SINGG survey Meurer et al. (2006), and can be important in polluting intergalactic space with the nucleosynthetic products of starbursts.

A large number of theoretical models have been constructed in the attempt to explain the structure of such starburst-driven winds, notably Tomisaka & Ikeuchi (1988), Tomisaka & Bregman (1993), Suchkov et al. (1994), Tenorio-Tagle & Muños-Tuñòn (1998) and Strickland et al. (2000). Nearly all of these are 2-D simulations. Recently Cooper et al. (2008) (see also this volume) have made full radiative 3-D radiative hydrodynamical simulations which better capture the physical complexity of starburst-driven winds. They find that the hot gas escapes in channels between the denser and cooler gas that is itself entrained in the flow, eventually forming long radial filaments. The general features of this flow are shown in figure 2. The zones familiar from the theory of mass-loss driven bubbles e.g. Castor, McCray & Weaver (1975) or Weaver et al. (1977) are clearly seen, although the bipolar structure and the entrainment zone at the base of the wind distinguishes these models from simple 1-D mass loss bubbles.

The limitation of the Cooper et al. (2008) and indeed, all other models is one of resolution. In order to properly resolve the cooling and entrainment regions a very high spatial resolution is required. Inadequate spatial resolution leads to over-cooling in the vicinity of the filaments, and an overestimate of the mass in the filaments. The highly non-spherical and non-uniform nature of the flow precludes the use of adaptive mesh

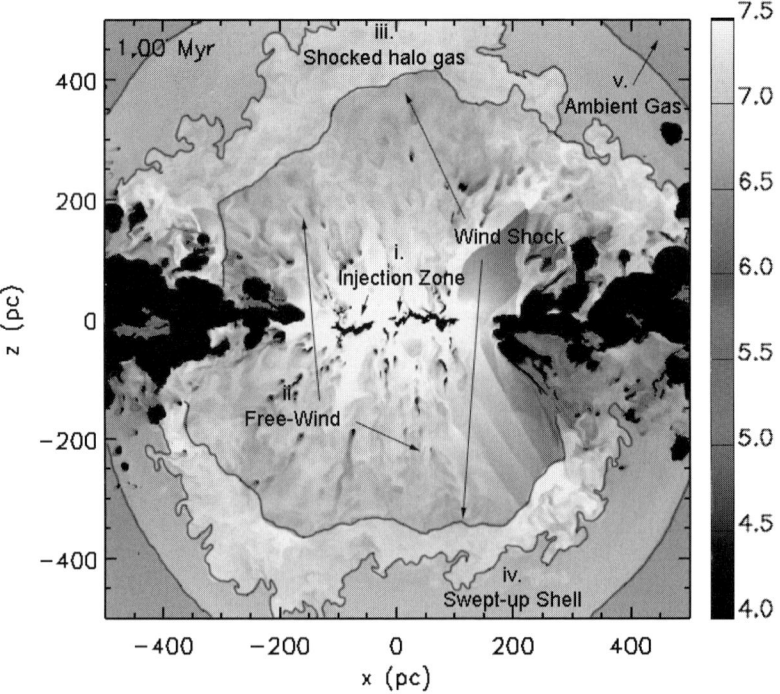

Figure 2. A slice through the 3-dimensional simulation of a galactic wind from Cooper et al. (2008), showing the various zones of the flow. The color coding gives the electron temperature of the plasma.

techniques, so there is little choice other than using as large a grid as possible, and as large a supercomputer as is available.

Nonetheless, despite these limitations, the general features of the superwind displayed in the simulations are valid. In particular, the rapid outflow velocities predicted in the entrained cooled gas are observed in the luminous starburst galaxies. For example, Martin (2005) find that the NaI $\lambda\lambda 5890, 5896$Å absorption lines indicate mean outflow velocities of over 300 km s^{-1} for the ultraluminous IR Galaxies (ULIRGs). She also infers that the rate of outflow of gas is comparable to the rate of consumption of gas through star formation, and that maximum outflow velocity is correlated with the star formation rate (SFR); $v \propto \mathrm{SFR}^{1/3}$. Even higher outflow velocities (between 490 and 2020 km s^{-1}) have been found by Tremonti, Moustakas & Diamond-Stanic (2007) in the Mg II $\lambda\lambda 2796, 2803$Å absorption lines associated with massive post-starburst galaxies as $z \sim 0.6$. These velocities are intermediate between ULIRGs and the Low-ionization Broad Absorption Line Quasars (LoBALs) Trump *et al.* (2006), suggesting that wind-driving by a buried AGN might be a rather significant contributor to the total kinetic energy budget of these objects.

5. Conclusions

In conclusion, it is fair to say that the confluence of high-quality spectrophotometric and imaging data obtained throughout the electromagnetic spectrum, coupled with theoretical results from the new generation of 3-D radiative hydrodynamical codes is finally furnishing a coherent picture of starburst-powered galactic winds throughout cosmic time. The mass budget of these winds is a significant fraction of the total baryonic content of galaxies, and the winds provide much of the hot gas in galactic clusters and in the intra-cluster medium, as well as providing the source of the chemical enrichment of the intergalactic medium.

References

Abel, T., Anninos, P., Zhang, Y. & Norman, M. L. 1997, *New Ast.*, 2, 181
Alvarez, M. A., Bromm, V. & Shapiro, P. R. 2006, *ApJ*, 639, 621
Blain, A. W., Chapman, S. C., Smail, I. & Ivison, R. 2004, *ApJ*, 611, 52
Booth, C. M., Theuns, T. & Okamoto, T. 2007, *MNRAS*, 376, 1588
Bromm, V. & Larson, R. B. 2004, *ARAA*, 42, 79
Castor, J., McCray, R. & Weaver, R. 1975, *ApJ*, 200, 107
Chevalier, R. A., & Oegerle, W. R. 1979, *ApJ*, 277, 398
Chu, Y.-H. 1999, *Ap&SS*, 269, 441.
Ciardi, B. & Ferrara, A. 2005, *Space Sci. Rev.*, 116, 625
Ciardi, B. 2008, in: B. O'Shea, A. Heger & T. Abel (eds.). *First Stars III* (New York: AIP) in press (arXiv:0709.1367)
Cooper, J. L., Bicknell, G. V., Sutherland, R. S., & Bland-Hawthorn, J. 2008, *ApJ*, 674, 157
De Breuck, C., van Breugel, W., Stanford, S. A. *et al.* 2002, *AJ*, 123, 637
Dopita, M. A., Groves, B. A., Fischera, J. *et al.* 2005, *ApJ*, 619, 755
Dopita, M. A., Fischera, J., Sutherland, R. S. *et al.* 2006, *ApJ*, 647, 244
Fardal, M. A., Katz, N., Weinberg, D. H. & Davé, R. 2007, *MNRAS*, 379, 985
Hanish, D. J. Meurer, G. R., Ferguson, H. C. *et al.* 2006, *ApJ*, 649, 150
Heiles, C. 1979, *ApJ*, 229, 533
Heiles, C. 1979, *PASP*, 91, 611
Johnson, J. L., Greif, T. H. & Bromm, V. 2007, *ApJ*, 665, 85
Kennicutt, R. C. Jr. 1998, *ApJ*, 498, 541
Kobayashi, C., Springel, V. & White, S. M. 2007, *MNRAS*, 376, 1465

Lilly, S. J., Le Fevre, O., Hammer, F. & Crampton, D. 1996, *ApJL*, 460, L1
McCarthy, P. J.; van Breugel, W. & Heckman, Ti. 1987, *AJ*, 93, 264
McClure-Griffiths, N. M., Dickey, J. M., Gaensler, B. M. & Green, A. J. 2002, *ApJ*, 578, 176
McClure-Griffiths, N. M., Dickey, J. M., Gaensler, B. M. & Green, A. J. 2003, *ApJ*, 594, 833
McClure-Griffiths, N. M. Ford, A., Pisano, D. J. *et al.* 2006, *ApJ*, 638, 196
Mac Low, M.-M. 2008, in: H. Beuther *et al.* (eds.), *Massive Star Formation: Observations Confront Theory* (San Francisco: ASP), *ASP Conf Ser.* in press (arXiv:0711.4071)
Madau, P., Ferguson, H. C., Dickinson, M. E. *et al.* 1996, *MNRAS*, 283, 138
Martin, C. 2005, *ApJ*, 621, 227
Mesinger, A.,Bryan, G. L. & Haiman, Z. 2006, *ApJ*, 648, 835
Meurer, G. R., Hanish, D. J., Ferguson, H. C. *et al.* 2006, *ApJS*, 165, 307
Mutchler, M. Bond, H. E., Christian, C. A. *et al.* . 2007, *PASP*, 119, 1
Naoz, S., Noter, S. & Barkana, R. 2006, *MNRAS*, 373, 98
OShea, B. W., Abel, T., Whalen, D., & Norman, M. L. 2005, *ApJ*, 628, L5
Oey, M. S. & Clarke, C. J. 2008, in: M. Livio (ed.), *Massive Stars: From Pop III and GRBs to the Milky Way*, (Cambridge: CUP) in press (astro-ph/0703036)
Read, J. I, Pontzen, A P. & Viel, M. 2006, *MNRAS*, 371, 885
Relaño, M., & Beckman, J. E. 2005, *A&A*, 430, 911
Reuland, M. van Breugel, W., de Vries, W. *et al.* 2007, *AJ*, 133, 2607
Rozas, M., Richter, M. G., López, J. A., *et al.* 2006, *A&A*, 455, 549
Silk, J. Rees, M. J. 1998, *A&A*, 331, L1
Stecher T. P, Williams D. A. 1967, *ApJ*, 149, L29
Strickland, D. K., Heckman, T. M, Weaver, K. A. & Dahlem, M. 2000, *AJ*, 120, 2965
Suchkov, A. A., Balsara, D. S., Heckman, T. M. & Leitherer, C. 1994, *ApJ*, 430, 511
Tenorio-Tagle, G. & Muños-Tuñòn, C. 1998, *MNRAS*, 293, 299
Tremonti, C. A., Moustakas, J. & Diamond-Stanic, A. M. 2007, *ApJ*, 663, L77
Trump, J. R., Hall, P. B., Reichard, T. A. *et al.* 2006, *ApJS*, 165, 1
Tomisaka, K. & Ikeuchi, S. 1988, *ApJ*, 330, 695
Tomisaka, K. & Bregman, J. N. 1993, *PASJ*, 45, 513
Weaver, R., McCray, R., Castor, J., Shapiro, P. & Moore, R. 1977, *ApJ*, 218, 377
Wise, J. H. & Abel, T. 2008a, *ApJ*, in press (arXiv:0710.4328)
Wise, J. H. & Abel, T. 2008b, *ApJ*, in press (arXiv:0710.3160)
Yoshida, N., Sokasian, A., Henquist, L. & Springel, V. 2003a, *ApJ*, 598, 73
Yoshida N, Abel T, Hernquist L, Sugiyama N. 2003b, *ApJ*, 592, 645
Yoshida, N. 2006, *New Ast. Revs.*, 50, 19
Yoshida, N., Oh, S. P., Kitayama, T., & Hernquist, L. 2007, *ApJ*, 663, 687

Discussion

STANEK: We have evidence that the Mg II "clouds" at $z \sim 2$ have small sizes, which would indicate that they are short-lived. Do you think that these super-winds could give rise to the Mg II complexes seen in QSO and GRB spectra?

DOPITA: This could well be. Crystal Martin, for example, finds outflows of roughly 300 km s^{-1} in absorption in the high-z ULIRGS she has looked at. *See ApJ 621, 227 (2005)*. The mass loaded outflows we model certainly are fragile, and rather transitory objects. The GRBs seem so be associated with low-mass galaxies, so outflows can be driven with quite modest rates of star formation.

STANEK: What is the fraction of total star formation that is in these starburst galaxies?

DOPITA: In the local universe, the fraction is rather small. However, by $z \sim 2$, the star formation is dominated by starburst galaxies.

SONNEBORN: Do the galactic fountain models you describe produce a hot $T > 3 \times 10^5$ K component in the outflowing gas?

DOPITA: Yes. These models are powered by the energy injection of shocked stellar winds and supernova explosions. The outflowing medium definitely forms a two-phase medium with one phase at coronal temperatures. However the final temperature in this phase depends on the details of the PdV work done in escaping the potential of the galaxy and on the degree of mixing and mass entrainment that has occurred.

KUDRITZKI: You referred to the simulation of Kobayashi, which showed the the IGM was very quickly enriched to $\log[Z/Z_\odot] \sim -2$ at high redshift. If this is true, how does one explain the existence of very metal poor stars in the Galactic Halo?

DOPITA: The ISM is really quite poorly mixed, and the IGM enrichment occurs in the low density phase. In the simulations of Wise & Abel (2007), the dense phase of the ISM is enriched only up to $\log Z/Z_\odot \sim -3$ while the lowest density phases are enriched up to solar metallicity. Thus, IGM enrichment does not preclude the formation of low metallicity stars. It is more relevant in relation to our efforts to determine the metallicity floor of the local universe, since this metal enriched IGM gas is likely to have only recently been captured by dwarf star forming galaxies in the Local Universe.

ZINNECKER: Can you give us a general definition of a "starburst" *i.e.* in terms of star formation rate, per unit area, per unit time or what?

DOPITA: The term "starburst" has been much used and abused, and there is currently no generally accepted definition. However, personally I would emphasize the importance of concentration, in terms of star formation per unit area, gas surface density or pressure in the star forming region. I suggest that a good working definition would be that $\log P/k \gtrsim 6$ cm^{-3}K in the star-forming region. Star formation regions characterized by such high pressures in the galactic plane would then stand a good chance to be able to blow gas out into their galactic haloes as a galactic wind.

Gloria Koenigsberger (left), Marco Limongi (center) and Mike Dopita (right).

Alfredo Sota (left) and Jesús Maíz Apellániz (right).

Jennifer Hoffman.

Gemini/IFU Observations of Galactic Outflows in Starburst Galaxies

Linda J. Smith[1,2] and Mark S. Westmoquette[2]

[1] Space Telescope Science Institute and European Space Agency, 3700 San Martin Drive, Baltimore, MD 21218, USA

[2] Department of Physics and Astronomy, University College London, Gower St., London WC1E 6BT, UK

Abstract. We present Gemini/IFU observations that sample the roots of the galactic wind outflows in the starburst galaxies NGC 1569 and M82. The good spatial and spectral resolutions of these observations allow us to probe the interactions of cluster winds with their environments on small scales. For both galaxies, we find a ubiquitous broad (200–300 km s^{-1}) Hα component underlying a brighter narrower component. By mapping the properties of the individual line components, we find correlations that suggest that the broad component results from powerful cluster wind-gas clump interactions. For NGC 1569, there is little evidence for organised gas flows within the central zone and we suggest that the flow-dominated wind must form well beyond the region containing the massive star clusters. For M82, we find that the kinematics of the wind base are very complex; the width of the broad component reaches values of > 350 km s^{-1} at the base of the wind, and the outflow itself rapidly becomes chaotic in the inner wind region.

Keywords. ISM: jets and outflows – ISM: kinematics and dynamics – galaxies: ISM – galaxies: starburst

1. Introduction

Massive stars feed back energy into the interstellar medium (ISM) via stellar winds and supernova explosions. This feedback mechanism is thought to play a fundamental role in the formation and evolution of galaxies. In starburst galaxies, where star formation is intense and young massive star clusters are formed, galaxy-wide superwinds are seen. The onset and development of these outflows can be studied by observing local starburst galaxies. In particular, we wish to understand how and where the individual winds from star clusters interact with each other, and combine to produce large-scale outflows. Past studies of galactic outflows have tended to concentrate on their large-scale aspects; here we present a summary of our recent work aimed at understanding the small-scale characteristics of the cluster wind flows in the two nearby starburst galaxies NGC 1569 and M82.

2. NGC 1569

NGC 1569 is a nearby (2.2 ± 0.6 Mpc) dwarf irregular galaxy that has recently undergone a galaxy-wide burst of star formation that peaked between 100 Myr and 5–10 Myr ago (Greggio *et al.* 1998). The two well-known super star clusters (SSCs) A and B are prominent products of the starburst episode, and probably dominate the energetics of the central region. An extended system of Hα filaments, indicative of a bipolar outflow is seen (Heckman *et al.* 1995; Martin 1998), and X-ray observations show that the wind emanates from the full extent of the Hα disk (Martin, Kobulnicky & Heckman 2002).

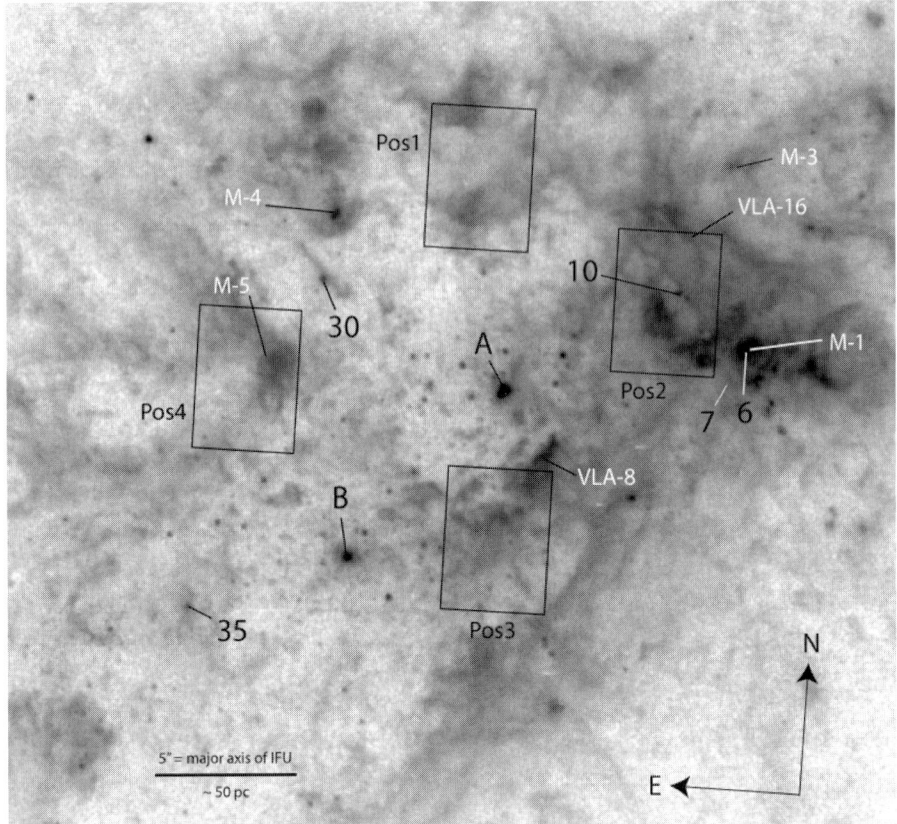

Figure 1. *HST*/WFPC2 F656N image of the central region of NGC 1569 showing the positions of the four IFU fields. A number of the most prominent star clusters (Hunter *et al.* 2000) and radio continuum sources (Greve *et al.* 2002) are labelled in black and white respectively.

2.1. *Observations and Analysis*

To probe the roots of the wind outflow from NGC 1569, we obtained observations of four regions near the centre of the galaxy with the Gillett Gemini North 8.1-metre telescope and the Gemini Multi-Object Spectrograph (GMOS) Integral Field Unit (IFU; Allington-Smith *et al.* 2002). In Fig. 1, we show the positions of the four IFU fields on an archive *Hubble Space Telescope (HST)* F656N image obtained with the Wide Field Planetary Camera 2 (WFPC2). The GMOS/IFU consists of 500 object fibres (each of diameter 0.2 arcsec) arranged in a rectangular array of size 5×3.5 arcsec^2 (or 50×35 pc^2 for NGC 1569). Our chosen spectral setup gave us a wavelength coverage of 4740–6860 Å with a dispersion of 0.34 Å pix^{-1}. The data were reduced using the standard Gemini pipeline; procedures included flat-fielding, cosmic-ray cleaning, sky subtraction, throughput correction, flux calibration and differential atmospheric correction. Full details can be found in Westmoquette *et al.* (2007a,b).

For each IFU position, we have 500 spectra covering the important nebular diagnostic emission lines of [O III], Hβ, [N II], Hα and [S II]. The signal-to-noise and spectral resolution (60–75 km s^{-1}) of these data are sufficiently high to resolve multiple Gaussian components to each emission line. In order to quantify and compare the properties of the ionized gas, we have fitted multiple Gaussian components to all of the spectral line profiles. To do this, we used an interactive χ^2 minimisation curve-fitting program written

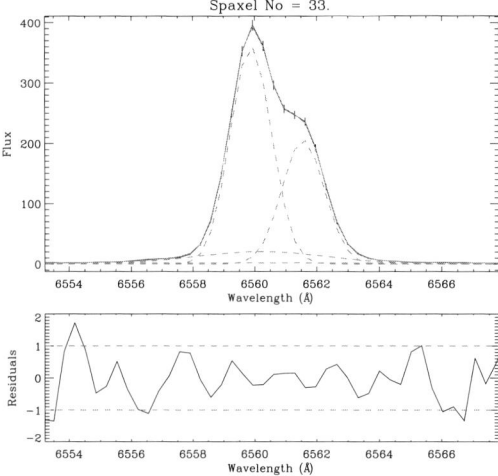

Figure 2. Example of a three component Gaussian fit to the Hα emission line profile for one spaxel in IFU position 1. The solid line is the observed line profile and the dashed lines represent the individual Gaussian fits. The normalised residuals to the fit are shown below.

in IDL (PAN; Dimeo 2005) to automate the line profile fitting. To determine how many Gaussian components best fit an observed profile, we applied the statistical F-test to the χ^2 ratios, together with a number of additional physical tests and filters (see Westmoquette *et al.* (2007a) for a detailed description of our fitting procedure and its associated uncertainties). Decomposing each compound line profile meant that we could analyse the properties of each component separately, and this allowed us to disentangle flux, velocity and line width information.

In general, we find the line profiles to be composed of a bright, narrow component (hereafter C1) overlaid on fainter, broad emission (hereafter C2). We find bright C1 emission across all four regions, well characterised by a Gaussian function with a distinct lower limit to its FWHM of $\sim$35 km s^{-1}. We also measure a fainter line component (C2) over all four regions with line widths between FWHM$\sim$100–400 km s^{-1}. The total radial velocity spread between these components is only $\sim$ 70 km s^{-1}. We also identify a third emission line component in some regions which is of higher velocity and can be associated with expanding shells of gas. An example of a line profile showing all three components is shown in Fig. 2.

By plotting maps of the fit parameters as a function of spatial position, we find a number of correlations. In particular, we find that the width of the broad underlying emission (C2) is both highly correlated with the flux of the narrow component (C1) and anti-correlated with its own flux. The former of these two correlations is shown in Fig. 3, where we plot the C1 line flux in contours over the C2 line width for IFU position 2.

2.2. *Interpretation*

We first discuss the origin of the bright "narrow" C1 component. The integrated H I velocity dispersion for the disk of NGC 1569 is $\sigma \approx 15$ km s^{-1} indicating that the neutral ISM is very disturbed (Mühle *et al.* 2005), presumably due to the combined effects of stellar winds and supernovae. We would thus expect the H II gas in NGC 1569 to have similar or higher velocity dispersions to the H I. Indeed, we find that there is a distinct $\sim$35 km s^{-1} lower limit to the C1 FWHM. After correction for thermal broadening, this becomes $\sigma \approx 12$ km s^{-1}. Line widths of this order for the main ionized gas component have been observed in many young star-forming regions e.g. 30 Dor or

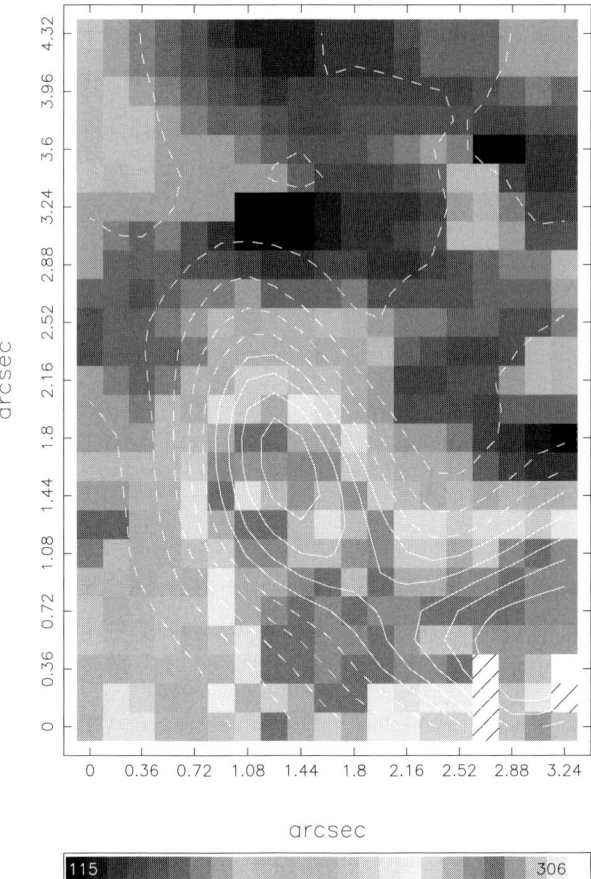

Figure 3. Contours of C1 flux overlaid on the C2 FWHM map for IFU position 2. Here the correlation between the bright Hα knots seen in C1 and the broadest C2 lines can be seen clearly.

NGC 604 (Chu & Kennicutt 1994; Yang *et al.* 1996). Various explanations have been put forward to explain its width, including multiple unresolved expanding shells integrated along the line-of-sight, and gravitational broadening through virial motions. We find that unresolved expanding shells cannot contribute significantly to the observed widths of C1, since the point at which coherent shells break up is when their expansion velocity equals the ambient ISM turbulent velocity (i.e. the $\sim$35 km s^{-1} lower limit). Simple calculations for the expected level of gravitational broadening show that it may contribute a significant fraction of the observed width. Thus, we interpret C1 as representing the turbulent ISM of NGC 1569, where its motions (width) result from a convolution of the general stirring effects of the starburst and gravitational virial motions (thus giving rise to the clear FWHM lower limit). Additional minor contributions may result from unresolved expanding shell components along the line-of-sight.

C2 cannot result from any of the effects described above since none are able to produce such broad line widths, or such strong correlations between the emission line maps as shown for example in Fig. 3. However, we expect that C2 must somehow be associated with the starburst environment i.e. the interactions of high-energy photons and hot, fast-flowing winds from the nearby clusters with the cold, ambient ISM. An exchange of energy and mass between the hot and cold phases of the ISM can occur through: photoevaporation; conductively-driven thermal evaporation; hydrodynamic ablation; and turbulent mixing (for a review see Pittard (2007)). When the hot, tenuous medium is fast-flowing, as in the case of stellar or cluster winds, shearing between the high velocity flow and the cloud surface can set up a turbulent mixing layer that can drive further evaporation. The impacting wind can also physically strip or ablate material off the clump surface, and entrain it into the flow (Scalo 1987).

Begelman & Fabian (1990) described a model for a turbulent mixing layer that forms between the hot, inter-cloud gas and the surface layers of cool or warm clouds in near-pressure equilibrium. The UV radiation emitted from the mixing layer photoionizes the cold gas in the cloud surface giving rise to optical emission lines with characteristic velocities of a fraction of the hot phase sound speed. Slavin *et al.* (1993) expanded on this idea by modelling the cloud interface in detail and considering the hot gas flow. Their model predicts strong optical [O III] and Hα emission with non-Gaussian (broad-winged) line profiles caused by the high levels of turbulence, clearly in good agreement with what we observe.

We therefore conclude the most likely explanation of the origin of C2 is emission from a turbulent mixing layer formed at the surface of dense clouds resulting from the viscous coupling between the cool clump material and the hot, fast, winds from the surrounding clusters. Material is also removed from this layer through thermal evaporation and/or mechanical ablation, resulting in the mass-loading of the flow. Our observations suggest that this mass-loading occurs close to or within the star-forming disk. In addition, we find from the lack of significant emission line velocity variations, that there is little evidence for organised gas flows within the central zone of NGC 1569. The flow-dominated wind must therefore form well beyond the region containing the massive star clusters; the wind sonic point probably lies at a distance of > 100–200 pc from the starburst.

3. M 82

Recently, we have obtained Gemini-North GMOS observations of the inner wind in the starburst galaxy M82 covering six IFU positions. The pointings were chosen to sample the roots of the wind close to the bright star-forming regions A and C.

We find that the Hα emission line profiles are similar to those observed for NGC 1569 with a narrow bright component superimposed on a broad underlying component. While the broad component is present throughout the starburst clumps (with widths of $\sim$200 km s^{-1}), it reaches maxima of $>$350 km s^{-1} at the base of the wind outflow region, following the linear 'streamer' morphology seen in *HST* imaging. In many regions we also identify a second narrow component offset in velocity. The kinematics of the wind base are clearly very complex.

Furthermore, for the first time we are able to measure the gas density (via the [S II] ratio) in all three line components, and find that the distribution differs in each. The density in the main, bright, narrow component peaks to the north of both regions A and C (in agreement with Westmoquette *et al.* 2007c), however the densities found in the broad underlying component peak within the starburst clumps themselves.

By extracting position–velocity diagrams along the major axis and along parallel lines stepping out into the inner-wind, we are able to examine the gas kinematics in more detail. We find that the outflow rapidly becomes chaotic. Interestingly, the broad component follows a shallower rotation curve, meaning that the motion of this gas may be de-coupled from that of the gas emitting the bright, narrow component. Clear evidence for expanding structures become visible at distances $>3''$ from the major axis, with line-splitting of $>$100 km s^{-1}.

Acknowledgements

We would like to thank and acknowledge our collaborators: Jay Gallagher, Katrina Exter (NGC 1569) and Jay Gallagher, Gelys Trancho, Nate Bastian and Iraklis Konstantopoulos (M82).

References

Allington-Smith, J., Murray, G., Content, R., et al. 2002, *PASP*, 114, 892
Begelman, M. C. & Fabian, A. C. 1990, *MNRAS*, 244, 26P
Chu, Y.-H. & Kennicutt, R. C. 1994, *ApJ*, 425, 720
Dimeo, R. 2005, PAN User Guide, ftp://ftp.ncnr.nist.gov/pub/staff/dimeo/ pandoc.pdf/
Greggio, L., Tosi, M., Clampin, M., et al. 1998, *ApJ*, 504, 725
Greve, A., Tarchi, A., Hüttemeister, S., et al. 2002, *A&A*, 381, 825
Heckman, T. M., Dahlem, M., Lehnert, M. D., et al. 1995, *ApJ*, 448, 98
Hunter, D. A., O'Connell, R. W., Gallagher, J. S., & Smecker-Hane, T. A. 2000, *AJ*, 120, 2383
Martin, C. L. 1998, *ApJ*, 506, 222
Martin, C. L., Kobulnicky, H. A., & Heckman, T. M., 2002, *ApJ*, 574, 663
Mühle, S., Klein, U., Wilcots, E. M., & Hüttemeister, S., 2005, *AJ*, 130, 524
Pittard, J. M. 2007, in T. W. Hartquist, J. M. Pittard, S. A. E. G. Falle, (eds.), *Diffuse Matter from Star Forming Regions to Active Galaxies.* (Dordrecht: Springer), *Astrophysics & Space Science* 245
Scalo, J. M., 1987, in D. J. Hollenbach, H. A. Thronson, Jr, (eds.), *Interstellar Processes*, (Dordrecht: Reidel), *Astrophys Space Sci Library* 134, 349
Slavin, J. D., Shull, J. M., & Begelman, M. C. 1993, *ApJ*, 407, 83
Westmoquette, M. S., Exter, K. M., Smith, L. J., & Gallagher, J. S., III, 2007a, *MNRAS*, 381, 894
Westmoquette, M. S., Smith, L. J., Gallagher, J. S., III, & Exter, K. M. 2007b, *MNRAS*, 381, 913
Westmoquette, M. S., Smith, L. J., Gallagher, J. S., III, et al. 2007c, *ApJ*, 671, 358
Yang, H., Chu, Y.-H., Skillman, E. D., & Terlevich, R. 1996, *AJ*, 112, 146

Claus Leitherer (left) and Linda Smith (right).

Radiative Feedback in Galaxies

M. S. Oey[1], E. S. Voges[2], R. A. M. Walterbos[2], G. R. Meurer,[3] S. Yelda[4] and E. Furst[5]

[1] Department of Astronomy, 830 Dennison Building, University of Michigan, Ann Arbor, MI 48109-1042, U.S.A.

[2] Department of Astronomy, MSC 4500, New Mexico State University, P.O. Box 30001, Las Cruces, NM 88003, U.S.A.

[3] Johns Hopkins University, Department of Physics and Astronomy, 3400 North Charles Street, Baltimore, MD 21218-2686, U.S.A.

[4] University of California, Department of Physics and Astronomy, P.O. Box 951547, Los Angeles, CA 90095-1547, U.S.A.

[5] 344 Greenlow Road, Catonsville, MD 21228, U.S.A.

Abstract. We examine the fate of ionizing radiation from massive stars on global scales. First, we compare the observed Hα luminosities of LMC H II regions with those predicted by the latest generation of stellar atmosphere models. Our results imply that classical H II regions are on average radiation-bounded, rather than density-bounded, as we found a decade ago. This is likely to necessitate an additional ionizing source for the diffuse, warm ionized medium (WIM) in galaxies. Secondly, we present new results from the SINGG Hα galaxy survey, showing that starburst galaxies have a lower fraction of WIM emission than normal star-forming galaxies. The most intriguing and consequential possible cause for this effect is the escape of ionizing radiation from starbursts. We show that the observations are also consistent with our predictions for the escape of ionizing radiation. Nevertheless, other observations do not necessarily support this scenario and other possible explanations must be considered.

Keywords. stars: atmospheres – stars: early-type – galaxies: evolution – galaxies: ISM – galaxies: starburst – H II regions – intergalactic medium – Magellanic Clouds

1. Introduction

The preceding speakers have described how mechanical and radiative feedback are intrinsically linked. Here, we will focus on the radiative feedback from massive stars, which refers specifically to the resulting photoionization of the gaseous environment. Classical H II regions are the most obvious manifestation, but another consequence is the photoionization of the interstellar medium (ISM) itself, resulting in the diffuse, warm ionized medium (WIM). In the case of starbursts or extreme star-formation intensity, ionizing radiation could escape from the galaxies altogether, ionizing the galactic halo, IGM, or even the cosmos itself.

2. Implications of Recent Stellar Atmosphere Models

Over the past decade, we have understood the ionization of the WIM in terms of a coherent picture: About half of the power is due to photoionization by field OB stars (Hoopes & Walterbos 2000; Oey *et al.* 2004), and the other half due to UV radiation escaping from density-bounded H II regions (Oey & Kennicutt 1997; Hoopes *et al.* 2000). Some of the strongest evidence pointing to the role of the H II regions was our work from ten years ago, in which we compared the observed Hα luminosities of a sample of nebulae in the Large Magellanic Cloud (LMC) with the values predicted by model atmospheres for the individual, spectroscopically classified stars observed in these regions (Oey & Kennicutt 1997). As is well-known in the massive star community, the more recent stellar atmosphere models are softer and predict less Lyman continuum radiation. We therefore decided to revisit the issue of radiative feedback in light of these newer models.

We take the same sample of 14 LMC H II regions from Oey & Kennicutt (1997) and now use the modern predictions from stellar atmosphere models by Smith, Norris, & Crowther (2002) and Martins, Schaerer, and Hillier (2005). The Smith *et al.* values are the most widely-used since they are incorporated into the STARBURST99 population synthesis code. The results are compared to those reported by Oey & Kennicutt (1997), which were based on ionizing emission rates predicted by Schaerer & de Koter (1997), Vacca, Garmany, & Shull (1996), and Panagia (1973).

The newer models predict roughly half the ionizing photon emission rates Q^0 relative to the older models. Figure 1 compares the observed Hα luminosities with the predicted values for the sample nebulae. Diamonds and squares correspond to predictions based on the Smith *et al.* (2002) and Martins *et al.* (2005) models, respectively. Points with crosses represent objects that are partly shock-ionized, and so their Hα luminosities are higher than expected simply from photoionization. Ignoring these shock-ionized objects, Figure 1 shows excellent agreement between the observed and predicted Hα luminosities. The deviation from the identity relation in this logarithmic plot, excluding the shock-ionized objects, is –0.031 and +0.122 for the Smith *et al.* (2002) and Martins *et al.* (2005) predictions, respectively. These correspond to linear observed-to-predicted ratios of 0.92 and 1.3, respectively, for the two sets of predictions. We furthermore examined a set of 39 LMC H II regions with stellar spectral types estimated photometrically. These estimates are less reliable than actual classifications, but the results are consistent with Figure 1 (Voges *et al.* 2008).

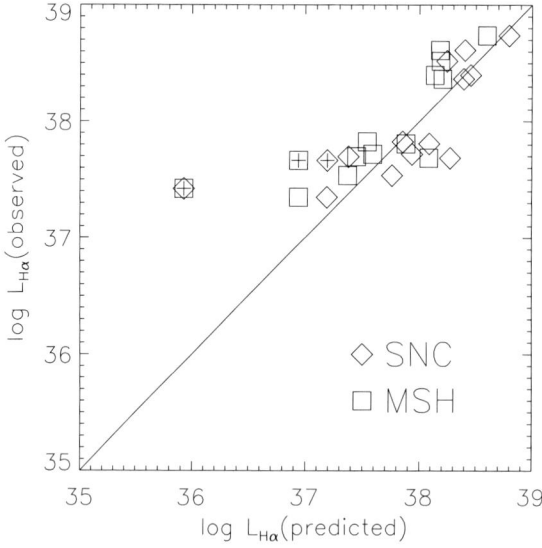

Figure 1. Comparison of observed and predicted Hα luminosities for the LMC H II regions. Diamonds and squares show the comparisons for the Smith *et al.* (2002) and Martins *et al.* (2005) predictions, respectively. Objects identified with '+' signs have significant additional contributions from shock ionization. (From Voges *et al.* 2008.)

While these newer models solve some outstanding problems for understanding massive stars, this however causes a new problem for understanding the ISM: These results imply that density-bounded H II regions may not be able to account for the required 50% ionization of the WIM. Figure 1 does show a large scatter, a factor of ~1.5, and so many objects are still likely to be density-bounded. However, the bulk reduction in available ionizing radiation by this mechanism requires other processes to help power the WIM.

3. Ionizing Radiation from Starbursts

We now turn to radiative feedback from galaxies, an issue of vital importance in understanding the reionization of the universe and galactic energy budgets. The Survey of Ionization in Neutral-Gas Galaxies (SINGG; Meurer *et al.* 2006) is an Hα and R-band survey of H I-selected galaxies aimed at understanding the local star-formation rate density and star-forming properties of galaxies in the local universe. The galaxies are selected to uniformly sample H I gas masses in the range 10^7 to 10^{11} $M_\odot$. The results reported here and by Oey *et al.* (2007) are based on the first data release of 109 galaxies.

We categorized the galaxies according to star-formation intensity (SFI), defined as the Hα surface brightness $\Sigma_{H\alpha}$ within the Hα half-light radius. Galaxies having $\log \Sigma_{H\alpha} > 39.4$ are categorized as "starburst" galaxies, and those having $\log \Sigma_{H\alpha} \leqslant 38.4$ are denoted as galaxies having "sparse" star formation; galaxies with $\Sigma_{H\alpha}$ between these boundaries are considered "normal". Furthermore, many galaxies have almost all of their star formation and Hα emission concentrated in the nuclear regions, and so we noted such galaxies in addition. We used the HIIphot code (Thilker *et al.* 2000) to identify the boundaries of the individual H II regions and thereby define the allocation of Hα emission between the classical H II regions and the WIM. Note that HIIphot does account for background diffuse emission associated with the identified H II regions.

Our study (Oey *et al.* 2007) examines a variety of parameters, but the most notable result is an anti-correlation between the fraction of total Hα luminosity occupied by the WIM and the star-formation intensity, shown in Figure 2. The relation is apparent for starburst galaxies in particular, recalling that these are systems having $\log \Sigma_{H\alpha} > 39.4$. The effect is also seen in the Hα surface brightness distributions: Our starburst galaxies

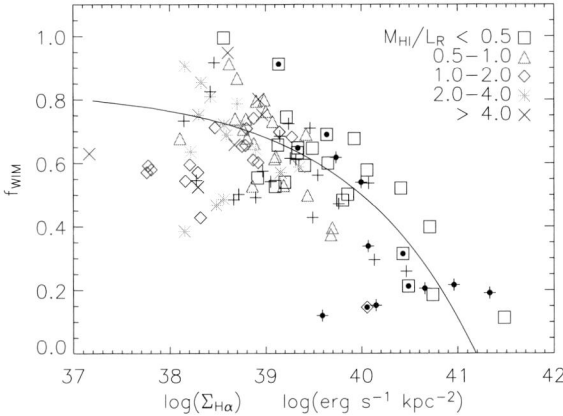

Figure 2. Fraction of diffuse Hα emission as a function of Hα surface brightness, a measure of star-formation intensity. Symbols show H I mass relative to R-band luminosity, a measure of H I gas fraction, as shown. (From Oey *et al.* 2007.)

show much shallower Hα surface brightness distributions than those for the "normal" and "sparse" star-formation categories (see Oey et al. 2007).

What is the origin of this intriguing decrease in WIM fraction with SFI for starbursts? There are two general possibilities: *a*) The sources of ionizing radiation are systematically reduced; or *b*) The starbursts are fully occupying the ISM, thereby leaving progressively less neutral, diffuse ISM to be ionized. The latter implies that the ISM is essentially density-bounded, allowing excess ionizing radiation to escape from the galaxy disks. It would be an important consequence if this radiation escapes from the galaxies altogether, allowing ionization of the intergalactic environment, and the cosmos itself.

In evaluating these two possibilities, note that starbursts have a morphological appearance that is consistent with this second scenario. The giant star-formation regions are closely packed together, leaving little external ISM that could serve as diffuse WIM (Figure 3). Clarke & Oey (2002) predicted a threshold critical star-formation rate (SFR) above which the mechanical feedback shreds the ISM, allowing the escape of ionizing radiation and hot, metal-laden superwinds. This is simply a balance of the ISM weight against the force of mechanical feedback. In terms of the global galaxy parameters, our rough estimate of this criterion is:

$$\mathrm{SFR}_{\mathrm{crit}} \sim 0.15 \, (M_{\mathrm{ISM},10} \, \tilde{v}_{10}^2 / f_d) \quad \mathrm{M}_\odot \, \mathrm{yr}^{-1} \,, \qquad (3.1)$$

where $M_{\mathrm{ISM},10}$ is the total ISM mass in units of 10^{10} M$_\odot$, $\tilde{v}_{10}$ is the thermal/turbulent velocity dispersion in units of 10 km s^{-1}, and f_d is a geometric correction factor for disk galaxies of order unity. Applying equation 3.1 to local starbursts, we find that typical critical SFR's are $\lesssim 1$ M$_\odot$ yr^{-1}, whereas actual SFR's are more like 1 – 20 M$_\odot$ yr^{-1}. This implies that in principle, local starbursts should largely be meeting this criterion, and allowing the escape of outflows and ionizing radiation. Figure 4 shows the Hα luminosities, relative to the value corresponding to the critical SFR, as a function of SFI for our sample galaxies. It is apparent that the galaxies with the highest SFI all exceed the threshold

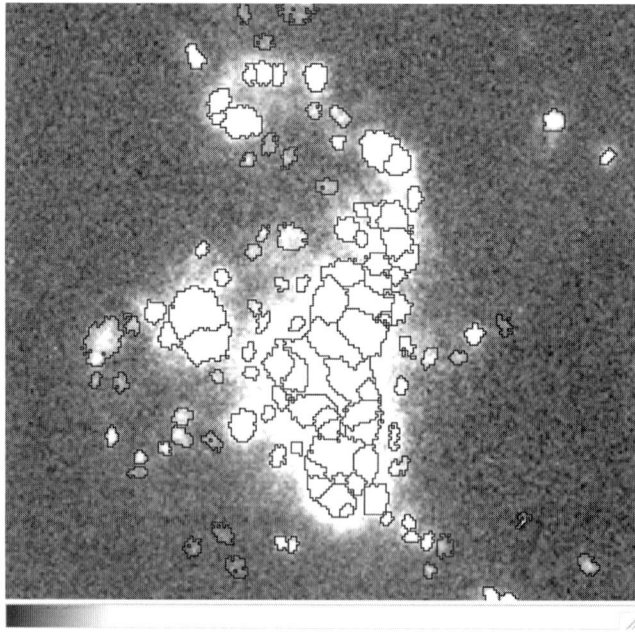

Figure 3. Continuum-subtracted Hα image of starburst galaxy J0355-42, showing H II region boundaries identified by the HIIphot software. (From Oey et al. 2007)

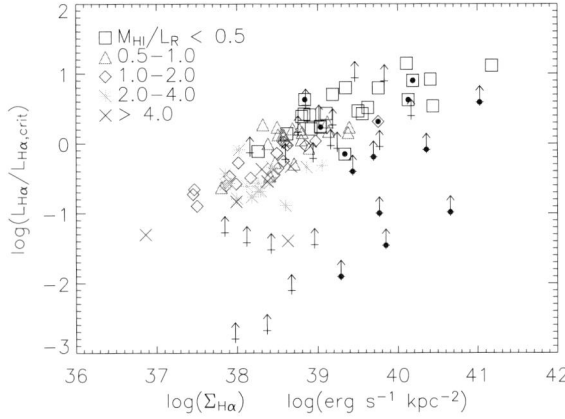

Figure 4. Total Hα luminosity, a measure of star-formation rate, relative to the critical threshold for dominating feedback. Symbols are as in Figure 2. (From Oey et al. 2007.)

criterion. We do caution that equation 3.1 is a crude parameterization, which is likely to need refinement and calibration. However, the general trend is fully consistent with an expectation that starburst galaxies should allow the escape of ionizing radiation. Moreover, both Figures 2 and 4 demonstrate that galaxies with low fractions of diffuse emission are also those that have the lowest H I gas fractions, which is also consistent with their ISM being largely density-bounded. Figure 2 shows a general analytic relation corresponding to greater ISM density-bounding as the SFR increases.

On the other hand, while outflows are widely seen in starbursts (e.g., Martin 2003), searches for evidence of escaping ionizing radiation are puzzlingly low or absent (e.g., Leitherer et al. 1995; Heckman et al. 2001). Recently, direct detections of Lyman continuum emission on the order of a few percent have been reported in the blue compact dwarf galaxy Haro 11 (Bergvall et al. 2006), and a sample of Lyman break galaxies (Shapley et al. 2006). These observations are at odds with the indirect evidence presented here. One way to reconcile these results is by invoking a limited opening angle for the escape of this radiation (see comment by Pettini below).

The other possibility to explain the trend of lower WIM fraction with SFI, is that the ionizing sources are reduced, as mentioned above. We discuss this further in Oey et al. (2007), where we argue that a reduction in the field star population is unlikely. It may also be that greater extinction traps ionizing radiation near their sources. We have initiated an investigation using *Spitzer* and *GALEX* observations to evaluate this possibility.

4. Summary

In summary, we find that the softer, current generation of hot star atmospheres imply that our LMC H II regions are largely radiation-bounded. If this result can be generalized to other star-forming galaxies, it will now likely require an additional source(s) to power the diffuse, warm ionized medium. On the other hand, our results from the SINGG survey suggest that the entire ISM of starburst galaxies are at least partly density-bounded, thereby allowing ionizing radiation to escape into their halos, and perhaps beyond, into the intergalactic medium.

Acknowledgements

We gratefully acknowledge travel support from the IAU. This work was supported by NSF grant AST-0448893 and by NASA grant NAG5-10768.

References

Bergvall, N., Zackrisson, E., Andersson, B.-G., et al. 2006, A&A, 448, 513
Clarke, C. J. & Oey, M. S. 2002, MNRAS, 337, 1299
Heckman, T. M., Sembach, K. R., Meurer, G. R., et al. 2001, ApJ, 558, 56
Hoopes, C. G., & Walterbos, R. A.M. 2000, ApJ, 541, 597
Leitherer, C., Ferguson, H. C., Heckman, T. M., & Lowenthal, J. 1995, ApJ, 454, L19
Martin, C. L. 2003, in: J. L. Rosenberg & M. E. Putman (eds.), *The IGM/Galaxy Connection: The Distribution of Baryons at z=0*, (Dordrecht: Kluwer), 205
Martins, F., Schaerer, D., & Hillier, D. J. 2005, A&A, 436, 1049
Meurer, G. R., Hanish, D. J., Ferguson, H. C. et al. 2006, ApJS, 165, 307
Oey, M. S., & Kennicutt, R. C., Jr. 1997, MNRAS, 291, 827
Oey, M. S., Meurer, G. R., Yelda, S. et al. 2007, ApJ, 661, 801
Panagia, N. 1973, AJ, 78, 929
Schaerer, D., & de Koter, A. 1997, A&A, 322, 598
Shapley, A. E., Steidel, C. C., Pettini, M., et al. 2006, ApJ, 651, 688
Smith, L. J., Norris, R. P.F., & Crowther, P. A. 2002, MNRAS, 337, 1309
Thilker, D. A., Braun, R., & Walterbos, R. A. M., 2000, AJ, 120, 2070
Vacca, W. D., Garmany, C. D., & Shull, J. M. 1996, ApJ, 460, 914
Voges, E. S., Oey, M. S., Walterbos, R. A. M., & Wilkinson, T. M. 2008, AJ, submitted

Discussion

GALLAGHER: Your results are consistent with our data for individual starbursts, that show little in the way of Hα emission beyond the starburst zone and associated wind. The mechanism, however, still isn't clear; preferential escape vs. dust blocking or something else. Do you have ideas for sharp tests?

OEY: We do have *Spitzer* and *GALEX* proposals in, to examine the SED's, including mid/far IR emission. Obvious tests to search for direct emission have been done by Leitherer and others, which are mostly negative.

LEITHERER: There is a rather straightforward test for probing the opacity of the diffuse ISM in these galaxies. It involves taking a UV longslit spectrum with STIS, which will hopefully be available in the next cycle. Since the Lyman continuum cannot be measured directly in these low-z galaxies, take a heavy element like C or Si as a proxy. The STIS spectrum should give you a column density of the respective ion. With an assumption on ionization and abundance, you will get the opacity.

PETTINI: A comment that the Shapley *et al.* (2006) results seem consistent with your suggestion that a narrow opening angle could be the way the radiation escapes. In their sample, most of the galaxies showed nothing, but in the two that were detected in the Lyman continuum, it looked like pretty much all of the radiation is escaping. Also, a second comment that the Bergvall results have been questioned recently by Grimes *et al.* (2007, *ApJ* 668, 891)

The Role of Massive Stars in Galactic Chemical Evolution

Francesca Matteucci

Dipartimento di Astronomia, Osservatorio Astronomica di Trieste (INAF)
Universita di Trieste, Via G.B. Tiepolo, 11
I-34124 Trieste, Italy
email: matteucci@oats.inaf.it

Abstract. I will review the role of massive stars in galactic evolution both from the nucleosynthesis and energetics point of view. In particular, I will highlight some important observational facts explained by means of massive stars in galaxies of different morphological type: the Milky Way, ellipticals and dwarf spheroidals. I will describe first the time-delay model and its interpretation in terms of abundance ratios in galaxies, then I will discuss the importance of mass loss in massive stars to reproduce the data in the Galactic bulge and disk. I will discuss also how massive stars can be important producers of primary nitrogen if rotation in stellar models is taken into account. Concerning elliptical galaxies, I will show that to reproduce the observed [Mg/Fe] versus Mass relation in these galaxies it is necessary to assume a more important role of massive stars in more massive galaxies and that this can be achieved by means of downsizing in star formation. I will discuss how massive stars are responsible in triggering galactic winds both in ellipticals and dwarf spheroidals. These latter systems show a low overabundance of α-elements relative to Fe with respect to Galactic stars of the same [Fe/H]: this is interpreted as due to a slow star formation coupled with very efficient galactic winds. Finally, I will show a comparison between the predicted Type Ib/c rates in galaxies and the observed GRB rate and how we can impose constraints on the mechanism of galaxy formation by studying the GRB rate at high redshift.

Keywords. galaxies: evolution – stars: early-type

1. Introduction

We call massive stars all stars with Main Sequence masses $M > 8 M_\odot$, namely those stars which do not develop a degenerate carbon-oxygen core. If their mass is lower than $10\, M_\odot$, they will explode as e-capture supernovae (SNe): the explosion is triggered by e-capture which destabilizes the star and then by the ignition of oxygen in a degenerate O-Ne-Mg core. All the stars with mass between 8 and $10\, M_\odot$ will instead ignite all the nuclear fuels up to Si-burning which produces ^{56}Ni which then decays into ^{56}Fe. They die as core-collapse SNe which include both Type II and Ib/c SNe. The massive stars are responsible for the production of the bulk of α-elements (O, Mg, Ne, Si, S, Ca and Ti) plus some Fe, originating either from the inner core or during the explosion by means of explosive Si-burning. However, the bulk of Fe should be produced by Type Ia SNe which are believed to originate from C-O white dwarfs (WDs) in binary systems, which explode by C-deflagration when the WD reaches the Chandrasekhar mass ($\sim 1.4 M_\odot$). Two main paths have been identified for Type Ia SNe: i) the single-degenerate scenario, made of a WD plus a normal star, where the WD explodes after reaching the limiting mass as a consequence of accreting mass from the companion, ii) the double-degenerate scenario, where two WDs of roughly $0.7\, M_\odot$ merge after losing angular momentum caused by gravitational wave emission. When they merge, the Chandrasekhar limiting mass is reached and the C-deflagration, producing mainly Fe, occurs as in the single-degenerate case.

Yields from massive stars have been calculated by many authors, here I will recall some of the most recent calculations: Nomoto *et al.* (2006) provided detailed yields of many isotopes including explosive nucleosynthesis for massive stars without mass loss as functions of the stellar metallicity. On the other hand, Hirschi (2007) presented yields of a limited number of elements without explosive nucleosynthesis but taking into account mass loss and rotation. In particular, mass loss in massive stars mainly affects the yields of He, C and O as already pointed out by Maeder (1992): stellar models with mass loss predict a larger He and C production at expenses of oxygen production in massive stars. Another interesting aspect is that stellar rotation, particularly important at low metallicities, can produce a considerable amount of primary ^{14}N from massive stars. We define "primary" a chemical element which is produced directly from H and He inside the stars, whereas we define "secondary" any chemical species which is produced by means of heavy elements already present in the star at birth. A typical example of secondary element is represented by ^{14}N which originates during the CNO cycle from the original C and O present in the star. However, ^{14}N can be primary if the C and O used to form it are produced by the star in situ and this is the case during the third dredge-up acting in conjunction with hot-bottom burning in Asymptotic Giant Branch (AGB) stars (e.g. Renzini & Voli 1981). Rotation in massive stars can also produce primary ^{14}N, as shown by Meynet & Maeder (2002).

When stellar yields are included in a chemical evolution model, namely a model which is aimed at predicting the temporal and spatial evolution of the abundances of the most abundant isotopes in the interstellar gas, we can compare the model results with detailed and precise abundance determinations. From this comparison we can then infer important constraints both on stellar nucleosynthesis, initial mass function (IMF), history of star formation (SF) and mechanisms of galaxy formation. In this paper, we will show how we can successfully interpret abundances in galaxies of various morphological type (Milky Way, ellipticals, dwarf spheroidals) by means of detailed chemical evolution models. In particular, we will highlight the role of massive stars in galactic chemical evolution both from the nucleosynthesis and energetics point of view. In fact, massive stars, besides producing the majority of heavy elements, inject large quantities of energy into the interstellar medium (ISM) by means of SN explosions but also by means of stellar winds. Such a feedback is an extremely important ingredient in studying galaxy formation and evolution since it can trigger galactic winds which in turn eject the heavy elements into the intergalactic and intracluster medium. Finally, since some Gamma Ray Bursts (GRBs) have been associated with Type Ib/c SNe, we will show a comparison between the rates of Type Ib/c SNe in galaxies and the observed GRB rate. Type Ib SNe are probably the result of the explosion of single Wolf-Rayet stars ($M > 25 M_\odot$), whereas Type Ic SNe should arise from the explosion of massive stars (12-20 $M_\odot$) in binary systems (e.g. Baron 1992). In particular, we will discuss predictions relative to the cosmic Type Ib/c SN rate in different scenarios of galaxy formation: monolithic and hierarchical, and show how the rate of GRBs at high redshift can be used to impose constraints on galaxy formation models.

2. Yields from massive stars

In Fig. 1 left panel we show the effect of mass loss in massive stars on the yield of carbon by comparing the yields of Maeder (1992) with a high rate of mass loss with those computed by Woosley & Weaver (1995) without mass loss. Both sets of yields are

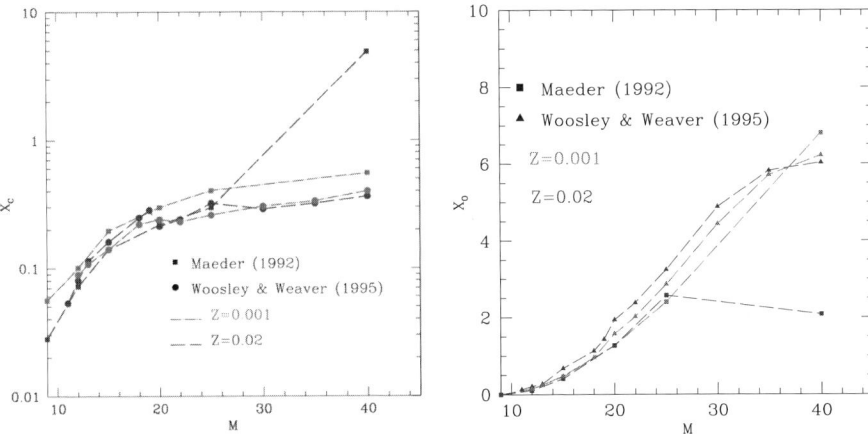

Figure 1. Left panel: the effect of mass loss on stellar yields: the yields of carbon from massive stars in presence of mass loss and as functions of metallicity from Maeder (1992). The yields of carbon from the conservative models of Woosley & Weaver (1995) are shown for comparison. Right panel: the effect of mass loss on stellar yields. The yields of oxygen from massive stars in presence of mass loss and as functions of metallicity from Maeder (1992). The yields of oxygen from the conservative models of Woosley & Weaver(1995) are shown for comparison.

computed for two different initial stellar metallicities. As one can see, the effect of mass loss (increase in the C production) becomes evident only for a metallicity $Z \geqslant Z_\odot$ and for stars with mass $M > 25 M_\odot$. This mass limit separates stars which end their lives as Type II SNe from those which become Wolf-Rayet stars and die as Type Ib SNe. For metallicities below the solar one the yields with and without mass loss are very similar. In Fig. 1 lower panel we show the effect of mass loss on the O yields: in this case, the O yields is severely decreased for stars with $M > 25 M_\odot$ and metallicities $Z \geqslant Z_\odot$. In fact, mass loss subtracts carbon to further processing through the $^{12}C(\alpha,\gamma)^{16}O$ reaction and increases with metallicity, thus its effect is evident mainly at high metallicities.

3. Galactic chemical evolution

The main ingredients to build models of galactic chemical evolution are:
- Initial conditions: open or closed model, primordial or pre-enriched gas.
- The birthrate function, in other words the star formation rate (SFR) and the initial mass function (IMF).
- Stellar yields: newly processed and unprocessed material restored into the ISM at the star death.
- Infall, outflow and inflow of gas.
- Equations including all of that (e.g. Tinsley 1980, Matteucci 2001).

3.1. The Milky Way

To describe our Galaxy we will assume the two-infall model proposed by Chiappini et al. (1997). In this model the Milky Way forms mainly during two main gas accretion episodes: during the first episode the halo, the central bulge and part of the thick disk form on a timescale not longer than 1-2 Gyr, whereas during a much longer second infall episode the thin disk formed inside-out. The timescale suggested for the formation of the thin disk at the solar neighbourhood is 6-8 Gyr (Chiappini et al. 1997; Boissier & Prantzos 1999). This model assumes the Scalo (1986) IMF and a Schmidt (1959) law for the SFR with exponent $k = 1.5$ and also the existence of a threshold density for the SF of $7 M_\odot pc^{-2}$ in the thin disk. This model reproduces the [X/Fe] vs. [Fe/H] relations found for halo and disk stars, the present time gas surface density, SFR, infall rate and SN rates. It can explain also the abundance gradients along the thin disk (Chiappini et al. 2001; Cescutti et al. 2007).

It is important to recall that the [X/Fe] vs. [Fe/H] relations depend mainly on the assumed nucleosynthesis, IMF and SFR. From the point of view of nucleosynthesis it is particularly important the role played by different SNe in the chemical enrichment. In particular, the delay with which Type Ia SNe restore the bulk of Fe relative to the fast production of α-elements by the core-collapse SNe. This interpretation first proposed by Tinsley (1979) and then developed by Greggio & Renzini (1983) and Matteucci & Greggio (1986) is known as *time-delay model*. To illustrate the time-delay model we show in Fig. 2 upper panel the predicted [α/Fe] vs. [Fe/H] for different histories of SF. If one assumes that O is mainly produced by Type II SNe and that 2/3 of the total Fe is produced by Type Ia SNe whereas the remaining 1/3 is formed in massive stars, one obtains a very good fit of the data relative to the stars of the halo and disk in the solar vicinity (central curve in the figure) as well as for the bulge (upper curve) and irregular galaxies (lowest curve). On the other hand, if one assumes that Fe is either produced entirely by Type Ia SNe or entirely by Type II SNe, the agreement with the data is lost. This simply means that both SN Types should contribute to the Fe production and that Type Ia SNe restore Fe into the ISM with a delay relative to the Fe produced by Type II SNe. In particular, the bulk of Fe production in the solar neighbourhood occurred with a delay of ~ 1 Gyr. This does not mean that the first Type Ia SNe occurred after 1Gyr, since the most massive binary systems giving rise to Type Ia SNe live no longer than 30-40 Myr. These prompt Type Ia SNe do exist, as shown by Mannucci et al. (2005;2006). However,

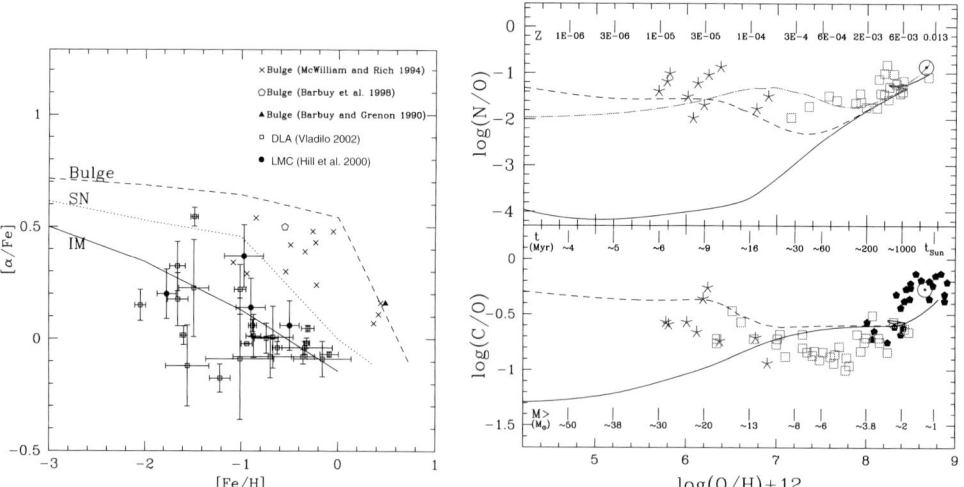

Figure 2. Left panel: illustration of the time-delay model for different histories of star formation. Right panel: C and N evolution in Galactic stars. Figure and references to the data in Chiappini et al. (2006). The dashed lines represent the predictions of a model including yields from massive stars with a strong primary N production due to stellar rotation in extremely metal poor stars.

the [α/Fe] vs. [Fe/H] relation is not the same everywhere: in fact, even if the roles of the two different Types of SNe in producing O and Fe are likely to be the same and the

IMF is not too different in different galaxies, the SFR is instead very different in different galaxies. In particular, the SFR must have been much faster in spheroids and ellipticals which have processed their gas in stars very quickly and at high redshift, as opposed to the slow and gasping SFR occurring in dwarf irregulars. Spirals like the Milky Way must have had a SFR intermediate and continuous. The different SFRs in galaxies influence the age-metallicity relation, producing a very fast increase of the [Fe/H] in spheroids due mostly to Type II SNe. The opposite occurs if the SFR is slow. This effect is clearly shown in Fig. 2 (left panel, lowest curve).

Another important aspect of the Galactic abundance patterns is related to the N production, as shown in Fig. 2 right panel. The most recent data seem to indicate that there should be a non negligible primary N production from massive stars. Models with stellar rotation produce N yields which can explain the observations in Galactic stars.

The Galactic Bulge is a spheroid and its properties are more similar to those of a small elliptical than to those of the Galactic disk. Successful models for the Galactic bulge suggest that it formed very quickly during the collapse of the inner halo and that its SFR was very high like in a star-burst (Matteucci & Brocato, 1990; Ballero *et al.* 2007). In particular, the model of Ballero *et al.* (2007) predicts that the bulge formed on a timescale not longer than 0.1-0.5 Gyr, that the SFR was roughly 20 times more efficient than in the Galactic disk and that the IMF was flatter than that in the disk in order to reproduce the bulge stellar metallicity distribution. In Fig. 3 (left panel) we show the predicted [O/Fe] vs. [Fe/H] in the bulge compared with the most recent and accurate data. As explained before, the strong SFR acts in a way such that a solar [Fe/H] is reached in the gas before the bulk of Fe is restored from Type Ia SNe. This produces a long plateau in the [α/Fe] ratio extending from low metallicities up to over-solar metallicities. The good agreement between predictions and observations strongly supports the time-delay model. In Fig. 3 left panel are shown two model predictions: one refers to the yields of Woosley & Weaver (1995) for massive stars whereas the other refers

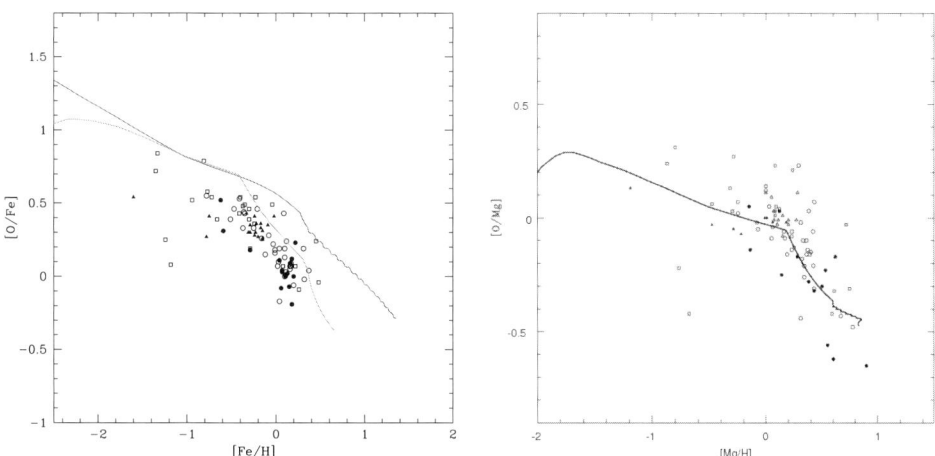

Figure 3. Left panel: observed and predicted [O/Fe] vs. [Fe/H]. The upper curve refers to the model with the yields of Woosley & Weaver (1995) in massive stars, whereas the lower one to the yields of Maeder (1992). For the references on the data see McWilliam *et al.* (2008). Right panel: predicted and observed [O/Mg] vs. [Mg/H] in the Galactic bulge. The model is that with the yields of Maeder (1992) for massive stars. Models adopting yields with no mass loss predict a flat [O/Mg] ratio (see McWilliam *et al.* 2008).

to a model adopting the yields of O from Maeder (1992) where mass loss depending on metallicity is taken into account. As one can see, the agreement is much better in this second case. However, as McWilliam *et al.* (2008) have pointed out, there is a diagram which illustrates even better the necessity of mass loss in massive stars. This is the plot in Fig. 3 right panel, where we show the [O/Mg] vs. [Mg/H]. Here the time-delay model does not work, since both Mg and O are produced mainly in massive stars. In spite of this, the [O/Mg] ratio declines strongly for [Mg/H] > 0, a rather unexpected result which can be explained only if the yields of O have a strong dependence on metallicity, whereas those of Mg do not, as it is in the case of massive stars with mass loss. It is important to note that the same behaviour of the [O/Mg] ratio is shown by galactic stars (Bensby *et al.* 2005 ; Mc William *et al.* 2007).

3.2. *Elliptical galaxies*

Early type galaxies and spheroids in general, where no gas is present now, are likely to have suffered galactic winds and/or stripping phenomena in high density environments. Galactic winds should be triggered by the energy injected by SNe and stellar winds from massive stars into the ISM. Unfortunately, there not exists a precise recipe for the feedback and in modelling galaxy evolution one is forced to assume that a certain fraction of the initial blast wave energy is transferred into the ISM as thermal energy. Pipino & Matteucci (2004) have shown that if a 20 % of the total blast wave energy is transferred from SNe into the ISM, galactic winds can occur even in massive ellipticals. However, their model does not take into account the gas cooling which can change drastically the situation. When cooling is assumed, then it is difficult to obtain galactic winds without assuming a contribution from the central AGN (e.g. Granato *et al.* 2001). In any case the situation is still unclear except for the fact that we know that to reproduce the observed features of local ellipticals one has to assume that their stars formed quickly at high redshift and that some mechanism must have kept the galaxies free of gas for several Gyrs. In the model of Pipino & Matteucci (2004) the galactic winds occur quite early in the life of ellipticals and earlier in the most massive ones. In other words, in their model the most massive ellipticals form stars more intensively and for a shorter time than the less massive ones. This is the "inverse wind scenario" proposed by Matteucci (1994) which produces a downsizing in the SF. This downsizing is responsible for the growth of the [Mg/Fe] ratio with total galactic mass observed in ellipticals and impossible to reproduce in the framework of classic hierarchical models for galaxy formation, as illustrated in Fig.4. The effect of massive stars on the chemical evolution of ellipticals is therefore evident: it is more important in more massive objects where SF lasts for a shorter period thus creating a higher [Mg/Fe] than in less massive objects, as a consequence of the time-delay model.

3.3. *Dwarf Spheroidals*

A great deal of observational work concerning dwarf spheroidals of the Local Group has appeared in the last few years. High resolution abundance determinations allow us to compare the chemical evolution of these objects with that of the Milky Way. In Fig. 5 we show a comparison between [α/Fe] vs. [Fe/H] relations in the dwarf spheroidals and in the Galaxy. As one can see, the observed patterns are different in the sense that except for a small overlapping of the [α/Fe] ratios at low metallicities, generally dwarf spheroidals show lower [α/Fe] ratios at the same [Fe/H] relative to Galactic stars. The behavior of the [α/Fe] ratios in these objects resembles that shown in Fig. 2 left panel for a galaxy with low star formation efficiency. By adopting a low star formation efficiency relative to the Galaxy and strong galactic winds, Lanfranchi & Matteucci (2003, 2004)

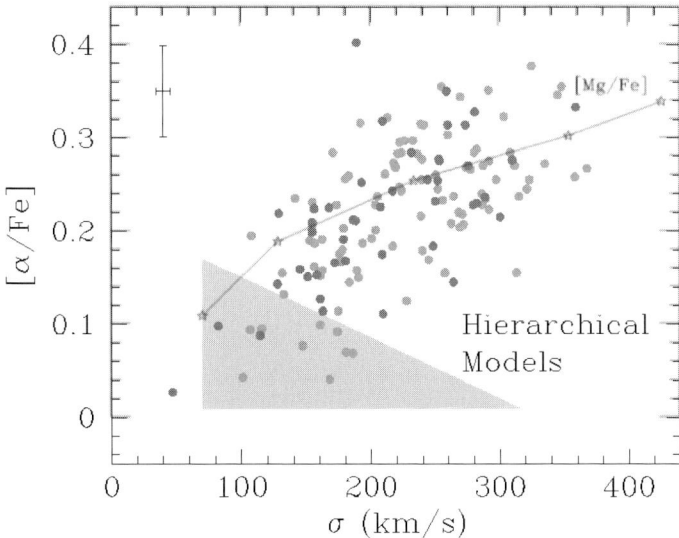

Figure 4. Predicted and observed [α/Fe] ratios in ellipticals. The continuous line represents the prediction of the model by Pipino & Matteucci (2004). The shaded area represents the prediction of hierarchical models for the formation of ellipticals. The symbols are observational data. Figure adapted from Thomas *et al.* (2002).

and Lanfranchi *et al.* (2006) reproduced the abundance patterns of six dwarf spheroidals of the Local Group both for α-, *s*- and *r*-process elements. They also reproduced very well the observed stellar metallicity distribution of the Carina galaxy.

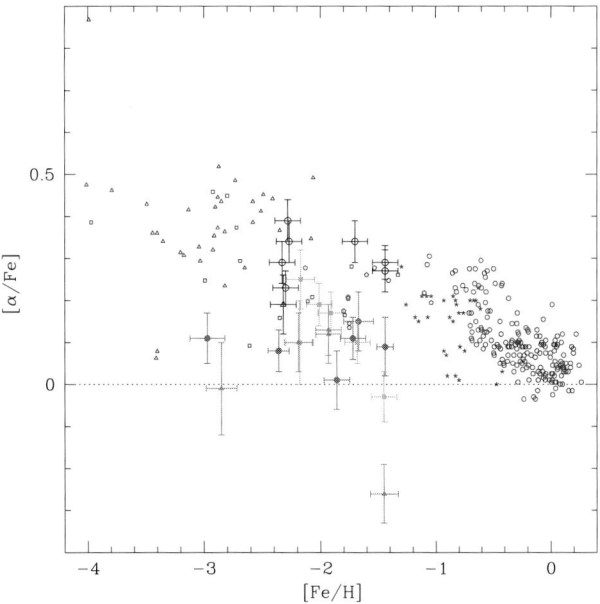

Figure 5. Observed [α/Fe] vs. [Fe/H] in the Galaxy and in dwarf spheroidals (data with error bars). The figure is from Shetrone *et al.* (2001).

4. Connection between Type Ib/c SNe and GRBs

Some long GRBs have been found to be associated with Type Ib/c SNe, therefore it is interesting to check whether the Type Ib/c SN rates in galaxies are compatible with the observed GRB rate. Bissaldi *et al.* (2007) assumed that Type Ib/c SNe arise either from : i) single WR stars with masses $M > 25 M_\odot$ or from ii) massive stars (12-20 $M_\odot$) in binary systems. In both cases, in fact, a massive star explodes after having lost its H mantle thus resulting into a Type b/c SN. They also assumed different SFRs in different galaxies going from a short and intense burst in ellipticals to a low and continuous SF in irregulars. The agreement with the observed rate was found to be good for spirals and irregulars (ellipticals do not show Type Ib/c SNe). As a second step, they computed the cosmic Type Ib/c SN rate by assuming several cosmic SFRs. In particular, they tested the cosmic SFR of Calura (2004) which assumes that ellipticals form very quickly and at high redshift as well as cosmic SFRs derived in the context of the hierarchical clustering scenario for galaxy formation, where ellipticals, especially the most massive ones formed last and until recently. Clearly the cosmic SFR of Calura predicts a high Type Ib/c rate at high redshift due to massive ellipticals, whereas the hierarchical cosmic rate is strongly decreasing at high redshift. In Fig. 6 we show the predicted cosmic Type Ib/c SN rates, under different assumptions about the cosmic SFR, compared with the observed GRB rate.

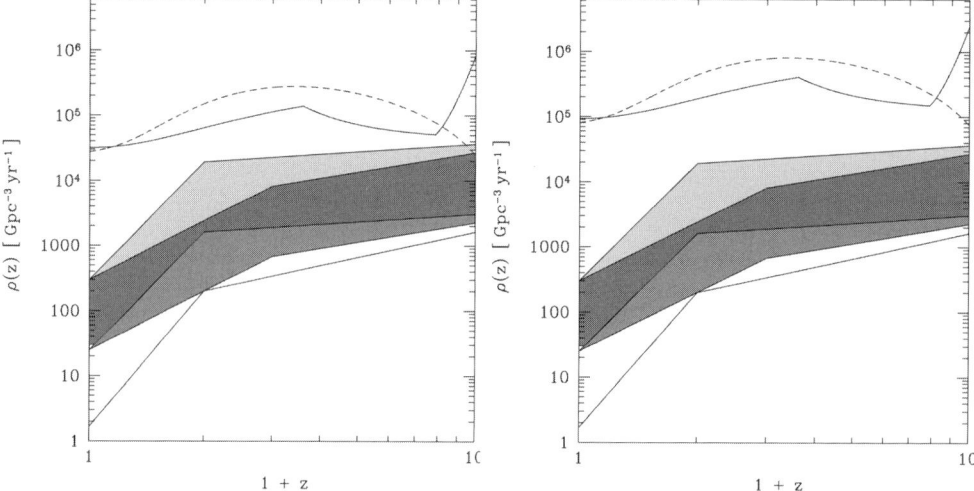

Figure 6. Comparison between the observed cosmic GRB rate and the predicted cosmic Type Ib/c SN rates as functions of redshift. Upper continuous line: predictions from the cosmic SFR favoring high star formation in ellipticals at high redshift (Calura 2004). Dashed line: predictions from a hierarchical cosmic SFR. Shaded area and lowest continuous line represents the data relative to the GRB rate. Right panel: predicted Type Ib/c SN rates assuming a Salpeter (1955) IMF. Left panel: predicted Type Ib/c SN rates assuming a top heavy IMF. Figure and references can be found in Bissaldi *et al.* (2007).

5. Conclusions

We have discussed the importance of the time-delay model in interpreting the abundance patterns in galaxies. In particular, we have discussed the role of core-collapse SNe (Type II, Ib/c) relative to the thermonuclear SNe (Type Ia) in the galactic chemical enrichment. We can conclude the following:

- abundances in metal poor stars do not always show the signature of massive stars, it rather depends on the history of star formation. In fact, some low metallicity objects show the signature of Type Ia SNe (e.g. dwarf spheroidals) and this is interpreted as due to slow star formation. On the contrary, the stars in the Galactic bulge and spheroids are metal rich and do not show the signature of Type Ia SNe: this is interpreted as due to a very fast star formation process.
- The [O/Mg] vs. [Mg/H] relation, both in the bulge and in the disk of the Milky Way, strongly favors a metallicity dependent mass loss in massive stars. The plot [N/O] vs. [O/H] in the Galaxy strongly favors the production of primary N from massive stars. This can be achieved by means of stellar rotation in massive stars.
- Elliptical galaxies show an increasing average [Mg/Fe] ratio in stars with galactic mass: this strongly supports the idea that the most massive ellipticals formed stars for a shorter time than the less massive ones (downsizing).
- By assuming that Type Ib/c SNe arise from either single massive stars ($M > 25 M_\odot$) or from massive stars in binary systems, we compared the cosmic Type Ib/c SN rate with the observed GRB rate. The Type Ib/c SN rate is much higher than the GRB rate which is only a fraction varying from 0.1 to 1% of the Type Ib/c rate, in agreement with observational determinations (e.g. Della Valle 2005). We also predict that if we accept downsizing in star formation in ellipticals, then the number of GRBs at high redshift should be much higher than predicted in hierarchical models of galaxy formation.

References

Ballero, S., Matteucci, F., Origlia, l. & Rich, R. M., 2007, *A&A*, 467, 123
Baron, E., 1992, *MNRAS*, 255, 267
Bensby, T., Feltzing, S., Lundstroem, I., & Ilyn, I., 2005, *A&A*, 185, 203
Bissaldi, E., Calura, F., Matteucci, F., *et al.* 2007, *A&A*, 471, 585
Boissier, S. & Prantzos, N., 1999, *MNRAS*, 307, 857
Calura, F., 2004, PhD Thesis, Trieste University
Cescutti, G., Matteucci, F., François, P. & Chiappini, C., 2007, *A&A*, 462, 943
Chiappini, C., Matteucci, F., & Gratton, R., 1997, *ApJ*, 477, 765
Chiappini, C., Hirschi, R., Meynet, G., *et al.* 2006, *A&A*, 449, L27
Chiappini, C., Matteucci, F. & Romano, D., 2001, *ApJ*, 554, 1044
Della Valle, M., 2005, in *Il Nuovo Cimento C*, 28, 563
Granato, G. L., Silva, L., Monaco, P., *et al.* 2001, *MNRAS*, 324, 757
Greggio, L. & Renzini, A., 1983, *Mem. SaIt*, 54, 311
Hirschi, R., 2007, *A&A*, 461, 571
Lanfranchi, G. & Matteucci, F. 2003, *MNRAS*, 345, 71
Lanfranchi, G. & Matteucci, F., 2004, *MNRAS*, 351, 1338
Lanfranchi, G., Matteucci, F. & Cescutti, G., 2006, *A&A*, 453, 67
Maeder, A., 1992, *A&A*, 264, 105
Mannucci, F., Della Valle, M., Panagia, N., *et al.* 2005, A&A 433, 807
Mannucci, F., Della Valle, M.& Panagia, N., 2006, *MNRAS*, 370, 773
Matteucci, F. & Brocato, E., 1990, *ApJ*, 365, 539
Matteucci, F., 2001, *The Chemical Evolution of the Galaxy* (Dordrecht: Kluwer)
Matteucci, F., 2003, *Ap&SS*, 284, 539
Matteucci, F., 1994, *A&A*, 288, 57
Matteucci, F. & Greggio, L., 1986, *A&A*, 154, 279
McWilliam, A., Matteucci, F., Ballero, S., *et al.* 2008, *AJ* submitted (arXiv:0708.4026)
Meynet, G. & Maeder, A., 2002, *A&A*, 390, 561
Nomoto, K., Tominaga, N., Umeda, H., *et al.* 2006, *Nucl. Phys.* A777, 424
Pipino, A. & Matteucci, F., 2004, *MNRAS*, 347, 968

Renzini, A. & Voli, M., 1981, *A&A*, 94, 175
Salpeter, E. E., 1955, *ApJ*, 121, 161
Scalo, J. M. *Fund. Cosmic Phys.* 11, 1
Schmidt, M., 1959, *ApJ*, 129, 243
Shetrone, M. D., Coté, P.& Sargent, W. L. W., 2001, *ApJ*, 548, 592
Thomas, D., Maraston, C., Bender, R., 2002, in: R. E. Schielicke (ed.), *JENAM 2001: Astronomy with Large Telescopes*, (Schielicke: Wiley), *Reviews in Modern Astronomy*, 15, 219
Tinsley, B. M, 1979, *ApJ*, 229, 1046
Tinsley, B. M., 1980, *Fund. Cosmic Phys.*, 5, 287
Woosley, S. E. & Weaver, T. A., 1995, *ApJS*, 101, 181

Discussion

HIRSCHI: Can you differentiate between a lower minimum mass limit for single stars to form Wolf-Rayet stars and a larger fraction of binaries to reproduce the number of SNe Ib/c?

MATTEUCCI: Yes, we can in principle. There should be a lower limit for single WR stars which can reproduce the observed Type Ib/c SN rate but in this case one has also to check that other constraints are not violated. If we find a too small limiting mass for WR stars, incompatible with stellar evolution calculations, then we have a constraint. We will test this point in the future.

Detailed Nucleosynthesis Yields from the Explosion of Massive Stars

Carla Fröhlich[1], T. Fischer[2], M. Liebendörfer[2], F.-K. Thielemann[2] and J.W. Truran[3,1]

[1] Enrico Fermi Institute, University of Chicago,
5640 South Ellis Avenue, Chicago IL 60637, USA
email: frohlich@uchicago.edu

[2] Dept. of Physics, University of Basel,
Klingelbergstr. 82, CH-4056 Basel, Switzerland

[3] Department of Astronomy and Astrophysics and ASC Flash Center,
5640 S. Ellis Ave, Chicago, IL 60637, USA

Abstract. Despite the complexity and uncertainties of core collapse supernova simulations there is a need to provide correct nucleosynthesis abundances for the progressing field of galactic evolution and observations of low metallicity stars. Especially the innermost ejecta are directly affected by the explosion mechanism, i.e. most strongly the yields of Fe-group nuclei for which an induced piston or thermal bomb treatment will not provide the correct yields because the effect of neutrino interactions is not included.

Recent observations of metal-poor halo stars support the suggested existence of a *lighter element primary process* (LEPP) which operates very early in the galaxy and is independent of the r-process. We present a candidate for the LEPP, the so-called νp-process.

Keywords. nuclear reactions, nucleosynthesis, abundances – supernovae: general – neutrinos

1. Introduction

Massive stars end their life as core collapse supernovae. At the end of their hydrostatic evolution stars with main sequence masses $M \gtrsim 9\,M_\odot$ produce a core massive enough to undergo core collapse. This is the only fate for stars above $10\,M_\odot$ (see e.g. Heger *et al.* 2003), whereas in between stars can either collapse and form a neutron star or lose their envelope and become a white dwarf. The final stellar fate is determined by the size of the CO core. However, the relation of the size of the CO core to the progenitor mass depends on the metallicity. At higher metallicities mass loss becomes more important, producing smaller He and CO cores for a given initial mass. For very massive stars this can result in such a strong mass loss with increasing metallicity that black hole formation is excluded (leaving neutron stars as the only possible type of remnant). At low metallicities the final fate of a massive star depends on its initial mass: for initial masses of ~ 10–$25\,M_\odot$ as neutron star, for initial masses between $\sim 25\,M_\odot$ and $\sim 40\,M_\odot$ as black hole through fallback on the neutron star, or directly as black hole for initial masses above $\sim 40\,M_\odot$ (Hirschi *et al.* 2006, Nomoto *et al.* 2006, Blinnikov 2006). Core collapse with neutron star formation leads to supernovae (divided into subclasses, type II and type Ib/Ic, depending on the observation of H- and/or He-lines).

Such core collapse supernova events produce intermediate mass elements Si – Ca and Fe and neighboring nuclei. The production of elements beyond Fe has long been postulated by three processes: the r-process, the s-process, and the p-process (or γ-process). The former two are caused by *r*apid or *s*low neutron captures. The latter stands for proton capture or alternative means to produce heavy neutron deficient, stable isotopes. The s-process takes place during stellar evolution and acts through neutron captures on Fe produced in previous stellar generations. The s-process is thus a "secondary" process. For the r-process and the p-process the location, operation, and uniqueness in astrophysical sites are still under debate. The r-process is required to be a primary process (Sneden and Cowan 2003); the production of such elements is independent of the initial heavy element content of the star. Most of the p-nuclei are thought to the produced in hot (supernova) environments through photodisintegration of preexisting heavy nuclei. This can account for the heavy p-nuclei but underproduces the light ones (see e.g. Arnould & Goriely 2003, Costa et al. 2000, Hayakawa et al. 2004). The production mechanism for the light p-nuclei ^{92,94}Mo and ^{96,98}Ru is currently unknown. From chemical evolution studies of the cosmochronometer ^{92}Nb (Dauphas et al. 2003) a primary (supernova) origin is inferred.

The enrichment of the interstellar medium with these elements heavier than H, He, and Li (which originate from the Big Bang) can be traced via the surface composition of low mass stars of different ages. These stars are unaltered since their formation and therefore measure the composition in the interstellar medium at the time of their birth. Observations of such "metal-poor" stars provide information about the nucleosynthesis processes at the earliest times in the evolution of our Galaxy. The recently discovered hyper-metal-poor stars in the Milky Way may witness the chemical enrichment of the first generation of massive stars (the fastest evolving species).

2. Core Collapse Supernova Nucleosynthesis

Observations of supernova remnants shows typical kinetic energies of 10^{51} erg. Introducing a shock of appropriate energy in the pre-collapse stellar model (Woosley & Weaver 1995, Thielemann et al. 1996, Nomoto et al. 1997, Hoffman et al. 1999, Nakamura et al. 1999, Rauscher et al. 2002, Nomoto et al. 2006) — either through a piston or through a thermal bomb — allows to perform nucleosynthesis calculations. However, such induced nucleosynthesis calculations are not self-consistent: they miss detailed knowledge of the explosion mechanism and of the location of the mass cut between the neutron star and the supernova ejecta. Therefore, they cannot predict the ejected ^{56}Ni masses from the innermost layers (undergoing explosive Si-burning) which powers the supernova light curves via the decay chain ^{56}Ni – ^{56}Co – ^{56}Fe. The situation is different for the intermediate mass elements Si – Ca. They only depend on the explosion energy and the stellar structure of the progenitor star. Even lighter elements such as O and Mg are determined by the stellar evolution of the progenitor. Thus, when moving in from the outermost to the innermost ejecta of SN II explosion, we see an increase in the complexity, depending (a) only on stellar evolution, (b) on the stellar evolution and explosion energy, and (c) on stellar evolution and the complete explosion mechanism.

The correct prediction of the amount of Fe-group nuclei ejected (which includes also one of the so-called alpha-elements, i.e. Ti) and their relative composition depends directly on the explosion mechanism and the size of the Fe core. Three types of uncertainties are inherent in the Fe-group ejecta, related to (i) the total amount of Fe(group) nuclei ejected and the mass cut between neutron star and ejecta, mostly measured by ^{56}Ni decaying to ^{56}Fe, (ii) the total explosion energy which influences the entropy of the ejecta and with it the amount of radioactive ^{44}Ti as well as ^{48}Cr (decaying to ^{48}Ti and being responsible for elemental Ti), and (iii) finally the neutron richness or $Y_e = <Z/A>$ of the ejecta, dependent on stellar structure, electron captures, and neutrino interactions (Fröhlich et al. 2006a). The electron fraction Y_e influences strongly the overall Ni/Fe ratio.

An example for the composition after explosive processing due to an (induced) shock wave is discussed in detail in Thielemann *et al.* 1996. The outer ejected layers ($M(r) > 2\,\mathrm{M}_\odot$) are unprocessed by the explosion and contain results of prior H-, He-, C-, and Ne-burning in stellar evolution. The interior parts of SNe II contain products of explosive Si, O, and Ne burning. In the inner ejecta, which experience explosive Si-burning, Y_e changes from 0.4989 to 0.494. The Y_e originates from beta-decays and electron captures in the pre-explosive hydrostatic fuel in these layers. Neutrino reactions during the explosion were not yet included in these induced explosion calculations, utilizing a thermal bomb prescription. Huge changes occur in the Fe-group composition for mass zones below $M(r) = 1.63\,\mathrm{M}_\odot$. There the abundances of ^{58}Ni and ^{56}Ni become comparable. All neutron-rich isotopes (^{57}Ni, ^{58}Ni, ^{59}Cu, ^{61}Zn, ^{62}Zn) increase, the even-mass isotopes (^{58}Ni, ^{62}Zn) show the strongest effect. One can also recognize the increase of ^{40}Ca, ^{44}Ti, ^{48}Cr, and ^{52}Fe for the inner high entropy zones, but a reduction of the $N = Z$ nuclei in the more neutron-rich layers. More details can be found in extended discussions (Thielemann *et al.* 1996, Nakamura *et al.* 1999).

2.1. *The νp-process*

While the influence of neutrino interactions on supernova nucleosynthesis has been emphasized for many years only recently a new nucleosynthesis mechanism involving neutrinos has been identified to operate in core collapse supernovae. Recent core collapse supernova simulations with accurate neutrino transport (Liebendörfer *et al.* 2001, Buras *et al.* 2003, Thompson *et al.* 2005) show the presence of proton-rich neutrino heated matter, both in the inner ejecta (Liebendörfer *et al.* 2001, Buras *et al.* 2003) and the early neutrino wind from the proto-neutron star (Buras *et al.* 2003). This matter, part of the initially shock heated material located between the surface of the proto-neutron star and the shock front expanding through the outer layers, is subject to a large neutrino energy deposition heating the matter. This and the expansion, lifting the electron degeneracy, make it possible for the reactions $\nu_e + n \leftrightarrow p + e^-$ and $p + \bar{\nu}_e \leftrightarrow n + e^+$ (i.e. neutrino and antineutrino captures on free nucleons and their inverse reactions, electron and positron capture) to drive the composition proton-rich (Fröhlich *et al.* 2005, Pruet *et al.* 2005, Fröhlich *et al.* 2006a), i.e. the electron fraction $Y_e > 0.5$. This effect will always be present in successful explosion with ejected matter irradiated by a strong neutrino flux, independent of the details of the explosion. While this matter expands and cools, nuclei can form. This results in a composition dominated by $N = Z$ nuclei, mainly ^{56}Ni and ^{4}He, and protons. Without the further inclusion of neutrino and antineutrino reactions the composition of this matter will finally consist of protons, alpha-particles, and heavy (Fe-group) nuclei, i.e. a proton- and alpha-rich freeze-out that results in enhanced abundances of ^{45}Sc, ^{49}Ti, and ^{64}Zn (Fröhlich *et al.* 2005, Pruet *et al.* 2005, Fröhlich *et al.* 2006a).

Traditional explosive (supernova) nucleosynthesis calculations did not include interactions with neutrinos and antineutrinos. The heaviest nuclei synthesized in these calculations have a mass number $A = 64$. The matter flow stops at the nucleus ^{64}Ge which has a small proton capture probability and a beta-decay half-life (64s) that is much longer than the expansion time scale (10s) (Pruet *et al.* 2005). When reactions with neutrinos and antineutrinos are considered for both free and bound nucleons the situation becomes dramatically different (Fröhlich *et al.* 2006b, Pruet *et al.* 2006, Wanajo 2006).

$N \sim Z$ nuclei are practically inert to neutrino capture (converting a neutron into a proton) because such reactions are endoergic for neutron-deficient nuclei located away from the valley of stability. The situation is different for antineutrinos that are captured in a typical time of a few seconds, both on protons and on nuclei, at the distances at which nuclei form (~ 1000 km). As protons are more abundant than heavy nuclei, antineutrino capture occurs predominantly on protons, causing a residual density of free neutrons of $10^{14} - 10^{15}$ cm^{-3} for several seconds when the temperatures are in the range 1–3×10^9 K. This effect is clearly seen in Figure 1 of Fröhlich et al. 2006a where the time evolution of the abundances of protons, neutrons, alpha-particles, and ^{56}Ni is shown for a trajectory of the model B07. The solid (dashed) lines display the nucleosynthesis results which include (omit) neutrino and antineutrino absorption interactions after nuclei are formed. ^{56}Ni serves to illustrate when nuclei are formed. The difference in proton abundances between both calculations is due to antineutrino captures on protons, producing neutrons which drive the νp-process. Without the inclusion of antineutrino captures the neutron abundance soon becomes too small to allow for any capture on heavy nuclei.

The neutrons produced via antineutrino absorption on protons can easily be captured by neutron-deficient $N \sim Z$ nuclei (for example ^{64}Ge) which have large neutron capture cross sections. While proton capture, (p,γ), on ^{64}Ge takes too long or is impossible, the (n,p) reaction dominates, permitting the matter flow to continue to nuclei heavier than ^{64}Ge via subsequent proton captures with freeze-out close to 1×10^9 K.

Figure 2 of Fröhlich et al. 2006a shows the results for the composition of supernova ejecta from one hydrodynamical model which includes neutrino absorption reactions in the nucleosynthesis calculations (filled circles) that lead initially to proton-rich conditions in the innermost zones, experiencing afterwards the νp-process. These abundances are compared to an older set of nucleosynthesis calculations (open circles, Thielemann et al. 1996) that did not include neutrino interactions and therefore did not produce the proton-rich matter resulting in models with accurate neutrino transport (Liebendörfer et al. 2001, Buras et al. 2003, Thompson et al. 2005). In later phases of the cooling proto-neutron star neutrino interactions will cause neutron-rich ejecta. Whether this permits weak or strong r-process is still debated (Thompson 2003).

3. Evidence for a Lighter Element Primary Process (LEPP)

Metal-poor stars in the Galactic halo provide us with a laboratory to study the earliest Galactic nucleosynthesis processes (Cowan & Sneden 2007 and references therein). Their chemical composition also provides us with hints about the nature of the very first generation of stars. Early observations (e.g. Spite and Spite 1978, Gilroy et al. 1988, Gratton & Sneden 1994, Burris et al. 2000) focussed on stars with metallicites of [Fe/H]< −1 and rare earth elements which are easily detectable with ground-based telescopes. However, with the latest generation of telescopes and with space based telescopes, such as the *Hubble Space Telescope* (HST), a much larger range of n-capture elements can be examined. The most metal-poor stars observed to date, HE0107-5240 (Christlieb et al. 2002) and HE1327-2326 (Frebel et al. 2005), with metallicities below [Fe/H]< −5 are enriched in C, N, and O but very poor in n-capture elements. However, the detection of Sr/Fe (exceeding 10 times the solar ratio) in the most metal-poor star (HE1327-2326, Frebel et al. 2005) suggests the existence of a primary process producing elements beyond Fe and Zn.

Based on abundance data from HST observations of metal-poor galactic halo stars (Cowan et al. 2005) analyze the behavior of Ge versus metallicity [Fe/H]. The observed trend (see Figs. 4 and 5 of Cowan et al. 2005) is consistent with an explosive (or charged particle) synthesis for Ge. A similar comparison for Zr is somewhat less conclusive but nevertheless seems to indicate a different synthesis origin for these two elements.

François et al. 2007 obtained abundances for 16 n-capture elements from a sample of 32 extremely metal-poor stars. Their measurements imply that not all n-capture elements in metal-poor stars were produced by a single r-process. From this they conclude that an additional process must contribute mainly to the production of the first peak elements in very metal-poor stars and extremely metal-poor stars.

Recent galactic chemical evolution studies (Travaglio *et al.* 2004a) of the Galactic enrichment of Sr, Y, and Zr (using homogeneous chemical models) suggest the existence of additional primary process, denoted *lighter element primary process* (LEPP), to explain the observed abundances. This process is independent of the r-process and operates very early in the Galaxy. Travaglio *et al.* 2004a point to massive stars as likely site for this process. The mean residuals of Sr, Y, and Zr in the François *et al.* 2007 sample (based on the Solar-system r-process abundances by Arlandini *et al.* 1999) show that this LEPP process is responsible for 90–95% of the total abundance of these elements at [Ba/H]$\simeq -4.3$. In a recent paper by Montes *et al.* 2007 the authors derive an abundance pattern of the LEPP and explore the possibility of a neutron-capture process.

A candidate for the LEPP is the νp-process (Fröhlich *et al.* 2006b). This nucleosynthesis process will occur in all core collapse supernovae and could explain the existence of Sr and other elements beyond Fe in the very early stage of galactic evolution. This process can also contribute to the nucleosynthesis of light p-process elements.

References

Arlandini, C, Käppeler, F., Wishak, K., *et al.* 1999 *ApJ*, 525, 886
Arnould, M. & Goriely, S. 2003, *Phys. Rep.*, 384, 1
Blinnikov, S. 2006, *Surv. High Energy Phys.*, 20, 89
Buras, R., Rampp, M., Janka, H.-T., *et al.* 2003, *Phys. Rev. Letters*, 90, 24, 241101
Burris D. L., Pilachowski, C. A., Armandroff, T. E. *et al.* 2000, *ApJ*, 544, 302
Cayrel, R., Depagne, E., Spite, M. *et al.* 2004 *A&A*, 416, 1117
Christlieb, N., Bessell, M. S., Beers, T. C. *et al.* 2002, *Nature*, 419, 904
Costa, V. Rayet, M. Zappala, R. A. *et al.* 2000, *A&A*, 358, L67
Cowan, J. J., & Sneden, C. 2007, *Nature*, 440, 1151
Cowan, J. J., Sneden, C., Beers, T. C. *et al.* 2005, *ApJ*, 627, 238
Dauphas, N., Rauscher, T., Marty, B. *et al.* 2003, *Nucl. Phys. A*, 719, C287
Domínguez, I., Höflich, P., and Straniero, O. 2001, *ApJ*, 557, 279
François, P., Depagne, E., Hill, V. *et al.* 2007, *A&A*, 476, 935
François, P., Matteucci, F. & Cayrel, R. *et al.* 2004, *A&A*, 421, 613
Frebel, A., Aoki, W., Christlieb, N., *et al.* 2005, *Nature*, 434, 871
Fröhlich, C., Hauser, P., Liebendörfer, M. *et al.* 2005, *Nucl. Phys. A*, 758, 27
Fröhlich, C., Hauser, P., Liebendörfer, M. *et al.* 2006, *ApJ*, 637, 415
Fröhlich, C., Martínez-Pinedo, G., Liebendörfer, M. *et al.* 2006, *Phys. Rev. Letters*, 96, 14, 142502
Gilroy, K. K., Sneden, C., Pilachowski, C. A., & Cowan, J. J. 1988, *ApJ*, 327, 298
Gratton, R. G. & Sneden, C. 1994, *A&A*, 287, 927
Gratton, R. G. & Sneden, C. 1991, *A&A*, 241, 501
Hayakawa, T., Iwamoto, N., Shizuma, T. *et al.* 2004, *Phys. Rev. Lett.*, 93, 16, 161102
Heger, A., Fryer, C. L., Woosley, S. E. *et al.* 2003, *ApJ*, 591, 288
Hirschi, R., Meynet, G., and Maeder, A. 2006,
Hoffman, R. D., Woosley, S. E., Weaver *et al.* 1999, *ApJ*, 521, 735
Khokhlov, A. M., Höflich, P. A., Oran, E. S. *et al.* 1999, *ApJ Lett.*, 524, L107
Langanke, K. & Martínez-Pinedo, G. 2000, *Nucl. Phys. A*, 673, 481
Liebendörfer, M., Mezzacappa, A., & Thielemann *et al.* 2001, *Phys. Review D*, 6310, 103004
Liebendörfer, M., Mezzacappa, A., Messer, O. E. B. *et al.* 2003, *Nucl. Phys. A*, 719, 144
Livio, M., Panagia, N., & Sahu, K. (eds.), 2001, *Supernovae and gamma-ray bursts: the greatest explosions since the Big Bang* (Cambridge: CUP)
Lodders, K. 2003, *ApJ*, 591, 1220

Montes, F. Beers, T. C., Cowan, J. *et al.* 2007, *ApJ*, 671, 1685
Nakamura, T., Umeda, H., Iwamoto, K. *et al.* 2001, *ApJ*, 555, 880
Nakamura, T., Umeda, H., Nomoto, K. *et al.* 1999, *ApJ*, 517, 193
Nomoto, K., Tominaga, N., Umeda, H. *et al.* 2006, *Nucl. Phys. A*, 777, 424
Nomoto, K., Hashimoto, M., Tsujimoto, T. *et al.* 1997, *Nucl. Phys. A*, 616, 79
Pruet, J., Hoffman, R. D., Woosley, S. E. *et al.* 2006, *ApJ*, 644, 1028
Pruet, J., Woosley, S. E., Buras, R. *et al.* 2005, *ApJ*, 623, 325
Rauscher, T., Heger, A., Hoffman, R. D. *et al.* 2002, *ApJ*, 576, 323
Sneden, C. & Cowan, J. J. 2003, *Science*, 299, 70
Spite, M. & Spite, F. 1978, *A&A*, 67, 23
Thielemann, F.-K., Nomoto, K., & Hashimoto, M. 1996, *ApJ*, 460, 408
Thompson, T. A. 2003, *ApJ*, 585, L33
Thompson, T. A., Quataert, E., & Burrows, A. 2005, *ApJ*, 620, 861
Timmes, F. X., Woosley, S. E., & Weaver, T. A. 1995, *ApJ Suppl.*, 98, 617
Travaglio, C., Gallino, R., Arnone, E. *et al.* 2004, *ApJ*, 601, 864
Travaglio, C., Hillebrandt, W., Reinecke, M. *et al.* 2004, *A&A*, 425, 1029
Wanajo, S. 2006, *ApJ*, 647, 1323
Woosley, S. E. 1993, *ApJ*, 405, 273
Woosley, S. E. & Weaver, T. A. 1994, in: Bludman, S. A., Mochkovitch, R., & Zinn-Justin, J. (eds.), *Supernovae*, (Amsterdam: Elsevier Science), *NATO Advanced Sci Inst Ser C*, 63
Woosley, S. E. & Weaver, T. A. 1995, *ApJ Suppl.*, 101, 181

Discussion

LIMONGI: Which is the main reason for the increase of [Sc/Fe] and [Cu/Fe] in your models?

FRÖHLICH: The [Sr/Fe] and [Cu/Fe] ratios in our models are increased compared to earlier calculations, e.g. Thielemann *et al.* (1996), due to the proton-rich material which emerges from the core collapse simulation. Even if we switch off the neutrino interactions in the proton-rich ejecta we still get these increased ratios for Sc and Cu which are consistent with observations of metal-poor stars.

MAEDER: The elements Sr, Y, Zr are typical s-elements which can be created by the s-process during the He-burning phase of low Z massive stars and not necessarily in the supernova explosions. In my opinion, these excesses of Sr,Y,Zr are related to the N-excesses (as well as C,O) prominently observed in this extreme star (Frebel star).

FRÖHLICH: Recent observations of extremely metal-poor stars e.g. by François *et al.* show clear indications for the need of an additional primary process (independent of the r-process) to explain the observed abundance pattern in these stars. The question is down to which metallicity the s-process, being a secondary process, can still operate.

Evidence for a Mass Outflow from Our Galactic Center

Casey Law

Astronomical Institute "Anton Pannekoek",
University of Amsterdam,
Kruislaan 403, Amsterdam, 1098 SJ, Netherlands
email: claw@science.uva.nl

Abstract. We discuss the nature of the Galactic center lobe (GCL), a degree-tall, loop-like structure apparently erupting from the central few hundred parsecs of our Galaxy. Although its coincidence with the Galactic center has inspired diverse models for its origin, the observational evidence connecting this structure to the GC region has been thin. We describe a multiwavelength observing campaign with the VLA, GBT, *Spitzer*, and other telescopes that finds compelling evidence that the structure is likely formed by a mass outflow from the central tens of parsecs of our Galaxy. The size and mass of the putative outflow is consistent with that expected from the observed supernova rate and gas pressure in the GC region. If the GCL is a mass outflow, its relative proximity offers a unique opportunity for studying these structures in unprecedented detail.

Keywords. Galaxy: center – ISM: jets and outflows – galaxies: starburst

1. Introduction

Dynamical processes make galactic nuclei host to massive black holes, stellar clusters, and dense molecular gas. Energetic jets and bursts of star formation are associated with massive outflows of gas and dust out of galaxies (e.g., NGC 3079; Cecil *et al.* 2001). Understanding the effect of these outflows impacts upon a wide range of topics, from galaxy evolution to the enrichment of the intergalactic medium.

While massive outflows are typically seen in more massive and active galaxies, our own Galactic center (GC) may host mass outflows, judging from the presence of several unusual properties. Massive stellar clusters ($M_{cl} \sim 10^4 M_\odot$) have been found there with properties similar to those seen in starburst galaxies (Figer *et al.* 1999). The ionized gas excitation conditions are similar to that seen in minor starbursts (Rodríguez-Fernández & Martín-Pintado 2004). The center of the Galaxy is known to have a massive ($\sim 4 \times 10^6 M_\odot$) black hole, such as found in active galaxies (Ghez et al. 2005). If the GC is host to an outflow, it would provide us with the closest view of this phenomenon, helping us understand them.

One possible signature of an outflow from our GC is the Galactic center lobe (GCL). The GCL was first discovered in radio continuum observations (Sofue & Handa 1984). As shown in Figure 1, the GCL is a $\sim 1°$ tall shell that spans the central ~ 100 pc of the GC region. Based on the coincidence of the shell with the GC and its morphology, it was suggested that the GCL is a sign of a mass outflow from the GC region. However, these ideas were founded mostly on morphology; the early observations allowed for the possibility that the east and west sides were unrelated to each other (Uchida *et al.* 1994). Bland-Hawthorn & Cohen (2003) resurrected the idea that the GCL is caused by a mass outflow by noting the presence of dust emission surrounding the structure in MSX 8 μm data.

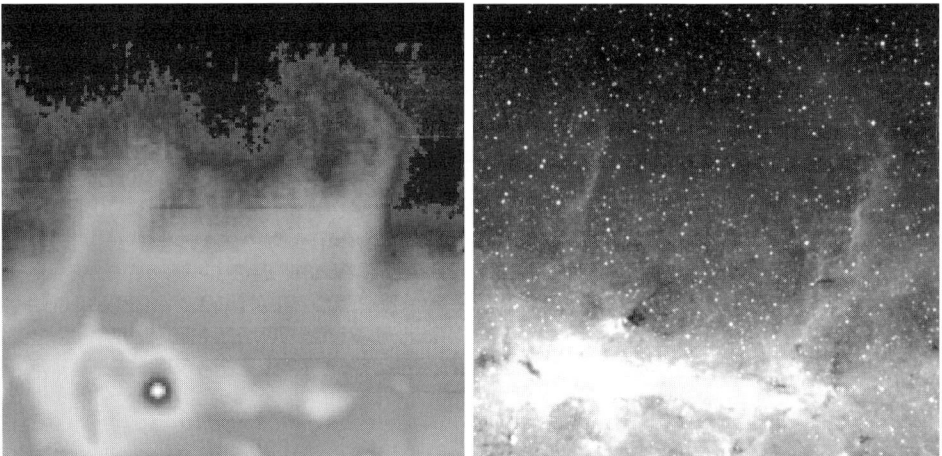

Figure 1. *Left:* Green Bank Telescope (GBT) 6 cm radio continuum image of the central two degrees of the GC region. The east and west sides of the GCL are seen here. *Right:* Midcourse Space Experiment (MSX) 8 µm continuum image of the same region.

To address earlier questions about the nature of the GCL and quantify its physical conditions, we conducted a series of observations. This work summarizes the results of this multiwavelength study of the GCL and presents evidence that it is a signature of a GC mass outflow.

2. New Results

We have collected and compared observations of the GCL by conducting new observations and searching public data archives. This study has revealed new aspects of the structure of the GCL. Here we summarize the available observations of the GCL; a more complete description will be given elsewhere (Law *et al.* 2008).

2.1. *GBT Radio Continuum at 3.6, 6 and 20 cm*

The radio continuum of the central few degrees of the GC region was surveyed at 3.6, 6, and 20 cm wavelengths with the GBT. These observations resolved the radio continuum shell of the GCL that was first noted by Sofue & Handa (1984). The east and west sides of the shell are connected about 1.4° north of the Galactic plane (not visible in Fig. 1).

A key result of the radio continuum observations is a clear measurement of the 3.6/6 cm and 6/20 cm radio spectral indices. Figure 2 shows how these spectral indices change with Galactic latitude for the east and west sides of the GCL. The values are generally nonthermal and steepen away from the plane. The similarity in the values for the two sides shows that they are very likely created by similar phenomena. The contiguous change in spectral index and shape also connects the GCL to the nonthermal filament known as the Radio Arc (Yusef-Zadeh, *et al.* 1984), an object known to be in the central few hundred parsecs of our Galaxy.

2.2. *GBT and Hat Creek Radio Observatory recombination lines near 6 cm*

The ionized gas structure was studied with two separate observations of radio recombination lines (HCRO data courtesy D. Backer). The observations find a shell-like morphology that is similar to radio continuum and mid-IR shells. The GBT observations show that the density and temperature of the ionized gas is similar throughout the GCL. The

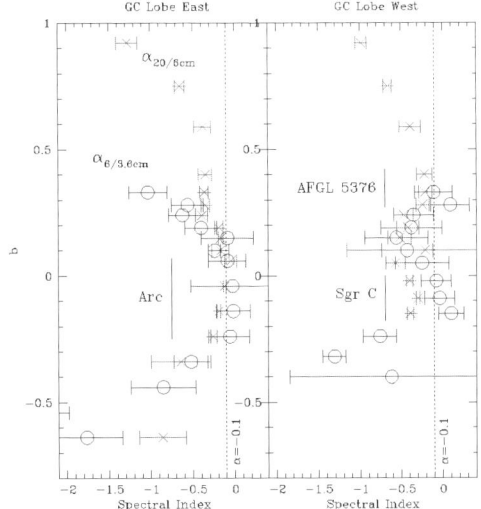

Figure 2. The plots show the dependence of the 6/20 cm and 3.6/6 cm spectral indices on Galactic latitude across the GCL. The left and right panel show the east and west sides of the GCL, respectively.

electron temperature in the ionized gas is unusually low, indicating a high metallicity, which is consistent with the enriched medium in the GC region.

2.3. SHASSA Hα image

A different view of the ionized gas is provided by the SHASSA Hα survey (Gaustad et al. 2001). Figure 3 shows a combined view of that survey with the HCRO radio recombination line survey, which covers lower latitudes. These maps show that there is a ionized shell of gas that matches the radio continuum and mid-IR shells over the entire structure.

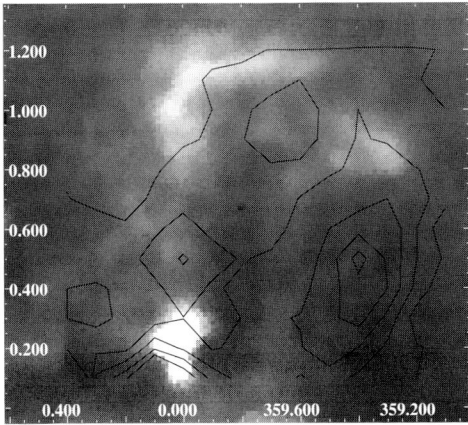

Figure 3. Two surveys of the ionized gas in the GCL. The gray scale shows Hα emission (SHASSA; Gaustad et al. 2001) and contours show 6 cm radio recombination line intensity observed by the HCRO (D. Backer, private communication). The two surveys show ionized gas, but are both biased differently by absorption and incomplete coverage. When combined, they provide a more complete view of the ionized gas in the GCL.

2.4. Spitzer/IRAC 8 μm images

The GCL was surveyed in part by a *Spitzer* survey of the GC region (Stolovy *et al.* 2008 in prep.) and the GLIMPSE survey (Benjamin *et al.* 2005). These surveys show that the mid-IR emission at arcsecond resolution traces a complete shell, much like that seen in radio continuum and recombination lines. This emission is dominated by PAH emission, which is caused by irradiation of dust.

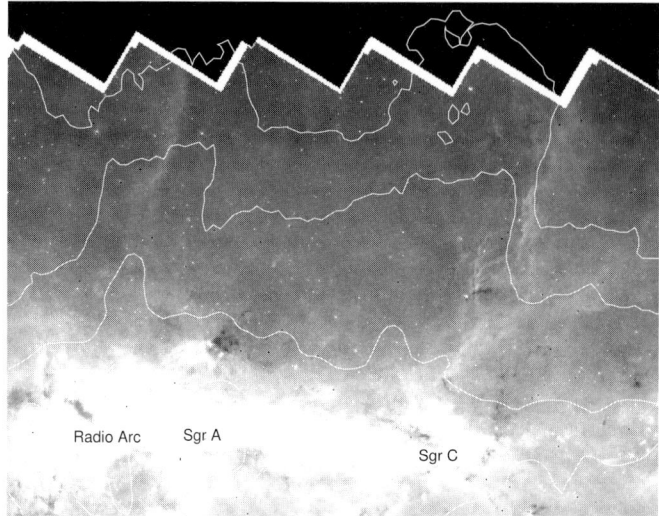

Figure 4. The 8 μm emission in the region of the GCL, as observed by *Spitzer* (from Stolovy *et al.* 2008).

2.5. Nature of GCL

The observations show that the GCL is composed of nested shells of gas and dust. There are three, morphologically-distinct components to the GCL, tracing synchrotron-emitting, magnetized gas (radio continuum), ionized gas (radio recombination and Hα lines), and PAH-emitting dust (8 μm continuum). Figure 5 shows how these three components align relative to each other.

The full analysis leads us to a few basic conclusions about the GCL:

• The GCL is a single object: The distinct morphology of the shell and the connection of the east and west sides at the top of the shell argues that the GCL has a single origin.

• The GCL is composed of layered shells: The multiple components are clearly distinct at arcminute size scales.

• The GCL is in the GC region: The connection of the GCL to the Radio Arc (Yusef-Zadeh *et al.* 1984) and the low electron temperature argue that the entire structure is in the central few hundred parsecs of the Galaxy.

Assuming the GCL is in the GC region (at a distance of roughly 8 kpc), the pressure measured by thermal x-rays, radio recombination line, and others means implies a formation energy $E \approx 4 \times 10^{52}$ ergs, assuming adiabatic expansion. Upper limits on the expansion velocity from radio recombination line observations show that the shell is not expanding along the line of sight faster than roughly 10 km s^{-1}.

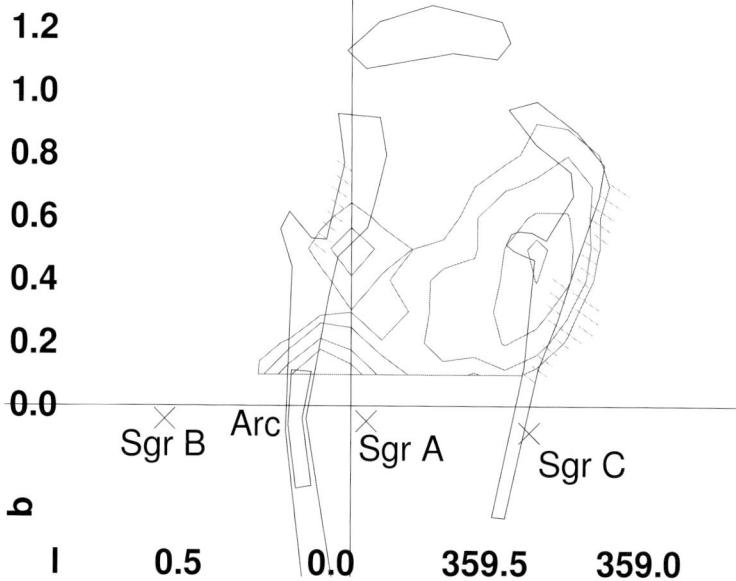

Figure 5. A schematic view of the morphology of the GCL. Red contours show radio recombination line emission. Blue line schematically shows the radio continuum emission. The green hatching schematically shows where the 8 μm emission lies.

3. Formation Models for the GCL

The physical structure of the GCL suggests that it is an outflow from the center of the Galaxy. Starting with this assumption, we can test models for its formation. The four standard models presented for the formation of the GCL are:

• **A small-scale starburst** (Chevalier 1992, Bland-Hawthorn & Cohen 2003) — Stellar winds and supernovae create a hot, buoyant bubble of gas that rises from the disk.

• **A jet from Sgr A*** (Melia & Falcke 2001) — Jets from accreting massive black holes are often found to create mass outflows.

• **Escape of 8 keV gas** (Muno et al. 2004, Belmont et al. 2005) — The GC region is believed to be filled with a hot gas of mysterious origin that may buoyantly escape.

• **A magnetodynamic effect** (Uchida et al. 1985, Shibata & Uchida 1987) — The inward migration of gas may twist the GC magnetosphere, possibly forcing ionized gas away from the plane.

We find that the morphology and physical conditions in the GCL are most consistent with the starburst model. The observed supernova rate in the central 20 pc ($\sim 10^{-5}$ yr^{-1}; Muno et al 2004) could power the GCL after 10^7 yr (for $\chi \approx 0.2$). The GCL is similar to the outflow seen in dwarf galaxy IC 10, which has layered nonthermal and thermal shells with no apparent expansion (Thurow & Wilcots 2005). In general, the formation energy ($\sim 4 \times 10^{52}$ ergs) and ionized mass (3×10^5 M$_\odot$) are similar that seen in dwarf starbursts (Veilleux et al. 2005). Furthermore, the upper limit on the expansion of the GCL is consistent with predictions of scaling relations, given the moderate star formation rate in the GC region and the GCL's small size (Martin 2005).

4. Conclusions

This work has presented new observations that show that the Galactic center lobe is a layered, shell-like structure in our Galactic center. The energy required to form it is consistent with the current supernova rate in the GC region, which suggests that it could be formed by the current star formation there. If so, the small size of the GCL and moderate star formation rate that powers it suggest that this kind of outflow is common. Furthermore, such an outflow would not easily be detected in the nearest spiral galaxies. Thus, the proximity of the GCL provides a unique perspective of this exciting phenomenon.

References

Belmont, R., Tagger, M., Muno, M. et al. 2005, *ApJ*, 631, L53
Benjamin, R. A., Churchwell, E., Babler, B. L. et al. 2005, *ApJ*, 630, L149
Bland-Hawthorn, J. & Cohen, M. 2003, *ApJ*, 582, 246
Cecil, G., Bland-Hawthorn, J., Veilleux, S. et al. 2001, *ApJ*, 555, 338
Chevalier, R. A. 1992, *ApJ*, 397, L39
Eckart, A. & Genzel, R. 1996, *Nature*, 383, 415
Figer, D. F., McLean, I. S., & Morris, M. 1999, *ApJ*, 514, 202
Figer, D. F., Rich, R. M., Kim, S. S. et al. 2004, *ApJ*, 601, 319
Gaustad, J. E., McCullough, P. R., Rosing, W., & Van Buren, D. 2001, *PASP*, 113, 1326
Ghez, A., Salim, S., Hornstein, S. D. et al. 2005, *ApJ*, 620, 744
Law, C. J., et al. 2008, *in preparation*
Martin, C. L., 2005 *ApJ*, 621, 227
Melia, F. & Falcke, H., 2001 *ARA&A*, 39, 309
Muno, M. P. Baganoff, F. K., & Bautz, M. W., 2004 *ApJ*, 613, 326
Rodríguez-Fernández, N. J. & Martín-Pintado, J., 2005 *A&A*, 429, 923
Shibata, K. & Uchida, Y., 1987 *PASJ*, 39, 559
Sofue, Y. & Handa, T., 1984 *Nature*, 310, 568
Thurow, J. C. & Wilcots, E. M., 2005 *AJ*, 129, 745
Uchida, K. I., Morris, M. R., Serabyn, E. & Bally, J., 1994 *ApJ*, 421, 505
Uchida, Y., Sofue, Y. & Shibata, K., 1985 *Nature*, 317, 699
Veilleux, S., Cecil, G., & Bland-Hawthorn, J., 2005 *ARA&A*, 43, 769
Yusef-Zadeh, F., Morris, M., & Chance, D., 1984 *Nature*, 310, 557

Session V

Massive Stars as Probes of the Early Universe

Massive Stars at High Redshifts

Max Pettini

Institute of Astronomy, University of Cambridge, Cambridge, CB3 0HA, UK

Abstract. The five years that have passed since the last IAU Symposium devoted to massive stars have seen a veritable explosion of data on the high redshift universe. The tools developed to study massive stars in nearby galaxies are finding increasing application to the analysis of the spectra of star-forming regions at redshifts as high as $z = 7$. In this brief review, I consider three topics of relevance to this symposium: the determination of the metallicities of galaxies at high redshifts from consideration of their ultraviolet stellar spectra; constraints on the initial mass function of massive stars in galaxies at $z = 2 - 3$; and new clues to the nucleosynthesis of carbon and nitrogen in massive stars of low metallicity. The review concludes with a look ahead at some of the questions that may occupy us for the next five years (at least!).

Keywords. galaxies: abundances – quasars: absorption lines – stars: abundances – stars: Wolf-Rayet

1. Introduction

The fact that a whole session of IAU Symposium 250 is devoted to the high redshift universe is an eloquent demonstration of how much our observational capabilities have improved in recent years. When we record the spectra of galaxies at redshifts $z > 1$ with our ground-based optical and infrared telescopes we of course look directly at their rest-frame ultraviolet (UV) emission that is dominated by massive stars. Thus, many of the ideas and techniques which we have discussed in the last four days find immediate application at high redshift, thereby reinforcing the already strong link between the stellar and extragalactic communities of astronomers.

In the last five years, since the last IAU Symposium devoted to massive stars in Lanzarote, studies of high redshift galaxies have proliferated. A number of different large-scale surveys have given us samples of hundreds or thousands of galaxies at redshifts from zero all the way to the end of the 'Dark Ages' at $z \sim 7$. A clear illustration of such progress is provided by Figure 1. The small size of the error bars in each bin of the luminosity functions attests to the improvements in the number statistics afforded by the large scale of galaxy surveys.

Out of this considerable body of recent work on high redshift galaxies I shall consider three main topics of particular relevance to this meeting: (i) determinations of element abundances; (ii) the initial mass function of massive stars; and (iii) new clues to the nucleosynthesis of C, N, and O in the first few episodes of star formation.

2. Abundance Determinations in High Redshift Galaxies

The degree to which galaxies at high redshifts have processed their interstellar media into stars, and the attendant detailed pattern of element abundances, are key physical characteristics that give us the means to assess their evolutionary status and link them to today's galaxy populations. To this end, a great deal of effort has been devoted to developing different metallicity diagnostics applicable at high redshifts—a review is given by Pettini (2006).

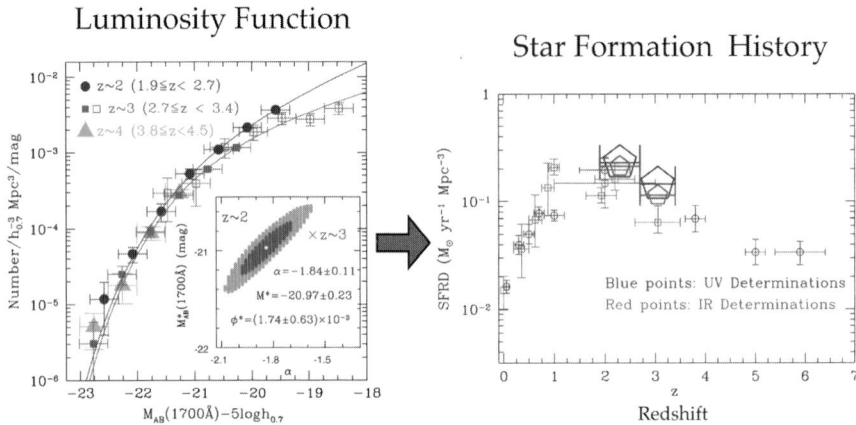

Figure 1. Figures reproduced from Reddy *et al.* (2008) showing the most recent and comprehensive determinations of: *(left)* the galaxy luminosity function at redshifts $z = 2-4$, and *(right)* the redshift evolution of the comoving star formation rate density.

Metallicities of galaxies at $z > 1$ have been deduced from consideration of *(i)* nebular emission lines formed in ionised gas at the sites of star formation; *(ii)* interstellar absorption lines which probe the more widely distributed interstellar medium; and *(iii)* spectral features from the photospheres and winds of massive stars. It is reassuring that these different approaches give concordant answers, to within a factor of about two, in the few cases where it has so far been possible to cross-check them in the same galaxy.

Of these methods, the one which has been found the widest application up to now is the first. For practical reasons, the ratio of two closely spaced emission lines, Hα and [N II] $\lambda 6583$, (Pettini & Pagel 2004) has proved to be particularly useful for obtaining a first, approximate measure of the abundance of oxygen in galaxies at $z \simeq 2$. In conjunction with estimates of the assembled stellar mass deduced from consideration of the rest-frame ultraviolet to near-infrared spectral energy distribution, this has led to the first determination of a mass-metallicity relation in galaxies at these redshifts, illustrated in Figure 2.

A general conclusion seems to be that UV-bright, star-forming, galaxies at $z \simeq 2$ were already at an advanced stage of chemical evolution: 85% of the galaxies in the Erb *et al.* (2006) sample have an oxygen abundance greater than, or equal to, 2/5 of that of the Orion nebula. There appears to have been only mild redshift evolution in the mass-metallicity relation over the last 3/4 of the age of the universe: some 10 billion years ago, galaxies of a given stellar mass were less metal-rich by only a factor of about two compared to the galaxies around us today as mapped by the Sloan Digital Sky Survey.

One of the difficulties in deriving abundances from nebular emission lines is that the strongest and best calibrated features all occur at rest-frame optical wavelengths which are redshifted into the near-infrared at $z > 1$, a wavelength regime where observations are complicated by a high (and variable) sky background and by the lack (until now) of multi-object spectrographs. There is therefore a strong incentive to try and develop abundance measures based on the spectral features from massive stars which dominate the rest-frame ultraviolet spectra of high-z galaxies—readily accessible from the ground at redshifts $z > 1$. This is at the approach adopted by Rix *et al.* (2004), partly stimulated by discussions at the Lanzarote meeting five years ago (see Figure 3).

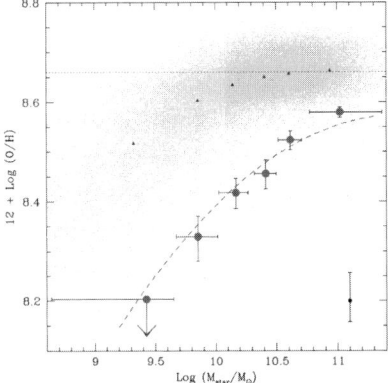

Figure 2. Stellar mass-metallicity relation for star-forming galaxies at $z \sim 2$ (large red circles with error bars) and in the nearby ($z \sim 0.1$) universe from the SDSS survey (small grey points). The blue triangles show the mean metallicity of the SDSS galaxies in the same mass bins as the $z = 2$ galaxies. The vertical error bar in the bottom right-hand corner indicates the uncertainty in the calibration of the $N2$ index (Pettini & Pagel 2004) used to deduce the oxygen abundance of both sets of galaxies. The horizontal dotted line is drawn at the solar abundance of oxygen. (Figure reproduced from Erb *et al.* 2006).

As pointed out by Leitherer *et al.* (2001), the spectral synthesis code *Starburst99* provides the machinery required to assess which stellar spectral features are most sensitive to changes in metallicity. However, its usefulness in this respect is limited by the lack of observations of OB stars of different metallicities—the currently available empirical libraries come in only two flavours: 'Milky Way' and 'Magellanic Cloud' (a hybrid of Large and Small Magellanic Cloud stars). To overcome this problem, Rix *et al.* (2004) turned to the fully theoretical massive star spectra generated by the *W-M Basic* code developed in Munich by Pauldrach *et al.* (2001) and used them to synthesise fully theoretical composite spectra of star-forming galaxies spanning a wide range of metallicities. Some examples are reproduced in Figure 4.

This approach turned out to be very successful. The *W-M Basic* galaxy spectra match closely those generated with the empirical libraries at the metallicities and wavelengths common to the two versions of *Starburst99*, giving us confidence in the extrapolations to

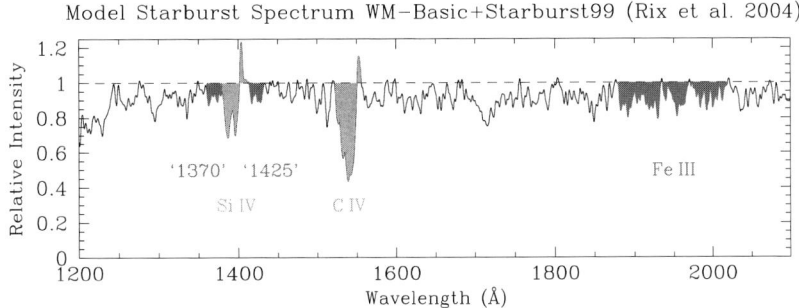

Figure 3. Fully theoretical UV spectrum of a starburst galaxy computed by Rix *et al.* (2004) for the case of continuous star formation with a Salpeter initial mass function (IMF) and solar metallicity. The shaded regions indicate stellar wind lines (green) and blends of photospheric lines (red) found by Rix *et al.* to be sensitive to metallicity.

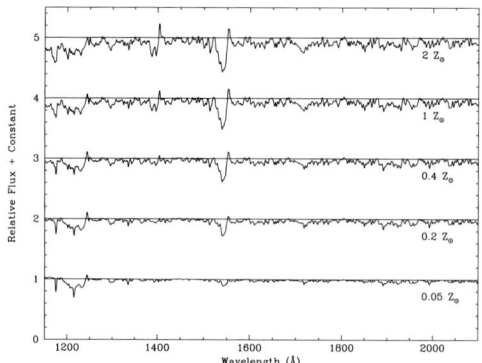

Figure 4. Fully synthetic UV spectra of star-forming galaxies with metallicities $Z = 2-0.05 Z_\odot$, from the work by Rix *et al.* (2004). All the examples shown here are for the 'standard' case of continuous star formation (at age 100 Myr) with a Salpeter slope of the IMF between 1 and 100 M$_\odot$. The spectra have been convolved with a Gaussian of FWHM = 2.5 Å (to match the typical rest-frame resolution of observed spectra of high-z galaxies) and normalised to the underlying stellar pseudo-continuum. See Rix *et al.* (2004) for additional details.

other metallicities and wavelength ranges (the empirical libraries are limited to a rather small portion of the UV spectrum). When compared to the best available spectra of

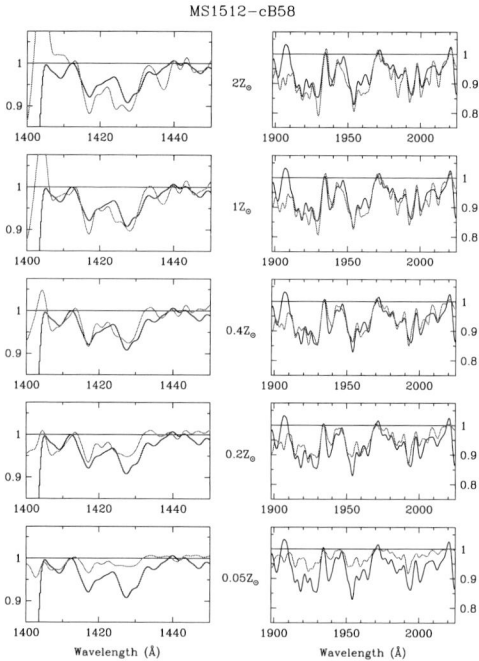

Figure 5. Comparison of the observed spectrum of the $z = 2.7276$ Lyman break galaxy MS1512-cB58 (*black*) with fully synthetic spectra produced by the *Starburst99+W-M Basic* combination. The left-hand set of panels shows the region near 1425 Å which includes a blend of Si, C, and Fe photospheric lines, while the right-hand set is for the blend of Fe III lines near 1978 Å produced in the photospheres of late O- and early B-type stars. The y-axis is residual intensity. (Figure reproduced from Rix *et al.* 2004).

high-z star-forming galaxies (Figure 5), the *W-M Basic* version of *Starburst99* scores very highly, in being able to reproduce the fine detail of blends of photospheric features centred near 1425 Å and 1978 Å. As can be seen from Figure 5, the closest match between model and observed spectra of MS1512-cB58 (a strongly lensed Lyman break galaxy at $z = 2.7276$) in these wavelength regions is obtained for a stellar metallicity of $\sim 2/5$ solar (middle panels). This value is in excellent agreement with the metallicity of the interstellar medium of this galaxy, as determined from absorption lines from H I regions (Pettini *et al.* 2002) and nebular lines from H II regions (Teplitz *et al.* 2000).

These results are very encouraging in offering an additional avenue to determining the metallicity of high-z galaxies, particularly at redshifts $z > 4$ where other diagnostics may be difficult to apply. On a less optimistic note, it is clear from these initial trials that data of high signal-to-noise ratio are essential in order to correctly interpret these low-contrast blends of stellar photospheric lines (e.g. de Mello *et al.* 2004). With current instrumentation such high quality data can only be secured for a handful of strongly lensed high redshift galaxies (whose number is now increasing thanks to the *Sloan Digital Sky Survey*). However, looking further ahead, it is easy to foresee that this technique will be fully exploited with the advent of 30-m class telescopes in the next decade (see Sandro D'Odorico's contribution to this volume).

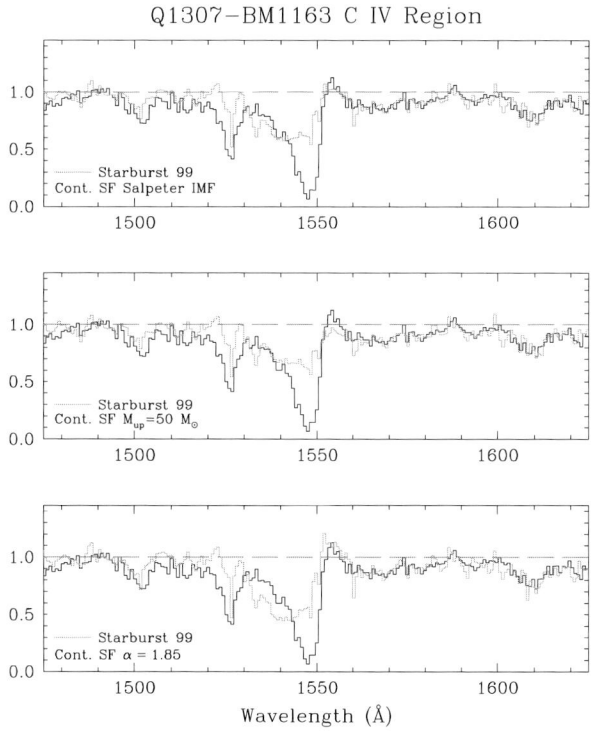

Figure 6. Sensitivity of the C IV P-Cygni emission-absorption profile to the upper end of the IMF. Black histogram: portion of the observed spectrum of Q1307-BM1163, a bright star-forming galaxy at $z = 1.409$; green histograms: model spectra produced by the *Starburst99* spectral synthesis code with, respectively, a standard Salpeter IMF (top panel), an IMF lacking stars more massive than $50 M_\odot$ (middle panel), and an IMF flatter than Salpeter (bottom panel). The changes were deliberately chosen to be relatively small to illustrate the sensitivity of the UV spectral features to the mix of stellar masses. (Figure reproduced from Steidel *et al.* 2004).

3. The Initial Mass Function in Star-Forming Galaxies at Redshifts $z = 2 - 3$

Between ~ 1000 and ~ 2000 Å, the UV spectrum of a star-forming galaxy is dominated by the light from stars more massive than about $10\,M_\odot$. Consequently, the details of such spectra—particularly the contrast of spectral features due to the most massive stars over the underlying integrated continuum—offer a direct test of the IMF of massive stars. A fair assessment of the situation so far would be to say that no evidence has yet been found for departures from a Salpeter slope at the high mass end of the IMF. An illustration is provided in Figure 6 showing a portion of the UV spectrum of the bright $z = 1.409$ galaxy Q1307-BM1163 from Steidel et al. (2004). After excluding the narrow interstellar core of the C IV line (Crowther et al. 2006), the best fit to the P-Cygni emission-absorption profile is achieved with a Salpeter IMF from 1 to $120\,M_\odot$ (top panel). Even minor changes, such as truncating the IMF at $50\,M_\odot$ (middle panel), or increasing the proportion of massive stars by adopting a flatter slope ($\alpha = 1.85$ instead of 2.35—bottom panel) result in synthetic C IV profiles that are evidently poorer matches to the observations.

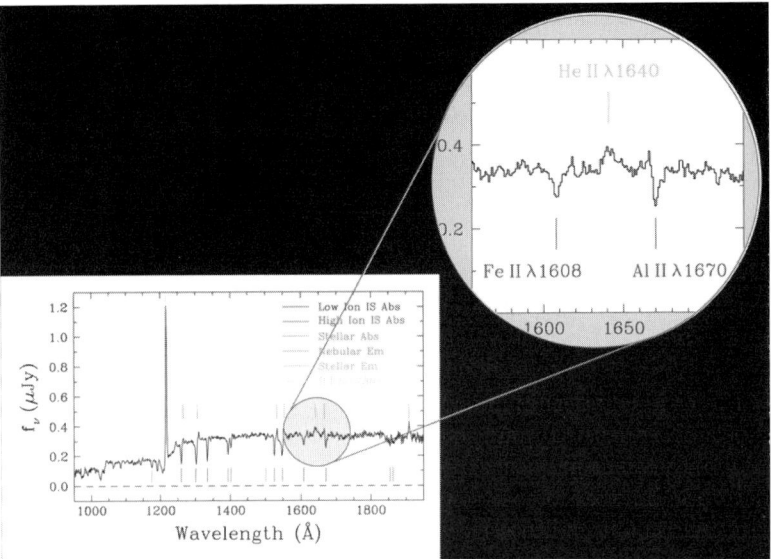

Figure 7. Composite rest-frame UV spectrum of 811 Lyman break galaxies at $\langle z \rangle = 3.0 \pm 0.3$ from Shapley et al. (2003). The co-addition of so many spectra reveals the wealth of spectral features available for analysis. The inset shows the broad He II $\lambda 1640$ emission line produced by Wolf-Rayet stars, superimposed on the integrated stellar continuum with an equivalent width EW$(\lambda 1640) = 1.3 \pm 0.3$ Å.

Another spectral feature that is regularly recognised in the spectra of Lyman break galaxies (at least when recorded at sufficiently high signal-to-noise ratio) is a broad He II $\lambda 1640$ emission line from Wolf-Rayet stars. The ubiquitous presence of this line is demonstrated by the fact that it can be readily discerned in the composite spectrum of 811 $z \simeq 3$ galaxies constructed by Shapley et al. (2003—see Figure 7). It has been suggested that a top-heavy IMF, or even a population of primordial stars (Jimenez & Haiman 2006), may be required to explain the strength of this feature. In reality, the recent analysis by Brinchmann, Pettini, & Charlot (2008), which uses up-to-date calibrations of the luminosity of this line in W-R stars of different sub-classes and different

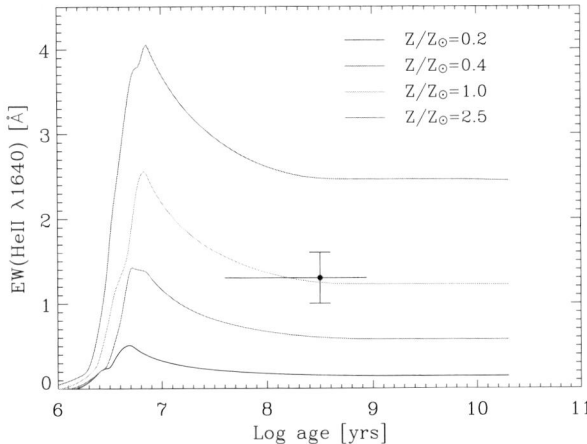

Figure 8. Time evolution of the equivalent width of the He II λ1640 emission line for four difference metallicities (from the models by Brinchmann et al. 2008). All four curves refer to a model with a constant star formation rate and Salpeter IMF. The dot at Log age [yrs] = 8.5 is the value of EW(He II λ1640) measured from the composite spectrum of 811 Lyman break galaxies published by Shapley et al. (2003—see Figure 7), while the error bar shows the inter-quartile range of their ages (Shapley et al. 2001). The measured value of EW(He II λ1640) is reproduced by the models with metallicities $Z = 0.75 - 1.5 \, Z_\odot$, somewhat higher than, but still consistent with, current estimates of the metallicity of LBGs.

metallicities (see the contributions to this volume by Paul Crowther and Lucy Hadfield), has shown that the observed EW(λ1640) = 1.3 ± 0.3 Å in the composite in Figure 7 can be naturally understood in terms of a Salpeter IMF and metallicities typical of LBGs, i.e. $Z_{\rm LBG} \simeq 0.3 - 1 Z_\odot$, as in Figure 2 (see Figure 8).

4. C, N, O Nucleosynthesis by Metal-Poor Stars

One of the topics of discussion at this meeting has been the recently recognised importance of massive stars to the nucleosynthesis not only of oxygen, but also of carbon and nitrogen, in the low metallicity regime (see the contributions to this volume by Raphael Hirschi, André Maeder, and Georges Meynet). Theoretical efforts to calculate the yields of these elements as a function of metallicity in stellar models that include the effects of rotation have been motivated by recent observations of elevated ratios of C/O and N/O in some of the most metal-poor stars in the halo of the Milky Way (Akerman et al. 2004; Spite et al. 2005).

Damped Lyα systems, or DLAs, (a class of absorption line systems in the spectra of high redshift quasars distinguished by their high column densities of neutral hydrogen) offer a complementary way to verify such trends in the relative abundances of C, N, and O (Pettini, Lipman, & Hunstead 1995). The measurements of element abundances are often more straightforward from the interstellar lines in DLAs than from the photospheres of halo stars, requiring fewer assumptions about the physical conditions and geometry of the gas in which they are formed. The main limitation is the restricted dynamic range of interstellar absorption lines—once a line becomes saturated it is no longer useful for abundance determinations.

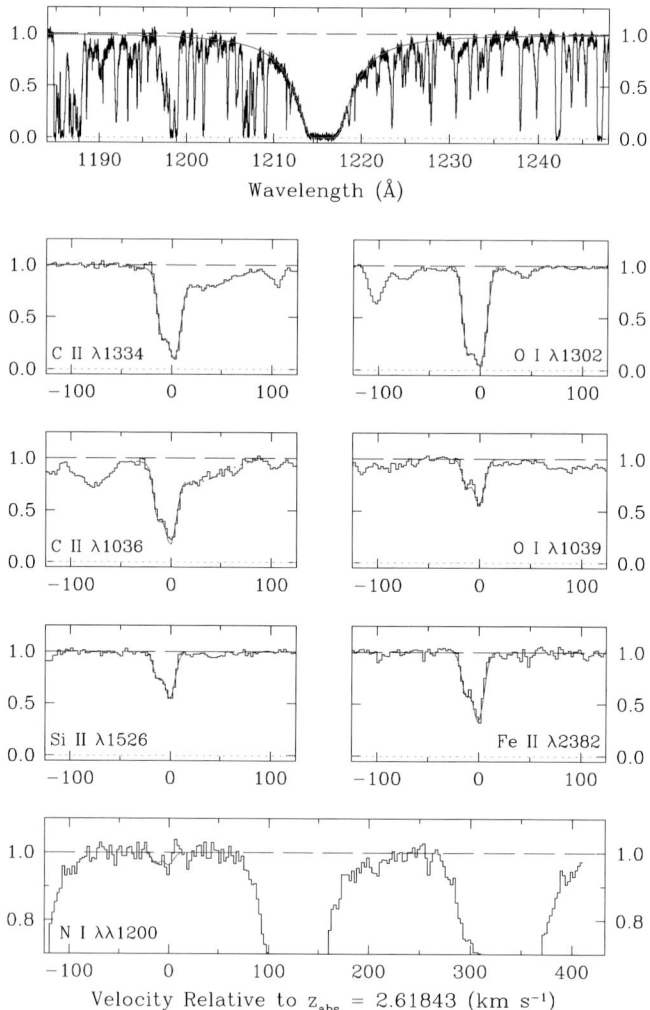

Figure 9. Observed profiles (black histograms) and fitted Voigt profiles (continuous red lines) of selected absorption lines in the $z_{\rm abs} = 2.61843$ DLA in the QSO Q0913+072. The y-axes of the plots show residual intensity; note the expanded y-scale of the bottom plot. The top panel shows the damped Lyα line, indicative of a column density of neutral hydrogen $N({\rm H\,{\sc i}}) = 2.2 \times 10^{20}$ atoms cm^{-2}. With an oxygen abundance of only 1/250 of solar, this is one of the most metal-poor DLAs known. The data shown here were obtained with UVES on the VLT and are reproduced from Pettini et al. (2008).

4.1. Carbon

The problem of line saturation has been particularly acute for C: the high abundance of this element (the fourth most abundant in the periodic table after H, He, and O), combined with the high transition probabilities of the resonance lines of its first ion (its dominant ion stage in H I regions), has been an obstacle to a precise determination of the interstellar abundance of C even in the nearby interstellar medium (e.g. Hobbs et al. 1982).

Recently, Pettini et al. (2008) have attempted to circumvent these difficulties by searching for the most metal-poor DLAs in the growing sample of such absorbers discovered by

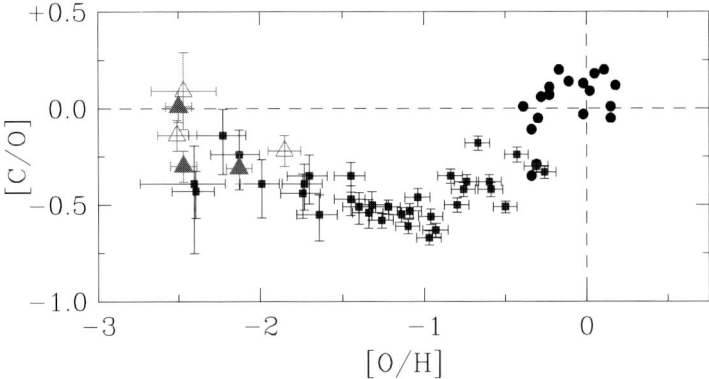

Figure 10. The values of [C/O] deduced by Pettini *et al.* (2008) for DLAs (red filled triangles) and sub-DLAs (open triangles) broadly agree with the stellar abundances (dots) measured at similar metallicities, supporting the suggestion by Akerman *et al.* (2004) that carbon and oxygen may be produced in near-solar proportions in the earliest stages of galactic chemical evolution. Stellar data from Akerman *et al.* (2004) and references therein (black dots: disk stars; blue squares: halo stars).

Prochaska, Herbert-Fort, & Wolfe (2005) in the spectra of QSOs from the *Sloan Digital Sky Survey*. An example is reproduced in Figure 9. At metallicities $Z_{\rm DLA} < 1/100 Z_\odot$, the resonance lines of C II are only mildly saturated and the velocity structure of the absorbing gas is generally simple, allowing for a more straightforward determination of the C/O and O/H ratios.

Figure 10 shows [C/O] vs. [O/H] in Galactic stars and the few DLAs in which the abundances of C and O have been measured to date; as usual, the square bracket notation denotes logarithmic abundances relative to the solar values (Asplund, Grevesse, & Sauval 2005). In halo stars [C/O] ≈ -0.5, while at the higher metallicities of most disk stars [C/O] rises to solar proportions. In the Galactic chemical evolution models considered by Akerman *et al.* (2004) this rise is interpreted as the additional contribution to carbon enrichment of the interstellar medium from the winds of Wolf-Rayet stars —whose mass loss rates are known to increase with metallicity—and, to a lesser extent, from the delayed evolution of low and intermediate mass stars.

More puzzling is the apparent upturn in [C/O] with *decreasing* metallicity when [O/H] $\lesssim -1$, since conventional stellar yields would predict the opposite behaviour: the expected time lag in the production of C relative to O in the first episodes of star formation should cause the [C/O] ratio to decrease dramatically as we move to the lowest metallicities in Figure 10. Akerman *et al.* (2004) speculated that the trend towards higher, rather than lower, values of [C/O] when [O/H] < -1 may be due to remaining traces of high carbon production by the first stars to form in the halo of the Milky Way proto-galaxy—high carbon yields (relative to oxygen) are indeed predicted by calculations of nucleosynthesis by Population III stars (e.g. Chieffi & Limongi 2002). More recently, Chiappini *et al.* (2006—see also Georges Meynet's and Raphael Hirschi's contributions to this volume) proposed that the observed behaviour can also be explained if stars of lower metallicity rotate faster: the higher rotation speeds greatly increase the yields of C (and N) by the massive stars which are the main source of O.

Some uncertainty in the interpretation of the stellar data is the use by Akerman *et al.* (2004) of one-dimensional, local thermodynamical equilibrium (LTE), stellar atmosphere models. The authors pointed out themselves the possibility that the neglect of non-LTE corrections to the high excitation C I and O I lines used for abundance determinations may mimic (or at least enhance) the rise in [C/O] with decreasing [O/H]. However, when the values of [C/O] measured in DLAs are compared with those of Galactic stars as in Figure 10, it appears that both sets of data paint a consistent picture.

The DLA measurements corroborate the conclusion by Akerman *et al.* (2004) in showing relatively high values of C/O in the low metallicity regime and complement the still very limited statistics of this ratio in stars of metallicity [O/H] < −2. It must be borne in mind here that, although the high redshift DLAs and the halo stars appear to constitute a continuous sequence in Figure 10, this does not imply that the chemical enrichment histories of the Milky Way halo and of the galaxies giving rise to DLAs were the same. It may well be that, as chemical evolution progresses, different DLAs follow different paths in the [C/O] vs. [O/H] plane, some similar to that of the Milky Way stellar populations while others may be different. Such a variety of behaviours has already been seen (in other elements) in Local Group galaxies (e.g. Pritzl *et al.* 2005). What our data show is that at metallicities below 1/100 of solar, DLAs and Galactic stars concur in showing elevated ratios of C/O. Akerman *et al.* (2004) speculated that an approximately solar value of C/O might be recovered when [O/H] $\lesssim -3$; the new data presented here are certainly consistent with this extrapolation of the Galactic measurements.

4.2. *Nitrogen*

The metal-poor DLAs targeted by Pettini *et al.* (2008) also help assess the contribution (if any) by massive stars to the primary production of N. The dual channels for the nucleosynthesis of N ('primary' or 'secondary', depending on whether the seed carbon and oxygen required for the CNO cycle are those manufactured by the star during helium burning, or were already present when the star first condensed out of the interstellar medium) are revealed by plots of N/O vs. O/H in extragalactic H II regions, such as the one reproduced in Figure 11. When the oxygen abundance is greater than about half solar (i.e. when $\log{(O/H)} + 12 \gtrsim 8.3$) N/O rises steeply with increasing O/H; this is the regime where N is predominantly secondary. At lower metallicities on the other hand ($\log{(O/H)} + 12 \lesssim 8.0$), N is mostly primary and its abundance tracks that of O; this results in the flat plateau at $\log{(N/O)} \simeq -1.5$ evident in Figure 11.

Primary N is thought to be synthesised most effectively by intermediate mass stars on the asymptotic giant branch; consequently, its release into the interstellar medium presumably takes place some time after the massive stars that are the main source of oxygen have exploded as Type II supernovae. Henry, Edmunds, & Köppen (2000) calculated the delay to be $\Delta t \sim 250 \, \text{Myr}$ for a 'standard' stellar initial mass function (IMF) and published stellar yields.

Damped Lyα systems add information that can be of value in assessing the validity of this overall picture. First of all, they offer measures of the abundances of O and N in entirely different environments from nearby star-forming galaxies. Second, they sample low metallicity regimes which are rare today. And third, by virtue of their being at high redshift, they can in principle provide a better sampling of the time delay in the production of primary N relative to O—the canonical $\Delta t \sim 250 \, \text{Myr}$ is a much larger fraction of the time available for star formation at $z = 2 - 3$ than today. In other words, the chances of 'catching' a galaxy in the interim period following a burst of star formation, when the O from Type II supernovae has already dispersed into the interstellar medium but lower mass stars have yet to release their primary N, are considerably higher in DLAs than in metal-poor star-forming galaxies in the local universe. Such a situation should manifest itself as a displacement below the primary plateau in the $\log{(N/O)}$ vs. $\log{(O/H)}$ plane in Figure 11.

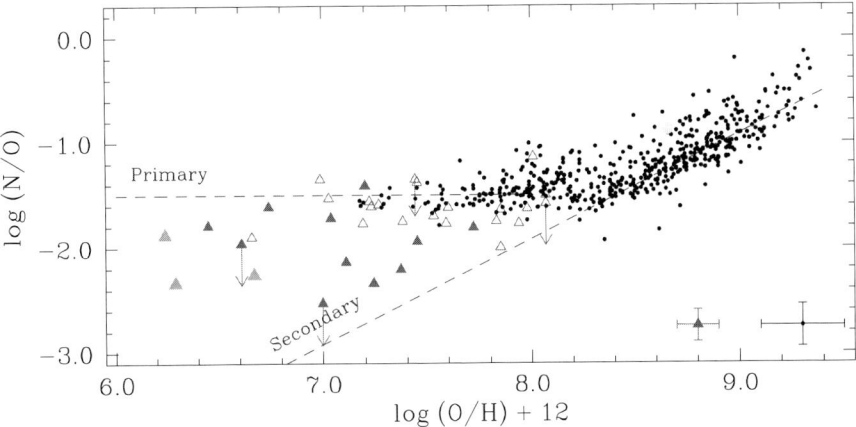

Figure 11. Abundances of N and O in extragalactic H II regions (small dots) and high redshift damped Lyα systems (triangles). See Pettini et al. (2008) for the sources of the original measurements. Filled triangles denote DLAs where the abundance of O could be measured directly, while open triangles are cases where S was used as a proxy for O. The error bars in the bottom right-hand corner give an indication of the typical uncertainties; the large yellow dot corresponds to the solar abundances of N and O. The dashed lines are approximate representations of the primary and secondary levels of N production (see text).

When DLAs are compared with present-day galaxies, as in Figure 11, it can be seen that in most cases their values of N/O fall close to the plateau defined by metal-poor H II regions. If primary N and O are synthesised by stars of different masses, as generally believed, the good agreement between local and distant galaxies attests to the universality of the IMF—a top-heavy IMF in DLAs would result in lower values of N/O, contrary to observations.

The finding that $\sim 1/3$ of DLAs in the current sample have values of N/O lower than the primary plateau is consistent with current ideas of a time delay by a few 10^8 years in the release of primary N from intermediate mass stars. But Figure 11 also emphasises the importance of extending measurements of N/O to the lowest metallicity DLAs: as the gap between secondary and primary N production widens, it becomes easier to recognise additional sources of primary nitrogen, if they exist. In particular, if massive stars were able to synthesise *some* primary N, we might expect to see a minimum value of N/O, corresponding to the N ejected into the ISM at the same time as O (Centurión et al. 2003). While the statistics are still limited, the available data do hint at the possibility of a 'floor' in the value of N/O at $\log{(\rm N/O)} \simeq -2.3$. If this does indeed reflect the primary N yield by massive stars, it would imply that at these metallicities massive stars can account for $\sim 15\%$ of the total primary N production.

We have heard much at this meeting about the importance of stellar rotation in facilitating the nucleosynthesis of primary N in massive stars (Hirschi et al. 2007—see also Georges Meynet's and Raphael Hirschi's contributions to this volume). It is therefore worthwhile pointing out here that the level of primary N production by massive stars suggested by the DLA measurements is lower than that apparently implied by the generally higher values of N/O derived by Spite et al. (2005) in metal-poor halo stars. It is the latter that has motivated much of the theoretical nucleosynthesis calculations so far. Unfortunately, the derivation of N abundances in very metal-poor stars remains one of the most difficult problems in stellar abundance measurements, beset with systematic uncertainties as Spite *et al.* (2005) themselves point out (see also Israelian *et al.* 2004). It may be, then, that we need to look to the still-increasing samples of DLAs for further clues to the origin of N in the earliest episodes of star formation.

5. Concluding Remarks

It was encouraging to see at this meeting a continuing trend of increasing interaction between stellar and extragalactic astronomers. The former can teach us (the latter) a great deal in our common quest to unravel the physical processes that led to the formation of the first stars, the assembly of galaxies and their evolution to the present time. It is also clear from even this brief review that the tremendous progress we have witnessed in the last five years has only just started us on many paths of enquiry. Returning to the individual topics discussed, I would single out the following as areas where we will be focusing our efforts in the years ahead:

1. The different metallicity diagnostics—from massive stars, H II regions, and the more widespread interstellar medium—need to be brought together, particularly at the high metallicity end of the scale. Do we understand why we have yet to find clear examples of supersolar metallicities at high redshifts, if the most luminous galaxies we observe then are the progenitors of today's massive ellipticals, as suggested by their masses, clustering and other properties?

2. What are the metallicities and other physical characteristics of star-forming galaxies at the faint end of the luminosity function? How do they overlap with other classes of high redshift objects, such as DLAs and the host galaxies of γ-ray bursts?

3. We have found no convincing evidence so far for departures from a Salpeter slope of the IMF in galaxies at $z = 2 - 3$. Is this also the case at higher redshifts, or is there a redshift where we can begin to see evidence for the top-heavy IMF generally expected to apply to the first stars? And what of stars less massive than $10\,\mathrm{M}_\odot$? The UV spectra described here give us no clues to their numbers, nor to the characteristic stellar mass M^* of the star-formation episodes we observe.

4. Determinations of element abundances in different classes of QSO absorbers continue to complement effectively the large body of stellar work in this area. Stellar rotation is now firmly established as an essential ingredient for a meaningful interpretation of this growing body of data. It will be of great interest to see if physically motivated metallicity-dependent C, N, and O yields can reproduce the relative abundances of these elements measured in DLAs, as well as Galactic stars, when incorporated into appropriate chemical evolution models.

Acknowledgements

It is a pleasure to acknowledge the major contributions by my collaborators in the various projects described in this review. Sam Rix also kindly offered valuable comments on the manuscript, and Amanda Smith helped with the preparation of some of the figures. I am indebted to the late Bernard Pagel for numerous inspiring conversations over the years. Finally, I should like to thank the organisers of the Symposium for inviting me to take part in this instructive and stimulating meeting on the beautiful island of Kaua'i.

References

Akerman, C. J., Carigi, L., Nissen, P. E., *et al.* 2004, *A&A*, 414, 931
Asplund, M., Grevesse, N., & Sauval, A. J. 2005, in T. G. Barnes III & F. N. Bash (eds.),

Cosmic Abundances as Records of Stellar Evolution and Nucleosynthesis (San Francisco: Astronomical Society of the Pacific), *ASP Conf. Ser.* 336, 25

Brinchmann, J., Pettini, M., & Charlot, S. 2008, *MNRAS*, 385, 679

Centurión, M., Molaro, P., Vladilo, G., et al. 2003, *A&A*, 403, 55

Chiappini, C., Hirschi, R., Meynet, G., et al. 2006, *A&A*, 449, L27

Chieffi, A. & Limongi, M. 2002, *ApJ*, 577, 281

Crowther, P. A., Prinja, R. K., Pettini, M., & Steidel, C. C. 2006, *MNRAS*, 368, 895

de Mello, D. F., Daddi, E., Renzini, A., et al. 2004, *ApJ*, 608, L29

Erb, D. K., Shapley, A. E., Pettini, M., et al. 2006, *ApJ*, 644, 813

Henry, R. B. C., Edmunds, M. G., & Köppen, J. 2000, *ApJ*, 541, 660

Hirschi, R., Chiappini, C., Meynet, G., et al. 2007, in: R.J. Stancliffe, G. Houdek, R. G. Martin & C. A. Tout (eds.), *Unsolved Problems in Stellar Physics*, (New York: AIP) *AIP Conf Proc.* 948, 397

Hobbs, L. M., York, D. G., & Oegerle, W. 1982, *ApJ*, 252, L21

Israelian, G., Ecuvillon, A., Rebolo, R., et al. 2004, *A&A*, 421, 649

Jimenez, R. & Haiman, Z. 2006, *Nature*, 440, 501

Leitherer, C., Leão, J. R. S., Heckman, T. M., et al. 2001, *ApJ*, 550, 724

Pauldrach, A. W. A., Hoffmann, T. L., & Lennon, M. 2001, *A&A*, 375, 161

Pettini, M. 2006, in: V. LeBrun, A. Mazure, S. Arnouts & D. Burgarella (eds.), *The Fabulous Destiny of Galaxies: Bridging Past and Present* (Paris: Frontier Group), 319

Pettini, M., Lipman, K., & Hunstead, R. W. 1995, *ApJ*, 451, 100

Pettini, M. & Pagel, B. E. J. 2004, *MNRAS*, 348, L59

Pettini, M., Rix, S. A., Steidel, C. C., et al. 2002, *ApJ*, 569, 742

Pettini, M., Zych, B. J., Steidel, C. C., & Chaffee, F. H. 2008, *MNRAS*, 385, 2011

Pritzl, B. J., Venn, K. A., & Irwin, M. 2005, *AJ*, 130, 2140

Prochaska, J. X., Herbert-Fort, S., & Wolfe, A. M. 2005, *ApJ*, 635, 123

Reddy, N. A., Steidel, C. C., Pettini, M., et al. 2008, *ApJS*, 175, 48

Rix, S. A., Pettini, M., Leitherer, C., et al. 2004, *ApJ*, 615, 98

Shapley, A. E., Steidel, C. C., Adelberger, K. L., et al. 2001, *ApJ*, 562, 95

Shapley, A. E., Steidel, C. C., Pettini, M., & Adelberger, K. L. 2003, *ApJ*, 588, 65

Spite, M., Cayrel, R., Plez, B. et al. 2005, *A&A*, 430, 655

Steidel, C. C., Shapley, A. E., Pettini, M., et al., 2004, *ApJ*, 604, 534

Teplitz, H. I., McLean, I. S., Becklin, E. E., et al., 2000, *ApJ*, 533, L65

Discussion

LEITHERER: The galaxies you discussed are UV/optically selected. Another significant component of the star formation census at redshifts $z = 2-3$ are sub-mm bright galaxies detected with SCUBA. Could they have higher metallicities than LBGs?

PETTINI: This is an interesting possibility, since SCUBA sources are often touted as being the most massive star-forming galaxies at these redshifts. However, the short answer seems to be that their metallicities are no higher than solar, and comparable to those of the most massive LBGs. Of course, the fact that SCUBA galaxies are often very faint at rest-frame UV wavelengths, and the preponderance of AGN amongst them, make this conclusion still rather tentative. However, the work by the Durham group, for example (Swinbank et al. 2004, *ApJ*, 617, 64), has shown that the [N II]/Hα ratio of composite spectra of SCUBA galaxies is indicative of slightly sub-solar abundances, consistent with the strengths of the UV stellar features. If galaxies with clearly supersolar abundances exist at high z, we have yet to find them. Part of the problem might well be that in most cases we only obtain a luminosity-weighted, single average value of Z for a given galaxy, with no information on any internal metallicity gradient which may be in place (but see Förster Schreiber et al. 2006, *ApJ*, 645, 1062).

CROWTHER: In Hadfield & Crowther (2006) we found that *Starburst99* with the Geneva or Padova stellar evolution tracks underestimates the ratio of Wolf-Rayet to O-type stars by about a factor of two, when compared with observations of starburst clusters in nearby galaxies whose metallicities are similar to that of the Large Magellanic Cloud. If we were to account for this factor of two, then the metallicity of the $z = 3$ LBGs implied by the equivalent width of the He II $\lambda 1640$ emission line measured in the Shapley *et al.* (2003) composite would fall from solar to near-LMC values (see Figure 8).

PETTINI: Thank you for pointing this out. An LMC-type metallicity (somewhere between half and one-third solar) is in fact more likely for the composite of hundreds of LBGs constructed by Shapley *et al.*, based on other abundance indicators. There may well be a number of such systematic offsets in the calibration of the luminosity of the He II $\lambda 1640$ emission line with metallicity. That's why I think that the studies of Wolf-Rayet stars in external galaxies which you and your collaborators are leading are so important for the correct interpretation of the spectra of high-z galaxies.

BURBIDGE: You have shown some very beautiful results. To obtain them you use the brightest sources. Do they tend to be average sources, i.e. typical objects in the redshift range, or are they to some extent anomalous in the sense that they are much brighter and easier to observe than average objects at the same redshift?

PETTINI: The observations I have presented refer to galaxies spanning a range of luminosities, from the bright end of the luminosity function down to one-to-two magnitudes (depending on redshift) *below* the fiducial luminosity L^*. Whether they are typical of the galaxy population at these redshifts depends on your definition of 'typical'. I would contend that they are not just the 'tip of the iceberg', as it is sometimes claimed. Of course, it would be of great interest to extend the work I have described to the more numerous galaxies at the faint end of the luminosity function. Such observations will have to wait until the advent of the next generation of giant optical-infrared telescopes, although we may get glimpses ahead of time aided by the boost of gravitational lensing.

MEYNET: Primary N production is accompanied by the production of primary ^{13}C. Thus, if signs of primary nitrogen are visible, one may expect low values of the ^{12}C/^{13}C ratio in the interstellar medium and at the surfaces of non-evolved stars. Is there any hope to check the ^{12}C/^{13}C values in DLAs?

PETTINI: There has already been one attempt to do so (Levshakov *et al.* 2006, *A&A*, 447, L21), although in that case the DLA considered was not among the most metal-poor, so that the lack of significant ^{13}C enhancement deduced by those authors may not be surprising. In general, this is a very difficult measurement to perform, because the isotope shifts involved are smaller than other sources of line broadening, even in the DLAs which are kinematically very quiescent, such as the examples I showed in my talk. I am not saying that it is an impossible measurement to perform, but it would require a great deal of careful preparatory work (to select the most likely cases), long exposure times, and a very rigorous analysis of the spectra.

Star Forming Galaxies at $z > 5$

Yoshi Taniguchi

Research Center for Space and Cosmic Evolution, Ehime University, 2-5 Bunkyo-cho,
Matsuyama 790-8577, Japan
email: tani@cosmos.ehime-u.ac.jp

Abstract. We present recent progress in searching for galaxies at redshift from $z \simeq 5$ to $z \simeq 10$. Wide-field and sensitive surveys with 8m class telescopes have been providing more than several hundreds of star forming galaxies at $z \simeq 5 - 7$ that are probed in the optical window. These galaxies are used to study the early cosmic star formation activity as well as the early structure formation in the universe. Moreover, near infrared deep imaging and spectroscopic surveys have found probable candidates of galaxies from $z \simeq 7$ to $z \simeq 10$. Although these candidates are too faint to be identified unambiguously, we human being are now going to the universe beyond 13 billion light years, close to the epoch of first-generations stars; i.e., Population III stars. We also mention about challenges to find Population III-dominated galaxies in the early universe.

Keywords. galaxies: formation – galaxies: evolution

1. Introduction

Massive stars are key ingredients in the early phase of galaxy formation. First stars (i.e., Population III stars) are considered to be very massive from theoretical aspects although the formation of low mass stars may not be ruled out. Once first massive stars were formed in subgalactic gas clumps within dark matter halos, they could begin to work as cosmic engines because of their huge number of ionizing photons. Then, since they could die after a few million years after their birth, they also supply both kinetic energy and heavy elements even into the intergalactic space. Therefore, massive stars also work as mechanical and chemical engines in the universe. All these issues suggest that the formation and early evolution of galaxies could be controlled by massive stars; how massive, how many massive stars (i.e., more generally, the initial mass function), and when did they begin to form? It is also worthwhile noting that massive stars in the phase of galaxy formation are considered to be related to the cosmic re-ionization of the universe.

In order to investigate these problems, it is necessary to carry out searches for galaxies at very high redshift. Such high-z galaxies have been identified mainly by the following survey methods; (1) optical narrow-band surveys for strong Lyα emitters (LAEs), and (2) optical broad-band surveys for Lyman break galaxies (LBGs) (i.e., the dropout technique; e.g., Steidel et al. 1999, 2003; Ouchi et al. 2004). Other methods are summarized in Taniguchi et al. (2003). Although both methods need follow-up optical spectroscopy to confirm that photometrically selected objects are real high-z galaxies, more than several hundreds of galaxies beyond $z = 5$ have been already found to date thanks to the great observational capability of 8m class telescope facilities (e.g., Dey et al. 1998; Rhodes et al. 2003; Hu et al. 2002, 2004; Ajiki et al. 2003; Kodaira et al 2003; Taniguchi et al. 2005; Kashikawa et al. 2006; Ouchi et al. 2008). To probe massive star formation or initial starbursts in early universe, hydrogen Lyα emission provides a powerful tool (Partridge and Peebles 1967; Haiman 2002). Therefore, in this review, we discuss the observational nature star-forming galaxies at high redshift found mainly by the optical narrow-band imaging surveys made so far (section 2).

In section 3, we give a brief summary of near infrared deep surveys for forming galaxies beyond $z = 7$ and discuss their implication. Another interesting issue is to search for evidence for Population III stars in early universe. Since theoretical consideration suggests that some galaxies at $z \gtrsim 2$ may be dominated by Population III stars because of low feedback from first-generation supernovae. Therefore, in section 4, we give a summary of recent trials for searching for such Population III-dominated galaxies. Finally, in section 5, we discuss future prospects in this research field briefly.

2. Surveys in the Optical Window; Star-forming galaxies at $z = 5 - 7$

When we use the optical window (e.g., 400 nm – 1000 nm), we are able to search for LAEs at $z \sim 2.5 - 7$. If we are interested in LAEs beyond $z = 5$, surveys should be made at wavelengths longer than 700 nm where OH airglow emission lines are bright. Therefore, we have to use some narrow windows where OH airglow is moderately weak; e.g., 815 nm, 920 nm, and so on. These atmospheric windows allow us to search for LAEs at $z \simeq 5.7$ and $z \simeq 6.6$, and so on. A summary of such LAE surveys for $z > 5$ is given in Table 1. We also give a list of the top ten of most distant galaxies found in the optical LAE surveys in Table 2; the most distant LAE known to date is IOK1 at $z = 6.96$ (Iye et al. 2006). Note that candidate galaxies around $z \sim 7$ have been found in the GLARE (Gemini Lyman-Alpha at Reionization Era) survey (Stanway et al. 2007).

We summarize the observational properties of LAEs at $z = 5 - 7$ as follows. (1) Stellar masses; $\sim 10^{9-10} M_\odot$, (2) Sizes (Lyα emission)); a few kpc, (3) Lyα luminosity; $L(\mathrm{Ly}\alpha) \sim 10^{42-43}$ erg s^{-1}; (4) Stellar ages; several million to several hundreds million years; (5) Star formation rates; a few to several tens $M_\odot$ yr^{-1}; (6) Star formation rate densities; $\sim 10^{-3}$ $M_\odot$ yr^{-1} Mpc^{-3}; (7) Morphology (Lyα emission); spatially extended, but only few information; and (8) Morphology (UV continuum); spatially extended (smaller than Lyα emission), but only few information (e.g., Rhoads et al. 2005; Taniguchi et al. 2008).

Note, however, that more massive galaxies tend to be found in high-z LBG samples with stellar masses of $\sim 10^{10-11} M_\odot$ (e.g., Egami et al. 2005; Mobasher et al. 2005; Yan et al. 2006; Eyles et al. 2007 and references therein). Moreover, the star formation rate densities for high-z LBGs is $\sim 10^{-2}$ $M_\odot$ yr^{-1} Mpc^{-3}, being higher by one order of magnitude than those of LAEs at similar redshifts (e.g., Taniguchi et al. 2005 and references therein).

The Lyα luminosity function for LAEs at $z > 5$ has been obtained for several samples of LAEs (e.g., Ajiki et al. 2004; Hu et al. 2004; Shimasaku et al. 2006; Kashikawa et al. 2006). One interesting point is that bright LAEs appear to be rarer at $z \approx 6.5$ than at $z \approx 5.7$ in the Subaru Deep Field (Kashikawa et al. 2006), suggesting a possible evolutionary effect from $z \approx 6.5$ to $z \approx 5.7$.

As for large-scale structures in early universe, there are lines of evidence for galaxy clustering below $z \approx 6$ [at $z \approx 5.7$ (Hu et al. 2004; Ouchi et al. 2005), and at $z \approx 4.9$ (Shimasaku et al. 2003)] although there is no strong evidence for such clustering at $z \approx 5.7$ (Ajiki et al. 2003; Murayama et al. 2007). Beyond $z = 6$, no such clustering feature has been found to date (Kashikawa et al. 2006). These observations suggest that the large-scale structure probed by LAEs could grow up from $z \sim 6$.

Table 1. Top ten of high-redshift galaxies found in the optical surveys

Window (nm)	Field	z	V (Mpc^3)	N(photo)	N(sp)	Ref.
973	SDF	6.96	3.2×10^5	2	1	a
921	SDF	6.5 – 6.6	3.2×10^5	58	34	b, c, d
912	6 fields	6.56	–	1	1	e
816	SDSS1044	5.7	2.0×10^5	20	2	f
815, 823	LALA	5.7 – 5.8	$\sim 2 \times 10^5$	18	2	g
816	SSA22	5.7	1.9×10^5	26	19	h
816	SDF	5.7	1.9×10^5	89	37	i
816	GOODS-N	5.7	4.0×10^4	10	–	j
816	GOODS-S	5.7	4.0×10^4	4	–	j
816	COSMOS	5.7	1.8×10^6	119	–	k
816	SXDS	5.7	9.2×10^5	401	17	l

Field: SDF = Subaru Deep Field (Kashikawa *et al.* 2004), SDSS1044 = SDSSp J104433.04-012522.2 (Fan *et al.* 2000), LALA = Large Area Lyman Alpha survey, SSA22 = Small Selected Area at RA = 22h, GOODS = Great Observatory Origins Deep Survey - North and South (Giavalisco *et al.* 2004), COSMOS = Cosmic Evolution Survey (Scoville *et al.* 2007), and SXDS = Subaru XMM-Newton Deep Survey (Sekiguchi *et al.* 2004)
Survey Volume (V): A flat universe with $\Omega_m = 0.3$, $\Omega_\Lambda = 0.7$, and $H_0 = 70$ km s^{-1} Mpc^{-1} is adopted.
N(photo): The number of LAEs identified by photometric conditions.
N(sp): The number of LAEs confirmed by follow-up spectroscopy. Note that follow-up spectroscopy is going on in some cases.
References: a = Iye *et al.* 2006, Ota *et al.* 2008; b = Kodaira *et al.* 2003; c = Taniguchi *et al.* 2005; d = Kashikawa *et al.* 2006; e = Hu *et al.* 2002; f = Ajiki *et al.* 2003 (see also Ajiki *et al.* 2004); g = Rhodes & Malhotra 2001, 2003; h = Hu *et al.* 2004; i = Shimasaku *et al.* 2005; j = Ajiki *et al.* 2006; k = Murayama *et al.* 2007; l = Ouchi *et al.* 2008

Table 2. Top ten of high-redshift galaxies found in the optical surveys

No.	Name	Redshift	Ref.
1	IOK1	6.96	a
2	SDF132522.3+273520	6.597	b
3	SDF132520.4+273459	6.596	c
4	SDF132357.1+272446	6.589	c
5	SDF132432.5+271647	6.580	b
6	SDF132528.8+273043	6.578	b
6	SDF132418.3+271455	6.578	b, d
8	HCM-6A	6.56	e
9	SDF132432.9+273124	6.557	c
10	SDF132408.3+271543	6.554	b

References: a = Iye *et al.* 2006, Ota *et al.* 2008; b = Taniguchi *et al.* 2005; c = Kashikawa *et al.* 2006; d = Kodaira *et al.* 2003; e = Hu *et al.* 2002

3. Surveys in the Near-Infrared Window; Star-forming galaxies at $z > 7$

The success of optical searches for high-z galaxies up to $z \approx 7$ urged us to carry out near-infrared (NIR) surveys for very high-z galaxies beyond $z = 7$. Such galaxies, if any, could be much fainter than galaxies with $z < 7$ because they could be in still in a mass assembly phase in their evolution. Moreover, the attenuation of H I atoms should be much severe for detecting Lyα emission although cosmological H II regions around LAEs could help for such detection (e.g., Haiman 2002). In this section, we give a summary of recent NIR surveys for forming galaxies.

[1] *Narrow-band Imaging Surveys*: The narrow-band imaging technique provides us a large number of LAEs at $2 < z < 7$. Therefore, this technique should be applied in the NIR window.

The first NIR trial was reported by Willis & Courbin (2005). Their project name is "ZEN (z equal nine)". They used a narrowband filter centered at 1.187 μm in J band with VLT/ISAAC (FOV = 2.5 arcmin $\times$ 2.5 arcmin). They found no LAE candidate with $L(\text{Ly}\alpha) > 10^{43}\ h^{-2}$ erg s^{-1} in the survey volume of 340 h^{-3} Mpc^{-3} where $h = H_0/100$. They also made a similar deep NIR narrowband imaging survey for LAEs in three lensing cluster areas (Abell 114, 1689, and 1835). Again, they found no LAE candidate with $L(\text{Ly}\alpha) > 10^{43}\ h^{-2}$ erg s^{-1} in the survey volume of 1270 h^{-3} Mpc^{-3}.

Using the same instrument and the NIR narrowband filter used in the ZEN survey, Cuby *et al.* (2007) made a survey for LAEs at $z \approx 9$ in seven fields, covering 31 arcmin^2 in total. Although their survey area is wider than that of ZEN, they found no LAE candidate.

[2] *Broad-band Imaging Surveys*: The broad-band imaging technique is also powerful to detect high-z galaxies. This is the same technique as the Lyman break method in the optical. The deepest NIR imaging survey data are provided by the Hubble Ultra Deep Field (Bouwens *et al.* 2005; see also Bouwens & Illingworth 2006). The survey depth is down to 28.6 in J_{110}(AB) and 28.5 in H_{160}(AB) with a 0.6 arcsec aperture. They found three probable galaxies at $z \sim 10$ with $H_{160} \simeq 28$, corresponding to $\approx 0.3 L^*_{z=3}$. Unfortunately, all these sources are too faint to be observed in spectroscopy. The presence of these objects gives a star formation rate density of $\sim 10^{-3}$ M$_\odot$ yr^{-1} Mpc^{-3}, being smaller by one order of magnitude than that at $z = 5 - 6$.

When we use ground-based telescopes, the survey depth is around 25 AB magnitude in J and H at best because of strong airglow emission. Therefore, NIR broadband imaging surveys for a blank field from the ground may not work well. Or, we need help of gravitational lensing. Richard *et al.* (2006) made deep NIR imaging survey for two lensing cluster regions (Abell 1835 and AC 114) by using VLT/ISAAC. Their survey depth is 25.5 – 25.6in J(AB), 24.7 in H(AB), and 24.3 – 24.7 in K_s(AB) for the two regions. They found 8 and 5 probable candidates at $7 < z < 10$ in Abell 1835 and AC 114, respectively. Given the magnification factor from 1.5 to 10, their star formation rate ranges from a few to 20 M$_\odot$ yr^{-1}.

[3] *Blind Spectroscopic Surveys along Caustic Lines*: This technique is very unique but very powerful in finding intrinsically faint objects at high redshift; see for optical trials, Ellis *et al.* (2001) and Santos *et al.* (2004). Stark *et al.* (2007) made NIR spectroscopic surveys for gravitationally-lensed LAEs around nine intermediate-z clusters of galaxies by using Keck/NIRSPEC. Since their spectroscopic survey is designed to sweep the caustic lines of the gravitational lensing, the amplification factor is as high as 10 – 50 for objects around $z = 10$. They found 6 probable LAE candidates at $z = 8.7 - 10.2$. Among them, two objects are the most likely candidates from their careful follow-up observations; Abell 68 c1 at $z = 9.32$ and Abell 2219 c1 at $z = 8.99$.

We have introduced five independent deep NIR surveys for high-z galaxies beyond $z = 7$. Generally speaking, it is more difficult to find candidates of forming galaxies in NIR than in optical. One reason is that there is no very wide-field NIR imager on 8m class telescopes. MOIRCS on the Subaru Telescope has FOV = 4 arcmin $\times$ 7 arcmin (http://www.subarutelescope.org/Observing/Instruments/MOIRCS/index.html) and HAWK-I on VLT has FOV = 7 arcmin $\times$ 7 arcmin (http://www.eso.org/instruments/hawki/).

Although these NIR cameras have widest FOV on such 8m class telescopes, their FOVs are still much smaller than those of optical cameras (> 30 arcmin $\times$ 30 arcmin). Note that both the stronger airglow emission and the much severe Tolman dimming [i.e., the surface brightness dimming as $(1+z)^{-4}$] also makes it difficult to detect LAEs at NIR. Also, forming galaxies at $z > 7$ could be faint intrinsically.

4. Surveys for Population III–dominated Galaxies

One interesting issue is to address on first stars in our universe: When were they made? How massive were they? How did they affect the nature of intergalactic medium, chemically and dynamically? Did only first stars contribute to the cosmic re-ionization? How were they related to the formation of supermassive black holes? In order to give firm answers to these questions, it is necessary to probe first stars or forming galaxies at very high redshift.

The major epoch of Pop III star formation may be at $z \sim 10 - 30$, when dark matter halos with mass of $M_{\rm halo} \sim 10^6$ $M_\odot$ could have grown up to form first stars (see for a review, Loeb & Barkana 2001). Shortly after the formation of Pop III stars ($\sim 10^6$ years after), first supernova explosions could occur and then pollute the gas inside and outside of dark matter halos. However, if this feedback process is inefficient, Pop III stars could be born in galaxies even at $z \sim 2$ or so (e.g., Scannapieco et al. 2003; Schneider et al. 2006; Jimenez & Haiman 2006; Tornatore et al. 2007; see also, for recent review, Norman 2008). If this is the case, it is possible to probe Pop III dominated forming galaxies in the optical window.

There are two important observational properties of such Pop III dominated galaxies. One is the large equivalent width (EW) of hydrogen Lyα emission. The other property is moderately strong He II λ1640 emission. These properties are attributed to the very high effective temperature ($\sim 10^5$ K) of Pop III stars (e.g., Tumlinson & Shull 2000; Tumlinson et al. 2001, 2003; Bromm et al. 2001; Oh et al. 2001; Schaerer 2002, 2003). Interestingly, it has been reported that some high-z LAEs have a large rest-frame equivalent width of Lyα emission with $EW_0 >$ a few 100 Å (e.g., Malhotra & Rhoads 2002; Nagao et al. 2004, 2005a, 2007; Shimasaku et al. 2006; Dijkstra & Wyithe 2007). These observations may not be explained with photoionization by usual massive stars.

For one of such LAEs with a large EW Lyα emission, Nagao et al. (2005a) made an ultra deep NIR spectroscopy to detect He II emission; SDF 132440.6+273607 with $EW_0({\rm Ly}\alpha) = 130$ Å and $L({\rm Ly}\alpha) = 1.8 \times 10^{43}$ erg s^{-1} at $z = 6.33$. However, they found no He II feature, suggesting $L({\rm HeII}) < 1.4 \times 10^{42}$ erg s^{-1}. Stacking analyses of spectra of high-z galaxies also found no evidence for Pop III-driven He II emission; (1) Shapley et al. (2003) examined the stacked spectrum of $\simeq 1000$ LBGs at $z \simeq 3$ and obtained evidence for He II emission. However, the authors suggested that this feature may be attributed to Pop I hot stars such as WR stars (see also the comment given by Max Pettini in the discussion; cf. Jimenez & Haiman 2006). (2) Ouchi et al. (2008) examined the stacked spectrum of $\simeq 50$ LBGs at $z = 3.7$ and found no He II emission. (3) See the comment given by Jon Eldridge in discussion.

Recently, Nagao et al. (2008) made a unique survey for Lyα-He II emitters by using combination of intermediate- and narrow-band filters in the optical window. They used (i) IA598 and NB816 filters for LAEs at $z \approx 4.0$ and (ii) IA 679 and NB921 for LAEs at $z \approx 4.6$, where the Subaru IA filter system consists of 20 filters covering from 4000 Å to 9500 Å with a spectroscopic resolution of $R = \lambda/\Delta\lambda = 23$ (Taniguchi 2001; see also Ajiki et al. 2004; Yamada et al. 2005). Their survey field is SDF. Although they found 10 dual-line emitters, they are not identified as Lyα/He II emitters (i.e., low-z dual-line emitters such as [O II]/[O III] ones. Their survey shows that there is no LAEs with Pop III star formation rate with $> 2 M_\odot$ yr^{-1} and the Pop III star formation rate density is lower than $5 \times 10^{-6} M_\odot$ yr^{-1} Mpc^{-3}, being smaller by a few orders of magnitudes than those of LAEs at similar redshifts.

5. Future Prospects

Finally, we give comments on future prospects. As shown in previous sections, optical deep surveys have been finding a large number of forming galaxies at $z \sim 5 - 7$. However, NIR deep surveys have been facing to the technical limit of the existing large telescope facilities including the Hubble Space Telescope. Therefore, we will have to wait for next-generation telescopes in order to improve our knowledge on massive stars at very high redshift (i.e., Pop III stars) and star formation history of in very young galaxies.

We will have two types of new-generation telescopes in near future; one is the James Webb Space Telescope (JWST: Sonneborn, this volume; see http://www.jwst.nasa.gov/) and the other is extreme large telescopes (ELTs) on the ground such as the European-ELT (D'Odorico, this volume; see http://www.eso.org/sci/facilities/eelt/) and the Thirty Meter Telescope (TMT: http://www.tmt.org/) led by Caltech and University of California. Extremely high observational capabilities of these telescopes will open the door to Pop III dominated universe. In particular, mid infrared observations with JWST will be able to go to $z \sim 30$.

Yet, we will have to do our best to find much younger, Pop III-dominated galaxies at high redshift up to $z \sim 30$ before next-generation, space and ground-based telescopes will come. For this effort, we have two nice helpers. One is the gravitational lensing, as demonstrated in section 3 (e.g., Stark *et al.* 2007; Richard *et al.* 2006). The other helper is very bright gamma-ray bursts at high redshift. After the gamma ray bursts (GRBs), very bright optical flashes have been often observed. The most distant GRB event known to date was found at $z = 6.33$ (Kawai *et al.* 2006). If such GRB events will occur at $z \sim 7$, we will be detect their rest-frame optical flashes in the NIR windows (see Fynbo, this volume).

As for the formation and evolution of galaxies, next-generation telescopes will allow us to explore what happened in galaxy formation and in early evolution phase. Then, we will be able to obtain a unified picture for various types of galaxies at high redshift, e.g., LAEs, LBGs, EROs (extremely red objects), DRGs (distant red galaxies), BzKs (galaxies selected from BzK photometry), SMGs (submillimeter galaxies), and so on. Finally, we will learn how massive stars have been working as cosmic engines for more than 10 billion years.

Acknowledgements

The author would like to thank all the members of both SOC and LOC of this wonderful symposium at Kauai/Hawaii. He also thanks his colleagues, in particular, Nobunari Kashikawa and Tohru Nagao.

References

Ajiki, M., Taniguchi, Y., Fujita, S. S., *et al.* 2003, *AJ*, 126, 2091
Ajiki, M., Taniguchi, Y., Fujita, S. S., *et al.* 2004, *PASJ*, 56, 597
Ajiki, M., Mobasher, B., Taniguchi, Y., *et al.* 2006, *ApJ*, 638, 596
Bouwens, R. J., Illingworth, G. D., Thompson, R. I., & Franx, M. 2005 *ApJ*, 624, L5
Bouwens, R. J. & Illingworth, G. D. 2006 *Nature*, 443, 189

Bromm, V., Kudritzki, R. P., & Loeb, A. 2001, *ApJ*, 552, 464
Cuby, J. G., Hibon, P., Le Febre, O., et al. 2007, *A&A*, 461, 911
Dey, A., Spinrad, H., Stern, D., et al. 1998, *ApJ*, 498, L93
Dijkstra, M. & Wyithe, J. S. B. 2006, *MNRAS*, 379, 1589
Egami, E., Kneib, J.-P., Rieke, G. H., et al. 2005, *ApJ*, 618, L5
Ellis, R. S., Santos, M. R., Kneib, J. -P., & Kuijken, K. 2001, *ApJ*, 560, L119
Eyles, L. P., Bunker, A. J., Ellis, R. S., et al. 2008, *MNRAS*, 374, 910
Fan, X., White, R. L., Davis, M., et al. 2000, *AJ*, 120, 1167
Giavalisco, M., Dickinson, M., Ferguson, H. C., et al. 2004, *ApJ*, 600, L93
Haiman, Z. 2002, *ApJ*, 576, L1
Hu, E. M., Cowie, L. L., McMahon, R. G., et al. 2002, *ApJ*, 568, L75 (Erratum, 576, L99)
Hu, E. M., Cowie, L. L., Capak, P., et al. 2004, *AJ*, 127, 563
Jimenez, R. & Haiman, Z. 2006, *Nature*, 441, 120
Kashikawa, N., Shimasaku, K., Yasuda, N., et al. 2004, *PASJ*, 56, 1011
Kashikawa, N., Shimasaku, K., Malkan, M. A., et al. 2006, *ApJ*, 648, 7
Kawai, N., Kogugi, G., Aoki, K., et al. 2006, *Nature*, 440, 184
Kodaira, K., Taniguchi, Y. Kashikawa, N., et al. 2003, *PASJ*, 55, L17
Loeb, A. & Barkana, R. 2001, *ARA&A*, 39, 19
Malhotra, S. & Rhoads, J. E. 2002, *ApJ*, 565, L71
Mobasher, B., Dickinson, M., Ferguson, H. C., et al. 2005, *ApJ*, 635, 832
Murayama, T., Taniguchi, Y., Scoville, N. Z., et al. 2007, *ApJS*, 172, 523
Nagao, T., Taniguchi, Y., Kashikawa, N., et al. 2004, *ApJ*, 613, L9
Nagao, T., Kashikawa, N., Malkan, M. A., et al. 2005a, *ApJ*, 634, 142
Nagao, T., Motohara, K., Maiolino, R., et al. 2005b, *ApJ*, 631, L5
Nagao, T., Murayama, T., Maiolino, R., et al. 2007, *A&A*, 468, 877
Nagao, T., Sasaki, S., Maiolino, R., et al. 2008, *ApJ*, in press (arXiv:0802.4123)
Norman, M. L. 2008, in: B. W. O'Shea, A. Heger & T. Abel (eds.), *First Stars III* (New York: AIP), *AIP Conference Ser* in press (arXiv:0801.4924)
Oh, S. P., Haiman, Z., & Rees, M. J. 2001, *ApJ*, 553, 73
Ota, K., Iye, M., Kashikawa, N., et al. 2008, *ApJ*, submitted (arXiv.0707.1561)
Ouchi, M., Shimasaku, K., Okamura, S., et al. 2004, *ApJ*, 611, 660
Ouchi, M., Shimasaku, K., Akiyama, M., et al. 2005, *ApJ*, 620, L1
Ouchi, M., Shimasaku, K., Akiyama, M., et al. 2008, *ApJS*, in press (arXiv:0707.3161)
Partridge, R. B. & Peebles, P. J. E. 1967, *ApJ*, 147, 868
Rhoads, J. E. & Malhotra, S. 2001, *ApJ*, 563, L5
Rhoads, J. E., Malhotra, S., Dey, A., et al. 2000, *ApJ*, 545, L85
Rhoads, J. E., Dey, A., Malhotra, S., et al. 2003, *AJ*, 125, 1006
Rhoads, J. E., Panagia, N., Windhorst, R. A., et al. 2005, *ApJ*, 621, 582
Richard, J., Pello, R., Schaerer, D., et al. 2006, *A&A*, 456, 861
Santos, M. R., Ellis, R. S., Kneib, J. -P., et al. 2004, *ApJ*, 606, 683
Scannapieco, E., Schneider, R., & Ferrara, A. 2003, *ApJ*, 589, 35
Schaerer, D. 2002, *A&A*, 382, 28
Schaerer, D. 2003, *A&A*, 397, 527
Schneider, R., Salvaterra, R., Ferrara, A., & Ciardi, B. 2006, *MNRAS*, 369, 825
Scoville, N., Aussel, H., Brusa, M., et al. 2007, *ApJS*, 172, 1
Sekiguchi, K., et al. 2004, *Bull. AAS*, Vol. 36, p. 1478
Shapley, A. E., Steidel, C. C., Pettini, M., & Adelberger, K. L. 2003, *ApJ*, 588, 65
Shimasaku, K., Ouchi, M., Okamura, S., et al. 2003, *ApJ*, 586, L111
Shimasaku, K., Kashikawa, N., Doi, M., et al. 2006, *PASJ*, 58, 313
Stanway, E. R., Bunker, A. J., Glazebrook, K., et al. 2007, *MNRAS*, 376, 727
Stark, D. P., Ellis, R. S., Richard, J., et al. 2007, *ApJ*, 663, 10
Steidel, C. C., Adelberger, K. L., Giavalisco, M., et al. 1999, *ApJ*, 519, 1
Steidel, C. C., Adelberger, K. L., Shapley, A. E., et al. 2003, *ApJ*, 592, 728

Taniguchi, Y. 2001, in: N. Arimoto & W. Duschul (eds.), *The Japan-German Workshop on Studies of Galaxies in the Young Universe with New Generation Telescopes* (astro-ph/0301097)
Taniguchi, Y., Shioya, Y., Ajiki, M., et al. 2003, *JKAS*, 36, 123 (Erratum, 36, 283)
Taniguchi, Y., Ajiki, M., Nagao, T., et al. 2005, *PASJ*, 57, 165 (Erratum, 59, 277)
Taniguchi, Y., et al. 2008, *ApJ*, to be submitted
Tornatore, L., Ferrara, A., & Schneider, R. 2007, *MNRAS*, 382, 945
Tumlinson, J., Giroux, M. L., & Shull, M. 2001, *ApJ*, 550, L1
Tumlinson, J., Shull, M., & Venkatesan, A. 2002, *ApJ*, 584, 608
Tumlinson, J. & Shull, M. 2000, *ApJ*, 528, L65
Willis, J. P. & Courbin, F. 2005, *MNRAS*, 357, 1348
Willis, J. P., Courbin, F., Kneib, J. -P., & Minniti, D. 2008, *MNRAS*, 384, 1039
Yamada, S. F., Sasaki, S. S., Sumiya, R., et al. 2005 *PASJ*, 57, 881
Yan, H., Dickinson, M. Giavalisco, M., et al. 2006 *ApJ*, 651, 24

Discussion

ELDRIDGE: We (myself and Elizabeth Stanway) also have looked for He II $\lambda1640$ at high redshift. We have a stacked spectrum of 50 galaxies with a mean redshift of 4.7 and we see no He II to a quite low limit.

TANIGUCHI: Thank you for your comment.

PETTINI: In our survey of UV-bright galaxies at $z = 2 - 3$, we have examples of galaxies with stellar populations which seem very young, with ages of only a few tens of million of years. These objects may be the lower redshift analogues of your galaxies at $z = 6 - 7$ and, if they are, they are of course much easier to study spectroscopically at the lower redshifts. Somewhat surprising perhaps, even these very young galaxies have already fairly high metallicities, between 1/5 and 1/3 of solar. It seems that once star-formation starts in these kinds of high redshift galaxies, it proceeds very fast, so that the traces of the first few generations of stars are quickly swamped. Thus, you may find that it is also very difficult to catch one of your z = 6 - 7 galaxies, which are also undergoing vigorous star formation (otherwise you would not see them), at such an early stage that the signatures of Pop III stars can be discerned.

TANIGUCHI: Now, one of the most important issues in this research field is to search for galaxies with Pop III stars at high redshift. Aslthough theoretical considereations suggest that such galaxies could be found at $z > 2.5$. However, any trials made so far failed to identify such galaxies. I think that we need more systematic search for Pop III dominated galaxies at $z > 2.5$ by using a large sample of galaxies (i.e., LAEs) in near future.

Core-Collapse Supernovae as Dust Producers

Rubina Kotak

Astrophysics Research Centre, Queen's University Belfast,
Belfast, BT7 1NN, United Kingdom
email: r.kotak@qub.ac.uk

Abstract. Although it has long been hypothesised that core-collapse supernovae may produce large quantities of dust, interest in this problem has recently been rekindled given the enormous dust masses inferred at very high redshifts ($z \gtrsim 6$), when conventional low-mass dust-producing stars would fail to contribute significantly to the universal dust budget. Emission due to warm dust peaks at mid-IR wavelengths. However, with the notable exception of SN 1987A, supernova studies in the mid-IR have been virtually non-existent until the advent of the *Spitzer Space Telescope*. On behalf of the Mid-Infrared Supernova Consortium, I briefly discuss recent exciting results from mid-IR studies of core-collapse supernovae using *Spitzer* and attempt to put the role of supernovae as major dust producers into perspective.

Keywords. supernovae: general – supernovae: individual (SN 2002hh, SN 2003gd, SN 2004dj, SN 2004et, SN 2005af) – dust, extinction – infrared: stars

1. Introduction

Recent years have witnessed a flurry of studies that have emphasised the important role that dust plays in our understanding of the near and distant Universe. Dust formation in the interstellar medium (ISM) has been shown to be extremely inefficient. The preferred sites for dust formation are the atmospheres of evolved, low-mass ($M < 8\,M_\odot$) stars from where it is transported into the ISM via stellar winds. This mechanism however, fails to explain the presence of dust at high redshifts as the evolutionary time-scales of these low-mass stars (up to 1 Gyr) begin to become comparable to the age of the Universe. Furthermore, the IR luminosities of z>6 quasars (e.g. Bertoldi *et al.* 2003) imply enormous dust masses ($10^8\,M_\odot$). The short time-scales required for dust enrichment make core-collapse supernovae rather natural candidates for dust producers in the early Universe.

It has long been hypothesized (Cernuschi *et al.* 1967, Hoyle & Wickramasinghe, 1970, Tielens *et al.* 1990) that the physical conditions in the ejecta core-collapse supernovae may lead to the condensation of large amounts of dust. This stems from a combination of factors: (i) core-collapse supernova ejecta contain large amounts of refractory elements from which dust grains could form; (ii) cooling of the ejecta occurs by adiabatic expansion and, in some cases, by molecular emission; (iii) dynamical instabilities in the ejecta results in regions of enhanced density which may further aid the growth of grains by self-shielding. On the modelling side, there seems to be no apparent problem in producing substantial amounts (0.1-1 $M_\odot$) of dust in supernova ejecta, even at high redshifts. This implies a dust condensation efficiency of about 0.2 (Morgan & Edmunds, 2003).

Yet, the observational support for the hypothesis that grains condense in sufficient amounts in core-collapse supernovae is remarkably meagre, and recent claims are highly controversial. Two of the most compelling ways of detecting dust are: (i) the attenuation of spectral lines at optical/near-IR wavelengths in the nebular phase; (ii) thermal emission from dust grains. Until recently, the strongest evidence for dust formation in supernova

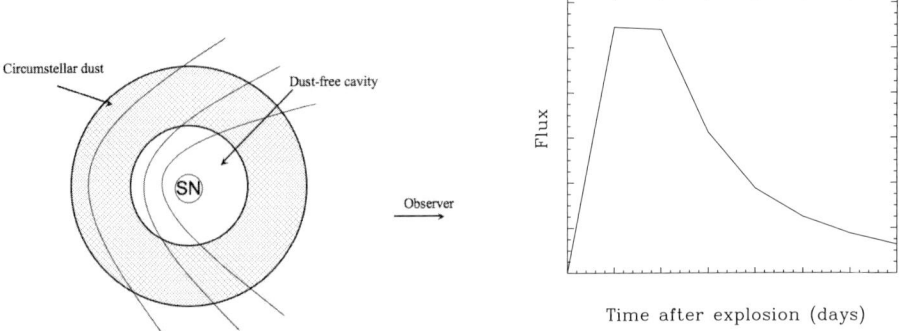

Figure 1. Left: Schematic illustration of an echo arising due to the explosion of a supernova in a dusty circumstellar medium. The series of paraboloids delineate the emitting volume which changes as a function of time. Adapted from Dwek (1983). **Right:** Example of a light curve resulting from a configuration such as that shown in the left-hand panel. Note the characteristic flat-top of the resulting light curve.

came from SN 1987A which showed a strong mid-IR excess that was accompanied by a decrease in optical emission and a blueward shift of emission line profiles (Lucy *et al.* 1989, Danziger *et al.* 1989, Wooden *et al.* 1991, Ercolano *et al.* 2007). However, even for this very well-studied object, the dust-mass estimates do not exceed 7.5×10^{-4} (Ercolano *et al.* 2007).

While the attenuation of spectral lines at late times is a relatively unambiguous signature of the presence of dust, it is difficult to derive quantitative measures of the amount and nature of the dust. As warm grains emit most strongly in the mid-IR, this is the ideal wavelength range for following dust condensation in real time. However, ground-based mid-IR observations are challenging – if not unfeasible – for the vast majority of supernovae. Even for SN 1987A (at only ~ 50 kpc), most of the mid-IR data came from the *Kuiper Airborne Observatory* (Wooden *et al.* 1991). Since the launch of the *Spitzer Space Telescope*, with vastly superior sensitivity and spatial resolution compared to previous instrumentation, this situation has been changing dramatically. With *Spitzer*, the Mid-Infrared Supernova Consortium (MISC)† has begun a vigorous programme of mid-IR supernova studies. Here, I will discuss recent exciting results, and attempt to put the role of core-collapse supernovae as dust producers into perspective.

2. New dust or circumstellar dust?

When studying the thermal emission from dust, it is important to bear in mind that even if a near- or mid-IR "excess" is detected, it might not necessarily be due to new dust that has condensed in the ejecta. Thermal emission may arise from pre-existing dust in the circumstellar medium e.g. due to a dusty wind from the progenitor star which has been heated by the flash from the supernova, resulting in an infrared echo (Bode & Evans, 1979, Dwek, 1983). Shock heating due to ejecta-circumstellar matter interaction may be another mechanism which gives rise to an echo. A schematic diagram of a configuration that would give rise to an echo is shown in Fig. 1.

Clearly, an infrared echo could potentially mask any signature of newly condensing dust, given that the magnitude of this effect. However, in general, emission due echoes

† http://star.pst.qub.ac.uk/~rk/misc.html

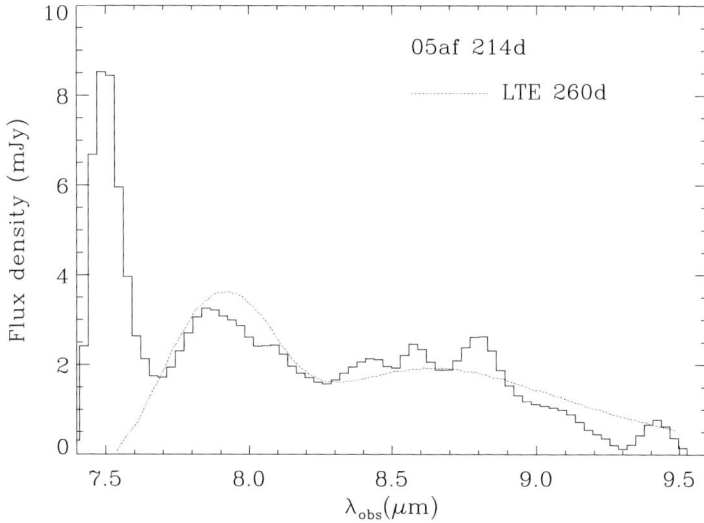

Figure 2. The 260 d LTE model from Liu & Dalgarno (1994), attenuated by a factor of 2.5, superimposed on the continuum-subtracted second-epoch spectrum of SN 2005af. From this, we deduce a SiO mass of $\sim 2 \times 10^{-4}$ $M_\odot$ (Kotak et al. 2006). The prominent line at 7.5 μm is due to [NiI].

tends to appear earlier in the evolution of a supernova, compared to dust formation, which tends to occur at epochs of several hundred days. A caveat, of course, is the geometry and spatial extent of the circumstellar matter. Given the characteristic light curves that arise from echoes, one way of potentially distinguishing between pre-existing and new dust is to monitor the light curve to determine its shape (see Fig. 1). In situations where there is contribution to the infrared luminosity from both an echo, and new dust, the situation is complicated, and detailed modelling is required. Meikle et al. (2006) report the first detection of an IR echo in the most common of supernova types, the type-IIP supernova SN 2002hh. However, given the complexity of the field around the supernova, it was not possible to conclusively establish whether the echo was due to dusty pre-existing circumstellar matter, or a dusty molecular cloud.

3. New mid-IR observations and models

The first mid-IR spectrum of any supernova obtained since SN 1987A was that of the type II-P supernova, SN 2004dj which, at distance of $\sim$3 Mpc, was also the nearest since SN 1987A. The spectrum showed a strong red-wing of the CO fundamental band, and prominent lines of [NiI] and [NiII] (see Fig. 2 in Kotak et al. (2005)). The features described above appear in all of the type II-P supernovae in our sample. Another example is shown in Fig. 3. SN 2005af, also a type II-P supernova, showed the clear development of the $\sim$7.5 μm SiO feature from day 67 to day 214 (Fig. 2) – again, the first time since SN 1987A. Using the SiO model of Liu & Dalgarno (1994), we estimate a SiO mass of $2 \times 10^{-4} M_\odot$.

The models shown in Fig. 3 are based on a spherical, uniform sphere of isothermal dust grains, following the escape probability treatment of Lucy et al. (1989). A typical grain size distribution is used for a mix of refractory materials as predicted by models of dust condensation in supernovae. In order to estimate the amount of dust, the mass

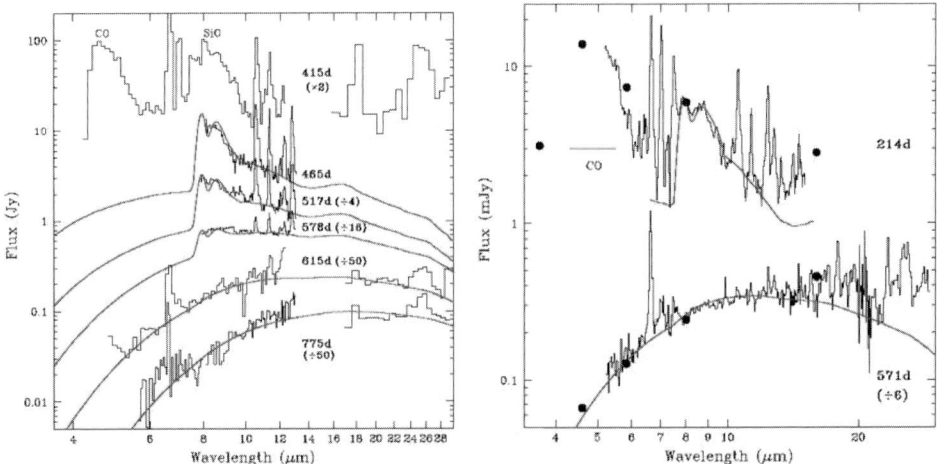

Figure 3. For both panels, the red lines show model fits to the data (see text) and include the SiO model of Liu & Dalgarno (1994). **Left panel:** Mid-IR spectra of SN 1987A, taken from Roche *et al.* (1993) and Wooden *et al.* (1993). **Right panel:** Spectra of the type II-P supernova, SN 2005af. The day 214 spectrum is taken from Kotak *et al.* (2006). Filled circles: 5-16 μm photometry.

of dust is increased until an adequate match to the spectrum is obtained. In order to be conservative, we model the spectra for the most optically thin case that will still provide an adequate fit to the spectra (see Meikle *et al.* (2007) for details).

This approach was first tested on SN 1987A (Fig. 3), which yields a maximum mass of $\sim 6 \times 10^{-4} M_\odot$. This is consistent with estimates using other methods e.g. $7.5 \times 10^{-4} M_\odot$ (Ercolano *et al.* 2007). As with SN 2004dj, and SN 2005af, emission due to SiO and CO is clearly evident in SN 1987A. Also, a cool dust continuum provides a good match to the spectra. The same pattern holds for SN 2005af, with strong molecular emission at earlier epochs (~ 200 d), which is replaced by a strong cool continuum at later times (~ 600 d). Our dust mass estimate for SN 2005af comes to $\sim 4 \times 10^{-4} M_\odot$. In the latter spectrum, there is a hint of an even cooler component, which might increase this estimate somewhat.

Clearly, the estimates above represent the amount of directly detected dust. A key point to remember is that these are only firm lower limits – there may be substantially more dust present in optically-thick clumps. This problem was already identified in the context of SN 1987A (Lucy *et al.* 1989, Wooden *et al.* 1993). The problems persists even at wavelengths as long as (24 μm, the extent of most of our data). In the event that clumps are optically-thick in the near-IR, but optically-thin in the mid-IR, it would be possible to obtain better constraints on the total dust mass by combining the data in both wavelength regions (Ercolano et al. 2007). However, for most – if not all – of our *Spitzer* targets, current indications are that the clumps are optically-thick in the mid-IR regime before significant dust condensation occurs.

Recently, Sugerman *et al.* (2006) claimed to have detected a significant amount of dust (0.02 $M_\odot$) in the type II-P supernova, SN 2003gd. However, reanalysis of the same data by Meikle *et al.* (2007) led them to conclude that SN 2003gd was not special in any way compared to the rest of the mid-IR II-P sample. They were forced to this conclusion for a number of reasons: Firstly, the large dust mass reported by Sugerman *et al.* (2006) rests entirely on their 24 μm flux measurement at an epoch of about 700 d. However, based on numerous tests of the sensitivity of the *Spitzer* images at 24 μm in the vicinity of the supernova, Meikle *et al.* (2007) showed that the Sugerman *et al.* (2006) flux is in error: it is too low by a factor of ~ 4. Secondly, their outer limit of their dust-forming zone lies at $8000\,\mathrm{km s}^{-1}$, which is conflict with the observed metal line velocities of $\sim 2000\,\mathrm{km s}^{-1}$ from late-time spectroscopy. Thus, the Sugerman *et al.* (2006) model implies a cooler dust component beyond $24\,\mu m$, for which there currently is no evidence. There is also a severe energy deficit in that the total radioactive decay energy deposited in the ejecta is a factor of ~ 4 too low to account for the observed luminosity. Even if one were to allow for a large uncertainty in the ^{56}Ni mass, the deficit would remain severe. Thus, on both energy and velocity grounds, most of the $24\,\mu m$ flux cannot be due to dust condensing in the ejecta.

4. Summary

From a sample of well-observed type II-P supernovae in the mid-IR, we find that all objects formed some dust. Prior to this work, evidence that the most common type of core-collapse supernova (the II-P subtype) form dust was meagre.

Grain formation models (e.g. Todini & Ferrara, 2001, Nozawa *et al.* 2003) predict that carbon, silicate, and magnetite grains should be present in substantial quantities, with the silicate grains probably dominating. Our sample of core-collapse supernovae all show evidence of strong emission due to CO, or SiO, or both at epochs as early as ~ 100 d. Thus, although our sample size remains small, all of the supernovae that showed evidence for dust condensation, also showed evidence at earlier epochs of strong molecular emission. Thus, molecules may well play a crucial role in the dust formation process.

Current estimates of the amount of dust remain small: 10^{-3} to $10^{-5} M_\odot$, i.e., 10–100 times lower than needed to account for the dust seen at high redshifts. However, the estimates presented here are only lower limits, and substantial amounts may well exist in optically-thick clumps.

Much work yet remains to be done in assessing the dust production as a function of supernova sub-type, and is limited mainly by current mid-IR facilities.

References

Bertoldi, F., Carilli, C. L., Cox, P., *et al.* 2003, *A&A*, 406, L55
Blair, W. P., Ghavamian, P., Long, K. S., *et al.* 2007 *ApJ*, 662, 998
Bode, M. & Evans, A. 1979, *A&A*, 73, 113
Cernuschi, F., Marsicano, F., & Codina, S. 1967, *Ann. d'Astrophys*, 30, 1039
Danziger, I. J., Gouiffes, C., Bouchet, P., & Lucy, L. B. 1989, *IAU Circ*, 4746, 1
Dwek, E. 1983, *ApJ*, 274, 175
Ercolano, B., Barlow, M. J., & Sugerman, B. E. K. 2007, *MNRAS*, 375, 753
Hoyle, F. & Wickramasinghe, N. C. 1970, *Nature*, 226, 62
Kotak, R., Meikle, P., van Dyk, S. D., *et al.* 2005, *ApJ*, 628, L123
Kotak, R., Meikle, P., Pozzo, M., *et al.* 2006, *ApJ*, 651, L117
Liu, W. & Dalgarno, A. 1994, *ApJ*, 428, L769
Lucy, L. B., Danziger, I. J., Gouiffes, C., & Bouchet, P. 1989, in: G. Tenorio-Tagle, M. Moles & J. Melnick (eds.), *Structure and Dynamics of the Interstellar Medium* (Berlin: Springer-Verlag), *Proc. IAU Coll 120*, 164
Meikle, W. P. S., Mattila, S., Gerardy, C. L., *et al.* 2006, *ApJ*, 649, 332
Meikle, W. P.S., Mattila, S., Pastorello, A., *et al.* 2007, *ApJ*, 665, 608
Morgan, H. L. & Edmunds, M. G., 2003, *MNRAS*, 343, 427
Nozawa, T., Kozasa, T., Umeda, H., *et al.* 2003, *ApJ*, 598, 785
Roche, P. F., Aitken, D. K., & Smith, C. H. 1993 *MNRAS*, 261, 522
Stanimirović, S., Bolatto, A. D., Sandstrom, K., *et al.* 2005, *ApJ* 632, L103

Sugerman, B. E. K., Ercolano, B., Barlow, M. J., *et al.* 2006, *Science*, 313, 196
Tielens, A. G. G. M., 1990, in *Carbon in the Galaxy: Studies from Earth and Space, NASA Conf. Publ.*, 3061, 59
Todini, P. & Ferrara, A. 2001, *MNRAS*, 325, 726
Wooden, D. H., Rank, D. M., Bregman, J. D., *et al.* 1993, *ApJS*, 88, 477

Discussion

MODJAZ: What about dust formation in supernova remnants?

KOTAK: Supernova remnants may be net creators of dust, but also net destroyers of dust. Currently, there is evidence on both sides: e.g. Rho *et al.* (this conference) report about $0.02\,M_\odot$ of dust in the Cassiopeia A supernova remnant, while Stanimirović *et al.* (2005) estimate in SNR 1E0102.2-7219 to be $\sim 8 \times 10^{-4} M_\odot$, and Blair *et al.* (2007) estimate $\sim 3 - 5 \times 10^{-4} M_\odot$ in the Kepler supernova remnant.

HIRSCHI: Do you expect dust formation in core-collapse supernovae to be metallicity-dependent?

KOTAK: Yes, one would naively expect this to be the case.

GRBs as Probes of Massive Stars Near and Far

Johan P. U. Fynbo and Daniele Malesani

Dark Cosmology Centre, Niels Bohr Institute, University of Copenhagen,
DK-2100 Copenhagen O, Denmark
email: jfynbo@dark-cosmology.dk

Abstract. Long-duration gamma-ray bursts are the manifestations of massive stellar death. Due to the immense energy release they are detectable from most of the observable universe. In this way they allow us to study the deaths of single (or binary) massive stars possibly throughout the full timespan massive stars have existed in the Universe. GRBs provide a means to infer information about the environments and typical galaxies in which massive stars are formed. Two main obstacles remain to be crossed before the full potential of GRBs as probes of massive stars can be harvested: *i)* we need to build more complete and well understood samples in order not to be fooled by biases, and *ii)* we need to understand to which extent GRBs may be intrinsically biased in the sense that they are only formed by a limited subset of massive stars defined by most likely a restricted metallicity interval. I describe the status of an ongoing effort to build a more complete sample of long-duration GRBs with measured redshifts. Already now we can conclude that the environments of GRB progenitors are very diverse with metallicities ranging from solar to a hundredth solar and extinction ranging from none to $A_V > 5$ mag. We have also identified a sightline with significant escape of Lyman continuum photons and another with a clear 2175 Å extinction bump.

Keywords. gamma rays: bursts – galaxies: distances and redshifts

1. Introduction

GRBs were discovered serendipitously in the late 1960's and first reported to the astrophysical community in the early 1970's (Klebesadel *et al.* 1973). For a long time their nature remained a mystery (e.g., Nemiroff 1994). For a review of the first decades of GRB research where only the prompt γ-ray emission was known see Fishman & Meegan (1995). The major breakthrough came in 1997 with the ability to determine celestial positions of GRBs thanks to the BeppoSAX satellite and the discovery of long-lived X-ray, optical and radio afterglows (e.g., Costa *et al.* 1997; van Paradijs *et al.* 1997; Frail *et al.* 1997). For a review of the early years of the so called afterglow era (after 1997) see van Paradijs *et al.* (2000).

GRBs come in at least two variants defined by their duration in the γ-ray band. The short bursts have duration less than about 2 se and the long bursts longer than this (Kouveliotou *et al.* 1993). In the reminder of this review we will only discuss the long duration bursts. The association between long-duration GRBs (LGRBs hereafter) and massive stars and hence the link between LGRBs and on-going massive star-formation found its first empirical basis with the detection of the first host galaxies (e.g., Hogg *et al.* 1999). Subsequently, the evidence was further strengthened with the discovery of supernovae (SNe) associated with LGRBs (Galama *et al.* 1998; Hjorth *et al.* 2003; Stanek *et al.* 2003; Malesani *et al.* 2004; Sollerman *et al.* 2006; Pian *et al.* 2006). For a recent review of the LGRB/SN association see Woosley & Bloom (2006).

Because of the link between LGRBs and massive stars and due to the fact that GRBs can be detected from both the most distant and the most dust obscured regions in the universe LGRBs were quickly identified to be very promising tracers of star-formation throughout cosmic history (e.g., Wijers *et al.* 1998). However, this potential has so far not really resulted in an improved census of the locations of massive stars due to complications discussed in the next section.

A major issue currently under discussion is if LGRBs are unbiased tracers of star formation. More precisely, it is not clear if LGRBs are caused by the same (small) fraction of all dying massive stars (unbiased tracers), or if LGRBs only trace a limited segment defined by parameters such as, e.g., metallicity or circumstellar density (biased tracers).

The currently operating *Swift* satellite (Gehrels *et al.* 2004) has revolutionized LGRB research with its frequent, rapid, and precise localization of LGRBs. Now it is for the first time possible in practice to use LGRBs as powerful probes. It is mandatory that this potential is exploited while *Swift* is still operating (at least until 2010).

2. Complications in the use of LGRBs as tracers of massive stars

2.1. Dark bursts and incomplete samples

A crucial issue when using LGRBs (or any other class of tracer) is sample selection. Whereas the detection of the LGRB itself poses no bias against dust obscured massive stars this is not the case for the softer afterglow emission which is crucial for obtaining the precise localization as well as measuring redshifts (see, e.g., Fiore *et al.* 2007).

In the samples of LGRBs detected with satellites prior to the currently operating *Swift* satellite the fraction of LGRBs with detected optical afterglows was only about 30% (Fynbo *et al.* 2001; Lazzati *et al.* 2002). Much of this incompleteness was caused by random factors such as weather or unfortunate celestial positions of the bursts, but some remained undetected despite both early and deep limits. It is possible that some of these so called "dark bursts" could be caused by LGRBs in very dusty environments (Groot *et al.* 1998) and hence the sample of LGRBs with detected optical afterglows could very well be systematically biased against dust obscured star formation (see also Jakobsson *et al.* 2004a; Rol *et al.* 2005; Rol *et al.* 2007 for recent discussions of the dark bursts).

In any case, such a high incompleteness imposes a large uncertainty on statistical studies based on LGRB host galaxies derived from these early missions. It should be stressed that the conclusions based on these samples may only be relevant for a minority of all LGRBs. Due to the much more precise and rapid localization capability of *Swift* it is now possible to build much more complete samples.

2.2. Are some LGRBs not associated with massive stellar death?

Recently, it has been found that some LGRBs are not associated with SNe, namely GRB 060505 (Fynbo *et al.* 2006a; Ofek *et al.* 2007) and GRB 060614 (Fynbo *et al.* 2006a; Della Valle *et al.* 2006; Gal-Yam *et al.* 2006). This means that either some massive stars die without producing SNe brighter than about $M_V=-13.5$ (Fynbo *et al.* 2006a, Della Valle *et al.* 2006) or, alternatively, some LGRBs are caused by other mechanisms than collapsing massive stars (Gal-Yam *et al.* 2006; Ofek *et al.* 2007).

At least for the case of GRB 060505 the evidence points to the former. The burst was located in a star-forming region in a relatively low metallicity region in the outer part of a spiral host (Fynbo *et al.* 2006a; Ofek *et al.* 2007; Thöne *et al.* 2008; see also Fig. 1). The burst itself displayed a significant spectral lag, which has so far never been seen for the short bursts that are believed to originate from merging compact objects (McBreen *et al.* 2008; Norris & Bonnell 2006). If some LGRBs indeed are caused by other progenitors than massive stars then a new classification that can distinguish between LGRBs from massive stars and those from other mechanisms is required. So far no such scheme has been found (see Gehrels *et al.* 2006).

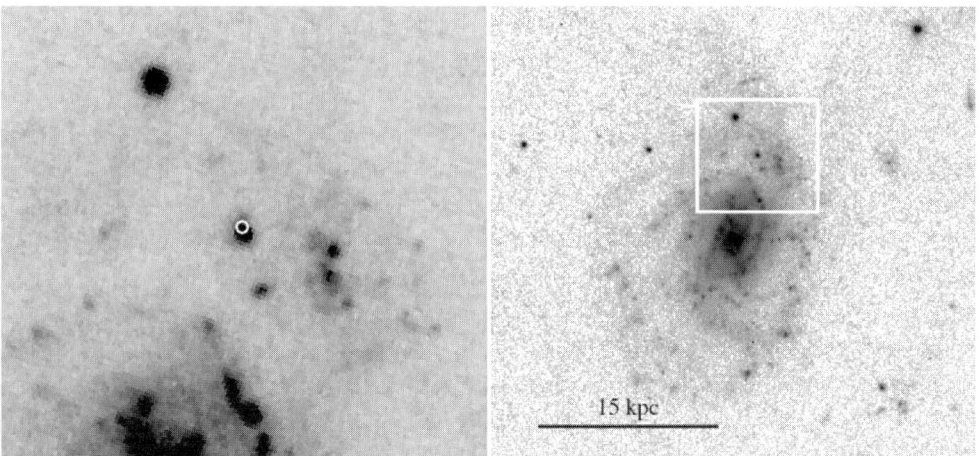

Figure 1. The host galaxy of GRB 060505 as observed with the HST (Ofek *et al.* 2007; Thöne *et al.* 2008). The white circle shows the error-circle of the burst consistent with a star-forming region in a spiral arm. This is strong evidence that the progenitor was a massive star. The properties of this host is also within the range found for other LGRBs. As an example, the host galaxy of GRB 050824 is very similar in terms of luminosity, R_{23} and star-formation rate (Sollerman *et al.* 2007), and the location within the host is similar to the location of other LGRBs in spiral hosts (Fynbo *et al.* 2000a; Sollerman *et al.* 2002; Le Floc'h *et al.* 2002; Jakobsson *et al.* 2005a).

2.3. The contamination from chance projections

The first important question to ask is: are LGRB host galaxies operationally well-defined as a class? In terms of an operational definition the case is not so clear. If we define the host galaxy of a particular burst to be the galaxy nearest to the line-of-sight, we need to worry about chance projection (Band & Hartmann 1998). In the majority of cases where an optical afterglow has been detected and localized with sub-arcsecond accuracy and where the field has been observed to deep limits a galaxy has been detected within an impact parameter less than 1 arcsec (see e.g., Bloom *et al.* 2002 and Fig. 2 for an example). The probability for this to happen by chance depends on the magnitude of the galaxy. The number of galaxies per arcmin^2 has been well determined to deep limits in the Hubble deep fields. In Fynbo *et al.* (2000b, their Fig. 2) the galaxy counts in the R and I bands based on the HDF-South can be found. To limits of R=24, 26 and 28 there are about 2, 6 and 13 galaxies arcmin^{-2}. Hence, the probability to find a R=24 galaxy by chance in an error circle with radius 0.5 arcsec is about 4×10^{-4}. For a R=28 galaxy the probability is about 3×10^{-3}. If the error circle is defined only by the X-ray afterglow with a radius of 2 arcsec in the best cases then we expect a random R=24 and R=28 galaxy in 0.6% and 5% of the error circles. For a sample of a few hundred LGRBs chance projection should hence not be a serious concern for LGRBs localized to sub-arcsecond precision, but for error-circles with radius of a few arcseconds we expect a few chance projections. In some cases it may be possible to eliminate the chance projects, e.g., based on conflicting redshift information from the afterglow and proposed host, but in general not.

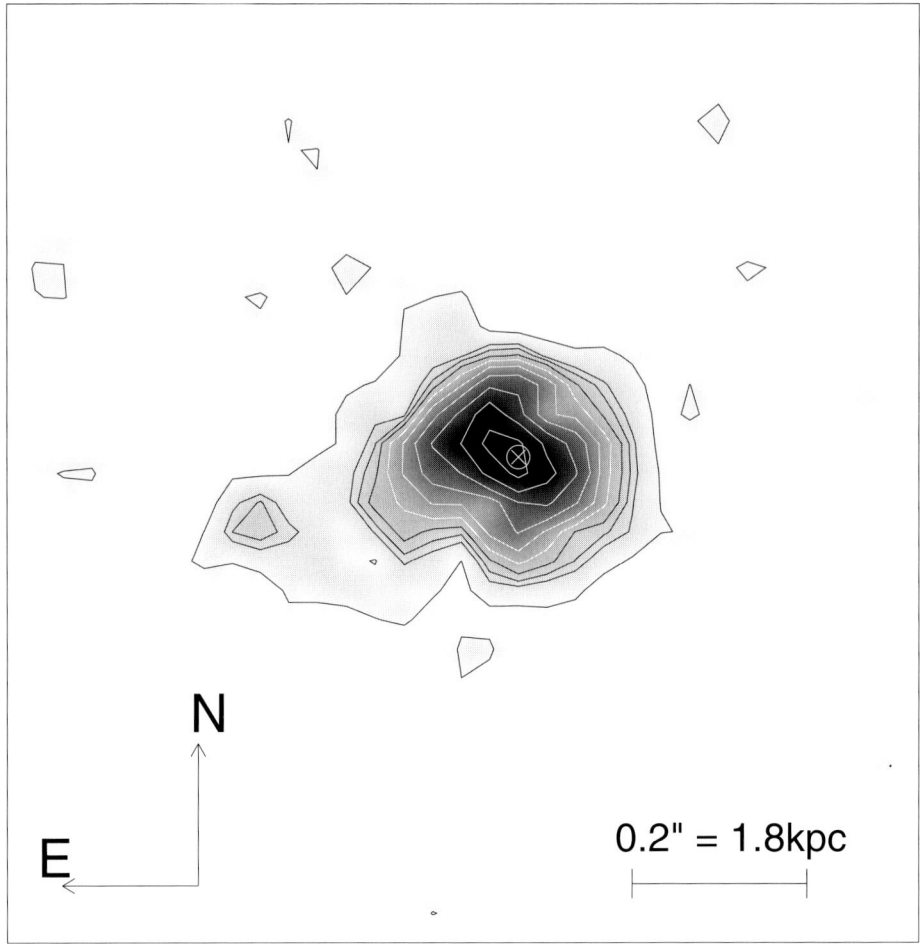

Figure 2. The 1×1 arcsec² field around the host galaxy of the $z = 2.33$ HETE-2 GRB 021004 observed with the *HST* (from Fynbo *et al.* 2005). The LGRB went off near the center of the galaxy. The position of the LGRB is marked with a cross and an error circle and it coincides with the centroid of the galaxy to within a few hundredths of an arcsec. In cases like this there is no problem in identifying the correct host galaxy. However, in cases of bursts localized to only a few arcsec accuracy chance projection needs to be considered.

3. Conclusions based on pre-Swift samples

Despite the complications mentioned above, the previous decade of LGRB host galaxy studies has after all taught us a lot about LGRBs and their link to massive star formation (see van Paradijs *et al.* 2000 and Djorgovski *et al.* 2003 for early reviews). LGRB hosts were early on found to be predominantly faint, blue star-forming galaxies (Hogg *et al.* 1999). The early studies found that these properties of LGRB hosts were consistent with the expectation if LGRBs are unbiased tracers of star-formation (Mao *et al.* 1998; Hogg *et al.* 1999). It was also realized early on that LGRBs offer a unique possibility to locate and study star-formation activity in dwarf galaxies at $z > 2$ (Jensen *et al.* 2001). This is basically impossible with any other currently existing method. The star-formation rates were found to be modest, but the specific star-formation rates among the highest ever found (Christensen *et al.* 2004). LGRB hosts are hence often in a starburst state.

Later evidence indicated that LGRBs maybe related only to massive stars with metallicity below a certain threshold. The first evidence for this came with the realization the LGRB hosts were fainter and bluer than expected according to certain models about the nature of the galaxies dominating the integrated star-formation activity (Le Floc'h et al. 2003, 2006; Tanvir et al. 2004). Nevertheless, it has recently been pointed out by Priddey et al. (2006) that "there is sufficient uncertainty in models and underlying assumptions, as yet poorly constrained by observation (e.g., the adopted dust temperature) that a correlation between massive, dust-enshrouded star formation and GRB production cannot be firmly ruled out." (see also Michałowski et al. 2008 concerning the issue of dust temperature). Further circumstantial evidence for a preference towards low metallicity came from the observation that Lyman-α emission seemed to be ubiquitous for LGRBs hosts (Fynbo et al. 2003; Jakobsson et al. 2005b).

Lately, evidence from hosts of more local LGRBs seems to point in the same direction. Several studies have found that local LGRBs hosts tend to be faint and metal poor although a caveat for some of these studies is the difficulty of using the strong line metallicity indicators like R_{23} to derive robust metallicities (Prochaska et al. 2004; Sollerman et al. 2005; Gorosabel et al. 2005; Stanek et al. 2006; Wiersema et al. 2007). However, nearly all of these studies targeted very incomplete pre-*Swift* samples and this raises the question whether the predominance of faint, metal poor hosts can be explained by, e.g., a bias against metal rich hosts and hence more dust obscured LGRB afterglows.

A very important result is that LGRBs and core-collapse SNe are found in different environments (Fruchter et al. 2006). The same study also found that LGRB host galaxies at $z < 1$ are fainter than the host galaxies of core-collapse SNe. This study is also based on incomplete pre-*Swift* samples, but as the SNe samples are if anything more biased against dusty regions than LGRBs this result does seem to be substantial evidence that LRBs are biased towards massive stars with relatively low metallicity. Wolf & Podsiadlowski (2007) however, find, based on an analysis of the Fruchter et al. (2006) data, that the metallicity threshold cannot be below half the solar metallicity. Concerning the different environments of core-collapse SNe and LGRBs it has recently been found that type Ic SNe have similar positions relative to their host galaxy light profiles as LGRBs, whereas all other SN types have a similar distribution, less centred on their host light than LGRBs and SN Ic's (Kelly et al. 2007). Larsson et al. (2007) find that the different distributions of different SN types relative to their host light can be naturally explained by assuming different mass ranges for the typical progenitor stars: $\gtrsim 8$ M$_\odot$ for typical core-collapse SNe and $\gtrsim 20$ M$_\odot$ for LGRB progenitors. The picture is complicated by the finding that type Ic SNe typically are found in substantially more metal rich environments than LGRBs (Modjaz et al. 2008, and in these proceedings). It is well established that WR-stars become more abundant with *increasing* metallicity - opposite to LGRBs that if anything are biased towards low metallicity. Taken together these findings suggest that progenitors of LGRBs and "normal" type Ic SNe are two different subsets of the $\gtrsim 20$ M$_\odot$ stars. For a thorough discussion of the relation between WR stars, SN Ic's and LGRBs we refer to Crowther (2007).

4. Building a complete sample of *Swift* LGRBs

The following is to a large extent based on Fynbo et al. (2007). The *Swift* satellite has been operating for about three years and is far superior to previous GRB missions. The reason for this is the combination of several factors: *i)* it detects LGRBs at a rate of about two bursts per week about an order of magnitude larger than the previous successful BeppoSAX and HETE-2 missions; *ii)* with its X-Ray Telescope (XRT) it localizes the bursts with a precision of about 5 arcsec also orders of magnitude better than previous missions; *iii)* it has a much shorter reaction time, allowing the study of the evolution of the afterglows literally seconds after the burst, sometimes during the prompt γ-ray emission itself. The *Swift* mission is funded at least until 2010. A crucial objective is to secure a large sample, as complete as possible, of LGRB afterglows while *Swift* is still operating. More concretely, rather than including all *Swift* detected GRBs, it is more optimal to concentrate on those LGRB afterglows with favourable observing conditions. Our group uses the following sample criteria:

(*a*) XRT afterglow detected within 12 hr
(*b*) Small foreground Galactic extinction: $A_V < 0.5$ mag
(*c*) Favourable declination: $-70 < \mathrm{Dec} < 70$
(*d*) Sun distance larger than 55°

By introducing these constraints, we are not biasing the sample towards optically bright afterglows, but we select a sample for which useful follow-up observations are likely to be secured.

About 50% of all *Swift* LGRBs do not fulfill these criteria, primarily because *Swift*, for technical reasons, has to point close to the Sun a significant fraction of the time. For bursts fulfilling the above criteria, we make every possible effort to detect optical and near-infrared afterglows and to measure their redshifts. As shown below, we have been very successful in this effort, using mainly the ESO VLT. Redshifts, or more generally spectroscopic observations, are crucial for almost all LGRB-related science. The most important science cases for which spectroscopy is critical are listed below:

• Determining the luminosity function for LGRBs (prompt emission as well as afterglows).
• Determining the redshift distribution of LGRBs and using LGRBs as tracers for the cosmic star-formation history (Jakobsson *et al.* 2006a; Fiore *et al.* 2007).
• Studying the host galaxies, in particular those faint, high-redshift galaxies that are unlikely to be found and studied with other methods (e.g., Vreeswijk *et al.* 2004).
• Studying LGRB-selected absorption-line systems (e.g., Jakobsson *et al.* 2004b; Prochter *et al.* 2006).
• Characterizing the dust extinction curves of high-z galaxies (e.g., Jensen *et al.* 2001; see also Fig. 4).
• Determining the Lyman continuum escape fraction from high-z galaxies (Chen *et al.* 2007, see also Fig. 5).
• Spotting very high redshift LGRBs (e.g., Kawai *et al.* 2006; Ruiz-Velasco *et al.* 2007).
• Probing cosmic chemical evolution with LGRBs (e.g., Savaglio 2006; Fynbo *et al.* 2006b; Prochaska *et al.* 2007).
• Studying if LGRBs can be used for cosmography (e.g., Ghirlanda *et al.* 2004).

Since the launch of *Swift*, we have had programmes running at the VLT with the aim of securing redshifts for *Swift* LGRBs. The status at the time of writing is that 109 *Swift* LGRBs fulfilled our selection criteria. For 57 of these a redshift measurement has been secured (see Fig. 6). The VLT has been the dominant single contributor in all LGRB redshift measurements, providing around 40% of the secure redshifts to date. The redshift of the most distant known GRB 050904 at $z = 6.295$ was measured with the Subaru telescope (Kawai *et al.* 2006). Most of the other redshifts have been measured using other 6–10 m telescopes (Keck, Gemini, Subaru, Magellan). This is contrary to the expectations prior to the launch of *Swift*, where it was suspected that *Swift* itself, or at least 2–4-m telescopes, would be able to measure most of the redshifts. However, optical afterglows turned out to be much fainter at early times than anticipated. As we show in Fig. 3, the majority of the afterglows are fainter than $R = 20$ when a slit can be placed

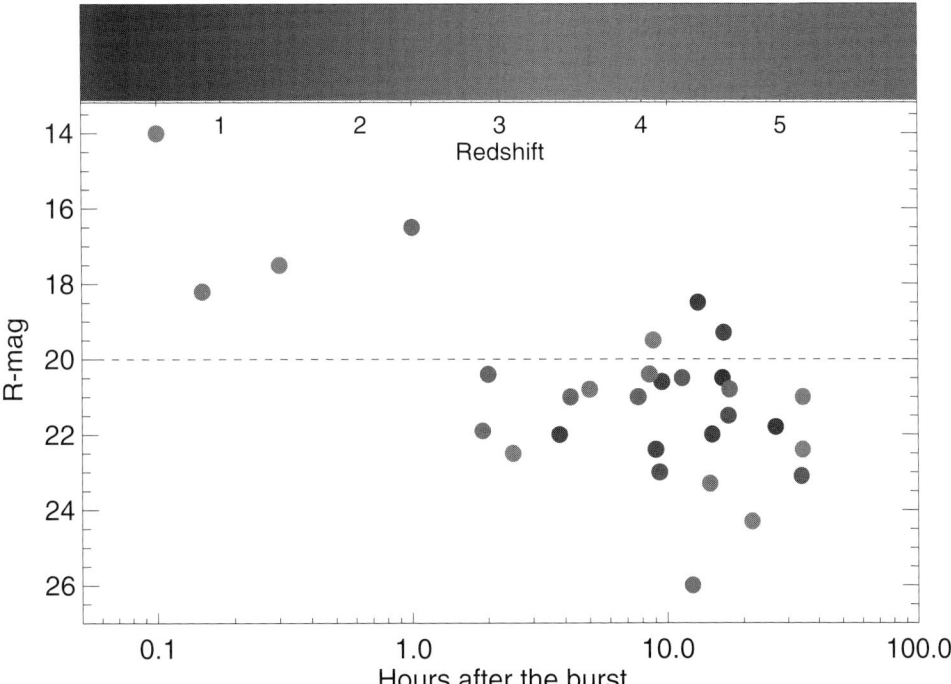

Figure 3. The R-band magnitude of the optical afterglows as a function of the time after the burst at which the spectroscopic observations were obtained. Only included are Swift bursts for which we have measured the redshift (using primarily the VLT, but also NOT, WHT and Gemini). The bar at the top indicates the code for the measured redshifts. The dashed line marks a magnitude of $R = 20$ which is roughly the spectroscopic limit for 2–4-m telescopes for detecting absorption lines. As seen, most afterglows are fainter than this limit when observable.

on them. $R = 20$ is, in our experience, the limit for spectroscopic redshift determination using 2–4-m telescopes (typically, no more than 1–2 hr exposure time is available for observing LGRB afterglows). Several optical afterglows are already fainter than $R = 22$ a few hours after the bursts. Hence, 6–10 m telescopes are crucial for securing redshifts for the majority of *Swift* LGRBs.

The fact that in particular the VLT, but also other 6-10 m telescopes, have made tremendous efforts to secure redshifts means that we now have a much higher redshift completion than for pre-*Swift* samples. But it is clear that we will not get redshifts for all bursts from spectroscopy of the afterglows for multiple reasons. In about 20-30% of the triggers we are not able to measure the redshift either due to lack of lines (probably bursts at redshifts between 1 and 2, see Fig. 6), bad weather or because the afterglow has faded too much before it is observable from Paranal. For these bursts our only chance of measuring the redshift is via spectroscopy of the host galaxy. We have also pursued this route extensively in an ESO large program (PI Hjorth). This is a challenging task due to the faintness of these systems, and the analysis of these data is still ongoing, but we have already determined a number of redshifts (included in Fig. 6).

4.1. The redshift distribution of Swift LGRBs: current status

The first conclusion from Fig. 6 is that *Swift* LGRBs are very distant. *Swift* LGRBs are more distant than LGRBs from previous missions due to the higher sensitivity of the satellite to the lower energies prevalent in the more distant events (Fiore *et al.* 2007).

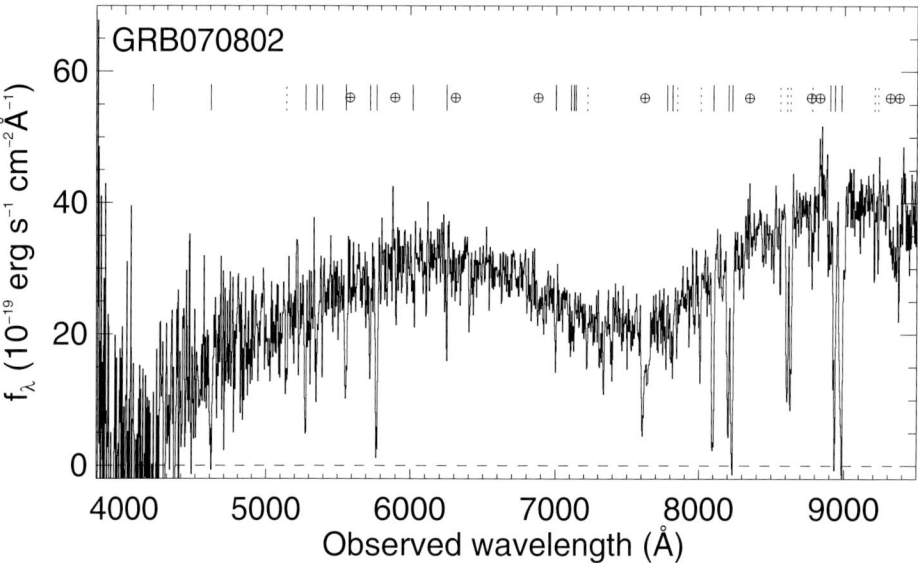

Figure 4. The VLT/FORS2 spectrum of the afterglow of GRB 070802 (Elíasdóttir *et al.*, in preparation). Plotted is the flux-calibrated spectrum against observed wavelength. Metal lines at the host redshift are marked with full-drawn lines whereas the lines from two intervening systems are marked with dotted lines. The broad depression centred around 7500 Å is caused by the 2175 Å extinction bump in the host system at $z_{\rm abs} = 2.4549$.

The median and mean redshift are now both 2.3, while for previous missions it was closer to 1 (Jakobsson *et al.* 2006a). The record holder is $z = 6.295$ (Kawai *et al.* 2006). It is striking how events at redshifts as large as 6 can be detected within such a small sample. For comparison, only a few QSOs are detected at similar distances out of a sample of hundred thousand QSOs. Remarkably, the redshift distribution, measured for just over 50% of all bursts, is consistent with the redshift distribution predicted if LGRBs are unbiased tracers of star formation (see, e.g., Jakobsson *et al.* 2006a and http://www.dark-cosmology.dk/~pallja/GRBsample.html for a regularly updated analysis).

4.2. *HI column densities*

The HI column density distribution for LGRB sightlines is extremely broad. It covers a range of about 5 orders of magnitude from $\sim 10^{17}$ cm^{-2} (Fig. 5, Chen *et al.* 2007) to nearly 10^{23} cm^{-2} (Jakobsson *et al.* 2006b). I still remain to be understood if this distribution is representative of the intrinsic distribution of HI column densities towards massive stars in galaxies or if the distribution is rather controlled by the ionizing emission from the afterglows themselves. In any case, as pointed out by Chen *et al.* (2007) the HI column density distribution provides an upper limit to the escape fraction of Lyman continuum emission from star-forming galaxies.

4.3. *Metallicities*

Afterglow spectroscopy often allows us to measure the metallicity of the line-of-sight in the host galaxy. In Fig. 7 we plot the metallicities along LGRB sightlines together with metallicities derived from QSO damped Lyman-α absorbers (QSO-DLAs). Here it can be seen that LGRBs are more metal rich than QSO-DLAs at similar redshifts. Some of the LGRB sightlines are almost as metal rich as the Lyman-break galaxies at similar redshifts (Pettini *et al.* 2001). The shift in metallicity relative to QSO-DLAs

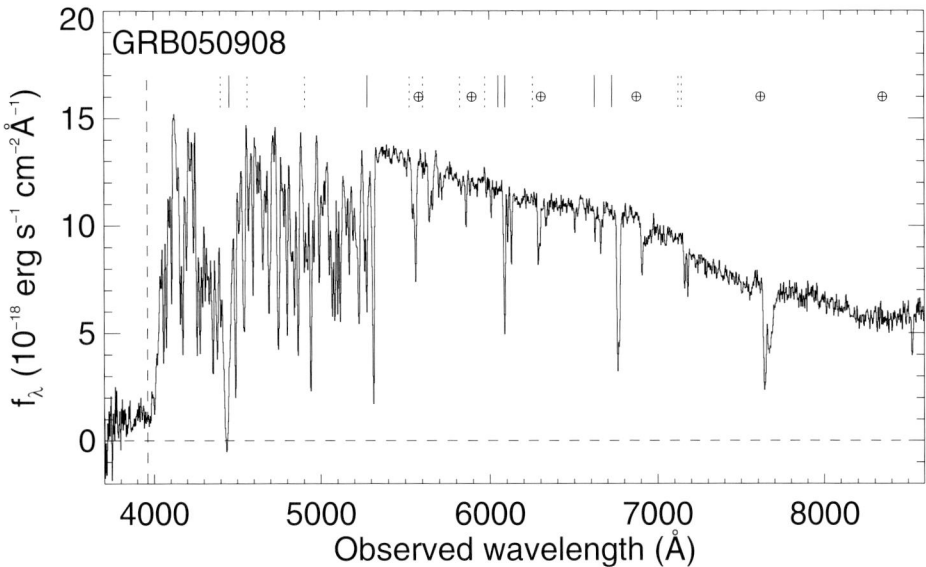

Figure 5. The VLT/FORS2 spectrum of the afterglow of GRB 050908 (Smette et al., in preparation). Plotted is the flux-calibrated 1-dimensional spectrum against observed wavelength. The vertical dashed line shows the position of the Lyman limit at the redshift of the GRB ($z = 3.343$). As seen, there is clear excess flux below the Lyman limit.

can be understood from the different selection functions of the (star-formation selected) LGRB-DLAs and the (HI cross-section selected) QSO-DLAs combined most likely with metallicity gradients in high-z galaxies (Fynbo et al. 2008). Hence, most likely LGRBs will give a reasonably unbiased census of where the massive stars are located, at least at $z > 2$ (Fynbo et al. 2006b).

4.4. Extinction

In addition to HI column densities, metal and molecular abundances and kinematics, the afterglow spectra also provide information of the extinction curves. The intrinsic spectrum of the afterglow is from theory predicted to be a power-law and therefore any curvature or other broad features in the spectrum can be interpreted as being due to features in the extinction curve. So far, almost all the extinction curves derived for LGRB host galaxy sightlines have been consistent with an extinction curve similar to that of the SMC. Recently, we obtained the clearest yet detection of the 2175 Å bump known from the Milky Way in a z=2.45 LGRB (Eliasdottir et al., in preparation and Fig. 4). This LGRB absorber also has unusually strong metal lines suggesting that the presence of the 2175 Å extinction bump is related to a high metallicity. However, we have examples of LGRBs with nearly solar metallicity for which the bump is not seen so it seems that metallicity is not the only parameter controlling the presence of the 2175 Å extinction bump. Concerning the amount of extinction the LGRB sightlines vary from no extinction (e.g., GRB050908, Fig. 5) to $A_V > 5$ mag (e.g., GRB070306, Jaunsen et al. 2008).

5. Conclusions and outlook

Spectroscopy of LGRB afterglows provides redshifts and information on the ISM properties for the population of galaxies containing the bulk of the high-redshift massive stars.

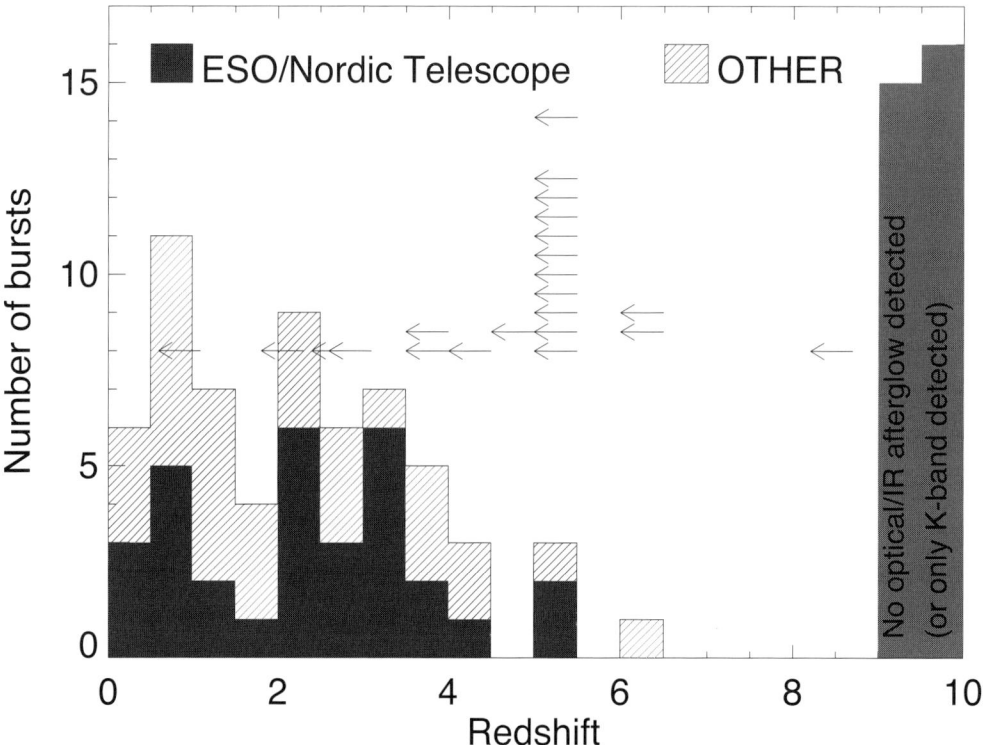

Figure 6. Redshift distribution (up to October 2007) of 109 *Swift* LGRBs localized with the X-ray telescope and with low foreground extinction $A_V < 0.5$. Of the 58 measured redshifts, our group has measured nearly half (25, shown in black). As shown, the VLT is the dominant source of redshifts in the *Swift* era (four of the black bursts in the histogram are also from the Nordic Optical Telescope). Bursts, for which only an upper limit on the redshift could be established so far, are indicated by arrows. Note that it is also difficult to secure redshifts for GRBs in the desert between $z = 1$ and $z = 2$. The grey histogram at the right indicates the 27 bursts for which no optical/J/H afterglow was detected and hence no redshift constraint could be inferred (see Ruiz-Velasco et al. 2007 for a full discussion).

We are currently working on securing this information for a complete sample of *Swift* LGRBs. Already now we can conclude that the environments of LGRB progenitors are very diverse with metallicities ranging from solar to a hundredth solar and extinction ranging from none to $A_V > 5$ mag. We have also identified a sightline with significant escape of Lyman continuum photons and another with a clear 2175 Å extinction bump.

Even though the completeness of the current *Swift* sample in terms of detections of optical afterglows (∼75%), redshift determinations (∼50%) and host galaxy detections (∼80%) is much higher than for LGRBs from previous missions we still need to do better - preferably all three fractions should be $\gtrsim 80\%$. We also need to improve the understanding of the link between LGRBs and massive stars. The SN-less GRBs GRB 060505 and GRB 060614 show that either some LGRBs are unrelated to massive stellar death or some massive stars die without causing SNe. The currently unfolding SN2008D/XRF080109 shows that there may be much more frequent bursts with softer prompt emission bridging the gap (in terms of both burst, SN and host properties) between GRBs and normal Ic SNe (e.g., Soderberg et al. 2008; Malesani et al., 2008).

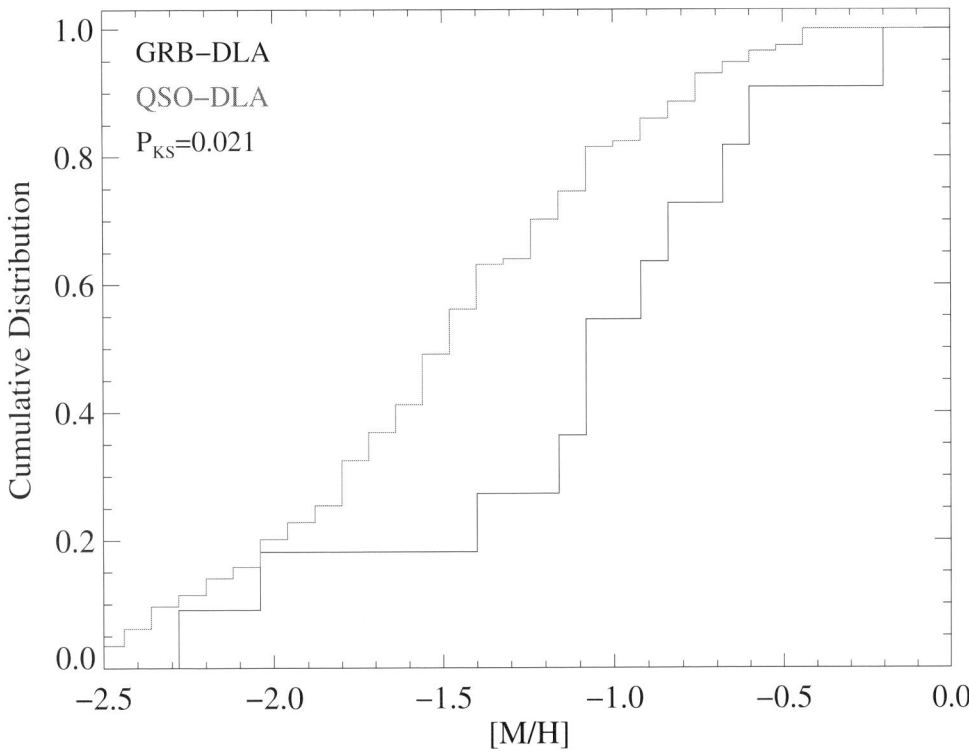

Figure 7. The histograms show the cumulative distribution of QSO-DLA and LGRB-DLA metallicities in the statistical samples compiled by Prochaska *et al.* (2003) and Prochaska *et al.* (2007). As seen, the LGRB-DLA metallicities are systematically higher than the QSO-DLA metallicities. See also Fynbo *et al.* (2008) for a simple model of these two distributions.

Acknowledgements

We thank our collaborators and the *Swift* team for carrying out such a wonderful experiment. The Dark Cosmology Centre is funded by the DNRF.

References

Band, D. L. & Hartmann, D. H. 1998 *ApJ*, 493, 555
Bloom, J. S., Kulkarni, S. R., & Djorgovski, S. G. 2002 *AJ*, 123, 1111
Chen, H.-W., Prochaska, J. X., & Gnedin, N. Y. 2007 *ApJ*, 667, L125
Christensen, L., Hjorth, J., & Gorosabel, J. 2004 *A&A*, 425, 913
Costa, E., Frontera, F., Heise, J., *et al.* 1997 *Nat*, 387, 783
Crowther, P. A. 2007 *ARA&A*, 45, 177
Della Valle, M., Chincarini, G., Panagia, N., *et al.* 2006 *Nat*, 444, 1050
Djorgovski, S. G., Kulkarni, S. R., Frail, D. A., *et al.* 2003, in: P. Guhathakurta (ed.), *Discoveries and Research Prospects from 6-10m Class Telescopes* (Bellingham: SPIE) *Proc. SPIE*, Vol. 4834, 238
Fiore, F., Guette, D., Piranomonte, S., *et al.* 2007 *A&A*, 470, 515
Fishman, G. J. & Meegan, C. A. 1995 *ARA&A*, 33, 415
Frail, D. A., Kulkarni, S. R., Nicastro, L., *et al.* 1997 *Nat*, 389, 261
Fruchter, A. S., Levan, A., Strolger, L., *et al.* 2006 *Nat*, 441, 463
Fynbo, J. P. U., Holland, S. T., Andersen, M. I., *et al.* 2000a *ApJ*, 542, L89
Fynbo, J. P. U., Freudling, W., & Møller, P. 2000b *A&A*, 355, 37

Fynbo, J. P. U., Jensen, B. L., Gorosabel, J., *et al.* 2001 *A&A*, 369, 373
Fynbo, J. P. U., Jakobsson, P., Møller, P., *et al.* 2003 *A&A*, 406, L63
Fynbo, J. P. U., Gorosabel, J., Smette, A., *et al.* 2005 *ApJ*, 633, 317
Fynbo, J. P. U., Starling, R. L. C., Ledoux, C., *et al.* 2006 *A&A*, 451, L47
Fynbo, J. P. U., Watson, D., Thöne, C. C., *et al.* 2006 *Nat*, 444, 1047
Fynbo, J. P. U., Vreeswijk, P., Jakobsson, P., *et al.* 2007 *ESO-Messenger*, 130, 43
Fynbo, J. P. U., Prochaska, J. X., Sommer-Larsen, J., *et al.* 2008 *ApJ*, submitted (arXiv:0801.3273)
Galama, T. J., Vreeswijk, P. M., van Paradijs, J., *et al.* 1998 *Nat*, 395, 670
Gal-Yam, A., Fox, D. B., Price, P. A., *et al.* 2006 *Nat*, 444, 1053
Gehrels, N., Chincarini, G., Giommi, P., *et al.* 2004 *ApJ*, 611, 1005
Gehrels, N., Norris, J. P., Barthelmy, S. D., *et al.* 2006 *Nat*, 444, 1044
Ghirlanda, G., Ghisellini, G., Lazzati, D., & Firmani, C. 2004 *ApJ*, 613, L13
Gorosabel, J., Pérez-Ramírez, D., Sollerman, J., *et al.* 2006 *A&A*, 444, 711
Groot, P., Galama, T. J., van Paradijs, J., *et al.* 1998 *ApJ*, 493, L27
Hjorth, J., Sollerman, J., Møller, P., *et al.* 2003 *Nat*, 423, 847
Hogg, D. W. & Fruchter, A. S. 1999 *ApJ*, 520, 54
Jakobsson, P., Hjorth, J., Fynbo, J. P. U., *et al.* 2004a *ApJ*, 617, L21
Jakobsson, P., Hjorth, J., Fynbo, J. P. U., *et al.* 2004b *A&A*, 427, 785
Jakobsson, P., Frail, D. A., Fox, D. B., *et al.* 2005a *ApJ*, 629, 45
Jakobsson, P., Björnsson, G., Fynbo, J. P. U., *et al.* 2005b *MNRAS*, 362, 245
Jakobsson, Levan, A., Fynbo, J. P. U., *et al.* 2006a *A&A*, 447, 897
Jakobsson, Fynbo, J. P. U., Ledoux, C., *et al.* 2006b *A&A*, 460, L13
Jaunsen, A. O., Rol, E., Watson, D. J., *et al.* 2008 *ApJ*, in press (arXiv:0803.4017)
Jensen, B. L., Fynbo, J. P. U., Gorosabel, J., *et al.* 2001 *A&A*, 370, 909
Kawai, N., Kosugi, G., Aoki, K., *et al.* 2006 *Nat*, 440, 184
Kelly, P. L., Kirshner, R. P., & Pahre, M. 2008, submitted (arXiv:0712.0430)
Klebesadel, R. W., Strong, I. B., & Olson, R. A. 1973, *ApJ*, 182, L85
Kouveliotou, C., Meegan. C. A., Fishman, G. J., *et al.* 1993, *ApJ*, 413, L101
Larsson, J., Levan, A. J., Davies, M. D., & Fruchter, A. S. 2007 *MNRAS*, 376, 1285
Lazzati, D., Covino, S., & Chisellini, G. 2002 *MNRAS*, 330, 583
Le Floc'h, E., Duc, P.-A., Mirabel, I. F., *et al.* 2002 *ApJ*, 581, L81
Le Floc'h, E., Duc, P.-A., Mirabel, I. F., *et al.* 2003 *A&A*, 400, 499
Le Floc'h, E., Charmandaris, V., Forrest, W. J., *et al.* 2006 *ApJ*, 642, L636
Mao, S. & Mo, H. J. 1998 *ApJ*, 339, L1
Malesani, D., Tagliaferri, G., Chincarrini, G., *et al.* 2004 *A&A*, 609, L5
Malaseni, D., Fynbo, J. P. U., Hjorth, J., *et al.* 2008, *ApJL*, submitted (arXiv:0805.1188)
McBreen, S., Foley S., Watson D., *et al.* 2008 *ApJL*, *ApJ*, 667, L85
Michałowski, M., Hjorth, J., Castro-Cerón, & J. M., Watson, D. 2008 *ApJ*, 672, 817
Modjaz, M., Kewley, L., Kirshner, R. P., *et al.* 2008 *AJ*, 135, 1136
Nemiroff, R. J. 1994, *Comments on Astrophysics*, 17, 189 (astro-ph/9402012)
Norris, J. P. & Bonnell, J. T. 2006 *ApJ*, 643, 266
Ofek, E. O., Cenko, S. B., Gal-Yam, A., *et al.* 2006 *ApJ*, 662, 1129
Pettini, M., Shapley, A. E., Steidel, C. C., *et al.* 2001 *ApJ*, 554, 981
Pian, E., Mazzali, P. A., Masetti, N., *et al.* 2006 *Nat*, 442, 1010
Priddey, R. S., Tanvir, N. R., Levan, A. J., *et al.* 2006 *MNRAS*, 369, 1189
Prochaska, J. X., Gawiser, E., Wolfe, A. M., *et al.* 2003 *ApJ*, 595, L9
Prochaska, J. X., Bloom, J. S., Chen, H.-W., *et al.* 2004 *ApJ*, 611, 200
Prochaska, J. X., Chen, H.-W., Dessauges-Zavadsky, M., & Bloom, J. S. 2007 *ApJ*, 666, 267
Prochter, G. E., Prochaska, J. X., Chen, H.-W., *et al.* 2006 *ApJ*, 698, L93
Rol, E., Wijers, R. A. M. J., Kouveliotou, C., *et al.* 2005 *ApJ*, 624, 868
Rol, E., van der Horst, A., Wiersema, K., *et al.* 2007 *ApJ*, 669, 1098
Ruiz-Velasco, A. E., Swan, H., Troja, E., *et al.* 2007 *ApJ*, 669, 1
Savaglio, S. 2006 *New Journal of Phys.*, 8, 195

Soderberg, A. M., Berger, E., Page, K. L. 2008, in press (arXiv:0802.1712)
Sollerman, J., Holland, S. T., Challis, P., et al. 2002 A&A, 386, 944
Sollerman, J., Östlin, G. Fynbo, J. P. U., et al. 2005 NewA, 11, 103
Sollerman, J., Jaunsen, A. O., Fynbo, J. P. U., et al. 2006 A&A, 454, 503
Sollerman, J., Fynbo, J. P. U., Gorosabel, J., et al. 2007 A&A, 466, 839
Stanek, K. Z., Matheson, T., Garnavich, P. M., et al. 2003 ApJ, 591, L17
Stanek, K. Z., Gnedin, O. Y., Beacom, J. F., et al. 2006 AcA, 56, 333
Tanvir, N., Barnard, V. E., Blain, A. W., et al. 2004 MNRAS, 352, 1073
Thöne, C. C., Fynbo, J. P. U., Östlin, G., et al. 2008 ApJ, 676, 1151
van Paradijs, J., Groot, P., Galama, T. J., et al. 1997 Nat, 386, 686
van Paradijs, J., Kouveliotou, C., & Wijers, R. A. M. J. 2000 ARA&A, 38, 379
Vreeswijk, P. M., Ellison, S. L., Ledoux, C., et al. 2004 A&A, 419, 927
Wiersema, K., Savaglio, S., Vreeswijk, P. M., et al. 2007 A&A, 464, 529
Wijers, R. A. M. J., Bloom, J. S., Bagla, J. S., & Natarajan, P. 1998 MNRAS, 294, L13
Wolf, C. & Podsiadlowski, P. 2007 MNRAS, 375, 1049
Woosley, S. & Bloom, J. S. 2006 ARA&A, 44, 507

Discussion

BURBIDGE: Some closeby -low redshift- GRBs are clearly associated with supernovae- we see traces of the SN light curves. But for most of the GRBs either no optical or radio object has been identified or only very faint afterglows are found. Thus the connection with galaxies- even star forming galaxies is much weaker. On the other hand, many optical spectra of GRB afterglows show the absorption features of QSOs. Thus I think that some attention should be paid to the connection of GRBs with QSOs. Here the MgII absorption seen in all of the afterglows with absorption but only in a fraction of QSOs presents a real problem for those who believe that the QSOs all have cosmological redshifts.

FYNBO: I would describe the situation as follows. For $z < 2$ GRBs we almost always detect a host galaxy at the position of the afterglows. For more distant GRBs the fraction of hosts detected to a detection limit of about $R = 27$ drops. The properties of the GRB absorption systems are most similar to the DLAs seen in QSO spectra although GRB systems on average are more metal rich. In no case has a GRB been associated directly with a galaxy hosting an AGN.

RAUW: There have been suggestions to use nuclear resonance absorption lines in the γ-ray spectrum to determine the redshifts of GRBs beyound z=6. That would require very large column densities, well in excess of the values you have presented. Can you comment on this?

FYNBO: I am afraid I do not have much insight into this issue. I can add that in some cases substantially higher column densities are inferred from X-ray absorption than from Hydrogen Lyman-α (e.g., Watson et al. 2007, ApJ, 660, L101). This is most likely due to ionized material close to the GRB progenitor. Still, the column densities inferred from X-ray absorption are so far all below 10^{23} cm^{-2} equivalent HI.

Max Pettini.

Yoshi Taniguchi.

Probing the Interstellar Medium and Stellar Environments of Long-Duration GRBs

Miroslava Dessauges-Zavadsky[1], Jason X. Prochaska[2] and Hsiao-Wen Chen[3]

[1] Geneva Observatory, University of Geneva, 51, Ch. des Maillettes, CH-1290 Sauverny, Switzerland
email: miroslava.dessauges@obs.unige.ch

[2] UCO/Lick Observatory, University of California, Santa Cruz, CA 95064, USA
email: xavier@ucolick.edu

[3] Department of Astronomy, University of Chicago, 5640 S. Ellis Ave., Chicago, IL 60637, USA
email: hchen@oddjob.uchicago.edu

Abstract. We review the properties of the gas surrounding high-redshift gamma-ray bursts (GRBs) assessed through the analysis of damped Lyman-alpha systems (DLAs) identified in their afterglow spectra. These GRB-DLAs are characterized by large H I column densities with a median of $N(\text{H I}) = 10^{21.7}$ cm^{-2}, no molecular gas signatures, metallicities ranging from 1/100 to nearly solar with a median exceeding 1/10 solar, and no anomalous abundance patterns. The detection of the atomic Mg lines and the time-variability of the fine-structure lines demonstrates that the majority of the neutral gas along the GRB sightlines is located between 50 pc and a few kpc from the GRB. This implies that this gas is presumably associated with the ambient interstellar medium of the host galaxy and that the derived properties from low-ionization lines do not directly constrain the local environment of the GRB progenitor. The highly ionized gas, traced by N V lines, which could result from a pre-existing H II region produced by the GRB progenitor and neighboring OB stars, appears on the other hand to be very local to the GRB at about 10 pc, yielding a snapshot of the medium's physical conditions at this radius.

Keywords. gamma rays: observations – galaxies: ISM – ISM: abundances

---

1. Introduction

Like quasars (QSOs), GRBs provide a bright – albeit transient – light beam. Two types of GRBs are now well established, the short and the long ones with durations < 2 s and > 2 s, respectively. The long-duration GRBs give rise to brighter optical afterglows and are of our interest in this work. Much of the recent advances was enabled by the launch of the Swift satellite. Thanks to its rapid and precise localization of GRBs and thanks to the 'target of opportunity' optical observations, high-quality and high-resolution spectroscopy was achieved for a number of afterglows on 10-m class telescopes. Imprinted in the quasi power-law spectrum of the afterglow are the signatures of the circumstellar material from the progenitor, the H II region produced by the progenitor and neighboring OB stars, the neutral interstellar medium (ISM) of the host galaxy, any diffuse gas within the galactic halo, and finally intergalactic material located on the GRB sightline. Unfortunately, even though these phases arise at distinct distances along the sightline, the observed spectrum resolves only the relative velocities of the gas. Therefore, for the analysis, one has to keep in mind that all these phases are potentially mixed in the afterglow spectrum.

It is expected that GRBs originate in star-forming regions and have massive stars as progenitors. Several observational evidences point toward this hypothesis. At low redshift,

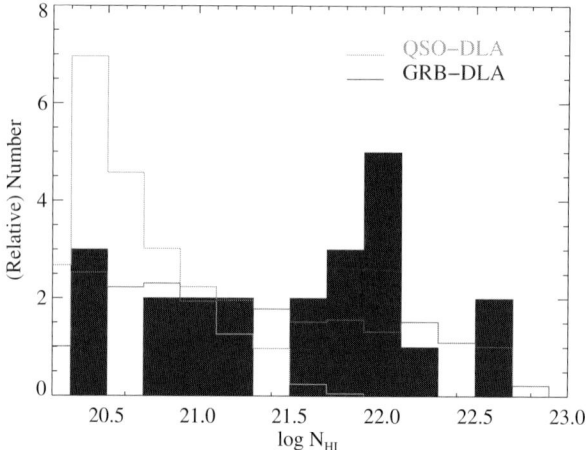

Figure 1. Distributions of H I column densities for both GRB-DLA (filled, solid histogram) and QSO-DLA (solid histogram). The GRB-DLA have a median value $\log N(\mathrm{H\,I}) = 21.7$, which exceeds all but a few QSO-DLA observed to date.

the association between the GRBs and massive star progenitors is well established by (i) Type Ic supernovae (SNe) identified at the positions of GRB events (Hjorth *et al.* 2003; Mirabal *et al.* 2006) and (ii) the location of GRBs in Wolf-Rayet galaxies (Hammer *et al.* 2006). At high redshift, the connection between GRBs and star-forming regions is inferred by (i) GRBs found in blue galaxies with nebular lines indicative of on-going star formation (Le Floc'h *et al.* 2003; Prochaska *et al.* 2004) and (ii) GRBs observed within a few kpc from the weighted flux centroid of their host (Fruchter *et al.* 2006).

What are the properties of the ISM within and near star-forming regions in the young Universe? What physical constraints does one place on the GRB surrounding gas (metallicity, density, temperature, dust and molecular content)? Where is the gas that we observe? Can we probe gas very local to the progenitor? These are the questions that motivate our work. Our GRB sample is composed of only afterglow spectra with (i) high signal-to-noise ratios and sufficient spectral resolution ($R > 2000$) to allow accurate column density measurements, and (ii) the presence of damped Lyα lines ($N(\mathrm{H\,I}) \geqslant 2 \times 10^{20}$ cm^{-2}) or of strong low-ionization lines. In total, our sample contains 16 afterglow spectra of GRB-DLA at redshifts between 1 and 6.5.

2. Neutral gas

2.1. *H I content: Fuel for star formation*

Figure 1 shows the distributions of H I column densities for both GRB-DLA and QSO-DLA. The GRB-DLA have large H I column densities, with a median value of $\log N(\mathrm{H\,I}) = 21.7$ (Prochaska *et al.* 2007). Out of about 1000 such measurements for QSO-DLA, not a single has $N(\mathrm{H\,I}) \geqslant 10^{22}$ cm^{-2}. This difference in $N(\mathrm{H\,I})$ between GRB-DLA and QSO-DLA can be naturally explained by sightline geometries: the GRB sightlines originate within the host galaxy and very certainly within a young star-forming region, while the QSO sightlines pass randomly through the galaxy and statistically more often through the outer regions, because of the higher cross-sections, and have lower $N(\mathrm{H\,I})$.

Massive stars are observed to form in molecular clouds, so we may wonder whether the high GRB-DLA H I column densities are consistent with that predicted for GRBs embedded within molecular clouds. Three observational results do not support the

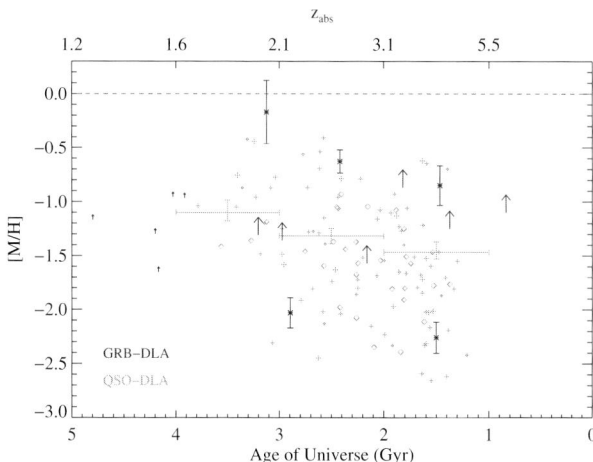

Figure 2. Metallicity [M/H] measurements (we considered the metal-lines which are unsaturated, so for most measurements [M/H] = [S/H]) for GRB-DLA (large symbols and arrows) and for QSO-DLA (small symbols) as a function of the age of the Universe. The GRB-DLA distribution spans two orders of magnitude and has a median value that exceeds 1/10 solar and is higher than the cosmic mean metallicity of H I in the ISM of QSO-DLA (large crosses).

presence of molecular gas: (i) there are too many GRB-DLA with H I column densities lower than $10^{21.5}$ cm^{-2} compared to the Galactic molecular clouds (Jakobsson et al. 2006); (ii) among the 5 sightlines for which H$_2$ lines are covered, none shows molecular line detection and sets a very low molecular fraction $f(H_2) < 10^{-6}$ (Tumlinson et al. 2007); and (iii) low C^0/C$^+$ ratios ($< 10^{-4}$) are observed favoring gas dominated by a warm, less dense phase (Prochaska et al. 2007).

As a first result, we can conclude that the observed H I gas traces more likely the ambient ISM gas, instead of gas associated with the GRB dense molecular cloud.

2.2. Metallicity

In the collapsar model, a large angular momentum is required to power the GRB. Since high-metallicity stars are expected to have significant mass-loss by winds, leading to the loss of their angular momentum, this suggests that GRB progenitors should have metallicities less than 1/10 solar (Hirschi et al. 2005; Langer & Norman 2006; Woosley & Heger 2006). The observations of GRB hosts at $z < 1$ do seem to show a bias toward low-metallicities. Indeed, the few metallicities derived from the nebular line analysis point to sub-solar metallicities (Sollerman et al. 2005). The GRB hosts also generally have sub-L$_*$ luminosities (Le Floc'h et al. 2003; Fruchter et al. 2006), consistent with low metallicities. Modjaz et al. (2008) has, in addition, shown that GRBs occur in lower metallicity galaxies than the ones hosting a random sample of Type Ic SN. If this is confirmed, it would offer compelling evidence that metallicity plays a role in the GRB progenitors.

What about the metallicities in GRB hosts at high redshift ($z > 2$)? Figure 2 shows the measured GRB-DLA metallicities as a function of the age of the Universe and compared with the QSO-DLA metallicities. GRB-DLA exhibit a large spread of metallicities ranging from 1/100 solar to nearby solar, with a median value of [M/H] exceeding 1/10 solar (Prochaska et al. 2007). Almost all the GRB-DLA metallicities are higher than the cosmic mean metallicity of H I in the ISM of QSO-DLA. This has two main implications. First, the gas near star-forming regions seems to have an enhanced metallicity. This can be a result of metallicity gradients within the galaxies, since as we have previously pointed out GRB-DLA more likely probe the inner regions of galaxies and QSO-DLA the outer ones. Second, there is little evidence that GRBs at high-redshift prefer low metallicities, i.e. lower than 1/10 solar.

2.3. Chemical properties

We have used the Zn/Fe ratio to examine the dust content in the GRB-DLA. They do show uniformly large [Zn/Fe] ratios (> 0.5 dex), suggesting a highly depleted gas with more than 50 % of Fe locked into dust grains. This is further supported by the sub-solar [Ti/Fe] ratios (Dessauges-Zavadsky *et al.* 2002). However, as pointed out by Savaglio (2006), the depletion pattern resembles more likely the warm Galactic ISM rather than the cold, dense clouds one associated with proto-molecular clouds.

The GRB-DLA show large [α/Fe] ratios, exceeding 0.5 dex and being systematically larger than in QSO-DLA. Is this α-enhancement a result of nucleosynthesis or dust depletion effects? On the nucleosynthetic point of view, an α-enhancement is expected, first because the GRB progenitors certainly are massive stars and second because high specific star formation rates are observed in the host galaxies implying young ages compared to the timescale for Type Ia SN enrichment. The contribution from differential dust depletion effects can be estimated to ~ 0.3 dex. This leaves place for a remaining α-enhancement larger than 0.3 dex, reflecting enrichment by massive stars and being consistent with GRBs tracing star formation.

> In summary, the GRB-DLA have a median metallicity larger than 1/10 solar, they show important dust depletion levels but still lower than those of cold/dense clouds, and they do not show any anomalous abundance patters which could occur if the gas represented partially mixed SN ejecta or circumstellar material. As a second result, we conclude that the derived metal abundances trace more likely the ambient ISM near the GRBs rather than the GRB circumstellar material.

3. Distance diagnostics of the neutral gas

Two distance diagnostics allow us to estimate the distance of the observed neutral gas. First, the detection of the atomic Mg lines. Indeed, we do observe large column densities of Mg^0 ($\approx 10^{14.7}$ cm^{-2}) with Mg I lines coincident in velocity with other low-ionization lines. With a very low ionization potential of 7.7 eV, this neutral gas is not shielded by the large H I columns, and therefore Mg^0 is fully ionized in less than 1000 s if the gas is located within 50 pc of the GRB. As a consequence, the detection of Mg I places the gas at more than 50 pc from the GRB.

Second, the transitions from the excited states of C^+, Si^+, O^0, Fe^+, and Ni^+. They appear to be generic features in the GRB afterglow spectra, with their absorption lines being coincident in velocity with the low-ionization lines. These transitions have not been detected in QSO-DLA so far. They have been observed in environments characterized by extreme densities and temperatures, such as the Broad Absorption Line quasars, η Carina and β Pictoris. Three excitation mechanisms allow to produce these fine-structure lines: indirect UV pumping, direct IR pumping, and collisional excitation. Prochaska *et al.* (2006) demonstrated that the UV pumping mechanism dominates out to 1 kpc. This has two main implications. First, the gas traced by the fine-structure lines is not a high-density circumstellar gas, and second the UV-pumping predicts line variability due to the decline of the afterglow ionizing flux as a function of time. The fine-structure line variability has been reported in two GRBs so far, GRB 020813 (Dessauges-Zavadsky *et al.*

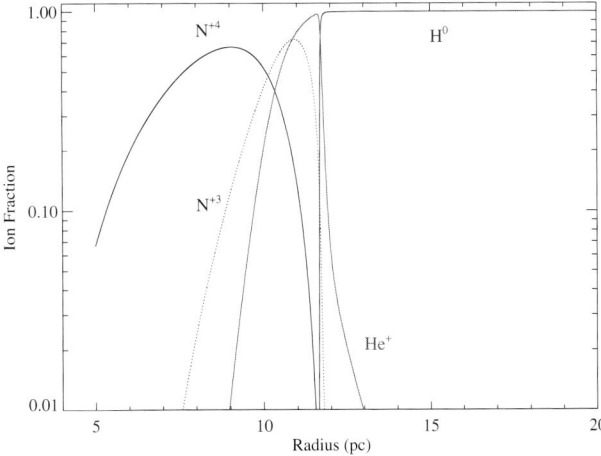

Figure 3. Ion fractions as a function of distance from the GRB afterglow. The calculation assumes a constant density medium ($n_H = 10$ cm^{-3}) and the total photon flux from an afterglow integrated from $t = +10$ s to $+1000$ s in the observer frame. The distance to the N^{4+} maximum ionization fraction is of about 10 pc.

2006) and GRB 060418 (Vreeswijk et al. 2007). These detections allow to constrain the distance of the neutral gas between 100 pc and 1.5 kpc.

> In summary and as a third result, the Mg I and fine-structure transitions demonstrate that the majority of neutral gas along the GRB sightline is located between 50 pc and a few kpc from the GRB. This is presumably beyond the immediate star-forming region of the GRB progenitor and therefore one associates this gas with the ambient ISM of the host galaxy. This implies that analysis related to the low-ionization transitions, e.g. metallicity measurements, chemical studies, does not directly constrain the local environment of the GRB progenitor.

4. Ionized gas

What have we learnt so far? The detected neutral gas in afterglow spectra traces the ambient ISM with most of this gas lying between 100 pc and 1.5 kpc from the GRB. There is no evidence for molecular gas, despite the observed large H I surface densities and the presence of dust. What about the ionized gas which may be present in pre-existing H II regions produced by the GRB progenitor and neighboring OB stars?

A number of GRB sightlines exhibit X-ray absorption with implied metal column densities that significantly exceed the neutral ISM column densities measured from optical spectra (Watson et al. 2007, Galama & Wijers, 2001). These observations provide hints for a large reservoir of highly ionized gas near the GRB. Therefore, we have initiated a survey for N V lines. Why N V? It is because N^{3+} has a high ionization potential (IP = 77 eV) that makes the production of N^{4+} difficult, especially using stellar radiation fields. N V is believed to trace collisionally ionized gas either in equilibrium at a high temperature ($T > 10^5$ K) or out of equilibrium due to a post-shocked gas cooling from $T > 10^6$ K.

We identify a high incidence (6/7) of spectra exhibiting N V gas (Prochaska et al. 2008). The N V lines show the following characteristics: (i) high detection rates; (ii) large optical depths with large integrated column densities $N(N^{4+}) \gtrsim 10^{14}$ cm^{-2}; (iii) narrow profiles ($\lesssim 50$ km s^{-1}); and (iv) small velocity offset from low-ionization and fine-structure lines

($\delta v \lesssim 20$ km s^{-1}). These characteristics are unlike the N V gas observed in the halo and disk of the Milky Way and high-redshift QSO-DLA, but best resemble the narrow absorption lines associated with QSOs and $z \sim 3$ starburst galaxies.

Scenarios related to material shock-heated by the progenitor's stellar wind are not the dominant mechanisms for producing the observed N^{4+} absorption, since then one would expect velocity offsets of the ionized material relatively to the neutral gas by several thousands of km s^{-1} (Starling et al. 2005; van Marle et al. 2005). The swept up ISM material by the stellar wind should on the other hand almost coincide in velocity with the neutral gas, but there the gas rapidly cools to $\approx 10^4$ K which is not sufficient to produce N^{4+} (Lamers & Cassinelli 1999). As a consequence, the GRB afterglow ionizing flux appears as the natural mechanism to produce kinematically cold N V lines, i.e. with no offset from the neutral gas and with narrow profiles. Figure 3 shows that the afterglow will photoionize nitrogen to N^{4+} at $r \approx 10$ pc. Within the photoionization scenario, the observations imply the progenitor's stellar wind is confined to $r < 10$ pc. This places strong constraints on the GRB progenitor and/or the ISM material close to the GRB: low mass-loss rates, short stellar lifetimes and/or high ISM densities of $n \geqslant 10^3$ cm^{-3} are required.

As a fourth result, we conclude that the highly ionized N V gas detected in the afterglow spectra traces the very local gas to the GRB at about 10 pc, yielding a snapshot of the medium's physical conditions at this radius.

References

Dessauges-Zavadsky, M., Prochaska, J. X., & D'Odorico, S. 2002, A&A, 391, 801
Dessauges-Zavadsky, M., Chen, H.-W., Prochaska, J. X., et al. 2006, ApJ, 648, L89
Fruchter, A. S., Levan, A. J., Strolger, L., et al. 2006, Nature, 441, 463
Galama, T. J. & Wijers, R. A. M. J. 2001, ApJ, 459, L209
Hammer, F., Flores, H., Schaerer, D., et al. 2006, A&A, 454, 103
Hirschi, R., Meynet, G., & Maeder, A. 2005, A&A, 443, 581
Hjorth, J., Sollerman, J., Moller, P., et al. 2003, Nature, 423, 847
Jakobsson, P., Fynbo, J. P.U., Ledoux, C., et al. 2006, A&A, 460, L13
Lamers, H. J. G. L. M. & Cassinelli, J. P. 1999, Introduction to Stellar Winds (Cambridge: CUP)
Langer, N. & Norman, C. A. 2006, ApJ, 638, L63
Le Floc'h, E., Duc, P.-A., Mirabel, I. F., et al. 2003, A&A, 400, 499
Modjaz, M., Kewley, L., Kirshner, R. P., et al. 2008, AJ, 135, 1136
Mirabal, N., Halpern, J. P., An, D., et al. 2006, ApJ, 643, L99
Prochaska, J. X., Bloom, J. S., & Chen, H.-W. et al. 2004, ApJ, 611, 200
Prochaska, J. X., Chen, H.-W., & Bloom, J. S. 2006, ApJ, 648, 95
Prochaska, J. X., Chen, H.-W., Dessauges-Zavadsky, M., & Bloom, J. S. 2007, ApJ, 666, 267
Prochaska, J. X., Dessauges-Zavadsky, M., Raminez-Ruiz, E. & Chen, H. W. 2008, ApJ, in press
Savaglio, S. 2006, New Journal of Phys., 8, 195
Sollerman, J., Östlin, G., Fynbo, J. P. U., et al. 2005, NewA, 11, 103
Starling, R. L. C., Wijers, R. A. M. J., Hughes, M. A., et al. 2003, MNRAS, 360, 305
Tumlinson, J., Prochaska, J. X., Chen, H.-W., et al. 2007, ApJ, 668, 667
van Marle, A. J., Langer, N., & Gracía-Segura, G. 2005, A&A, 444, 837
Vreeswijk, P. M., Ledoux, C., Smette, A., et al. 2007, A&A, 468, 83
Watson, D., Hjorth, J., Fynbo, J. P. U., et al. 2007, ApJ, 660, L101
Woosley, S. E. & Heger, A. 2006, ApJ, 637, 914

The Connection between Gamma-Ray Bursts and Extremely Metal-Poor Stars as Nucleosynthetic Probes of the Early Universe

K. Nomoto[1,2], N. Tominaga[1], M. Tanaka[1], K. Maeda[2], and H. Umeda[1]

[1]Department of Astronomy, University of Tokyo, Bunkyo-ku, Tokyo 113-0033, Japan
email: nomoto@astron.s.u-tokyo.ac.jp

[2]Institute for the Physics and Mathematics of the Universe, University of Tokyo, Kashiwa, Chiba 277-8582, Japan

Abstract. The connection between the long GRBs and Type Ic Supernovae (SNe) has revealed the interesting diversity: (i) GRB-SNe, (ii) Non-GRB Hypernovae (HNe), (iii) X-Ray Flash (XRF)-SNe, and (iv) Non-SN GRBs (or dark HNe). We show that nucleosynthetic properties found in the above diversity are connected to the variation of the abundance patterns of extremely-metal-poor (EMP) stars, such as the excess of C, Co, Zn relative to Fe. We explain such a connection in a unified manner as nucleosynthesis of hyper-aspherical (jet-induced) explosions of Pop III core-collapse SNe. We show that (1) the explosions with large energy deposition rate, $\dot{E}_{\rm dep}$, are observed as GRB-HNe and their yields can explain the abundances of normal EMP stars, and (2) the explosions with small $\dot{E}_{\rm dep}$ are observed as GRBs without bright SNe and can be responsible for the formation of the C-rich EMP (CEMP) and the hyper metal-poor (HMP) stars. We thus propose that GRB-HNe and the Non-SN GRBs (dark HNe) belong to a continuous series of BH-forming massive stellar deaths with the relativistic jets of different $\dot{E}_{\rm dep}$.

Keywords. Galaxy: halo – gamma rays: bursts – nuclear reactions, nucleosynthesis, abundances – stars: abundances – stars: Population II – supernovae: general

---

1. Introduction

Among the most interesting recent developments in the study of supernovae (SNe) is the establishment of the Gamma-Ray Burst (GRB)-Supernova Connection (Woosley & Bloom 2006). Three GRB-associated SNe have been observed so far: GRB 980425/ SN 1998bw (Galama et al. 1998, Iwamoto et al. 1998), GRB 030329/SN 2003dh (Stanek et al. 2003, Hjorth et al. 2003), and GRB 031203/SN 2003lw (Malesani et al. 2004). They are all very energetic supernovae, whose kinetic energy E exceeds 10^{52} erg, more than 10 times the kinetic energy of normal core-collapse SNe. In the present paper, we use the term 'Hypernova (HN)' to describe such a hyper-energetic supernova with $E_{51} = E/10^{51}$ erg $\gtrsim 10$ (Fig.1; Nomoto et al. 2004, Nomoto et al. 2006). The above three SNe associated with GRBs are called "GRB-HNe".

In contrast, "non-SN GRBs" (or *dark HNe*) have also been discovered (GRBs 060605 and 060614) (Fynbo et al. 2006, Gal-Yam et al. 2006, Della Valle et al. 2006, Gehrels et al. 2006). Upper limits to brightness of the possible SNe are about 100 times fainter than SN 1998bw. These correspond to upper limits to the ejected ^{56}Ni mass of $M(^{56}{\rm Ni}) \sim 10^{-3} M_\odot$ (see, e.g., Nomoto et al. 2007 for prediction).

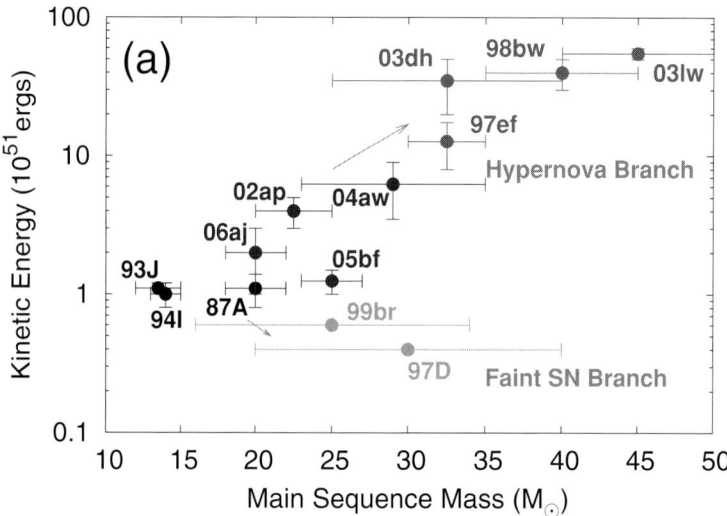

Figure 1. The kinetic explosion energy E as a function of the main sequence mass M of the progenitors for several supernovae/hypernovae. Hypernovae are the SNe with $E_{51} > 10$.

Such hypernovae and GRBs are also likely to be hyper-aspherical explosions induced by relativistic jet(s) as suggested from photometric and spectroscopic observations.

We calculate nucleosynthesis in such hyper-energetic and hyper-aspherical explosions and find that resultant abundance features show some important differences from normal supernova explosions (e.g., Maeda *et al.* 2002, Maeda & Nomoto 2003, Tominaga *et al.* 2007, Tominaga 2008). We show that such features can explain the peculiar abundance patterns observed in the extremely metal-poor (EMP), and hyper-metal-poor (HMP) halo stars (e.g., Hill, François, & Primas 2005, Beers & Christlieb 2005). This approach would lead to identifying the First Stars in the Universe, which is one of the important challenges of the current astronomy.

2. Nucleosynthesis in Jet-Induced Explosions

We calculate hydrodynamics and nucleosynthesis of the explosions induced by relativistic jets (*jet-induced explosions*) (Fig. 2) (Tominaga *et al.* 2007, Tominaga 2008). For the $40M_\odot$ stars (Umeda & Nomoto 2005, Tominaga, Umeda, & Nomoto 2007), the jets are injected at a radius $R_0 \sim 900$ km (corresponding to an enclosed mass of $M_0 \sim 1.4 M_\odot$). The most important parameter in our models is the energy deposition rate $\dot{E}_{\rm dep}$. We investigate the dependence of nucleosynthesis outcome on $\dot{E}_{\rm dep}$ for a range of $\dot{E}_{{\rm dep},51} \equiv \dot{E}_{\rm dep}/10^{51} {\rm erg\, s}^{-1} = 0.3 - 1500$. The diversity of $\dot{E}_{\rm dep}$ is consistent with the wide range of the observed isotropic equivalent γ-ray energies and timescales of GRBs (Amati *et al.* 2006 and references therein). Variations of activities of the central engines, possibly corresponding to different rotational velocities or magnetic fields, may well produce the variation of $\dot{E}_{\rm dep}$.

The upper panel of Figure 3 shows the dependence of the ejected $M(^{56}{\rm Ni})$ on $\dot{E}_{\rm dep}$. Generally, higher $\dot{E}_{\rm dep}$ leads to the synthesis of larger $M(^{56}{\rm Ni})$ in explosive nucleosynthesis because of higher post-shock densities and temperatures (e.g., Maeda & Nomoto 2003, Nagataki *et al.* 2006). If $\dot{E}_{{\rm dep},51} \gtrsim 60$, we obtain $M(^{56}{\rm Ni}) \gtrsim 0.1 M_\odot$, which is consistent with the brightness of GRB-HNe. Some C+O core materials are ejected along

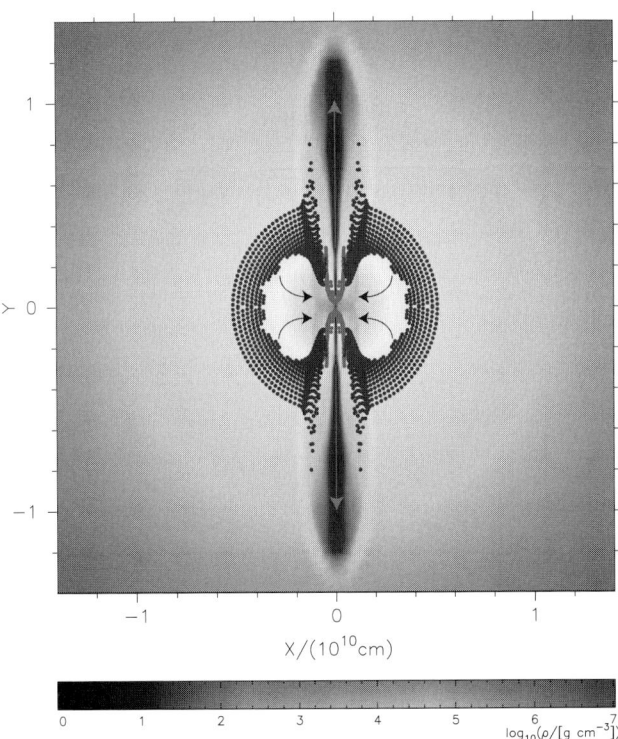

Figure 2. The density structure of the 40 $M_\odot$ Pop III star explosion model of $\dot{E}_{\rm dep,51} = 15$ at 1 sec after the start of the jet injection. The jets penetrate the stellar mantle (*grey arrows*) and material falls on the plane perpendicular to the jets (*black arrows*). The dots represent ejected Lagrangian elements dominated by Fe (^{56}Ni, *grey*) and by O (*black*).

the jet-direction, but a large amount of materials along the equatorial plane fall back (Fig. 2).

For $\dot{E}_{\rm dep,51} \gtrsim 60$, the remnant mass is initially $M_{\rm rem}^{\rm start} \sim 1.5 M_\odot$ and grows as materials are accreted from the equatorial plane (Fig. 2). The final BH mass is generally larger for smaller $\dot{E}_{\rm dep}$. The final BH masses range from $M_{\rm BH} = 10.8 M_\odot$ for $\dot{E}_{\rm dep,51} = 60$ to $M_{\rm BH} = 5.5 M_\odot$ for $\dot{E}_{\rm dep,51} = 1500$, which are consistent with the observed masses of stellar-mass BHs (Bailyn *et al.* 1998). The model with $\dot{E}_{\rm dep,51} = 300$ synthesizes $M(^{56}{\rm Ni}) \sim 0.4 M_\odot$ and the final mass of BH left after the explosion is $M_{\rm BH} = 6.4 M_\odot$.

For low energy deposition rates ($\dot{E}_{\rm dep,51} < 3$), in contrast, the ejected ^{56}Ni masses ($M(^{56}{\rm Ni}) < 10^{-3} M_\odot$) are smaller than the upper limits for GRBs 060505 and 060614.

If the explosion is viewed from the jet direction, we would observe GRB without SN re-brightening. This may be the situation for GRBs 060505 and 060614. In particular, for $\dot{E}_{\rm dep,51} < 1.5$, ^{56}Ni cannot be synthesized explosively and the jet component of the Fe-peak elements dominates the total yields (Fig. 4). The models eject very little $M(^{56}{\rm Ni})$ ($\sim 10^{-6} M_\odot$).

For intermediate energy deposition rates ($3 \lesssim \dot{E}_{\rm dep,51} < 60$), the explosions eject $10^{-3} M_\odot \lesssim M(^{56}{\rm Ni}) < 0.1 M_\odot$ and the final BH masses are $10.8 M_\odot \lesssim M_{\rm BH} < 15.1 M_\odot$. The resulting SN is faint ($M(^{56}{\rm Ni}) < 0.01 M_\odot$) or sub-luminous ($0.01 M_\odot \lesssim M(^{56}{\rm Ni}) < 0.1 M_\odot$).

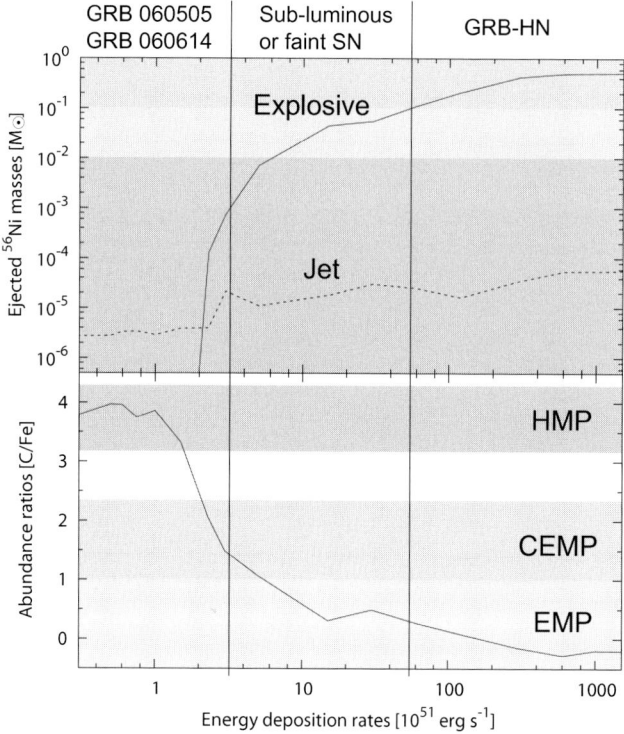

Figure 3. *Upper*: the ejected ^{56}Ni mass (*solid*: explosive nucleosynthesis products, *dashed*: the jet contribution) as a function of the energy deposition rate. The background shows the corresponding SNe (*top*: GRB-HNe, *top-middle*: sub-luminous SNe, *middle*: faint SNe, *bottom*: GRBs 060505 and 060614). Vertical lines divide the resulting SNe according to their brightness. *Lower*: the dependence of abundance ratio [C/Fe] on the energy deposition rate. The background color shows the corresponding metal-poor stars (*bottom*: EMP, *middle*: CEMP, *top*: HMP stars).

Table 1. Metal-poor stars.

Name	[Fe/H]	Features	Reference
HE 0107–5240	−5.3	C-rich, Co-rich?, [Mg/Fe]∼ 0	Christlieb *et al.* 2002
HE 1327–2326	−5.5	C, O, Mg-rich	Frebel *et al.* 2005, Aoki *et al.* 2006
HE 0557–4840	−4.8	C, Ca, Sc, Ti-rich, [Co/Fe]∼ 0	Norris *et al.* 2007
HE 1300+0157	−3.9	C, Si, Ca,Sc,Ti, Co-rich	Frebel *et al.* 2007
HE 1424–0241	−4.0	Co,Mn-rich, Si,Ca,Cu-poor	Cohen *et al.* 2007
CS 22949–37	−4.0	C,N,O,Mg,Co,Zn-rich	Depagne *et al.* 2002
CS 29498–43	−3.5	C,N,O,Mg-rich, [Co/Fe]∼ 0	Aoki *et al.* 2004
BS 16934–002	−2.8	O,Mg-rich, C-poor	Aoki *et al.* 2007

3. Abundance Patterns of Extremely Metal-Poor Stars

Table 1 summarizes the abundance features of various EMP stars. In addition to HMP stars, we focus on the recently discovered first Ultra Metal-Poor (UMP) star (Norris *et al.* 2007) and the very peculiar Si-poor star (Cohen *et al.* 2007). Many of these EMP stars have high [Co/Fe], suggesting the HN-connection.

3.1. *C-rich Metal-Poor Stars (CEMP)*

The lower panel of Figure 3 shows the dependence of the abundance ratio [C/Fe] on $\dot{E}_{\rm dep}$. Lower $\dot{E}_{\rm dep}$ yields larger $M_{\rm BH}$ and thus larger [C/Fe], because the infall decreases the

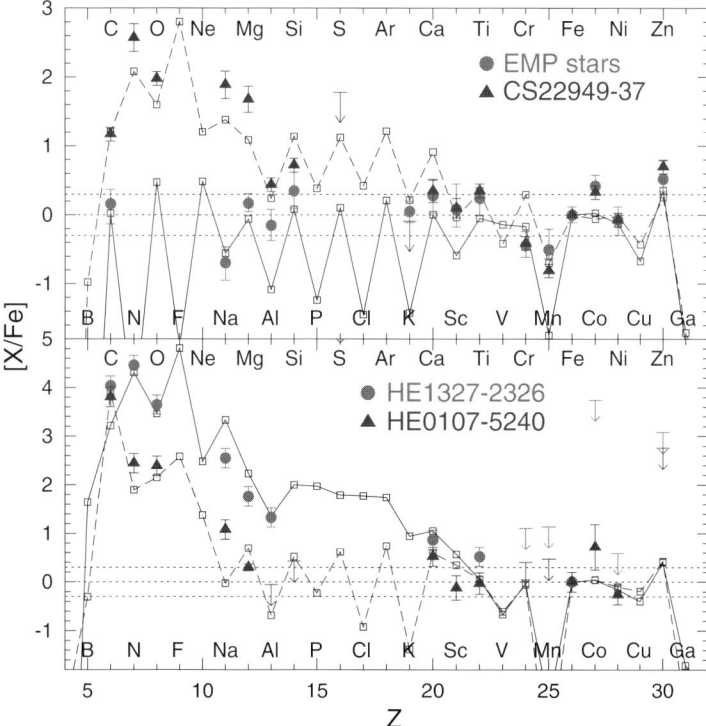

Figure 4. A comparison of the abundance patterns between the metal-poor stars and our models. *Upper*: typical EMP (*dots*, Cayrel *et al.* 2004) and CEMP (*triangles*, CS 22949–37, Depagne *et al.* 2002) stars and models with $\dot{E}_{\rm dep,51} = 120$ (*solid line*) and $= 3.0$ (*dashed line*). *Lower*: HMP stars: HE 1327–2326, (*dots*, e.g., Frebel *et al.* 2005), and HE 0107–5240, (*triangles*, Christlieb *et al.* 2002, Bessell & Christlieb 2005) and models with $\dot{E}_{\rm dep,51} = 1.5$ (*solid line*) and $= 0.5$ (*dashed line*).

amount of inner core material (Fe) relative to that of outer material (C) (see also Maeda & Nomoto 2003). As in the case of $M(^{56}{\rm Ni})$, [C/Fe] changes dramatically at $\dot{E}_{\rm dep,51} \sim 3$.

The abundance patterns of the EMP stars are good indicators of SN nucleosynthesis because the Galaxy was effectively unmixed at [Fe/H] < -3 (e.g., Tumlinson 2006). They are classified into three groups according to [C/Fe]:

(1) [C/Fe] ~ 0, normal EMP stars ($-4 <$ [Fe/H] < -3, e.g., Cayrel *et al.* 2004);

(2) [C/Fe] $\gtrsim +1$, Carbon-enhanced EMP (CEMP) stars ($-4 <$ [Fe/H] < -3, e.g., CS 22949–37, Depagne *et al.* 2002);

(3) [C/Fe] $\sim +4$, hyper metal-poor (HMP) stars ([Fe/H] < -5, e.g., HE 0107–5240, Christlieb *et al.* 2002, Bessell & Christlieb 2005; HE 1327–2326, Frebel *et al.* 2005).

Figure 4 shows that the abundance patterns of the averaged normal EMP stars, the CEMP star CS 22949–37, and the two HMP stars (HE 0107–5240 and HE 1327–2326) are well-reproduced by models with $\dot{E}_{\rm dep,51} = 120$, 3.0, 1.5, and 0.5, respectively. The model for the normal EMP stars ejects $M(^{56}{\rm Ni}) \sim 0.2 M_\odot$, i.e., a factor of 2 less than SN 1998bw. On the other hand, the models for the CEMP and the HMP stars eject $M(^{56}{\rm Ni}) \sim 8 \times 10^{-4} M_\odot$ and $4 \times 10^{-6} M_\odot$, respectively, which are always smaller than the upper limits for GRBs 060505 and 060614. The N/C ratio in the models for CS 22949–37 and HE 1327–2326 is enhanced by partial mixing between the He and H layers during pre-supernova evolution (Iwamoto *et al.* 2005).

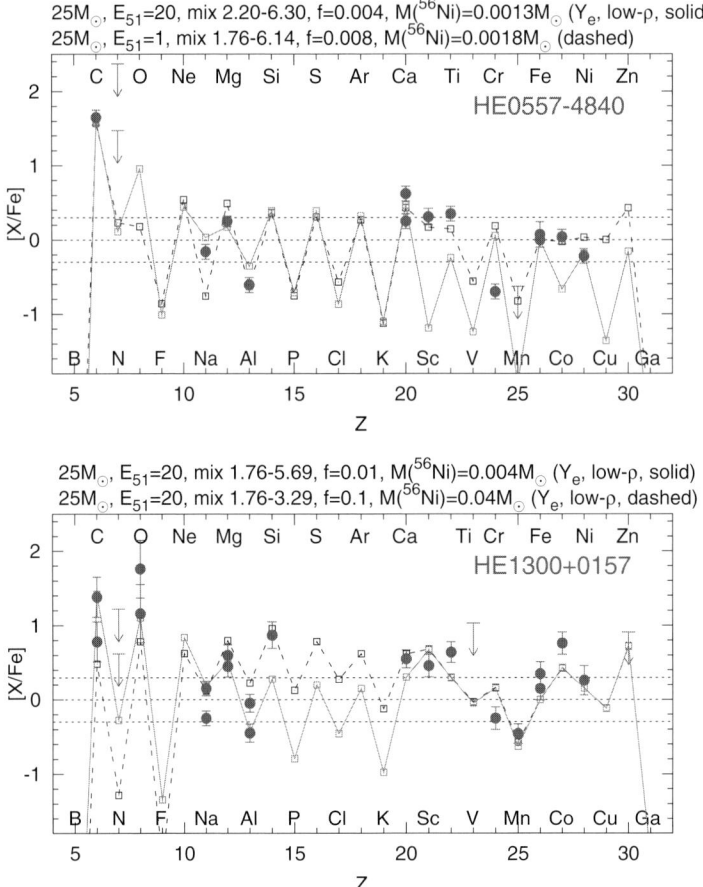

Figure 5. Comparisons of the abundance patterns between the mixing-fallback models and the UMP star HE0557–4840 (upper: Norris et al. 2007), and the CEP star HE1300+0157 (lower: Frebel et al. 2007).

3.2. *UMP Star HE 0557–4840 and CEMP-no Star HE 1300+0157*

The abundance pattern of the first Ultra Metal-Poor (UMP) star (HE 0557–4840: Norris et al. 2007) is shown in Figure 5 and compared with the HN ($E_{51} = 20$) and SN ($E_{51} = 1$) models of the $25 M_\odot$ stars. The Co/Fe ratio ([Co/Fe]~ 0) requires a high energy explosion and the high [Sc/Ti] and [Ti/Fe] ratios require a high-entropy explosion. As shown in Figure 5 (upper), a HN model with a "low-density" modification (Tominaga et al. 2007) is in a good agreement with the abundance pattern of HE 0557–4840. The model indicates $M(^{56}\mathrm{Ni}) \sim 10^{-3} M_\odot$ being similar to faint SN models for CEMP stars. The [Cr/Fe] ratio in the model is much higher than that of HE 0557–4840.

The abundance pattern of the CEMP-no star (i.e., CEMP with no neutron capture elements) HE 1300+0157 (Frebel et al. 2007) is shown in Figure 5 (lower) and marginally reproduced by the hypernova model with $M_{\mathrm{MS}} = 25 M_\odot$ and $E_{51} = 20$. The large [Co/Fe] particularly requires the high explosion energy.

3.3. *Si-Poor Star: HE 1424–0241*

The very peculiar Si-poor abundance pattern of HE 1424–0241 (Cohen et al. 2007) is shown in Figure 6 (upper) and compared with the model of $M_{\mathrm{MS}} = 50 M_\odot$ and $E_{51} = 40$.

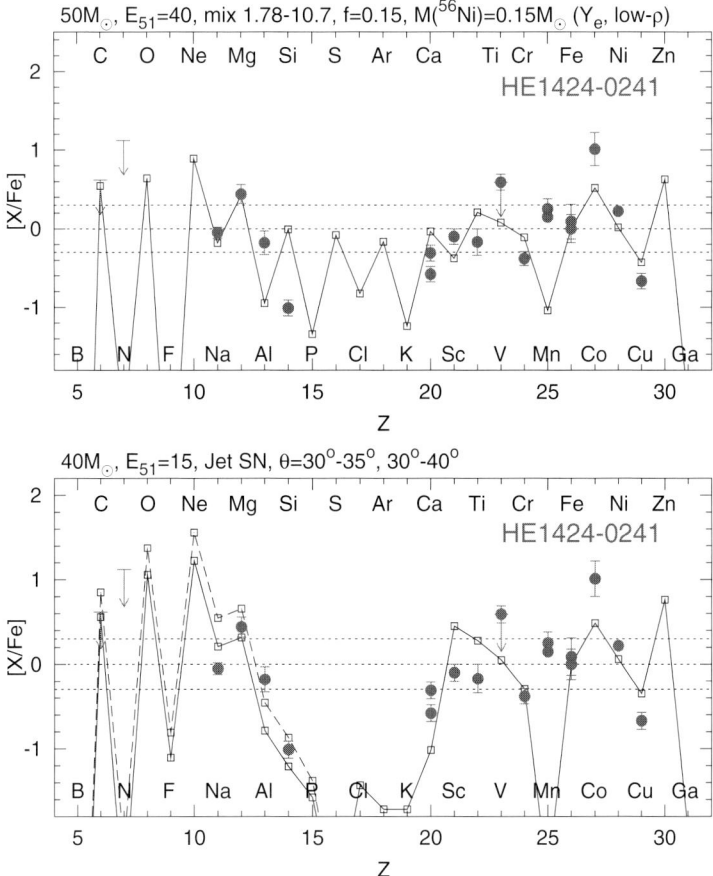

Figure 6. Comparisons between the abundance patterns of HE1424–0241 (Cohen *et al.* 2007) and the mixing-fallback model (upper), and the angle-delimited yields integrated over $30° - 40°$ (*dashed line*) and $30° - 35°$ (*solid line*) of the jet-induced SN model with $\dot{E}_{\rm dep} = 1.2 \times 10^{53}$ erg s^{-1} (lower).

The high [Mg/Si] ratio cannot be reproduced by this model. The peculiar abundance pattern of HE 1424–0241 is a challenge to the explosion models.

The angle-delimited yield provides a possibility to explain the high [Mg/Si] and normal [Mg/Fe]. Figure 6 (lower) shows that the yields integrated over $30° - 40°$ or $30° - 35°$ reproduce the abundance pattern of HE 1424–0241. The yields consist of Mg in the inner region and Fe in the outer region.

Thus the most difficult pattern can be reproduced by the angular dependence of the yield. The high [Mg/Si] and normal [Mg/Fe] are realized if the heavy elements penetrate into the stellar mantle and expand laterally (i.e., the duration of the jet injection is long) and if Mg along the equatorial plane is not accreted onto the central region (i.e., $\dot{E}_{\rm dep}$ is large).

4. Concluding Remarks

We show that (1) the explosions with large energy deposition rate, $\dot{E}_{\rm dep}$, are observed as GRB-HNe and their yields can explain the abundances of normal EMP stars, and (2) the explosions with small $\dot{E}_{\rm dep}$ are observed as GRBs without bright SNe and can be responsible for the formation of the CEMP and the HMP stars. We thus propose that GRB-HNe and GRBs without bright SNe belong to a continuous series of BH-forming massive stellar deaths with the relativistic jets of different $\dot{E}_{\rm dep}$. The very peculiar Si-poor EMP star can also be explained by the angle-delimited yield.

References

Amati, L., Della Valle, M., Frontera, F., et al. 2007, A&A, 463, 913
Aoki, W., Norris, J. E., Ryan, S. G., et al. 2004, ApJ, 608, 971
Aoki, W., Frebel A., Christlieb, N., et al. 2006, ApJ, 639, 897
Aoki, W., Honda, S., Beers, T. C., et al. 2007, ApJ, 660, 747
Bailyn, C. D., Jain, R. K., Coppi, P., & Orosz, J. A. 1998, ApJ, 499, 367
Beers, T. & Christlieb, N. 2005, ARA&A, 43, 531
Bessell, M. S. & Christlieb, N. 2005, in V. Hill, François, P., & F. Primas (eds.), *From Lithium to Uranium: Elemental Tracers of Early Cosmic Evolution*, Proc. IAU Symposium No. 228 (Cambridge: Cambridge Univ. Press), 237
Cayrel, R., Depagne, E., Spite. M., et al. 2004, A&A, 416, 1117
Christlieb, N., Bessell, M. S., Beers, T. C., et al. 2002, Nature, 419, 904
Cohen, J. G., McWilliam, A., Christlieb, N., et al. 2007, ApJ, 659, L161
Della Valle, M., Chincarini, G., Panagia, N., et al. 2006, Nature, 444, 1050
Depagne, E., Hill, V., Spite, M., et al. 2002, A&A, 390, 187
Fynbo, J. P. U., Watson, D., Thöne, C. C., et al. 2006, Nature, 444, 1047
Frebel, A., Aoki, W., Christlieb, N., et al. 2005, Nature, 434, 871
Frebel, A., Norris, J. E., Aoki, W., et al. 2007, ApJ, 658, 534
Galama, T., Vreeswijk, P. M., van Paradijs, J., et al. 1998, Nature, 395, 670
Gal-Yam, A., Fox, D. B., Price, P. A., et al. 2006, Nature, 444, 1053
Gehrels, N., Norris, J. P., Barthelmy, S. D., et al. 2006, Nature, 444, 1044
Hill, V., François, P., & Primas, F. (eds.) 2005, *From Lithium to Uranium: Elemental Tracers of Early Cosmic Evolution*, Proc. IAU Symp. No. 228 (Cambridge: Cambridge Univ. Press)
Hjorth, J., Sollerman, J., Møller, P., et al. 2003, Nature, 423, 847
Iwamoto, K., Mazzali, P. A., Nomoto, K., et al. 1998, Nature, 395, 672
Iwamoto, N., Umeda, H., Tominaga, N., et al. 2005, Science, 309, 451
Maeda, K., Nakamura, T., Nomoto, K., et al. 2002, ApJ, 565, 405
Maeda, K. & Nomoto, K. 2003, ApJ, 598, 1163
Malesani, J., Tagliaferri, G., Chincarini, G., et al. 2004, ApJ, 609, L5
Nagataki, S., Mizuta, A., & Sato, K. 2006, ApJ, 647, 1255
Nomoto, K., Maeda, K., Mazzali, P. A., et al., 2004, in: C. L. Fryer (ed.), *Stellar Collapse* (Dordrecht: Kluwer), Astrophys. Space Science Lib, 302, 277 (astro-ph/0308136)
Nomoto, K., Tominaga, N., Umeda, H., et al. 2006, Nuclear Phys A, 777, 424
Nomoto, K., Tominaga, N., Tanaka, M., et al. 2007, in: *Swift and GRBs: Unveiling the Relativistic Universe*, Nuovo Cimento B, 121, 1207 (astro-ph/0702472)
Norris, J. E., Crhistlieb, N., Korn, A. J., et al. 2007, ApJ, 670, 774
Stanek, K. Z., Matheson, T., Garnavich, P. M., et al. 2003, ApJ, 591, L17
Tominaga, N., Tanaka, M., Nomoto, K., et al. 2005, ApJ, 633, L97
Tominaga, N., Maeda, K., Umeda, H., et al. 2007, ApJ, 657, L77
Tominaga, N., Umeda, H., & Nomoto, K. 2007, ApJ, 660, 516
Tominaga, N. 2008, ApJ, submitted (arXiv:0711.4815)
Tumlinson, J. 2006, ApJ, 641, 1
Umeda, H. & Nomoto, K. 2002, ApJ, 565, 385
Umeda, H. & Nomoto, K. 2005, ApJ, 619, 427
Woosley, S. E. & Bloom, J. S. 2006, ARA&A, 44, 507

The First Stars

Jarrett L. Johnson[1], Thomas H. Greif[2] and Volker Bromm[1]

[1]Department of Astronomy, University of Texas, Austin, TX 78712
[2]Institut für Theoretische Astrophysik,
Albert-Ueberle Strasse 2, 69120 Heidelberg, Germany

Abstract. The formation of the first generations of stars at redshifts $z \geqslant 15-20$ signaled the transition from the simple initial state of the universe to one of ever increasing complexity. We here review recent progress in understanding the assembly process of the first galaxies, starting with cosmological initial conditions and modelling the detailed physics of star formation. In particular, we study the role of HD cooling in ionized primordial gas, the impact of UV radiation produced by the first stars, and the propagation of the supernova blast waves triggered at the end of their brief lives. We conclude by discussing how the chemical abundance patterns observed in extremely low-metallicity stars allow us to probe the properties of the first stars.

Keywords. cosmology: theory – early universe – galaxies: formation – HII regions – ISM: molecules — stars: formation – supernovae: general — hydrodynamics

1. Introduction

One of the key goals in modern cosmology is to study the formation of the first generations of stars and to understand the assembly process of the first galaxies. With the formation of the first stars, the so-called Population III (Pop III), the universe was rapidly transformed into an increasingly complex, hierarchical system, due to the energy and heavy element input from the first stars and accreting black holes (Barkana & Loeb 2001; Bromm & Larson 2004; Ciardi & Ferrara 2005; Miralda-Escudé 2003). Currently, we can directly probe the state of the universe roughly a million years after the Big Bang by detecting the anisotropies in the cosmic microwave background (CMB), thus providing us with the initial conditions for subsequent structure formation. Complementary to the CMB observations, we can probe cosmic history all the way from the present-day universe to roughly a billion years after the Big Bang, using the best available ground- and space-based telescopes. In between lies the remaining frontier, and the first stars and galaxies are the sign-posts of this early, formative epoch.

To simulate the build-up of the first stellar systems, we have to address the feedback from the very first stars on the surrounding intergalactic medium (IGM), and the formation of the second generation of stars out of material that was influenced by this feedback. There are a number of reasons why addressing the feedback from the first stars and understanding second-generation star formation is crucial:

(i) The first steps in the hierarchical build-up of structure provide us with a simplified laboratory for studying galaxy formation, which is one of the main outstanding problems in cosmology.

(ii) The initial burst of Pop III star formation may have been rather brief due to the strong negative feedback effects that likely acted to self-limit this formation mode (Greif & Bromm 2006; Yoshida *et al.* 2004). Second-generation star formation, therefore, might well have been cosmologically dominant compared to Pop III stars.

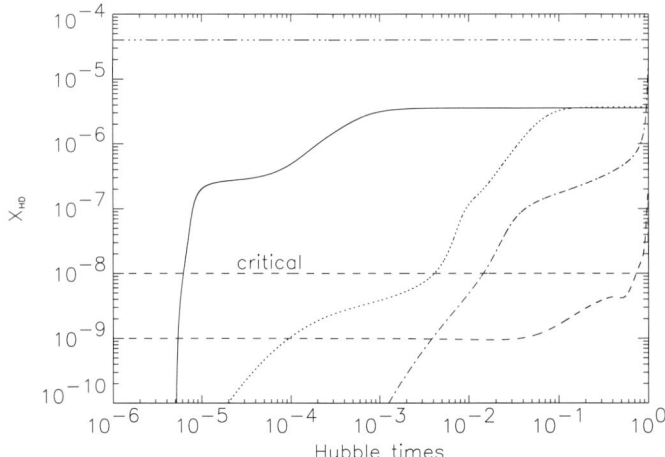

Figure 1. Evolution of the HD abundance, $X_{\rm HD}$, in primordial gas which cools in four distinct situations. The solid line corresponds to gas with an initial density of 100 cm^{-3}, which is compressed and heated by a SN shock with velocity $v_{\rm sh} = 100$ km s^{-1} at $z = 20$. The dotted line corresponds to gas at an initial density of 0.1 cm^{-3} shocked during the formation of a 3σ halo at $z = 15$. The dashed line corresponds to unshocked, un-ionized primordial gas with an initial density of 0.3 cm^{-3} collapsing inside a minihalo at $z = 20$. Finally, the dash-dotted line shows the HD fraction in primordial gas collapsing from an initial density of 0.3 cm^{-3} inside a relic H II region at $z = 20$. The horizontal line at the top denotes the cosmic abundance of deuterium. Primordial gas with an HD abundance above the critical value, $X_{\rm HD,crit}$, denoted by the bold dashed line, can cool to the CMB temperature within a Hubble time.

(iii) A subset of second-generation stars, those with masses below $\simeq 1$ M$_\odot$, would have survived to the present day. Surveys of extremely metal-poor Galactic halo stars therefore provide an indirect window into the Pop III era by scrutinizing their chemical abundance patterns, which reflect the enrichment from a single, or at most a small multiple of, Pop III SNe (Beers & Christlieb 2005; Frebel *et al.* 2007). Stellar archaeology thus provides unique empirical constraints for numerical simulations, from which one can derive theoretical abundance patterns to be compared with the data.

Existing and planned observatories, such as HST, Keck, VLT, and the *James Webb Space Telescope (JWST)*, planned for launch around 2013, yield data on stars and quasars less than a billion years after the Big Bang. The ongoing *Swift* gamma-ray burst (GRB) mission provides us with a possible window into massive star formation at the highest redshifts (Bromm & Loeb 2002, 2006; Lamb & Reichart 2000). Measurements of the near-IR cosmic background radiation, both in terms of the spectral energy distribution and the angular fluctuations provide additional constraints on the overall energy production due to the first stars (Dwek *et al.* 2005; Fernandez & Komatsu 2006; Kashlinsky *et al.* 2005; Magliocchetti *et al.* 2003; Santos *et al.* 2002). Understanding the formation of the first galaxies is thus of great interest to observational studies conducted both at high redshifts and in our local Galactic neighborhood.

2. Population III Star Formation

The first stars in the universe likely formed roughly 150 Myr after the Big Bang, when the primordial gas was first able to cool and collapse into dark matter minihalos with masses of the order of 10^6 M$_\odot$ (Bromm *et al.* 1999, 2002; Abel *et al.* 2002). These stars

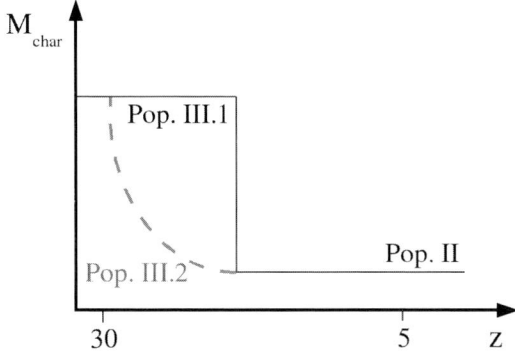

Figure 2. Characteristic stellar mass as a function of redshift. Pop III.1 stars, formed from unshocked, un-ionized primordial gas are characterized by masses of the order of 100 $M_\odot$. Pop II stars, formed in gas which is enriched with metals, emerged at lower redshifts and have characteristic masses of the order of 1 $M_\odot$. Pop III.2 stars, formed from ionized primordial gas, have characteristic masses reflecting the fact that they form from gas that has cooled to the temperature of the CMB. Thus, the characteristic mass of Pop III.2 stars is a function of redshift, but is typically of the order of 10 $M_\odot$.

are believed to have been very massive, with masses of the order of 100 $M_\odot$, owing to the limited cooling properties of primordial gas, which could only cool in minihalos through the radiation from H_2 molecules (e.g. Tegmark et al. 1997). While the initial conditions for the formation of these stars are, in principle, known from precision measurements of cosmological parameters (e.g. Spergel et al. 2007), Pop III star formation may have occurred in different environments which may allow for different modes of star formation. Indeed, it has become evident that Pop III star formation might actually consist of two distinct modes: one where the primordial gas collapses into a DM minihalo (see below), and one where the metal-free gas becomes significantly ionized prior to the onset of gravitational runaway collapse (Johnson & Bromm 2006). We had termed this latter mode of primordial star formation 'Pop II.5' (Greif & Bromm 2006; Johnson & Bromm 2006; Mackey et al. 2003). To more clearly indicate that both modes pertain to *metal-free* star formation, we here follow the new classification scheme suggested by Chris McKee (see McKee & Tan 2008; Johnson et al. 2008). Within this scheme, the minihalo Pop III mode is now termed Pop III.1, whereas the second mode (formerly 'Pop II.5') is now called Pop III.2. The hope is that McKee's terminology will gain wide acceptance.

While the very first Pop III stars (so-called Pop III.1), with masses of the order of 100 $M_\odot$, formed within DM minihalos in which primordial gas cools by H_2 molecules alone, the HD molecule can play an important role in the cooling of primordial gas in several situations, allowing the temperature to drop well below 200 K (Abel et al. 2002; Bromm et al. 2002). In turn, this efficient cooling may lead to the formation of primordial stars with masses of the order of 10 $M_\odot$ (so-called Pop III.2) (Johnson & Bromm 2006). In general, the formation of HD, and the concomitant cooling that it provides, is found to occur efficiently in primordial gas which is strongly ionized, owing ultimately to the high abundance of electrons which serve as catalyst for molecule formation in the early universe (Shapiro & Kang 1987).

Efficient cooling by HD can be triggered within the relic H II regions that surround Pop III.1 stars at the end of their brief lifetimes, owing to the high electron fraction that persists in the gas as it cools and recombines (Johnson et al. 2007; Nagakura & Omukai 2005; Yoshida et al. 2007). The efficient formation of HD can also take place

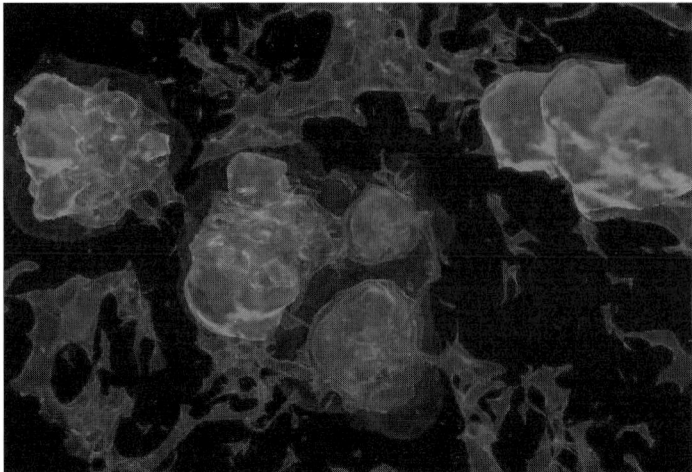

Figure 3. The chemical interplay in relic H II regions. While all molecules are destroyed in and around active H II regions, the high residual electron fraction in relic H II regions catalyzes the formation of an abundance of H_2 and HD molecules. The light and dark shades of blue denote regions with a free electron fraction of 5×10^{-3} and 5×10^{-4}, respectively, while the shades of green denote regions with an H_2 fraction of 10^{-4}, 10^{-5}, and 3×10^{-6}, in order of decreasing brightness. The regions with the highest molecule abundances lie within relic H II regions, which thus play an important role in subsequent star formation, allowing molecules to become shielded from photodissociating radiation and altering the cooling properties of the primordial gas.

when the primordial gas is collisionally ionized, such as behind the shocks driven by the first SNe or in the virialization of massive DM halos (Greif & Bromm 2006; Johnson & Bromm 2006; Machida et al. 2005; Shchekinov & Vasiliev 2006). In Figure 1, we show the HD fraction in primordial gas in four distinct situations: within a minihalo in which the gas is never strongly ionized, behind a 100 km s^{-1} shock wave driven by a SN, in the virialization of a 3σ DM halo at redshift $z = 15$, and in the relic H II region generated by a Pop III.1 star at $z \sim 20$ (Johnson & Bromm 2006). Also shown is the critical HD fraction necessary to allow the primordial gas to cool to the temperature floor set by the CMB at these redshifts. Except for the situation of the gas in the virtually un-ionized minihalo, the fraction of HD becomes large quickly enough to play an important role in the cooling of the gas, allowing the formation of Pop III.2 stars.

Figure 2 schematically shows the characteristic masses of the various stellar populations that form in the early universe. In the wake of Pop III.1 stars formed in DM minihalos, Pop III.2 star formation ensues in regions which have been previously ionized, typically associated with relic H II regions left over from massive Pop III.1 stars collapsing to black holes, while even later, when the primordial gas is locally enriched with metals, Pop II stars begin to form (Bromm & Loeb 2003; Greif & Bromm 2006). Recent simulations confirm this picture, as Pop III.2 star formation ensues in relic H II regions in well under a Hubble time, while the formation of Pop II stars after the first SN explosions is delayed by more than a Hubble time (Greif et al. 2007; Yoshida et al. 2007a,b; but see Whalen et al. 2008b).

3. Radiative Feedback from the First Stars

Due to their extreme mass scale, Pop III.1 stars emit copious amounts of ionizing radiation, as well as a strong flux of H_2-dissociating Lyman-Werner (LW) radiation (Bromm et al. 2001b; Schaerer 2002). Thus, the radiation from the first stars dramatically

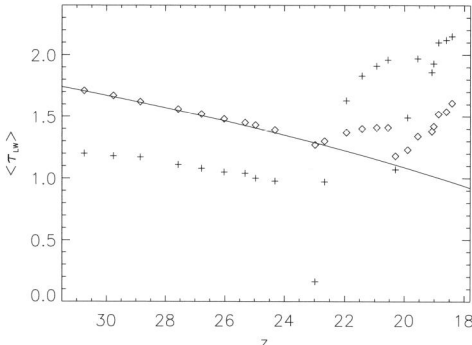

Figure 4. Optical depth to LW photons due to self-shielding, averaged over two different scales, as a function of redshift. The diamonds denote the optical depth averaged over the entire cosmological box of comoving length 660 kpc, while the plus signs denote the optical depth averaged over a cube of 220 kpc (comoving) per side, centered on the middle of the box. The solid line denotes the average optical depth that would be expected for a constant H_2 fraction of 2×10^{-6} (primordial gas), which changes only due to cosmic expansion.

influences their surroundings, heating and ionizing the gas within a few kpc (physical) around the progenitor, and destroying the H_2 and HD molecules locally within somewhat larger regions (Alvarez et al. 2006; Abel et al. 2007; Ferrara 1998; Johnson et al. 2007; Kitayama et al. 2004; Whalen et al. 2004). Additionally, the LW radiation emitted by the first stars could propagate across cosmological distances, allowing the build-up of a pervasive LW background radiation field (Haiman et al. 2000).

3.1. Local Radiative Effects

The impact of radiation from the first stars on their local surroundings has important implications for the numbers and types of Pop III stars that form. The photoheating of gas in the minihalos hosting Pop III.1 stars drives strong outflows, lowering the density of the primordial gas and delaying subsequent star formation by up to 100 Myr (Johnson et al. 2007; Whalen et al. 2004; Yoshida et al. 2007a). Furthermore, neighboring minihalos may be photoevaporated, delaying star formation in such systems as well (Ahn & Shapiro 2007; Greif et al. 2007; Shapiro et al. 2004; Susa & Umemura 2006; Whalen et al. 2008a). The photodissociation of molecules by LW photons emitted from local star-forming regions will, in general, act to delay star formation by destroying the main coolants that allow the gas to collapse and form stars.

The photoionization of primordial gas, however, can ultimately lead to the production of copious amounts of molecules within the relic H II regions surrounding the remnants of Pop III.1 stars (Johnson & Bromm 2007; Nagakura & Omukai 2005; Oh & Haiman 2002; Ricotti et al. 2001). Recent simulations tracking the formation of, and radiative feedback from, individual Pop III.1 stars in the early stages of the assembly of the first galaxies have demonstrated that the accumulation of relic H II regions has two important effects. First, the HD abundance that develops in relic H II regions allows the primordial gas to re-collapse and cool to the temperature of the CMB, possibly leading to the formation of Pop III.2 stars in these regions (Johnson et al. 2007; Yoshida et al. 2007b). Second, the molecule abundance in relic H II regions, along with their increasing volume-filling fraction, leads to a large optical depth to LW photons over physical distances of the order of several kpc. The development of a high optical depth to LW photons over such short length-scales suggests that the optical depth to LW photons over cosmological scales may

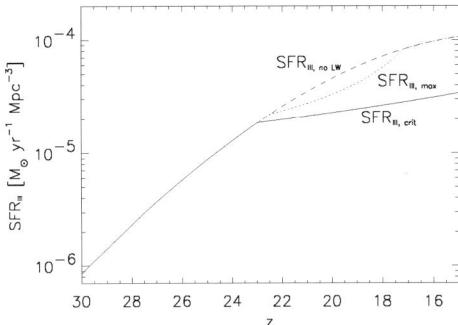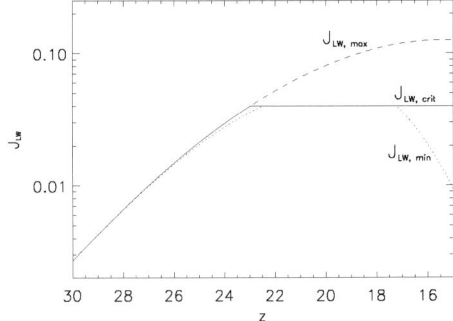

Figure 5. The Pop III star formation rates (*left panel*) and the corresponding LW background fluxes (*right panel*) for three models of the build-up of the LW background by Pop III.1 stars formed in minihalos. The maximum possible LW background, $J_{\rm LW,max}$, is generated for the case that every minihalo with a virial temperature $\geqslant 2 \times 10^3$ K hosts a Pop III star, without the LW background in turn diminishing the SFR, labeled here as SFR$_{\rm III,noLW}$. The self-regulated model considers the coupling between the star formation rate, SFR$_{\rm LW,crit}$, and the critical LW background that it produces, $J_{\rm LW,crit}$. The minimum value for the LW background, $J_{\rm LW,min}$, is produced for the case of a high opacity through the relic H II regions left by the first stars, in which case the self-consistent SFR, SFR$_{\rm III,max}$ can approach the undiminished SFR$_{\rm III,noLW}$.

be very high, acting to suppress the build-up of a background LW radiation field, and mitigating negative feedback on star formation.

Figure 3 shows the chemical composition of primordial gas in relic H II regions, in which the formation of H$_2$ molecules is catalyzed by the high residual electron fraction. Figure 4 shows the average optical depth to LW photons across the simulation box, which rises with time owing to the increasing number of relic H II regions.

3.2. Global Radiative Feedback

While the reionization of the universe is likely to have occurred at later times, as inferred from the *WMAP* third year results (Spergel *et al.* 2007), the process of primordial star formation can be affected by the build up of a LW background very soon after the formation of the first stars. This LW radiation, which acts to destroy H$_2$, the very coolant that enables the formation of the first stars, can, in principle, dramatically lower the formation rate of Pop III stars in minihalos (e.g. Haiman *et al.* 2000; Machacek *et al.* 2001; Yoshida *et al.* 2003; Mackey *et al.* 2003).

While star formation in more massive systems may proceed relatively unimpeded, through atomic line cooling, during the earliest epochs of star formation these atomic-cooling halos are rare compared to the minihalos which host individual Pop III stars. While the process of star formation in atomic-cooling halos is not well understood, for a broad range of models the dominant contribution to the LW background is from Pop III.1 stars formed in minihalos at $z \geqslant 15$-20 (Johnson *et al.* 2008). Therefore, at these redshifts the LW background radiation may be largely self-regulated, with Pop III.1 stars producing the very radiation which, in turn, suppresses their formation. Johnson *et al.* (2008) argue that there is a critical value for the LW flux, $J_{\rm LW,crit} \sim 0.04$, at which Pop III.1 star formation occurs self-consistently, with the implication that the PopIII.1 star formation rate in minihalos at $z > 15$ is decreased by only a factor of a few, as shown in Fig. 5. The star formation rate may be even higher if the cosmological average optical depth to LW photons through the relic H II regions left by the first stars is sufficiently high. An analytical model of this effect shows that the SFR may be only

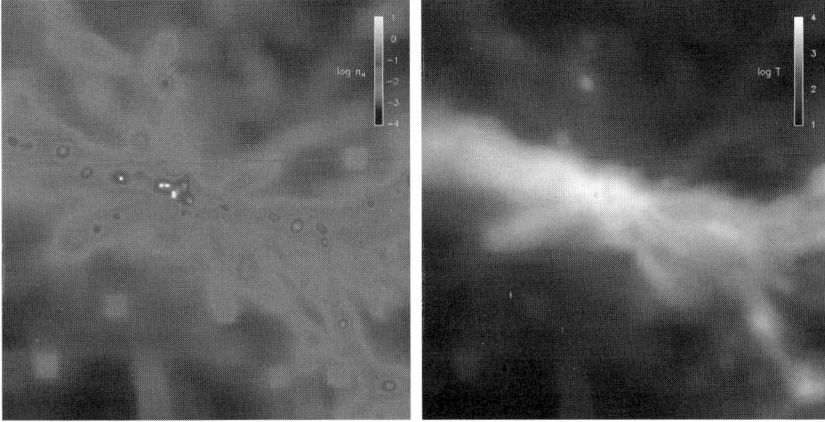

Figure 6. The hydrogen number density and temperature of the gas in the region of the forming galaxy at $z \sim 12.5$. The panels show the inner ~ 10.6 kpc (physical) of our cosmological box. The cluster of minihaloes harboring dense gas just left of the center in each panel is the site of the formation of the two Pop III.1 stars which are able to form in our simulation including the effects of the self-regulated LW background. The remaining minihaloes are not able to form stars by this redshift, largely due to the photodissociation of the coolant H_2. The main progenitor DM halo, which hosts the first star at $z \sim 16$, by $z \sim 12.5$ has accumulated a mass of 9×10^7 $M_\odot$ through mergers and accretion. Note that the gas in this halo has been heated to temperatures above 10^4 K, leading to a high free electron fraction and high molecule fractions in the collapsing gas. The high HD fraction that is generated likely leads to the formation of Pop III.2 stars in this system.

negligibly reduced once the volume-filling factor of relic H II regions becomes large, as is also shown in Fig. 5 (Johnson *et al.* 2008).

Simulations of the formation of a dwarf galaxy at $z \geqslant 10$ which take into account the effect of a LW background at $J_{\rm LW,crit}$ show that Pop III.1 star formation takes place before the galaxy is fully assembled, suggesting that the formation of metal-free galaxies may be a rare event in the early universe (Johnson *et al.* 2008). Figure 6 shows the temperature and density of the protogalaxy simulated by these authors at $z \sim 12.5$.

4. The First Supernova Explosions

Recent numerical simulations have indicated that primordial stars forming in DM minihalos typically attain 100 $M_\odot$ by efficient accretion, and might even become as massive as 500 $M_\odot$ (Bromm & Loeb 2004; O'Shea & Norman 2007; Omukai & Palla 2003; Yoshida *et al.* 2006). After their main-sequence lifetimes of typically $2-3$ Myr, stars with masses below $\simeq 100$ $M_\odot$ are thought to collapse directly to black holes without significant metal ejection, while in the range $140 - 260$ $M_\odot$ a pair-instability supernova (PISN) disrupts the entire progenitor, with explosion energies ranging from $10^{51} - 10^{53}$ ergs, and yields up to 0.5 (Heger *et al.* 2003; Heger & Woosley 2002).

The significant mechanical and chemical feedback effects exerted by such explosions have been investigated with a number of detailed calculations, but these were either performed in one dimension (Kitayama & Yoshida 2005; Machida *et al.* 2005; Salvaterra *et al.* 2004; Whalen *et al.* 2008b), or did not start from realistic initial conditions (Bromm *et al.* 2003; Norman *et al.* 2004). The most realistic, three-dimensional simulation to date took cosmological initial conditions into account, and followed the evolution of the gas until the formation of the first minihalo at $z \simeq 20$ (Greif *et al.* 2007). After the gas

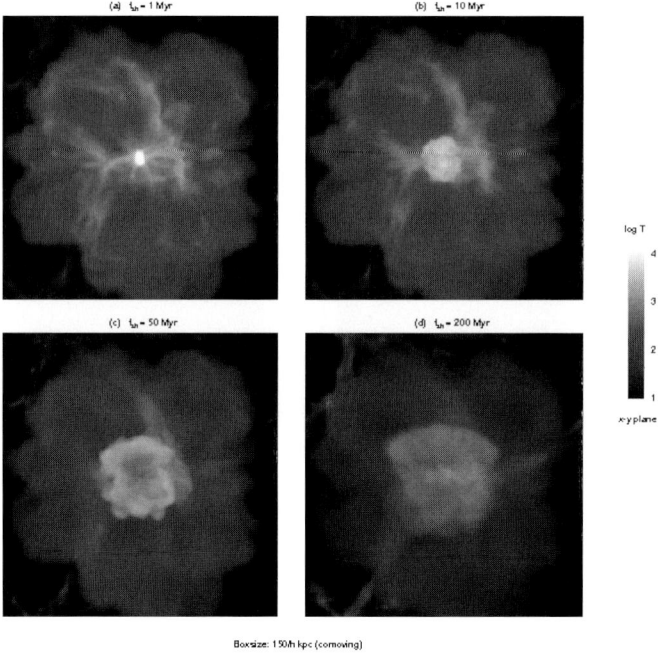

Figure 7. Temperature averaged along the line of sight in a slice of $10/h$ kpc (comoving) at 1, 10, 50, and 200 Myr after a Pop III PISN (Greif et al. 2007). In all four panels, the H II region and SN shock are clearly distinguishable, with the former occupying almost the entire simulation box, while the latter is confined to the central regions. *(a)*: The SN remnant has just left the host halo, but temperatures in the interior are still well above 10^4 K. *(b)*: After 10 Myr, the asymmetry of the SN shock becomes visible, while most of the interior has cooled to well below 10^4 K. *(c)*: The further evolution of the shocked gas is governed by adiabatic expansion. *(d)*: After 200 Myr, the shock velocity approaches the local sound speed and the SN remnant stalls. By this time the post-shock regions have cooled to roughly 10^3 K.

approached the 'loitering regime' at $n_{\rm H} \simeq 10^4$ cm^{-3}, the formation of a primordial star was assumed, and a photoheating and ray-tracing algorithm determined the size and structure of the resulting H II region (Johnson et al. 2007). An explosion energy of 10^{52} ergs was then injected as thermal energy into a small region around the progenitor, and the subsequent expansion of the SN remnant was followed until the blast wave effectively dissolved into the IGM. The cooling mechanisms responsible for radiating away the energy of the SN remnant, the temporal behavior of the shock, and its morphology could thus be investigated in great detail (see Figure 7).

The dispersal of metals by the first SN explosions transformed the IGM from a simple primordial gas to a highly complex medium in terms of chemistry and cooling, which ultimately enabled the formation of the first low-mass stars. However, this transition required at least a Hubble time, since the presence of metals became important only after the SN remnant had stalled and the enriched gas re-collapsed to high densities (Greif et al. 2007; but see Whalen et al. 2008b). Furthermore, the metal distribution was highly anisotropic, as the post-shock gas expanded into the voids in the shape of an 'hour-glass', with a maximum extent similar to the final mass-weighted mean shock radius (Greif et al. 2007).

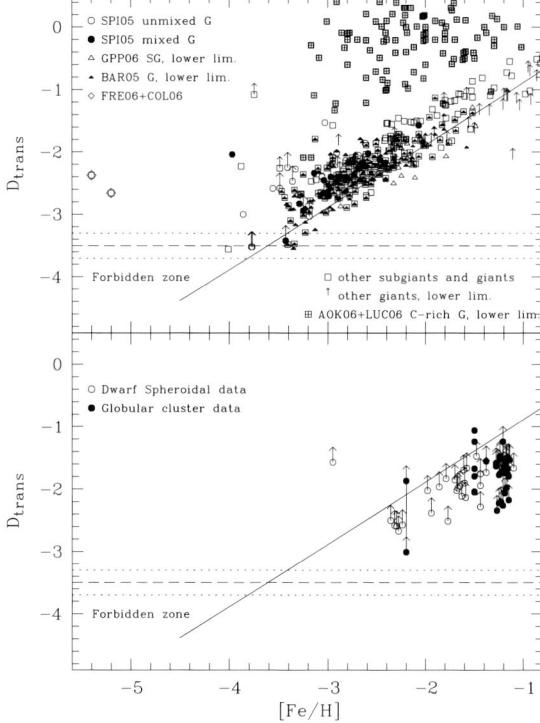

Figure 8. Transition discriminant, $D_{\rm trans}$, for metal-poor stars collected from the literature as a function of [Fe/H]. *Top panel*: Galactic halo stars. *Bottom panel*: Stars in dSph galaxies and globular clusters. G indicates giants, SG subgiants. The critical limit is marked with a dashed line. The dotted lines refer to the uncertainty. The detailed references for the various data sets can be found in Frebel *et al.* (2007).

To efficiently mix the metals with all components of the swept-up gas, a DM halo of at least $M_{\rm vir} \simeq 10^8$ M$_\odot$ had to be assembled (Greif *et al.* 2007), and with an initial yield of 0.1, the average metallicity of such a system would accumulate to $Z \simeq 10^{-2.5}$Z$_\odot$, well above any critical metallicity (Bromm & Loeb 2003; Bromm *et al.* 2001a; Schneider *et al.* 2006; see also Wise & Abel 2008). Thus, if energetic SNe were a common fate for the first stars, they would have deposited metals on large scales before massive galaxies formed and outflows were suppressed. Hints to such ubiquitous metal enrichment have been found in the low column density Lyα forest (Aguirre *et al.* 2005; Songaila & Cowie 1996; Songaila 2001), and in dwarf spheroidal satellites of the Milky Way (Helmi *et al.* 2006).

5. The Chemical Signature of the First Stars

The discovery of extremely metal-poor stars in the Galactic halo has made studies of the chemical composition of low-mass Pop II stars powerful probes of the conditions in which the first low-mass stars formed. While it is widely accepted that metals are required for the formation of low-mass stars, two general classes of competing models for the Pop III – Pop II transition are discussed in the literature: *(i)* atomic fine-structure line cooling (Bromm & Loeb 2003; Santoro & Shull 2006); and *(ii)* dust-induced fragmentation (Schneider *et al.* 2006). Within the fine-structure model, C II and O I have been suggested as main coolants (Bromm & Loeb 2003), such that low-mass star formation can occur in

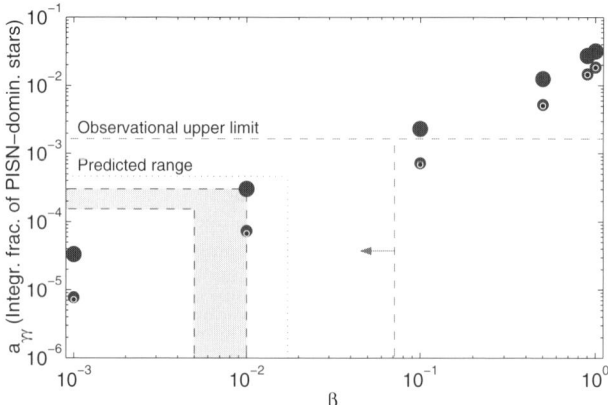

Figure 9. The predicted integrated (total) fraction of PISN-dominated stars below [Ca/H] = −2 as a function of β, corresponding to > 90% (big blue dots), > 99% (medium sized, blue dots), and > 99.9% (small, dark blue dots) PISN-enrichment. The dashed (red) lines indicate the observational upper limit of β, assuming that none of the ∼ 600 Galactic halo stars with [Ca/H] ⩽ −2 for which high-resolution spectroscopy is available show any signature of PISNe (N. Christlieb, priv. comm.). The dotted (blue) line and shaded (blue) area denote the predicted range of $a_{\gamma\gamma}$ anticipated from the calculations of Padoan *et al.* (2007) and Greif & Bromm (2006), respectively.

gas that is enriched beyond critical abundances of $[C/H]_{crit} \simeq -3.5 \pm 0.1$ and $[O/H]_{crit} \simeq -3 \pm 0.2$. The dust-cooling model, on the other hand, predicts critical abundances that are typically smaller by a factor of $10 - 100$.

Based on the theory of atomic line cooling (Bromm & Loeb 2003), a new function, the 'transition discriminant' has been introduced:

$$D_{\rm trans} \equiv \log_{10}\left(10^{[C/H]} + 0.3 \times 10^{[O/H]}\right), \qquad (5.1)$$

such that low-mass star formation requires $D_{\rm trans} > D_{\rm trans,crit} \simeq -3.5 \pm 0.2$ (Frebel *et al.* 2007). Figure 8 shows values of $D_{\rm trans}$ for a large number of the most metal-poor stars available in the literature. While theories based on dust cooling can be pushed to accommodate the lack of stars with $D_{\rm trans} < D_{\rm trans,crit}$, it appears that the atomic-cooling theory for the Pop III – Pop II transition naturally explains the existing data on metal-poor stars. Future surveys of Galactic halo stars will allow to further populate plots such as Figure 8, and will provide valuable insight into the conditions of the early universe in which the first low-mass stars formed.

The abundance patterns observed in the most metal-poor stars can also provide information about the types of SN that ended the lives of the first stars, as the metals that are emitted in these explosions will become incorporated into later generations of stars, some of which are observed in the halo of the Galaxy. Interestingly, while detailed numerical simulations of the formation of the first stars suggest that they were often massive enough to explode as PISN, no clear signature of such a PISN has yet been detected in a metal-poor star. Does this imply that the first stars did not explode as PISN, or that they were not very massive? (see also Ekström in this volume).

PISNe may have ejected enough mass in metals to enrich the IGM to a metallicity well above those of the most metal-poor stars. Therefore, one possible explanation for the apparent lack of Pop III PISNe is that the few stars which might have formed from gas dominantly enriched by a PISN may have relatively high metallicities, and so may have eluded surveys seeking such true second generation stars at lower metallicity (Karlsson et al. 2008). Karlsson et al. (2008) developed a model for the inhomogeneous chemical enrichment of the gas collapsing to become a dwarf galaxy at z ∼ 10 in which the formation of both Pop III stars from metal-free gas and the formation of Pop II stars from metal-enriched gas were tracked self-consistently. These authors find that the lack of the discovery of a metal-poor star showing signs of enrichment dominated by PISN yields in the existing catalog of metal poor stars is not inconsistent with theories predicting that the first stars were very massive, as shown in Figure 9. It is hoped that future surveys of stars in the Galactic halo will test this model by searching for PISN-enriched stars with metallicities [Fe/H] ⩾ -2.5.

6. Conclusion

Understanding the formation of the first galaxies marks the frontier of high-redshift structure formation. It is crucial to predict their properties in order to develop the optimal search and survey strategies for the *JWST*. Whereas *ab-initio* simulations of the very first stars can be carried out from first principles, and with virtually no free parameters, one faces a much more daunting challenge with the first galaxies. Now, the previous history of star formation has to be considered, leading to enhanced complexity in the assembly of the first galaxies. One by one, all the complex astrophysical processes that play a role in more recent galaxy formation appear back on the scene. Among them are external radiation fields, comprising UV and X-ray photons, and possibly cosmic rays produced in the wake of the first SNe (Stacy & Bromm 2007). There will be metal-enriched pockets of gas which could be pervaded by dynamically non-negligible magnetic fields, together with turbulent velocity fields built up during the virialization process. However, the goal of making useful predictions for the first galaxies is now clearly drawing within reach, and the pace of progress is likely to be rapid.

Acknowledgements

V.B. acknowledges support from NSF grant AST-0708795. The simulations presented here were carried out at the Texas Advanced Computing Center (TACC).

References

Abel, T., Bryan, G. L., & Norman, M. L. 2002, *Science*, 295, 93
Abel, T., Wise, J. H., & Bryan, G. L. 2007, *ApJ*, 659, L87
Aguirre, A., Schaye, J., Hernquist, L., *et al.* 2005, *ApJ*, 620, L13
Ahn, K. & Shapiro, P. R. 2007, *MNRAS*, 375, 881
Alvarez, M. A., Bromm, V., & Shapiro, P. R. 2006, *ApJ*, 639, 621
Barkana, R. & Loeb, A. 2001, *Phys. Rep.*, 349, 125
Beers, T. C. & Christlieb, N. 2005, *ARA&A*, 43, 531
Bromm, V., Coppi, P. S., & Larson, R. B. 1999, *ApJ*, 527, L5
Bromm, V., Coppi, P. S., & Larson, R. B. 2002, *ApJ*, 564, 23
Bromm, V., Ferrara, A., Coppi, P. S., & Larson, R. B. 2001a, *MNRAS*, 328, 969
Bromm, V., Kudritzki, R. P., & Loeb, A. 2001b, *ApJ*, 552, 464
Bromm, V. & Larson, R. B. 2004, *ARA&A*, 42, 79
Bromm, V. & Loeb, A. 2002, *ApJ*, 575, 111
Bromm, V. & Loeb, A. 2003, *Nature*, 425, 812
Bromm, V. & Loeb, A. 2004, *New Astronomy*, 9, 353
Bromm, V. & Loeb, A. 2006, *ApJ*, 642, 382
Bromm, V., Yoshida, N., & Hernquist, L. 2003, *ApJ*, 596, L135
Ciardi, B. & Ferrara, A. 2005, *Space Science Reviews*, 116, 625
Dwek, E., Arendt, R. G., & Krennrich, F. 2005, *ApJ*, 635, 784

Fernandez, E. R. & Komatsu, E. 2006, *ApJ*, 646, 703
Ferrara, A. 1998, *ApJ*, 499, L17
Frebel, A., Johnson, J. L., & Bromm, V. 2007, *MNRAS*, 380, L40
Greif, T. H. & Bromm, V. 2006, *MNRAS*, 373, 128
Greif, T. H., Johnson, J. L., Bromm, V., & Klessen, R. S. 2007, *ApJ*, 670, 1
Haiman, Z., Abel, T., & Rees, M. J. 2000, *ApJ*, 534, 11
Heger, A., Fryer, C. L., Woosley, S. E., et al. 2003, *ApJ*, 591, 288
Heger, A. & Woosley, S. E. 2002, *ApJ*, 567, 532
Helmi, A. et al. 2006, *ApJ*, 651, L121
Johnson, J. L. & Bromm, V. 2006, *MNRAS*, 366, 247
Johnson, J. L. & Bromm, V. 2007, *MNRAS*, 374, 1557
Johnson, J. L., Greif, T. H., & Bromm, V. 2007, *ApJ*, 665, 85
Johnson, J. L., Greif, T. H., & Bromm, V. 2008, *MNRAS*, submitted (astro-ph/0711.4622)
Karlsson, T., Johnson, J. L., & Bromm, V. 2008, *ApJ*, in press (astro-ph/0709.4025)
Kashlinsky, A., Arendt, R. G., Mather, J., & Moseley, S. H. 2005, *Nature*, 438, 45
Kitayama, T. & Yoshida, N. 2005, *ApJ*, 630, 675
Kitayama, T., Yoshida, N., Susa, H., & Umemura, M. 2004, *ApJ*, 613, 631
Lamb, D. Q. & Reichart, D. E. 2000, *ApJ*, 536, 1
Machacek, M. E., Bryan, G. L., Abel, T. 2001, *ApJ*, 548, 509
Machida, M. N., Tomisaka, K., Nakamura, F., & Fujimoto, M. Y. 2005, *ApJ*, 622, 39
Mackey, J., Bromm, V., & Hernquist, L. 2003, *ApJ*, 586, 1
Magliocchetti, M., Salvaterra, R., & Ferrara, A. 2003, *MNRAS*, 342, L25
McKee C. F. & Tan, J. C. 2008, *ApJ*, submitted (astro-ph/0711.1377)
Miralda-Escudé, J. 2003, *Science*, 300, 1904
Nagakura, T. & Omukai, K. 2005, *MNRAS*, 364, 1378
Oh, S. P. & Haiman, Z. 2002, *ApJ*, 569, 558
Omukai, K. & Palla, F. 2003, *ApJ*, 589, 677
O'Shea, B. W. & Norman, M. L. 2007, *ApJ*, 654, 66
Ricotti, M., Gnedin, N. Y., & Shull, J. M. 2001, *ApJ*, 560, 580
Salvaterra, R., Ferrara, A., & Schneider, R. 2004, *New Astronomy*, 10, 113
Santoro, F. & Shull, J. M. 2006, *ApJ*, 643, 26
Santos, M. R., Bromm, V., & Kamionkowski, M. 2002, *MNRAS*, 336, 1082
Schaerer, D. 2002, *A&A*, 382, 28
Schneider, R., Omukai, K., Inoue, A. K., & Ferrara, A. 2006, *MNRAS*, 369, 1437
Shapiro, P. R., Iliev, I. T., & Raga, A. C. 2004, *MNRAS*, 348, 753
Shapiro, P. R. & Kang, H. 1987, *ApJ*, 318, 32
Shchekinov, Y. A. & Vasiliev, E. O. 2006, *MNRAS*, 368, 454
Songaila, A. 2001, *ApJ*, 561, L153
Songaila, A. & Cowie, L. L. 1996, *AJ*, 112, 335
Spergel, D. N., Bean, R., Doré, O. et al. 2007, *ApJS*, 170, 377
Stacy, A. & Bromm, V. 2007, *MNRAS*, 382, 229
Susa, H. & Umemura, M. 2006, *ApJ*, 645, L93
Tegmark, M., Silk, J., Rees, M. J., et al. 1997, *ApJ*, 474, 1
Whalen, D., Abel, T., & Norman, M. L. 2004, *ApJ*, 610, 14
Whalen, D., O'Shea, B. W., Smidt, J., & Norman, M. L. 2008a, in: B. O'Shea, A. Heger & T. Abel (eds.), *First Stars III* (New York: AIP) in press (astro-ph/0708.3466)
Whalen, D., van Veelen, B., O'Shea, B. W., & Norman, M. L. 2008b, *ApJ*, submitted (arXiv:0801.3698)
Wise, J. H. & Abel, T. 2008, *ApJ*, submitted (astro-ph/0710.3160)
Yoshida, N., Abel, T., Hernquist, L., & Sugiyama, N. 2003, *ApJ*, 592, 645
Yoshida, N., Bromm, V., & Hernquist, L. 2004, *ApJ*, 605, 579
Yoshida, N., Oh, S. P., Kitayama, T., & Hernquist, L. 2007a, *ApJ*, 663, 687
Yoshida, N., Omukai, K., & Hernquist, L. 2007b, *ApJ*, 667, L117
Yoshida, N., Omukai, K., Hernquist, L., & Abel, T. 2006, *ApJ*, 652, 6

Imprint of First Stars Era in the Cosmic Infrared Background Fluctuations

Alexander Kashlinsky

SSAI and Observational Cosmology Lab, Code 665, Goddard Space Flight Center, Greenbelt, MD 20771, U.S.A.
email: `Alexander.Kashlinsky.1@nasa.gov`

Abstract. We present the latest results on Cosmic Infrared Background (CIB) fluctuations from early epochs from deep Spitzer data. The results show the existence of significant CIB fluctuations at the IRAC wavelengths (3.6 to 8 μm) which remain after removing galaxies down to very faint levels. These fluctuations must arise from populations with a significant clustering component, but only low levels of the shot noise. There are no correlations between the source-subtracted IRAC maps and the corresponding fields observed with the *HST* ACS at optical wavelengths. Taken together, these data imply that 1) the sources producing the CIB fluctuations are individually faint with $S_\nu <$ a few nJy at 3.6 and 4.5 μm; 2) have different clustering pattern than the more recent galaxy populations; 3) are located within the first 0.7 Gyr (unless these fluctuations can somehow be produced by - so far unobserved - local galaxies of extremely low luminosity and with the unusual for local populations clustering pattern), 4) produce contribution to the net CIB flux of at least 1-2 nW m^{-2} sr^{-1} at 3.6 and 4.5 μm and must have mass-to-light ratio significantly below the present-day populations, and 5) they have angular density of $\sim$ a few per arcsec^2 and are in the confusion of the present day instruments, but can be individually observable with *JWST*.

Keywords. diffuse radiation – early universe – large-scale structure of universe

1. Introduction

The cosmic infrared background (CIB) is a repository of emissions throughout the entire history of the Universe. The recent years have seen significant progress in CIB studies, both in identifying and/or constraining its mean level (isotropic component) and fluctuations (see Kashlinsky 2005 for a recent review). The CIB contains emissions also from objects inaccessible to current (or even future) telescopic studies and can, therefore, provide unique information on the history of the Universe at very early times. One particularly important example of such objects, of particular reference to this conference, concerns Population III stars (hereafter Pop III), the still elusive zero-metallicity stars expected to have been much more massive than the present stellar populations (see Bromm & Larson 2004 for a recent review). Here I will use the term "era of the first stars", or "Pop III era", with the understanding that the actual era may be composed of objects of various nature from purely zero-metallicity stars, to low- metallicity stars to even possibly mini-quasars whose contribution to the CIB is driven by energy released by gravitational accretion, as opposed to stellar nucleosynthesis.

Extensive numerical investigations of collapse and fragmentation of the first objects forming out of density fluctuations specified by the standard ΛCDM model suggest that Pop III stars were quite massive and lived at $z > 10$, well within the first Gyr of the Universe's evolution. If predominantly massive, they are expected to have left a significant level of diffuse radiation redshifted today into the IR, and it has been suggested that the CIB contains a detectable contribution from Pop III in the near-IR, manifest in both its mean level and its anisotropies (e.g. Bond *et al.* 1986, Santos *et al.* 2002, Salvaterra & Ferrara 2003, Cooray *et al.* 2004, Kashlinsky *et al.* 2004).

In the past several years a group of us (Kashlinsky, Arendt, Mather & Moseley 2005, 2007a,b,c - hereafter KAMM1, KAMM2, KAMM3, KAMM4) have used deep-integration *Spitzer* data to measure the CIB fluctuations component arising from early populations. These provide first observational insights into the global evolution of the Universe at early cosmic epochs. Our measurements revealed significant CIB fluctuations at the IRAC wavelengths (3.6 to 8 μm) which remain after removing galaxies down to very faint levels (KAMM1, KAMM2). These fluctuations must arise from populations that have a significant clustering component, but only low levels of the shot noise (KAMM3). Furthermore, there are no correlations between the source-subtracted IRAC maps and the corresponding fields observed with the *HST* ACS at optical wavelengths (KAMM4). Taken together, these data imply that 1) the sources producing the CIB fluctuations have a very different clustering pattern than galaxies at intermediate redshifts and are individually faint with $S_\nu <$ a few nJy at 3.6 and 4.5 μm; 2) are located within the first $\simeq 0.7$ Gyr (unless these fluctuations can somehow be produced by - so far unobserved - local galaxies of extremely low luminosity and with the unusual for local populations clustering pattern), 3) they produce contribution to the net CIB flux of at least 1-2 nW m^{-2} sr^{-1} at 3.6 and 4.5 μm and must have mass-to-light ratio significantly below the present-day populations, and 4) their angular density is $\sim$(a few) arcsec^{-2}, so they are in the confusion of the current instruments, but can be individually observable with *JWST*.

Below, I will discuss the latest measurements of the fluctuations in the CIB by our team (Kashlinsky, Arendt, Mather & Moseley - KAMM) and explore their implications for the nature of the sources contributing to these anisotropies, specifically in the era of the first stars. Following the Introduction, Sec. 2 reviews the current measurements at both near-IR (IRAC) and optical (ACS) wavelengths and Sec. 3 discusses the nature of the cosmological populations producing these CIB anisotropies.

2. Source subtracted CIB fluctuations vs optical galaxies

Before we discuss the interpretation of the KAMM measurements, it is important to review the steps done in the analysis leading to the measured CIB fluctuations. The data have been assembled from the individual AORs using the self-calibration method from Fixsen *et al.* (2000). The exposure times ranged from $\sim 8-9$ hours/pixel for the initial $5' \times 10'$ QSO 1700 field (KAMM1) to $\sim 23-25$ hours/pixel in each of the four GOODS fields of $10' \times 15'$ (KAMM2) The latter have been observed at two different epochs separated by ~ 6 months allowing us to better handle zodiacal gradients and possible instrument systematics. The images have been cleaned of foreground galaxies and stars in two steps: 1) all pixels with flux fluctuations exceeding a fixed number of standard deviations of the image were iteratively blanked, and 2) the residual faint parts of the sources were removed iteratively using a modified CLEAN algorithm (Hogbom 1974) where the maximum pixel intensity is located and the wide PSF is then scaled to half of this intensity and subtracted from the image. In the first step, which defines the final mask, it is important to remain in the regime when the fraction of the removed pixels is small enough to allow a reliable computation of the power spectrum using FFTs. The second step allows us removal of progressively fainter foreground populations and enables a better characterization of the remaining (and removed) populations.

Left panels of Fig. 1 show the source-subtracted CIB fluctuations (the instrument noise has been subtracted) for four GOODS fields at 3.6 and 4.5 μm adopted from

KAMM2. The detected signal is significantly higher than the instrument noise and the various systematics effects cannot account for it. Similarly, fluctuations due to emissions in the Solar System and the Galaxy are too weak, except at 8 μm where Galactic cirrus may contribute to the measured signal. There was a statistically significant correlation between the channels for the regions of overlap meaning that the same population is responsible for the fluctuations. The correlation function at deeper clipping cuts, when too few pixels were left for Fourier analysis, remains the same and is consistent with the power spectrum numbers (KAMM1). The signal is to a good accuracy isotropic on the sky, as required by its extragalactic origin, and must thus contain contributions from the "ordinary" galaxies and from unresolved populations at high z.

The extragalactic signal is made of two components: 1) shot noise from the remaining faint galaxies (shown with dotted lines in Fig. 1), and 2) on arcminute scales the fluctuations are produced by clustering of the emitters. It is important to emphasize that as fainter foreground galaxies are removed so that the remaining shot noise is reduced the details of the fluctuations change. This is due to the varying contribution from the remaining foreground galaxies. The large-scale part of the fluctuations remains as the foreground sources are removed down to the lowest shot-noise levels. Left panels of Fig. 2 show the decrease in the shot-noise power, $P_{\rm SN}$, as progressively higher iterations in the source removal are reached. The final shot noise reached by us with the GOODS data is shown with horizontal lines and is a factor of ~ 2 lower than in the QSO 1700 data (KAMM1). The right panels of Fig. 2 compare the final shot noise with that produced by the observed galaxy counts (which at these wavelengths are confusion limited at $m_{\rm AB} > 21-22$ for IRAC beam). The figure shows that galaxy removal is efficient to $m_{\rm AB} > 26-27$ and the signal in Fig. 1 comes from very faint sources.

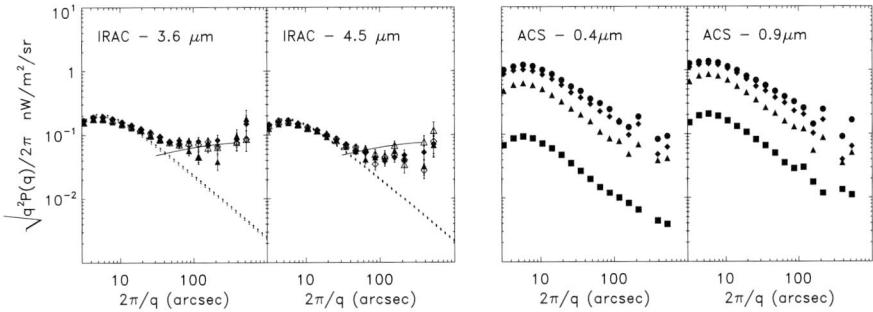

Figure 1. *Left*: Source-subtracted CIB fluctuations from KAMM2 at 3.6 and 4.5 μm. Four sets of symbols correspond to the four GOODS fields. Dotted lines show the shot-noise contribution. Solid line shows the slope of sources at high-z with the ΛCDM model spectrum of the same amplitude at 3.6 and 4.5 μm. *Right*: CIB fluctuations due to ACS galaxies for the 972×972 0.6″ pixel field at HDFN-Epoch2 region for the ACS B and z-bands. Filled circles correspond to ACS galaxies fainter than $m_0 = 21$ with the mask defined by the clipping. Filled diamonds, triangles and squares correspond to fluctuations produced by sources fainter than $m_0 + 2, m_0 + 4, m_0 + 6$.

GOODS fields have also been observed at optical wavelengths with the *Hubble* ACS instruments reaching source detection levels fainter than 28 AB mag. This allowed us to further test the origin of the source-subtracted CIB fluctuations. If the latter come from local populations there should be a strong correlation between the source-subtracted IRAC maps and the ACS sources. Conversely, there should be no such correlations if the CIB signal arises at at epochs where the Lyman break (at rest $\sim 0.1\mu$m) gets shifted passed the longest ACS z-band at $\simeq 0.9\mu$m. To test for this in KAMM4 we have constructed synthetic maps, overlapping with the GOODS fields, using sources in the ACS

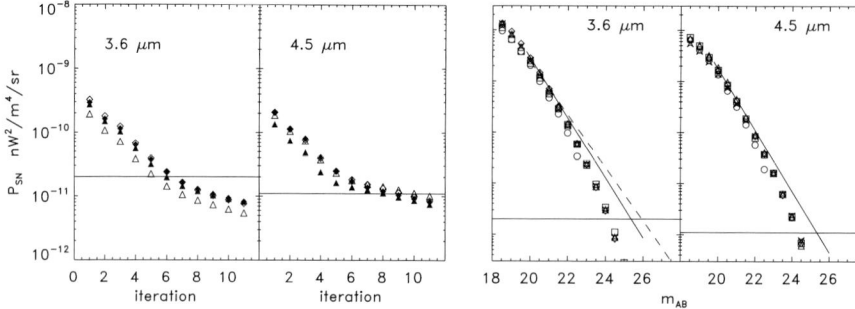

Figure 2. *Left*: Decrease of the shot noise power vs the iteration number of the source cleaning. Horizontal lines show the levels reached in Kashlinsky *et al.* (2007), which are a factor of ~ 2 below those in Kashlinsky *et al.* (2005). Four sets of symbols correspond to the four GOODS fields. Horizontal lines mark the shot noise levels of KAMM2. *Right*: Shot noise power $P_{\rm SN}$ estimated by integrating the counts. Symbols identical to those on the left denote the four GOODS fields; open circles correspond to counts for the QS1700 field. Solid line shows $P_{\rm SN}$ according to the fit to IRAC counts of Fazio *et al.* (2004) used in KAMM1.

B, V, i, z bands from the ACS sources catalog of Giavalisco *et al.* (2004). These maps were then convolved with the IRAC 3.6 and 4.5 μm beams. Finally, we applied the clipping mask from the IRAC maps and computed the fluctuations spectrum produced by the ACS sources and their correlations with the IRAC-based maps.

The fluctuations in the diffuse light produced by the ACS galaxies are shown in the right panels of Fig. 1. The contrast between the spectrum of the source-subtracted CIB fluctuations and those produced by the optical galaxies is obvious. The former has the power spectrum such that the fluctuations are flat to slowly rising with increasing angular scale, whereas the latter have power spectrum with fluctuation amplitude decreasing with increasing scale in agreement with CIB measurements from deep 2MASS data arising from galaxies at $z \sim$1-2 (Kashlinsky et al 2002). This by itself shows that the populations producing the source-subtracted CIB fluctuations are not in the ACS source catalog.

More importantly, Fig. 3 shows that the correlations between the ACS galaxies and the source-subtracted CIB maps are very small and on arcminute scales are within the statistical noise. Thus, at most, the remaining ACS sources contribute to the shot-noise levels in the residual KAMM maps, but not to the large scale correlations. At the same time, there are excellent correlations (shown with open symbols) between the ACS source maps and the sources *removed* by KAMM prior to computing the remaining CIB fluctuations.

The excess source-subtracted CIB fluctuation on arcminute scales in the 3.6 μm channel is ~ 0.1 nW m^{-2} sr^{-1}; KAMM measure a similar amplitude in the longer IRAC bands indicating that the energy spectrum of the arcminute scale fluctuations is flat to slowly rising with increasing wavelength at least over the IRAC range of wavelengths. This is illustrated with the solid line in the left panels of Fig. 1.

3. Cosmological implications

Any interpretation of the KAMM results must reproduce three major aspects:
- The sources in the KAMM data were removed to a certain (faint) flux limit, so the CIB fluctuations arise in populations with magnitudes fainter than the corresponding magnitude limit, $m_{\rm lim}$. Furthermore, these sources are not present among the optical ACS galaxies as demonstrated by the absence of correlations between these galaxies and the IRAC source-subtracted CIB maps.

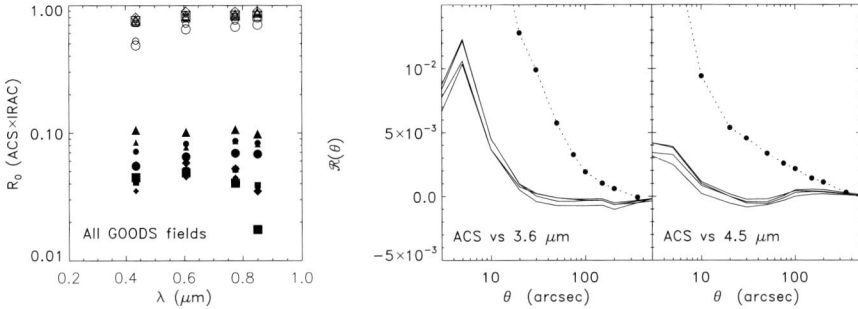

Figure 3. *Left*: Correlation coefficient between clipped/masked ACS and KAMM data. Large and small symbols correspond to the IRAC Ch 1 and Ch 2; the four sets of symbols correspond to the four GOODS fields. Open symbols correspond to correlations with the maps of the removed sources and filled symbols with the residual KAMM maps which contain the fluctuations shown in Fig. 1. *Right*: Solid lines show the dimensionless correlation function between the diffuse light in the ACS and KAMM maps for B, V, i, z-bands in order of increasing thickness. Dotted line shows the dimensionless correlation function of the KAMM maps, $C_{\rm KAMM}(\theta)/\sigma^2_{\rm KAMM}$, which remains positive out to $\sim 100''$ and is better viewed in log-log plots as in Fig. SI-4 of KAMM1.

- These sources must reproduce the excess CIB fluctuations by KAMM on scales $> 0.5'$. They must also reproduce the measured spectrum of the CIB fluctuations, which is different from the observed clustering pattern of ordinary galaxies at intermediate z.
- Lastly, the populations fainter than the above magnitude limit must account not only for the correlated part of the CIB, but - equally important - they must reproduce the (low) shot-noise component of the KAMM signal, which dominates the power at $<0.5'$. The discussion below by-and-large follows KAMM3:

1) **Magnitude limits and epochs**. Since the ACS galaxies do not contribute to the source-subtracted CIB fluctuations, the latter must arise at $z > 7$ as is required by the Lyman break at rest $\sim 0.1\mu$m getting redshifted past the ACS z-band of peak wavelength $\simeq 0.9\mu$m. This would place the sources producing the KAMM signal within the first 0.7 Gyr. If the KAMM signal were to originate in lower z galaxies which escaped the ACS GOODS source catalog because they are below the catalog flux threshold, they would have to be extremely low-luminosity systems ($< 2 \times 10^7 h^{-2} L_\odot$ at $z=1$) and these galaxies would also have to cluster very differently from their ACS counterparts.

2) **Clustering component**. Solid lines in Fig. 1 show the expected CIB fluctuations from sources with the (biased) concordance ΛCDM power spectrum at $z > 5$. The fit is reasonably good making such sources a plausible candidate for producing the observed signal. At the same time, the observed galaxy populations out to $z \sim 2 - 3$ cluster very differently. Thus any model attempting to assign the KAMM signal to more recent sources will have to account for this *observed* difference in the clustering patterns.

3) **Net CIB levels from the new sources**: The angle of $1'$ in the concordance cosmology subtends comoving scales of 2.2-3 Mpc at $5 \leqslant z \leqslant 20$. For ΛCDM density fields with reasonable biasing one can reach relative arcminute-scale fluctuations of $\sim$5-10% meaning that the net CIB from sources contributing to the KAMM signal at 3.6 μm is at least 1-2 nW m^{-2} sr^{-1}, which is well within the uncertainties of the recent CIB measurements of Thompson *et al.* (2007a,b).

4) **Shot noise constraints**. The amplitude of the shot-noise power gives a particularly strong indication of the epochs of the sources contributing to the KAMM signal. This

can be seen from the expressions for the shot noise (Kashlinsky 2005a):

$$P_{\rm SN} = \int_{m_{\rm AB} > m_{\rm lim}} f(m)dF(m) \equiv f(\bar{m})F_{\rm tot}(m_{\rm AB} > m_{\rm lim}) \qquad (3.1)$$

where $f(m)$ is the flux in Jy of a source of magnitude m and $F_{\rm tot}(m_{\rm AB} > m_{\rm lim})$ is the net CIB flux produced by the remaining sources. Above it was shown that the sources contributing to the fluctuations must have CIB flux greater than a few nW m^{-2} sr^{-1} and combining this with the values for $P_{\rm SN} \sim 10^{-11}$ nW^2 m^{-4} sr^{-1}, reached in the KAMM2 analysis, leads via eq. 3.1 to these sources having typical magnitudes $m_{\rm AB} < 29-30$ or individual fluxes < 4 nJy. *Such faint sources are expected to lie at very high z.*

5) **Mass/light ratio of the new populations**: This information on the nature of the populations responsible for these CIB fluctuations, can be obtained from the fact that the significant amount of flux (>1-2 nW m^{-2} sr^{-1}) required to explain the amplitude of the fluctuations must be produced within the short time available at these high z (cosmic times <0.5-1 Gyr). The implied comoving luminosity density associated with these populations is related to the fraction of baryons locked in these objects with the additional assumption of their $\Gamma \equiv M/L$. The smaller the value of Γ, the fewer baryons are required to explain the CIB fluctuations detected in the KAMM studies. It turns out that in order not to exceed the baryon fraction observed in stars, the populations producing these CIB fluctuations had to have Γ much less than the solar value, typical of the present-day populations (KAMM3). This is consistent with the general expectations of the first stars being very massive.

6) **Resolving the new sources**: In order to directly detect the faint sources responsible for the CIB fluctuations with fluxes below a few nJy, their individual flux must exceed the confusion limit. If such sources were to contribute to the CIB required by KAMM data, at 3.6 and 4.5 μm they had to have the average surface density of $\bar{n} \sim F_{\rm CIB}^2/P_{\rm SN} \sim 5$ arcsec^{-2}. In order to avoid the confusion limit and resolve these sources individually at, say, 5-sigma level ($\alpha = 5$) one would need a beam of the area $\omega_{\rm beam} \leqslant \alpha^{-2}/\bar{n} \sim 5 \times 10^{-3}$ arcsec^2 or of circular radius below ~ 0.04 arcsec. This is clearly not in the realm of the currently operated instruments, but the *JWST* could be able to resolve these objects given its sensitivity and resolution.

Acknowledgements

I thank my collaborators, Rick Arendt, John Mather and Harvey Moseley for many contributions to the KAMM results, and the NSF AST-0406587 grant for support.

References

Bond, J. R., Carr, B. J., & Hogan, C. J. 1986, *ApJ*, 306, 428
Bromm, V. & Larson, R. 2004, *ARA&A*, 42, 79
Cooray, A., Bock, J. J., Keatin, B., et al. 2004, *ApJ*, 606, 611
Fazio, G., Ashby, M. L. N., Barnby, P., et al. 2004, *ApJS*, 154, 39
Fixsen, D., Moseley, S. H. & Arendt, R. G. 2000, *ApJS*, 128, 651
Giavilisco, M., Dickinson, M., Ferguson, H. C., et al. 2004, *ApJ*, 600, L93
Hogbom, J. 1974, *A&AS*, 15, 417
Kashlinsky, A. 2005, *Phys. Rep.*, 409, 361
Kashlinsky, A. 2005, *ApJ*, 633 L5
Kashlinsky, A., Odenwald, S., Mather, J., et al. 2002, *ApJ*, 579, L53
Kashlinsky, A., Arendt, R. G., Gardner, J. P., et al. 2004, *ApJ*, 608, 1
Kashlinsky, A., Arendt, R. G., Mather, J., & Moseley, S. H. 2005, *Nature*, 438, 45 (KAMM1)
Kashlinsky, A., Arendt, R. G., Mather, J., & Moseley, S. H. 2007a, *ApJ*, 654, L5 (KAMM2)

Kashlinsky, A., Arendt, R. G., Mather, J., & Moseley, S. H. 2007b, *ApJ*, 654, L1 (KAMM3)
Kashlinsky, A., Arendt, R. G., Mather, J., & Moseley, S. H. 2007c, *ApJ*, 666, L1 (KAMM4)
Santos, M., Bromm, V., & Kamionkowski, M. 2002, *MNRAS*, 336, 1082
Thompson, R., Eisenstein, D., Fan, X., *et al.* 2007a, *ApJ*, 659, 667
Thompson, R., Eisenstein, D., Fan, X., *et al.* 2007b, *ApJ*, 666, 658

Discussion

FYNBO: In Chary *et al.* (2007, *ApJL* submitted, ArXiv:0711.4099) a significant fraction of the fluctuation signal is associated with low luminosity galaxies at $<z>=2.5$. Could you please comment on that?

KASHLINSKY: We have demonstrated in Kashlinsky *et al.* (2007c) that there are completely negligible correlations between the ACS galaxies, such as considered by Chary *et al.*, and our source-subtracted IRAC maps from which we derive the CIB fluctuations at 3.6 μm and 4.5 μm. This means that the ACS-detected galaxies contribute negligibly to the CIB fluctuations measured by us. Additionally, the clustering pattern of the ACS galaxies is such that their diffuse light fluctuations decrease strongly with angular scale, whereas our CIB fluctuations are flat to slowly rising with increasing angular scale. This further demonstrated that populations producing the source-subtracted CIB fluctuations are of a different nature than those in the ACS measurements.

Ken Nomoto.

George Sonneborn.

Claus Leitherer.

Imaging and Spectroscopy with the James Webb Space Telescope

George Sonneborn

Laboratory for Observational Cosmology, Code 665
NASA Goddard Space Flight Center
Greenbelt, MD 20771, USA
email: george.sonneborn@nasa.gov

Abstract. The James Webb Space Telescope (JWST) is a large, infrared-optimized space telescope scheduled for launch in 2013. JWST will find the first stars and galaxies that formed in the early universe, connecting the Big Bang to our own Milky Way galaxy. JWST will peer through dusty clouds to see stars forming planetary systems, connecting the Milky Way to our own Solar System. JWST's instruments are designed to work primarily in the infrared range of 1 - 28 μm, with some capability in the visible range. JWST will have a large segmented mirror, $\sim$6.5 m in diameter, and will be diffraction-limited at 2 μm ($<$ 0.1 arcsec resolution). JWST will be placed in an L2 orbit about 1.5 million km from the Earth. The instruments will provide imaging, coronography, and multi-object and integral-field spectroscopy across the 1 - 28 μm wavelength range. The breakthrough capabilities of JWST will enable new studies of massive stars from the Milky Way to the early universe.

Keywords. telescopes – space vehicles: instruments – instrumentation: high angular resolution – instrumentation: spectrographs

1. Mission Overview

The James Webb Space Telescope (JWST) is a large (6.5 m), cold ($T \sim 40$ K), infrared-optimized space observatory being built for launch in 2013 on an Ariane 5 ESA rocket into orbit around the Sun-Earth Lagrange point L2. The observatory will have four instruments: a camera, a multi-object spectrograph, and a tunable filter imager will cover the near-infrared spectrum ($0.6 < \lambda < 5.0 \mu$m). A mid-infrared instrument will provide imaging and spectroscopy from $5.0 < \lambda < 28$ μm. Coronography will also be possible across the entire JWST bandpass. JWST is a cooperative project between NASA and the European and Canadian Space Agencies (ESA and CSA). The mission design and development is led and managed by NASA's Goddard Space Flight Center. JWST will be operated for NASA and the scientific community by the Space Telescope SCience Institute.

The JWST science goals, observatory design, operational concept, and expected performance are described in detail by Gardner *et al.* (2006, see the link on the JWST web site www.jwst.nasa.gov/science.html) to download the reprint of this large paper). The present contribution is a highly abbreviated version of the information contained in Gardner *et al.*

The JWST science goals are divided into four themes. The End of the Dark Ages: First Light and Reionization theme is to identify the first luminous sources to form and to determine the ionization history of the early universe. The Assembly of Galaxies theme is to determine how galaxies and the dark matter, gas, stars, metals, morphological structures, and active nuclei within them evolved from the epoch of reionization to the present day. The Birth of Stars and Protoplanetary Systems theme is to unravel the birth and early evolution of stars, from infall on to dust-enshrouded protostars to the genesis of planetary systems. The Planetary Systems and the Origins of Life theme is to determine the physical and chemical properties of planetary systems, including our own, and investigate the potential for the origins of life in those systems.

To enable these observations, JWST consists of a telescope, an instrument package, a spacecraft and a sunshield. The instrument package contains the four science instruments (SIs) and a fine guidance sensor. The Sun, Earth, and Moon are always behind the sunshield, so that the telescope and instruments are passively cooled to $T \sim 40$ K. The spacecraft, located on the sunward side of the sunshield, provides pointing, power, orbit maintenance and communications. The JWST operations plan is based on that used for previous space observatories, and the majority of JWST observing time will be allocated to the international astronomical community through annual peer-reviewed proposal opportunities.

The telescope is a deployable optical system consisting of 18 hexagonal beryllium segments that provide diffraction-limited performance at 2 μm. The mirror segments are adjustable in orbit in six degrees of freedom. Image quality is maintained by periodic monitoring of the wavefront errors by NIRCam (Section 2.1) and biweekly updates to the mirror positions. The wavelength range of JWST and the SIs spans 0.6 to 28 μm, limited at the short end by the gold coatings on the primary mirror and at the long end by the detector technology.

JWST is designed to provide instantaneous sky coverage over the solar elongation range of 85° to 135° (95° to 45° from the anti-solar direction). There s a continuous viewing zone within 5° of both the north and south ecliptic poles. Thirty percent of the sky can be viewed continuously for at least 197 continuous days. All regions of the sky have at least 51 days of continuous visibility per year. The JWST architecture provides an instantaneous visibility of ∼40% of the sky.

The sunshield attenuates the incident solar radiation by a factor of $\sim 10^5$. This solar attenuation is a result of the five-layer configuration of the sunshield. Its physical size and shape determine the instantaneous sky coverage for the observatory. By reducing the solar radiation to the milliwatt level, the observatory has an intrinsically stable point spread function at all solar orientations within the allowed range.

Many JWST observations will be background limited. The background is a combination of in-field zodiacal light, scattered thermal emission from the sunshield and telescope, scattered starlight, and scattered zodiacal light. Over most of the sky, the zodiacal light dominates at wavelengths $\lambda < 10\mu m$. The sensitivities of the JWST instruments are given in Table 1.

2. JWST Science Instruments

JWST has four science instruments: NIRCam, NIRSpec, TFI, and MIRI. A cryo-cooler will be used for cooling the MIRI focal plane and its Si:As detectors. The near-infrared detector arrays in the other three instruments are passively cooled HgCdTe of HAWAII II heritage. The instrument module also houses the Fine Guidance Sensor (FGS, provided by CSA) and a computer that directs the daily science observations based on plans received from the ground. The science instruments and FGS have non-overlapping FOVs. Simultaneous operation of all science instruments is possible. This capability will be used for parallel calibration, including darks and possibly sky flats. FGS is used for guide star acquisition and fine pointing. Its FOV and sensitivity are sufficient to provide a greater than 95% probability of acquiring a guide star for any valid pointing direction and roll angle.

Table 1. JWST Instrument Sensitivities

Instrument/Mode	λ (μm)	Bandwidth	Sensitivity
NIRCam	1.1	R=4	12.1 nJy, AB=28.7
NIRCam	2.0	R=4	10.4 nJy, AB=28.9
TFI	3.5	R=100	126 nJy, AB=26.1
NIRSpec/Low Res	3.0	R=100	120 nJy, AB=26.2
NIRspec/Med Res	2.0	R=1000	1.64×10^{-18} erg s^{-1} cm^{-2}
MIRI/Broad-Band	10.0	R=5	700 nJy, AB=24.3
MIRI/Broad-Band	21.0	R=4.2	7.3 μJy, AB=21.7
MIRI/Spect.	9.2	R=2400	1.0×10^{-17} erg s^{-1} cm^{-2}
MIRI/Spect.	22.5	R=1200	5.6×10^{-17} erg s^{-1} cm^{-2}

NOTE – Sensitivity is evaluated for a point source detected at 10 σ in 10000 s.

2.1. Near-Infrared Camera (NIRCam)

NIRCam provides filter imaging and coronography in the 0.6 to 5.0 μm range with wavelength multiplexing. It includes the ability to sense the wavefront errors of the observatory. NIRCam uses a dichroic to simultaneously observe the short (0.6 to 2.3 μm) and long (2.4 to 5.0 μm.) wavelength light paths from the same field of view.

The instrument contains a total of ten 2K×2K detectors, including those in the identical redundant optical trains. The short wavelength arm in each optical train contains four detectors, optimized for the 0.6 - 2.3 μm wavelength range, with a small gap (~3 mm = ~5 arcsec) between adjacent detectors. The detectors will all have thinned substrates to avoid cosmic ray scintillation issues, as well as to extend their sensitivity below 0.85 μm. Each optical train contains filter and pupil wheels that hold a range of filters and wavefront sensing optics.

To enable the coronagraphic imaging, each of the two identical optical trains in the instrument contains a traditional focal plane coronagraphic mask plate held at a fixed distance from the detectors. This mounting ensures that the coronagraph spots are always in focus at the detector plane. Each coronagraphic plate is transmissive, and contains a series of spots of different sizes, including linear and radia-sinc occulters, to block the light from a bright object.

2.2. Near-Infrared Spectrograph (NIRSpec)

NIRSpec is a near infrared multi-object dispersive spectrograph provided by ESA that is capable of simultaneously observing up to ~100 sources over a field-of-view (FOV) larger than 3 × 3 arcmin. In addition to the multi-object capability, it includes fixed slits and an integral field unit (IFU) for imaging spectroscopy. Six gratings will yield resolving powers of R ~ 1000 and ~ 2700 in three spectral bands, spanning the range 1.0 to 5.0 μm. A prism will yield R ~ 100 over 0.6 to 5.0 μm.

Targets in the FOV are normally selected by opening groups of shutters in a micro-shutter assembly to form multiple apertures. The micro-shutter assembly itself consists of a mosaic of 4 subunits producing a final array of approximately 750 (spectral) by 350 (spatial) individually addressable shutters with 200 × 450 milliarcsec (mas) openings and 250 × 500 mas pitch. The minimum aperture size is 1 shutter (spectral) by 1 shutter (spatial) at all wavelengths. Multiple pointings may be required to avoid placing targets near the edge of a shutter and to observe targets with spectra that would overlap if observed simultaneously at the requested roll angle. The nominal slit height perpendicular to the dispersion direction is three shutters in all wavebands. These provide background spectra adjacent to the science target in the central shutter. In the open configuration, a shutter passes light from the fore-optics to the collimator. A slitless mode can be configured by opening all of the micro shutters. As the shutters are individually addressable, long slits, diagonal slits, Hadamard transform masks, and other patterns can also be configured.

In addition to the slits defined by the micro-shutter assembly, NIRSpec also includes five fixed slits that can be used for high-contrast spectroscopy. They are placed in a central strip of the aperture focal plane between sub-units of the micro-shutter assembly. Three fixed slits are 3.5 arcsec long and 200 mas wide. One fixed slit is 4 arcsec long and 400 mas wide for increased throughput at the expense of spectral resolution. One fixed slit is 2 arcsec long and 100 mas wide for brighter targets.

2.3. Tunable Filter Imager (TFI)

The TFI, built by CSA, provides narrow-band near-infrared imaging over a field of view of 2.2×2.2 arcmin^2 with a spectral resolution R $\sim$ 100. The etalon design allows observations at wavelengths of 1.6 μm to 2.6 μm and 3.1 μm to 4.9 μm. The gap in wavelength coverage allows the single channel to reach more than one octave in wavelength.

The TFI incorporates four coronagraphic occulting spots permanently to one side of the field of view, and occupying a region 20 by 80 arcsec. A set of selectable apodization masks is located at the internal pupil images of each channel by the filter wheels. The coronagraph is designed to deliver a contrast ratio of $\sim 10^4$ (10σ) at 1 arcsec separation. The sensitivity is limited by speckle noise. Contrast ratios of 10^5 may be achievable at sub-arcsec scales using roll or spectral deconvolution techniques.

2.4. Mid-Infrared Instrument (MIRI)

MIRI, provided by a consortium of European and U.S. institutions, is designed to obtain imaging, coronography, and spectroscopic measurements over the wavelength range 5 to 28 μm. A cryo-cooler will keep the MIRI Si:As detectors at $T \sim 6$ K. The optical bench contains two actively cooled subcomponents, an imager and IFU spectrograph, plus an on-board calibration unit. The imager module provides broad-band imaging, coronography, and low-resolution (R$\sim$100, 5-10 μm) slit spectroscopy using a single 1024$\times$1024 pixel Raytheon Si:As detector with 25 μm pixels. The coronagraphic masks include three phase masks for a quadrant-phase coronagraph and one opaque spot for a Lyot coronagraph. The coronagraphic masks each have a square field of view of 26 $\times$ 26 arcsec and are optimized for particular wavelengths.

The IFU obtains simultaneous spectral and spatial data on a small region of sky. The spectrograph field of view is next to that of the imager so that accurate positioning of targets will be possible by locating the image with the imager channel and off-setting to the spectrograph. The light is divided into four spectral ranges by dichroics, and two of these ranges are imaged onto each of two detector arrays. A full spectrum is obtained by taking exposures at each of three settings of the grating wheel. The spectrograph uses four image slicers to produce dispersed images of the sky on two 1024 $\times$ 1024 detectors, providing R $\sim$ 3000 integral field spectroscopy over the 5 to 29 μm wavelength range, although the sensitivity of the detectors drops longward of 28 μm.

References

Gardner, J. P., Mather, J. C., Clampin, M., *et al.* 2006, *Sp. Sci. Rev.*, 123, 485

The Impact of Extremely Large Telescopes on the Study of the Most Luminous Stellar Objects

Sandro D'Odorico

European Southern Observatory
K. Schwarzschild Str. 2, 85748 Garching bei Mnchen, Germany
email: sdodoric@eso.org

Abstract. The potential advantages of the new generation of Extremely Large Telescopes are briefly summarized. When used in combination with advanced adaptive optics modules which can substantially remove the effect of atmospheric turbulence at infrared wavelengths, these telescopes will provide unique capabilities both in terms of photon collecting power ($\rightarrow$2-4 magnitude advantage) and angular resolution (4-5 times higher than with current 8-10m telescopes). The instruments under study for the TMT and E-ELT projects are presented and compared. I discuss the impact of the ELTs on three major science topics: stellar populations in galaxies to the Virgo distance, chemical abundances of the brighter stars in nearby galaxies and high redshift SN and GRBs.

Keywords. instrumentation: adaptive optics – galaxies: stellar content – supernovae: general – gamma rays: bursts – telescopes

1. ELTs: context and potential advantages

In the last 25 years as many as 15 new 8-10m telescopes have entered or are about to enter into operation. This gigantic increase in photon collecting power has had already a major impact on many field of astrophysics. It has pushed the brightness limit of the objects which can be studied 2-3 magnitudes fainter and has led to much higher S/N data in other types of observations. Beside the increase in the area of the telescopes, two other parallel developments had an equally positive influence. First, array detectors for both UV-Visual and NIR wavelengths have become regularly available in size up to 2K × 2K pixels (for CCD up to 4K × 2K) and with QE above 60% over the whole range of sensitivity. Secondly, through the introduction of active optics pioneered by ESO at the NTT in the late 80's, the optical image quality of the telescope has matched the best seeing achievable at the high quality sites (that is down to 0.3″). Finally, in the last decade an additional observing capability has been demonstrated at the telescope and is now operating on a regular basis: the acquisition of images corrected for atmospheric turbulence by use of adaptive mirrors (e.g. NACO, SINFONI at the VLT, OSIRIS at Keck). These instruments are competitive in the NIR with the observing modes offered by the Hubble Space Telescope (HST) which operates above the atmosphere. Extremely Large Telescopes (ELTs) with diameters of the primary mirror larger than 20m have now become the subject of detailed feasibility studies. The new telescopes will offer an order of magnitude increase in collecting area and they are going to rely on the use of adaptive optics to perform at the diffraction limit ($\theta = 1.22\lambda/D$) over a moderate field with high Strehl ratios (concentration of light in the diffraction peak with respect to the total light from the source) at infrared wavelengths. The large diameter of the primary mirror of the ELTs will boast the angular resolution capability with respect to the space-based James Webb Space Telescope (JWST), which is expected to start to operate in 2015, has the advantage of the absence of a blurring atmosphere but has a diameter of 6.5m only. Many classes of scientific objectives which rely on observations at the highest angular resolution(e.g. planetary discs, regions of star formation, active galactic nuclei) will benefit from this unique advantage. Beside the gain of the unique angular resolution when operating close to the diffraction limit of the telescope, the advantages of the ELTs with respect to the 8-10 m class ground-based telescopes can be quantified as follows, depending on the nature of the targets and the type of observations:

For photon-noise dominated observations, the faintness limit (at a given integration time and for a chosen S/N) and the speed (defined as the reciprocal of the time needed to reach a given S/N for a given magnitude) are proportional to D^2. When comparing the European Extremely Large Telescope (now nominally 42m) with the VLT UT (8m) this translates into a gain of 3.5 magnitudes. This regime corresponds e.g. to high time resolution or high spectroscopy resolution observations of bright sources.

For detector noise dominated observations, the faintness limit is proportional to D^2, the speed to D^4. This is a regime where the advantage of the ELTs is more relevant but with the improvement in the r.o.n. of modern detectors, this case is confined to the observations at high resolution spectroscopy at low S/N.

For sky background limited observations, at natural seeing image quality, the faintness limit is proportional to D, the speed to D^2 . This points out that for this type of observations (e.g. large field imaging at red or NIR wavelength) the gain from the VLT 8m UT to a 42m telescope is less than two magnitudes for a fixed exposure time. An imaging survey with an 8m telescope of large field (1 square degree of larger) can be more efficient than one carried out at the ELT with a field of a few square minutes. For operation at the telescope with an adaptive optics module which delivers images at the diffraction limit with a significant fraction of the flux of point-like sources within the Airy diffraction disc (high Strehl ratio), then the advantage of the 42 m in sky background-limited observations becomes more relevant. The faintness limit is again proportional to D^2 and the speed is proportional to D^4. This regime corresponds to the observation of faint stellar sources like stars in external galaxies, GRBs and supernovae. By combining the intervals of best natural seeing with the power of AO, it becomes possible to get observations of stellar fields which in limiting magnitudes are as deep as expected from the JWST but with an higher angular resolution. The gain in the observations of sources like high z galaxies which are not stellar but can be made up by sub-seeing knots will be lower but still relevant. It is important to note however that the AO modules of first generation will deliver DL performance at IR wavelengths only and over a field of less than 1 arcmin.

2. The European ELT

ESO is carrying out a Phase B study of a 42m segmented primary telescope with the goal to prepare a construction proposal by the beginning of 2010. The European ELT current baseline foresees a 5 mirror optical design which delivers a 10' field at the Nasmyth foci with diffraction-limited image quality and 9% vignetting (Fig. 1, Delabre 2008). The optical train includes one adaptive mirror, M4, and a fast tip-tilt correcting mirror, M5, which can provide a correction of the ground-layer atmospheric turbulence and of wind buffeting on the telescope at the observatory site. Instruments at the E-ELT will be used either directly at the telescope focus (providing just a ground-layer correction with M4-GLAO mode) or in combination with additional adaptive mirrors in a post-focal module

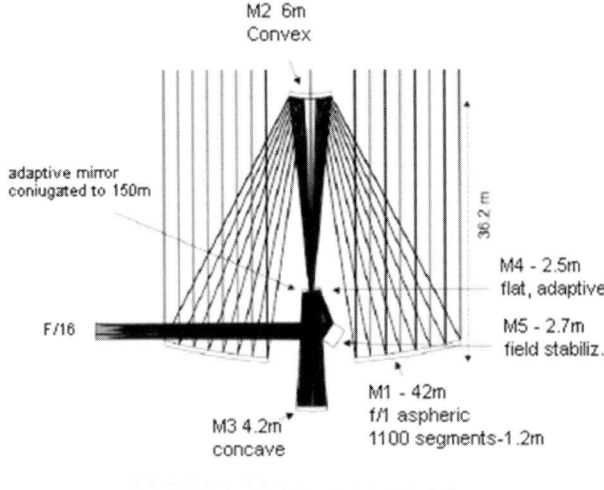

Figure 1. Basic Optical Parameters of the E-ELT baseline configuration

to obtain a diffraction limited performance with high light concentration over fields up to 1-2' at the Nasmyth foci. Fig. 2 and 3 show possible layout of instruments on the two Nasmyth platforms and on the additional coudè focus, which provides a protected location for instruments requiring a large volume and a controlled environment. This simulation shows that we could have up to 7 instruments permanently mounted at the telescope, offering a wide range of observing capabilities on line. This will permit the implementation of a flexible scheduling at the telescope to optimize the use of observing time and the scientific output. In the current Study Phase B of the project, ESO has launched 6 instruments and two post focal adaptive optics studies with institutes or consortia of institutes in the ESO member states. The choice of the instruments was based on the science case for the 42m telescope as initially identified in the report by the E-ELT Science Working group (http://www.eso.org/sci/facilities/eelt/publ/ELT-SWG-apr30-1.pdf) and on the basis of concept studies of instruments carried out in 2005-2006 at ESO and in other institutes under the European Community FP6 ELT Design Study program.

3. A comparison of the first light TMT versus E-ELT Instrumentation

The Thirty Meter Telescope (TMT) is a project of an optical-infrared telescope carried out in collaboration between CALTECH, the University of California and Association of Canadian Universities for Research in Astronomy (ACURA). Among the ELTs projects, the TMT is the more advanced in terms of elaboration of the science case and of the technical studies. A construction proposal and the supporting science case have been recently released (http://www.tmt.org/foundation-docs/index.html). The TMT design foresees a 3 mirror Ritchey Chretien telescope, with two wide Nasmyth platforms which can host a suite of permanently mounted instruments. Unlike the E-ELT the TMT has no AO mirrors in its baseline optical train but includes a first light AO module (NFI-RAOS) which will feed DL images to two of the first light instruments. Table 1 shows a

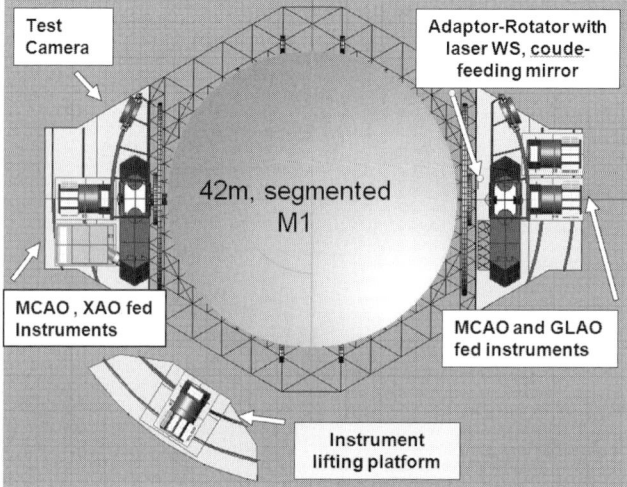

Figure 2. A top view of the two Nasmyth platforms of the European ELT showing a possible arrangement of 4 instrument and the adaptive optics modules.

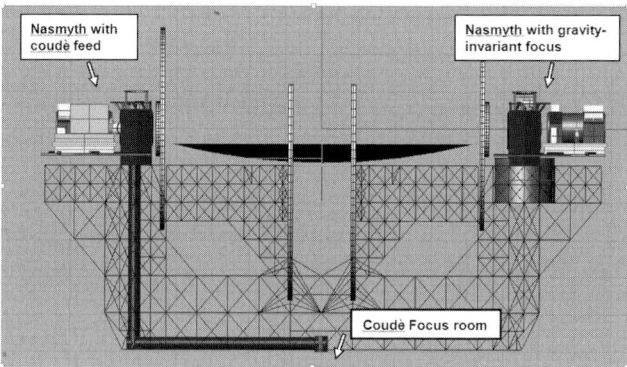

Figure 3. A side view of the European ELT showing the adaptors, AO modules and instruments mounted on the Nasmyth focus and the path to the coudè focus room. The gravity-invariant focus below the platform is foreseen for large field instrument which need to rotate with the sky but need to minimize flexures.

comparison of the initial requirement of the TMT and E-ELT instruments of first generation. In the case of the TMT, the project has identified two categories of instruments: first light instruments and instruments to be installed in the first decade of operation. For the E-ELT the 6 instruments now under detailed studies are candidates for first generation. Two additional instrument concepts will be selected in 2008 for advanced study and a selection will take place at the beginning of 2010 only, when the E-ELT construction proposal will be presented. There are many similarities and a few differences between the two suites of instruments. Both include a multi conjugate adaptive optics module to provide diffraction limited images in the NIR over a moderate field ($30''$-$2'$). This module can be coupled with an imager (IRIS at the TMT, MICADO at the E-ELT) and a spectrograph (again IRIS, HARMONI at the E-ELT) to take advantage of the unique size of the telescope from the early days of operation. Both telescopes foresee multi-object spectrographs working in the spectral range 800-2400 nm with AO , with the observations of galaxies at intermediate and high redshift as main driver (IRMS, EAGLE). Both

Table 1. E-ELT and TMT Instrumentation

	E-ELT INSTRUMENT		TMT INSTRUMENT
EAGLE (1st gen candidate)	Wide-field, multi IFU NIR spectrograph +AO Wavelength range: 0.8-2.5 mm; FoV >=5'; Multiplexing >20; Spectral R=5000 (R>15000 option); IQ: >30% EE in 100mas	IRMS (early light)	NIR multi-slit spectrometer-imager (clone of MOSFIRE @ Keck); FoV: 2.27 arcmin Multiplexing: 46; Spectral R=3270 -4500; Spectral coverage: all of Y to K
MICADO (1st gen candidate)	High angular resolution camera; Wavelength range: 0.8- 2.4 mm; FoV:=>30''; operation with GLAO, LTAO, MCAO	IRIS (early light)	IFU NIR spectrometer and imager; Wavelength range: 0.8-2.5 mm; FoV<2'' for IFU ,10×10 for Imaging; Spectral R=4000 (J,H,K)
HARMONI (1st gen candidate)	Single field wide band spectrograph; Wavelength range: 0.8-2.4 mm; FoV: tbd; Spectral R: R∼4000 (R∼20000); GLAO, MCAO modes	WFOS (early light)	Wide-field Optical MOS - seeing limited; Wavelength range: 0.34-1.0 mm; FoV: 40.5 arcmin2; Spectral R=500-5000
CODEX (1st gen candidate)	High-resolution visual spectrograph; Wavelength range: 0.40-0.69mm; Spectral Res > 120000; Stability: < 5cm/sec	HROS (1st decade)	High-resolution visual spectrograph; Wav. range goal: 0.31-1.0 mm; Spectral resolution: R=50000 (1'' slit)
MIDIR (1st gen candidate)	Mid-IR imager and spectrograph + AO; Wavelength range: 7-20 mm (goal 3.5-27 mm); FoV: => 30'' diameter; Spectral R: tbd	MIRES (1st decade)	Mid-IR imager and spectrograph +AO; Wavelength range: 8-18 mm; FoV: 10''; Spectral R: 5000<R<100000
EPICS (1st gen candidate)	Planet imager and spectrograph; Coupled to XAO, coronograph; Wavelength range: 0.6-1.8 mm; Spectral R>50 in Y to H bands, Polarimetry	PFI (1st decade)	Planet imager and spectrograph; Coupled to XAO, coronograph; Wavelength range: 1-2.5 mm; Spectral R=50 full FOV, R=500 partial FOV

include an instrument dedicated to imaging/spectroscopy of planets close to bright stars (PFI, EPICS) and MID Infrared instruments (MIRES and MIDIR). High resolution visual spectrographs have been studied for both projects: HROS - a first decade TMT instrument- is a classical echelle aiming at a resolution of 50000 while CODEX for the E-ELT is optimized for high stability, high accuracy radial velocity measurements. TMT is the only project to include a blue-visual, wide field MOS spectrograph optimized for operation with natural seeing. Among the first decade TMT instruments, but not included in Table 1 are also an Infrared Multi Object Spectrometer (IRMOS, a scaled up version of IRMS) and a Near Infrared Echelle Spectrometer (NIRES).

4. The impact of ELTs on the study of the most luminous stellar objects

In the previous sections the potential advantages of the new generation of ELTs have been presented and the current concepts of the first generation instruments for the two more advanced projects, the TMT and the European ELT have been compared. Most of the instrument studies have not yet been finalized. Those for the E-ELT are in their initial phase of the detailed design and their impact on specific scientific areas has not yet been accurately quantified. We consider below three topics related to the subject of this Symposium where the predicted advantages of observations with the ELTs are outstanding:

Color-Magnitude diagrams of stars in external galaxies. The angular resolving power of an ELT used in combination with AO will be uniquely suited to do accurate photometry

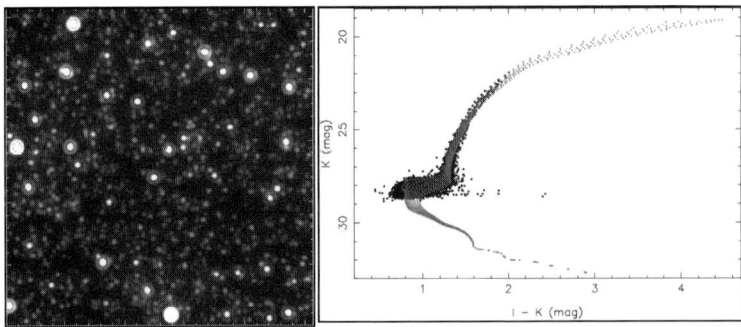

Figure 4. Left: Simulated K image of a stellar population of an elliptical galaxy at DM=26, with a V surface brightness of 22mag/arcsec^2. The image area is $\sim$ 10 arcsec^2, observed with the 42m E-ELT for 100hrs. **R**ight: input stellar population (light circles) and extracted CMD (darker circles). From Liske (2007).

of stellar populations in external galaxies. It will provide for the first time both the depth and the angular resolution to build up color magnitude diagrams well beyond the Local Group. These could be used to reconstruct the history of star formation in galaxies of different morphological type than of our own galaxy. Measurements of the bright part of the CMD of elliptical galaxies as far as Virgo (DM=31) and the HB to Centaurus A (DM=26) are among the science drivers of instruments like IRIS at the TMT and MICADO at the E-ELT. Liske (2007) at ESO has carried out a first set of simulations of crowded stellar fields in external galaxies using the predicted PSF of the telescope with an adaptive optics module and an assumed stellar populations. Different photometry packages have been used on the simulated images to extract the photometric measurements from which the CMD has been reconstructed (Fig. 4). This exercise is essential to investigate the ultimate capabilities of the telescope and to optimize both the instrument parameters and the data reduction software.

Chemical abundances in stars of different stellar populations. High and intermediate resolution spectrographs at 8-10m telescopes (e.g. UVES and FLAMES+GIRAFFE at the ESO VLT, HIRES at Keck) have been used very effectively in the last decade to extend our knowledge of the abundances in stellar populations in our Galaxy and in other members of the Local Group (Tolstoy 2007 and references therein). The observations do hint to a complex star formation history even for galaxies which have a relatively simple morphological structure like the Local Group dwarfs. This type of observations is essentially photon-limited and has been confined to the most luminous stars in each system. They will benefit from the improvement of more than one order of magnitude in collecting area provided by the ELTs and in any improvement in angular resolution to resolve the brightest stars in the crowded regions as it could be obtained with a small field AO system. The expected gain is of 3-4 magnitudes, making possible measurements for RGB in Local Group galaxies. Of the instruments under study for the TMT and the E-ELT, HROS appears as the best option for the high resolution work. The NIR multi-object spectrographs like IRMS and EAGLE could be used very effectively in the statistical study of the Ca II triplet as metallicity indicator (Fig. 5) in a large sample of extragalactic stars like it has been done at brighter magnitudes with 8-10m telescopes (Battaglia *et al.* 2008). Here it is essential that the spectral range of the ELT NIR spectrographs will include the I band. The gain with the ELT at these wavelengths can be as high as 4-5 magnitudes because the observations will benefit from the light concentration provided by a wide field GLAO or MOAO system associated to the telescope.

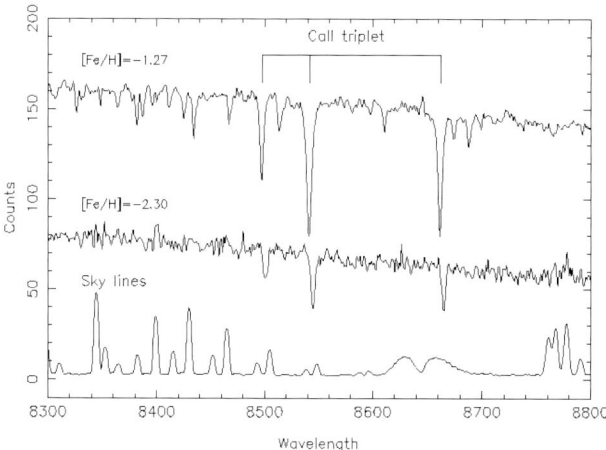

Figure 5. Observations of the Ca II triplet at intermediate resolution have been shown to be effective in measuring the metallicity of stars in different stellar populations (Battaglia *et al.* 2008).

The WFOS instrument at the TMT is potentially a powerful instrument to extend the work carried out on supergiants stars in the Local Group spirals and dwarf irregulars (Urbaneja *et al.* 2008, Bresolin *et al.* 2008, and references therein) to fainter magnitudes and to galaxies beyond the Local Group, up to 10 Mpc. It has been demonstrated that by comparing intermediate resolution spectroscopy with model simulations, it is possible to derive both the physical parameters and the metallicity of the brightest stars. These can be complemented by the gaseous phase abundances from the emission lines in the H II regions. A multi object spectroscopy capability is a strong requirement for this program. The observations would also greatly benefit from any improvements in the angular resolution but a modest gain can be expected at the visual wavelengths of this instrument from AO. Observations with WFOS will have to rely on the best natural seeing of the site.

Near-Infrared Spectroscopy of faint stellar sources. Both the TMT and the E-ELT instrument suites feature faint object spectrographs which will be operated in combination with AO modules pushing the limit of spectroscopy in the NIR to magnitude as faint as 28. Stellar sources like SN up to z=2 and GRBs at redshifts up to z=10 are prime targets for single field spectrographs like IRIS (TMT) and HARMONI (E-ELT). These observations are expected to have a major impact on a large spectrum of fields of astrophysics, from the last phases of stellar evolution to star formation, from cosmology to the properties of the IGM.

References

Battaglia, G., Irwin, M., Tolstoy, E., *et al.* 2008, *MNRAS*, 383, 183
Bresolin, F., Urbaneja, M. A., Gieren, W., *et al.* 2008, *ApJ*, 691, 2028
Delabre, B. 2008, *A&A*, in press
Liske J. 2007, http://www.eso.org/sci/facilities/eelt/science/drm/
Tolstoy E. 2007, in: A. Vazdekis & R. F. Peletier (eds.), *Stellar Populations as Building Blocks of Galaxies* (Cambridge: CUP), *Proc. IAU Symp 241*, 279
Urbaneja, M. A., Kudritzki, R. P., & Bresolin F. 2008, in: S. Koribalski & H. Jerjen (eds.), *Galaxies in the Local Volume*, (Berlin: Springer), *Astrophys. Space Sci.*, in press

Discussion

WALBORN: Has the exact size of the E-ELT been selected yet?

D'ODORICO: In the present Phase B study to be completed by the end of 2009 the baseline concept foresees a 42m segmented primary mirror.

MARTAYAN: I did not see in your talk 2 instruments for the E-ELT initially foreseen for the first generation and described in earlier ESO documents: a multi-object spectrograph like a super VLT FLAMES and a spectro-polarimeter . These instruments, according to me, seem to be extremely important to obtain information and statistics on stars of the galaxies beyond the local group. Are they definitely removed from the 1st generation instruments if the E-ELT or could we influence yet ESO?

D'ODORICO: The two instruments you quote are indeed not included in the six currently under study. However they could be proposed in response to the Call for two additional instrument concept studies that ESO will release in February 2008. A decision on the first generation E-ELT instrumentation will not be taken before the beginning of 2010.

Metallicities at the Sites of Nearby SN and Implications for the SN-GRB Connection

Maryam Modjaz[1,2] L. Kewley[3], R. P. Kirshner[2], K. Z. Stanek[4],
P. Challis[2], P. M. Garnavich[5], J. E. Greene[6], P. L. Kelly[7], &
J. L. Prieto[4]

[1] UC Berkeley Astronomy Department
email: mmodjaz@astro.berkeley.edu

[2] Harvard-Smithsonian CfA
email: kirshner,pchallis@cfa.harvard.edu

[3] Department of Astronomy, Univerisity of Hawaii
email: kewley@IfA.Hawaii.Edu

[4] Department of Astronomy, The Ohio State University
email: kstanek,prieto@astronomy.ohio-state.edu

[5] Department of Physics, University of Notre Dame
email: pgarnavi@nd.edu

[6] Department of Astrophysical Sciences, Princeton University
email: jgreene@astro.princeton.edu

[7] Kavli Institute for Particle Astrophysics and Cosmology
email: pkelly3@stanford.edu

Abstract. While the broad-lined Type Ic supernovae (SN Ic-bl) associated with long-duration gamma-ray bursts (GRBs) have been studied, we do not fully understand the conditions that lead to each kind of explosion in a massive star. Here we show clues as to the production mechanism of GRBs by comparing the chemical abundances at the sites of 5 nearby ($z < 0.25$) broad-lined SN Ic that accompany nearby GRBs with those of 12 nearby ($z < 0.14$) broad-lined SN Ic that have no observed GRBs. We show that the oxygen abundances at the GRB sites are systematically lower than those found near ordinary broad-lined SN Ic. A unique feature of this analysis is that we present new spectra of the host galaxies and analyze the measurements of both samples in the same set of ways, using 3 independent metallicity diagnostics. We demonstrate that neither SN selection effects (SN found via targeted vs. non-targeted surveys) nor the choice of strong-line metallicity diagnostic can cause the observed trend. Though our sample size is small, the observations are consistent with the hypothesis that low metal abundance is the cause of some massive stars becoming SN-GRB. We derive a cut-off metallicity of 0.2−0.6 $Z_\odot$, with the exact value depending on the adopted metallicity scale and solar abundance value.

Keywords. gamma rays: bursts– supernovae: individual – galaxies: abundances

1. Introduction

We seek clues as to the production mechanism of long-durations gamma-ray bursts (GRBs) by comparing the chemical abundances at the sites of nearby broad-lined SN Ic (SN Ic) that accompany some GRBs with the broad-lined SN Ic that have no observed GRBs. In nearby ($z < 0.25$) long-duration GRBs, after the afterglow has faded, if the spectrum of the underlying event is observed, it is that of a broad-lined SN Ic (Galama et al. 1998, Stanek et al. 2003, Hjorth et al. 2003, Modjaz et al. 2006). These are SN Ic whose optical spectra show no hydrogen or helium lines and whose line widths approach 30,000 km s^{-1} (2-3 times larger than normal SN Ic, see Figure 1), also known as

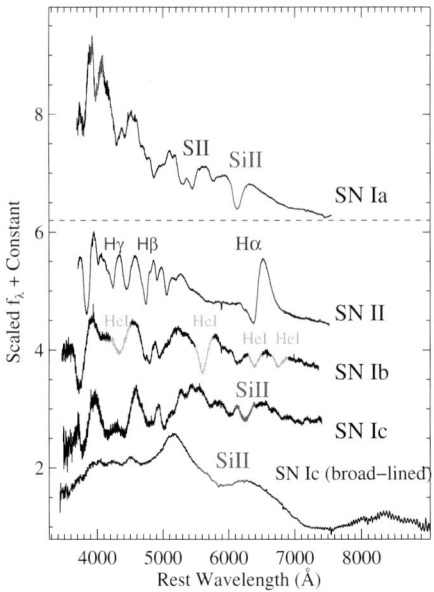

Figure 1. Representative sample of SN spectra illustrating the main SN classifications (but not all of them), which are based on the presence or absence of certain elements in the SN spectrum at maximum light. The main line identifications are marked. Broad-lined SN Ic (bottom spectrum) are the SN spectroscopically identified in the afterglow of nearby long GRBs. The SN shown are: SN 1994D (SN Ia; Meikle *et al.* 1996); SN 1999em (SN II; Leonard *et al.* 2002), SN 2004gq (SN Ib; Modjaz 2007), SN 2004gk (SN Ic; Modjaz 2007), and SN 1998bw (broad-lined SN Ic; Galama *et al.* 1998).

"Hypernovae". However, a growing number of broad-lined SN Ic are also seen without accompanying GRBs. If low metallicity is necessary for producing conditions suitable for GRBs, as proposed by observational studies (e.g., Stanek *et al.* 2006) and theoretical work (e.g., Yoon & Langer 2005, Woosely & Heger 2006), then we would expect broad-lined SN Ic intrinsically without GRBs to erupt in higher metallicity environments. This is the hypothesis we test here, using the best available list of SN and our own uniform determinations of the local metal abundance. Details of this work are presented in Modjaz *et al.* (2008a).

2. Sample and Observations

2.1. SN Host Galaxy Samples

For our sample of SN connected with GRBs, we considered the four nearby secure cases of direct, and spectroscopically determined, SN-GRB associations, namely GRB 980425 / SN 1998bw, GRB 030329/SN 2003dh, GRB 031203/SN 2003lw and GRB/XRF 060218 / SN 2006aj. We also include the host of XRF 020903 that had a clear supernova signature in its light curve and in its afterglow spectrum, but we do note that this SN confirmation has a lower degree of certainty than the other GRB-SN associations. In order to derive oxygen abundances, we used the published emission-line fluxes for the sites of the SN-GRBs, when possible, and otherwise the light-integrated host galaxy spectra (see Modjaz *et al.* 2008b for a full list of references).

We obtained and analyzed host galaxy spectra of a total of 12 broad-lined SN Ic without observed GRBs. A few of these broad-lined SN Ic are well-documented in the literature, and to increase our sample, we searched for additional ones in the announcements of the International Astronomical Union Circulars. Half of our sample consists of host galaxies of SN that were found via traditional galaxy-targeted SN surveys (e.g., Lick Observatory SN Search), whereas the other half consists of galaxies in which the SN were found via large field-of-view or rolling surveys, i.e., where the host galaxy was not targeted (e.g., Texas SN survey, SDSS-II SN survey). We are aware that selection effects can be introduced by different methods of discovery. To mitigate these selection effects, we worked hard to include SN that were found in host galaxies that had not been targeted for search. We believe they provide a better match to the host galaxies that are selected by the appearance of a GRB. For completeness, we analyze and discuss spectra of all broad-lined SN Ic. But when directly comparing to the sample of SN Ic connected with GRBs, we identify the SN Ic found in the same, non-targeted fashion in order to minimize discovery selection effects. The sample size of broad-lined SN Ic that are free of galaxy-selection effects is comparable to that of SN-GRB. More details about the sample are given in Modjaz *et al.* (2008a).

2.2. *Observations and Metallicity Measurements*

We obtained spectra of the SN sites and of their galaxy nuclei via our own observations (with the 6.5m Magellan, the 6.5m MMT, and 1.5m Mt. Hopkins telescopes) and archival data (SDSS DR5). Using the galaxy emission-line measurements corrected for stellar absorption and extinction, we derive central oxygen abundances and abundances at the SN position, while explicitly including the statistical uncertainties in the line measurements. We compute oxygen abundances via strong-line diagnostics using three independent and well-known calibrations: Kewley & Dopita (2002), McGaugh (1991), and Pettini & Pagel (2004), the latter being effectively in the direct electron temperature (T_e) scale. Furthermore, we compute local star formation rates, and draw from the literature the host galaxy B-band luminosities, M_B. We compare the properties of our sample with properties of the 5 nearby SN-GRB hosts for which we derive chemical abundances using published emission line fluxes via the same metallicity diagnostics as for the SN without GRBs in our sample. Furthermore, we include the sample of local ($0.005 < z < 0.2$) SDSS star forming galaxies (Tremonti *et al.* 2004), whose abundances we computed in the same three scales using their published line fluxes.

3. Discussion

Figure 2 shows the main result of our comparison: we plot host galaxy metallicity (as expressed in terms of oxygen abundance $12+\log(O/H)$ with the KD02 calibration) and host galaxy luminosity (M_B) of broad-lined SN Ic without observed GRBs (called "SN Ic (broad)", circles) and with GRBs (called "SN Ic (broad & GRB)", squares). Objects whose host galaxies had not been targeted during the discovery have an extra circle (for SN without GRBs) or an extra square (for SN with GRBs) around their plotted symbol. We note that three of the five broad-lined SN Ic found in the lower luminosity galaxies ($M_B > -19$ mag) were discovered by the SDSS-II SN Survey, which is a galaxy-impartial survey. For the broad-lined SN Ic we only plot abundances measured at or extrapolated to the SN position. These plotted values are lower than the values we measure for the center of the same host galaxies presumably due to metallicity gradients. The SN-GRB host abundances also reflect the metallicity at the SN-GRB position, since they were either measured specifically at the SN position (for GRB 980425/SN 1998bw), or the SN reside in the nucleus of their dwarf-galaxy hosts that are chemically homogeneous (Kobulnicky *et al.* 1997). Due to the short life times of the massive SN progenitor stars

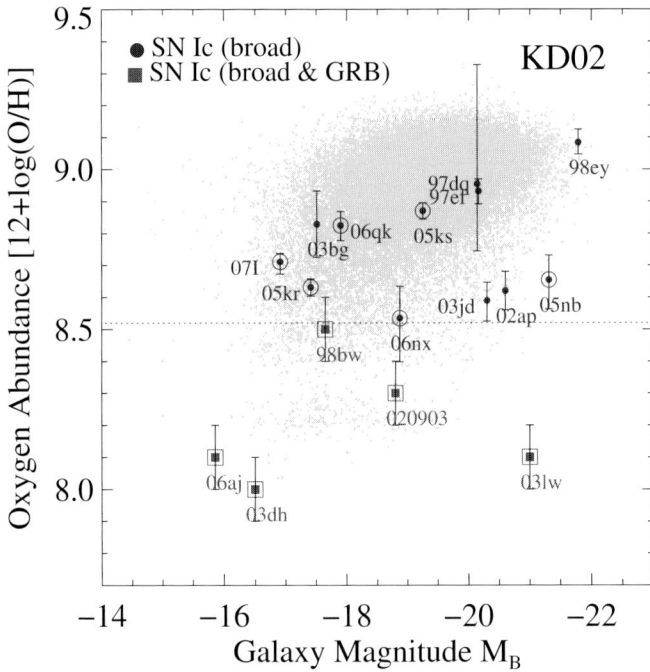

Figure 2. Host galaxy luminosity (M_B) and host galaxy metallicity (in terms of oxygen abundance) at the sites of nearby broad-lined SN Ic ("SN Ic (broad)": filled circles) and broad-lined SN Ic connected with GRBs ("SN (broad & GRB)": filled squares; see also Stanek et al. (2006), Kewley et al. (2007)). Extra circles and squares designate SN which were found in a non-targeted fashion. The oxygen abundances are in the Kewley & Dopita (2002) (KD02) scale and represent the abundance at the SN position. Due to radial metallicity gradients, the gas abundance at the SN position is lower than the central galaxy abundance for some SN. Labels represent the SN names while one ("020903") refers to its associated GRB. Yellow points are values for local star-forming galaxies in SDSS Tremonti et al. (2004), re-calculated in the Kewley & Dopita (2002) scale for consistency, and illustrate the empirical luminosity-metallicity ($L-Z$) relationship for galaxies. Host environments of GRBs are systematically less metal-rich than host environments of broad-lined SN Ic where no GRB was observed, even for the same range of host luminosities. The dotted line at $12+\log(\mathrm{O/H})_{KD02} \sim 8.5$ designates the apparent dividing line between SN with and without observed GRBs.

($<$10 Million years for $M_{\mathrm{ZAMS}} >20 M_\odot$), we regard the metallicities at the SN position as natal metallicities.

Moreover, we plot in Figure 3 the comparison between the two host samples and the local SDSS galaxies in the metallicity scales of M91 and of PP04-O3N2, with the latter being effectively in the T_e scale. While the absolute values of the abundances are different in different scales, as expected, the bimodal distribution persists in each scale and thus, is independent of the choice of metallicity diagnostic. The K-S Test applied to the host abundances of SN Ic (broad & GRB) and SN Ic (broad) found in a non-targeted fashion yields low probabilities of 4% (M91-based abundances) and 3% (T_e-based abundances) that they are drawn from the same population. For each scale, we plot as a dotted line the boundary that separates the two samples: $12+\log(\mathrm{O/H})_{M91} \sim 8.4$ and $12+\log(\mathrm{O/H})_{T_e} \sim 8.1$. Although our sample is small our findings are consistent with the hypothesis that low metal abundance is the cause of some very massive stars

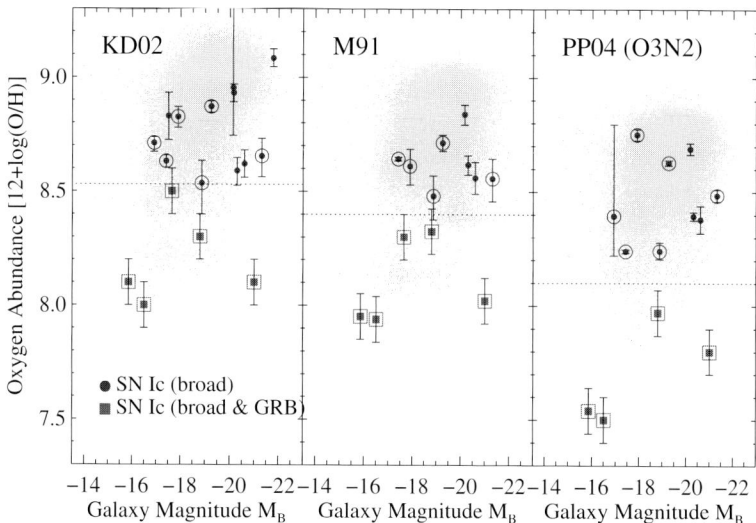

Figure 3. Similar to Figure 2, but using different metallicity diagnostics; Kewley & Dopita (2002) (KD02) as in Figure 2 (*left*); McGaugh (1991) (M91, *middle*); and Pettini & Pagel (2004) (PP04-O3N2), which is effectively on the T_e scale, (*right*). Host environments of GRBs are more metal-poor than environments of broad-lined SN Ic where no GRB was observed, independent of the abundance scale used. Yellow dots designate the SDSS galaxy values calculated in the respective metallicity scales.

becoming SN-GRB. The absolute value for the cut-off metallicity ranges between 0.2–0.6 $Z_\odot$ depending on the adopted metallicity scale and solar abundance value.

At face value, our results differ from various studies in the literature, which have concluded that high-z GRB hosts are not necessarily low-metallicity systems. However, there might be intrinsic differences in GRB population at low and high z (e.g., Guetta & Della Valle 2006): the low-z GRBs might constitute a different class of low-luminosity GRBs that are not detected at higher z. Moreover, most of the techniques for measuring abundances at higher z are different from our direct approach, as they use absorption techniques in GRB afterglows (Prochaska *et al.* 2007 and references therein).

Radio observations do not support off-axis GRBs with broad-lined SN Ic in half of our sample (Soderberg *et al.* 2006), while we cannot exclude the possibility of off-axis GRBs in the rest of the SN sample. Recently, Mazzali *et al.* (2005) argued for an off-axis jet in one SN of our sample, SN 2003jd (see Figure 2), since they interpreted double-peaked oxygen profiles observed in its late-time spectra as signs of an aspherical, axisymmetric explosion caused by a jet. However, a large set of SN with nebular spectra exhibit double-peaked oxygen profiles, which indicates that asphericities are common in normal SN Ib/c (Modjaz *et al.* 2008b, Maeda *et al.* 2008) and not necessarily a sign of an off-axis GRB.

4. Conclusions

We presented spectroscopic data of a statistically significant set of host galaxies of 12 nearby ($z < 0.14$) broad-lined SN Ic with no observed GRBs. Using the galaxy emission-line measurements corrected for stellar absorption and extinction, we derived central oxygen abundances and abundances at the SN position based on strong-line diagnostics in three independent scales. We compared the properties of our host sample with the properties of five nearby SN-GRB hosts, for which we derived chemical abundances using the same three metallicity diagnostics as for SN without observed GRBs. Broad-lined SN Ic without GRBs tend to consistently inhabit more metal-rich environments, and their host galaxies, for the *same* luminosity range ($-17 < M_B < -21$ mag), are systematically more metal-rich than corresponding GRB host galaxies. The trend is independent of the choice of diagnostic and cannot be due to selection effects as we include six SN found in a similar non-targeted manner as GRB-SN.The boundary between broad-lined SN Ic that have a GRB accompanying them and broad-lined SN Ic without a GRB lies at an oxygen abundance of $\sim 12+\log(O/H)_{KD02} \sim 8.5$, which corresponds to $0.2-0.6$ $Z_\odot$ depending on the adopted metallicity scale and solar abundance value.

M.M. acknowledges support from the Miller Foundation for the time during which part of this study was conducted and thanks the organizers of IAUS 250 for an enjoyable and enlightening conference. Supernova research at Harvard University has been supported in part by the National Science Foundation grant AST06-06772.

References

Galama, T. J., Vreeswijk, P. M., van Paradijs, J. et al. 1998, *Nat*, 395, 670
Guetta, D. & Della Valle, M. 2006, *ApJ*, 657, L73
Hjorth, J., Sollerman, J., Møller, P. et al. 2003, *Nat*, 423, 847
Kewley, L. J. & Dopita, M. A. 2002, *ApJS*, 142, 35
Kewley, L. J., Brown, W. R., Geller, M. J. et al. 2007, *AJ*, 133, 882
Leonard, D. C., Filippenko, A. V., Gates, E. L. et al. 2002, *PASP*, 114, 35
McGaugh, S. S. 1991, *ApJ*, 380, 140
Maeda, K., Kawabata, K., Mazzali, P. A. et al. 2008, *Sci*, 319, 1220
Meikle, W. P. S., Cummings, R. J., Geballe, T. R. et al. 1996, *MNRAS*, 281, 263
Modjaz, M., Stanek, K. Z., Garnavich, P. M. et al. 2006, *ApJ*, 645, L21
Modjaz, M. 2007, Ph. D. Thesis, Harvard University
Modjaz, M., Kewley, L., Kurshner, R. P. et al. 2008a, *AJ*, 135, 1136
Modjaz, M., Kirshner, R. P., & Challis, P. 2008b, *ApJL*, submitted (arXiv:0801.0221)
Pettini, M. & Pagel, B. E. J. 2004, *MNRAS*, 348, L59
Soderberg, A. M., Nakar, E., Berger, E., & Kulkarni, S. R. 2006, *ApJ*, 638, 930
Stanek, K. Z., Matheson, T., Garnavisch, P. M. et al. 2003, *ApJ* 591, L17
Stanek, K. Z., Gnedin, O. Y., Beacom, J. F. et al. 2006, *Acta Astronomica*, 56, 333
Tremonti, C. A., Heckman, T. M., Kauffmann, G. et al. 2004, *ApJ*, 613, 898
Woosley, S. E. & Heger, A. 2006, *ApJ*, 637, 914
Yoon, S.-C. & Langer, N. 2005, *A&A*, 443, 643

Discussion

HIRSCHI: Very interesting! Where would the LMC and SMC oxygen abundances lie on your plots?

MODJAZ: The values for the LMC and the SMC are $\log(O/H)+12=8.3$ and 8.0 in the Kewley & Dopita scale, respectively. So exactly where the GRB-SN Hosts lie.

MACFADYEN: So there must be something fundamentally important about the impact of low metallicity on the GRB progenitors?

MODJAZ: Yes. I hope I have convinced you that, observationally, we tried to be very thorough by taking selection effects and metallicity systematics into account, and this is what the data are telling us.

ABSTRACTS OF ADDITIONAL ORAL TALKS

We collect here the abstracts of five oral presentations for which the written contribution did not reach the editors in time for publication.

Magnetars and Their Massive Star Progenitors

Bryan M. Gaensler (The University of Sydney)

Magnetars are a small group of young neutron stars powered by extreme magnetic fields, with surface strengths in excess of 10^{15} gauss. Now that magnetars have been established as a signicant fraction of the overall neutron star population, we need to understand why some core-collapse supernovae make "normal" radio pulsars, while other supernovae make exotic magnetars. I will review a series of recent multi-wavelength results on the environments and birth-sites of magnetars, which as an ensemble provide strong evidence that magnetars originate from unusually massive stars. I will discuss the resultant implications for the formation and birthrate of magnetars, for the magnetic fields of high-mass stars, and for the rate of long gamma-ray bursts in high-metallicity galaxies.

Long Gamma-Ray Bursts – Core-Collapse SN Connection

Andrew MacFadyen (New York University)

I will review the GRB-SN connection in the context of the collapse and asymmetric explosion of massive rotating stars.

The Nature of Gamma-Ray Burst Progenitors: Observational Constraints

Kris Stanek (Ohio State University)

I will discuss the observational constraints we have on the GRB progenitors. I will concentrate on the properties of the hosts of nearby GRBs, contrasting them with what we know about hosts of various types of supernovae.

Massive Stars in Dwarf and Irregular Galaxies

Eva K. Grebel (University of Heidelberg)

I will discuss our knowledge of massive stars in resolved dwarf and irregular galaxies and the impact of these stars on the evolution of their host galaxies.

Massive Stars, Super Star Clusters, and Feedback in Starbursts

Jay Gallagher (University of Wisconsin), Linda J. Smith (ESA/STScI), & Mark S. Westmoquette (University College London)

The majority of stars now in existence likely formed in conditions more similar to those in starbursts than to the situation in the present-day solar neighborhood. Unique features in the spatial and temporal patterns of massive star formation in starbursts thus must be taken into account in understanding how feedback shaped galaxies. This talk considers how the concentration of massive stars into compact star clusters with sizes of ~ 10 pc, which in turn often are found in multi-100-pc-scale starburst clumps, affect the host galaxy. M82 provides a particularly accessible example of the starburst clump evolutionary mode. HST and ground-based observations show that the resulting high power densities can energize large-scale galactic winds and also support long-term photoionization over kpc scales. Working from the case of M82 we consider how changes in patterns of star formation can inuence galaxy evolution through the redistribution of baryons and newly synthesized metals.

Conclusion

Symposium Summary

Claus Leitherer

Space Telescope Science Institute, 3700 San Martin Drive, Baltimore, MD 21218, USA
email: leitherer@stsci.edu

Abstract. I summarize the highlights of the conference. First I provide a brief history of the beach symposia series our massive star community has been organizing. Then I use most of my allocated space discussing what I believe are the main answered and open questions in the field. Finally I conclude with a perspective of the future of massive star research.

Keywords. sociology of astronomy – radiative transfer – telescopes – stars: early-type – ISM: evolution – open clusters and associations: general – galaxies: stellar content – early universe – gamma rays: bursts

1. Past

This is the ninth meeting in what has become the traditional beach symposia series of the massive star community. The series had its origins in 1968 at a workshop on Wolf-Rayet (W-R) stars in Boulder (definitely not a beach location). At that time, the cosmic significance of W-R stars was largely unknown but understanding their nature was deemed important enough for follow-up meetings. This then led to IAU Symp. 49 in Argentina in 1971, and to the subsequent tradition of holding massive star symposia at roughly five-year intervals. For those who like statistics, here is the complete tally: IAU Symp. 83 (1978, Canada), IAU Symp. 99 (1981, Mexico), IAU Symp. 116 (1985, Greece), IAU Symp. 143 (1990, Indonesia), IAU Symp. 163 (1994, Italy), IAU Symp. 193 (1998, Mexico), IAU Symp. 212 (2002, Spain), and finally IAU Symp. 250 (2007, USA).

Each meeting had its distinct flavor and theme. For instance, IAU Symp. 163 had the emphasis on binary stars, and IAU Symp. 116 allocated a large fraction of time to massive stars in Local Group galaxies. Overall, a clear trend is apparent: the median distance of the astronomical objects at each meeting has been steadily increasing. Distances were expressed in kpc in the early symposia, then became Mpc, and during this meeting redshifts larger than 3 were commonly mentioned.

Those who attended IAU Symp. 49 would have been astounded had they known how the field would have progressed by the time of IAU Symp. 250. Even as recently as during IAU Symp. 193 (which I attended as well), it would have been preposterous to discuss massive stars in connection with gamma-ray bursts, Lyman-break galaxies, and Population III stars.

I will finish my historical notes with a bit of trivia. Three participants of this symposium have witnessed this rapid development from the very beginning. They already participated in the 1971 meeting: Peter Conti, Lindsay Smith†, and Nolan Walborn. Peter Conti distinguished himself by an impressive feat: he attended *all* nine beach symposia. His record would have been tied were it not for Virpi Niemela's untimely death in 2006.

† I apologize to Lindsay Smith whom I failed to name in my original talk.

2. Present

The symposium *Massive Stars as Cosmic Engines* was structured along the five major science themes Atmospheres and Winds, Stellar Evolution, Nearby Populations, Feedback, and Early Universe. I will follow these themes and try highlighting the major issues that have emerged during the meeting.

2.1. *Atmospheres and Winds*

40 years after the construction of the first non-LTE model atmospheres and 30 years after the development of the radiatively driven wind theory the surfaces of OB stars are basically understood, and models and observations agree quite well (*Puls*). Independent atmosphere codes used by different groups give consistent results. We can routinely generate fully blanketed non-LTE models and compare them to observations. This has led to a major revision of the relation between spectral type and effective temperature (T_{eff}), the latter being 10 – 20% lower than previously thought. The ramifications for the ionizing photon output or the bolometric corrections are significant. The long-standing discrepancy between stellar spectroscopic and evolutionary masses still persists, although the new atmospheres with their higher mass-luminosity ratio improve the agreement.

The dependence of the stellar mass-loss rate ($\dot{M}$) on heavy-element abundance (Z) is of particular interest to those who need to extrapolate to metal-free Population III stars. Empirical studies of hot stars in the solar neighborhood, the Large (LMC) and the Small (SMC) Magellanic Cloud suggest a smooth progress to lower rates in agreement with the wind theory (*de Koter*). The small lever in Z makes this conclusion a challenge: one would really like to perform such a study in truly metal-poor galaxies such as, e.g., I Zw 18 whose oxygen abundance is 1/20 that of the LMC.

Of course, the devil is in the detail. While we understand the overall physics and derived parameters of OB stars, the effects of micro- and macroturbulence, variability, porosity, and wind clumping are still poorly understood (*Hillier*). Wind clumping, if not taken into account properly, leads to an overestimate of $\dot{M}$. Back in the 1980s and 1990s we were overly confident in our ability to determine mass-loss rates. It has become clear that many of the published values (including those by myself) need to be revised downward.

Wind instabilities can be traced with thermal X-ray emission (*Cohen*). Interestingly, two classes of X-ray emitters are found among (single) massive stars: those with soft and with hard keV spectra. The former are associated with shocks in the wind whereas the latter are related to magnetic fields. Soft X-ray spectra are found in evolved O stars. In contrast, the hard X-ray emitters tend to be located closer to the zero-age main-sequence. The caveat is the small sample size of the observed stars, which precludes studying dependencies other than on age.

It was gratifying to learn that the atmospheric modeling of W-R stars has reached the state-of-the-art of that for O stars (*Crowther*). We believe we understand the W-R parameters as well (or as badly) as those of O stars. This is quite an improvement in comparison with the 1980's when W-R luminosities were uncertain by an order of magnitude. One of the key points we should all take away from this meeting is $WN \approx O \neq WC$. Most WN's (in particular those of the latest types) are core-hydrogen burning O stars in disguise. They may still be close to the main-sequence. WC stars are truly devoid of hydrogen.

Empirical hydrodynamical models of W-R winds allow one to determine mass and mass-loss-rate (*Gräfener*). This opens up the possibility of using $\dot{M}$ as a proxy for the W-R mass. In cases when the most massive stars in a populous star cluster, like the

Arches cluster in the Galactic Center, are dominated by WN-like objects, one can utilize the W-R mass-loss rates to probe the initial mass function (IMF).

Luminous Blue Variables (LBVs) have made a strong come-back at this meeting. LBVs were a hot commodity in the 1980s and 1990s, when they were thought to be an essential phase in the evolution of a massive star. However, they took the backseat – quite undeservedly – at the last two symposia. Now they are back in full force. One reason is the revived interest in an evolutionary phase with strong mass loss. Since $\dot{M}$ in all pre- and post-LBV phases has been recognized to be lower than assumed, an additional source for removing stellar material is needed. Needless to say that LBVs are fascinating objects which deserve to be studied in their own right.

The physical mechanism responsible for the giant LBV outbursts differs from radiation pressure. The key is the location of LBVs close to the Eddington limit. In this case, the mechanical luminosity of a continuum driving wind can in principle reach the radiative luminosity limit (*Owocki*). The Eddington limit may even be temporarily exceeded when pulsations are taken into account (*Onifer*). Such pulsations are triggered by a build-up of luminosity when the interior reaches a temperature of ∼200,000 K, corresponding to the Fe opacity peak. A special case may be η Car. The fact that the star is a binary may help understand some of the properties of the circumstellar material (*Okazaki*). The audience was divided as to whether binarity can explain the entire LBV phenomenon.

The cool side of the uppermost Hertzsprung-Russell (HR) diagram often looks rather un-cool to the massive star community: red supergiants (RSGs) are rarely covered in this symposium series. I was glad to learn about a fresh analysis of an RSG sample with MARCS atmospheres that result in substantially higher T_{eff} and lower luminosity (L). When placed on the HR diagram, the RSGs are now in excellent agreement with stellar evolution models (*Massey*). The most luminous RSGs are surrounded by resolved material. They offer one of the few occasions when stellar astronomers can directly observe the 2-dimensional structure of a star, rather than inferring it from point source spectroscopy. The morphology of the circumstellar material is direct evidence of stellar magnetic fields and photospheric convection (*Humphreys*).

The royal method of determining stellar masses utilizes binary systems. Binaries hosting the most luminous, and therefore the most massive stars are of particular interest. There are only a few of them but their analysis uniformly suggests a mass of about 100 $M_\odot$ for the most massive stars (*Moffat*). This is consistent with the upper mass limit found for the IMF in rich clusters. Interestingly, mid/late WN stars turn out to be the most massive stars known. All of them are core-hydrogen burning.

2.2. *Stellar Evolution*

The stellar evolution modelers delivered another message we all should take home: *stellar rotation is a fundamental channel leading to the formation of W-R stars* (*Maeder*). Originally, all W-R stars were suggested to be components of binaries, with orbital forces being responsible for the mass loss. In 1976, Peter Conti recognized that stellar winds in single stars could remove mass as efficiently, and a 2nd W-R channel was established. 30 years later, rotation becomes the 3rd channel. At low Z, rotation may even be the dominant formation channel (*Meynet*).

Many model predictions of stellar evolution must rely on sometimes rather daring extrapolations. This is particularly true when it comes to evolution to very low Z. I was impressed about the diligence of the modelers who try to test their predictions using all angles of observational evidence: not only do they perform the classical HR diagram comparisons, but they also seek guidance from the WN/WC ratio, the nitrogen surface abundance, or the pulsar rotation velocities. The fact that so vastly different constraints lead to consistent answers gives confidence in the basic soundness of the models.

We can even dare to make predictions for stellar evolution at essentially zero metal abundance (*Hirschi*). What matters is rotation, rotation, rotation.... Centrifugal forces lead to stronger mechanical and radiatively driven winds, the latter being enhanced via chemically enhanced material that is brought to the surface. Since even the first stellar generations effectively switch from the pp to the CN cycle and synthesize metals, one may expect significant mass loss already early on.

Stellar pulsations are ubiquitous across the HR diagram but evolution models generally ignore them. Exploratory calculations suggest that pulsations do in fact modify rotation and trigger additional mixing (*Townsend*). Perhaps the buzz word at the next beach symposium will be pulsation, pulsation, pulsation....

Binary evolution can play a crucial role for understanding the observed properties of individual stars and of massive populations as a whole. Mass transfer affects evolution in many ways, one of them being increased spin-up and therefore even higher rotation velocity. Could all rapid rotators be binary stars (*Langer*)? Population models accounting for binary evolution appear to improve the agreement between predicted and observed stellar number ratios (*Eldridge*). Regrettably, the stellar census in Local Group galaxies is still incomplete and/or subject to large statistical errors because small number statistics. Clearly, a push for more extensive surveys is needed.

Most (all?) massive stars end their lives as core-collapse supernovae (SNe). It has been a long-standing challenge to revive the shock during the core-collapse (*Burrows*). Traditional models need an ad hoc mechanism to prevent the shock from stalling, such as an artificial "piston" or "bomb". The key to making the core-collapse self-consistent is the geometry: once the explosion is assumed to be non-spherical, instabilities will set in, and the shock will not stall.

Being affiliated with STScI, I was delighted to see the good use of the Hubble archive in the search for SN progenitors. Many potential host galaxies of core-collapse SNe have been observed with one of the HST imagers. Chances are, the SN progenitor has been captured by HST and its properties can be extracted from the HST archive (*Smartt*). The typical SN II progenitor is a RSG with a mass of about 10 $M_\odot$. This is the progenitor canonically expected for a type II SN, suggesting that SN1987A's blue progenitor really was an oddball confusing the issue. SN2006gy seems to be yet another oddball. Its properties hint at a dense circumstellar environment (*N. Smith*). One possible explanation is an LBV progenitor. Could this occur more commonly than previously thought?

Magnetars are neutron stars with "insanely" high magnetic fields of order 10^{15} G. A variety of independent arguments suggest an association with very massive progenitors (*Gaensler*). We now believe that 10% of all massive stars end their lives as magnetars. While this is not the dominant end phase, it is certainly not insignificant, either. Extremely massive stars are predicted to evolve into pair-instability SNe. Since Population III stars are expected to be strongly biased towards very high masses, there should be many such pair-instability SNe in the early universe. The absence of any evidence in favor of their existence (e.g., from abundance patterns) may indicate that pair-instability SNe do not form in large numbers. How can we avoid their formation (*Ekström*)? Rotation again leads the way: primordial massive stars may not enter the pair-instability SN phase if they suffer from rotation induced mass loss and have strong magnetic fields.

Gamma-ray bursts (GRBs) were one of the major themes of this conference. There is now convincing evidence that long GRBs are located in star-forming galaxies and are associated with SNe of type Ibc (*MacFadyan*). Lifetime arguments favor a small star with a hydrogen-free envelope. This then leads to a progenitor model with a massive, rotating star. Both single and binary models are being discussed for GRB progenitors (*Yoon*). The observed diversity of GRBs may be accounted for by magnetic fields with a range of properties. It is truly remarkable to witness the progress we have made on this subject since the last massive star symposium.

The host galaxies of nearby long GRBs can be studied in some detail. The chemical composition is of particular interest. GRBs are preferentially found in relatively small, metal-poor galaxies (*Stanek*). The GRB hosts, when plotted on a metallicity vs. luminosity diagram, tend to fall below the mean relation. The implication is that GRBs trace only low-metallicity star formation. Obviously, life on Earth benefits from this bias. It is unlikely that a long GRB will be found in our own Milky Way, something we should not be too keen about.

2.3. *Nearby Populations*

We witnessed a major change of guard in the session on massive-star populations. 30 Doradus used to be the gold standard and Rosetta Stone, yet we heard very little about this region. Rather, the heavily dust-obscured star clusters in our own Galaxy attracted a lot of the attention during this meeting. There are currently 10 massive ($\sim 10^5 M_\odot$) star clusters known in our Galaxy, although this is just the tip of the iceberg (*Figer*). Notable examples are the Arches and Quintuplet clusters near the Galactic Center. The Arches cluster is dense and rich enough to allow a full sampling of the IMF well in excess of 100 $M_\odot$ if such massive stars existed. However, no stars more massive than about 150 $M_\odot$ are found. This is interpreted as a genuine upper limit to the IMF.

The Galactic Center is the birth site of 30% of the W-R stars known in our Galaxy. Many of these W-R stars are still hydrogen-rich and should actually be considered O stars (*Martins*). The W-R stars near the Galactic Center serve as a warning for those using the presence of such stars as a cluster age indicator. Finding such "pseudo-W-R" stars does not necessarily indicate an evolved cluster age. The corresponding ages are well below 2 Myr. The same caveat applies to the well-known Galactic cluster NGC 3603.

The large reddening towards the Galactic Center requires the infrared (IR) spectral region for atmospheric analyses. Non-LTE models are now available that allow us to determine abundances of both α- and Fe-group elements (*Najarro*). The abundances found in hot and cool supergiants as well as in the surrounding ionized gas are consistent and suggest roughly solar composition. These techniques are highly relevant for current and upcoming observing facilities which emphasize the near-IR window.

Westerlund 1 is the most massive young star cluster known in our Galaxy. Will it become the new Rosetta Stone of star clusters? Since Westerlund 1 is heavily dust obscured, the IR is the wavelength of choice. The cluster has a mass of order $10^5 M_\odot$ and an age of 4.5 Myr (*Negueruela*). There is a significant binary population as indicated by numerous discrete X-ray sources. The most pressing issue for resolution is the uncertain distance. Like most Galactic clusters we do not know the distance to better than about 20%, which is a serious limitation in comparison with extragalactic clusters or the clusters near the Galactic Center.

Westerlund 1 should have many brethren in the Galaxy. Where are they? Extrapolating a standard cluster luminosity function for our Galaxy leads to an expected number of about 100 such clusters (*Hanson*). Most of them would have been missed in previous surveys. New searches are underway to complete the census.

Ultracompact HII (UCHII) regions are at the lower end of the HII luminosity function. The recently completed GLIMPSE survey with the Spitzer Space Telescope turns out to be a treasure trove for identifying UCHII regions (*Conti*). UCHII regions are in a pre-Orion evolutionary state with ages of a few 10^5 yr. Their ionizing stellar population is tiny by Westerlund 1 standards. There are typically only just a few ionizing stars. An important result is the trend to find no isolated ionizing stars. They all come in small clusters, which is consistent with the suggestion that most (all?) massive stars are not born in isolation.

If massive stars live most of their lives in clusters, it becomes important to address the interplay between stellar evolution and the dynamical evolution of the cluster (*Vanbeveren*). Close stellar encounters between single and multiple stars affect both stellar and cluster evolution. Some stars will merge and/or be ejected from the cluster. As a result, runaway stars would be the products of stellar mergers.

Several recently completed surveys of nearby star-forming galaxies with, e.g., GALEX and Spitzer have provided us with a panchromatic view of extragalactic massive star populations (*Bresolin*). Comparison of IR and ultraviolet (UV) luminosities suggests there is no major hidden star formation in *normal* galaxies. The stellar light is attenuated by about $A_V = 1$ but otherwise no significant part of the stellar population is missed in the UV/optical. We can now measure stellar and nebular abundances in a variety of nearby galaxies. The comparison indicates agreement to within a factor of 2.

We were reminded that star formation can occur in unexpected environments, such as in the dSp galaxy Fornax and in the dE galaxy NGC 205 (*Grebel*). Significant star formation occurred in both galaxies as recently as ∼100 Myr ago. Asymptotic giant branch (AGB) stars are the sign posts of that star-formation episode. While AGB stars are not massive stars by most definitions, they are close to RSGs in many of their properties. In particular, AGB stars and RSGs have similar stellar wind physics since their winds are both thought to be dust driven.

Spectroscopists do it better (Kudritzki). "Extragalactic stellar astrophysics" was another buzzword at this meeting. Owing to new instrumentation *and because of the vision and drive of dedicated astronomers* we are able to obtain stellar spectra in galaxies beyond the Local Group whose quality rivals those obtained in the Galaxy. Large surveys like FLAMES have been mentioned numerous times at this meeting. Its impact cannot be overemphasized. Targeted observations of individual stars out to a distance of about 7 Mpc via spectroscopy provides us with reddening independent distances. We were told that H_0 may not be as well known as we thought...

Some 25 years ago Phil Massey and Tony Moffat and their groups pioneered the systematic search for W-R stars in Local Group galaxies. In 2007, such surveys can be done in NGC 1313 at a distance of 4.1 Mpc (*Hadfield*). Pushing beyond the Local Group is highly complementary to the more detailed studies in, e.g., M33 or M31. We sacrifice individuality for completeness.

2.4. Feedback

The term "feedback" kept appearing throughout this symposium. This is one theme which crosses the boundaries of traditional hot star research and embraces topics such as the interstellar medium (ISM) dynamics, galaxy evolution, and the first stellar generations.

We observe nebular structures around stars on scales ranging from $10^1 - 10^3$ pc and with ages of 10^4–10^7 yr (*Chu*). Depending on their sizes and origination, these structures are commonly referred to as bubbles, superbubbles, and supergiant shells. 30 years ago, Weaver *et al.* published their classic paper describing the evolution of a circumstellar bubble with simple analytical scaling relations. These relations are still commonly used. Unfortunately, real bubbles tend not to follow the relations very well. Only when the detailed radiation-hydrodynamics are taken into account, modeled and observed bubbles agree (*Arthur*). In particular, observations clearly show complex substructure, like clumps, whose presence turns out to be crucial for the modeling.

HD 316285 is an extreme P Cygni star which has been proposed as an LBV candidate. A recent Spitzer mid-IR survey of circumstellar matter around Galactic star detected circumstellar emission around this star. Surprisingly, the distance to the mid-IR nebula turns out to be ∼8 kpc, much farther than the canonical (uncertain) distance to HD 316285. If correct, this distance would lift HD 316285 into the status of the most luminous star known in our Galaxy, essentially identical to the Pistol star. Since the uncertainties are still large, we may not see a press release anytime soon.

Feedback is fundamental at any redshift and at any time. We were given beautiful examples of the regulatory forces of feedback at redshifts 30, 2 and 0 (*Dopita*). Molecular hydrogen is the major coolant on the primordial ISM. Star formation sets in once the gas has cooled down sufficiently. When the newly formed stars inject wind and SN material, the H_2 may be removed, thus effectively curbing the cooling and star formation process. Redshift 2 is the main epoch of massive galaxy formation. These galaxies host massive central black holes which can trigger galaxy-wide outflows. The outflows remove material from the galaxy centers, inhibiting further growth. Finally, local starbursts inject matter and energy into the ISM and may eventually pollute the surrounding intergalactic medium (IGM).

M82 is for the starburst community what ζ Pup, η Car, or 30 Dor are for those working on individual hot stars. Its numerous super star cluster exemplify the very nature of a starburst: *it is the intensity that makes a starburst (Gallagher)*. Super star clusters have masses of order $10^6 M_\odot$ and form rapidly over 10^6 yr. The corresponding star-formation rate of $1 M_\odot$ yr^{-1} is comparable to that of the Milky Way, with the notable difference that the super star cluster has a size of ∼1 pc. The term "socialized photoionization" was introduced, meaning that starburst regions in a galaxy do not exist in isolation but may mutually share their photon output and affect each other's evolution.

The dwarf galaxy NGC 1569 is one of the closest starbursts and consequently offers a detailed view of its ISM morphology (*Linda Smith*). Spectroscopy of the bottom of the outflow of NGC 1569 leads to a picture where the outflow consists of a superposition of individual superbubbles. There is no evidence yet for a highly organized galactic superwind as observed farther out in X-rays.

What is the ionization source of the diffuse ISM (*Oey*)? This long-standing open question has been brought back on the agenda owing to the revised relation between stellar spectral type and ionizing photon output (see above). Previously, there seemed to be an excess of photons in HII regions that could leak out and heat the diffuse ISM. The new calibration essentially eliminates this excess, and the ionizing source of the diffuse ISM is again an open issue.

Massive star yields are often considered as being determined by Type II supernova yields alone. While it is true that SNe are often the dominant nucleosynthetic source, some subtle but important observations can only be understood when stellar wind yields are taken into account (*Matteucci*). For instance, we observe an excess of nitrogen at very low chemical composition. This excess is most readily understood as due to primary nitrogen produced in massive stars. Another example is [O/Mg], which plotted vs. O/H is not constant. Intuitively (at least to my intuition), one would expect the ratio of two α-elements to be constant. However, pre-SN stellar winds in massive stars modify the oxygen abundance prior to the SN explosion.

Despite a general understanding of the core-collapse SN explosion mechanism, the predicted yields for heavy elements like ^{45}Sc, ^{49}Ti, and ^{64}Zn are lower than indicated by observations of metal-poor stars (*Fröhlich*). These (and other elements) are produced in the core of the SN where the details of the explosion mechanism matter. The fact that we know that a supernova explodes in principle, is not sufficient for computing the corresponding yields!

2.5. *Early Universe*

The conference organizers devoted a full session to the high-redshift universe — the largest allocation in the history of the beach symposium series. Spectroscopic features of massive stars in high-z galaxies (the so-called Lyman-break galaxies) were first reported in the mid-1990s. By the time of IAU Symp. 212 the quality and quantity of the spectra had reached a state to permit IMF studies using the strongest lines. Now, we have reached the stage when faint absorption features are used to study abundances (*Pettini*). Starforming galaxies at $z \approx 3$ have oxygen abundances already close to solar. This means galaxies observed at a lookback time of about 10 Gyr have almost the same metal content as galaxies in the local universe.

Lyman-α emitters are another class of star-forming galaxies at cosmological redshift (*Taniguchi*). They are the lower luminosity extension of Lyman-break galaxies. Lyman-α emitters can be found at higher redshift and lower luminosity because a few percent of the entire starburst power is concentrated just in the Lyman-α line. These galaxies are now routinely found out to redshift 7. The Subaru telescope has turned out to be one of the most prolific discovery machines. Galaxies harboring genuine Population III stars are predicted to have a strong nebular He II 1640 line. Searches for this line are underway, so far with no firm detections. Since the prediction entirely hinges on the correct radiative transfer prediction in the stellar atmosphere below 228 Å, one would definitely like to perform some independent local tests of the models before allocating too many resources to such searches.

The early universe shows ample evidence of dust. For instance, the spectra of quasars at high redshift imply enormous dust masses. What is the source of this dust? Core-collapse SNe are likely candidates, and Spitzer was used to search for dust features around SNe in the mid-IR (*Kotak*). Surprisingly little dust was detected leading to the speculation that a significant amount of dust may be hidden in clumps.

GRB host galaxies have by now been detected out to the highest redshifts. Thanks to the rapid response of the Swift satellite, the host galaxies of tens of GRBs have been identified, and many of them could be studied in great detail (*Fynbo*). The hosts tend to be blue and irregular but display an enormous diversity in parameters such as reddening, Lyman continuum escape, or hydrogen column density. So far, most of the information on the hosts refers to the galaxy as a whole. Obviously one would like to make the direct association between the GRB and the progenitor and its birth place. One route that is being followed is the analysis of damped Lyman-α absorbers in GRB afterglow spectra (*Dessauges-Zavadasky*). However, the spectra analyzed so far do not exhibit circumstellar material but are dominated by the general galaxy ISM. The direct signatures of GRB progenitors are still elusive.

Hypernovae are a particular subclass of SNe whose kinetic energies are an order of magnitude or more above the canonical value of 10^{51} erg (*Nomoto*). There is fairly good evidence that hypernovae originate from progenitors with masses more massive than about 40 $M_\odot$. Morphologically they are associated with SNe of type Ic. Several long-duration GRBs have been firmly linked to these objects. The nucleosynthesis in hypernovae differs from that of standard supernovae by generating elevated carbon and diminished iron yields. This trend may in fact by observed in very metal-poor stars, emphasizing the importance of hypernovae in the early universe.

There is a fairly strong conviction among theorists that the first generation of stars was born with a top-heavy IMF (*Bromm*). The absence of dust and a higher equilibrium temperature favor higher accretion during the star-formation process. As a result, the characteristic masses of stars are an order of magnitude higher than in the present-day universe, and the masses of the most massive stars may be as high as 500 M$_\odot$. This is a research area that is begging for close collaborations between star formation experts, atmosphere modelers and cosmologists: such extremely massive and hot stars are ideal testbeds for the predictions of, e.g., the ionizing radiation in the extreme UV and the mass loss at zero metallicity.

We heard about a rather subtle imprint of the first stellar generation in the cosmic IR background (*Kashlinsky*). In principle, existing Spitzer mid-IR spectra carry the information of this early star formation. However, extracting the signal and separating it from other effects is a formidable task, and final verification may have to await the James Webb Space Telescope.

3. Future

I am sure all participants will agree with me that this symposium was a resounding success. The field of massive stars is healthy and shaping many other "hot" topics in contemporary astronomy. For everyone's entertainment, I am presenting in Table 1 the result of a little exercise prior to the symposium. I was asking myself these questions: What is the impact of massive stars? How did the impact evolve over time? How does it compare to other fields of astronomy? My approach was very simplistic and would not withstand a truly rigorous analysis. Nevertheless I hope one can recognize a few trends that some of you may find useful.

The entries in Table 1 were generated as follows: I searched the ADS for the keywords in the first column, binned in five-year intervals from 1975 until 2005. The last column of the table gives the ratio of the entries in the 2000–2005 over the 1975–1980 columns. For instance, "galaxies" was found in 6812 articles between 1975 and 1980, and 41,087 times between 2000 and 2005. The increase is the result of more publications with time, but also the increasing completeness of the ADS, large number of preprints, etc. Therefore the numbers should be looked at only in a relative sense.

The first three entries are the generic keywords "galaxies", "interstellar", and "stars". Their frequency increases by factors of a few, with galaxies showing the steepest rise. These numbers are the benchmarks for comparison. The next seven entries are related to stellar astronomy. My reading of these numbers is as follows. Keywords traditionally associated with hot, massive stars are essentially flat (i.e., just mirroring stellar astronomy). This does not suggest all stellar subcategories are flat: "brown" and "AGB" show a significant increase.

The next six entries are typically associated with extragalactic publications. They display a clear upward trend, even neglecting the 1975–1980 column where statistics are small. Note, for instance the increase of "survey" in the past 15 years, reflecting publications related to the major extragalactic surveys.

The divergent trends of "ultraviolet" and "infrared" are quite illuminating. They trace the increasing numbers of IR facilities and related results and the relative decline of the field of UV astronomy. The latter has traditionally been more closely connected with massive stars than the former. The final two entries compare "Hertzsprung-Russell" and "color-magnitude". This again signals a shift towards color-magnitude diagrams using photometric techniques with wide-field detectors as opposed to HR diagrams constructed from spectroscopy of individual stars.

Here is my reading of the lessons learned from this little exercise.

• The number of papers dealing with "classical" topics related to massive stars is lagging behind some other subjects in stellar astrophysics, such as AGB stars. The hot

Table 1. Popularity of some keywords over time as returned by ADS.

Keyword	75–80	80–85	85–90	90–95	95–00	00–05	(00–05)/(75–80)
galaxies	6812	9928	13556	17649	32400	41087	6.0
interstellar	6458	8091	9433	10017	9337	10234	1.6
stars	16862	23227	26788	29264	42483	52792	3.1
AGB	1	82	344	883	2100	2796	2796
atmosphere	7949	8228	8278	8871	9641	14621	1.8
brown	121	135	326	707	1742	3000	24.8
non-LTE	179	217	355	435	644	765	4.3
OB	369	631	618	747	1075	1826	4.9
supernova	2099	2747	5221	5203	7890	12735	6.1
Wolf-Rayet	297	760	869	952	1135	1120	3.8
AGN	1	103	632	1706	3723	6514	6514
bulge	219	408	658	1215	2297	2944	13.4
redshift	1527	2170	3090	5000	11066	16010	10.5
reionization	3	3	17	75	287	1049	350
starburst	1	97	737	1454	3236	4702	4702
survey	1752	2812	3870	6501	12631	22836	13.0
ultraviolet	3808	5759	5419	6141	6520	7473	2.0
infrared	4872	6439	9987	11770	16975	23487	4.8
Hertzsprung-Russell	370	683	547	435	237	282	0.8
color-magnitude	144	344	730	916	1164	1326	9.2

star community should be more aggressive in leveraging their scientific expertise and take ownership of fields where stellar astrophysics of massive stars adds value. The subject of Population III stars in the early universe is an excellent example. We saw this happening at this meeting but stronger efforts are needed.

• On average, the growth of many extragalactic/cosmological topics exceeds that of stellar topics. While this may seem disconcerting to some of us, this should not necessarily be the case. The separation between stellar and extragalactic is becoming increasingly fuzzy and meaningless. For instance, "starbursts" are more and more becoming embraced by the massive star community. At the same time, the extragalactic community is studying local template stars for comparison with Lyman-break galaxies.

• The growing emphasis on the IR is clear and present. This community should take advantage of this wavelength regime in areas where progress is lacking: understanding the formation of massive stars, dissecting heavily obscured regions like the Galactic Center, and developing diagnostic techniques to utilize spectral information at these wavelengths. These efforts will pay off when the next generation of large grand-based and space telescopes will be functional.

• I am speculating that one of the reasons for the waning popularity of the HR diagram may be the trend towards surveys and wide-field observation techniques. If so, we may see a revival of spectroscopically determined stellar properties with the advent of several multi-object/integral field unit spectrographs over the next few years. The results of the FLAMES project are an example of the enormous scientific returns to be expected.

The field of massive stars will greatly benefit from the new generation of large ground-based telescopes (*D'Odorico*) and JWST (*Sonneborn*). I am excited to learn about the results at future beach symposia of the massive star community.

Poster Abstracts

Edited by F. Bresolin, P. Crowther and J. Puls

(Grouped by sessions in alphabetical order by first author)

SESSION I: Atmospheres of Massive Stars

Pulsational Seeding of Structure in a Line-Driven Stellar Wind

Nurdan Anilmis & Stan Owocki (University of Delaware)

Massive stars often exhibit signatures of radial or non-radial pulsation, and in principal these can play a key role in seeding structure in their radiatively driven stellar wind. We have been carrying out time-dependent hydrodynamical simulations of such winds with time-variable surface brightness and lower boundary conditions that are intended to mimic the forms expected from stellar pulsation. We present sample results for a strong radial pulsation, using also an SEI (Sobolev with Exact Integration) line-transfer code to derive characteristic line-profile signatures of the resulting wind structure. Future work will compare these with observed signatures in a variety of specific stars known to be radial and non-radial pulsators.

Wind and Photospheric Variability in Late-B Supergiants

Matt Austin (University College London), Nevyana Markova (National Astronomical Observatory, Bulgaria) & Raman Prinja (University College London)

There is currently a growing realisation that the time-variable properties of massive stars can have a fundamental influence in the determination of key parameters. Specifically, the fact that the winds may be highly clumped and structured can lead to significant downward revision in the mass-loss rates of OB stars. While wind clumping is generally well studied in O-type stars, it is by contrast poorly understood in B stars.

In this study we present the analysis of optical data of the B8 Iae star HD 199478. Data collected primarily from the Bulgarian NAO 2 m telescope is used to probe the temporal behaviour of the Hα line, and any potential connections to deep-seated photospheric changes. Of some interest is a rare "high velocity absorption" episode that was monitored during 2000. The role of pulsations and ordered magnetic fields are discussed in understanding the Hα changes.

Modeling Ultraviolet Wind Line Variability in Massive Hot Stars

Ronny Blomme & Alex Lobel (Royal Observatory of Belgium)

We model the detailed time-evolution of Discrete Absorption Components (DACs) observed in P Cygni profiles of the fast-rotating supergiant HD 64760 (B0.5 Ib). We assume that the DACs are caused by co-rotating interaction regions (CIRs) in the stellar wind. We use the Zeus3D code to calculate hydrodynamic models of these CIRs (limited to the equatorial plane). Our model assumes that the CIRs are due to "spots" on the stellar surface. We then perform 3-D radiative transfer to calculate the resulting DACs in the Si IV ultraviolet resonance line.

Our best-fit model for HD 64760 consists of two spots of unequal brightness and size on opposite sides of the equator. The additional mass-loss rate due to these spots is less than 1% of the smooth-wind mass-loss rate. The recurrence time of the observed DACs compared to the estimated rotational period shows that the spots move 5 times slower than the rotational velocity. This strongly suggests that the "spots" are due to a beat pattern of non-radial oscillations. The fact that DACs are observed in a large number of hot stars constrains the amount of clumping that can be present in their winds, as substantial amounts of clumping would tend to destroy the CIRs.

Clumping and Rotation in ζ Puppis

Jean-Claude Bouret (Laboratoire d'Astrophysique de Marseille), John D. Hillier (University of Pittsburgh) & Thierry Lanz (University of Maryland)

We have analyzed the spectrum of the Galactic O4 supergiant, ζ Puppis, with a special emphasis on its clumping and rotation properties. The wind has been modeled using the NLTE code CMFGEN. Synthetic spectra have been computed with a code recently developed for CMFGEN to compute synthetic spectra in two-dimensional geometry allowing for the effects of rotation. As already showed in recent studies, introducing clumping deep in the wind improves the agreement to FUV/UV lines compared to homogeneous models. The fit to Hα is also improved but requires that clumping starts at larger distances in the wind. We demonstrate that it is possible to have clumping occur close to the star while still achieving an excellent fit to Hα by consistently treating the wind's rotation in the spectral modeling. An excellent agreement to other important optical lines such as HeII 4686 Å and NIII 4634–4640 Å is also obtained for the first time.

First Estimation of the Rotation Rates of Wolf-Rayet Stars

André-Nicolas Chené (Herzberg Institute of Astrophysics-Canadian Gemini Office) & Nicole St-Louis (Université de Montréal)

See p. 139 for full article.

Shock Variations in η Carinae

Michael Corcoran, Kenji Hamaguchi (Center for Research and Exploration in Space Science & Technology, University of Maryland/Universities Space Research Association/Goddard Space Flight Center), David Henley (University of Georgia), *et al.*

η Carinae is a binary system composed of an extremely massive luminous blue variable star and a bright companion. Thermal X-ray emission generated where the wind from η Car collides with the companion's wind provides unique diagnostics of the mass loss phenomena in both stars. We use variations in resolved X-ray emission lines from HETGS spectra taken around the orbit and concentrating on the approach to periastron passage to constrain the wind-wind interaction, the flow of hot gas along the colliding wind boundary, and the orbital elements of the system.

Propagating Waves in Hot-Star Winds: Leakage of Long-Period Pulsations

Steven R. Cranmer (Harvard-Smithsonian Center for Astrophysics)

Massive stars have strong stellar winds that exhibit variability on timescales ranging from hours to years. Many classes of these stars are also seen, via photometric or line-profile variability, to pulsate radially or non-radially. It has been suspected for some time that these oscillations can induce periodic modulations in the surrounding stellar wind and produce observational signatures in line profiles or clumping effects in other diagnostics. The goal of this work is to investigate the detailed response of a line-driven wind to a given photospheric pulsation mode and amplitude. We ignore the short-wavelength radiative instability and utilize the Sobolev approximation, but we use a complete form of the momentum equation with finite-disk irradiation and finite gas pressure effects.

For large-scale perturbations appropriate for the Sobolev approximation, though, the standard WKB theory of stable "Abbott waves" is found to be inapplicable. The long periods corresponding to stellar pulsation modes (hours to days) excite wavelengths in the stellar wind that are large compared with the macroscopic scale heights. Thus, both non-WKB analytic techniques and numerical simulations are employed to study the evolution of fluctuations in the accelerating stellar wind. This progress report describes models computed with one-dimensional (radial) isothermal motions only. However, even this simple case produces a quite surprising complexity in the phases and amplitudes of velocity and density, as well as in the distribution of outward and inward propagating waves throughout the wind.

Empirical Calibration of Mid-IR Fine Structure Lines Based on O Stars

P.A. Crowther, J.P. Furness (University of Sheffield) & P.W. Morris (IPAC)

We present mid-IR Spitzer IRS spectroscopy, plus VLT ISAAC H- and K-band spectroscopy of O stars identified by Blum *et al.* (2001) within the W31 cluster. We provide an empirical calibration of [NeII/III] and [SIII/IV] fine-structure lines versus effective temperature for early- to mid-O stars, as derived from spectroscopic near-IR analysis. Comparisons with photoionization models are presented, suggesting a discrepancy between empirical and predicted mid-IR line ratios, as previously identified by Morisset *et al.* (2004). We also discuss the embedded timescales for high-mass stars, since W31 hosts O stars, massive young stellar objects, and UCHII regions.

Rotating Radiation Driven Winds of Massive Stars

M. Curé (Departamento de Física y Astronomía, Universidad de Valparaíso, Chile), L. Cidale & R. Venero (Facultad de Ciencias Astronómicas y Geofísicas, Universidad Nacional de La Plata, Argentina)

Radiation driven wind theory, taking into account the centrifugal force due to the stellar rotation, has been re-examined. For rotational speeds above about 75% of the critical velocity, the wind solution can switch to an alternative mode (slow solution), characterized by a much slower outward acceleration. Together with a moderately enhanced mass flux, the resulting lower speed outflow then implies a substantial enhancement in density, relative to the standard (or fast) solution. We present different applications of these slow solutions, that may explain wind features, as the outflowing disk from B[e] supergiants and the winds of A supergiants and its influence in the wind momentum luminosity (WML) relationship.

A Multispectral View of the Periodic Events in η Carinae

Augusto Damineli (Instituto de Astronomia, Geofísica e Ciências Atmosféricas, Universidade de São Paulo)

We present a multi-spectral analysis, ranging from X-rays to radio, of the 5.5-yr events in η Carinae. We show that the events are bimodal, composed by a short duration core that hides an eclipse and by a "main body" that encompasses the whole cycle, due to the immersion of the secondary star in the wind of the primary. We show anti-correlated radial velocity curves, giving indications of the orbital parameters. We make predictions to be tested in a monitoring campaign to cover the 2009 event.

η Carinae: The Central Counter-Paradigm for Very Massive Stars
Kris Davidson (University of Minnesota)

η Car is the only star above 120 M$_\odot$ that has been observed in great detail; it is the most accessible supernova impostor; and it has repeatedly exposed gaps and errors in existing theory. Two high-priority questions will be featured here. The first concerns η Car's 5.5-year spectroscopic cycle. Intensive HST spectroscopy indicates that each recurring "spectroscopic event" is a disruption in the latitude-dependent wind, probably triggered by a companion star's periastron approach. This indicates a surface instability that was not predicted and is stronger than one would normally have expected based on tidal and radiative effects. The hypothetical companion object continuously alters the primary star's surface parameters, revealing information that would otherwise have been hidden. HST data 1991–2007 also show that η Car recently accelerated its secular rate of change. This presumably involves the thermal/rotational recovery process following the supernova impostor event seen 160 years ago. Major variations in the recovery rate convey unique clues to the interior structure.

For almost 30 years we have recognized that very luminous stars can lose mass chiefly in eruptions rather than steady winds. Some theorists assume that this is metallicity-dependent, but the contrary has not been disproven. Thus it is premature to say that Pop III stars are immune to LBV-like eruptions.

Colliding-Wind Binaries As Non-Thermal X-ray Emitters: An Observational Investigation from XMM-Newton to Next Generation X-Ray Observatories
Michaël De Becker (Institut d'Astrophysique et Géophysique, Université de Liège)

The investigation of massive stars in the radio domain revealed about 25 years ago that some of them are synchrotron emitters, showing that these objects are able to accelerate particles up to relativistic velocities. In this context, non-thermal emission processes such as inverse Compton scattering are expected to be at work in the high-energy domain. For this reason, an observational campaign devoted to the X-ray investigation of non-thermal radio emitters has been carried out with XMM-Newton. However, considering the rather strong thermal X-ray emission from these systems below 10 keV, XMM-Newton does not appear to be the ideal observatory to detect their putative non-thermal X-rays. As a consequence, the advent of next generation X-ray observatories with a bandpass extending significantly above 10 keV (SIMBOL-X, XEUS, or NEXT) is expected to provide important results related to the non-thermal high-energy emission from colliding-wind binaries and their capability to accelerate particles.

Magnetic Fields in Wolf-Rayet Stars
Antoine de la Chevrotière, Nicole St-Louis & Anthony F. J. Moffat (Université de Montréal)

Independently of the possibility of internally generated fields, there are good reasons to believe that massive stars actually harbour significant magnetic fields of fossil origin. As the ISM itself possesses organized, large-scale magnetic fields, it is unavoidable because of magnetic flux conservation and flux freezing, that stars that result from cloud collapse have a magnetic field, at least at first. Magnetic fields have been observed and measured in a variety of hot stars, now including O stars. Furthermore, neutron stars (NS), the descendants of many massive stars including Wolf-Rayet stars (WR), display large magnetic fields. In the case of core He-burning WR stars, the hydrostatic core radii (R_*)

are typically an order of magnitude smaller than those of their O progenitors, so the NS-extrapolated surface magnetic fields in WR stars could be ∼100 times higher than in O stars, i.e. $B \sim 200$–2000 G. However, the strong WR winds only allow us to see emission lines down to $\sim 4R_*$, where the magnetic fields vary as R^{-2} and will thus be smaller than at the hydrostatic core, assuming the likely scenario of a split monopole to best simulate the fields, pulled radially outward by the strong flows. This implies fields of ∼10–100 G expected in the inner observable parts of WR winds. The Zeeman effect is the most direct way to measure the strength of magnetic fields as more observations are needed.

Are WC9 as Violent as WN8?

Remi Fahed, Anthony F. J. Moffat (Université de Montréal) & Alceste Z. Bonanos (Carnegie Institution of Washington)

Do some WR stars owe their strong winds to something else besides radiation pressure? The answer to this question is still not obvious, especially in certain subclasses, mainly WN8 and WC9. Both of these types of WR stars are known to be highly variable, possibly due to pulsations. However, only the WN8 stars have so far been vigorously and systematically investigated for variability. We present here some preliminary results of a systematic survey during 3 consecutive weeks in 2007 June at Las Campanas Observatory, Chile, of 20 Galactic WC9 stars for photometric variability in two optical bands, V and I. Of particular interest are the variations in colour index in the context of carbon dust formation, which occurs frequently in WC9 stars.

Weak Winds in Orion?

Miriam Garcia, Artemio Herrero (Instituto de Astrofísica de Canarias, ULL) & Sergio Simón-Díaz (Observatoire de Genève)

The theory of radiatively driven winds apparently fails to predict the wind momentum of low luminosity ($\log L/L_\odot < 5.2$) early-type stars from metal poor environments like the SMC, but there are also some Galactic cases. The reason of this alleged theoretical breakdown is still unknown. While it may hint a metallicity-dependent threshold luminosity to initiate the wind, it may also relate to age, the discrepant objects being too young to have turned it on.

Starting from previous quantitative analyses of their optical spectra, we study the ultraviolet spectra of a sample of OB-type stars in Orion, to constrain their wind terminal velocity and mass-loss rate. Orion is a young star-forming region; thus our results will contribute to ascertain whether there is a "weak wind-young object" connection.

Analytical Wide and Thin Colliding-Wind Bow-Shock Asymptotics

Kenneth Gayley (University of Iowa), Stan Owocki (University of Delaware) & Peter Tuthill (University of Sydney)

Bow shocks from colliding winds and runaway hot stars are typically either modeled with detailed numerical simulations, or are approximated analytically in the thin-shell approximation. The latter allows for an elegant analytical treatment based on momentum conservation, but requires rapid radiative cooling to be valid. Many types of wind collisions yield low enough densities, high enough temperatures, or rapid enough flow times that the response is expected to be largely adiabatic. The shock heating and wind re-acceleration makes for a potentially wide interaction region. This poster considers simplifying assumptions that allow an analytic determination of the bow-shock angle far from the star, and argues that adiabaticity can significantly widen a bow shock only when the winds are fairly equal. However, adiabaticity may strongly constrain the fraction of a wind that avoids being asymptotically embroiled in the interaction zone.

Time-Dependent Effects in the Wind of Luminous Blue Variables

Jose H. Groh (Max-Planck-Institut für Radioastronomie), D. John Hillier (University of Pittsburgh) & Augusto Damineli (University of São Paulo)

LBVs are characterized by strong photometric and spectroscopic variability on timescales from days to decades, arising from changes in stellar and wind parameters. Therefore, the common assumption of a steady-state outflow is invalid for LBVs. We present a newly developed method to include the effects of time variability in the radiative transfer code CMFGEN. We show how time-dependent effects significantly change the velocity law and density structure of the wind, affecting the derivation of the mass-loss rate, volume filling factor, wind terminal velocity, and luminosity. We quantitatively show that striking features seen in the spectrum of LBVs are due to time variability, such as the presence of multiple absorption components in Fe II and Balmer lines, and high-velocity absorption in the UV resonance lines. The results of this work are directly applicable to all active LBVs, such as AG Car, S Dor, R 127, and HD5980, and can result in a revision of their stellar and wind parameters.

Ejecta of η Carinae: Clues to Mass Ejection

Theodore R. Gull (Goddard Space Flight Center)

We have used the spatially resolved spectra of HST/STIS and VLT/UVES to gain insight on the structure of the current extended interacting binary wind and the material ejected in the 1840s and 1890s. Much has been learned about the very N-rich, O- and C-poor ejecta, the abundant metals not ordinarily seen in the ISM (for example, V, Sr, Sc, Ti) and some molecules that formed. The dust is very peculiar, likely formed of SiO, AlO, and metal complexes (such as TiN, TiC, FeN). Abundances of these metals and the chemistry in N-rich, O- and C-starved gas will be discussed.

Time-Resolved FUSE Observations of the Mysterious WR46

Vincent Hénault-Brunet, Nicole St-Louis (Université de Montréal), Sergey V. Marchenko (Western Kentucky University), *et al.*

The Wolf-Rayet star WR46 (of spectral type WN3p) is known to exhibit a very complex variability pattern on relatively short timescales of a few hours. Eight-hour periodic radial velocity variations in the highest ionization lines have been found but seem to intermittently disappear. In addition, multiple photometric periods have been claimed by various authors. Nonradial oscillations of the Wolf-Rayet star or the presence of a yet-to-be-confirmed compact companion have been proposed to explain the observed short-term behavior, but the true nature of this system is still a mystery. In an effort to better constrain the above scenarios, we present time-resolved *FUSE* observations of WR46 extending over approximately 50 hours and covering several variability cycles. A TVS analysis performed on our time series of spectra confirms a significant variability in the blue wing of the absorption trough of the shock-sensitive O VI doublet (1032,1038 Å) P Cygni profile. We also obtain the far-UV continuum light curve and find a dominant photometric period close to 8 hours. The blue wing of the absorption trough of the O VI

P Cygni profile varies on the same timescale. We discuss the implication of these new results in terms of the different hypotheses for the cause of the short-term variability.

Is WR 104 Really a Face-on Colliding-Wind Binary?

Grant M. Hill (Keck Observatory)

WR 104 is the prototype for a small but growing group of stars that present the remarkably striking appearance of pinwheels. Many of them appear to be nearly face-on spirals. The assumption that these objects are dust producing, low-inclination, colliding-wind binaries has been very successful in modeling the imaging. This assumption remains largely untested by spectroscopy, though.

Confrontation of the colliding-wind binary model with six years of spectroscopy, offering full phase coverage of WR 104, is presented. A number of predictions are examined, and the results imply there is more to these systems than currently understood.

The Magnetic O Star HD191612

Ian Howarth (University College London), Nolan R. Walborn (Space Telescope Science Institute), Danny Lennon (The Isaac Newton Group of Telescopes, La Palma), et al.

We present extensive optical spectroscopy of the early-type magnetic star HD 191612 (O6.5f?pe–O8fp). The Balmer and He I lines show strongly variable emission which is highly reproducible on a well-determined 538-d period. Metal lines and He II absorptions (including many selective emission lines but excluding He II $\lambda 4686$ Å emission) are essentially constant in line strength, but are variable in velocity, establishing a double-lined binary orbit with $P_{\rm orb} = 1542$d, $e = 0.45$. We conduct a model-atmosphere analysis of the spectrum, and find that the system is consistent with a $\sim$O8 giant with a $\sim$B1 main-sequence secondary. Since the periodic 538-d changes are unrelated to orbital motion, rotational modulation of a magnetically constrained plasma is strongly favoured as the most likely underlying 'clock'. An upper limit on the equatorial rotation is consistent with this hypothesis, but is too weak to provide a strong constraint.

A Study of the Peculiar B0 Star θ Carinae.

S. Hubrig (European Southern Observatory), M. Briquet, T. Morel (Universiteit Leuven), et al.

Massive stars end their evolution, with a final supernova explosion, as neutron stars or black holes. The initial masses of these stars range from 8–10 $M_\odot$ to 100 $M_\odot$ or more, which corresponds to spectral types earlier than about B2. While magnetic fields in the Sun and solar-like stars have been studied intensively, very little is known about their existence, origin, and role in massive stars. We present the results of our recent study of the magnetic field and abundances in the peculiar B0 star θ Car. This star was already "preselected" as a suitable candidate for a magnetic field study by Walborn (2006, *IAU Joint Discussion*, 4E, 19) because it has a peculiar, variable spectrum both in the optical and UV, a high $L_X/L_{\rm bol}$ ratio, and enhanced nitrogen (Walborn 1976, *ApJ*, 205, 419).

Phase Resolved FUV Spectra of Massive Binaries

Rosina C. Iping (Goddard Space Flight Center & CUA), George Sonneborn (GSFC) & Doug Gies (Georgia State University)

We present FUV observations of massive binary stars in the Galaxy, LMC, and SMC. A large sample of close, massive binaries including detached and semi-detached systems that are at pre- and post-Roche lobe overflow evolutionary stages is presented. The binaries are generally double-line spectroscopic binaries, many are eclipsing systems, with well-determined orbits and periods in the range 1.6–12 days. The FUV spectra are used to determine stellar wind mass-loss rates and terminal velocities from species tracing a range of wind ionization states. Each system was observed more than once to sample different orbital phases and spectral variability. The spectral features are modeled to study photospheric abundances. We searched for evidence for CNO enhancements as the result of mass transfer and measured projected rotational velocities.

The Wind Properties of O-type Stars Constrained with Millimeter Observations

Eric Josselin, Fabrice Martins (GRAAL, Université Montpellier) & Jean-Claude Bouret (Laboratoire d'Astrophysique de Marseille)

Clumping in hot stars winds is currently one of the main questions in massive stars astrophysics. In particular, the variation of clumping with distance is not well constrained. A first step towards a determination of this variation was made by Puls *et al.* (2006, *A&A*, 454, 625). Here, we present millimeter observation of a sample of late-type O supergiants conducted at IRAM aiming at better constraining the clumping behavior in those objects. Complemented by UV-optical-IR data already obtained, these data will probe different parts of the atmospheres and clumping factors in those regions will be analysed with CMFGEN models.

New Massive Stars in Cyg OB2

Daniel C. Kiminki, Henry A. Kobulnicky (University of Wyoming) & M. Virginia McSwain (Lehigh University)

As part of an ongoing study to determine the distribution of orbital parameters for massive binaries in the Cygnus OB2 association, we present the orbital solutions for two new single-lined spectroscopic binaries, MT059 (O8V) & MT258 (O8V), and one double-lined eclipsing binary (Schulte 3). We also constrain the orbital elements of three additional double-lined systems (MT252, MT720, MT771). Periods for all systems range from 1.5–19 days and eccentricities range from 0–0.11. The six new OB binary systems bring the total number of multiple systems within the core region of Cyg OB2 to 11.

Energy Dissipation Variations and Periastron Passage Events

Gloria Koenigsberger, Edmundo Moreno (Universidad Nacional Autonoma de Mexico) & Andrea Avena (UAEM)

Numerous examples are now emerging of binary systems in which activity levels significantly increase around periastron passage. In some cases, the activity is so intense that it is classified as an eruptive event, as in WR 140 and other such systems. This raises the question of how the increasing tidal perturbations at periastron passage and the observational manifestations of stellar activity may be linked. We have constructed a model that computes how the tidal shear energy dissipation rates change over the orbital cycle, and in this poster we illustrate the results for particular binary systems. A noteworthy feature of our model is that it predicts maximum dissipation rates after periastron passage, similar to what is observed in many systems where periastron passage events are reported. We speculate that the degree of activity may correlate with the rate of change in the magnitude of the tidal interactions over the orbital cycle and, thus, depends on orbital period and separation, as well as stellar radius and equatorial rotational velocity.

Resonance Scattering in the X-ray Emission-Line Doppler Profiles of ζ Puppis

Maurice A. Leutenegger (Goddard Space Flight Center), Stanley P. Owocki (Bartol Research Institute, University of Delaware), Steven M. Kahn (KIPA/SLAC/Stanford University), et al.

We present XMM-Newton Reflection Grating Spectrometer observations of pairs of X-ray emission line profiles from the O star ζ Pup that originate from the same He-like ion. The two profiles in each pair have different shapes and cannot both be consistently fit by models assuming the same wind parameters. We show that the differences in profile shape can be accounted for in a model including the effects of resonance scattering, which affects the resonance line in the pair but not the intercombination line. This implies that resonance scattering is also important in single resonance lines, where its effect is difficult to distinguish from a low effective continuum optical depth in the wind. Thus, resonance scattering may help reconcile X-ray line profile shapes with literature mass-loss rates.

The Coolest Stars in the Clouds: Late-Type Red Supergiants in the Magellanic Clouds

Emily M. Levesque (Institute for Astronomy, University of Hawaii), Philip Massey (Lowell Observatory), K. A. G. Olsen (NOAO), et al.

Red supergiants (RSGs) are a He-burning phase in the evolution of moderately high-mass stars (10–25 solar masses). The physical properties of these stars continue to challenge our understanding of their evolutionary theory, particularly at low metallicities. The latest-type RSGs in the LMC and SMC are cooler than the current evolutionary tracks allow, occupying the "forbidden" region to the right of the Hayashi limit, which shifts to warmer temperatures at lower metallicities. Among these outliers, we have discovered four Cloud RSGs that display remarkably similar and unusual variations in their physical properties, varying dramatically in their V magnitudes and effective temperatures (and hence their spectral types). When these stars are warmer, they are also brighter, more luminous, and show an increased amount of extinction, with these substantial physical changes happening in timescales on the order of months. At their greatest, the amount of extinction is characteristic of that due to circumstellar dust around other RSGs, and thus suggests that we are seeing sporadic dust production from these stars while they are in their cooler states. Two of the SMC RSGs, HV 11423 and [M2002] SMC 055188, have been observed in an M4.5 I state, making them considerably later and cooler than any other supergiant in the SMC. We believe that this unusual behavior is indicative of a unstable, short-lived, and previously unobserved evolutionary phase in RSGs, and consider the implications such behavior could have for our understanding of the latest stages of massive star evolution in low-metallicity environments.

Hydrodynamic Modeling of UV Line Profiles in the η Carinae System: The Search for η Car's Missing Companion

Thomas I. Madura (Department of Physics & Astronomy, University of Delaware), Theodore Gull (Goddard Space Flight Center, Astrophysics Science Division), Stanley Owocki (Bartol Research Institute, University of Delaware), *et al.*

The extremely massive and luminous star η Carinae, with its bipolar Homunculus nebula, comprises one of the most remarkable and intensely observed stellar systems in the galaxy. Observed X-ray variations are interpreted as being due to the interaction of a massive wind from the primary star with the fast, less dense wind from a hot companion star. There are still no direct detections of the companion, but UV ionization signatures from the "Weigelt blobs" and other regions of the surrounding nebula generally support this binary scenario. This poster presents recent work aimed at modeling UV line profiles of η Car as a function of orbital phase by analyzing the results of smoothed particle hydrodynamics (SPH) simulations of the binary interactions in the system and comparing these to spectra obtained with the Hubble Space Telescope Space Telescope Imaging Spectrograph (HST STIS). The primary goals are to further constrain the parameters of the binary orbit (including the stellar mass ratio), determine how/where UV light is escaping in the system, and ascertain what, if any, direct signatures of the companion are present in the HST spectra.

Disentangling the Radio Emission Nature in Wolf Rayet Stars

Gabriela Montes, Miguel A. Perez-Torres & Antonio Alberdi (Instituto de Astrofísica de Andalucía, Consejo Superior de Investigaciones Científicas)

We present simultaneous, multiwavelength (at 1.3, 3.6, and 6 cm) VLA observations in D configuration of a sample of WR stars, aimed at disentangling the nature of their radio emission. Wolf-Rayet (WR) stars display high-mass loss rates and terminal velocities that result in strong winds. These winds are expected to have a positive spectral index at radio wavelengths. However, several WR sources have been found to present variable emission and negative spectral indices characteristics of non-thermal emission. Our sample sources have been detected at least at one wavelength; however, due to the variable nature of their radio emission, simultaneous multiwavelength observations are the best way to determine spectral indices and therefore disentangle their radio emission nature from thermal or non-thermal. We have found variability in the radio emission and the spectral indices, suggesting the presence of an emission process in addition to the thermal radio emission described for ionized stellar winds in single stars.

Systematic Uncertainties in OB Star Analysis

M. Fernanda Nieva & Norbert Przybilla (Bamberg Observatory, University of Erlangen-Nuremberg)

Precise stellar parameters and chemical abundances are crucial for constraining theories of stellar and galactochemical evolution. These constraints can be inferred through quantitative spectroscopic analyses where the sources of systematic errors are not always fully recognised. We show that (a) accurate input atomic data for the non-LTE modelling are often underestimated, which may lead to systematic biases when only a few strongly non-LTE affected lines are used for the analysis; and (b) accurate atmospheric parameters are also often underestimated, which may give larger biases than the neglect of non-LTE effects on many strategic lines. We present an improved spectral modelling and a self-consistent analysis methodology for OB-type dwarfs and giants that help us to reduce the major sources of systematic uncertainties. The spectrum synthesis is based on model atoms built from precise atomic data, and it employs a hybrid non-LTE approach, appropriate for this kind of star. The analysis technique is able to reproduce the entire H and He spectra simultaneously from the visual to the near-IR. Spectral energy distributions, Stark-broadened lines, and multiple ionization equilibria are employed to derive precise atmospheric parameters, resulting in a simultaneous tight agreement from 6 independent spectral indicators and therefore demonstrating a large reduction of systematic errors. The metal abundances (CNO + alpha elements) of our programme stars in the solar vicinity also show high accuracy from the consideration of the entire line spectra. This allows stellar and Galactic chemical evolution models to be tested in unprecedented detail.

A 3-D Geometric Model of the Colliding Winds in η Carinae

E. Ross Parkin, Julian M. Pittard (School of Physics and Astronomy, University of Leeds), Mike F. Corcoran (Exploration of the Universe Division, Goddard Space Flight Center), *et al.*

We have developed a 3-D geometric model of the wind collision region (WCR) in a colliding wind binary (CWB), where the curvature of the WCR due to the orbital motion of the stars is accounted for. We have used this model to simulate the X-ray light curve and spectra from the massive CWB η Carinae, and are able to constrain the orbital orientation and wind parameters of the system. A notable success is the ability to match the long duration of the X-ray minimum without recourse to additional mass ejection during periastron passage, though the exact shape of the minimum remains to be explained.

On the X-ray Emission and the Incidence of Magnetic Fields in Massive Stars of the Orion Nebula Cluster

Véronique Petit (Université Laval), Gregg A. Wade (Royal Military College of Canada) & Laurent Drissen (Université Laval)

Magnetic fields have been frequently proposed as a likely source of variability and confinement of the winds of massive stars. Recently, Stelzer *et al.* (2005, *ApJS*, 160, 557) found significant X-ray emission from all massive stars in the Orion Nebula Cluster (ONC). Periodic rotational modulation in X-rays and other indicators suggested that there might be many magnetic B- and O-type stars in this star-forming region. We have carried out sensitive ESPaDOnS observations to search for direct evidence of such fields, detecting unambiguous Zeeman signatures in three objects. We also obtained dipole field upper limits for the remaining stars with a state-of-the-art Bayesian analysis, resulting in a precise magnetic characterisation of all ONC massive stars. This allows us to explore for the first time the connections between fields, winds, and X-rays in a complete, co-eval and co-environmental sample of massive stars. These remarkable results bring forth new challenges for understanding the processes leading to X-ray emission in massive stars. We also expect to provide unique data regarding the incidence of magnetic fields in massive stars with which to confront models of magnetic field origin in neutron stars and magnetars, such as that proposed by Ferrario & Wickramasinghe (2006, *MNRAS*, 367, 1323).

Colliding Wind Binaries: Mass-Loss Rates and Particle Acceleration

Julian M. Pittard (University of Leeds)

Clumping in hot star winds can significantly affect estimates of mass-loss rates, the inferred evolution of the star, and the environmental impact of the wind. I present hydrodynamical simulations of colliding wind binary systems with clumpy winds, and demonstrate that in many cases the clumps are rapidly destroyed in the wind-wind collision region. X-ray emission from this region is thus a clumping-independent measure of the mass-loss rates. I also discuss the implications for a variety of other phenomena, including particle acceleration, non-equilibrium ionization, and electron heating.

New Observations of the Non-radial Pulsator HD93521

Gregor Rauw (Université de Liège), *et al.*

We present the results of long-term spectroscopic monitoring, as well as of an intensive photometric campaign on the runaway O9.5Vp star HD93521. This star has a very high rotational velocity and is known to display line profile variability probably due to non-radial pulsations. Our observations indicate that the amplitude of the dominant pulsation modes with periods of 1.75 (l = 8) and 2.89 hr (l = 4) change with time. Whilst light variations are detected, they are apparently not related to these periodicities.

Comprehensive Analysis of the WN Stars in the LMC

Ute Ruehling, Götz Gräfener & Wolf-Rainer Hamann (Universität Potsdam)

We present preliminary results from an analysis of almost all known WN stars in the LMC. We fit $\sim$100 archival spectra from the UV to the IR range with PoWR atmosphere models, taking Fe-group line-blanketing and wind clumping into account. In this way we are able to determine reliable stellar parameters like luminosities, effective temperatures, and mass-loss rates for all stars of the sample. This allows us to perform a statistical study of the WN properties in the LMC without selection bias. Among the objects, we find candidates for fast rotators, binaries, and stars with peculiar surface abundances. To investigate the impact of the low LMC metallicity, we compare our results to previous analyses of the Galactic WN population.

The X-Ray Light Curve of η Carinae from a 3-D SPH Binary Colliding Wind

Christopher M. P. Russell (Department of Physics and Astronomy, University of Delaware), Stanley P. Owocki (Bartol Research Institute, University of Delaware) & Atsuo T. Okazaki (Department of Architecture and Building Engineering, Hokkai-Gakuen University)

We model the RXTE X-ray light curve for η Carinae using a 3-D smoothed particle hydrodynamics (SPH) simulation of the collision of the strong wind from the primary star with a weaker but faster wind of an assumed secondary star. For a reasonable choice of stellar, wind, and orbital parameters, the SPH simulations provide a dynamical model of the relatively low-density cavity carved out by the secondary wind, and how this varies with orbital phase. Assuming then the main X-ray emission occurs near the head of the wind-wind interaction cone, and varies in intensity with the inverse of the binary separation at any given orbital phase, we generate trial X-ray light curves by computing the phase variation of absorption to observers at various assumed lines of sight.

Comparison with the RXTE light curve suggests an optimal viewing angle approximately 36 degrees out of the orbital plane and 36 degrees from apastron in the prograde direction. Such a viewing angle is consistent with the orbit being in roughly the same plane as the equatorial skirt. Our derived synthetic light curve naturally reproduces many of the key features of the RXTE light curve, namely the increase in X-rays approaching periastron, the sudden decline into the X-ray eclipse, the appropriate duration of the X-ray eclipse, and the less sharp incline out of the eclipse. The naturalness of the fit provides strong evidence in favor of the basic wind-wind binary interaction model, and with further analysis of, e.g., X-ray and UV spectra, it should be possible to place further constraints on the basic stellar, wind, and orbital parameters.

Spectroscopic Analysis of Deneb: A Hybrid Non-LTE Approach

Florian Schiller & Norbert Przybilla (Dr. Remeis Sternwarte Bamberg)

Quantitative spectroscopy of luminous BA-type supergiants offers a high potential for modern astrophysics. The degree to which we can rely on quantitative studies of this class of stars as a whole depends on the quality of the analyses for benchmark objects. We constrain the basic atmospheric parameters and fundamental stellar parameters as well as chemical abundances of the prototype A-type supergiant Deneb to unprecedented accuracy ($T_{\text{eff}} = 8525 \pm 75\,\text{K}$, $\log g = 1.10 \pm 0.05\,\text{dex}$, $M_{\text{spec}} = 19 \pm 3\,M_\odot$, $L = 1.96 \pm 0.32 \cdot 10^5\,L_\odot$, $R = 203 \pm 17\,R_\odot$, enrichment with CN-processed matter) by applying a sophisticated hybrid NLTE spectrum synthesis technique which has recently been developed and tested. The study is based on a high-resolution and high-S/N spectrum obtained with the Echelle spectrograph FOCES on the Calar Alto 2.2 m telescope. Practically all inconsistencies reported in earlier studies are resolved. Multiple metal ionization equilibria and numerous hydrogen lines from the Balmer, Paschen, Brackett, and Pfund series are brought into match simultaneously for the stellar parameter determination. Stellar wind properties are derived from Hα line-profile fitting using line-blanketed hydrodynamic non-LTE models. A self-consistent view of Deneb is thus obtained, allowing us to discuss its evolutionary state in detail by comparison with the most recent generation of evolution models for massive stars.

Wind Clumping and Neon Abundances in Galactic WC Stars

Olivier Schnurr, Paul A. Crowther (University of Sheffield), Patrick W. Morris (IPAC/Caltech), *et al.*

We have obtained high-quality, flux-calibrated, mid-IR Spitzer/IRS spectroscopy of six Galactic WC4–8 Wolf-Rayet stars including the key fine-structure neon lines [Ne II] 12.8 μm and [Ne III] 15.5 μm. In addition, we carry out tailored line-blanketed, non-LTE model atmosphere analyses using UV/optical and near-IR spectroscopy, to obtain quantitative stellar and wind properties of our program stars, necessary to measure abundances of neon, which in WC stars is expected to be greatly enhanced. Wolf-Rayet stars display a strong free-free continuum emission at IR and radio wavelengths, whose slope can be used as a diagnostic of clumping in the outer wind. We find that, while UV/optical diagnostics confirm that the inner wind is clumped, there appears to be no difference in clumping for the outer wind, contrary to what is expected from hydrodynamical considerations. From our derived wind and chemical properties, we are able to accurately determine the neon abundance in our program WC4–8 stars. Taking into account the latest correction to the solar abundances by Asplund *et al.* (2004, *A&A*, 417, 751), we find that the neon abundances in our WC stars are in excellent agreement with predictions from evolutionary models.

Using HII Region Spectra to Probe the Ionizing Radiation from Massive Stars

Sergio Simón-Díaz (Observatoire de Genève), Jorge García-Rojas (Universidad Nacional Autonoma de Mexico), Grazyna Stasińska (LUTH, Observatoire de Paris-Meudon), *et al.*

We are performing a study of HII regions ionized by a single massive star to test the prediction of the new generation of stellar atmosphere codes in the H Lyman continuum (below 911 Å). The observations collected for this study comprise the optical spectra of the corresponding ionizing stars, along with imaging and long-slit spatially resolved nebular observations. The analysis of the stellar spectra allow us to obtain the stellar parameters of the ionizing star and $Q(H^0)$, while the nebular observations provide us constraints on the nebular gas distribution and nebular abundances. Finally, the ionized SEDs predicted by the stellar atmosphere codes are being tested by comparing various diagnostic line ratios predicted by a photoionization code with the observed line ratios across the nebulae. We will present some results on this on-going project.

A Complete Spectroscopic Survey of O stars

Alfredo Sota (Universidad Autónoma de Madrid), Jesús Maíz Apellániz (Instituto de Astrofísica de Andalucía), Rodolfo Barbá (Universidad de La Serena), *et al.*

We have recently compiled the most complete Galactic O star catalog with accurate spectral types (Maíz Apellániz *et al.* 2004, *ApJS*, 151, 103), and now we are conducting a spectroscopic survey to observe all known Galactic O stars with $B < 14$ based on v2.0 of the catalog. The survey will be used for a number of purposes, such as a precise determination of the IMF for massive stars, the measurement of radial velocities for Galactic kinematic studies, and the detection of unknown massive binaries. Results will be made available through a dedicated web server, will be incorporated into the virtual observatory, and will include the most complete spectral atlas of massive stars to date. The northern part of the survey is being carried out from the Sierra Nevada Observatory (Spain), and the southern part from Las Campanas (Chile), La Silla (Chile), and CASLEO (Argentina). As part of our early results we have discovered 13 new massive spectroscopic binary systems.

The Nature of WC9 stars–Hints from Variability

Nicole St-Louis (Université de Montréal) & André-Nicolas Chené (Herzberg Institute of Astrophysics-Canadian Gemini Office)

We have carried out a spectroscopic variability survey of all apparently single Galactic Wolf-Rayet stars brighter than $V \simeq 13$ (e.g. Chené & St-Louis 2007, *ASP Conf* 367, 117). One intriguing result from this project is that *all* WC9-type stars in our sample present large-scale (more than 6% of the line flux) variability. Among these are three of the rare (only 10%) WC9 stars for which evidence of dust formation has *not* been found: WR 81, WR 88, and WR 92. These stars have been monitored in photometry for more than two decades without any evidence for episodic dust formation events (Williams & van der Hucht 2000, *MNRAS*, 314, 23). The variability we have found for these three stars is in the form of rather large excess emissions on the top of the normal wind lines that move across the profile on relatively short timescales (on the order of days). These can either be caused by the presence of co-rotating interaction regions in the wind of the Wolf-Rayet star or by colliding winds. In the latter case, the relatively short timescale of the changes would indicate that the orbital period is much shorter than that of other dust-making WC9 stars in binaries such as WR 140 or WR 137, which have periods on the order of several years. This might be a clue as to why these stars are dust free.

Observations of a New Massive Binary in the Cygnus OB2 Association.

Vanessa Stroud (LCOGTN/The Open University), Ignacio Negueruela (Universidad de Alicante) & Simon Clark (The Open University)

We present optical spectroscopy and photometry of a newly discovered binary system in the Cygnus OB2 Association. The system is likely to contain two O-type supergiants, one of them perhaps in a transitional state. If correct, this would make it one of the most massive binaries in the Galaxy.

A Near-Infrared View of the Homunculus Nebula Around η Carinae

M. Teodoro, A. Damineli (Instituto de Astronomia, Geofísica e Ciências Atmosféricas, Universidade de São Paulo), R. Sharp (Anglo-Australian Observatory), *et al.*

We present a near-infrared view of the Homunculus nebula around η Car using the Gemini/CIRPASS spectrograph. We show the velocity maps in the light of [FeII] 12567, which confirm the presence of a hole in the polar region of both lobes. Using this result and a model for the shape of the Homunculus, we could derive its thickness at the polar region, which is consistent with values found by other authors. Furthermore, we confirm the presence of regions with intrinsic components of HeI 10830, which might be formed by UV photons coming from the central source. Considering the Little Homunculus as a small HII region, we could use the radio flux to estimate the nature of the hot companion of η Car.

The Inner Wind Structure of Supergiant Stars

Jose M. Torrejon-Vazquez (Massachusetts Institute of Technology), Ignacio Negueruela (Universidad de Alicante) & David M. Smith (University of California, Santa Cruz)

The newly discovered phenomenon of the fast X-ray transients in supergiants is straightforwardly explained if the wind is highly structured in clumps. The existence of these hypothetical clumps, however, could be at odds with the persistent emission displayed by classical supergiant X-ray binaries. In this paper we present a coherent picture that unifies the emission behavior of all supergiant X-ray binaries in terms of the porous winds in the supergiant stars. Within the framework of the porous wind model, we discuss the implications of the available X-ray observations on the model parameters.

Winds of Magellanic Cloud B Supergiants

Carrie Trundle (Queen's University Belfast)

The analysis of radiation driven winds in B-type supergiants has highlighted discrepancies between the predictions of these massive star winds and observational results. These inconsistencies will have an important impact on the later stages of stellar evolution due to stellar winds being drivers of the evolution of massive stars during the core hydrogen burning phase. The discrepancies between observations and theory will be reviewed here along with some new results from the Magellanic Cloud B star population.

The Effects of Field-Aligned Rotation on Magnetically Channeled Line-Driven Stellar Winds

Asif ud-Doula (University of Delaware)

Based on a MHD simulation study of magnetic channeling in radiatively driven stellar winds, we examine here the dynamical effects of stellar rotation in the 2-D axisymmetric case of an aligned dipole surface field. We characterize the stellar rotation in terms of a parameter $W(=V_{\rm rot}/V_{\rm orb})$ (the ratio of the equatorial surface rotation speed to orbital speed). We find that rotation effects are weak for models with Alfven radius (RA) smaller than the Kepler co-rotation radius (RK), RA < RK, but can be substantial and even dominant for models with RA > RK. In particular, by extending our simulations to the very strong magnetic confinement case, we find that these do indeed show clear formation of the rigid-body disk predicted in previous analytic models, with however a rather complex, dynamic behavior characterized by both episodes of downward infall and outward breakout that limit the buildup of disk mass. Overall, the results provide an intriguing glimpse into the complex interplay between rotation and magnetic confinement, and form the basis for a full MHD description of the rigid-body disks expected in strongly magnetic Bp stars like σ Ori E.

Numerical Simulations of Continuum-Driven Winds of Super-Eddington Stars

Allard Jan van Marle, Stanley P. Owocki (University of Delaware) & Nir J. Shaviv (Hebrew University)

Continuum driving is an effective method to drive a strong stellar wind. It is governed by two limits: the Eddington limit and the photon-tiring limit. A star must exceed the effective Eddington limit for continuum driving to overcome the stellar gravity. The photon-tiring limit places an upper limit on the mass-loss rate that can be driven to infinity, based on the energy available in the radiation field of the star. Because continuum driving does not require the presence of metals in the stellar atmosphere, it is particularly suited to removing mass from low- and zero-metallicity stars and can play a crucial part in their evolution.

We compute numerical simulations of super-Eddington, continuum-driven winds using a porosity length formalism. We find that below the photon-tiring limit, continuum driving can produce a large, steady mass-loss rate at velocities on the order of the escape velocity.

If the star exceeds the photon-tiring limit, a steady solution is no longer possible. While the effective mass loss rate is still very large, the wind velocity is quite small. These objects will show a highly variable luminosity as radiation escapes from the dense circumstellar wind in short bursts. Since most of the radiative energy is used to maintain the high mass-loss rate, such a star would appear to be quite dim.

Since continuum driving can use a large fraction of the available energy, an accurate estimate of the luminosity of super-Eddington objects must include both mechanical and radiative luminosity.

High Angular Resolution Interferometric Observations of Evolved Massive Stars

Debra J. Wallace (College of Charleston), Douglas R. Gies (Georgia State University), William C. Danchi (Goddard Space Flight Center), *et al.*

Recent aperture-masking and interferometric observations of late-type WC Wolf-Rayet stars strongly support the theory that dust formation in these objects is a result of colliding winds in binary systems. To explore and quantify this possible explanation, and build on our high-resolution HST WFPC2 imaging results, we have conducted a high-resolution interferometric survey of late-type massive stars utilizing the VLTI, KI, IOTA, and FGS1r interferometers. We present here the first results from the MIDI instrument on the VLTI, and the KI and IOTA observations. Our VLTI study is aimed primarily at resolving and characterizing the dust around the WC9 star WR 85a and the LBV WR 122, both dust-producing but at different phases of massive star evolution. Our IOTA and KI interferometric observations resolve the WR star WR 137 into a dust-producing binary system.

The Progenitors of Gamma-Ray Bursts; A New Spectropolarimetric Survey of Galactic Wolf-Rayet Stars

G. Grant Williams (MMT Observatory), G. Schmidt & P. Smith (Steward Observatory)

We present results from a spectropolarimetric survey of 18 Galactic Wolf-Rayet (WR) stars that were not included in the Harries et al. (1998, *MNRAS*, 296, 1072) survey. Our observations increase the number of spectropolarimetrically studied Galactic Wolf-Rayet stars by more than 50%. The results from this survey are used to further characterize the progenitors of gamma-ray bursts (GRBs). The collapsar model is the most widely accepted model for producing GRBs. In this model, a massive star that has shed its hydrogen envelope undergoes core collapse, resulting in a black hole and an accretion disk. A preferred axis, which is generally attributed to rapid rotation, provides a path for the relativistic jet. This preferred axis may produce asymmetries in the geometrical mass-loss structure that can be measured with spectropolarimetry. Therefore, our results provide insight into the parameters and/or environment surrounding progenitors of GRBs.

The Massive LMC Binary [L72] LH 54-425

Stephen J. Williams, Douglas R. Gies, Todd Henry (Department of Physics and Astronomy, Georgia State University), *et al.*

We present results from an optical spectroscopic investigation of the massive binary system [L72] LH 54-425. We find an orbital period of 2.247409 ± 0.00001 days. We find spectral types of O3 V for the primary and O5 V for the secondary. We made a combined solution of the radial velocities and previously published V-band photometry to determine the inclination of the system, $i = 52^{+2}_{-3}$ degrees, and obtain radii and masses for each star in the system: $M_1 = 53^{+7}_{-4}\ M_\odot$ and $R_1 = 11.0^{+0.7}_{-0.3}\ R_\odot$ for the primary, and $M_2 = 30^{+4}_{-2}\ M_\odot$ and $R_2 = 9.7^{+1.0}_{-0.2}\ R_\odot$ for the secondary. Based on the position of the two stars plotted on a theoretical H-R diagram, we find the age of the system to match most closely with a ~ 2 Myr isochrone.

HD 45166: A Causal Connection between Photospheric and Wind Structure and Variability

Allan Willis (University College London)

What are the true mass-loss rates of O stars and WR stars? Currently, there are discrepancies of factors of 10–100 between UV and radio studies, with severe implications for stellar and galaxy evolution models. Inherent in modelling stellar wind emissions (UV P Cyg profiles, radio continua, etc.) is the realization that the winds are highly structured and variable. This is epitomized by the ubiquitous appearance of discrete absorption components (DACs) in OB and WR winds. However, the physical origin of DACs is unknown. HD 4516 is a low-mass, hot star, with $T_{\rm eff}$ recently determined from FUSE spectra to be 37000 K. IUE high-resolution spectra at many epochs show substantial DAC variability in CIV, NV, SiIV resonance P-Cyg profiles. In addition, the *photospheric* absorption lines in FeV show substantial strength changes that appear to be directly associated with those seen in the DACs. This is the first time that a direct observational link has been isolated between photospheric and wind activity—a facet probably linked to the relatively low mass-loss rate and wind density. It is suggested that this apparent photospheric-wind variation linkage in HD 45166 may also be operating in massive O and WR analogues. This may open up a route to explaining the physical origin of massive star wind structure and variability, and to a proper understanding of their true mass-loss rates.

Implementation of Plasma Emission Calculations into the Non-LTE Radiative Transfer Program, CMFGEN

Janos Zsargo & D. John Hillier (University of Pittsburgh)

We present preliminary X-ray emission results for ζ Puppis calculated by a stellar atmosphere code that combines CMFGEN (Hillier & Miller 1998, *ApJ*, 496, 407) and the Astrophysical Plasma Emission Code (APEC, Smith *et al.* 2001, *ApJ*, 556, L91). The merged codes will allow for the combined spectral analysis of all observable wavelengths (from X-ray to IR) and will provide stellar parameters, wind parameters, and the temperature and location of the X-ray emitting plasma. Furthermore, the attenuation of the X-ray emission by the cool wind will be self-consistently accounted for. Our progress, so far, includes the modification of APEC to take into account photo-excitation by UV radiation (calculated by CMFGEN) in the statistical equilibrium equations for the He-like ions. The resulting emission measures are then used to produce observed spectra.

SESSION II: Physics and Evolution of Massive Stars

Core Overshoot and Nonrigid Interior Rotation of Massive Stars: Current Status from Asteroseismology

Conny Aerts (Instituut voor Sterrenkunde, Katholieke Universiteit Leuven)

See p. 237 for full article.

Can Envelope Convection Zones in Hot Stars Cause Wind Clumping?

Matteo Cantiello (Astronomical Institute, Utrecht University), *et al.*

We study convection zones in the envelopes of hot massive stars. These regions are caused by opacity peaks associated with iron and helium ionization. Such convective regions can be very close to the stellar surface and we discuss possible implications. We argue that convection close to the surface may affect the stellar mass loss by triggering wind clumping. We present a simple model for this 'convection-driven clumping'.

End of Massive Stars and GRBs

Pascal Chardonnet (Université de Savoie)

In this poster, I will present a new scenario for gamma-ray bursts. I found observational evidence to support this theory. I will explain some enigmatic facts like the "Amati Relation," and I will draw some interesting perspectives in terms of global stellar evolution.

Homogeneous Evolving Stars in Close Binaries

S. E. de Mink, M. Cantiello & N. Langer (Utrecht University)

Rotational mixing is proposed to explain observed N and He enhancements in OB stars. If the rotation rate is of the order of 30% of the critical rotation rate or higher, rotational mixing can be so efficient that the star evolves chemically homogeneously: the star stays compact, gradually becoming an Wolf-Rayet star. This evolutionary scenario was proposed for the formation of long gamma-ray burst progenitors.

Rotation rates needed for homogeneous evolution can be achieved in close, massive, pre-mass transfer binaries (primary mass > 40 solar masses, periods 2–5 days), in which the tidal forces lock the rotation of the stars with the rotation of the orbit. Hence we consider the possibility of forming GRB progenitors in this way.

In less massive or wider binaries, current models predict N and He enhancements. These systems are suggested to constitute potentially stringent test cases for the physics of rotational mixing. We indicate 5 double lined eclipsing binaries in the SMC that may be particularly suitable for such a test.

Simulations of Magnetically Driven Supernova Explosions

Luc Dessart (Steward Observatory), *et al.*

I will present results from 2-D rotating, multi-group, radiation magneto-hydrodynamics simulations of supernova core collapse, bounce, and explosion. In the context of rapid rotation, magnetic stresses at the neutron-star surface lead to the creation and propagation of MHD jets that are powered by the energy extracted from the differentially rotating core. I will review the properties of the resulting ejecta and discuss the implications for the collapsar model of long-duration GRBs.

The Massive Star Newsletter

Philippe Eenens (Universidad de Guanajuato)

We present the newsletter of the IAU Working Group on Massive Stars. We retrace its history after 100 issues. We analyze its role and discuss its future.

Unveiling the Internal Structure of Massive Supergiants with Asteroseismology: Effect of Mass-loss

Mélanie Godart (Université de Liège), Marc-Antoine Dupret (Observatoire de Paris-Meudon), Arlette Noels (Université de Liège) et al.

Saio et al. (2006, *ApJ* 650, 1111) have detected p and g-modes in a B supergiant star HD 163899 with MOST. The presence of excited g-modes in a post-main sequence star is explained by the existence of a convective shell which prevents some modes from entering into the damping radiative core.

We show that this intermediate convective zone disappears when sufficient mass loss is included in the models. Hence, the non-radial p- and g-modes are not excited anymore. Our study is one of the first showing how asteroseismology allows us to get precise information on the physics of the very deep layers of massive stars.

New Measurements of Magnetic Fields in SPB and β Cephei Stars.

S. Hubrig (ESO, Chile), M. Briquet (Universiteit Leuven), M. Schöller (ESO, Chile), et al.

Our recent study of the evolutionary state of Bp and SPB stars indicates that Bp stars are younger that SPBs and stars with stronger magnetic fields have much lower pulsation amplitudes. We present the results of new magnetic field measurements of SPBs and beta Cephei stars which are important to establish the link between the presence of a magnetic field and other fundamental properties of pulsating stars.

Rotational Mixing in Rapidly Rotating Massive Stars

Ian Hunter (Queen's University Belfast), Ines Brott (Utrecht University), Danny Lennon (Isaac Newton Group of Telescopes), et al.

Rotation has become an important element in evolutionary models of massive stars, specifically the prediction of rotational mixing. Here detailed non-LTE surface chemical compositions are presented for 135 early B-type stars in the Large Magellanic Cloud. These objects have projected rotational velocities up to 300 $\mathrm{km\,s^{-1}}$ with over 40% having rotational velocities greater than 100 $\mathrm{km\,s^{-1}}$. This represents the largest sample to date of fast rotators with chemical abundance estimates and allows the effects of rotation on stellar evolution to be examined in detail. Specifically we find no significant evidence that the amount of material mixed into the surface is correlated with the rotational velocity. Indeed, we find both fast rotators that are not enriched and slow rotators that are highly enriched. These observations are in conflict with theoretical predictions. Additionally, the blue supergiants generally are more enriched than normal core hydrogen burning objects which suggests that their enrichment is not due to rotational mixing. We conclude that rotational mixing is not the dominant enrichment process in massive stars.

Black Hole Formation: Progenitors and Kicks

Vicky Kalogera, Tassos Fragos, Bart Willems (Northwestern University), *et al.*

In recent years, an increasing number of proper motions have been measured for Galactic X-ray binaries. When supplemented with accurate determinations of the component masses, orbital period, and donor luminosity and effective temperature, these kinematical constraints harbor a wealth of information on the systems' past evolution. We have developed an analysis that allows us to consider all this available information and reconstruct the full evolutionary history of X-ray binaries back to the time of core collapse and compact object formation. This analysis accounts for four evolutionary phases: mass transfer through the ongoing X-ray phase, tidal circularization before the onset of Roche-lobe overflow, motion through the Galactic potential after the formation of the compact object, and binary orbital dynamics at the time of core collapse. The constraints on compact object progenitors and kicks derived from this are of immense value for understanding compact object formation and exposing common threads and fundamental differences between black hole and neutron star formation. Here, we present the results of such an analysis for the black hole X-ray binary XTE J1118+480 (and also GRO J1655-40). Assuming that the system originated in the Galactic disk and the donor had solar metallicity, we find that a high-magnitude (>100 km s^{-1}) asymmetric natal kick is not only plausible but required for the formation of the system. We also investigate a globular cluster origin of XTE J1118+480 that would require a low-metallicity donor star. It turns out that such a scenario involves a lot of fine tuning and seems rather improbable.

Be Stars and Stellar Evolution

C. Martayan (Royal Observatory of Belgium, GEPI Observatoire de Paris), Y. Frémat (Royal Observatory of Belgium), A.-M. Hubert (GEPI Observatoire de Paris), *et al.*

In this poster we present the impact of the star-formation conditions and stellar evolution on the appearance of Be stars in environments with different metallicities. Using the observations obtained with the VLT-FLAMES, we focus on the incidence of the ZAMS rotational velocities and the metallicity of the environment on the evolution of B-type stars. Specifically, we show at what evolutionary status the Be phenomenon can appear in the Milky Way and in the Magellanic Clouds as a function of the stellar mass. To validate a given diagram of the evolutionary status of Be stars, it is important to observe very young clusters with emission line stars (ELS) present in environments with different metallicities and to distinguish clearly, which of them are classical Be stars and which are Herbig Be stars. To this purpose, we present new results on ELS in the very young open clusters NGC 6611, Trumpler 14, Trumpler 15, Trumpler 16, Collinder 232 (MW), and NGC 346 (SMC).

A Magnetosynthesis Model for Massive Stars

Mary E. Oksala & Richard Townsend (University of Delaware)

Magnetic fields in massive stars are often accompanied by inhomogeneous surface abundance distributions. It is well understood how rotational modulation of abundance inhomogeneities leads to spectroscopic and photometric variability. However, the possibility that the inhomogeneities can also impact magnetic field measurements has not yet been investigated. This motivated us to develop a new "magnetosynthesis" model for simulating the effective field strength of magnetic massive stars having arbitrary surface abundance distributions.

The effective field strength of a magnetic star is determined from spectropolarimetry as a weighted average of the field component along the line of sight. The weighting function is proportional to the local equivalent width (EW) of the absorption line used in the measurement, and therefore depends on the local surface abundance of the element responsible for the line. In our magnetosynthesis model, we specify the abundance distribution on a triangle-based mesh representing the stellar photosphere. For each triangular element, the EW is calculated from specific intensity data in a 5-dimensional (temperature, gravity, wavelength, angle, abundance) grid of pre-computed TLUSTY/SYNSPEC spectra. The effective field strength is then found by integrating the weighted longitudinal field component over all visible elements.

We present results from a preliminary application of the magnetosynthesis model to He-strong Bp stars. The principal findings are that (i) the shapes of effective field-strength curves depart from sinusoidal, and (ii) the amplitudes of these curves vary from element to element, and from line to line.

A Very Faint Core-Collapse SN in M85

A. Pastorello (Queen's University Belfast), M. Della Valle (INAF-Arcetri, Florence), S. J. Smartt (Queen's University Belfast), *et al.*

An anomalous transient in the early Hubble-type (S0) galaxy Messier 85 (M85) in the Virgo cluster was discovered by Kulkarni *et al.* (2007 *Nat*, 447, 458) on 7 January 2006 that had very low luminosity (peak absolute R-band magnitude M_R of about -12 mag) that was constant over more than 80 days, red colour and narrow spectral lines, which seem inconsistent with those observed in any known class of transient events. Kulkarni *et al.* (2007) suggest an exotic stellar merger as the possible origin. An alternative explanation is that the transient in M85 was a type II-plateau supernova of extremely low luminosity, exploding in a lenticular galaxy with residual star-forming activity. This intriguing transient might be the faintest supernova that has ever been discovered.

Massive Stars as Tracers for Stellar and Galactochemical Evolution

N. Przybilla, M. F. Nieva, M. Firnstein (Dr. Remeis Observatory Bamberg), *et al.*

Recent advances in the modelling of the atmospheres of massive stars, in particular by considering improved atomic data, allow stellar parameters and elemental abundances to be constrained with unprecedented accuracy: effective temperatures to $\sim$1%, surface gravity to $\sim$10–20%, and abundances to $\sim$10–20%, largely unbiased by systematic errors.

We discuss the implications of the modelling improvements for testing stellar evolution models observationally. This is done for a sample of stars from the OB-type main sequence to later evolution stages up to the BA-type supergiant phase in a homogeneous way. Particular emphasis is given to the light elements (He, CNO) as tracers of nuclear-processed material. The results imply that mixing appears to be more efficient than predicted by current evolution models accounting for mass loss and rotation, by a factor $\sim$2. This may favour scenarios with enhanced mixing due to the interaction of rotation and a magnetic field.

Abundances of the heavier elements, unchanged by nuclear burning, may be used for testing galactochemical evolution models. Our results imply a much higher homogeneity of elemental abundances from massive stars in the solar vicinity than reported before. This is finally in agreement with the chemical uniformity of the ISM and predictions of Galactic evolution models. Reference abundances for several astrophysically important elements are presented, which are often systematically higher than reported before. The study re-establishes the recently doubted status of massive stars as excellent proxies for present-day abundances. Implications for the derivation of Galactic abundance gradients are discussed.

SESSION III: Massive Star Populations in the Nearby Universe

W51 IRS2: A Compact HII Region in Detail

Cassio Barbosa (IP&D, Universidade do Vale do Paraíba, Brazil), Robert Blum (NOAO), Peter Conti (JILA), *et al.*

We present the first results of a recent campaign on W51 IRS2 carried out at Gemini Observatories. We derived mid-infrared fluxes and luminosities as well as line-of-sight extinction from images taken with T-ReCS at 7.7, 9.8, 12.3, and 24.5 μm. We report the detection of a new source at 24.5 μm. Near-infrared spectra taken with high spatial resolution show nebular lines on all selected sources. We report for the first time the detection of the first overtone CO band in emission in IRS2E, a typical signature of circumstellar disks.

A Downward Revision to the Distance of the 1806-20 Cluster and Associated Magnetar from Gemini Near-Infrared Spectroscopy

Joanne L. Bibby, Paul A. Crowther, James P. Furness (Department of Physics & Astronomy, University of Sheffield), *et al.*

We present H- and K-band GNIRS spectroscopy of OB and Wolf-Rayet (WR) members of the Milky Way cluster 1806-20, obtaining a revised cluster distance relevant to the 2004 giant flare from the SGR 1806-20 magnetar. We confirm four candidate OB stars as late O/early B supergiants and support previous mid WN and late WC classifications for two WR stars.

The distance modulus (DM) achieved from theoretical isochrone fitting using the age inferred by the stellar content is combined with the DM from Ks-band magnitude calibration for B supergiants and WR stars, to produce a cluster DM of 14.7 ± 0.4 mag ($8.7^{+1.8}_{-1.5}$ kpc). This is significantly lower than the 15 kpc to the magnetar and reduces the peak luminosity of the giant flare to 7×10^{46} erg s^{-1}, hence contamination of BATSE short gamma ray bursts from such events is reduced to 8%. We infer a magnetar progenitor mass of $\sim 48^{+20}_{-8}$ M$_\odot$, in agreement with the magnetar in Westerlund 1.

A Survey of the Most Massive Stars in the Local Universe

Alceste Z. Bonanos (Carnegie Institution of Washington, Department of Terrestrial Magnetism)

The physical parameters of very massive stars (>30 M$_\odot$) remain unexplored. The most accurate method for deriving masses, radii, and luminosities of such distant stars is to measure them in eclipsing binary systems. Currently, the most massive eclipsing binary known is WR20a, which consists of two 80 solar mass stars in a 3.7 day orbit. In total, only ~ 20 very massive stars (>30 M$_\odot$) belonging to our Galaxy and Local Group galaxies have accurate determinations of their parameters. I will present the first results of a wide-ranging survey targeting the brightest and thus most massive stars in eclipsing binaries in both young massive clusters in the Milky Way and in nearby galaxies. The measurement of fundamental parameters for massive stars at a range of metallicities will provide much needed constraints on theories that model the formation and evolution of massive stars and will observationally probe the upper limit on the stellar mass.

Massive Blue Stars in NGC 55

Norberto Castro, Artemio Herrero (Instituto de Astrofísica de Canarias, ULL), Miriam Garcia (IAC), *et al.*

We present the first spectral census of hot massive stars in NGC55, a spiral galaxy of the Sculptor group. Using VLT spectra taken with FORS2 in MXU mode, we have produced spectral classifications for 200 objects spread throughout the galaxy. The resulting catalogue will be published shortly. The metallicity and rotational velocity curve of the galaxy were estimated in an approximate way. This work is part of a larger project to determine stellar parameters, evolutionary status and abundances of B-supergiants in NGC55, and to use the results to probe the chemical composition of the galaxy.

The Massive Galactic Red Supergiant Clusters

Ben Davies, Don Figer (Rochester Institute of Technology), Rolf-Peter Kudritzki (Institute for Astronomy, University of Hawaii), *et al.*

I present the recent discoveries of two Galactic massive young clusters, which together contain 40 red supergiants (RSGs)—20% of all those known in the Galaxy—and as many in the entire Large Magellanic Cloud. From observations and evolutionary synthesis models, we argue that the cluster masses are comparable to the other Galactic "super star clusters" such as Westerlund 1 and the Arches Cluster. The distinctly different ages of the clusters, uniform metallicity, and large number of RSGs mean that these objects now offer an unprecedented opportunity to study the pre-supernova evolution of massive stars. Further, their location at the point where the Scutum-Crux spiral arm meets the bulge allows us to study the metallicity gradient at this location in the Galaxy, key to the constraining of Galaxy evolution models.

Mapping the 'Intimate' Spectral Properties of Gas Flows in the Orion Nebula: HH 202

César Esteban, Adal Mesa-Delgado, Luis López-Martín (Instituto de Astrofísica de Canarias), *et al.*

We present preliminary results on low-resolution 2-D and high-resolution echelle spectrophotometry of the head of the Herbig-Haro object HH 202 in the central part of the Orion Nebula. The 2-D maps, obtained with the Postdam Multi-Aperture Spectrophotometer (PMAS) at the 3.5m telescope in Calar Alto, show the spatial distribution of a large number of nebular properties (line fluxes, reddening coefficient, electron density and temperature, chemical abundances) permitting to carry out a thorough study of the effects of gas flows and shocks onto the nebular ionized gas. In addition, the echelle spectrum, obtained with the Ultraviolet Echelle Spectrograph (UVES) at the VLT in Cerro Paranal and covering from 3500 to 10000 Å, permits to resolve the nebular and shock components of the gas and study the physical conditions and chemical abundances of each kinematical component. One of the main aims of this work has been the measurement of very faint OII recombination lines, in order to compare the abundances of O^{++} determined from OII and [OIII] lines and, therefore, determine the so-called abundance discrepancy.

Spectroscopy of Resolved Stellar Populations with the E-ELT

Chris Evans (UK Astronomy Technology Centre) & Miska Le Louarn (ESO)

We present results from new adaptive optics simulations to illustrate the performance of the European Extremely Large Telescope (E-ELT). We also introduce EAGLE, one of the Phase A E-ELT instrument studies now underway, and how it will contribute to studies of resolved stellar populations.

Circumnuclear Activity in NGC 7469 as Revealed by Integral Field Spectroscopy with Subaru-Kyoto3DII

Takashi Hattori (Subaru Telescope, National Astronomical Observatory of Japan), Hajime Sugai (Kyoto University) & Hiroshi Ohtani (Okayama Astrophysical Observatory, NAOJ)

We present the results of an optical spectroscopic study of the circumnuclear region of the Seyfert 1 galaxy NGC 7469. High spatial resolution integral field spectroscopy performed with Subaru-Kyoto3DII allows us to investigate the spectroscopic properties of the nuclear starburst ring without contamination from the bright nucleus.

For the first time in NGC 7469, we detect a strong Wolf-Rayet emission feature at 4650 Å from a small part of the ring, indicating the presence of a few thousand late-type WN stars. Spectral properties of this region, located at the southern edge of the ring and coincident with a bright MIR peak, are consistent with a moderately reddened, young (several Myr), metal-rich, instantaneous starburst population. We also identify a star-forming region that is characterized by metal absorption lines such as FeII and MgII, suggesting an older (several 10 Myr) stellar population.

The high spatial resolution and careful analysis of the data cube also allow us to investigate the spatial structure of emission lines inside of the starburst ring. As a result, we find a spatially extended, highly blueshifted (~ -1000 km s^{-1}) component in [SII] and [OIII] lines. In [OIII], the high velocity component is bright at the nucleus and extends toward the northeast, which corresponds to the direction of the minor axis of the nuclear gas disk. In contrast, it is distributed at the north to southwest side of the nucleus in [SII].

These observational results provide important clues to understanding the star-forming activity and outflow in this system, and demonstrate the importance of spatially resolved spectroscopy.

Star Formation in the Outer Disks of Dwarf Galaxies

Deidre Hunter, Bonnie Ludka (Lowell Observatory) & Bruce Elmegreen (IBM T. J. Watson Research Center)

Outer edges of dwarf galaxies present an extreme environment for star formation. Dwarf galaxies already challenge models of star formation because of their low gas densities even in the central regions. Outer parts of dwarfs, where the gas density is even lower, therefore, present a particularly difficult test of our understanding of the cloud/star formation process. Yet, we see that stars have formed in the outer parts to very low surface brightness levels. We are using UV images obtained with the GALEX satellite to trace and characterize star formation in the outer disks of dwarf galaxies where Hα may not be an effective tracer of recent star formation. Here we compare the far-UV surface brightness profiles to those of Hα and to those of the older stars.

On the Binary Characteristics of Massive Stars

Chip Kobulnicky (University of Wyoming) & Chris Fryer (Los Alamos National Laboratory)

We compare radial velocity survey data on 114 early-type stars in the 2–3 Myr old Cygnus OB2 Association with the expectations of Monte Carlo models based on several popular binary system prescriptions to constrain the properties of massive binaries. We explore a range of true binary fraction, F, a range of power-law slopes, alpha describing the distribution of companion masses, between the limits $q_{low} < 1$ and a range of power law slopes, beta, describing the distribution of orbital separations, between the limits r_{in} and r_{out}. We also consider distributions of secondary masses described by a Miller-Scalo type initial mass function (IMF) and by a two-component IMF which includes a substantial "twin" population with $M_2 \sim M_1$. If the distribution of orbital separations is not far from the canonical Opik's Law distribution (i.e., flat; beta=0), several seemingly disparate formulations of the massive binary characteristics can be reconciled by adopting carefully specified values for F, r_{in} and r_{out}. We show that binary fractions $F < 0.7$ are less probable than $F > 0.8$ for reasonable choices of r_{in} and r_{out}. The secondary star mass function cannot be drawn from a Miller-Scalo-like IMF unless the lower end of the mass function is truncated below $\sim$2–4 $M_\odot$. A Salpeter or Miller-Scalo IMF extending to low stellar masses can produce sufficiently large radial velocity variations consistent with the data if there exists a substantial "twin" population with $q \sim 1$ comprising $\sim$40% of all systems.

A Spectroscopic Study of G61.48+0.09

A. Lenorzer, A. Herrero, A. Marin (Instituto de Astrofísica de Canarias), *et al.*

We present a study of the obscured Galactic cluster G61.48+0.09, based on *JHK* photometry and low-resolution *HK* multiobject spectroscopy obtained with LIRIS attached to the WHT. Based on this information, we investigate whether G61.48+0.09 is a nearby cluster (at about 2.5 kpc) or a far one (at about 8 kpc). In the first case, its mass (assuming a Salpeter IMF) is relatively modest (1000 solar masses), while in the second case, it is a very massive cluster (20000–40000 solar masses). Other intriguing aspects, like the main ionization source of G61.48+0.09 are also explored.

The Quintuplet Cluster

Adriane Liermann, Wolf-Rainer Hamann & Lida M. Oskinova (Institute for Physics, Potsdam University)

We present new near-infrared integral-field spectroscopy data of the central region of the Quintuplet cluster that is located about 30 pc (projected distance) from the Galactic center. The Quintuplet cluster is young (3–5 Myr old) and one of the most massive star clusters in the Galaxy with a rich population of hot massive stars. The observations were obtained with the ESO VLT SINFONI-SPIFFI instrument covering the inner parts of the Quintuplet cluster with 22 FOVs of 8×8 arcsec in the spectral range of 1.94 to 2.45 μm (*K*-band). The 3-D data cubes from the observations are flux-calibrated with standard stars and combined into a contiguous cube from which the spectra of all detected point sources are extracted. The spectral atlas and catalog of the sources, with coordinates and spectral classification, will be published soon (Liermann *et al.*, *A&A*, in prep.). We report the identification of two new Wolf-Rayet star candidates, adding to the 11 already-known

WR stars in the Quintuplet cluster. These newly discovered WR stars are of spectral type WC9 or later.

The quantitative analysis of the extracted spectra is underway. Using the Potsdam Wolf-Rayet (PoWR) model atmosphere code, we will determine the parameters of all O-, B-, and WR-type stars in the cluster. This will shed light on the massive star population of the Quintuplet cluster, its formation and evolution.

Accurate Distances to Nearby Massive Stars with the New Reduction of the Hipparcos Raw Data

Jesús Maíz Apellániz, Emilio Alfaro (Instituto de Astrofísica de Andalucía-CSIC) & Alfredo Sota (Universidad Autónoma de Madrid)

The new reduction of the Hipparcos raw data (Van Leeuwen 2007, *A&A*, 474, 653) has reduced its parallax uncertainties up to a factor of 4 for bright stars. In this work we use the new data to recalculate the spatial distribution of massive stars in the solar neighborhood and to provide for the first time accurate trigonometric distances for several tens of massive stars. We will discuss alternative measurements and will show that previous issues with the Hipparcos parallaxes (e.g., the distance to the Pleiades) have been significantly reduced or altogether eliminated.

The Open Cluster Pismis 11 and Its Hypergiant

Amparo Marco & Ignacio Negueruela (University of Alicante)

We present photometry and spectroscopy of stars in the open cluster Pismis 11, which contains one of the brightest hypergiants in the Galaxy, the B2Ia$^+$ star HD80077. We calculate cluster parameters and hence, the distance and luminosity of the hypergiant.

A Systematic Search for Stellar Clusters in the Galactic Plane

Maria Messineo (Rochester Institute of Technology)

The identification of stellar clusters in the plane of our Galaxy is of primary importance for gaining a better understanding of the Galactic structure and current star formation rate as well as to identify and characterize the most massive stars in the Galaxy. However, such a study is strongly hampered by interstellar extinction. Several hundred new Galactic stellar clusters have been discovered in the last few years thanks to large near- and mid-infrared surveys, such as 2MASS and GLIMPSE. However, their census remains highly incomplete. I will give an overview of the total number of stellar clusters known and present a list of a dozen of candidate massive stellar clusters towards the inner Galaxy, one of them a new discovery. The clusters have been selected on the basis of their infrared color-magnitude diagrams. We are currently following up these results with NIR spectroscopy to obtain a spectral classification of the brightest and probably more massive stars of each cluster and therefore to better characterize the nature of the cluster.

The Massive Star Content of NGC 1140

Sarah Moll (University of Sheffield), Sabine Mengel (European Southern Observatory), Richard de Grijs (University of Sheffield), *et al.*

We use new UVES spectroscopy and HST imaging to study the massive stellar content of the Wolf-Rayet (WR) dwarf starburst galaxy NGC 1140. We obtain $12 + \log(O/H) = 8.3$

from a nebular analysis of the brightest knot of the central giant HII region. We find that cluster 1 within this knot has an age of ~5 Myr and a mass of 1×10^6 $M_\odot$, from which 6000 O stars are inferred. Fitting the blue WR bump with LMC metallicity WR templates, we estimate that cluster 1 contains 550 late WN and 200 early WC stars. In common with other studies of metal-poor WR clusters, we estimate N(WR)/N(O) ~ 0.1, a factor of two higher than instantaneous burst predictions from Geneva or Padova evolutionary models.

The Multi-Wavelength Picture of the Open Cluster Westerlund 2 and the Very Massive Binary WR 20a

Yaël Nazé & G. Rauw (Université de Liège)

Westerlund 2 is a young open cluster in a blowout region of the HII region RCW 49. The cluster contains a dozen O-type stars and WR20a, the most massive binary (83+84 $M_\odot$) known to date. Recently, we have undertaken a large, multiwavelength monitoring of the cluster. In the optical domain, the new photometric and spectroscopic data have led to a significant revision of the massive star content (with several stars shown to be as early as O3-4) and the discovery of additional eclipsing binaries. In addition, the deepest exposure in the X-ray domain was obtained with the Chandra satellite. This unprecedented X-ray observation reveals a wealth of sources in the cluster. In this presentation, we will focus on the very first spatially resolved X-ray data of the colliding wind binary WR20a, and we highlight the results obtained for the other massive objects in the cluster.

Formation of High-Mass Stars in Violent Cluster Environments

Dieter E. A. Nürnberger (European Southern Observatory, Chile)

High-mass stars are usually forming deeply embedded in their natal environment, which can be penetrated only at wavelengths beyond the mid IR. In my presentation, I will summarize our recent efforts to search for and to characterize high-mass protostars in interfaces between Galactic HII regions and their adjacent molecular clouds, e.g. in NGC 3603 (Nürnberger 2003, *A&A*, 404, 255) and in M17 (Chini et al. 2004, *A&A*, 427, 849; Chini *et al.* 2006, *ApJ*, 645, L61; Hoffmeister *et al.* 2006, *A&A*, 457, L29; Nielbock *et al.* 2007, *ApJ*, 656, L81; Nürnberger *et al.* 2007, *A&A*, 465, 931). Taking advantage of 'curtain-lifting' stellar winds and energetic photons from the central clusters of early-type main sequence stars and making use of sensitive, high angular resolution observations in the near and mid IR, we have identified promising candidates that play a decisive role in our understanding of the basic formation processes of high-mass stars. In particular, as we see strong evidence for the existence of (accretion) disks around these sources, one has to favour the accretion scenario against the collision (coalescence) scenario.

LBT Discovery of a Yellow Supergiant Eclipsing Binary in the Dwarf Galaxy Holmberg IX

J. L. Prieto (Ohio State University), *et al.*

See p. 333 for full article.

Massive Stars in Young Dense Clusters

Fred Rasio (Northwestern University), *et al.*

We will present results from recent N-body simulations of massive stars in young dense clusters, focusing on the role of stellar collisions, binaries, and the possibility of runaway growth through successive collisions.

The Molecular Gas in High-Density Environments: From Submillimeter (mid-J) to Far-Infrared (high-J) Observations

J. Ricardo Rizzo (Laboratorio de Astrofísica Espacial y Fisica Fundamental-INTA, Spain) & F. M. Jiménez-Esteban (Observatorio Astronómico Nacional, Spain)

The study of molecular line emission surrounding evolved massive stars (mostly Wolf-Rayet and LBV) has been developed in the last decade. The role played by the molecules in the whole feedback between massive stars and their surroundings has changed from merely testimonial to becoming a significant fraction of it.

In this contribution we report the first detection of mid-J CO lines surrounding LBVs and Wolf-Rayets ever recorded. The main advantages of these lines are the high critical density and the high energy of the lower level. We have detected high density (from 10^4 to several 10^5 cm^{-3}) and warm (above 70 K) molecular gas surrounding ionized nebulae. The detection are not only supported by the morphology, but also by the dynamics, including the identification of low-velocity shockfronts (less than 20 km s^{-1}). Masses span from a few solar to 60 solar masses.

Our first results may serve as templates for future studies in three directions: (1) they can help to improve the numerical models about the interplay between massive stars and their CSM; (2) they can provide inputs for high-J observations which will help us to learn about the excited molecular gas; (3) they can be used as guides for the search in other cases, particularly stars of similar evolutionary stages. The future in this field concerns the search for vibrationally excited H$_2$, high-J CO emission, complex molecules, and PAHs. The advent of new high-sensitivity IR and submillimeter instrumentation will give support to this ample set of new observational lines.

Massive Star Binary Fraction in Nearby Open Clusters

H. Sana (European Southern Observatory, Chile)

While it is generally accepted that most (if not all) massive stars form in a cluster environment, the relation (if any) between the properties of a cluster and those of the O star population it hosts remains unclear. We revise here the properties of the massive stars in nearby young open clusters, with an emphasis on the binary fraction and on the orbital parameters distribution. The aim pursued is to provide accurate observational constraints, that should serve as guidelines for the theories of massive star formation and evolution,

Spitzer and Near-Infrared Imaging of the Massive Protostar IRAS 20126+4104 and Its Shocked Outflow

Steve Skinner & John Bally (University of Colorado), Manuel Güdel (Paul Scherrer Institut), *et al.*

Previous observations of the massive protostar IRAS 20126+4104 have revealed complex structure including a candidate massive circumstellar disk, radio jets, an extended (and possibly precessing) bipolar outflow, and maser emission. High-resolution Gemini images reveal a tightly spaced cluster of 18.3 μm sources near the methanol maser positions (De Buizer 2007, ApJ, 654, L147). We present Spitzer IRAC observations of IRAS 20126+4104 and ground-based images of the outflow in the H_2 S(1) (2.12 μm) and [Fe II] (1.64 μm) lines. The IRAC images at 3.6, 4.5, and 5.8 μm show emission extending to the northwest and southeast of IRAS 20126+4104, but no peak is seen at the protostar. However, a dominant peak in the IRAC 8 μm image is offset by only about 1 arc-second from the methanol maser and 18.3 μm source group, and is likely associated with a heavily obscured protostar. The H_2 image traces shocks extending for over 2 arc-minutes from IRAS 20126+4104 in a S-shaped pattern. This S-shaped pattern was also noted in CO (3-2) maps obtained by Su et $al.$ (2007, ApJ, 671, 571) and was interpreted in terms of a collimated precessing outflow. However, we note that the symmetric placement of several H_2 knots about stars embedded in the extended IRAS 20126 core region indicates that multiple outflow sources may be present and could confuse the interpretation of the outflow morphology.

Distance Determination for the Brightest Cool Stars in Galaxies Using the Wilson-Bappu and Wing Emission Line Correlations

Robert Stencel (Denver University)

An empirical correlation between the FHWM of the emission core of the CaII K-Line at 3933 Å and the intrinsic luminosity among late-type dwarf, giant, and supergiant stars was published first by Wilson & Bappu (1957, ApJ, 125, 661). Later on, Stencel (1977, ApJ, 215,176) extended this luminosity calibrator by using so-called wing emission lines in the wings of the H and K lines. Efforts to extend these techniques to the brightest supergiants in Local Group galaxies were frustrated by the limits of photographic coudé spectra even on 4-meter telescopes at the time. With the advent of CCD spectra and S/N possible with 8-meter telescopes, I advocate the potential for extragalactic hypergiant star distance calibration.

Using the Paranal Observatory library of high-resolution spectra (www.sc.eso.org/-santiago/uvespop/) obtained from the UVES instrument at an ESO Very Large Telescope, we measure the line widths of the CaII H and K lines and the wing emission lines in late-type stars. By plotting the measured FWHM and absolute magnitude, we re-evaluate the Wilson-Bappu line-width-to-luminosity correlation for the K core emission and $H - K$ core wing emission lines. Because the $H - K$ wing emission lines remain visible in very luminous stars cooler than F1 whereas circumstellar absorption obliterates the core emission, the line-width-to-luminosity correlation may be useful in estimating intrinsic luminosity for these stars, particularly in comparison with new angular diameters possible with interferometry.

Massive Stars in the Galactic Center as Traced by Compact HII Regions

Susan Stolovy (Spitzer Science Center/CalTech), et $al.$

We investigate massive star formation in the Galactic Center with a study of compact HII region candidates identified in a Spitzer/IRAC survey. The survey covers approximately 3 square degrees in 4 channels at 3.6, 4.5, 5.8, and 8.0 μm, with a spatial resolution of 2 arcsec. These candidate compact HII regions exhibit strong 8 μm emission, are compact (typically <20′), have a variety of morphologies, and are strongly clustered toward the Galactic plane. A search is conducted for counterparts at other wavelengths.

Quantitative Spectral Analysis of BA Supergiants in M33

Vivian U, Rolf P. Kudritzki, Miguel A. Urbaneja (Institute for Astronomy, University of Hawaii), *et al.*

We present atmospheric parameters and metallicities for a sample of 10 late-B and early-A supergiants in M33. High-resolution spectra have been obtained with DEIMOS on Keck II during a 2003 observing campaign. Spectral types, initially estimated by a qualitative comparison between models and observed spectra, were confirmed by numerical analysis of several spectral lines; luminosity class was obtained from measuring the equivalent width of Hγ. We measure the radial and rotational velocities empirically and provide well-characterized spectra for our objects. We use a model grid of 1600 detailed non-LTE spectra varying in T_{eff}, g, and metallicity parameter space to determine stellar parameters and metallicity, and then we discuss the evolutionary status of our objects and the metallicity gradient in M33. Finally, we use the results to investigate the relationship between stellar luminosity and flux-weighted gravity and compare them with the findings from Kudritzki *et al.* (2003, *ApJ*, 582, L83; Kudritzki 2007, *AAS*, 210, 40.03).

Discovery of Two Dust Pillars near the Galactic Plane

Leonardo Ubeda (Université Laval) & Anne Pellerin (Johns Hopkins University)

We report the discovery of two dust pillars using GLIMPSE archival images obtained with the Infrared Array Camera on board the Spitzer Space Telescope. They are located close to the Galactic molecular cloud GRSMC45.453+0.060, and they appear to be aligned with the ionizing region associated with GRSMC45.478+0.131. Our three colour mosaics show that these stellar incubators present different morphologies as seen from planet Earth. One of them shows the unquestionable existence of young stellar objects in its head, whose influence on the original cocoon is evident, while the other presents a well defined bright-rimmed ionizing front. We argue that second-generation star formation has been triggered in these dust pillars by the action of massive stars present in the nearby H II regions.

FGLR Distance to WLM

Miguel A. Urbaneja, Rolf Kudritzki, Fabio Bresolin (Institute for Astronomy, University of Hawaii), *et al.*

This contribution presents the first practical application of the flux-weighted gravity–luminosity relation of BA Supergiants (Kudritzki, Bresolin & Przybilla 2003, *ApJ*, 582, L83), to determine the distance to the Local Group galaxy WLM. The FGLR distance is consistent with the recent result obtained by the ARAUCARIA project based on Cepheids. The potential application of the FGLR to other galaxies is also discussed.

The Open Cluster Berkeley 90

Ana Ursúa, Amparo Marco, Ignacio Negueruela (University of Alicante), *et al.*

We present a deep optical and infrared photometric study of the area around the young open cluster Berkeley 90 and spectroscopy of its OB population. In spite of the presence of two early O-type stars, we find a low number of members earlier than B3. We find evidence for triggered star formation in the vicinity of the cluster.

A Near-IR Imaging Survey of Intermediate- and High-Mass Young Stellar Outflow Candidates

Watson P. Varricatt, Christopher J. Davis (Joint Astronomy Centre, Hilo, Hawaii), Suzanne K. Ramsay Howat (Institute of Astronomy, Royal Observatory, Edinburgh), *et al.*

We have carried out a near-infrared imaging survey of luminous young stellar outflow candidates using the United Kingdom Infrared Telescope. Observations were obtained in the near-infrared K band (2.2 μm) and at the wavelengths of H_2 (2.122 μm) and Br_γ (2.166 μm) lines. Fifty regions were imaged with a field of view of $2.2' \times 2.2'$. Seventy-four percent of the objects exhibited H_2 emission, and 50% exhibited aligned H_2 emission features implying collimated outflows, many of which are new discoveries. These observations imply that accretion is probably the leading mechanism in the formation of stars at least up to early B and even late O type. The YSOs responsible for many of these outflows are positively identified in our near-IR images based on their locations with respect to the outflow lobes, 2MASS colours, and association with MSX, IRAS, millimetre, and radio sources. The close association of the molecular outflows detected in CO with the H_2 emission features produced due to shock excitation by the jets from the YSOs implies that the outflows in these objects are jet driven. Br_γ emission is not detected in any of the outflows; it is therefore a poor tracer of outflows.

Aligned Circumstellar Disk Systems in Young Clusters

John P. Wisniewski (Goddard Space Flight Center), Karen S. Bjorkman (University of Toledo), Antonio M. Magalhaes (University of São Paulo), *et al.*

We have obtained and analyzed intrinsic polarization observations of Small Magellanic Cloud and Large Magellanic Cloud classical Be circumstellar disk systems. We found evidence of a single, cluster-wide preferential orientation of disk rotational axes in 2 of 11 (18%) clusters. The statistical significance of these trends has been confirmed via use of the Kuiper statistical test. For NGC 1948, the common orientation of disk rotational axes is parallel to the projected direction of the cluster's local magnetic field, whereas aligned disk systems in NGC 2100 do not have their rotational axes oriented parallel to the direction of the local magnetic field. We discuss the mechanisms that might be responsible for producing the observed alignment of disks in these two cluster environments.

The Physics of the Stellar Upper Mass Limit

Hans Zinnecker (Astrophysikalisches Institut Potsdam)

Is there an upper mass limit to the stellar IMF? And if yes, is it determined by stellar evolution physics (mass-loss mechanisms) or by star formation physics (mass accretion despite radiation pressure)?

Observational evidence suggests that there is a physical rather than a statistical upper limit to the mass of massive stars in rich young clusters such as R136 in the LMC (Weidner & Kroupa 2004, *MNRAS*, 348, 187; Oey & Clarke 2005, *ApJ*, 620, L43; Koen 2006, *MNRAS*, 365, 590) and the Arches cluster near the Galactic center (Figer 2005, *Nat*, 434, 192).

While in the past, pulsational instabilities near the Eddington limit were blamed for an upper mass limit (e.g., Smith & Owocki 2006, *ApJ*, 645, L45, for a review of η Car and other LBV stars), the current discussion focusses more on the mass accretion physics in the face of radiation pressure and the fragmentation or photo-evaporation of massive circumstellar disks (Keto & Wood 2006, *ApJ*, 637, 850; Kratter & Matzner 2006, *MNRAS*, 373, 1563; Krumholz 2006, *ApJ*, 641, L45). Cooperative rather than competitive accretion in dense ionized embedded OB clusters (Keto 2007, *ApJ*, 666, 976) and the merging of young massive close binaries (Bonnell & Bate 2005, *MNRAS*, 362, 915; Bally & Zinnecker 2005 *AJ*, 129, 2281) have also been suggested as important processes to limit the potentially infinite growth of massive stars.

The literature and the physical arguments are reviewed in a recent *Annual Reviews* article by Zinnecker & Yorke (2007, *ARA&A*, 45, 481).

In this contribution, I will present a summary of the physics of the upper mass limit of star formation, based on the *Annual Reviews* article, and also discuss future observational tests of the theoretical predictions (e.g., metallicity-dependence).

SESSION IV: Hydrodynamics and Feedback from Massive Stars in Galaxy Evolution

Winds of Embedded O Star Clusters

Sara Beck (Tel Aviv University)

Single O stars and WR stars drive powerful winds. How, then, do the small, dense supernebulae of embedded O star clusters persist long enough to be observed? Is it overpressure, gravitation, slow mass-loaded winds, or a combination? What does the IR and radio data on clusters suggest about this?

SPH Simulations of Star Formation Triggered by Expanding HII Regions

Thomas G. Bisbas, Anthony P. Whitworth & Richard Wunsch (School of Physics and Astronomy, Cardiff University)

We introduce a new 3-D SPH algorithm to study the evolution of HII regions. The algorithm constructs a set of rays around the source of ionizing photons using HEALPix (Górksi *et al.*, 2005, *ApJ*, 622, 759). Along each ray we locate the ionization front by applying the condition for photoionization equilibrium. We split the rays adaptively so that their angular separation is always comparable to the particle sizes in their neighbourhood. This ensures the necessary resolution of ray tracing while keeping the computational costs low.

We have used this new algorithm to study the properties of fragments formed by self gravity in the dense shells swept up by expanding HII regions. These simulations demonstrate that under suitable conditions the shell breaks up into massive ($\gtrsim 10\,M_\odot$) fragments some of which are likely to collapse to form massive stars. Therefore this is a viable mechanism for sequential self-propagating star formation.

NLTE Hydrostatic Equilibrium Solutions for Viscous Keplerian Disks

J. E. Bjorkman (University of Toledo) & A. C. Carciofi (University of São Paulo)

We investigate the interplay between the temperature structure and the geometrical structure of the disks around hot stars. Observational evidence suggests that these disks are Keplerian (rotationally supported) gaseous disks. The essential physics that determines the geometrical structure of Keplerian disks is reasonably well understood in the case of pre-main-sequence stars. The primary result is that the disks are hydrostatically supported in the vertical direction, while the radial structure is governed by the viscous transport. Since the disk is pressure-supported in the vertical direction, the geometrical structure (flaring) of the disk is determined by the radiative equilibrium temperature. Similarly the viscous transport of material is temperature dependent, so the radial density structure also depends on the radiative equilibrium temperature. To investigate the coupling between the temperature and density structures, we performed 3-D NLTE Monte Carlo simulations of the radiative transfer and solved self-consistently for the temperature and density structure of the disk. These simulations also calculate the emergent spectrum and its polarization. We find that the hydrostatic solution for the disk departs significantly from the often-assumed isothermal structure and simple power laws, with significant observational effects on the emergent spectrum.

Probing Variable Circumstellar Disks with Contemporaneous Optical and IR Spectroscopy

Karen S. Bjorkman, Erica N. Hesselbach (University of Toledo), John P. Wisniewski (Goddard Space Flight Center), *et al.*

Asymmetric double-peaked hydrogen emission line profiles in classical Be stars have been interpreted as evidence of one-armed density waves in the circumstellar disks. Contemporaneous optical and IR spectroscopy can aid in mapping the density structure of these putative one-armed waves as a function of radius. Variability has been detected in these stars over both short (days to weeks) and longer (months) timescales. We present preliminary results from contemporaneous Ritter Observatory (Hα) and IRTF SpeX (0.8–5.4 μm) spectroscopy of a selection of classical Be stars observed between 2004 and 2007. The data illustrate a range of line profiles common in Be stars and show significant variability, present in both the optical and IR lines. By combining with detailed models, these observations can be used to investigate the physical density distribution, temperature structure, and variability of the circumstellar disks.

The Carina Nebula: A Laboratory for Feedback and Triggered Star Formation

Kate Brooks (CSIRO Australia Telescope National Facility) & Nathan Smith (Astronomy Department, University of California, Berkeley)

The Carina Nebula (NGC 3372) is our richest nearby laboratory in which to study feedback through UV radiation and stellar winds from very massive stars during the formation of an OB association, before supernova explosions have disrupted the environment. In Carina, this feedback is triggering successive generations of new star formation around the periphery of the nebula, while simultaneously evaporating the gas and dust reservoirs out of which young stars are trying to accrete material. Carina is currently powered by UV radiation from 64 O stars and 3 WNL stars but for most of its lifetime when its most massive star (η Carinae) was on the main-sequence, the Carina Nebula was powered by 69 O stars that produced a hydrogen ionizing luminosity 200 times stronger than the Orion Nebula. At a distance of only 2.3 kpc, Carina has the most extreme stellar population within a few kpc of the Sun, and suffers little interstellar extinction. In this poster I will present a census of the Carina Nebula.

Circumstellar Medium around Massive Stars

Sabina Chita & Norbert Langer (Astronomical Institute Utrecht)

Massive stars interact with their surroundings by emitting winds and ionizing photons. Here we simulate the evolution of the circumstellar medium around stars of 12 solar masses from their birth up to the supernova stage. These stars are expected to expand at least twice into red supergiants with intermediate blue stages where fast winds are emitted. Stellar wind anisotropies expected during the blue loop will give rise to a latitudinal dependence of the shell structures as observed in Sher 25 and SN 1987A. We utilize the stellar parameters as function of time from detailed stellar evolution calculations as input for our hydrodynamic models.

Filament Formation and X-ray Emission in Starburst Winds

Jackie L. Cooper (Research School of Astronomy and Astrophysics, The Australian National University), *et al.*

Galactic winds are an important component of feedback processes in galaxy formation and the enrichment of the intergalactic medium. We have performed a series of three-dimensional simulations of a starburst-driven galactic wind in an inhomogeneous interstellar medium. We find that the emission-line filaments, which are a spectacular feature of starburst winds, are formed from disk gas that has been accelerated into the outflow by the ram-pressure of the wind. Chandra observations of starburst winds have revealed a close spatial relationship between the filamentary gas and the soft X-ray emission. While higher resolution simulations are required in order to determine the importance of mixing processes, we propose four mechanisms that give rise to soft X-ray emission that is also naturally correlated with the filaments: (i) Mass-loading from ablated clouds, (ii) the intermediate temperature interface between the hot wind and cool filaments, (iii) bow shocks upstream of clouds accelerated into the outflow, and (iv) interactions between these bow shocks.

Kinematics of Small Starbursts: NGC 2363 and NGC 5253

Laurent Drissen, Leonardo Ubeda, Maxime Charlebois (Université Laval), *et al.*

I will present GMOS/IFU observations of young bursts of star formation in the nearby dwarf galaxies NGC 2366 and NGC 5253.

A High-Resolution Study of SN Ejecta and Circumstellar Mass-Loss Interactions

Robert Fesen & Jordan Zastrow (Dartmouth College)

We present HST images of the outer edges of the young, core-collapse supernova remnant Cas A which reveal dozens of small (<1 arcsec) knots of a SN's metal-rich, high-speed ejecta interacting with a clumpy, N-rich circumstellar mass loss medium. Such SN ejecta–CSM interactions are seen to give rise to substantial ejecta knot brightness variations on timescales as short as one year, along with trailing ablation emission "tails" extending along the direction of a knot's motion. We will primarily present and discuss SN ejecta–CSM interactions occurring along the remnant's eastern limb and in the northeastern "jet" containing especially high-velocity ejecta.

BLAST Observations of the Cassiopeia A Supernova Remnant

Peter Hargrave (Department of Physics & Astronomy, Cardiff University), *et al.*

We present BLAST observations at 250, 350, and 500 μm of the SNR Cassiopeia A (Cas A), and its surroundings. We find that the SED in the direction of Cas A is best fit by a two-temperature modified black-body function, whilst the surrounding cloud region SED can be fit by a single temperature. The temperature of the cloud region is found to be the same as the temperature of the cold component of the fit in the direction of Cas A. This suggests that the cold component of the Cas A SED may be associated with an extended foreground cloud structure, rather than with the remnant. This result does not exclude the possibility of cold dust associated with the remnant, but sets upper limits on the mass of any dust which may be present.

The Circumstellar Structure and Massive Progenitors of Interacting Supernovae

Jennifer L. Hoffman (University of Denver)

In the past few years, more and more supernovae whose spectra show signatures of interaction with circumstellar material ejected by their evolved progenitor stars have been discovered. Studying the environments of these "interacting supernovae," which include members of the Type IIn subclass, can yield important information about core-collapse progenitors and the role of mass loss in the end stages of massive stellar evolution. One obstacle to understanding these supernovae has been their heterogeneity as a group; the IIn supernovae alone span a broad range of spectral characteristics, light-curve morphologies, intrinsic brightnesses, and many other properties. Spectropolarimetric observations provide a way to break the degeneracies often inherent in supernova spectra; analysis of polarized spectra may thus hold the key to subdividing the category of interacting supernovae and thereby constraining the properties of their progenitors.

One common signature of interacting supernovae is the presence of a strong narrow Hα line, often consisting of several superposed components and often intrinsically polarized by scattering in the circumstellar envelope. I will present results from numerical modeling of Hα line profiles in direct and scattered light that provide clues to the geometrical structure of the circumstellar material around interacting supernovae. I will also review the observed line profile behavior of Type IIn supernovae and its correlations with other supernova characteristics. Finally, I will discuss what these results can tell us about the massive stellar progenitors of interacting supernovae.

High-Resolution Observations of Obscured LBV Nebulae

Cornelia Lang (Department of Physics & Astronomy, University of Iowa), *et al.*

Recent progress has been made on understanding the physical properties of LBVs and their circumstellar nebulae (LBVNe) in obscured regions of the Galaxy by using radio and infrared observations. We present high-resolution radio and infrared data on 7 new LBV sources in our Galaxy (including the Pistol Star, AFGL 2298, NaSt1, LBV 1806-20, and others). Multi-frequency radio data provide measurements of the stellar wind source and nebular spectral indices, the mass-loss rates for the central stellar winds, and parameters of the physical conditions of the LBVNe. We present interpretation of the radio data along with high-resolution infrared spectroscopy that provides new insight into the processes by which these stars are ejecting mass, and the interstellar environments into which they are expanding.

Evidence for a Mass Outflow from Our Galactic Center

Casey J. Law (Astronomical Institute, University of Amsterdam)

See p. 407 for full article.

SNR 4449-1: A Very Young Supernova Remnant from a Massive Progenitor Star

Dan Milisavljevic & Robert Fesen (Dartmouth College)

A young (age ∼100 yr) and highly luminous O-rich supernova remnant in NGC 4449 shares many properties with the Galactic supernova remnant Cas A. Ground-based and

HST optical images and spectra show 6000 km s^{-1} expanding O-rich ejecta. The remnant's high luminosity is likely due to a recent and strong interaction of the ejecta with dense, pre-SN circumstellar material. This is indicated by 500 km s^{-1} Hα and [N II] emissions, with line ratios suggesting a nitrogen overabundance around 10 times solar. The surrounding CSM appears clumpy ($n \sim 10^5$) with some portion possibly distributed in a ring. The remnant lies near the center of a rich OB + WR cluster, and we estimate a progenitor mass of at least 20 $M_\odot$. This object may represent a much more luminous version of the Cas A remnant seen at a considerably earlier stage of CSM interaction.

Spitzer Observations of the Young Core-Collapse Supernova Remnant E0102: Infrared Ejecta Emission and Dust Formation

Jeonghee Rho, William Reach (Spitzer Science Center/Caltech), Achim Tappe (Harvard-Smithsonian Center for Astrophysics), *et al.*

We present Spitzer IRS and IRAC observations of the young supernova remnant E0102 (SNR 0102.2-7219) in the Small Magellanic Cloud. E0102 has some notable similarities to Cas A: both had massive progenitors of comparable masses and show optically emission from highly enriched oxygen ejecta as high velocity knots. The infrared spectra of E0102 showed ejecta lines of Ne, Si, S, and O. Among these lines, Ne lines are dominant: two [Ne III] lines at 15.5 and 36.0 μm, and two [Ne V] lines of 14.3 and 24.3 μm. The main element difference in E0102 is lack of Ar and Fe lines, compared to those of Cas A. It implies different nucleosynthesis and/or emitting conditions between the two SNRs. The [Ne II] line at 12.8 μm shows high-velocity dispersion of $\sim$3000 km s^{-1}, showing fast moving ejecta material. The continuum emission is from the same places as the ejecta lines, suggesting that dust forms in the supernova ejecta. The spectra also show a broad dust feature at 17 μm. IRAC 8 μm emission is detected from the strong optical ejecta knots. We will discuss the distribution and physical conditions of ejecta elements and the inferred total dust mass from freshly formed dust, and compare dust mass and composition with those of Cas A.

Kinematics of Superbubbles in Irregular Galaxies

Margarita Rosado & Patricia Ambrocio-Cruz (Instituto de Astronomia, Universidad Nacional Autonoma de Mexico)

This work presents three examples of previous kinematical studies on superbubbles in irregular galaxies: (i) the non-thermal superbubble in IC 10, possibly formed by a hypernova (Yang & Skillman 1993, *AJ*, 106, 1448; Bullejos & Rosado 2002, *Rev. Mex. A&A*, 12, 254; Lozinskaya & Moiseev, 2007, *MNRAS*, 381, L26), (ii) the network of supernova-plus wind-driven superbubbles that have perforated the ISM in IC 1613 (Valdez-Gutiérrez *et al.* 2001, *A&A*, 366, 35); (iii) the S3 nebula, hosting a WO star also in IC 1613 (Rosado *et al.* 2001; *AJ*, 122, 194; Borissova et al. 2004, *A&A*, 423, 97). These examples illustrate how massive stars of irregular galaxies have a profound influence in shaping the ISM in their host galaxies. The Fabry-Perot interferometer PUMA (Rosado *et al.* 1995, *Rev. Mex A&A*, 3, 263) was used to carry out the Hα and S II line observations.

Investigating the Circumstellar Environments of the Cool Hypergiants

Michael T. Schuster (University of Minnesota), Massimo Marengo (Harvard-Smithsonian Center for Astrophysics) & Roberta Humphreys (University of Minnesota)

The cool hypergiants are among a few highly unstable, very massive stars that lie on or near the empirical upper luminosity boundary in the H-R diagram. As a consequence of their very high mass-loss rates many of these stars are expected to have extensive circumstellar (CS) nebulosity. We investigate the cool hypergiants' CS environments and discuss how the presence and extent of CS nebulosity provide a record of their evolutionary histories. Each star's local interstellar environment can also play an important role in shaping the CS nebulosity and determining what we observe. The extremely luminous OH/IR M-type hypergiant NML Cyg is surrounded by an inverted HII region, where gas is ionized externally by Lyman continuum radiation from Cyg OB2. High angular resolution, high-contrast HST and mid-IR AO images of NML Cyg reveal an enigmatic asymmetric CS cocoon likely shaped through photo-dissociation and grain destruction by the near-UV radiation from the massive, hot stars within Cyg OB2. High-resolution HST and mid-IR AO images of the peculiar F-type hypergiant ρ Cas, famous for its shell ejections, show no CS material. The G-type hypergiant HR5171A also has no discernible CS material, but Spitzer/IRAC images reveal that it once dominated its local environment by creating a large HII and photo-dissociation region.

SESSION V: Massive Stars as Probes of the Early Universe

The Nucleosynthetic Products of the First Massive Stars

Ann Merchant Boesgaard, Emily M. Levesque & Jeffrey A. Rich (Institute for Astronomy, University of Hawaii)

The extremely metal-poor (EMP) stars ($-4.0 <$ [Fe/H] < -3.0) were formed in the earliest days of the Galaxy. Such stars contain the products of the early generations of massive stars. During stellar evolution of the massive stars and through their supernovae explosions, atoms of C, N, and O are created. The production of CNO, in turn, produces the rare light elements, Li, Be, and B through spallation reactions, either in the vicinity of the supernovae or in the ambient interstellar gas. Both Be and O are tied to the rate of supernovae; the abundance of Be in EMP stars is a tracer of massive star formation within our Galaxy. We have obtained high-resolution, high signal-to-noise spectra of about 15 stars with [Fe/H] < -2.8 using HIRES on the Keck I telescope. We have determined abundances of Be from the resonance lines of Be II at 3130 and 3131 Å and O from 3 OH lines in the UV in order to investigate the early nucleosynthesis processes in the Galaxy. We have found the stellar effective temperatures and gravities of our sample spectroscopically, using equivalent widths of Fe I, Fe II, Ti I, and Ti II. Our results indicate that there is a linear relationship between the logarithmic abundances of Be and O. As O increases in the Galaxy by two orders of magnitude, Be increases by a factor of 170. The SNe type II processes can produce both the Be and the O. This is supported by NSF AST05-05899.

Type Ib/c and II SN Rates and the Hubble Sequence

Francesco Calura (INAF-Osservatorio Astronomico di Trieste)

We compute the type Ib/c and II supernova (SN) rates as functions of the cosmic time for galaxies of different morphological types. We use four different chemical evolution models, each one reproducing the features of a particular galactic morphological type: E/S0, S0a/b, Sbc/d, and Irr galaxies. These models are used to study the evolution of the SN rates per unit luminosity and per unit mass as functions of cosmic time and as functions of the Hubble type. We explain the increase of the Ibc and II SN rate per unit mass observed in local galaxies as due to the higher star formation rates per unit mass of the latest Hubble types, in agreement with the popular downsizing scenario for galaxy formation.

Massive Stars in Hickson Compact Groups as a Tracer of the Early Universe

Jane C. Charlton, Jason E. Young (Pennsylvania State University), Patrick R. Durrell (Youngstown State University, Ohio), *et al.*

Compact groups of galaxies have some important similarities to protogalactic groups in the early universe. They have frequent interactions, the velocities of encounters are similar, and their gas is often stripped from the galaxies. Thus the formation of massive stars in compact groups, particularly those that form outside of galaxies, and their influence on their environment could provide clues to how these processes work at high redshift. We have embarked on a multi-wavelength study of twelve Hickson compact groups.

We will summarize the results of our Spitzer study of these groups. We find that the groups that are most gas-rich have the most active star formation. About half of the giant galaxies in the groups show red, mid-infrared colors in their nuclei, characteristic of AGNs and/or star formation, and that there is more such activity in the gas-rich groups.

We also present an analysis of the star clusters populations and the extended tidal sources in the contrasting examples of HCG 7 and HCG 31. This analysis is based on our HST/ACS BVI images of these groups. HCG 31 has widespread extremely recent star formation in the debris outside of the galaxies while HCG 7 is relatively quiescent.

Two- and Three-Dimensional Simulations of Mixing in Type II Supernovae

Candace Church Joggerst (University of California, Santa Cruz)

We present 2- and 3-dimensional simulations of zero- and solar-metallicity stars in the last stages of their lives and estimate their resulting nucleosynthetic output. When primordial stars with masses less than about 100 solar masses explode as type II supernovae, some portion of the star falls back onto the black hole, while the rest escapes to enrich the next generation of stars. The composition of the escaped gas depends on processes, such as Rayleigh-Taylor-induced mixing, that cannot be adequately modeled in one dimension. Multidimensional simulations are needed to capture the inherently asymmetric processes that enrich the outer layers of the star and determine its final yield. We find that Rayleigh-Taylor-induced mixing operates more efficiently in solar-metallicity stars than in their zero-metallicity counterparts. Our nucleosynthetic yields for zero-metallicity stars reproduce the high [C/Fe] and [O/Fe] ratios observed in the most metal poor stars known.

Detecting $z > 2$ Type IIn Supernovae

Jeff Cooke (University of California, Irvine), Mark Sullivan (Oxford University) & Elizabeth Barton (UC Irvine)

Luminous blue variables ($M \gtrsim 80\,M_\odot$) have been identified as the likely progenitors of Type IIn SNe (SNe IIn). The intrinsic bright and blue continua of SNe IIn enable photometric detection at $z > 2$ in existing deep optical surveys. In addition, observations of the bright emission lines in low-redshift SNe IIn indicate that the emission lines of $z > 2$ events are above the spectroscopic thresholds of existing facilities. Detections of $z > 2$ SNe IIn will measure the supernova rate and trace the sites of recent star formation during this early epoch. In addition, detections will provide a statistical means to quantify the SN contribution to galactic feedback, mass loss, and enrichment of the IGM and constrain the high-mass end of the IMF for targeted $z > 2$ galaxy populations. We present results from Phase I of our program to detect $z > 2$ SNe IIn monitoring color-selected galaxies in the CFHTLS Deep Synoptic survey. Our technique exploits (1) the efficiency of $z > 2$ galaxy photometric color selection, (2) the sensitivities and wide-field capabilities of existing optical facilities, and (3) the intrinsic brightness and dispersion ($= -19.0 \pm 0.9$) of Type IIn SNe that extends well into the FUV (1200–2000 Å).

Modelling Dust Formation of Supernovae Ejecta

Christa Gall, Anja C. Andersen (Dark Cosmology Center, University of Copenhagen), Ernst A. Dorfi (Institute of Astronomy, University of Vienna), *et al.*

Dust grains play a crucial role in the formation and evolution history of stars and galaxies in the early Universe. The presence of dust at high redshift seems to require efficient condensation of dust grains in supernova ejecta. Yet, observations of a few well-studied supernova remnants imply condensation efficiencies, which are several orders of magnitude below what is predicted by some theoretical models.

The aim of this project is the development of a self-consistent model, which treats the evolution of core-collapse supernova ejecta, arising from the first population of stars in the Universe. The constructed model is based on radiation hydrodynamic and will simulate the dust nucleation and grain growth scenario and the possible survival of these grains during the forward and so-called "reverse shock," which develops out of the code in succession of the interaction with the ambient interstellar medium. Of particular interests are the abundance of the formed dust, the time of the whole dust forming process, the grain temperature and the grain-size distribution of the particles, the possible crystal structure, and the different mineral types, as well as the dust production efficiency.

Radiative Transport Using SimpleX

Chael Kruip, Vincent Icke & Jan-Pieter Paardekooper (Sterrewacht Leiden)

SimpleX is a new, versatile radiative transfer code applicable to a wide variety of problems. It describes the transport of radiation as a random walk on an unstructured grid which represents the physical problem. Using such a physical grid makes the method computationally very fast and allows for high spatial resolution. Furthermore, the computational cost does not scale with the number of sources, making SimpleX ideal for simulations of, for example, re-ionization. Because the number of sources is irrelevant, diffuse photons can be incorporated trivially. We are in the process of coupling SimpleX to the Flash hydro-code in order to model radiation driven outflows in stellar atmospheres.

Metallicities at the Sites of Nearby SNe Ic and Implications for the SN-GRB Connection

Maryam Modjaz (University of California, Berkeley), *et al.*

See p. 503 for full article.

BRIght Target Explorer Constellation

A. F. J. Moffat (Département de Physique, Université de Montréal), W.W. Weiss (Institut für Astronomie, Universität Wien), S. M. Rucinski (Department of Astronomy and Astrophysics, University of Toronto), *et al.*

The primary goal of BRITE-Constellation is to constrain the basic properties of intrinsically luminous stars—stars that most affect the ecology of the Universe—by measuring their oscillations in both brightness and temperature with precision down to 10's of micromagnitudes on hour to month timescales, based on dual-broadband photometric time-series from space. BRITE-Constellation is a fleet of 4 independently free-flying nanosatellites to be launched starting in 2009. Each nanosat contains a 30 mm telescope with either a blue or red optical filter and CCD detector to image the sky in a 24-degree field of view. This will enable mmag-precision photometry or better on timescales of $\sim$15 minutes for all the $\sim$300 stars in the sky brighter than $V \sim 3.5$. As it turns out, most of these stars are also among the intrinsically brightest stars, which fall into two broad classes: massive stars during their whole lifetimes and intermediate-mass stars at the end of their nuclear-burning phases. It may appear strange that the brightest naked-eye stars in the sky have not generally been photometrically scrutinized with the same high precision as many of their fainter cousins. The reason for this is that the brightest stars are relatively sparsely distributed across the sky and thus difficult to measure properly from the ground. Because its targets are the brightest naked-eye stars in the sky, BRITE Constellation will also provide a special appeal to the public. The three-axis pointing performance (1 arcminute RMS stability) of each BRITE satellite is a significant advancement over anything that has ever flown before on a nanosatellite and is an important factor that enables this relatively inexpensive high-precision photometry mission.

SN Less Long Duration GRBs: Clues to their Nature from Their Environment?

Christina C. Thöne, Johan P. U. Fynbo, Paul M. Vreeswijk (Dark Cosmology Centre), et al.

The connection between long-duration GRBs and supernovae (SNe) seemed to be an established fact when last year, two long-duration GRBs with no SN signature in their light curves were found. One of them, GRB 060505 was an especially interesting case to study as it was hosted in a star-forming region in a rather nearby spiral galaxy. The properties of the burst site are rather different from the rest of the galaxy and resemble the properties of the more usual long GRB hosts which are highly star-forming, low-metallicity dwarf galaxies. Similar studies can be done with a few other spiral galaxies that hosted long-duration GRBs of which one, GRB 980425, was clearly connected to a SN. Comparing these hosts and the conditions at the GRB site, we try to get a conclusion about the nature of the SN-less long GRBs.

Galaxy-wide Census of Massive Dust-Producing Stars

Earl S. Wood (Western Kentucky University)

Massive stars can produce copious amounts of dust. We attempt to define the relevant yields and pathways of dust-production and extrapolate them to the environment of the early Universe. For this purpose, we use the 2MASS database and conduct a Galaxy-wide search for relatively massive ($M_i > 2 M_\odot$) dust-producing stars, with ∼1100 sources detected so far. Among them, ∼75% have no previous classification. Combining all the available optical-IR measurements and constructing optical-IR spectral energy distributions, we separate the known sources into four major categories: [extreme] carbon, OH/IR, Wolf-Rayet, and B[e]wd stars. As for the unknown sources, we have discovered that while [extreme] carbon stars and OH/IR stars likely comprise 85% of them, there is a handful of objects that could be classified as Wolf-Rayet WC-binary pinwheel candidates. We also single out ∼50 unclassifiable objects with rather exotic (extreme reddening, low envelope temperatures) spectral energy distributions.

Reports on Special Sessions

Evolution of Massive Stars at Low Metallicity

Georges Meynet[1], Nolan R. Walborn[2], Ian Hunter[3], Christophe Martayan[4], Allard Jan van Marle[5], Sergey Marchenko[6], Jorick S. Vink[7], Marco Limongi[8], Emily M. Levesque[9], and Maryam Modjaz[10]

[1]Observatory of Geneva University, Switzerland
email: georges.meynet@obs.unige.ch

[2]Space Telescope Science Institute, USA
email: walborn@stsci.edu

[3]Queen's University of Belfast, UK
email: i.hunter@qub.ac.uk

[4]Observatoire Royal de Belgique, Belgium, and, Observatoire de Paris-Meudon, France
email: martayan@oma.be

[5]Bartol Research Institute, University of Delaware, USA
email: marle@udel.edu

[6]Western Kentucky University, USA
email: sergey.marchenko@wku.edu

[7]Armagh Observatory, UK
email: jsv@arm.ac.uk

[8]Istituto Nazionale di Astrofisica, Italy, and Monash University, Australia
email: marco@oa-roma.inaf.it

[9]University of Hawaii, USA
email: emsque@IfA.Hawaii.Edu

[10]UC Berkeley Astronomy Department, USA
email: mmodjaz@astro.berkeley.edu

Abstract. This paper reports the contributions made on the occasion of the Special Session entitled "Evolution of Massive Stars at Low Metallicity" which was held on Sunday, December 9, 2007 in Kauai (USA).

Keywords. stars: abundances, early-type, evolution, mass loss, emission-line, Be, rotation; gamma rays: bursts

1. Introduction

Georges Meynet

Understanding the evolution of massive stars at low and very low metallicity is a requirement to address questions such as the nature of the sources of the reionization in the early Universe, the evolution of the interstellar abundances during the early phases of the evolution of galaxies, for finding possible signatures of primordial stellar populations in the integrated light of very distant galaxies and for discovering which objects are the progenitors of the long soft Gamma Ray Bursts. At present, the most "iron" poor objects known in the Universe are not very far from us since they are galactic halo stars. Provided these stars are trustworthy very metal poor stars (a view recently challenged by Venn & Lambert 2008), these objects offer a unique opportunity to study the yields of the first generations of stars and supernovae. Very interestingly, many observed features are unexpected and will probably require some revision of the classical ideas on how very metal poor massive stars evolve. To cite a few of them, let us recall that in the abundance patterns of the most metal poor halo stars, 1) no sign of Pair Instability Supernovae has been observed (Cayrel *et al.* 2004), 2) a high plateau of the N/O ratios (as a function of O/H) are obtained requiring the activity of efficient sources of primary nitrogen (see e.g. Spite *et al.* 2005), 3) simultaneously the C/O ratio as a function of O/H shows an upturn at a metallicity of [O/H]$\sim$-2 (Spite *et al.* 2005, see also Pettini, this volume), 4) a significant fraction of very metal poor stars are C-rich stars showing very peculiar abundance pattern at their surface (Beers & Christlieb 2005), 5) very helium-rich stars and 6) stars with high abundance of sodium and low abundance of oxygen are detected in globular clusters (Gratton *et al.* 2004; Piotto *et al.* 2005). In order to go further in the understanding of these very interesting questions, we have to obtain detailed and solid observations of massive stars at low metallicity in our neighborhood, produce models able to reproduce these observed features and then, using the same successful physics, explore the consequences for primordial and very metal poor stars.

The contributions presented during this special session brought new and very interesting results in the following areas of research: a first series of three contributions presented recent observations of massive stars at low metallicity, a second series of three contributions discussed the important question of the variation of the stellar winds with metallicity, then Pop III interior models have been presented and finally two presentations were devoted to the study of Gamma Ray Burst hosts.

2. The Onfp Class in the Magellanic Clouds
Nolan R. Walborn

The primary defining characteristic of Onfp spectra is the presence of an absorption reversal in the He II λ4686 emission line. Most members of the class also have substantially broadened absorption and emission lines. These properties suggest rapid rotation and perhaps disk structures—but see Bouret, Hillier, & Lanz, this volume, who reproduce this λ4686 profile with just wind clumping and rotation. The Galactic prototypes are the two brightest Of stars in the sky, ζ Puppis and λ Cephei. Luminosity classification of these objects is problematic, since the line profile of the primary He II criterion is peculiar. Recent spectroscopic surveys of OB stars in the Magellanic Clouds have found increasing numbers of the Onfp class, currently numbering about two dozen. Their spectral types span the entire O-type range, while their reliable absolute magnitudes range between those of classes V and I. Significant spectral variations are seen in all objects with multiple observations, including single-peaked λ4686 profiles, and two of them have been found to be spectroscopic binaries. Most of them are located near the peripheries of clusters or associations, with a few exceptions in compact clusters. The Onfp line-broadening distribution is distinct and shifted toward larger values from those of normal O dwarfs and supergiants with >99.99% confidence (analysis by Ian Howarth). These objects are of interest in the contexts of massive star formation, binary mergers, and GRB progenitors, the last particularly at the SMC metallicity.

3. Rotational Mixing in Magellanic Cloud Massive Stars
Ian Hunter

The prediction that core processed material is rotationally mixed into the photosphere is now considered a fundamental property in evolutionary models of massive stars. However, such theories have never been observationally tested, for example, one would expect the fastest rotators to be the most mixed but analyses of such fast rotators have never before been carried out. The analysis of over 100 LMC B-type massive stars has been carried out and surface nitrogen enrichments are used as an indicator of the efficiency of the mixing mechanism. Models including rotation have been generated to best fit the observed data and while the models can reproduce a significant fraction of the data there are a number of discrepancies between these theoretical models and the observations. In particular several fast rotators with little evidence of rotational mixing are observed despite being close to the end of their hydrogen burning lifetimes. Additionally both populations of main-sequence slow rotators and evolved blue supergiants with significant chemical enrichments are seen. These populations violate the current theories of rotational mixing and challenge our understanding of massive star evolution.

4. The Low Metallicity Effects on the B and Be Stars
Christophe Martayan

We present new observational results of the low metallicity effects on B and Be star populations of the Magellanic Clouds (MC). First, we show the results obtained with the VLT-GIRAFFE about the rotational velocities during the main sequence and at the ZAMS. These results indicate that the lower the metallicity, the higher the rotational velocities of B and Be stars. The ZAMS rotational velocities for Be stars we found are mass and metallicity-dependent. Second, we present new clues on the appearance of Be stars and explain the difference of evolution for the more massive of them between the Milky Way (MW) and the Small Magellanic cloud (SMC). Third, with a cross-matching between the 520 B-type stars observed with the VLT and the large photometric survey MACHO, we report on the detection of binaries and for the first time of short-term multi-periodicity in $\sim$30 Be stars and $\sim$10 B stars in the SMC. This result plaids in favor of pulsations at low metallicity, which was not foreseen by theory. Fourth, thanks to a Hα survey conducted with the ESO-WFI in slitless spectroscopic mode, we obtained 8 millions of spectra in the MC, and we found that the proportion of Be stars to B stars in 85 SMC open clusters is higher than in the MW.

5. Continuum Driven Winds from Super-Eddington Massive Stars
Allard Jan van Marle

If a star exceeds the Eddington limit, continuum scattering becomes a very efficient mechanism for removing mass from a star, producing mass loss rates that are several orders of magnitude larger than those achieved by line driving. This, coupled to the fact that continuum driving does not depend on the metallicity of the star makes it particularly well suited to removing mass from zero- and low-metallicity stars, which is necessary to prevent them from exploding as pair instability supernovae. Numerical simulations using the porosity length formalism derived by Owocki *et al.* (2004), show that super-Eddington stars can drive steady winds. However, if the mass loss rate becomes so large that the mechanical luminosity of the wind exceeds the radiative luminosity of the star, then the wind becomes unstable as part of the material will fall back toward the star. See van Marle *et al.* (2008).

6. Wind-Clumping in Low-Z Environment: Theory and Observations
Sergey Marchenko

We briefly review theoretical expectations for the wind-clumping factors in the environments with a low heavy-metal content. We show that, based on the recent observations of a small sample of Population I Wolf-Rayet stars in the Small Magellanic Cloud, general properties of the detected wind-clumps closely match the corresponding characteristics of small-scale inhomogeneities in the winds of Galactic Wolf-Rayet stars.

7. Mass Loss at Very Low Z
Jorick Vink

Stellar wind models for OB and WR stars as a function of Z are presented. Since the mass-loss rate is determined by the Fe lines in the inner wind, the theoretical dependence is found to be relatively steep, with $\dot{M} \propto Z^{0.7-0.8}$ down to $Z = 10^{-2} Z_\odot$ (Vink et al. 2001, Vink & de Koter 2005). The new prediction of Z-dependent mass-loss for WR winds may be particularly relevant for the progenitor evolution of long duration gamma-ray bursts (GRBs), which are widely believed to be rapidly rotating WR stars. Testing the rotational properties and asymmetries of WR stars at low Z using recent spectropolarimetric data, Vink (2007) discussed the threshold metallicity for making a long-duration GRB.

Turning attention to the early Universe at extremely low Z, we emphasize a flattening of the $\dot{M}$-Z dependence below $Z = 10^{-3} Z_\odot$ for late type WC stars. This might suggest that mass loss could still play a relevant role for massive stars in the early Universe, especially if these stars turn out to be close to their Eddington limit, e.g. undergoing LBV outbursts (Vink & de Koter 2005). It has also been suggested that characteristic LBV S Doradus mass-loss behavior in the present-day Universe might indicate that some LBVs could be direct precursors of supernovae (Kotak & Vink 2006).

8. Pop III Stellar Models
Marco Limongi

We presented our latest set of massive star models with initial zero metallicity (in the mass range 13-80 solar masses) and their associated explosive yields. The presupernova models have been computed by means of the FRANEC code (050419) which is described in Limongi & Chieffi (2006). The explosive yields are computed by means of a hydro code with an induced explosion in the framework of the kinetic bomb. First, we described the main differences in the evolutionary properties between zero and solar metallicity models. Among these, the most relevant one being the substantial primary ^{14}N production in zero metallicity models. This production is due to the fact that protons, in the H shell tail, are ingested by the underlying He convective shell. Such an ingestion activates a hot CNO cycle leading to a substantial production of primary ^{14}N. Such a partial mixing between the He convective shell and the overlying H-rich zone is a rather common occurrence in stellar models of initial zero metallicity (see e.g. Woosley & Weaver 1982) because of the low-entropy barrier that develops at the H-He interface in these stars. The second difference between solar and zero metallicity models is the well known odd-even effect, i.e., the difference between the production factors of the odd (Na-Sc) and the even (Ne-Ca) nuclei is much larger in the zero metallicity models compared to their corresponding solar metallicity ones. Finally, we presented a comparison between the explosive yields obtained for the zero metallicity models and the element abundance ratios observed in extremely metal poor stars (EMPS). These stars, in fact, probably formed in the very early epochs of Galaxy formation by gas clouds chemically enriched by the first stellar generations (POP III). Hence, whether or not they are associated to single supernovae or single burst events, they provide very useful constraints to test presupernova models, supernova explosion and nucleosynthesis theories as well as they can be used to infer the nature of the first generations of stars and supernovae. We showed that the element abundance pattern of both the NORMAL and the C-RICH EMPS can be explained in terms of enrichment of a population of STANDARD POP III core collapse supernovae (CC-SNe) with fluctuations of the IMF. In particular we showed that clouds dominated by the ejecta of high mass SNe are characterized by a chemical composition typical of the C-RICH EMPS, while clouds dominated by the ejecta of the lower mass SNe are characterized by a composition typical of the NORMAL EMPS.

9. Metal-Poor Galaxies and Gamma-Ray Burst Hosts

Emily Levesque

We have used the newest generation of the Starburst99/Mappings code to generate an extensive suite of models covering a wide range of physical parameters suitable for metal-poor galaxies and gamma-ray burst hosts. We use our theoretical models with optical emission line diagnostic diagrams to constrain the star-formation histories and ISM properties of metal-poor galaxies and gamma-ray burst hosts. Our comparisons reveal several important short-comings of the current model atmospheres and stellar evolutionary tracks in metal-poor environments – for more discussion, see Levesque *et al.* (2008, in prep). We have applied our diagnostics to the host galaxy of GRB 060505, one of the most hotly contested GRBs to date. We do not find compelling evidence to suggest that GRB 060505 originated in a long-duration core collapse progenitor. Our emission line diagnostic analysis suggests that the environment of GRB 060505 is more consistent with the host environments of compact-object-merger GRB progenitors (Levesque & Kewley 2007). In the future, a detailed understanding of stellar evolution at low metallicities will provide a deeper understanding of the unique and unusual environments of GRB progenitors.

10. Metallicities at the Sites of Nearby SN Ic

M. Modjaz

We show clues as to the production conditions of long-duration Gamma-ray Bursts (GRBs) by comparing the chemical abundances at the sites of five broad-lined SN Ic that accompany nearby GRBs with those of twelve nearby broad-lined SN Ic with no observed GRBs. We show that the oxygen abundances at the GRB sites are significantly lower than those found near ordinary broad-lined SN Ic. We demonstrate that neither SN selection effects (SN found via targeted vs. non-targeted surveys) nor the choice of the three strong-line metallicity diagnostic are convincing causes of the observed trend. Though the sample size is small, the observations are consistent with the hypothesis that low metal abundance is the main cause of some massive stars becoming SN-GRB, with a cut-off metallicity of 0.2-0.6 $Z_\odot$, depending on the adopted metallicity scale and solar abundance value (Modjaz *et al.* 2008).

References

Beers, T. C. & Christlieb, N. 2005, *ARA&A*, 43, 531
Cayrel, R., Depagne, E., & Spite, M. 2004, *A&A* 416, 1117
Chieffi, A. & Limongi, M. 2006, *ApJ*, 647, 483
Gratton, R., Sneden, C., & Carretta, E. 2004, *ARA&A*, 42, 385
Kotak, R. & Vink, J. S. 2006, *A&A*, 460, L5
Levesque, E. M. & Kewley, L. J. 2007, *ApJ*, 667, L121
Modjaz, M., Kewley, L., Kirshner, R. P. *et al.* 2008, *AJ*, 135, 1136
Owocki, S. P., Gayley, K. G., & Shaviv, N. J. 2004, *ApJ*, 616, 525
Piotto, G., Villanova, S. & Bedin, L. R. 2005, *ApJ*, 621, 777
Spite, M., Cayrel, R., & Plez, B. 2005, *A&A*, 430, 655
van Marle, A. J., Owocki, S. P., & Shaviv, N. J. 2008, in: B. O'Shea, A. Heger & T. Abel (eds.), *First Stars III* (New York: AIP), in press (arXiv:0708.4207)
Venn, K. A. & Lambert, D. L. 2008, *ApJ*, in press, (arXiv:0801.0752)
Vink, J. S., 2007, *A&A* 469, 707
Vink, J. S., & de Koter, A. 2005, *A&A* 442, 587
Vink, J. S., de Koter, A., & Lamers, H. J. G. L. M. 2001, *A&A* 369, 574
Woosley, S. E. & Weaver, T. A., 1982, in: M.J. Rees & R.J. Stoneham (eds.), *Supernovae: A Survey of Current Research*, (Dordrecht: Reidel), 79

Massive Stars as Cosmic Engines
Proceedings IAU Symposium No. 250, 2007
F. Bresolin, P.A. Crowther & J. Puls, eds.

© 2008 International Astronomical Union
doi:10.1017/S1743921308020966

Magnetic Massive Stars

Rich Townsend[1], David H. Cohen[2], Luc Dessart[3], Swetlana Hubrig[4], Yaël Nazé[5], Véronique Petit[6], Asif ud-Doula[1] and Nolan R. Walborn[7]

[1] Bartol Research Institute, University of Delaware, Newark, DE 19716, USA

[2] Swarthmore College, Department of Physics and Astronomy, 500 College Ave., Swarthmore College, Swarthmore, PA 19081, USA

[3] Department of Astronomy and Steward Observatory, University of Arizona, Tucson, AZ 85721, USA

[4] European Southern Observatory, Casilla 19001, Santiago, Chile

[5] Institut d'Astrophysique et de Géophysique, Université de Liège, Bât. B5c, Allée du VI Août 17, B-4000 Liège, Belgium

[6] Département de Physique, de Génie Physique et d'Optique and Observatoire du mont Mégantic, Université Laval, Québec, QC G1K 7P4, Canada

[7] Space Telescope Science Institute, 3700 San Martin Drive, Baltimore, MD 21218, USA

Abstract. Magnetic fields are unexpected in massive stars, due to the absence of a sub-surface convective dynamo. However, advances in instrumentation over the past three decades have led to their detection in a small but growing subset of these stars. Moreover, complementary theoretical developments have highlighted their potentially significant influence over the structure, evolution and circumstellar environments of massive stars. Here, we summarize a special session convened prior to the main conference, focused on presenting recent developments in the study of massive-star magnetic fields.

Keywords. stars: early-type, magnetic fields, stars: winds, stars: mass-loss, hydrodynamics, MHD, techniques: spectroscopic, techniques: polarimetric, X-rays: stars, supernovae: general

---

1. Optical, Ultraviolet, and X-Ray Spectral Morphology of Hot Magnetic Stars
Nolan R. Walborn

It is noteworthy that the four hottest massive magnetic stars were all isolated as peculiar from their spectral morphology and/or variability prior to the field detections. They are the mid-O stars θ^1 Orionis C and HD 191612, and the early-B stars τ Scorpii and ξ^1 Canis Majoris. These peculiarities extend to the optical, UV, and X-ray domains, in comparison with the corresponding normal spectral sequences. Indeed, it is essential to recognize and isolate the peculiar objects, which otherwise obscure the normal trends. The existence of the latter in X-rays is a surprising new development, emerging from purely morphological analysis of *Chandra* data as a function of the optical spectral types, just as the UV wind-profile systematics did from *IUE* data during the early 1980's. Figure 1 shows a normal X-ray spectral-type sequence; the progression of the strongest lines toward longer wavelengths with advancing spectral type, and the correlated H-like/He-like line ratios of Si, Ne, and O, are readily seen. The Ne behavior is most useful, since the ratio reverses through this domain (note interference from an Fe XVII line at the later types). In contrast, Figure 2 shows the drastic departures of some magnetic and rapidly rotating stars from the normal trends. (Figures are courtesy of Wayne Waldron.)

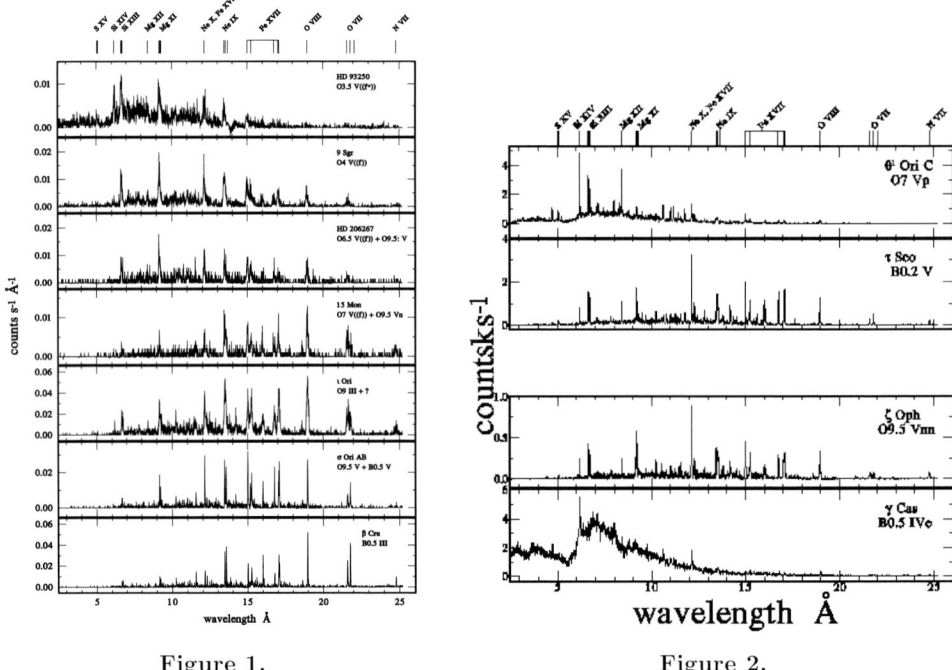

Figure 1.

Figure 2.

These results have two important implications: (1) other hot, massive stars with unexplained spectral peculiarities or variations are strong magnetic candidates that should be observed for fields (HD 108, HD 36879, HD 148937, θ Car, 15 Mon); and (2) the rarity of these phenomena indicates that the fields are likely primordial, rather than generated by the intrinsic stellar structure and evolution. (Acknowledgements: NASA/*FUSE* grant NNG06G179G).

2. The Mysterious Of?p Class and the Magnetic O-Star θ^1 Ori C

Yaël Nazé

In our Galaxy, there are only 3 stars in the Of?p category: HD 108, HD 148937 and HD 191612. Their peculiar properties (varying line profiles, X-ray over-luminosity, and photometric changes) were recently discovered. A magnetic field was even detected for HD 191612, leading to the suggestion that this object was a magnetic oblique rotator, similar to θ^1 Ori C but older. We review here the physical properties of these objects and re-examine the evidence leading to that conclusion.

• **Spectral types**: That of θ^1 Ori C is still uncertain (O5.5–O7V), but it is obvious that the star does not share the defining characteristics of the Of?p category, i.e. strong CIII emission at 4650Å. In fact, the types of HD 108 and HD 191612 are not even fixed: due to contamination by varying emission, they alternate between early Of?p (O4–5.5) with strong CIII emission and late O (O8–8.5) with a much weaker CIII emission.

• **Variability**: H and HeII 4686 are variable for all objects of our sample, but the CIII and HeI lines also vary strongly for HD 108 and HD 191612. Note that the changes in HD 148937 are of very weak amplitude.

• **Periodicity**: The line profile changes are periodic with a timescale of 15.4d for θ^1 Ori C, 537.6d for HD 191612, ∼55yrs for HD 108 and possibly 7d for HD 148937.

- **Binarity**: Both θ^1 Ori C and HD 191612 are eccentric binaries with $M_1/M_2 \sim 0.5$. The orbital period is much longer than that of the line profile changes: 10.9yrs for the former and 1542d for the latter. The binary status of HD 108 and HD 148937 is not yet fully ascertained: no clear evidence for orbital motion was detected, but very long-term and/or low-amplitude velocity variations can not be excluded.
- **Physical parameters**: The Fourier method for determining rotation rate shows that all of these objects present low $v\sin(i)$. Atmosphere model fits yields similar values of the luminosity and temperature for HD 108, HD 191612 and θ^1 Ori C: these stars appear very close to each other in the HR diagram while HD 148937 is twice more luminous.
- **Other characteristics**: A magnetic field was clearly detected for θ^1 Ori C and HD 191612, but observations are needed to check if the other two Of?p objects are magnetic. HD 148937 is surrounded by a nebula, thought to have been ejected during a LBV-like event; the star and its nebula are both clearly enriched in nitrogen, suggesting that HD 148937 might be more evolved than the other 3 stars.
- **Photometry**: Both HD 108 and HD 191612 appear brighter when the emission lines are stronger and the spectral type earlier. No large variation of the photometry of HD 148937 was found; photometric changes in θ^1 Ori C have been proposed to explain the small variations of the EW of the photospheric lines, but have not been confirmed yet.
- **X-ray properties**: At high energies, θ^1 Ori C displays a large overluminosity, a very hard spectrum (dominated by a temperature of $\sim$3keV), phase-locked variations (simultaneous brightening of the X-ray and visible emissions), and very narrow lines. The X-ray spectrum of the three Of?p is clearly different: though a large overluminosity is present, their spectrum is soft (dominated by a temperature of $\sim$0.3keV) and displays rather broad lines. For HD 191612, flux variations are also present, but their phase-locked nature is uncertain.

It is now clear that θ^1 Ori C and the Of?p stars share many similarities. However, there are also crucial differences that should not be discarded, especially in the X-ray domain. It seems thus premature to simply identify these objects with θ^1 Ori C. Most probably, what is observed is a combination of effects due to the magnetic field with a completely different phenomenon (still unknown). (Acknowledgement: FRS-FNRS for grant and support).

3. Observations of Magnetic Fields of Massive Stars with FORS 1 at the VLT

Swetlana Hubrig

Our recent studies using FORS 1 in spectropolarimetric mode included observations of magnetic fields in a sample of massive stars with spectral types earlier than B2. The excellent potential of FORS 1 for measuring magnetic fields in massive stars was demonstrated by observations of the mean longitudinal magnetic field in θ^1 Ori C in 2007 (Hubrig *et al.* 2008). This star was the first O-type star with a detected weak magnetic field varying with a rotation period of 15.4 days. The FORS 1 measurements show a sinusoidal curve in spite of the phase gap between 0.60 and 0.88. However, our observations determined a magnetic geometry distinct from the one deduced by Wade *et al.* (2006). The maxima and minima of the measured longitudinal field as well as the phases of the field extrema appeared to be significantly different. Assuming an inclination of the rotation axis to the line-of-sight of $i = 45°$, our modeling of the longitudinal field variation constrains the dipole magnetic field geometry of θ^1 Ori C to $B_{\rm d} \approx 900$ G and $\beta \approx 80°$, where $B_{\rm d}$ is the dipole intensity and β is the obliquity angle.

A longitudinal magnetic field at a level larger than 3σ ($\langle B_z \rangle = -115 \pm 37$ G) has been detected by us in another O-type star, HD 155806, with a spectral type O7.5IIIe, which is currently the third O-type star with a diagnosed magnetic field. Clear variations of Si IV, He I and other lines have been detected in FEROS and UVES spectra retrieved from the ESO archive (Hubrig et al. in preparation).

Our search of magnetic fields in a sample of 13 Be stars revealed the presence of weak magnetic fields in two stars, HD 56014 and HD 148184 (Hubrig et al. 2007a). For two other stars in the studied sample, HD 58011 and HD 11735, we detected distinctive circular polarization signatures in the Stokes V spectra of the Ca II H&K lines. The profiles of these Ca lines in the FORS 1 spectra taken in integral light are deeper than predicted by synthetic spectra. Additional high signal-to-noise spectroscopic observations are needed to study the Ca line profiles to be able to decide whether they are formed in circumstellar disks around these stars.

Finally, we report the discovery of weak magnetic fields in a few β Cephei stars and in about 20 slowly pulsating B (SPB) stars (Hubrig et al. 2006; Hubrig et al. in preparation) implying that massive pulsating stars can no longer be considered as classes of non-magnetic pulsators. However, the effect of the fields on the oscillation properties remains to be studied.

Our spectropolarimetric studies using FORS 1 at the VLT demonstrate that magnetic fields are indeed present in massive stars. For the case of magnetic fields in massive stars that are weaker in strength and likely more complex in their geometry than classical magnetic Ap and Bp stars, progress in their study may potentially come from detailed studies of polarized line profiles. At present, it is not obvious to what extent magnetic fields can be directly discovered in circumstellar material. Previous detections of magnetic fields in circumstellar material include a detection of magnetic fields in the circumstellar disk of FU Ori (Donati et al. 2005) and in circumstellar Ca lines of Herbig stars (Hubrig et al. 2006; Hubrig et al. 2007). However, modeling diagnostics of magnetic fields in these environments are still under development (Ignace & Gayley 2007).

4. X-ray Emission and the Incidence of Magnetic Fields in Massive Stars of the Orion Nebula Cluster

Véronique Petit

Magnetic fields have been frequently proposed as a likely source of variability and confinement of the winds of massive stars. Recently, Stelzer et al. (2005) found significant X-ray emission from all massive stars in the Orion Nebula Cluster (ONC). Possibly periodic rotational modulation in X-rays and other indicators suggested that there might be many magnetic B- and O-type stars in this star-forming region.

Magnetic fields can be directly detected in stellar atmospheres by the means of the Zeeman effect. If the field is strong enough, and the spectral lines narrow enough, one can directly see the Zeeman splitting of lines in the intensity spectrum. However, if the field is weaker, and the lines broadened either intrinsically or by fast rotation, the splitting is much more difficult to detect, even at high spectral resolution. In that case, the most effective way to detect magnetic fields is to look for circular polarization signatures across photospheric spectral lines, generated by the longitudinal Zeeman effect. During the past decade, a powerful multi-line analysis procedure called 'Least Squares Deconvolution' (LSD; Donati et al. 1997) has been developed and applied for extracting a mean Stokes

V profile from a stellar spectrum, simultaneously exploiting the signal contained in all the lines present. This allows for significantly improved sensitivity to magnetic fields, as it substantially increases the signal-to-noise ratio.

Notwithstanding the gains provided by these sophisticated tools, magnetic fields in hotter OB stars remain a challenge to detect. The few photospheric lines present in the optical spectrum and the large intrinsic width of these lines, compounded by the usual rapid rotation of these stars, require large-bandwidth and high signal-to-noise ratio observation to start with, even using LSD. The advent of a new generation of spectropolarimeters such as ESPaDOnS at the Canada-France-Hawaii Telescope (CFHT) and its twin NARVAL at the Télescope Bernard-Lyot (TBL) now allows a new level of investigation of magnetic fields in massive OB stars. ESPaDOnS consists of a polarimetric module located at the Cassegrain focus of the CFHT, linked by optical fibers to the high-resolution echelle spectrometer. A resolution of 65,000 for a spectral range covering 360 nm to 1 μm can be achieved in a single observation. The high spectral resolution enables the resolution of the Zeeman signature across essentially all individual spectral lines, which provides a qualitative advantage over lower resolution instruments.

We have carried out sensitive ESPaDOnS observations to search for direct evidence of magnetic fields in the massive stars of the ONC. We used the statistical test described by Donati *et al.* (1997) to diagnose the presence of a signal in LSD mean Stokes V profiles. A signal is unambiguously detected whenever the associated detection probability is larger than 99.999 per cent (corresponding to a false alarm probability smaller than 10^{-5}), and when no signature is detected in the associated diagnostic null (N) spectrum. The associated longitudinal field can be inferred from the Stokes I and V profiles. It is important to note that the longitudinal field is not the primary diagnostic for inferring the presence of a magnetic filed, because a magnetic configuration with a null longitudinal component still usually produces a non-zero Stokes V signature.

We report the detection of two new massive magnetic stars in the Orion Nebula Cluster: Par 1772 (HD 36982) and NU Ori (HD 37061), for which the estimated dipole polar field strengths, with 1σ error bars, are 1150^{+320}_{-200} G and 650^{+220}_{-170} G respectively. We also obtain dipole field upper limits for the remaining stars with a state-of-the-art Bayesian analysis, resulting in a precise magnetic characterization of all ONC massive stars. This allows us to explore for the first time the connections between fields, winds and X-rays in a complete, co-eval and co-environmental sample of massive stars. These remarkable results bring forth new challenges for understanding the processes leading to X-ray emission in massive stars. We also expect to provide unique data regarding the incidence of magnetic fields in massive stars with which to confront models of magnetic field origin in neutron stars and magnetars, such as that proposed by Ferrario & Wickramasinghe (2006).

5. X-rays from Magnetically Channeled Winds of OB Stars

David Cohen

OB stars with strong radiation-driven stellar winds and large-scale magnetic fields generate strong and hard X-ray emission via the Magnetically Channeled Wind Shock (MCWS) mechanism (Shore & Brown 1990; Babel & Montmerle 1997; ud-Doula & Owocki 2002). There are four separate X-ray diagnostics that confirm the MCWS scenario for the young, magnetized O star that illuminates the Orion Nebula, θ^1 Ori C, and constrain the physical properties of its X-ray emitting magnetosphere:

1. High X-ray temperatures, determined from thermal spectral model fitting. The differential emission measure of θ^1 Ori C peaks at temperatures above 10 MK, which

is in contrast to the few million K peak temperatures in mature, unmagnetized O stars (Wojdowski & Schulz 2005), and which is well reproduced by MHD simulations of the MCWS mechanism (Gagné et al. 2005).

2. Relatively narrow X-ray emission lines. The X-ray emitting plasma in the MCWS scenario is predominantly in the closed magnetic field regions and thus the plasma velocity is relatively low and the associated Doppler line broadening is modest. This is seen in the MHD simulations and confirmed by the *Chandra* grating observations.

3. The rotational modulation of the X-ray emission is consistent with part of the magnetosphere being eclipsed near phase 0.5, when the viewing orientation is magnetic equator-on. The depth of the eclipse provides information about the location of the X-ray emitting plasma (deeper eclipses imply more plasma close to the star). The observed eclipse depth for θ^1 Ori C implies that the bulk of the plasma is within a stellar radius of the photosphere. This is a somewhat closer than the MHD simulations predict (e.g. Gagné et al. 2005).

4. The ratio of the forbidden to intercombination line strengths in helium-like ions also puts a constraint on the location of the X-ray emitting plasma, via the sensitivity of these line ratios to the local UV mean intensity. The closer the hot plasma is to the photosphere, the stronger the UV photoexcitation of electrons from the upper level of the forbidden line to the upper level of the intercombination line, and the smaller the f/i ratio. This is demonstrated in Fig. 3 for the Mg XI complex in the co-added (over four observations) *Chandra* grating spectrum of θ^1 Ori C, where we see that the very weak forbidden line requires a plasma location below $r \approx 2$ R$_*$.

6. MHD Simulation of Magnetic Channeling and Spindown in Rotating Hot-Star Winds

Asif ud-Doula

Building upon our previous MHD simulation studies of magnetic channeling in radiatively driven stellar winds, this talk examines the dynamical effects of stellar rotation in the 2-D axisymmetric case of an aligned dipole surface field. In addition to the magnetic confinement parameter $\eta_* \equiv B_{\rm eq}^2 R_*^2 / \dot{M} v_\infty$ introduced in ud-Doula & Owocki (2002), we now add a rotational parameter $W \equiv V_{\rm rot}/V_{\rm orb}$ (the ratio of the equatorial surface rotation speed to orbital speed), examining specifically models with moderately strong rotation W = 0.25 and 0.5, and comparing these to analogous nonrotating cases. Defining the associated Alfvén radius $R_{\rm A} \approx \eta_*^{1/4} R_*$ and Kepler corotation radius $R_{\rm K} \approx W^{-2/3} R_*$, we find rotation effects are weak for models with $R_{\rm A} < R_{\rm K}$, but can be substantial and even dominant for models with $R_{\rm A} > R_{\rm K}$. In particular, by extending our simulations to magnetic confinement parameters (up to $\eta_* = 1000$) that are well above those ($\eta_* = 10$) considered in ud-Doula & Owocki (2002), we are able to study cases with $R_{\rm A} \gg R_{\rm K}$; we find that these do indeed show clear formation of the rigid-body disk predicted in previous analytic models (Townsend and Owocki 2005), with however a rather complex, dynamic behavior characterized by both episodes of downward infall and outward breakout that limit the buildup of disk mass. Overall, the results provide an intriguing glimpse into the complex interplay between rotation and magnetic confinement, and form the basis for a full MHD description of the rigid-body disks expected in strongly magnetic Bp stars like σ Ori E.

This simulation study also allows us to examine the role of such rotation-aligned dipole filed in enhancing the net loss of angular momentum from the wind. Compared to the 1D analytic model derived by Weber and Davis (1967; WD) for the effect of a monopole

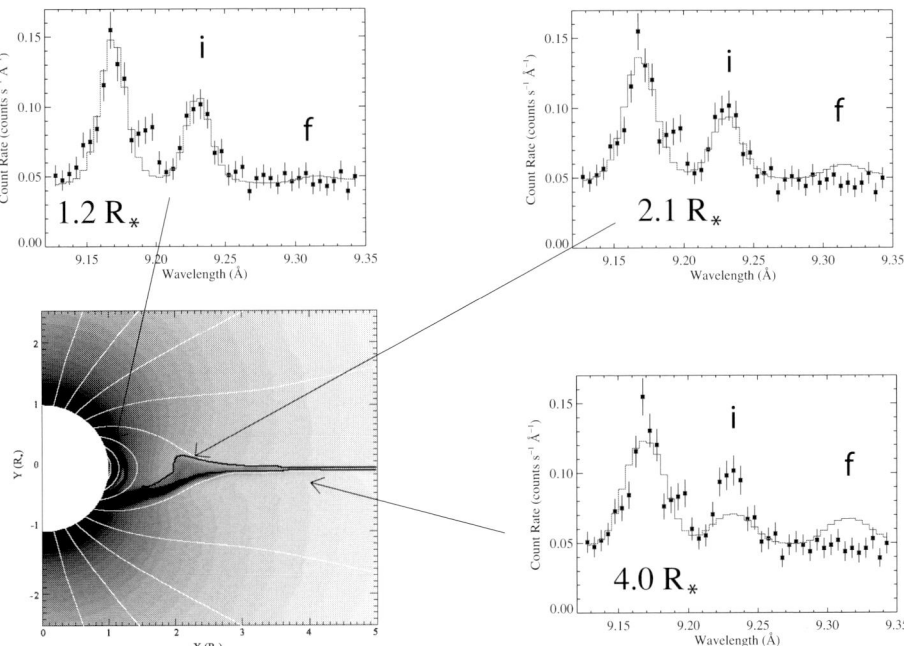

Figure 3. A snapshot from a 2-D MHD simulation of θ^1 Ori C, taken from Gagné et al. (2005), showing emission measure in grayscale, magnetic field lines as white contours, and with a thick, black contour enclosing plasma with temperature above 10^6 K (lower left). The three other panels show the *Chandra* spectrum in the vicinity of the helium-like Mg XI complex, with a model of the resonance, intercombination (i), and forbidden (f) lines overplotted. The relative strengths of the f and i lines are different in each of the three panels, as the three models were calculated assuming a source location of $1.2\,\mathrm{R}_*$, $2.1\,\mathrm{R}_*$, and $4.0\,\mathrm{R}_*$, respectively, starting at the top left and moving clockwise. The arrows indicate the approximate location in each case. The intermediate case – $r = 2.1\,\mathrm{R}_*$, which seems to agree with the MHD simulation – is marginally consistent with the data (68% confidence limit).

field on the equatorial plane outflow of the gas-pressure-driven solar, our 2D simulations of rotation-aligned dipole in a line-driven stellar wind show much more complex, time-dependent flows; but the overall scaling for the net loss of angular momentum still can be cast in terms similar to the classic WD form, $\dot{J} \approx -(2/3)\dot{M}\Omega R_A^2$, where Ω the stellar rotation frequency. As in the WD model, only ca. 20% of this angular momentum is carried by wind material itself, with the bulk of the loss instead carried by the field Poynting stress. However, while the WD monopole model implies a scaling $R_A \sim \eta_*^{1/2} R_*$, our more realistic dipole simulation models give a smaller Alfvén radius $R_A \sim \eta_*^{1/4} R_*$. The associated stellar rotation spindown time then has the scaling $t_J \equiv J/\dot{J} \approx 0.05 t_M/\sqrt{\eta_*}$, where $t_M \equiv M/\dot{M}$ is the characteristic mass loss time. For strong magnetic rotators like σ Ori E, the expected spindown time is of order a million years or shorter.

7. Rigid-Field Hydrodynamics Models for the Magnetospheres of Strongly Magnetic Massive Stars

Rich Townsend

Among the differing classes of magnetic massive star, the 'He-strong' stars — characterized by elevated and spatially inhomogeneous photospheric helium abundances — are the most challenging to model using standard magnetohydrodynamical (MHD) approaches. Their strong surface fields ($B_{\rm pole} \sim 10,000\,{\rm kG}$), paired with radiatively driven winds having modest mass-loss rates ($\dot{M} \sim 10^{-9}\,{\rm M_\odot\,yr^{-1}}$), imply magnetic confinement parameters on the order of $\eta_* \sim 10^5 - 10^7$. This parameter characterizes the relative energy densities of the magnetic field and the wind outflow, with the field dominating the wind in the limit $\eta_* \gg 1$, and vice versa in the limit $\eta_* \ll 1$ (for details, see ud-Doula & Owocki 2002). In the large-η_* limit, the field lines are almost rigid, and the Alfvén speed is very large; this makes MHD simulations very computationally expensive. (Asif ud-Doula's contribution above reports on simulations at the current upper limit of feasibility, $\eta_* = 1,000$; this is still orders-of-magnitude smaller than the He-strong domain).

To tackle the strong-field limit, Townsend & Owocki (2005) employed the simplifying *Ansatz* that the field lines are *completely rigid*; together with the frozen flux condition of ideal MHD, the result is that the field lines behave simply as fixed conduits (in the co-rotating frame of the star) for the wind material flowing along them. For closed field lines, it is inevitable that the wind streams from opposing footpoints will collide with each other, producing strong shocks in the manner envisaged by Babel & Montmerle (1997a,1997b). As it cools down by radiative emission, the post-shock wind material is acted on by the gravitational and centrifugal forces. Beyond the Kepler co-rotation radius, the centrifugal force is sufficient to overcome gravity, and supports the cooled material near the tops of the rigid field loops in a warped, thin, co-rotating disk.

This *Rigidly Rotating Magnetosphere* model (RRM; Townsend & Owocki 2005) has proven very successful in modeling the optical variability of the archetypal σ Ori E (Townsend *et al.* 2005). However, the RRM model does not encompass the wind streams or shocks feeding the rigid magnetosphere, and therefore cannot be used for analysis of, e.g., X-ray emission. To incorporate these phenomena, we have developed a new *Rigid-Field Hydrodynamics* (RFHD) approach that models the wind flow along each field line using a 1-D hydrodynamical code. By piecing together many independent 1-D models, we can build up a 3-D dynamical simulation for a complete magnetosphere in the strong-field limit, at a tiny fraction of the cost of an equivalent MHD simulation. Initial results from applying the RFHD approach to σ Ori E confirm the analytical findings of the RRM formalism, but a number of novel results have also emerged. For instance, downstream of the shocks, the post-shock material is heated by centrifugal compression, to temperatures $\sim 10^8$ K where it emits hard X-rays; moreover, instabilities in the radiative cooling process excite periodic, dipole-mode oscillations in the disk material. Full details can be found in Townsend, Owocki & ud-Doula (2007). (Acknowledgement: NASA/*LTSA* grant NNG05GC36G).

8. Radiation Magneto-Hydrodynamics Simulations of Core-Collapse Supernovae

Luc Dessart

Recent stellar evolutionary calculations of low-metallicity massive fast-rotating main-sequence stars yield iron cores at collapse that are endowed with high angular momentum. It is thought that high angular momentum and black hole (BH) formation are critical ingredients of the collapsar model (Woosley 1993) of long-soft γ-ray bursts (GRBs). We recently performed (Dessart *et al.* 2008) 2D multi-group, flux-limited-diffusion MHD simulations of the collapse, bounce, and immediate post-bounce phases of a 35 $M_\odot$ collapsar-candidate model of Woosley & Heger (2006). We find that, provided the magneto-rotational instability (MRI) operates in the differentially-rotating surface layers of the millisecond-period neutron star, a magnetically-driven explosion ensues during the proto-neutron star phase, in the form of a baryon-loaded non-relativistic jet, and that a BH, central to the collapsar model, does not form. Paradoxically, and although much uncertainty surrounds stellar mass loss, angular momentum transport, magnetic fields, and the MRI, current models of chemically homogeneous evolution at low metalicity yield massive stars with iron cores that may have *too much* angular momentum to avoid a magnetically-driven, hypernova-like, explosion in the immediate post-bounce phase. We surmise that fast rotation in the iron core may inhibit, rather than enable, collapsar formation, which requires a large angular momentum not in the core but *above* it.

We conclude that variations in the angular momentum distribution of pre-collapse massive stars may lead to different post-bounce scenarios. Non- or slowly-rotating progenitors may explode with weak/moderate energy ($\lesssim 1$ B; 10^{51} erg $\equiv 1$ Bethe [1 B]) through a neutrino or an acoustic mechanism $\lesssim 1$ s after bounce, or may collapse to a BH. Objects with large angular momentum in the envelope, but little in the core, may proceed through the proto-neutron star (PNS) phase, transition to a BH and form a collapsar with a GRB signature. Owing to the modest magnetic-field amplification above the PNS, a weak precursor polar jet may be launched, soon overtaken by a baryon-free, collimated relativistic jet. At the same time, the progenitor envelope is exploded by a disk wind, resulting in a hypernova-like SN with a large luminosity (large ^{56}Ni mass). Finally, and this is what we conclude here, objects with large angular momentum in the core may not transition to a BH. Instead, and fueled by core-rotation energy, a magnetically-driven baryon-loaded non-relativistic jet is obtained without any GRB signature. The explosion has the potential of reaching energies of a few B to 10 B, and for viewers along the poles of looking like a Type Ic hypernova-like SN with broad lines. For a viewer at lower latitudes, the delayed and less energetic explosion nearer the equator may look more like a standard Type Ic SN (Höflich *et al.* 1999). This volume-restricted jet-like explosion is dimmer, as the amount of processed ^{56}Ni may be significantly less than the $\sim 0.5\,M_\odot$ obtained in the collapsar context (MacFadyen & Woosley 1999). Hence, magnetic processes during the post-bounce phase of fast-rotating iron cores offer a potential alternative to collapsar formation and long-soft GRBs by producing non-relativistic non-Poynting-flux-dominated baryon-loaded hypernova-like explosions without any GRB signature. Importantly, while our study narrows the range over which the collapsar model may exist, it also offers additional routes to explain the existence of GRB/SN-hypernova events like SN 1998bw (Woosley *et al.* 1999), and hypernova events like SN 2002ap without a GRB signature (Mazzali *et al.* 2002).

References

Babel, J. & Montmerle, T. 1997a, *A&A*, 323, 121
Babel, J. & Montmerle, T. 1997b, *ApJ*, 485, L29
Dessart, L., Burrows, A., Livne, E., & Ott, C. 2007, *ApJ*, 673, L43
Donati, J.-F., Semel, M., Carter, B. D., *et al.* 1997, *MNRAS*, 291, 658
Donati, J.-F., Paletou, F., Bouvier, J., & Ferreira, J. 2005, *Nature*, 438, 466
Ferrario, L. & Wickramasinghe, D. 2006, *MNRAS*, 367, 1323
Gagné, M., Oksala, M., Cohen, D. H., *et al.* 2005, *ApJ*, 628, 986
Höflich, P., Wheeler, J. C., & Wang, L. 1999, *ApJ*, 521, 179
Hubrig, S., Briquet, M., Schöller, M., *et al.* 2006a, *MNRAS*, 369, L61
Hubrig, S., Yudin, R. V., Schöller, M., & Pogodin, M. A. 2006b, *A&A*, 446, 1089
Hubrig, S., North, P., & Schöller, M. 2007a, *AN*, 328, 475
Hubrig, S., Pogodin, M. A., Yudin, R. V., *et al.* 2007b, *A&A*, 463, 1039

Hubrig, S., Schöller M., Briquet, M. *et al.* 2008, in: *CP#AP Workshop, Contrib. Astron. Obs. Skalnaté Pleso* (arXiv:0712.0191)
Ignace, R. & Gayley, K. G. 2007, *arXiv:0708.1942*
MacFadyen, A. I., & Woosley, S. E. 1999, *ApJ*, 524, 262
Mazzali, P. A., Deng, J., Maeda, K. *et al.* 2002, *ApJ*, 572, L61
Shore, S. N. & Brown, D. N. 1990, *ApJ*, 365, 665
Stelzer, B., Flaccomio, E., Montmerle, T., *et al.* 2005, *ApJS*, 160, 557
Townsend, R. H. D. & Owocki, S. P. 2005, *MNRAS*, 357, 251
Townsend, R. H. D., Owocki, S. P., & Groote, D. 2005, *ApJ*, 630, L81
Townsend, R. H. D., Owocki, S. P., & Ud-Doula, A. 2007, *MNRAS*, 382, 139
ud-Doula, A. & Owocki, S. P. 2002, *ApJ*, 576, 413
Wade, G. A., Fullerton, A. W., Donati, J.-F., *et al.* 2006, *A&A*, 451, 195
Weber, E. J. & Davis, L. J. 1967, *ApJ*, 148, 217
Wojdowski, P. & Schulz, N. S. 2005, *ApJ*, 627, 953
Woosley, S. E. 1993, *ApJ*, 405, 273
Woosley, S. E., Eastman, R. G., & Schmidt, B. P. 1999, *ApJ*, 516, 788
Woosley, S. E. & Heger, A. 2006, *ApJ*, 637, 914

Author Index

Aerts, C. – **237**
Alberdi, A. – 534
Alfaro, E. – 552
Ambrocio-Cruz, P. – 563
Anderson, A.C. – 566
Anilmis, N. – **525**
Arthur, S.J. – **355**
Austin, N. – **525**
Avena, A. – 532

Bally, J. – 554
Barbá, R. – 538
Barbosa, C. – **548**
Barton, E. – 566
Beck, S. – **559**
Beffa, C. – 147
Belkus, H. – 293
Bibby, J.L. – **548**
Bisbas, T.G. – **559**
Bjorkman, J.E. – **559**
Bjorkman, K.S. – 557, **560**
Blomme, R. – **525**
Blum, R. – 548
Boesgaard, A.M. – **565**
Bonanos, A. – 529, **548**
Bouret, J.-C. – **526**, 532
Bresolin, F. – **273**, 313, 556
Briquet, M. – 531, 544
Bromm, V. – 471
Brooks, K. – **560**
Brott, I. – 167, 544
Burrows, A. – **185**

Calura, F. – **565**
Cantiello, M. – 167, 231, **543**, 543
Carciofi, A.C. – 559
Castro, N. – **549**
Challis, P. – 503
Chardonnet, P. – **543**
Charlebois, M. – 561
Charlton, J.C. – **565**
Chen, H.-W. – 457
Chené, A.-N. – **139**, 538
Chiappini, C. – 217
Chita, S. – **560**
Chu, Y.-H. – **341**
Cidale, L. – 527
Clark, J.S. – 301, 539
Cohen, D.H. – **17**, 577
Conti, P.S. – **285**, 548
Cooke, J. – **566**
Cooper, J.L. – **561**

Corcoran, M.F. – 133, **526**, 535
Cranmer, S.R. – **526**
Crockett, R.M. – 201
Crowther, P.A. – **47**, 285, 301, 327, **527**, 537, 548
Curé, M. – **527**

Damineli, A. – **527**, 530, 539
Danchi, W.C. – 541
Davidson, K. – **528**
Davies, B. – **549**
Davis, C.J. – 557
De Becker, M. – **528**
de Grijs, R. – 552
de Koter, A. – **39**
de la Chevrotière, A. – **528**
de Mink, S. – 167, **543**
Della Valle, M. – 546
Dessart, L. – 185, **543**, 577
Dessauges-Zavadsky, M. – **457**
D'Odorico, S. – **495**
Dopita, M.A. – **367**
Dorfi, E.A. – 566
Drissen, L. – 535, **561**
Dupret, M.-A. – 544
Durrell, P.R. – 565

Eenens, P. – **544**
Eisenhauer, F. – 257
Ekström, S. – 3, 147, **209**, 217
Eldridge, J.J. – **179**, 201
Elmegreen, B. – 550
Esteban, C. – **549**
Evans, C.J. – **550**

Fahed, R. – **529**
Fesen, R. – **561**, 562
Figer, D.F. – **247**, 549
Firnstein, M. – 546
Fischer, T. – 401
Fragos, T. – 545
Frémat, Y. – 545
Fröhlich, C. – **401**
Fryer, C. – 551
Furness, J.P. – 285, 527, 548
Furst, E. – 385
Fynbo, J.P.U. – **443**, 568

Gaensler, B.M. – **509**
Gallagher, J.S. – **510**
Gall, C. – **566**
Garcia, M. – **529**, 549

García-Rojas, J. – 538
Garnavich, P.M. – 503
Gayley, K. – **529**
Genzel, R. – 257
Georgy, C. – 3, 147
Gies, D. – 532, 541
Gillessen, S. – 257
Glatzmaier, G.A. – 231
Godart, M. – **544**
Gräfener, G. – **63**, 536
Grebel, E.K. – **509**
Greene, E. – 503
Greif, T.H. – 471
Groh, J.H. – **530**
Güdel, M. – 554
Gull, T.R. – **530**, 534
Guzik, J.A. – 83

Hadfield, L. – 301, **327**
Hamaguchi, K. – 526
Hamann, W.-R. – 63, 536, 551
Hanson, M.M. – **307**
Hargrave, P. – **561**
Hattori, T. – **550**
Hénault-Brunet, V. – **530**
Henley, D. – 526
Henry, T. – 541
Herrero, A. – 529, 549, 551
Hesselbach, E.N. – 560
Hirschi, R. – 3, 147, **217**
Hill, G.M. – **531**
Hillier, D.J. – **89**, 257, 526, 530, 542
Hoffman, J.L. – **562**
Howarth, I.D. – **531**
Hubert, A.-M. – 545
Hubrig, S. – **531**, **544**, 577
Humphreys, R.M. – **111**, 564
Hunter, D. - **550**
Hunter, I. – 167, **544**, 571,

Icke, V. – 567
Iping, R.C. – **532**
Izzard, R.G. – 179

Jiménez-Esteban, F.M. – 554
Joggerst, C.C. – **566**
Johnson, J.L. – **471**
Josselin, E. – **532**

Kahn, S.M. – 533
Kalogera, V. – **545**
Kashlinsky, A. – **483**
Kelly, P.L. – 503
Kewley, L. – 503
Kiminki, D.C. – **532**

Kirshner, R.P. – 503
Kobulnicky, H.A. – 532, **551**
Kochanek, C.S – 333
Koenigsberger, G. – 532
Kotak, R. – **437**
Kruip, C. – **567**
Kudritzki, R.-P. – **313**, 549, 556

Lang, C. – **562**
Langer, N. – **167**, 231, 543, 560
Lanz, T. – 526
Law, C. – **407**
Le Louarn, M. – 550
Leitherer, C. – **513**
Lennon, D. – 167, 531, 544
Lenorzer, A. – **551**
Leutenegger, M.A. – **533**
Levesque, E.M. – 97, **533**, 565, 571,
Liebendörfer, M. – 401
Liermann, A. – **551**
Limongi, M. – 571
Livne, E. – 185
Lobel, A. – 525
López-Martín, L. – 549
Ludka, B. – 550

MacDonald, J. – 161
MacFadyen, A.I. – **509**
Madura, T.I. – **534**
Maeda, K. – 463
Maeder, A. – **3**, 147, 209, 217
Maíz Apellániz, J. – 538, **552**
Magalhaes, A.M. – 557
Malesani, D. – 443
Marchenko, S. – 530, 571
Marco, A. – **552**, 556
Marengo, M. – 564
Markova, N. – 525
Marin, A. – 551
Martayan, C. – **545**, 571
Martins, F. – **257**, 532
Massey, P. – **97**, 533
Matteucci, F. – **391**
Maund, J.R. – 201
McSwain, M.V. – 532
Mengel, S. – 552
Mennekens, N. – 293
Mesa-Delgado, A. – 549
Messineo, M. – **552**
Meurer, G.R. – 385
Meynet, G. – 3, **147**, 209, 217, **571**
Milisavljevic, D. – **562**
Modjaz, M. – **503**, 571
Moffat, A.F.J. – **119**, 528, 529, **567**
Moll, S.L. – **552**
Montes, G. – **534**
Morel, T. – 531

Moreno, E. – 532
Morris, P.W. – **361**, 527, 537
Murphy, J. – 185

Najarro, F. – **265**
Nazé, Y. – **553**, 577
Negueruela, I. – **301**, 539, 552, 556
Nieva, M.F. – **534**, 546
Noels, A. – 544
Nomoto, K. – **463**
Nürnberger, D.E.A. – **553**

Oey, M.S. – **385**
Ohtani, H. – 550
Okazaki, A.T. – **133**, 536
Oksala, M.E. – **545**
Olsen, K.A.G. – 97, 533
Onifer, A.J. – **83**
Oskinova, L.M. – 551
Ott, C.D. – 185, 257
Owocki, S.P. – **71**, 133, 525, 529, 533, 534, 536, 540

Paar de Kooper, J.-P. – 567
Parkin, E.R. – **535**
Pastorello, A. – **546**
Pellerin, A. – 556
Perez-Torres, M.A. – 534
Petit, V. – **535**, 577
Pettini, M. – **415**
Pittard, J.M. – 535, **536**
Plez, B. – 97
Popescu, B. – 307
Prieto, J.L. – **333**, 503
Prinja, R. – 525
Prochaska, J.X. – 457
Przybilla, N. – 313, 534, 537, **546**
Puls, J. – **25**

Ramsay Howat, S.K. – 557
Rasio, F. – **554**
Rauw, G. – **536**, 553
Reach, W. – 563
Rho, J. – 285, **563**
Rich, J.A. – 565
Rizzo, J.R. – **554**
Rosado, M. – **563**
Rucinski, S.M. – 567
Ruehling, U. – **536**
Russell, C.M.P. – 133, **536**

Sana, H. – **554**
Schiller, F. – **537**
Schmidt, G. – 541
Schnurr, O. – **537**
Schöller, M. – 544
Schuster, M.T. – **564**

Sharp, R. – 539
Shaviv, N. – 540
Simón-Dìaz, S. – 529, **538**
Skinner, S. – **554**
Smartt, S.J. – **201**, 546
Smith, D.M. – 539
Smith, L.J. – **379**, 510
Smith, N. – **193**, 560
Smith, P. – 541
Sonneborn, G. – **491**, 532
Sota, A. – **538**, 552
St Louis, N. – 139, 528, 530, **538**
Stanek, K.Z. – 333, 503, **509**
Stasińska, G. – 538
Stencel, R. – **555**
Stolovy, S. – **555**
Stroud, V. – **539**
Sugai, H. – 550
Sullivan, M. – 566

Tanaka, M. – 463
Taniguchi, Y. – **429**
Tappe, A. – 563
Teodoro, M. – **539**
Thielemann, F.-K. – 401
Thöne, C.C. – **568**
Tominaga, N. – 463
Torrejon-Vazquez, J.M. – **539**
Tout, C.A. – 179
Townsend, R. – **161**, 545, **577**
Trippe, S. – 257
Trundle, C. – **539**
Truran, J.W. – 401
Tuthill, P. – 529

U, V. – **556**
Ubeda, L. – **556**, 561
ud-Doula, A. – **540**, 577
Umeda, H. – 463
Urbaneja, M.A. – 313, 556, **556**
Ursúa, A. – **556**

Vanbeveren, D. – **293**
Van Bever, J. – 293
van Marle, A.J. – 71, **540**, 571
Varricatt, W.P. – **557**
Venero, R. – 527
Verheijdt, M. – 167
Vink, J.S. – 571
Voges, E.S. – 385
Vreeswijk, P.M. – 568

Wade, G.A. – 535
Walborn, N.R. – 531, 571, 577
Wallace, D.J. – **541**
Walterbos, R.A.M. – 385

Weiss, W.W. – 567
Weisz, D.R. – 333
Westmoquette, M.S. – 379, 510
Whitworth, A.P. – 559
Willems, B. – 545
Williams, G.G. – **541**
Williams, S.J. – **541**
Willis, A.J. – **542**
Wisniewski, J.P. – **557**, 560
Wood, E.S. – **568**

Woosley, S.E. – 231
Wunsch, R. – 559

Yelda, S. – 385
Yoon, S.-C. – 167, **231**
Young, J.E. – 565

Zastrow, J. – 561
Zinnecker, H. – **557**
Zsargo, J. – **542**